Oceanography and Marine Biology

An Annual Review

Volume 56

OCEANOGRAPHY
and
MARINE BIOLOGY
AN ANNUAL REVIEW

Volume 56

Editors

S. J. Hawkins
Ocean and Earth Science, University of Southampton,
National Oceanography Centre, UK
and
The Marine Biological Association of the UK, The Laboratory, Plymouth, UK

A. J. Evans
Ocean and Earth Science, University of Southampton,
National Oceanography Centre, UK
and
The Marine Biological Association of the UK, The Laboratory, Plymouth, UK

A. C. Dale
Scottish Association for Marine Science, Argyll, UK

L. B. Firth
School of Biological and Marine Sciences, Plymouth University, UK

I. P. Smith
School of Biological Sciences, University of Aberdeen, United Kingdom

CRC Press
Taylor & Francis Group
Boca Raton London New York

CRC Press is an imprint of the
Taylor & Francis Group, an **informa** business

International Standard Serial Number: 0078-3218

CRC Press
Taylor & Francis Group
6000 Broken Sound Parkway NW, Suite 300
Boca Raton, FL 33487-2742

International Standard Book Number-13: 978-1-138-31862-5 (Hardback)

Contents

Preface

The 56th volume of *Oceanography and Marine Biology: An Annual Review* (OMBAR) contains six reviews that cover a range of topics, reflecting the wide readership of the series. OMBAR welcomes suggestions from potential authors for topics that could form the basis of appropriate reviews. Contributions from physical, chemical and biological oceanographers that seek to inform both oceanographers and marine biologists are especially welcome. Because the annual publication schedule constrains the timetable for submission, evaluation and acceptance of manuscripts, potential contributors are advised to contact the editors at an early stage of manuscript preparation. Contact details are listed on the title page of this volume.

The editors gratefully acknowledge the willingness and speed with which authors complied with the editors' suggestions and requests and the efficiency of CRC Press, especially Jennifer Blaise, Alice Oven and Marsha Hecht, in ensuring the timely appearance of this volume. The editors also thank Dr. Anaelle Lemasson of the Marine Biological Association, Plymouth for her help with proofreading.

Oceanography and Marine Biology: An Annual Review, 2018, **56**, 1-72
© S. J. Hawkins, A. J. Evans, A. C. Dale, L. B. Firth, and I. P. Smith, Editors
Taylor & Francis

IMPLICATIONS OF LONG-TERM CLIMATE CHANGE FOR BIOGEOGRAPHY AND ECOLOGICAL PROCESSES IN THE SOUTHERN OCEAN

CHRISTOPHER D. MCQUAID*

Department of Zoology and Entomology, Rhodes University, Grahamstown 6140, South Africa
*Corresponding author: Christopher D. McQuaid
e-mail: c.mcquaid@ru.ac.za

Abstract

Understanding the long-term consequences of climate change for Southern Ocean ecosystems is important because this is one of the last remaining wildernesses on the planet and because the Southern Ocean is a major driver of global climate. The Southern Ocean is roughly the size of Africa and experiences exceptional seasonality. Its many habitats include the permanently open ocean, sea ice, frontal systems and neritic waters, and different zonal (east-west) and meridional (north-south) regions are on different trajectories in terms of climate, sea ice cover and biological populations. The Western Antarctic Peninsula has experienced substantial warming, loss of sea ice and declining Adélie penguin populations, while eastern Antarctica has cooled, shown increased ice cover and increasing numbers of Adélies. In the ocean itself, warming seems to be concentrated north of the Antarctic Circumpolar Current and at depth rather than in surface waters. Densities of Antarctic krill are correlated with ice cover in the previous winter and in the south-west Atlantic have decreased over the last century while salps have shown increasing numbers south of the Antarctic Circumpolar Current. Even in this example, the mechanisms involved are uncertain, making predictions difficult.

The historic loss of an enormous biomass of consumers through fisheries, led to top-down ecological effects including competitive release among predators. These pelagic food webs are, however, strongly physically forced, making them particularly vulnerable to changes in environmental conditions. The size and species composition of the primary producers affect food chain length and the efficiency of the biological pump. Critically, the primary producer community is profoundly shaped by factors influencing the availability of light (e.g. season, ice melt and water column stability), micro and macronutrients. Changes in these will have deeply important bottom-up effects, and this brings us to defining biogeography. In these pelagic systems, biogeographic provinces are defined by the frontal systems that delineate sharp discontinuities in conditions in the water column and the taxa that dominate primary production. Because these are not geographically fixed, changes in biogeography in this context describe the expansion, contraction or simple displacement of biomes. The associated food webs revolve around a small number of key species that differ among habitats and biomes. They are not simple, exhibit considerable flexibility and include a number of taxa, particularly the cephalopods, that are difficult to sample and remain poorly studied. A major difficulty in understanding how climate change is likely to manifest is the brevity of relevant datasets. We have few physical or biological benchmarks to use in separating short-term noise from long-term signal. As a physical example, Southern Ocean fronts can exhibit short-term meridional shifts of >100 km in a matter of weeks. Biologically, regional differences in trajectories of Adélie penguin numbers need to be viewed against the background of substantial variability over the last 45,000 years. Except for ice or sediment cores, such data are available for few variables or species. In addition, research efforts are geographically unbalanced for logistic reasons. Remote sensing and the Argo float programme reduce this problem by increasing spatial coverage enormously for some variables, but not others, and even then offer relatively new time series.

Important variables that are undergoing, or will undergo, change include: sea ice cover (essentially habitat loss), sea temperatures, wind and mixing regimes, the positions of fronts, ultraviolet levels and ocean pH. Many of these will have interacting effects, and the areas likely to be exposed to multiple environmental changes far exceed those already experiencing important changes. Species are potentially vulnerable to stressors at all ontogenetic stages and in many cases sublethal effects (e.g. on reproductive success) are likely to affect population dynamics before stressors reach lethal levels. Likely, and complicated, will be possible changes to the microbial plankton, including increased microbe deposition by aeolian dust and greater susceptibility of potential hosts to viral infection because of other stresses. All these stressors can have direct and interacting effects as well as indirect effects. For example, meridional shifts in frontal positions will alter the foraging ranges of top predators that are land-based during the breeding season. The ecosystem-level consequences will depend on individual species reactions, including the rates at which they can respond and stressors to which they respond, with the potential disruption of species interactions through different phenological responses.

Perhaps the key characteristic of predictions for the effects of climate change on the Southern Ocean lies in an even greater degree of uncertainty than the norm.

Introduction

The responses of biological and physical processes in the Southern Ocean to long-term climate change carry implications on a global scale, and there are three primary reasons for a focus on this part of the planet. First, the Southern Ocean is an important driver of global climate and source of deep bottom water (White & Peterson 1996, Orsi et al. 1999, Marinov et al. 2006, Mayewski et al. 2009, Jaccard et al. 2013, Hansen et al. 2016), and there is an increasing recognition of its role in driving the meridional overturning circulation of the ocean (Marshall & Speer 2012). Second, Southern Ocean ecosystems are exceptionally productive, strongly shaped by physical conditions and unusually sensitive to climate change (Moore et al. 2000, Constable et al. 2003, Nicol et al. 2000, Smetacek & Nicol 2005). Modelling studies suggest that this is exacerbated by the fact that the Southern Ocean is responsible for an inordinate proportion ($75 \pm 22\%$) of oceanic uptake of anthropogenic heat (Frölicher et al. 2015). Third, the Southern Ocean is one of the key places on the globe where signal-to-noise ratios in surface and subsurface temperatures are high, making it a good region to detect or monitor climate change (Banks & Wood 2002), though a recent study indicates that there may be a delay of 50–60 years before it is possible to separate background climate variability from an anthropogenic signal (Hawkins & Sutton 2012). More recently still, modelling studies by Armour et al. (2016) suggest that the Southern Ocean is warming very slowly because of cooling by upwelled water that originates in the North Atlantic so that a long-term response will be delayed for some centuries until these upwelled waters are themselves warmed.

It is not sensible to consider the Southern Ocean in isolation from Antarctica because the two influence each other so much, so I will inevitably refer to the continent especially in terms of understanding past and future climate change. Additionally, the Subantarctic differs in important ways from the high Antarctic. Recent palaeoceanographic literature indicates that different parts of the Southern Ocean have responded differently to climate change in the past (Kohfeld et al. 2005, Marinov et al. 2006, Sigman et al. 2010), reflecting the efficiency of carbon sequestration by the biological pump (Jaccard et al. 2013, Turner 2015) and increased iron fertilisation of the Subantarctic by dust during glacial periods (Kumar et al. 1995, Robinson et al. 2005). Thus, the northern and southern regions of the Southern Ocean are expected to exhibit differences in the way that properties such as salinity, wind stress and rate of warming will alter in the future (Boyd et al. 2015).

Definitions and scope

The Southern Ocean covers a vast area and includes a wide range of biomes, habitats and interconnected ecosystems. I will not include the terrestrial ecosystems or biogeography of the Subantarctic islands and the continent itself (see Chown et al. 1998, Convey et al. 2014). Nor will I include benthic systems, though benthic-pelagic links can be very important in neritic waters, for example, the indirect effects of ice thickness on the benthos (Lohrer et al. 2013) or benthic feeding by krill (Schmidt et al. 2011). The benthic fauna of the Southern Ocean has been recently reviewed and exhibits a high level of endemism that has evolved in the context of a relatively stable environment (Kaiser et al. 2013). There have been major recent advances in our understanding of these systems, such as the inclusion of molecular techniques (Allcock & Strugnell 2012), the identification of new habitats, including rather inactive vent systems beneath the former position of the Larsen ice shelf (Domack et al. 2005a,b), and the development of international collaborative initiatives (Griffiths et al. 2011). Nevertheless, the study of benthic systems in the Southern Ocean is still at an early stage with strong geographic biases in sampling effort (see Figure 3 in De Broyer & Danis 2011 and Figure 3 in Kaiser et al. 2013) plus the problem that at least some 'species' are genetically segregated geographically and by bathymetry (e.g. Brandão et al. 2010).

Two components of the Southern Ocean deserve particular attention. The first comprises the oceanic frontal systems. These are important in themselves as feeding sites for top predators and because they act as biogeographic and ecological boundaries. Transitions across fronts in the size composition of the phytoplankton community have important implications for food web function. The second component is the sea–ice habitat because of its influence on ocean/atmosphere CO_2 exchange and on primary production during the annual ice retreat, and because it drives krill populations.

The Southern Ocean includes the Antarctic Circumpolar Current (ACC), the Antarctic Coastal Current, the marginal ice zone (MIZ) and the Subantarctic waters that lie between the Antarctic Convergence and the Subtropical Convergence, or very roughly between about 46 and 60°S (Figure 1). The idea that these represent a single marine ecosystem is long outdated, and there is clear recognition that there are major differences in biology and the physical environment among regions, including how biology is forced by physical factors, while models indicate that the balance of factors within regions can change periodically (Constable et al. 2003). Similarly, the perception that this vast oceanic ecosystem is highly biologically isolated from the waters farther north is also under revision with evidence of the rafting of kelps (and their associated fauna) across the Antarctic Polar Front (Fraser et al. 2016) and eddy transport of propagules and plankton across the ACC (Clarke et al. 2005).

With a surface area of approximately $30 \times 10^6 \, km^2$ (El-Sayed 2005), the Southern Ocean forms over 20% of the world ocean (Tomczak & Godfrey 2013) and is one of the most important drivers of world climate (Gille 2002), particularly through its role in the formation of Antarctic bottom water. Water dense enough to form the bottom water of the ocean interior can only be formed where surface evaporation caused by cold dry winds combines with ice formation and brine rejection to produce very cold, saline water (Longhurst 1998). This occurs in few places, including the Labrador and East Greenland Sea, and is a major driving force in global thermohaline circulation (Lutjeharms 1985). Bottom water is also formed in Antarctica at the Coastal Convergence Zone, where very dense ice-shelf water is transported beneath the warmer circumpolar deep water that upwells at the Antarctic Divergence. It has been argued that physical processes in the Southern Ocean largely control nutrient distribution to the rest of the world ocean (Ribbe 2004), and models suggest that nutrients supplied by the Southern Ocean may be responsible for three-quarters of biological production north of 30°S (Sarmiento et al. 2004a,b). As one of the three major high nitrogen low chlorophyll (HNLC) regions of the ocean, the Southern Ocean is also a huge potential sink for atmospheric CO_2 (Falkowski et al. 1998), though ocean-atmosphere models of global warming suggest that this may be modified as increased

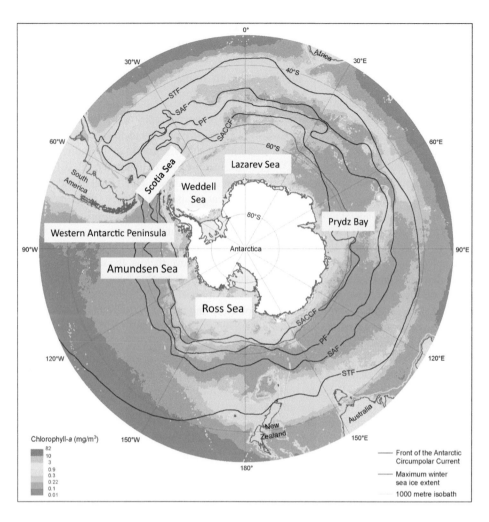

Figure 1 Seas and frontal systems of the Southern Ocean. STF, subtropical front; SAF, sub-Antarctic front; PF, Polar front; SACCF, Southern Antarctic Circumpolar Current front. Chlorophyll values are means for summer 2002–2016. (Modified from Deppeler, S.L. & Davidson, A.T. 2017. *Frontiers in Marine Science* **4**, 40, originally produced by the Australian Antarctic Division Data Centre.)

rainfall would lead to greater stratification and a reduction in the downward flux of carbon and loss of heat to the atmosphere, both effects reducing oceanic uptake of CO_2 (Sarmiento et al. 1998).

Biogeography

Ainley et al. (2007) argue strongly that research in the Southern Ocean has undergone an unacknowledged paradigm shift. They argue that a top-down perspective linking changes in the system, including populations of krill eaters, to the loss of whale biomass and consequent competitive release has been replaced by a bottom-up understanding. In the latter view, the system is seen to be driven almost exclusively by changes in environmental conditions, linked to climate change and alterations to sea ice cover. In their view, this shift in our collective perspective or understanding began in the 1990s, coinciding with the striking increase in access to large-scale environmental data through remote sensing. They list a number of cases in which environmental change cannot be the sole explanation for changes in the distributions and abundances of animals and point out that environmental effects will have interacted with biology as highlighted by overviews such as that of Croxall (1992). Similarly, Ducklow et al. (2007), while recognising the role of ocean circulation in 'translating climate warming into ecosystem changes' (p. 74), acknowledge the difficulty of incorporating biological interactions, particularly trophic interactions, into a framework in which physical factors such as ice extent have a clear influence on population dynamics. There are also complex indirect effects of removing vast biomass of whales from the system. For example, models by Branch and Williams (2006) indicate that it is possible that the removal of large whales led to a switch in the diets of killer whales to include more fur seals and southern elephant seals, driving their numbers down, although this would not explain postulated declines in Minke whale populations. Essentially Ainley et al. (2007) plead for a recognition that the Southern Ocean ecosystem has changed through a mixture of both environmental change and through direct and indirect consequences of the removal of huge numbers of predators, not only whales but also fish (Pauly et al. 1998, 2005).

While this is almost certainly true, it is also true that pelagic ecosystems are particularly strongly forced by the physical environment. Although grazers may have a crucial role in structuring the pelagic community, for example in the Weddell Sea (Bathmann 1996), physical forcing appears to be especially important in the Southern Ocean. Nicol et al. (2000) provide striking evidence, based on a large-scale quasi-synoptic survey of the coast of eastern Antarctica, of the importance of physical conditions (primarily the extent of sea ice) in structuring marine ecosystems in the Antarctic, with clear links between oceanographic conditions and ecosystem structure all the way from primary production to the abundances of krill, salps, birds and whales. Thus, ironically, the Southern Ocean not only helps drive global climate, it is also especially vulnerable to changes in climate because of the degree to which ecological processes in this vast ecosystem are driven by physical factors.

In considering biogeography and ecological processes in the ocean, it is important to acknowledge fundamental differences between life in water and life on land. For instance, on land nutrient recycling happens more or less *in situ*, whereas in the ocean, nutrient uptake and remineralisation are profoundly uncoupled. Similarly, prey advection can allow predator populations to persist in the absence of large prey standing stocks (e.g. Barkai & McQuaid 1988), and advection is especially important in certain parts of the Southern Ocean, such as the Scotia Sea (Murphy et al. 1998, 2007a). Consequently, our understanding of time and space is fundamentally different for marine and terrestrial systems (McQuaid 2010). This is not trivial. For example, Crampton et al. (2016) recently published a paper on turnover of phytoplankton species in the Southern Ocean over millions of years. Their data were derived from core samples, and they rightly point out that this does not allow them to differentiate between global speciation/extinction and local effects or biogeographic changes. Samples collected from cores reflect the phytoplankton community in the overlying water, and this can undergo substantial changes as frontal systems undergo latitudinal fluctuations. In other words, species may go extinct from a perspective fixed in space but may simply have moved from the perspective of the water column. Changes in core samples are related to changes in overlying water masses rather than changes within those water masses. Likewise, the difference between the marginal ice zone (the part of the ocean seasonally covered by ice) and the permanently open ocean is not geographic but the presence or absence of ice, the extent of which differs among years.

Longhurst (1998) concludes that the processes that force stratification of the surface layers also determine characteristically different phytoplankton regimes, which are the primary biomes of pelagic ecology and comparable to terrestrial biomes such as savannah, grassland, and so forth. On this basis, he recognises four primary biomes, defined by the factors that control mixed layer depth. Three of these occur in the Southern Ocean:

1. Coastal biome, where diverse coastal processes force the mixed layer depth.
2. Westerlies biome, where mixed layer depth is forced largely by local winds and irradiance.
3. Polar biome, where the mixed layer depth is constrained by a surface brackish layer that forms each spring in the marginal ice zone.

Both wind regimes and ice melt are likely to be strongly affected by long-term climate change, altering the processes that define these biomes. Within the Southern Ocean, Longhurst identifies four major biogeographic provinces. The Subtropical Convergence (STC) is the only frontal system in the ocean recognised as a biome in its own right; the others are defined primarily by the positions of various frontal systems.

Antarctic Westerly Winds Biome:
The South Subtropical Convergence (STC) province at approximately 35°S.
The Subantarctic water ring province, bounded by the STC and the Antarctic polar front (APF 50–55°S).
Antarctic Polar Biome:
This comprises the Antarctic province, between the APF and the Antarctic Divergence (60–65°S) and the Austral Polar province, the seasonally ice-covered seas from the Antarctic Divergence to the coast of Antarctica.

The point here is that these biogeographic provinces are defined primarily by the positions of oceanic fronts.

Grant et al. (2006) took a different approach. They used hierarchical classification to produce a bioregional scheme for the Southern Ocean that identified regions that are relatively homogenous in terms of their physical and ecological conditions but differ from other regions. This was primarily based on physical characteristics: bathymetry, sea surface temperature, silicate and nitrate. They then included the proportion of the year that areas were covered by sea ice and satellite derived chlorophyll values. The basal divisions were among coastal Antarctica, the marginal ice zone and the open waters farther north, separating these into four primary clusters: Antarctic waters, the Polar plus Subantarctic Fronts and temperate waters, which grouped out as shallow, including the neritic waters of the islands and continental shelf, and deep, including the Subtropical Front (Figure 2). This scheme broadly conforms to the concept of the Southern Ocean as nested rings defined by fronts but introduces a higher level of detail. Importantly, their classification identified the Weddell Gyre as a large, standout region grouped with the banks of the Ross Sea.

Ecological processes

Marine ecosystems are likely to be more vulnerable to long-term climate change than terrestrial ecosystems. Partly this is because, with the exception of coastal macroalgal communities, the primary producers in marine ecosystems are microscopic organisms that do not accumulate biomass and must undergo an annual recovery of standing stocks from winter minima. A consequence of this is that grazers must depend on new production each year, while terrestrial grazers can utilise primary producer biomass accumulated over years. Thus, although trophic cascades have been observed even in speciose terrestrial ecosystems, implying the importance of top-down effects (Pace et al. 1999, Schmitz et al. 2000), the land is essentially green because primary producers are only rarely controlled by grazing (Hairston et al. 1960). In the sea, phytoplankton carbon turnover times are of the order of a week (Falkowski et al. 1998), and unutilised primary production largely undergoes senescence and sedimentation. A second consequence is that, with no accumulation of biomass, marine primary producers do not have the indirect effects of their terrestrial counterparts. They do not provide habitat complexity, nor do they moderate the effects of physical stress through shading, amelioration of wind effects, increasing humidity and so forth. Longhurst (1998) notes the further effect that on land ecotones shift over very long timescales (centuries or millennia), so that biomes are relatively stable in their geographic distribution, while in the sea, ecological boundaries are more ephemeral and unstable, and respond rapidly to changes in global climate and ocean circulation.

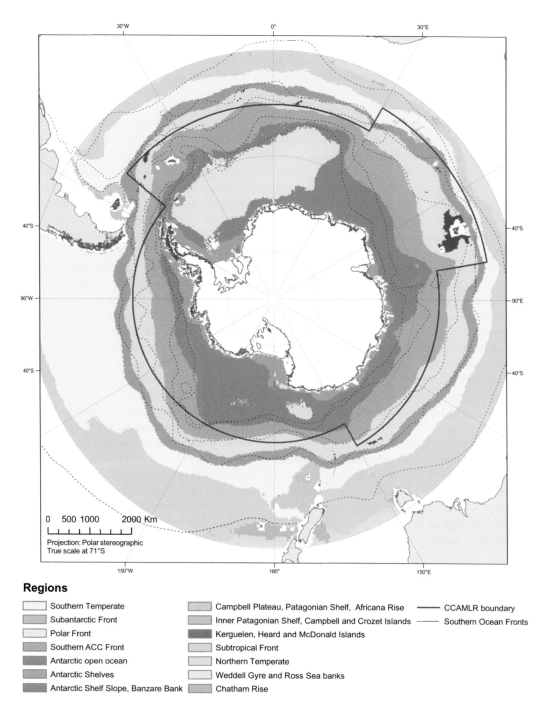

Regions

Southern Temperate	Campbell Plateau, Patagonian Shelf, Africana Rise	—— CCAMLR boundary
Subantarctic Front	Inner Patagonian Shelf, Campbell and Crozet Islands	—— Southern Ocean Fronts
Polar Front	Kerguelen, Heard and McDonald Islands	
Southern ACC Front	Subtropical Front	
Antarctic open ocean	Northern Temperate	
Antarctic Shelves	Weddell Gyre and Ross Sea banks	
Antarctic Shelf Slope, Banzare Bank	Chatham Rise	

Figure 2 Bioregions of the Southern Ocean. Regions are based primarily based on bathymetry, sea surface temperature, silicate and nitrate, the proportion of the year that areas were covered by sea ice and satellite derived chlorophyll values. (Reproduced with kind permission from Grant, S. et al. 2006. *Bioregionalisation of the Southern Ocean: Report of experts workshop*, Hobart, September 2006. WWF-Australia and ACE CRC.)

Given this, interannual variability in factors that affect primary production is likely to have more immediate and more dramatic effects on ecosystems in the sea than on land. This is implicit in Longhurst's (1998) approach to defining marine biogeography. He postulates that general phytoplankton–herbivore–predator relationships differ among pelagic habitats and are predictable, at least in outline, given information on the way in which algal growth is forced environmentally. This in turn influences biogeochemical cycling (Lampitt & Antia 1997, Lutz et al. 2007). Effectively, understanding the seasonal phytoplankton cycle for a region is equivalent to describing the climax community of a terrestrial system and allows us to make inferences about the ecology of that region.

A key element to understanding marine ecosystems in general and perhaps Southern Ocean systems in particular is a recognition of the importance of advection. This is important because it means that the factors shaping the observed system may have been applied far away. Thus, predator populations can be maintained in areas with low productivity and in the absence of obvious prey standing stocks, while population dynamics can reflect the success of ontogenetic stages that occur in very different places from the adult population. Advection even applies to the transport of habitat in the form of sea ice.

Consequently, different parts of the Southern Ocean are characterised by different assemblages of species, largely in response to a combination of the type of physical environment and the composition of the primary producers. Probably the key factors shaping Southern Ocean food webs are the presence of ice and factors that influence the size composition of the primary producer community, including seasonality. To generalise in the face of strong spatial variability, and bearing in mind that most of our data are collected during the summer season, to the north of the marginal ice zone and within the Subantarctic away from frontal systems, nano and picoplankton dominate the phytoplankton, and characteristic primary and secondary consumers include copepods and mesopelagic fish, particularly myctophids. In seasonally ice-free waters farther south, diatoms are seasonally dominant, and Antarctic krill and myctophids are characteristic. In the southerly ice-covered shelf areas of the Western Antarctic Peninsula (WAP) where there is a seasonal succession from diatoms to cryptomonads and flagellates (Ducklow et al. 2007), the dominant consumers are ice krill and notothenoid fish. These different habitats and groups of animals, described by Murphy et al. (2013) as indicators of ecosystem structure, are associated with different higher predators. In other words, the signature of primary producer composition is driven by abiotic conditions and is transmitted to the higher predators. Importantly, however, there is very strong seasonality. There can also be strong connections among the ecosystems of different regions through both advection of plankton, particularly krill, by currents, and through migration of larger predators, many of which can forage over enormous ranges. Consequently, some regions are at least partially allochthonous, being driven by production imported from elsewhere. For example, nearshore feeding birds at the Prince Edward Islands depend on food brought to the islands by the Antarctic Circumpolar Current, and predation can be so intense that there is a characteristic 'hole' in the zooplankton downstream of the islands where zooplankton standing stocks are markedly lower than upstream (Perissinotto & McQuaid 1992). Similarly, there is no local recruitment of krill at South Georgia; instead they are carried there from the WAP. In fact, coastal circulation results in complex mixing of populations in western Antarctica, including the Bellingshausen, Amundsen, Weddell and Scotia Seas (Murphy et al. 2007a, Thorpe et al. 2007, Piñones et al. 2011).

Additionally, there is non-seasonal variation. Variation on scales of years or decades in the strength of the El Niño-Southern Oscillation (ENSO) and the Southern Annular Mode (SAM) impose another layer of variability on the system by changing conditions at the base of the system. Modelling studies by Hall and Visbeck (2002) indicate that the SAM shows weak seasonality but is closely correlated with zonal wind stress and has significant effects on the Southern Ocean south of 30°S, influencing surface current variability and sea ice cover. Loeb et al. (2009) found significant correlations between the Southern Oscillation and phytoplankton, zooplankton and sea ice cover that were consistent with Antarctic Dipole forcing. These and many other studies illustrate the strength of physical forcing on these ecosystems.

Food webs

There is a clear consensus that major change is coming to the Southern Ocean and the ecosystems it supports, but there is considerable uncertainty over exactly how that change will manifest. We have a much better understanding of the ecosystems of some regions than others, but given the considerable differences among regions, both in how they function and how environmental conditions are changing, it is simply not possible to extrapolate among regions (e.g. Melbourne-Thomas et al. 2013). How things will change and indeed how they are already changing depends on where you look.

There have been two main alterations to our understanding of the Southern Ocean: (a) the system is not uniform, and (b) although food webs involve fewer species than those at lower latitudes, they are not simple. While international organisations such as Commission for the Conservation of Antarctic Marine Living Resources (CCAMLR) exist, research agendas are often set nationally and are not always optimally coordinated, with research groups from different countries focussing on different geographical areas largely due to logistical constraints or regional interests. Except for aspects that can be measured using remote sensing, the result is geographically patchy understanding of these systems, with the south-west Atlantic region being particularly well studied.

This section gives a general overview of food webs, but it is important to recognise the high degree of flexibility that exists both in broad patterns of energy flow and at finer, even interindividual levels. Where diets have been examined in detail, there can be important differences between conspecific populations living in different places (cf. Adams & Brown 1989 with Piatkowski & Pütz 1994). There can also be differences in diet or foraging behaviour between the sexes, for example shags (*Phalacrocorax* spp.), Gentoo penguins (Bearhop et al. 2006) and elephant seals (McIntyre et al. 2010), and of course there are also seasonal differences as the larger predators such as elephant seals can forage in quite different places in different seasons (e.g. Bradshaw et al. 2003). This is particularly the case for the birds, which breed on land and must feed their chicks until they fledge. During this time, their foraging range is restricted, and additionally, they can feed themselves and their chicks on different diets (e.g. Connan et al. 2014). Some species, including sperm whales, macaroni penguins and Antarctic fur seals, are known to exhibit specialisation of sympatric same-sex individuals that feed on different prey or in different locations (e.g. Evans & Hindell 2004, Cherel et al. 2007).

The very close links between biology and physics mean that coupling biogeochemical, physical and biological models is essential. Murphy et al. (2012) offer an extremely useful modelling framework (Figure 3) and emphasise the need to understand the links between food webs and the physical drivers that shape them across multiple scales. Detailed food web descriptions exist but only for specific, relatively well studied regions, particularly the Western Antarctic Peninsula (WAP), the Scotia and Ross Seas, and Eastern Antarctica. These tend to highlight differences rather than commonalities among regions. For example, the WAP has exhibited a southerly extension of the warmer moist climate of the northern peninsula into the colder, drier south, resulting in ice loss and changes in community composition ranging from phytoplankton to the abundances of different species of penguins (Ducklow et al. 2007), though this trend seems to have reversed since the 1990s (see below). The Ross Sea, on the other hand, has been described as supporting neritic food webs that differ dramatically from others in the Antarctic, with stronger benthic-pelagic coupling (Smith et al. 2007).

Given all these reservations, it is useful to consider broad patterns within the Southern Ocean, and I have taken an approach that is similar to that of Grant et al. (2006), considering the following five major habitat types: permanently open ocean, fronts, marginal ice zone, islands, coastal waters. But first, central to understanding how food webs may differ among these broad habitats is the role of body size, particularly the size of the primary producers. To a large degree, this reflects the tendency of different phytoplankton taxa to feed into one of the three energetic pathways described by Clarke et al. (2007): grazing by zooplankton, sedimentation, the microbial loop.

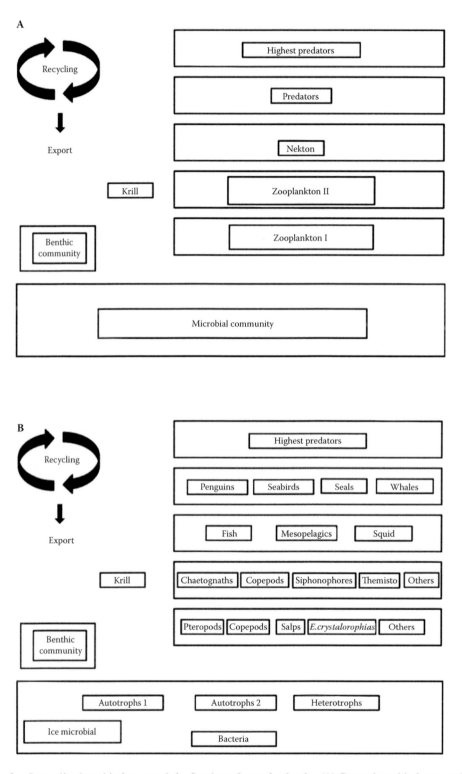

Figure 3 Generalised trophic framework for Southern Ocean food webs. (A) General trophic framework. (B) Key species and functional groups, emphasising the intermediate zooplankton groups and alternative energy pathways. (*Continued*)

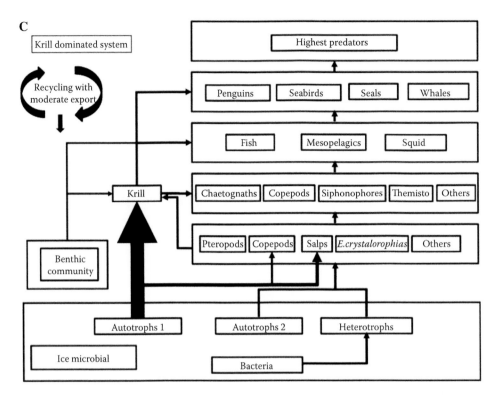

Figure 3 (Continued) Generalised trophic framework for Southern Ocean food webs. (C) Application of the generic structure to a summer, krill-dominated system. Arrow thickness indicates the magnitude of energy flow. The degree of recycling and export is indicated by the size of the arrows on the left. Autotrophs 1 = large autotrophs, Autotrophs 2 = small autotrophs. (Reproduced from Murphy, E. et al. 2012. *Progress in Oceanography* **102**, 74–92.)

Primary production

Trophic position in pelagic systems is closely linked to body size, and Ainley et al. (1991) offer the interesting concept that zooplankton essentially package phytoplankton and microheterotrophs as larger particles, making them available to micronekton, which in turn repackage energy as particles that are useable by larger predators and so on. This is reflected in the fact that phytoplankton size is critical to an understanding of these food webs. At the base of the food web, rates of primary production show large-scale differences among regions and habitats, with generally low levels of productivity but hotspots of periodically intense primary production, and it is useful to envisage comparatively stable levels of background production by small cells on top of which are superimposed increases in biomass and productivity due to blooms of diatoms and *Phaeocystis* (Smetacek et al. 2004). Thomalla et al. (2011) used satellite data to characterise the seasonal cycle of primary production in the Southern Ocean, producing a scheme of four primary regions representing the combinations of high versus low average chlorophyll levels with either high or low seasonal cycle reproducibility, that is, a high or low percentage of total variance explained by the seasonal cycle. The physical mechanisms leading to enhanced chlorophyll levels remain unclear and varied, however, and there were marked differences among zonal regions (Figure 4). While iron-limited regions of the Southern Ocean are characterised by high levels of microbial recycling and consumption by copepods and salps, blooms of larger cells underpin the diatom–krill–whale food chain and fuel the carbon pump (Smetacek et al. 2004).

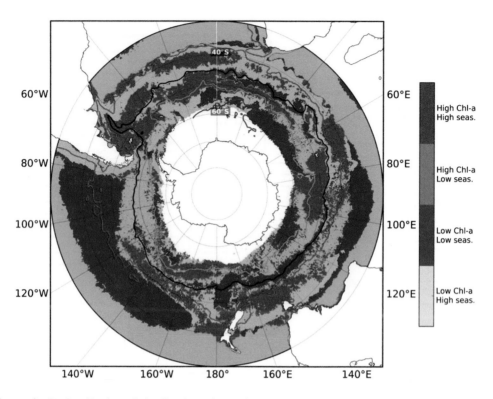

Figure 4 Regionalisation of the Southern Ocean based on the response of phytoplankton biomass to seasonality. Regions are characterised as having high (green) or low (blue) average chlorophyll levels with either a high (high seas.) or a low (low seas.) percentage of total variability explained by the seasonal cycle. Mean (1998–2007) frontal positions are shown for the subtropical front (STF – red), the sub-Antarctic front (SAF – black), the polar front (PF – orange) and the Southern Antarctic Circumpolar Current front (SACCF – blue). (Reproduced with permission from Thomalla, S. et al. 2011. *Biogeosciences* **8**, 2849.)

As predicted by Longhurst's (1998) model, water column stability seems to have a major influence on phytoplankton community structure (e.g. Kang & Lee 1995, Garibotti et al. 2003). Chlorophyll levels over most of the Southern Ocean are generally <0.03 mg m^{-3}, mainly comprising nano and picophytoplankton (Detmer & Bathmann 1997, Gervais et al. 2002), but with blooms of >1 mg m^{-3} in coastal waters of the continent and islands, in the marginal ice zone and at fronts (Moore & Abbott 2000). Bloom development is by no means a given, even at the edge of retreating ice where water column stability, although critical (Lancelot et al. 1993), is not alone enough to predict bloom development (e.g. Smith & Nelson 1990, Smetacek et al. 1997). Blooms are generally dominated by a few medium-sized species of around 10–15 µm, usually diatoms or *Phaeocystis* (Smetacek et al. 2004), with the species differing among individual blooms (e.g. Bathmann et al. 1997, Smetacek et al. 1997). Blooms can be limited by iron availability, with experimental addition of iron stimulating a response in the larger size classes, particularly diatoms (Boyd et al. 2000, Hoffmann et al. 2006), and there are multiple natural sources of iron, including resuspension of benthic material, terrestrial dust, sea ice retreat and the melting of icebergs (Boyd et al. 2012). The relative importance of these mechanisms, of course, depends on place and time (de Jong et al. 2012).

Arrigo et al. (2008a) analysed data for the period 1997–2006 and found that, although rates of primary production were highest over the continental shelf, the low productivity open ocean accounted for 90% of total primary production because of its enormous area. Similarly, although the marginal ice zone (MIZ) can show intense plankton blooms because ice retreat in spring seeds the water column with ice plankton and iron as well as stabilising the water column (Smith & Nelson 1985), Arrigo et al. found that over their 10-year study period, the MIZ was only slightly more productive than the open pelagic. They attributed this to the fact that the MIZ periodically shows exceptionally high rates of production but also minimum values that are lower than those of the open ocean. Generally, chlorophyll and ice retreat are positively correlated, reflecting stabilisation and seeding of the water column by melting ice, but the relationship is not consistent and differs considerably among regions. Regionally, the Ross and Weddell Seas show high levels of chlorophyll but also high variability, whereas productivity in the open ocean, while low, is more stable (Constable et al. 2003).

Just as important as rates of primary production is the size composition of the primary producer community, as this fundamentally influences carbon drawdown and patterns of energy flow through the system. Primary production by large and small cells leads into food chains of different lengths, with implications for the efficiency of the biological pump and the transfer of energy to top predators (Fenchel 1988). In situations characterised by small phytoplankton and high levels of microzooplankton grazing, tight coupling of small grazers and the microbial loop results in a high proportion of carbon being retained and recycled within the euphotic zone so that there is relatively little drawdown of atmospheric carbon (Azam et al. 1983, Longhurst & Harrison 1989). In contrast, meso and macrozooplankton grazing of larger phytoplankton tends to lead to high efficiency in the transfer of carbon to depth (Roman et al. 1993). Even here the details become important. Copepods, salps and krill are the major pelagic consumers of phytoplankton but contribute with different effectiveness to the transfer of carbon to depth. Copepods dominate in terms of dry biomass and production (Voronina 1968), but the rate of sinking of zooplankton faeces is closely related to their size (Small et al. 1979, Komar et al. 1981), and copepods produce small faeces that are rarely found in sediment traps, presumably because they are channelled through the microbial loop and largely recycled within the water column. As a result, copepods are inefficient at transferring carbon to depth (Turner 2002, Michels et al. 2008). In contrast, both salps and krill produce large faeces that sink relatively quickly (Yoon et al. 2001, Turner 2002), giving them an important role in the transfer of carbon to depth (Le Fèvre et al. 1998).

Although the open ocean is typically characterised by small-celled nano and picophytoplankton, there are marked seasonal changes in the size composition of the phytoplankton with a much higher contribution of netphytoplankton in summer (Froneman & Perissinotto 1996a,b). As a result, grazing is primarily due to microzooplankton in winter and to meso and macrozooplankton in summer, with implications for the seasonal efficiency of the biological pump (Froneman & Perissinotto 1996a,b). Larger netphytoplankton also tend to dominate south of the Antarctic polar front and at frontal systems, which can exhibit phytoplankton blooms that are as dense as those of the marginal ice zone (Froneman & Perissinotto 1996a,b). This makes fronts important as foraging grounds for both birds and mammals (Bost et al. 2009). These blooms tend to be dominated by specific size classes and species of phytoplankton, indicating the importance of *in situ* production rather than advection and accumulation. Again, these show a seasonal shift in dominance from large (>20 μm) netphytoplankton such as *Chaetoceros* spp. and *Nitzschia* spp. in early summer to smaller nanoplankton in late summer (Laubscher et al. 1993).

Food web structure

Although there are substantial differences among geographic (east-west) regions, these are superimposed on broad differences in food web structure among the within-region habitats or biomes formed by the permanently open ocean, the marginal ice zone and the coastal waters of the continent, while embedded within and different from the open ocean are fronts and a variety of island systems. The last are particularly important as they are the breeding and moulting grounds for many of the air-breathing top predators.

The largest of these major biomes is the ice-free open ocean, where primary production is normally by small-size classes of phytoplankton, and grazing is generally dominated by copepods, salps and small euphausiids. The marginal ice zone periodically exhibits particularly high levels of biological activity fuelled by the input of ice algae, and diatom blooms and grazing here is largely attributable to krill, which is taken in turn by mammals and birds.

Importantly, the difference between the permanently open ocean and the marginal ice zone is not geographic; it is the presence or absence of sea ice that is the key factor. Thus, the pelagic can be free of the influence of ice in space, in other words north of the maximum extension of winter sea ice (the permanently open ocean), or in time, reflecting seasonal fluctuations in the extent of sea ice (the marginal ice zone).

Permanently open ocean

Primary production in ice-free waters away from fronts is primarily due to small-size classes of phytoplankton. Broadly, phytoplankton mortality due to grazing tends to decrease with cell size (Smetacek et al. 2004), and grazing of small cells by microzooplankton can be very intensive. For example, using the dilution technique, Burkill et al. (1995) recorded grazing by microzooplankton (mostly protozoans) of over 270% of phytoplankton daily production in open waters, as opposed to 27% within the pack ice. Overall, there tends to be a major transition in the relative importance of grazing by microzooplankton versus meso and macrozooplankton at the Antarctic polar front, with the latter assuming the greater importance to the south where diatoms dominate primary production (Froneman & Perissinotto 1996a,b). Quéguiner (2013) described the seasonal development of the phytoplankton of the permanently open ocean, dividing the diatoms into two groups. The first comprised lightly silicified species showing a succession of dominant species in response to conditions of light, temperature and nutrients. The second comprised slow-growing, heavily silicified species concentrated near the pycnocline and supplied with nutrients by mixing. Neither group suffers strong grazing, the first showing mortality through senescence and bacterial activity, while the second exhibits mass mortality and sedimentation following decreases in light and nutrients in autumn.

Salps or krill?

Krill densities can fluctuate markedly among years (Siegel and Loeb 1995, Brierley et al. 1999a,b, Saunders et al. 2007). This is linked to the occurrence of so-called salp years (Huntley et al. 1989, Nishikawa et al. 1995) and can be linked to the influence of sea ice on krill population dynamics (Hewitt et al. 2003, Flores et al. 2012; see below). Krill and salps show marked spatial segregation that was once suggested as being due to competition (Loeb et al. 1997, Perissinotto & Pakhomov 1998a) but is now believed to reflect their confinement to water bodies with different properties (Pakhomov et al. 2002), particularly in terms of phytoplankton. Krill needs high densities of chlorophyll, while salps are found in areas with low chlorophyll values (Nicol et al. 2000). Consequently, we see the replacement of krill by salps as the phytoplankton community shifts from large diatoms to small cryptophytes during summer in coastal waters of the Antarctic Peninsula (Moline et al. 2004) and over decades in the south-west Atlantic sector as a whole (Atkinson et al. 2004). While salps are eaten by a range of other species and appear not to be the trophic dead end previously assumed (Pakhomov et al. 2002), they do not occupy the sort of central trophic position occupied by krill (e.g. Laws 1985), so that there are obvious direct and indirect implications for top predators in which group dominates.

The contributions of these species to vertical carbon flux are determined not only by their abundances and grazing rates, but also by the details of their biology. Salps are non-selective feeders that use a mucous net to trap particles. They are exceptionally efficient at filtering very small particles, including those in the size range 0.1–1 μm, which includes bacteria (Sutherland et al. 2010), so that grazing pressure by salp blooms can be extremely high (e.g. Bernard et al. 2012). In contrast, *Euphausia superba* shows selective feeding in mixed phytoplankton assemblages, favouring diatoms over *Phaeocystis*, even if the cells are of similar sizes (Haberman et al. 2003). *Euphausia superba* uses the long setae on the legs to form a filter basket that expands as water is sucked in from the front. Food particles are retained as the water is squeezed out of the sides of the basket (Riisgård & Larsen 2010); this is not efficient for small, round particles (Quetin & Ross 1985). As a result, *E. superba* is less effective at transferring carbon to depth, though this may be mitigated by preying on microzooplankton and repackaging them as faeces (Perissinotto et al. 2000).

The comparison is aggravated by their vertical migratory behaviour. Adult *E. superba* exhibit a reduction in metabolic rate in winter (Torres et al. 1994) when they show a seasonal increase in the depth of vertical migration (Siegel 2005) in response to changes in the light regime (Meyer 2012). Even in winter, however, migration is generally shallower than 300 m (Siegel 2005), while salps undergo daily migration to depths of 800 m or more (Wiebe et al. 1979). Defaecation at depth takes salp faeces well below the zone of regeneration, greatly increasing the efficiency of the vertical transfer of carbon (Phillips et al. 2009).

Both krill and salps are key grazers, sometimes removing total daily primary production entirely (e.g. Pakhomov et al. 2002), and, based mainly on biochemical analyses, it has been suggested that consumption of salps may have been underestimated in the past in the Southern Ocean and elsewhere (Cardona et al. 2012, Dubischar et al. 2012). Nevertheless, senescent salp blooms can be a major source of organic input to the benthos (e.g. Smith JLK et al. 2014, Smith WO et al. 2014), and it is krill that are the focal prey species for a wide range of top predators (e.g. Murphy et al. 2012). Because Antarctic krill are long-lived ($\geq$6 years), like fish, they can exhibit strong and weak year classes (Nicol 1990, Kilada et al 2017), potentially minimising the effects of serial predator–prey mismatching. Despite this, there is good evidence linking the reproductive success of top predators to variations in krill abundance (e.g. Croxall et al. 1999, Reid et al. 2005, Nicol et al. 2008). Braithwaite et al. (2015) go so far as to link condition in baleen whales to sea-ice cover via the abundance of krill.

These considerations led to an early understanding that Southern Ocean food webs were relatively simple, with krill as the dominant prey species, but it has inevitably become clear that the situation is more complex with a variety of possible energy pathways and flexibility on the part of consumers. In many ways, this ultimately reflects spatial and temporal variability in levels of primary production and the species composition of the phytoplankton. For example, Montes-Hugo et al. (2009) note that changes in water column stability in the Western Antarctic Peninsula have resulted in a southward shift in the centres of primary production, specifically the occurrence of diatom-driven blooms that is mirrored by changes in populations of krill and penguins. Similarly, Morgenthaler et al. (2018) tentatively linked mass mortality of Southern Rockhopper penguins to the occurrence of low levels of chlorophyll immediately prior to the peak moulting period.

Actual food webs are complex largely because of their flexibility, but there are essentially two major routes for energy flow, via krill and its direct use by top predators where large phytoplankton dominate and via grazing by other zooplankton and then fish (especially myctophids) where primary producers are small. The second pathway involves more steps and is less efficient at transferring energy to the top predators. The framework proposed by Murphy et al. (2012) for modelling Southern Ocean ecosystems (see Figure 3) emphasises the need to link physical, biogeochemical and biological models. Their generalised trophic model illustrates the alternative pathways that connect the primary producers to top predators, indicating how the balance among pathways can shift. Fundamentally, it is the size composition of the primary producers that is critical so that, for example, there will conceptually be no krill in the absence of large autotrophs, though of course there may be no krill anyway for other reasons. Given the role of the phytoplankton community in shaping patterns of energy flow, we might expect strong seasonality in ecosystem function. Hunt et al. (2011) showed that the Lazarev Sea exhibits seasonal shifts in the predominance of different macrozooplankton species and rates of primary production. They propose that the balance among primary production, grazing and predation alters so that the epipelagic system exhibits a transition from bottom-up control in summer to top-down control during winter when the system is supported mainly by copepod prey. This is in agreement with the observation of Pilskaln et al. (2004) that sediment trap particles collected during summer in Prydz Bay in Antarctica suggest a simple food chain with diatoms grazed by large grazers producing fast-sinking macrozooplankton faeces and algal-rich aggregates. In contrast, winter-sinking material had a completely different composition: no diatom aggregates, smaller faecal pellets and many crustacean filter feeders, suggesting a longer winter food chain with particle reprocessing. Hunt GL et al. (2002) proposed something similar for cold and warm years in the Bering Sea.

The middle trophic levels of the Southern Ocean, which link primary production with the top predators, are occupied primarily by krill, fish and squid. The role of krill is relatively well studied, but those of squid and fish are less well understood, not just in the Southern Ocean but globally (e.g. Young et al. 2015). Particularly important are pteropods, myctophid fish and squid. I will briefly discuss fish and squid.

Myctophids, icefish and krill all have high energetic values as prey, reflecting their particularly high lipid content (Barrera-Oro 2002). Fish are the most important prey for top predators after krill (Barrera-Oro 2002, Barrera-Oro & Casaux 2008), and Flores et al. (2008) have suggested that myctophids may be even more important than krill in the transfer of energy through the food web, at least in the Lazarev Sea. Although they are critical components of an array of Southern Ocean ecosystems, the fish fauna is relatively depauperate in terms of species. While the Southern Ocean comprises about 10% of the ocean's surface, the 322 species identified there so far form only about 0.01% of known fish diversity (Eastman 2005). Different parts of the Southern Ocean are dominated by different groups of pelagic fish with, broadly, the myctophids dominating oceanic waters and nototheniids, which probably arose in the Antarctic (Briggs & Bowen 2012), dominating the neritic waters (Koubbi et al. 2011). Fish form an intermediate level within these food webs being important as both prey and predators, and vertically migrating mesopelagic species have an important role in enhancing the efficiency of the biological pump (Flores et al. 2008, Kock et al. 2012). Fish, especially myctophids, are critically important, not only to many of the flying seabirds, but also to king penguins and some mammals (e.g. Adams & Klages 1987, Kozlov 1995, Cherel et al. 1996, Guinet et al. 1996, Raclot et al. 1998, Connan et al. 2007), offering an alternative energy pathway that may be as important as the krill–predator energy route (Murphy et al. 2007a, Flores et al. 2008), but that appears also to be more complex, involving multiple, flexible predator–prey interactions (La Mesa et al. 2004). In situations where the energy demands of individual predators are especially high, such as during lactation or egg incubation, the high energy content of fish can make them preferred as prey to krill, even if krill are available (Ichii et al. 2007).

Myctophids feed opportunistically on mesozooplankton, particularly copepods, pteropods, hyperiid amphipods and euphausiids. Stomach contents suggest a high degree of dietary overlap among species (Pakhomov et al. 1996), but stable isotope analysis indicates strong niche partitioning, with larger species occupying a higher trophic level (Cherel et al. 2010). Krill are important in the diets of many Antarctic fish, especially the larger myctophids, though their consumption is limited by fish gape size (Cherel et al. 2010). Myctophids exhibit stable isotope values between those of large crustaceans and the top predators, with some larger species having a trophic signature similar to those of seabirds that depend on crustaceans (Cherel et al. 2010). In the southern Ross Sea, the nototheniid *Pleuragramma antarcticum* dominates the fish community and is preyed on by a variety of mammals and birds. Together with *Euphausia crystallorophias*, it plays a similar role to that of myctophids and *E. superba* in other areas (La Mesa et al. 2004). Estimates of consumption of krill by fish are very approximate due to lack of data for some abundant species and for large areas of the Indian and Pacific sectors but are probably of the order of 20–30 million tonnes of krill and other prey per annum by demersal fish, and 5–32 million tonnes by mesopelagic species in the Atlantic sector (Kock et al. 2012). Like krill, fish occupy a key position as both prey and predators, and climate-driven alterations to fish populations would clearly have enormous effects on Southern Ocean ecosystems. While myctophids are unlikely to be directly affected by ice retreat, species that use sea ice as spawning or nursery grounds are expected to be heavily influenced. Unfortunately, Antarctic fish show such a wide range of life histories and reproductive strategies that, even among species with close links to ice cover, population-level responses will almost certainly be species-specific (Moline et al. 2008), making the prediction of likely ecosystem-level consequences fraught.

Underemphasised in many models are the cephalopods that are known from diet analyses to be very important to both birds and mammals but that are extremely difficult to sample. Squid in particular have high consumption rates, show very rapid growth and are likely to be particularly sensitive to climate change (Coll et al. 2013). Imber (1978) suggested that our understanding of the role of squid has been impaired partly because seabirds are better at catching some species of cephalopods while trawlers catch more of others, and the idea of a polar frontal zone food chain running from copepods to myctophids to top predators through squid emerged relatively recently (Rodhouse et al. 1992a,b). Fortunately, Southern Ocean cephalopods have recently been reviewed by Collins and Rodhouse (2006). Their diets are not well known, but the squid are pelagic feeders, often feeding on other squid (González & Rodhouse 1998). Many live near the surface when young and in deeper waters as adults (Clarke 1966), moving to higher trophic levels as they age and exhibiting an ontogenetic shift from crustacean to fish prey (Rodhouse & Nigmatullin 1996, Phillips et al. 2003). Species or age classes in surface waters feed primarily on euphausiids, hyperid amphipods and mysids, while the larger meso and bathypelagic species feed on migratory fish, particularly myctophids (Phillips et al. 2001, Dickson et al. 2004). The importance of krill in squid diets is unclear, though krill are certainly eaten (Kear 1992; Nemoto et al. 1985, 1988; Collins et al. 2004). Squid are highly opportunistic (Nixon 1987), and Dickson et al. (2004) found that interannual differences in the contribution of krill to the diet of *Martialia hyadesi* was not explained by variation in krill abundances. Instead, they suggested that encounter rates might be important.

The role of cephalopods as prey comes out clearly in Figure 5, taken from Collins and Rodhouse (2006), with different predators predominating at different depths. Cephalopods are taken by all

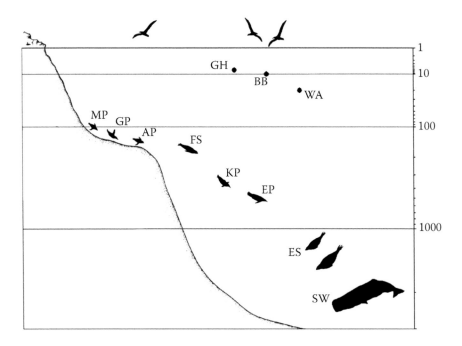

Figure 5   Foraging depths of Southern Ocean cephalopod predators. MP: macaroni penguin; GP: gentoo penguin; Ap: Adélie penguin; GH: grey-headed albatross; BB: black-browed albatross; WA: wandering albatross; FS: fur seal; KP: king penguin; EP: emperor penguin; ES: elephant seal; SW: sperm whale. Depths are in metres. (Reproduced from Collins, M.A. & Rodhouse, P.G. 2006. *Advances in Marine Biology* **50**, 191–265.)

major groups of predators, including other cephalopods, birds, mammals and fish, and squid can be the main prey of many predators such as the grey-headed albatross (Xavier et al. 2003, Richoux et al. 2010), although the diet shifts with the season and differs between self- and chick-provisioning (Connan et al. 2014). Estimates of total consumption of squid by Southern Ocean predators, though crude, are enormous (e.g. Laws 1977, Clarke 1983, cited in Collins & Rodhouse 2006). Difficulties with such estimates include the fact that they are often based on the presence of squid beaks in predator stomachs, and these can be retained for prolonged periods, raising the risk of seriously overestimating consumption rates (Piatkowski & Pütz 1994).

A wide variety of flying seabirds prey on squid, though albatrosses may feed on them largely by scavenging. Lipinski and Jackson (1989) analysed the diets of 14 species of seabirds, dividing their cephalopod prey into those that sink and those that float on dying and concluded that most were taken after death, rather than alive near the surface at night time. A wide range of penguins take live squid (Collins & Rodhouse 2006), but only king and emperor penguins take them in substantial numbers. Even for these species, the diet can differ markedly among locations. At the Prince Edward Islands, king penguins feed predominantly on myctophid fish, and squid are a small part of the diet (Adams & Brown 1989, Perissinotto & McQuaid 1992), while in the Weddell Sea, squid form over 50% of the diet (Piatkowski & Pütz 1994). Similarly, when rearing chicks, fish dominated the stomach contents of king penguins at South Georgia, while at the Crozet archipelago, squid were more important (Cherel et al. 1996). Squid are important in the diets of pinnipeds, such as southern elephant seals (Rodhouse et al. 1992a,b), particularly when feeding farther offshore in summer (Bradshaw et al. 2003), and cetaceans, including sperm whales, which show individual differences in prey species and size (Evans & Hindell 2004). The only fishes known to take cephalopods in numbers are the toothfishes, though other fish are their main prey (García de la Rosa et al. 1997, Xavier et al. 2002). Sharks are poorly studied in the Southern Ocean, but Cherel and Duhamel (2004) found that cephalopods were important in the diets of the three species they examined, with giant squids forming a significant proportion by mass of the diets of sleeper sharks (*Somniosus* cf. *microcephalus*).

Although Southern Ocean ecosystems are relatively poor in species with comparatively simple food webs, the flexibility in diets described previously is a clear indication of complex species interactions that are poorly understood. For example, within the plankton, protozoan grazing can promote diatom populations by reducing competition between them and flagellates (Walsh et al. 2001), while abundances of krill and copepods around South Georgia are negatively correlated as krill can switch to feeding on copepods when phytoplankton biomass is low (Atkinson et al. 2001). At intermediate scales, Ainley et al. (2006) describe a summer trophic cascade involving penguins, whales and their prey, crystal krill and icefish. At still larger scales, Laws (1977) estimated the degree to which hunting of baleen whales reduced their biomass, releasing resources in the form of up to 150 million tonnes of krill per annum for their competitors. This led to substantial increases in the populations of krill-eating birds and seals and has been described by Mori and Butterworth (2004) as the greatest ever human-induced perturbation of a marine system. Smaller minke whales also showed rapid increases in populations as the large whales disappeared, but more recently krill populations have declined from the 1970s to 1990s (Reid & Croxall 2001) with negative effects on minke whales, while blue whales appear to be better at tolerating declines in krill availability (Mori & Butterworth 2004), presumably making them better exploitation competitors. Similarly, Waluda et al. (2017) suggest that the different responses to krill availability of co-occurring gentoo penguins, which are generalist predators, and macaroni penguins, which are krill specialists, is important in differentiating their ecological niches. Generalising for such a vast ecosystem necessarily ignores details that may be critical to predicting responses to long-term change accurately.

Fronts

Frontal systems have long been recognised as barriers to species distributions, though they differ dramatically in their effectiveness (Ward & Shreeve 2001), and the convergent frontal zones of the Southern Ocean form particularly distinct ecotones. Southern Ocean fronts exhibit a degree of topographic steering (Marshall & Speer 2012) and are often characterised by high biomass of plankton (Moore & Abbott 2000) and increased primary production driven by eddy shedding (Kahru et al. 2007). Eddy shedding is particularly prolific at the Antarctic Circumpolar Current, and mesoscale eddies influence local atmospheric conditions including cloud cover and rainfall (Frenger et al. 2013). More importantly here, this generally (not always) leads to enhanced overall biological activity (Laubscher et al. 1993, Pakhomov et al. 1996). Frontal systems often support food webs that are quite different from the open waters of the Southern Ocean, but, although Longhurst (1998) considers the Subtropical Convergence to be a frontal system that forms a distinct biogeographic province, there are generally no biota that are endemic to fronts. Southern Ocean fronts are not uniform longitudinally, and because the strengths of fronts vary with longitude, their effectiveness as biogeographic boundaries also varies. For example, the Subtropical Convergence has a much stronger biogeographic effect on microphytoplankton and mesozooplankton south of Africa than in the mid-Indian Ocean (Froneman et al. 1995a,b, 1998) or the mid-Atlantic (Barange et al., 1998). Nevertheless, the fronts generally form regions of relatively sharp transition in the physico-chemical quality of water masses and can delimit distinct biotopes (e.g. Froneman et al. 1995a,b, Grant et al. 2006). As expected, this effect tends to be more obvious for planktonic organisms, with weaker effects on more mobile species. Evidence from the composition of the microbial community reveals clear differences across the Antarctic polar front, with predominance of cyanobacteria to the north and of phytoplankton to the south (Wilkins et al. 2013). Apart from the effects of water column stability on light availability, temperature and nutrients are among the primary determinants of the distributions of coccolithophores and other phytoplankton (Takahashi & Okada, 2000, Boeckel et al. 2006, Gravalosa et al. 2008) and distinct species assemblages of diatoms characterise water masses with different properties that are separated by fronts south of Africa (Boden et al. 1988, Boden & Reid, 1989), an effect confirmed for a wide range of surface water phytoplankton and microzooplankton (Malinverno et al. 2016). Within biogeographic regions, there are also important habitat effects. Analysing diatoms from sediments in the Southern Ocean, Armand et al. (2005) made a primary division between those associated with sea ice and open-ocean species. They concluded that the former showed species-specific distributions related to sea-ice cover and surface temperature, while open-sea species formed three groupings: one associated with cool open-ocean conditions, particularly the winter sea-ice edge, a pelagic open-ocean group most abundant at the Antarctic polar front and a warm open-ocean assemblage with maximum abundances in the polar front zone (Crosta et al. 2005). Taxa such as euphausiids (Gibbons 1997, McLeod et al. 2010), ostracods (Chavtur & Kruk 2003) and copepods (Razouls et al. 2000) show clear biogeographic effects linked to water quality, and there are often marked differences in species composition, or at least species abundances, among water masses separated by fronts (e.g. Tarling et al. 1995, Hunt & Hosie 2003, Pakhomov & Froneman 2004). Thus, overall zooplankton community structure is well defined by the positions of the main frontal systems, with the Subantarctic front separating assemblages characterised by Antarctic and the Subantarctic/subtropical species, marking it out as a clear biogeographic boundary (Pakhomov et al. 2000b). This effect may be masked by cross-frontal mixing, eddy shedding and other forms of water exchange across the fronts as parcels of water crossing fronts carry biota characteristic of their origins (Hunt B et al. 2002). Consequently, while different species or groups of species may dominate zooplankton abundance or biomass on either side of a front, conspecific individuals can often be found on both sides. This is perhaps especially true in the polar frontal zone, which lies between the Subantarctic front and the Antarctic polar front (e.g. Hunt et al. 2001, Hunt B et al. 2002). For the Southern Ocean fish, the Subtropical Convergence and especially the Antarctic polar front act as strong biogeographic barriers, though with some leakage (Stankovic et al. 2002), while the Subantarctic front is much less effective at limiting species distributions (Lutjeharms 1990). For the still more mobile avifauna, there are strong latitudinal changes in species composition that are associated with the fronts but are not clearly defined by them (Abrams 1985, Pakhomov & McQuaid 1996). Nevertheless, taken overall, factors that affect the positions, structures or strength of fronts can have important biogeographic effects. The effects of frontal systems may be obvious over long time periods, and it has been suggested that, in causing isolation, Southern Ocean fronts may promote speciation (Bargelloni et al. 2000). This is debatable, however (Page & Linse 2002), and the effect may depend on the taxon considered.

As well as defining the limits of species distributions, the Southern Ocean fronts are critical feeding grounds for many top predators (Bost et al. 2009, Raymond et al. 2010) and define regions of the ocean in which different ecological processes predominate. This is implicit in Longhurst's approach as different size classes of phytoplankton dominate primary production in different regions and lead into different trophic pathways and food webs. This has consequences that reverberate to the top of the food chain and influence the efficiency of the biological pump. For example, the small-celled nano and picoplankton that dominate the phytoplankton in the permanently open ocean between the Antarctic Polar and Subantarctic fronts (Froneman & Perissinotto 1996a,b) are grazed primarily by microzooplankton, resulting in re-processing of material in the surface waters, relatively long food webs and an inefficient biological pump (Azam et al. 1983, Legendre & Le Fèvre 1995). In contrast, the food webs supporting the Southern Ocean giants, such as the great whales, exist south of the Antarctic polar front and are based on primary production by diatoms and *Phaeocystis* (Smetacek et al. 2004). So, shifts in the positions of frontal systems indicate changes in biological processes in different parts of the ocean as water masses with particular physico-chemical properties expand or contract. This can represent contraction or expansion of biomes with the sort of consequences observed for the expansion of salp and contraction of krill distributions reported by Atkinson et al. (2004), described previously.

Large predators do not necessarily feed where primary or secondary production are high; some focus on places where food accumulates. In other words, it is not just the factors that limit production that are important but also the circulation patterns that cause prey to accumulate, and this has implications for the use of fronts as foraging areas. At-sea observations more recently complemented by tracking studies highlight strong associations between top predators (birds and mammals) and the major frontal systems of the Southern Ocean. Diversity and abundances of seabirds are high at the main fronts, and high species richness of both whales and seabirds is consistently found at the Subtropical Convergence and the Subantarctic front (Bost et al. 2009). Birds particularly tend to prey on small species that form swarms, such as myctophids and krill, moving rapidly to the position of the fronts before slowing down and turning frequently to adopt a foraging mode once they reach areas where the probability of prey is higher (Fauchald et al. 2000, Pinaud & Weimerskirch 2005, Weimerskirch 2007). Similarly, the presence of fronts can affect foraging by seals. Arthur et al. (2016) found that female Antarctic fur seals, *Arctocephalus gazella*, dispersing for the winter after breeding on the Prince Edward Islands foraged differently north and south of the Antarctic polar front, with shallower, briefer foraging bouts to the south, presumably reflecting better prey availability.

Given that they mark sharp changes in the physical and biological properties of the water, it is perhaps unsurprising that fronts even mark sharp changes in stable isotopes. The Subtropical Convergence is a point of sharp change in a general latitudinal gradient in stable isotopes. The $\delta^{13}C$ signature of particulate organic matter drops markedly from north to south of the Convergence (Francois et al. 1993) so that the isotopic landscape differs markedly between subtropical waters and the Subantarctic zone. The latitudinal shift in the isotopic baseline of the system can be traced to the top predators. Penguin chicks living at different latitudes show marked differences in their stable isotope profiles (Cherel & Hobson 2007), and isotope studies of tracked albatrosses have shown strong correlations with the latitudes at which the birds foraged (Jaeger et al. 2010). The north-south decrease in $\delta^{13}C$ is less marked in predators than at lower trophic levels because theirs is an integrated dietary signal, however, and attempts to identify the natal sites of Antarctic terns based on their isotope signatures have proven unsuccessful (Connan et al. 2015). Importantly, the fronts themselves can be positioned isotopically (Jaeger et al. 2010).

Islands

Subantarctic islands present unique situations for several reasons. First, they frequently support phytoplankton populations that differ from those of the surrounding ocean (Armand et al. 2008, Salter et al. 2012). Being volcanic, they are also sites where phytoplankton communities are naturally fertilised by iron (Pollard et al. 2009, Salter et al. 2012) and are highly productive, exhibiting frequent plankton blooms that originate locally (Moore & Abbott 2002, Ward et al. 2007). Higher up the food chain, they are critical sites for reproduction for land-breeding animals (pinnipeds and birds) and for moulting for birds. Because the proportion of sea to land is so huge in the Southern Ocean, breeding localities are restricted to 23 island groups lying within a narrow band of latitudes, mostly in the Atlantic and Indian Ocean sectors (Figure 6). Consequently, these are sites where enormous aggregations of animals occur seasonally, and predation pressure is particularly high around these archipelagos (e.g. Siegfried 1985). Considerable effort has gone into understanding these food webs, for example at Kerguelen, Crozet, South Georgia and the Prince Edward islands among others.

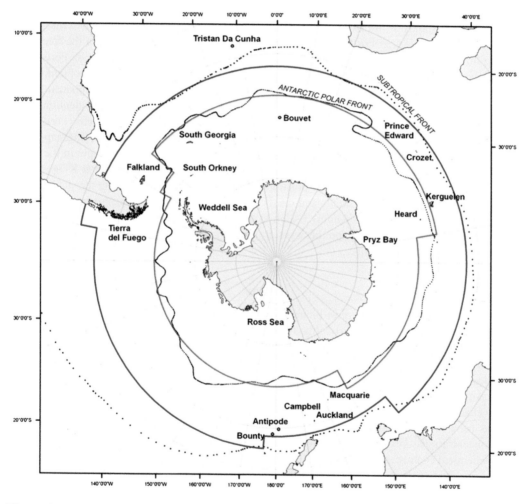

Figure 6 Positions of the sub-Antarcic islands. The northern limits to the areas of interest for the Marine Biodiversity Information Network of the Scientific Committee on Antarctic Research (SCAR-MarBIN) are shown for the Antarctic (inner line, in green, close to the Antarctic polar front) and for the sub-Antarctic (outer line, in blue, close to the subtropical front). (Reproduced from De Broyer, C. & Danis, B. 2011. *Deep Sea Research Part II: Topical Studies in Oceanography* **58**, 5–17.)

Because breeding and moulting are both energetically expensive, energy demands seasonally exceed local production, and elements of these systems can be highly allochthonous, depending on the import of prey by advection to supplement local production by diatom blooms (Perissinotto & McQuaid 1992, Croll et al. 1998, Murphy et al. 2004). This can lead to situations in which different groups of predators respond differently to long-term change because they depend to different degrees on allochthonous and autochthonous production (e.g. Allan et al. 2013).

Many predators necessarily function as central place foragers while breeding and can change foraging direction and distance once their young become independent (Blanchet et al. 2013, Lowther et al. 2014). Latitudinal shifts in the positions of the circumpolar fronts can alter circulation to and around these islands, with potentially dramatic consequences. The Subantarctic front forms the northern boundary of the Antarctic Circumpolar Current. It lies to the north of the Prince Edward Islands, and its proximity influences the speed with which water flows past the islands. When the front approaches the islands, water flows past and around them. This brings in prey from upstream of the islands, but there is little of the local water retention that leads to autochtonous diatom blooms. Such a 'through-flow' state (Pakhomov & Froneman 1999a, 2000) benefits birds that feed away from the islands at the front itself by bringing their feeding grounds closer to the islands. When the front lies farther north, there is more likelihood of trapping of water in the interisland plateau through the formation of a Taylor column-type eddy (Perissinotto et al. 1990, Perissinotto & Duncombe Rae 1990), resulting in the formation of diatom blooms. Nanophytoplankton are preferred over larger diatoms by all grazers, including pteropods, copepods and euphausiids, with only two euphausiids, *Euphausia vallentini* and *Thysanoessa vicina,* showing selection of the smaller netphytoplankton. Consequently, most of the blooms formed by large diatoms (Perissinotto 1992) suffer little grazing but contribute through senescence to bentho-pelagic coupling, supporting benthic populations of the shrimp *Nauticaris marionis* (Perissinotto 1992), a staple in the diet of nearshore feeding birds there such as macaroni and rockhopper penguins (Brown & Klages 1987). A long-term southward shift in the mean position of the front is detectable in both the stable isotope signatures of birds around the islands and, in the sizes of their populations, particularly declines in the populations of nearshore feeders (Allan et al. 2013). This is a telling example of possible indirect effects of climate change on these systems and the likely complexity of their responses to long-term environmental change.

Sea ice

Sea ice constitutes one of the largest biomes on the planet (Arrigo & Thomas 2004). When at its maximum, it has a surface area (roughly $19 \times 10^6 \, \text{km}^2$) that is greater than that of the Antarctic continent ($13 \times 10^6 \, \text{km}^2$; Bye et al. 2006). Southern hemisphere sea ice is unconstrained by land to the north and subject to strong circumpolar winds, resulting in more open water within the extent of the ice pack than in the Arctic (about 21% of total area; Gloersen et al. 1993). The average albedo of the Antarctic pack ice is relatively low largely because of this open water (Allison et al. 1993, Brandt et al. 2005), but it still makes an important contribution to planetary albedo (Rind et al. 1995) and has been implicated in reducing deep water ventilation and contributing to low atmospheric CO_2 concentrations during glacial periods (Stephens & Keeling 2000). In addition to its role in biogeochemical cycling and climate, Southern Ocean sea ice has enormous ecological importance. It includes a wide variety of habitats at different scales (Brierley & Thomas 2002), from the brine channels that are dominated by copepods (Arndt & Swadling 2006) and microhabitats formed during ice disintegration where irradiance and primary productivity are both high (Gleitz et al. 1996) to the polynyas, extensive areas of open water within the pack ice where biological productivity is high, even exceptionally high, in spring and summer (Smith & Gordon 1997, Arrigo & Van Dijken 2003). Polynyas form an important habitat for a variety of species and act as focal points for populations of top predators, such as Adélie penguins, but they vary considerably in their productivity

(Arrigo & Van Dijken 2003). Productivity underneath sea ice can be high as well. Arrigo et al. (2012) reported the occurrence of an enormous phytoplankton bloom underneath sea ice in the Arctic. Unfortunately, sea ice is extremely difficult to access logistically (Dieckmann & Hellmer 2010), so that it is difficult to interpret the relationships among fluctuations in ice cover and animal populations (Croxall et al. 2002). Little attention was given to the ecology of this habitat until recently, but there have now been a number of comprehensive and also more focussed reviews (Brierley & Thomas 2002, Arndt & Swadling 2006, Massom & Stammerjohn 2010, Vancoppenolle et al. 2013). Recent advances include the work of Flores et al. (2011), who used a new design of sampling gear to illustrate the ecological importance of the ice water interface in linking sea-ice biota and pelagic food webs. Sampling the top 2 m of the water column under the ice and in open water in summer, autumn and winter, they recovered nearly 40 species from eight animal phyla, with seasonal changes in community composition. Biomass was greatest under winter ice and lowest under autumn ice, with Antarctic krill generally dominating numbers and biomass, and diel vertical migration indicating grazing at night. This points to perhaps the most important attribute of the pack ice. Antarctic krill is associated with sea ice almost year-round, depending critically on sea ice at certain stages of its life cycle, particularly as postlarvae (Nicol 2006, Flores et al. 2012). High-density krill swarms have been found associated with under-ice topographic features within 13 km of the ice edge (Brierley et al. 2002), indicating the ice edge as an ecologically critical area.

Schools or swarms of Antarctic krill can be kilometres long (Nicol 2006) and can contain tens of thousands of individuals per cubic metre (Hamner et al. 1983), so that total numbers in swarms are vast. The distribution and abundance of *Euphausia superba* differs markedly among regions (e.g. Nicol et al. 2000), with the south-west Atlantic sector of the Southern Ocean supporting over 50% of total stocks (Atkinson et al. 2004). *Euphausia superba* seems to be associated with areas with high densities of chlorophyll (Nicol et al. 2000), and there is a broad relationship between krill abundance and productivity at the oceanic scale (Tynan 1998, Atkinson et al. 2004). In fact, Perissinotto et al. (1997, 2000) found that phytoplankton alone could not fulfil the energy requirements of krill during summer and was supplemented by micro and mesozooplankton.

Nicol (2006) has drawn on earlier studies to develop a conceptual model of the relationship of Antarctic krill with its physical environment, emphasising the relationships of the various ontogenetic stages (eggs, larvae, juveniles and adults) with their different environments as there is spatial separation of developmental stages (Siegel & Harm 1996). Simplified (see Figure 7), in summer gravid females migrate offshore, and the eggs are laid in December to March, sinking into deep water and hatching at depths of 700–1000 m. It is assumed that eggs reaching the benthos suffer high mortality, and it seems likely that the need for deep water for the laying of sinking eggs helps separate *E. superba* from the more coastal *E. crystallorophias*, which has buoyant eggs. The two also show separation by depth of their furcilia stages (Daly & Zimmerman 2004). After hatching, the larvae of *E. superba* migrate to the surface over the next 30 days. The larvae develop in the east-flowing Antarctic Circumpolar Current where the water becomes ice-covered in winter, and they drift with the pack ice. During winter, the adults and the juveniles occur in deeper water so that there is no competition between them and the developing larvae. Cyclonic gyres carry the ice and larvae inshore, and the juveniles are found inshore of the adults in summer, migrating offshore to the shelf break as they grow. Egg-laying females require food and deep water for egg laying, while the larvae rely on the under-ice community for food during winter and circulation to carry them to the juvenile habitat. The combination of these ontogenetic requirements will contribute to distribution patterns, heavily modified as we have seen by migration/advection of adults.

The key aspect of this life cycle is spatial separation of the different stages, and Melbourne-Thomas et al. (2013) suggest that stage-structured models are necessary to resolve krill metapopulation dynamics. Spatial separation reduces competition and also the danger of cannibalism as it separates the small, vulnerable eggs and larvae from the swarms of adult *E. superba*; adult clearance rates depend on the size of particles, not their identity (Quetin & Ross 1985). Thus, ice is critical for the

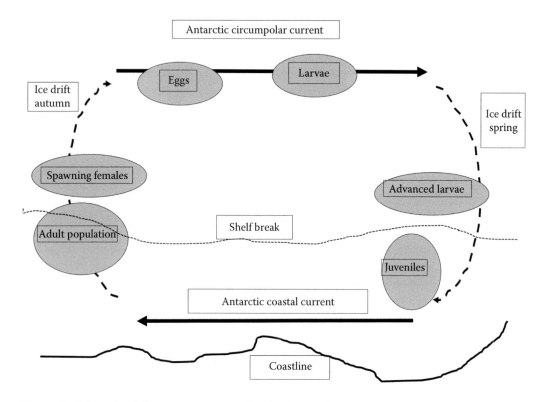

Figure 7 Schematic of the seasonal ontogenetic migration of *Euphausia superba*, showing the use of the gyral circulation that links the two major current systems. (Redrawn with permission from Nicol, S. 2006. *Bioscience* **56**, 111–120.)

overall life cycle, allowing krill to maintain huge stocks of large planktivores despite low oceanic productivity (Smetacek et al. 1990). Models by Wiedenmann et al. (2009) showed a nonlinear relationship between recruitment and sea-ice area from the previous season in the combined Bellingshausen and Amundsen Seas. Recruitment was low outside a narrow range of ice cover, and the authors suggested the possibility that successful recruitment requires a threshold level of ice cover. Despite the apparent relationship between ice and krill recruitment, the mechanisms driving it are not obvious. The model of Wiedenmann et al. produced results that did not conform with the assumption that ice provides food during winter, matching Daly's (2004) finding that low gut fluorescence values indicate that there is little food under winter ice and that krill show opportunistic feeding and slow growth there. The enzyme kinetics of *E. superba* indicate an unusual ability to adapt rapidly to patchy food sources and changing food types (Buchholz & Saborowski 2000), and heterotrophic organisms, mainly micro and mesozooplankton can be important in its diet, even in summer (Perissinotto et al. 1997, 2000). Wiedenmann et al. suggest an alternative interpretation of the krill/ice relationship with the mechanistic link being through the effects of ice melt on the spring phytoplankton bloom. Even in such a relatively well-studied system there are huge unknowns. The level of uncertainty is epitomised by the fact that Murphy et al. (2007b) predict dramatic reductions in krill numbers, with consequences for krill feeders, but the habitat usage models of Melbourne-Thomas et al. (2016) suggest that large-scale loss of sea ice will not have particularly severe consequences for krill abundances.

The effects of advection are particularly important in the case of Antarctic krill (e.g. Hofmann & Murphy 2004, Atkinson et al. 2008), and it is difficult to interpret krill population dynamics in the context of locally observed effects because the important interactions occurred elsewhere, particularly to the west, or 'upstream'. Large-scale teleconnections are critical in the Scotia Sea (e.g. Brierley et al. 1999a,b; Murphy et al. 2007a), but Nicol (2006) offers an alternative view of more isolated local stocks essentially maintained by retentive coastal gyres and suggests that these could conceivably represent distinct genetic stocks. At least in the case of *Euphausia crystallorophias*, there can be a high degree of intraregional genetic differentiation (Jarman et al. 2002). Earlier studies on *E. superba* using allozymes at relatively small spatial scales (e.g. Schneppenheim & MacDonald 1984) were equivocal, and a study by Zane et al. (1998) using a single mitochondrial gene suggested that oceanographic barriers could drive genetic differentiation in *E. superba*. More recently, although Batta-Lona et al. (2011) concluded that the genetic signatures of Antarctic krill from several areas of the Western Antarctic Peninsula region indicated multiple centres of reproduction rather than a single panmictic population, while a study by Bortolotto et al. (2011) using microsatellites indicated a lack of genetic structure at large geographic scales from the Weddell Sea eastwards to the Ross Sea. Overall, it is important to remember that genetic and ecological connectivity are not the same as even low rates of individual migration/immigration can lead to genetic homogeneity. At this stage, it is not possible to determine the degree to which krill population dynamics and processes can be explained by local rather than allochthonous effects, and it seems probable that both concepts apply but to different degrees in different regions. Alternatively, Nicol (2006) suggests that the Weddell Gyre can be interpreted as a single, larger system with a longer cycle than the coastal gyres of the rest of the continent.

Coastal waters

The coastal or neritic waters of the continental shelf are relatively deep in Antarctica, with the shelf break at around 800 m in the Ross Sea. This embayment constitutes the largest expanse of shelf around the continent (Smith et al. 2007), but neritic waters are also extensive from the Amundsen Sea to the Weddell Sea. As usual, some regions have been more intensively studied than others, particularly the Western Antarctic Peninsula (WAP) and the Ross Sea. While these two areas show commonalities, they also show important differences, highlighting the difficulty in generalising for the waters surrounding a continent twice the size of Australia.

Given that physical conditions differ between neritic and oceanic waters, it should be no surprise that there are differences in community composition and function all the way from the primary producers to the top predators. The transition from oceanic to neritic waters around the shelf break generally sees a shift in the abundances and importance of the major players in these food webs. There is a distinct shift from a system dominated by Antarctic krill and myctophids to one in which the Antarctic silverfish, *Pleuragramma antarcticum*, and the crystal krill, *Euphausia crystallorophias* dominate, both of which are characteristic of cold shelf waters (Thomas &Green 1988, Sala et al. 2002, Vacchi et al. 2004). These two species form the major link between predators and primary production in neritic waters (e.g. Pakhomov & Perissinotto 1996, 1997), and in the Ross Sea, they are preyed on by all top predators (Ainley 2002).

The Antarctic continental shelf is globally important as the site of the formation of Antarctic bottom water (Orsi et al. 1999, Gordon et al. 2004), which has shown signs of warming over the last 30 years (Purkey & Johnson 2012, 2013). The shelf waters of the Southern Ocean are also critically important in carbon sequestration, and Arrigo et al. (2008b) estimate that the continental shelf of the Ross Sea alone accounts for perhaps 27% of total carbon uptake by the Southern Ocean as a whole. This is partly because of the high levels of primary production in shallow shelf waters made possible by high levels of iron from either ice melt or sediment resuspension (Fitzwater et al. 2000).

While primary production in the WAP is dominated by diatoms, the Ross Sea is different. The spring phytoplankton bloom develops earlier there than elsewhere in Antarctica (Smith & Asper 2001), and pigment concentrations are particularly high (Sullivan et al. 1993), especially in coastal polynyas (Comiso et al. 1993), with growth regulated by the availability of light and iron (Smith JLK et al. 2014, Smith WO et al. 2014). Diatoms are common, and cryophytes can form large blooms (Smith et al. 2007), but a key organism is the prymnesiophyte *Phaeocystis antarctica*. While diatoms dominate in highly stratified waters, *P. antarctica* is the dominant organism where mixing is deeper (Arrigo et al. 1999). *Phaeocystis* has a polymorphic life cycle, with both flagellated solitary and gelatinous colonial forms (Verity & Smetacek 1996). Grazing pressure on *Phaeocystis* colonies is low (Hansen & Van Boekel 1991), and export of material to the benthos is highly efficient where *P. antarctica* dominates primary production (DiTullio et al. 2000). *P. antarctica* can repair photodamage due to high light intensities when carried near the surface by mixing, allowing it to develop while mixing is still deep and average light exposure is low during early spring (Kropuenske et al. 2009). This results in a succession with *Phaeocystis* forming the early spring bloom, declining after about two weeks, probably due to iron limitation (DiTullio et al. 2000), to be replaced by diatoms as water column stratification increases (Smith JLK et al. 2014, Smith WO et al. 2014).

In coastal waters up to 100 km off the coast of the Antarctic Peninsula, seasonal changes in phytoplankton species composition as meltwater run-off reduces salinity involve a shift from diatoms to cryptophytes, algae that favour brackish water (Moline et al. 2004). The much smaller size of these organisms (around 8 μm) inevitably alters food web structure as such small particles are not easily grazed by krill. Moline et al. (2004) suggest a link between this shift in the primary producer community, reductions in krill abundances and the occurrence of large populations of salps. Given the central position of krill in the food web, this of course has implications for energy transfer through the entire system, including the top predators. Similar correlations between low salinity, water column stability and dominance by cryptophytes have been reported by Kang and Lee (1995) and Garibotti et al. (2003), again around the Antarctic Peninsula, and in eastern Antarctica by McMinn and Hodgson (1993).

Boysen-Ennen and Piatkowski (1988) identified distinctly different zooplankton communities in the Weddell Sea, with clearly separate oceanic and neritic communities. The latter included many larvae and juveniles of benthic species and showed a geographic effect, with a southern and a north-eastern grouping in a cluster analysis. The oceanic and neritic communities differed in the replacement of *Euphausia superba* by *E. crystallorophias* in shallower waters and also in trophic structure, with a much higher proportion of carnivores in oceanic waters and more filter-feeders in the coastal waters, reflecting the high levels of primary production measured during the same survey (von Bröckel 1985). Similarly, Siegel and Piatkowski (1990) identified distinct oceanic and neritic mesozooplankton communities in the WAP, though the co-occurrence of salps and Antarctic krill there in summer was unusual (Ducklow et al. 2007).

Hopkins (1985, 1987) analysed the stomach contents of mesopelagic species (i.e. to depths of c. 1000 m) in both the WAP and the Ross Sea and found that most species fed on detrital particles, including both phytoplankton and krill moults, with the carnivores preying mostly on copepods and coelenterates. While shared species had essentially the same diets in the two areas, krill were only occasionally taken by larger predators in the Peninsula region, while krill furcilia were taken as live prey in the Ross Sea, though in the comparison, geography was confounded with differences in timing with respect to phytoplankton blooms.

Bentho-pelagic coupling in the Southern Ocean is believed to occur primarily through the agency of zooplankton, particularly the production and sinking of faecal matter, so that the vertical migratory behaviour of zooplankton has a strong influence on energy flux to the benthos as does the nature of the phytoplankton community (Schnack-Schiel & Isla 2005). Coupling is important as benthic biomass is limited by the availability of food (Smith JLK et al. 2014, Smith WO et al. 2014), and coupling is closer in neritic than deep waters. This is because material reaches the benthos more directly (i.e. with less opportunity for lateral advection) and in a comparatively unprocessed state. Coupling is also closer because of the production of planktonic larvae by benthic animals (e.g. Boysen-Ennen & Piatkowski 1988) and the contribution of demersal fish, which feed on both the benthos and zooplankton (Barrera-Oro 2002). In inshore waters of the WAP, demersal fish have greater ecological importance than krill, while in deeper waters farther off shore, they become inaccessible to predators and pelagic fish form the link between zooplankton and predators (Barrera-Oro & Casaux 2008).

Neritic waters are most obviously characterised by very high densities of top predators (e.g. Ducklow et al. 2007, Smith et al. 2007). These include both ice-dependent species such as Weddell and crabeater seals that are resident year-round and those that use neritic waters as summer feeding grounds, such as fur and elephant seals. The two groups can show interesting differences in population trends. There has been a long-term loss of sea ice within the Bellingshausen Sea/ Amundsen Sea region, including the WPA, which is in contrast to the Ross Sea where there has been an increase in ice coverage and the length of the sea-ice season (Parkinson 2002, Zwally et al. 2002, Ducklow et al. 2007). In the WAP, the loss of ice has had contrasting effects on top predators that show different affinities for ice. For example, while populations of Adélie penguins have decreased, those of ice-intolerant gentoo and chinstrap penguins have increased (Ducklow et al. 2007). Large-scale effects such as sea-ice loss can, however, interact with smaller-scale effects. Some populations of Adélie penguins have been declining more rapidly than others, and this appears to be driven by local landscape effects on the accumulation of snow. Despite warming temperatures (Thompson et al. 1994), the Antarctic Peninsula has experienced increasing precipitation (Thomas et al. 2008) and populations of Adélie penguins nesting where snow accumulates have declined faster than those nesting where the effects of wind mitigate snow accumulation (Ducklow et al. 2007). In contrast, and fittingly, in Adélie Land in eastern Antarctica, Adélie penguins have shown increases in their populations (Micol & Jouventin 2001).

The presence of particularly high abundances of large predators in the Ross Sea and the fact that all are consumers of silverfish, *Pleuragramma antarcticum,* and crystal krill, *Euphausia crystallorophias*, suggests the possibility of top-down regulation of these systems. Ballard et al. (2012) found some spatial partitioning among nine mesopredators (penguins, flying birds, seals and whales) in the Ross Sea but substantial dietary overlap. This links with the suggestion by Ainley et al. (2007) that Southern Ocean biologists have lately tended to favour bottom-up processes in explaining population changes, while discounting top-down regulation. For example, Adélie penguins foraging over the Ross Sea shelf exhibit prey switching and prey depletion (Ainley 2002), as do Weddell seals on local scales in McMurdo Sound (Testa et al. 1985), while minke whales in the Ross Sea have unusually small stomach contents and low body mass (Ichii et al. 1998). These can all be interpreted as important indications of top-down effects. Ultimately such interactions can reveal themselves through trophic cascades. Ainley et al. (2006) suggest that such a cascade occurs at Ross Island where heavy predation by Adélie penguins, minke whales and fish-eating killer whales depletes populations of crystal krill, and eventually silverfish, reducing grazing of the diatom community by the krill. Such prey depletion in areas with dense predator populations are not well documented elsewhere in the Southern Ocean (Smith et al. 2007) and presumably reflects the history of the area. Given the comparatively low levels of human exploitation in the Ross Sea, apart from the universal loss of the large baleen whales, it may be the closest we have to an unspoilt Antarctic neritic ecosystem (Ainley 2002).

Evidence for long-term change in Southern Ocean ecosystems

Understanding change in the Southern Ocean is an area of intense research, but for both biology and the physical environment, a central problem is separating short-term fluctuations from long-term trends. A good example here is the use of radiocarbon dating by Emslie et al. (2007) to establish that Adélie penguins have shown substantial expansion and contraction of their distribution in the Ross Sea over the last 45,000 years, with the present colonies being relatively recently established within the last couple of thousand years. In both the physical and biological domains, we are constrained by the lack of long-term datasets. This is particularly true for the Southern Ocean itself, and there has necessarily been a major focus on the use of modelling approaches.

The use of stable isotopes, pollen records and other techniques provides evidence from both the Antarctic Peninsula and Subantarctic Campbell Island that marine and terrestrial temperatures can show strikingly different trends over periods of tens of thousands of years that may be linked to meridional shifts in the position and intensity of westerly winds (Bentley et al. 2009, McGlone et al. 2010).

Physical evidence

Evidence of change comes from air temperatures, changes in fast ice and sea ice, and the ocean itself, the last being the most difficult.

Changes in surface temperatures in Antarctica have been complex over the last 50 years with patterns varying enormously among regions (Vaughan et al. 2001, Kwok & Comiso 2002). Rates and even directions of temperature change have been spatially variable and complex, and stations with long-term records differ in showing significant or non-significant warming or cooling depending on the season (see Figure 2 in Turner et al. 2005). There seems to have been an overall warming, particularly in the Antarctic Peninsula, where warming has been faster than anywhere else on the planet (Clarke et al. 2007, Bromwich et al. 2013), but cooling elsewhere (Vaughan et al. 2001, 2003), and the closest we can come to a generalised picture is one of significant warming in western Antarctica but not the rest of the continent (Turner et al. 2005, Mayewski et al. 2009). Unfortunately, even this turns out to be an oversimplification. There is very considerable natural variability in these systems, and Turner et al. (2016), while acknowledging the brevity of their records, show that temperatures in the Antarctic Peninsula have actually cooled since the 1990s.

There has also been shrinkage and breakup of the ice masses around the continent, again especially in western Antarctica (Mercer 1978, Doake & Vaughan 1991, Vaughan & Doake 1996, Rott et al. 1996, Domack et al. 2005a,b). However, the situation is different for the marginal ice zone. While there has been a dramatic decrease in Arctic sea ice in the last century, both through thinning and a 33% decrease in cover (e.g. Stroeve et al. 2012, Dobricic et al. 2016), the evidence for Antarctic sea ice is less clear (Vinje 2001, Wadhams & Munk 2004). Earlier perceptions of massive shrinkage have been found to be incorrect (Ackley et al. 2003), and most measurements indicate high regional variability (e.g. Stammerjohn & Smith 1997) with a slight, non-significant overall increase in sea-ice cover (Zwally et al. 2002), while indirect measures correlating methanesulphonic acid concentrations in ice cores from the Law Dome with satellite imagery suggest a 20% decline since about 1950 (Curran et al. 2003). Obviously, the situation remains unclear, but the consensus seems to be that, since the late 1970s, there has been an overall increase in ice cover in the Ross Seas, with ice loss in the Bellingshausen and western Weddell Seas (Liu et al. 2004, Parkinson 2004, Moline et al. 2008), and there is no evidence of significant changes in total sea-ice cover since satellite monitoring began (Mayewski et al. 2009). This long-term trend closely resembles the patterns in year-to-year variability in ice and surface air temperature described as the Antarctic dipole, with ice extent being out of phase between the central/eastern Pacific and the Atlantic sectors of the Antarctic (Yuan & Martinson 2001). Modelling exercises suggest that this may be because increases in temperature gradients in one ocean basin strengthen storms in the other (Rind et al. 2001), while Li et al. (2014) make a clear link with temperature changes in the Atlantic. Again, there are reservations about the brevity of the time series, but, as with temperature, there are clearly strong differences in regional effects with regards to ice cover. At this stage, there are no reliable methods for monitoring ice thickness over large scales, though ship-based estimates indicate considerable spatial as well as seasonal variability (Worby et al. 2008).

Regarding the ocean itself, many of the purely oceanographic signals from the Southern Ocean are equivocal or poorly understood because the lack of long-term data makes it difficult to separate trends from cyclical or stochastic change. The Southern Ocean has a globally dominant role in heat uptake.

Partly, this is because heat is rapidly transported northwards so that surface temperatures remain relatively low, allowing heat uptake to continue (Morrison et al. 2016). Gille (2002) provides evidence that mid-depth (700–1100 m) temperatures in the Southern Ocean, particularly along the Subantarctic front, have risen faster than warming in the global ocean, but such direct measurements are rare, and their interpretation difficult. Nevertheless, there are some indications of large-scale change. These include warming and freshening of Antarctic bottom water, which may be driven by different mechanisms in different regions, including erosion of the ice sheet in western Antarctica where surface warming has been especially rapid (see Meredith & King 2005, Mayewski et al. 2009). Gille (2008) later showed that Southern Ocean warming was concentrated within the Antarctic Circumpolar Current (ACC). The data could result from poleward migration of the ACC, though different basins exhibit different patterns of meridional shifts of fronts (e.g. Sallée et al. 2008) but could also be interpreted as overall heat gain within the ocean, possibly linked to increased eddy flux driven by intensified westerly winds (Hogg et al. 2008). Meridional (i.e. north/south) shifts in the positions of frontal systems provide indirect measures of the effects of global warming but are difficult to substantiate because frontal positions can be inherently variable. For example, the position of the Subantarctic front can exhibit short-lived meridional shifts or meanders of 120 km in two weeks (Hunt et al. 2001).

Part of the difficulty in determining the response of the Southern Ocean to global warming is the difficulty in separating climate variability due to intrinsic interactions among ocean, atmosphere, land and sea ice from external drivers such as increased greenhouse gases (Latif et al. 2013). In addition, there is the problem of the lack of historical datasets with dense coverage in time and space. In that sense, the transition from datasets relying on expendable bathythermographs to the development of the Argo float programme in the early twenty-first century has been described by Lyman et al. (2010) as a revolution in ocean observation. The enormous capacity of the ocean for absorbing and storing heat comes partly from its ability to remove heat from the sea surface through mixing and the action of currents (Roemmich et al. 2012). Most heat gain by the ocean is in the southern hemisphere, with important differences among the main ocean basins and regional patterns of heat gain reflecting the effects of ocean dynamics (Roemmich et al. 2012, 2015). Belkin (2009) examined surface warming in the world's 63 large marine ecosystems and identified no warming in Antarctica, in stark contrast to all other systems except the Humboldt and California current systems, which showed mild cooling. Yin et al. (2011) ascribed the lack of warming in the Southern Ocean to the combined effects of upwelling and the ACC in preventing the intrusion of warmer water towards the pole, and indeed Armour et al. (2016) identified marked warming of surface waters immediately to the north of the ACC, while Llovel and Terray (2016) identify maximum warming at about 40°S. Armour et al. (2016) went on to suggest that Southern Ocean surface warming will be driven by the transmission of heat by warmed North Atlantic deep water that ultimately upwells in Antarctica so that the effects of warming will only be seen over timescales of multiple decades. Upwelling in turn is driven by the westerly winds (Marshall & Speer 2012), linking Southern Ocean warming to meridional overturning circulation. This highlights the importance of the Southern Ocean in contemporary and historical climate variability (Marshall & Speer 2012).

Pedro et al. (2016) link deep convection with multicentennial-scale warming events in Antarctica, and De Lavergne et al. (2014) suggest that the recent lack of deep ocean convection following the closure of the Weddell polynya in the late 1970s may indicate a pre/postindustrial era regime shift in the deep ventilation of the Southern Ocean. They regard this as a possible response to anthropogenic effects that could contribute to slow surface warming and enhanced subsurface warming. In contrast to surface waters, the deeper waters of the Southern Ocean have shown significant warming, especially in the deep waters south of the Subantarctic front in basins that are ventilated by Antarctic bottom water (Purkey & Johnson 2010). Gille (2008) found that the upper 1000 m of the Southern Ocean have warmed, especially within the ACC, and attributed this to poleward migration of the ACC, possibly driven by a poleward shift in the westerlies which strengthened by at least 20% between 1980 and 2010 (Gent 2016). Desbruyères et al. (2016) found that the strongest warming has occurred in abyssal waters of the Southern and Pacific Oceans. By comparing pre to post 2000 estimates, they concluded that rates of warming are increasing in the Southern Ocean, with particularly enhanced warming in the Amundsen/Bellingshausen Sea sector. This is presumably due to the zonal inequities described by Landschützer et al. (2015). They describe a tendency towards the development of high pressure systems over the Atlantic/Indian sector of the Southern Ocean and low pressures over the Pacific sector, resulting in stronger meridional warm (Atlantic) or cold (Pacific) winds and increased upwelling in the Pacific sector (see their Figure 3).

Essentially, it appears that heat is accumulating within the Southern Ocean in deep, rather than surface waters and that the Southern Ocean may respond to anthropogenic warming over decadal-centennial timescales (Armour et al. 2016).

Biological evidence

As with physical evidence of change, long-term datasets are necessary to separate the noise of natural variability from possible signal. The best candidates here are probably chlorophyll, which is comparatively easy to monitor across large scales using satellites, and seabirds. The birds have more or less fixed breeding sites, making them much easier to follow than say pelagic invertebrates, and they have been monitored for relatively long periods. Birds also have the advantage of being top predators, which is where we might see long-term trends amplified. An important point here is that polar species may suffer from genetic disadvantages when it comes to adapting to environmental change.

Somero (2010) suggests that ectotherms endemic to the Southern Ocean are likely to be particularly vulnerable to warming. Tewksbury et al. (2008) point out that, like tropical animals, marine polar animals have evolved in a context of little temperature variation. As a result, they are remarkably stenothermal and likely to display DNA decay, the loss of the genetic information required to adapt to increasing temperatures (Somero 2010). Nototheniid fish are an excellent example of this problem, exhibiting such extreme stenothermy that many species die at temperatures only marginally higher than those they normally experience (Beers & Jayasundara 2015). For southern rockhopper penguins, and possibly other long-lived seabirds, there is the additional problem that they display little phenotypic plasticity in breeding phenology, reducing their capacity for microevolutionary adaptation (Dehnhard et al. 2015). In the longer term, this leaves them with the need to respond by dispersal, rather than adaptation (Forcada & Trathan 2009). Interestingly, this contrasts with the findings of Mock et al. (2017) for the polar diatom *Fragilariopsis cylindrus*. This species has an exceptionally high level of allelic divergence, with different alleles being expressed, depending on environmental conditions. Mock et al. (2017) suggest that such a high level of allelic divergence may be an adaptation for the extreme variability of Antarctic conditions, as experienced by phytoplankton.

Boyce et al. (2010) calculated that phytoplankton concentrations have declined globally over the last century. They attribute this largely to increases in sea surface temperature. These have had a negative effect on chlorophyll levels over most of the world's oceans but, surprisingly, a strong positive effect in the Southern Ocean. At similar global scales, Cheung et al. (2013) relate changes in the species composition of fishery catches to the changes in sea surface temperatures in the world's large marine ecosystems described by Belkin (2009). This is unsurprising but indicates the degree to which community composition and ecosystem function track environmental conditions. Several lines of biological evidence of long-term change in Southern Ocean ecosystems have emerged. These include: changes in the phytoplankton community composition in the Western Antarctic Peninsular (WAP); the biogeographic affinities of zooplankton species in the polar frontal zone; the stable carbon isotope signature of benthic shrimps; long-term changes in populations of seabirds of different types (e.g. nearshore versus offshore feeding or ice-loving versus ice-avoiding); extension of the salp (*Salpa thompsoni*) biotope to the south. These lines of evidence cover a wide range of temporal and spatial scales, but all point to dramatic changes in the biogeography of the Southern Ocean and the processes that support Southern Ocean food webs.

Moline et al. (2008) recorded a short-term (5 year) increase in chlorophyll towards higher latitudes in the Southern Ocean, which they suggest may be linked to melting of coastal glaciers and the ice shelf. At smaller scales, Montes-Hugo et al. (2009) found dramatic changes in chlorophyll levels around the WAP over a 30-year period. In this case, the directions of change differed between the north, where chlorophyll levels have decreased, and southerly waters where they increased. The authors linked this to changes in both sea ice and climate in the form of wind-driven mixing and cloud cover leading to a southerly shift in the occurrence of large phytoplankton such as diatoms.

There is evidence of crucial changes in Southern Ocean ecosystems gathered from studies covering large, whole-ocean scales. *Salpa thompsoni* and *Euphausia superba* are characteristic of much of the Southern Ocean. The two generally show strong spatial separation at all physical scales (Nicol et al. 2000, Pakhomov et al. 2002, Pakhomov 2004, but see Ducklow et al. 2007) and are characteristic of different biotopes, effectively representing different modes for the ecosystem. Krill can show enormous interannual variability in abundance. For example, Brierley et al. (1997) found a 20-fold increase between 1994 and 1996 at South Georgia and correlated this with the breeding success of most krill-eating predators. Furthermore, these fluctuations are synchronised over large spatial scales and appear directly linked to environmental conditions including the extent of sea ice, air and sea surface temperatures (Brierley et al. 1999a,b). Although there is such huge background variability in the abundances of planktonic species, there are strong long-term trends. An examination of salp distribution in the 1950s and in the late twentieth century (1980–1998) indicated a pronounced southward shift in the centre of gravity for salp populations, suggesting a warming trend in the high Antarctic (Pakhomov et al. 2002). This trend was confirmed in an analysis of krill and salp populations from 1926 to 2003 by Atkinson et al. (2004). Their study showed increases in salp populations and concomitant deceases in krill populations over very large scales. As salps prefer warmer water than krill, and krill recruitment is linked to sea-ice extent in the previous year, this is indirect evidence of a strong warming trend. This has major implications for regional carbon flux (Perissinotto & Pakhomov 1998b) but also, importantly, for Southern Ocean ecosystems. Such a change is linked with a decrease in the extent of the krill biotope, with enormous implications for the entire Antarctic marine food web (Loeb et al. 1997).

While many Southern Ocean predators feed on other species, especially fish (Barrera-Oro 2002), krill have long been recognised as central to Antarctic food webs, and we would expect reductions in krill abundance to be reflected in populations of their predators. There is evidence of this (Loeb et al. 1997, Murphy et al. 1998), particularly from studies of conspicuous species for which we have relatively long-term population data, especially birds. For example, based on stable isotope analysis of old-ancient eggshells, Emslie and Patterson (2007) concluded that Adélie penguins exhibited a shift to the use of krill as a major part of their diet within the last 200 years as populations of large krill eaters declined. Some authors refer to a 'regime shift' within the Southern Ocean ecosystem, occurring in the 1980s and involving changes in the sea-ice environment, the availability of prey and the population dynamics of species such as penguins, albatrosses, fulmars, fur and elephant seals (Weimerskirch et al. 2003, Jenouvrier et al. 2005a). Apart from the absolute availability of prey, change to the sea-ice regime can disrupt predator–prey relationships by altering the timing of prey availability (Moline et al. 2008). Fraser and Hofmann (2003) suggest direct links among variability in ice cover, krill recruitment and foraging behaviour of Adélie penguins. Such effects may take effect by influencing overwinter survival or the ease with which birds regain breeding condition (Trathan et al. 1996). Broadscale changes in Antarctic seabird populations are thus seen as likely to be direct or indirect responses to climate change (e.g. Cunningham & Moors 1994), but such responses can be complex (Croxall et al. 2002). The effects of changes in sea-ice extent may differ dramatically among populations of the same species (e.g. Jenouvrier et al. 2005b). This is partly because the biological response to such changes may differ between genders or among age classes. Biological responses may also differ among locations and may not be linear, making the population-level consequences of changes in sea ice very difficult to predict. For example, too much sea ice may reduce survival of subadult Adélie penguins in the Ross Sea (Wilson et al. 2001), but too little ice leads to population decreases of the same species in the Antarctic Peninsula (Smith et al. 1999). Trivelpiece et al. (2011) argue that the effects of ice loss are not direct, through loss or gain of winter habitat, but indirect, with changes in penguin populations tracking changes in krill populations. Hinke et al. (2007) conclude that juvenile survival is particularly sensitive to krill availability. The extent of the pack ice also has two opposing effects on emperor penguins. In winter, extensive sea ice increases the distance between the colony and the feeding grounds, reducing hatching success, while in summer, high ice cover increases food availability and improves adult survival. Even then the consequences are not simple, as past reductions in sea ice reduced survival of male emperor penguins more than females (Barbraud & Weimerskirch 2001a). Similarly, Barbraud et al. (2000) showed that 44% of the variation in survival of adult snow petrels was inversely associated with sea-ice extent in July, yet breeding success improved when ice cover was high in the previous November and July–September, so that the same ice conditions can have different effects on performance of the same species (Barbraud & Weimerskirch 2001b). The overall conclusion must be that direct and indirect alterations to sea-ice extent are crucially important but highly complex and variable.

To make matters worse, there are strong indications of the overriding importance of stochastic effects. Roberts et al. (2017) recently published evidence from geochemical analyses of lake sediments that disruption of nesting and foraging by ash deposited by volcanic eruptions has been the key driver of gentoo penguin population fluctuations in the Antarctic Peninsula. They suggest that colonies took hundreds of years to recover following eruptions and found no links between guano deposits (used to estimate population size) and air temperatures, sea temperatures or sea ice across the Holocene.

Reid and Croxall (2001) provide an analysis of long-term trends in populations of krill eaters at South Georgia that highlights the difficulty of separating competing hypotheses on population regulation, and hence on making predictions on future change. The reproductive performance of top predators at South Georgia reflects variability in the regional abundance of krill (Croxall et al. 1999), and Reid and Croxall analysed population data from 1980 to 2000 for fur seals and several birds (macaroni and gentoo penguins and blackbrowed albatross). All four predators showed relatively stable populations during the 1980s, followed by a decline during the 1990s. For blackbrowed albatross and macaroni penguins, this entailed about a 50% decline. There are many possible contributing factors to such declines, but the data provided indicate causes operating on large scales of time and space. A concurrent decline in the predominant size classes of krill in the diets of these predators suggests that the biomass of the larger krill size classes was insufficient to support the needs of the predators during the 1990s. Krill numbers at South Georgia depend on influx from the Antarctic Peninsula, growth and local mortality, and Reid and Croxall suggest that changes in mortality rates of krill are responsible for changes in its population structure, as mortality within the foraging range of the predators is much higher than the overall average. The authors suggest that there was a period when high krill availability following the removal of the great whales supported large populations of other krill eaters, but that this period may be coming to an end, while changes in krill populations around South Georgia are amplified by the more local effects of predation. Krill abundance correlates well with physical factors, especially ice cover and sea temperature (Brierley et al. 1999a,b, Trathan et al. 2003), and physical conditions must ultimately define habitable conditions for any species, but it is clear that biological interactions produce important complications and provide feedback mechanisms in the regulation of both predator and prey populations. In fact, simple abundance of prey may not be the only relevant factor. Ruck et al. (2014) examined the quality of Antarctic krill and fish as prey around the WAP. They found that *Euphausia antarctica* had the highest lipid content of the five species they studied, but they also found that *E. superba* from more southerly waters had markedly higher lipid content than those from farther north. They suggest that this may be related to higher microzooplankton biomass in inshore waters (Garzio & Steinberg 2013). The results indicate clear geographic differences in the quality of this major prey species but also indicate that prey quality may alter over long and short timescales.

Several other lines of evidence emerge from smaller-scale studies at the Subantarctic Prince Edward Islands. The impact of global climate change on the terrestrial ecosystem of these islands and its interaction with invasive species are relatively well documented (Smith 1991, 2002, Smith & Steenkamp 1990, Chown & Smith 1993, Bergstrom & Chown 1999, Greve et al. 2017), but evidence of change in the marine ecosystem is more recent and is largely linked to a southward shift in the average latitude of the Subantarctic front. The strongest evidence concerns long-term correlations between changes in the average meridional position of the Subantarctic front and changes in seabird populations.

The Prince Edward Islands lie within the polar frontal zone (PFZ), the region between the Antarctic polar front and the Subantarctic front. Longhurst (1998) places the PFZ firmly within his Subantarctic water ring province, with its boundaries at the Subtropical Convergence (35°S) and the Antarctic polar front (50–55°S). Nevertheless, there is good reason to consider the PFZ separately as a region of transition between the low-productivity waters of the Subantarctic and the more productive Antarctic surface waters, so that the PFZ effectively acts as an ecotone, with no endemic biota (Froneman & Pakhomov 1998). Spatial variability in species composition of the zooplankton in this zone is often extreme and reflects the periodic intrusion of subtropical surface waters from the north and of Antarctic surface waters from the south (Pakhomov & Froneman 1999b, Hunt et al. 2001, Bernard & Froneman 2003, Froneman & Bernard 2004), resulting in the advection of tropical and Antarctic species into the region. As a result, the zooplankton community includes animals with different biogeographic affinities such as subantarctic, subtropical and antarctic crustaceans (Pakhomov et al. 2000a, Hunt et al. 2001, Bernard & Froneman 2003, Froneman & Bernard 2004).

Over the past two decades, there has been a change in the biogeographic affinities of the zooplankton species found near the islands. The contribution of Antarctic species to total zooplankton species richness has decreased by some 20%, while the contribution of subtropical species has increased from 6 to 26% (Pakhomov et al. 2000a). This shift in species composition of the zooplankton is believed to be associated with an increase in the frequency of intrusions of warmer waters into the PFZ (Pakhomov et al. 2000a). In addition to changes in the zooplankton species composition, Hunt et al. (2001) suggested that more frequent intrusion of Subantarctic surface waters into the PFZ will coincide with a decrease in the total zooplankton biomass in the vicinity of the islands, as Subantarctic surface waters are generally characterised by low productivity (Hunt et al. 2001, Bernard & Froneman 2003). The implication is of changes in biotope extent as this frontal system loses its effectiveness as a biogeographic boundary.

Looking more closely at the Subantarctic front, it represents a high-speed region of the Antarctic Circumpolar Current (ACC) (Hofmann 1985). The geographic position of the front is, of course, not fixed, but it has shown a gradual drift to the south since the 1970s (Allan et al. 2013), and a change in its average position alters hydrodynamics around the Prince Edward Islands. The high chlorophyll values associated with Southern Ocean Islands are partly explicable in terms of iron fertilisation (Longhurst 1998), but phytoplankton production can be limited by multiple, interacting factors (Boyd 2002), and at the Prince Edward Islands, there is evidence that local phytoplankton blooms are also partly driven by the interaction of the ACC with bottom topography. Water column stabilisation and the retention of guano run-off from the islands that is rich in reduced nitrogen both contribute to high productivity (Perissinotto et al. 1990, Perissinotto & Duncombe-Rae 1990). A southward change in the mean position of the Subantarctic front suggests a shift in the balance between advective and frictional forces at the islands, with advective forces becoming more powerful. This should result in weakening of water retention over the shallow waters around the islands, so that local diatom blooms become rarer and overall primary production in the immediate vicinity of the islands decreases. Zooplankton grazing on these blooms tends to be weak, and most of their production sediments out into a shallow interisland plateau at about 200 m, contributing to the support of a diverse benthic community (Perissinotto 1992, Branch et al. 1993, McQuaid & Froneman 2008).

A key element in the nearshore ecosystems of these islands is the caridean shrimp *Nauticaris marionis*. *Nauticaris marionis* is an opportunistic predator feeding on a variety of prey including detritus, benthic organisms and plankton (Perissinotto & McQuaid 1992 Vumazonke et al. 2003), and it forms an important prey item for a number of land-based seabirds on the islands, so that it helps to couple the benthic and pelagic subsystems (Perissinotto & McQuaid 1990, Kaehler et al. 2000). The $\delta^{13}C$ signature of this shrimp decreased significantly over the period 1984–2009 (Pakhomov et al. 2004, Allan et al. 2013), suggesting a shift in the primary carbon sources it utilises. This is likely to reflect a reduction in productivity near the islands, particularly a reduction in localised diatom blooms (Pakhomov et al. 2004). Although there are no data on changes in the size of *N. marionis* populations, a reduction in diatom blooms is also likely to lead indirectly to a decrease in the biomass of *N. marionis*.

A southward shift of the Subantarctic front has different implications for the various land-based predators. Those living on the islands but foraging at the front should benefit because foraging distances decrease. In contrast, nearshore foragers, depend on allochthonous production carried towards the islands by the ACC, and indirectly on local production via *N. marionis*. These predators would experience a reduction in food availability because local productivity is expected to decrease, while advection past the islands increases. There is strong evidence from long-term changes in bird populations that exactly these ramifications are occurring towards the top of the food chain. During the 30-year period of southward migration of the front, offshore feeding birds such as grey-headed albatrosses and northern giant petrels, that are likely to feed at the front, have shown medium- to long-term population increases. At the same time, the inshore-feeding rockhopper penguins have decreased in numbers, and populations of the macaroni penguin, which has a mixed foraging strategy, have remained relatively stable (Allan et al. 2013). Similarly, Inchausti et al. (2003) found that changes in oceanographic conditions had opposite effects on population trends in seabirds with different feeding strategies in the Crozet/Kerguelen area, even between species that were closely related. Budd (2000) reports long-term recolonisation of Heard Island by both king penguins and fur seals, so that long-term shifts in the ecology of Subantarctic islands appear to be the norm, even if the mechanisms involved are not always well understood.

The observed changes in zooplankton community structure, isotopic signature of *N. marionis* and the dynamics of bird populations all suggest that global climate change has already influenced the biology of the region in quite profound ways. The consequences of a change in physical conditions, driven by latitudinal shifts in the position of the southern boundary of the ACC, have repercussions that reverberate all the way to the top of the food chain.

Future climatic change and consequences

There is clear consensus that change is coming to the Southern Ocean but considerable uncertainty about how exactly that change will manifest, and it is important to bear in mind the dangers of making predictions based on short-term studies of one or a few species (Clarke et al. 2007). For example, life-history characteristics will have important effects on species responses in the longer term; seabirds tend to evolve slowly (Barbraud et al. 2012), while salps show very rapid evolution (Jue et al. 2016).

Sigman and Boyle (2000) found that parts of the Southern Ocean have reacted very differently to climate change and glacial-interglacial cycles. Modelling studies by Marinov et al. (2006) highlight the consequences of regional differences in the efficiency of carbon sequestration by the biological pump and the production of different deep-water masses in different parts of the Southern Ocean. Low efficiency of the biological pump and the formation of deep water in the high-latitude Antarctic have a profound influence on the air–sea carbon balance, while levels of low-latitude export production are driven by the Subantarctic. Climate change is likely to affect the high Antarctic and the Subantarctic differently.

Ainley et al. (2007) argue that there has been an unchallenged paradigm shift concerning the functioning of Southern Ocean ecosystems from an understanding that top-down regulation was of key importance to an almost exclusive focus on bottom-up forcing. In fact, ecosystems are rarely controlled by one or the other but by the balance between them, which can shift in both time and space (e.g. see Menge (2000) for intertidal systems). Either form of forcing can dominate under the right circumstances, and the ecologically rapid removal of finfish and the great whales from Southern Ocean ecosystems was the sort of massive perturbation that would lead to major adjustment within the system. The types of environmental changes now taking place are another, and because their effects will be felt at the base of the food web, they will undoubtedly lead to major re-balancing of these systems. Croxall et al. (2002) reviewed the data for Southern Ocean seabirds and concluded that such re-balancing is likely to be rapid, rather than gradual.

One of the main problems with attempting to understand how ecosystems will respond to climate change is predicting how species interactions will be affected (e.g. Winder & Schindler 2004). In the case of highly seasonal systems such as the Southern Ocean, where annual primary production is largely limited to a narrow window of time, this is especially important in terms of mismatches between prey availability and predator needs (Moline et al. 2008). To a large degree, this will depend on whether the phenology of predator and prey populations respond to different environmental cues or respond at different rates to the same cues (Visser et al. 2004).

Even in terms of physical conditions, we have a much better understanding of some regions than others, but given the differences among regions (or even subregions) in terms of how their ecosystems function and how conditions are already changing, it is simply not possible to extrapolate among regions. Indeed, there is even uncertainty about what is driving observed changes in the biology of even the best-studied regions (Melbourne-Thomas et al. 2013). How things will change and indeed how they are already changing depends on where you are. Perhaps it is naïve to expect that we can anticipate or predict change in an ecosystem that is larger than the African continent. Where and when research is conducted in the Southern Ocean is profoundly shaped by logistics, and the key must be to develop a better understanding of the considerable extent of the Southern Ocean ecosystem(s) that are presently poorly studied. It is also important to recognise the rapidity with which our understanding of these systems is changing and that much remains controversial.

Climate change implies potentially critical alterations to physical conditions in the Southern Ocean with enormous implications for an ecosystem that is very strongly physically forced. Climate change alters several environmental parameters simultaneously in complex ways, and a major problem in understanding possible effects on ecosystems is that the various environmental stressors applied to any biological system do not operate in isolation from one another. Rose et al. (2009) tested the effects of iron availability and temperature together on the Southern Ocean phytoplankton and microzooplankton that lie at the base of the food web. The result was a synergistic interaction that magnified the effects of both. Importantly, this synergy could not have been predicted by examining either stressor in isolation. The situation is made more complicated because the ways in which environmental conditions are changing differs among regions, and the balance of synergistic and antagonistic effects is difficult to forecast. Boyd et al. (2015) took a modelling approach to reveal an enormous range of possible responses of global phytoplankton to anticipated environmental change. They illustrated the problem by noting differences in anticipated environmental changes in the north and south Southern Ocean and then predicting the responses of coccolithophores and diatoms in the north Southern Ocean. The results, based on complex interactions of silicate depletion, warming, higher CO_2 and increased iron availability, indicated a poleward increase in the coccolithophore biome. They highlight the fact that accurately predicting the biological consequences of climate change is not possible at this stage. Likewise, Deppeler and Davidson (2017) concluded that how phytoplankton communities will respond to long-term change will depend not only on the degree of environmental changes but also on the sequence in which stressors take effect so that responses will differ even within regions.

Gutt et al. (2015) avoided making predictions about biological responses but produced maps of where combinations of stressors are likely to occur in the Southern Ocean. Some of the environmental changes they considered, such as the effects of icebergs on pelagic communities, are likely to affect relatively restricted areas, while others such as acidification, seasonal ultraviolet (UV) increase and surface water warming will affect much of the Southern Ocean. They also acknowledge that, even in their simplest form, the relationships among environmental variables and the biota are complex, while in reality, biological interactions will be superimposed on this complexity. This would include not only interactions within a re-balanced community of native species but could also include the arrival of species that were previously excluded by low temperatures, such as the predatory king crabs reported in western Antarctica (Smith et al. 2011) or the eddy-mediated advection of the red tide flagellate *Noctiluca* into the Southern Ocean (McLeod et al 2012). The latter seems to have been a stochastic event rather than a range expansion, but the appearance of novel species has the potential for dramatic consequences.

Overall, the possible combinations of interacting environmental and biological effects are almost limitless. They range from the interaction of decreasing UV-B exposure due to reduction in the ozone hole (Newman et al. 2006) and long-term reductions in sea ice (Bracegirdle et al. 2008) to the possibility that changes in the transmission of sound through water caused by changes in pH (Brewer & Hester 2009) will affect echolocation and communication among marine mammals. Perhaps the most important conclusion from Gutt et al. (2015) is that the areas expected to experience multiple environmental changes are substantially larger than those that have already experienced important changes.

Nevertheless, with the proviso that events invariably turn out to be more complicated than anticipated, particularly when biological interactions are involved, several major changes can be predicted that will involve both large-scale and more localised effects. As with the animals, a few phytoplankton species play key roles in Southern Ocean ecosystems (Smetacek et al. 2004), and many of the anticipated changes involve alterations to the structure or productivity of the phytoplankton community, with subsequent effects on the entire pelagic ecosystem. Of even greater significance is that many of these effects have important implications for the efficiency of the biological pump and hence have the potential to form critical feedback loops among climate change, planktonic community structure and carbon drawdown.

Obvious effects are: a reduction in sea-ice cover; a rise in sea temperature; increased water column stability associated with warming changes in wind regime; decreasing pH. These and several other effects are considered below.

Reduction in sea-ice cover

Antarctic sea ice cover reached maximum levels for the last 50 years in 2012–2014, at virtually the same time that Arctic ice reached minimum values (http://www.nasa.gov/content/goddard/nasa-study-shows-global-sea-ice-diminishing-despite-antarctic-gains). Although the duration of the sea ice season shows contrasting patterns between decreasing in the Antarctic peninsula, while increasing in the Ross Sea (Stammerjohn et al. 2012), it is difficult to imagine that large-scale loss of Southern Ocean sea ice will not occur in the long term. Bracegirdle et al. (2008) project a loss of 2.6 million km^2 during the present century. Ice retreat would reduce the influence of the marginal ice zone, with major implications for krill biology and knock-on effects for the entire food web. Large-scale ice retreat could also involve a reduction or even loss of specialist biotopes such as the polynyas on which several species depend, including *Euphausia crystallorophias* (Pakhomov & Perissinotto 1996) and seabirds such as snow petrels, Antarctic petrels and emperor penguins (Croxall 2004). At this stage, there is no evidence of large-scale changes in total ice cover (Mayewski et al 2009). In fact, Parkinson (1990) found no evidence for a substantial difference in the ice edge between now and 200 years ago. There have, however, been the large changes within regions, described previously (Parkinson 2004).

The pack ice is a major characteristic of the Southern Ocean and undergoes enormous seasonal changes in extent, from a minimum of 4×10^6 km^2 in summer to about 20×10^6 km^2 in winter (Comiso et al. 1993), when it reaches to about 58°S in August–October (Longhurst 1998). The pack ice has important physical effects on the flux of heat and gases between the sea and the atmosphere and provides a unique habitat for a range of organisms including brine diatoms, krill and seals. Not only is the ice important in itself, it is important through its effects during ice retreat. The marginal ice zone represents an area of millions of square kilometres in which unique conditions of water column stability, salinity and epontic algae exist. Ice retreat triggers a spring bloom of production that follows water column stabilisation through the release of melt water. Ice melt also seeds the water column with algae released from the ice (El-Sayed 2005), and possibly releases iron (Sedwick et al. 2000). The net effect is to make a significant contribution to total annual primary production. Nevertheless, it has been suggested that a reduction in sea-ice cover would result in an overall increase in primary production (Holm-Hansen & Sakshaug 2002), and modelling exercises by both Sarmiento et al. (2004b) and Arrigo and Thomas (2004) support this. The models of Arrigo and Thomas predict that a drop of 25% in sea-ice cover would lead to an increase of 10% in total Southern Ocean primary production, as the relatively unproductive sea-ice habitat would be replaced by the more productive marginal ice zone habitat and permanently open ocean. However, sea-ice algae are important not only in terms of their contribution to total Southern Ocean production but because they provide food in places and at times when water column production is low (Lizotte 2001). A reduction in ice cover would have enormous consequences for species that depend on ice algae either directly or indirectly, which is effectively to say almost the entire food web.

The most significant ecological effect of a reduction in the extent of sea ice would be on krill abundance. Although they are invertebrates, krill are among the largest pelagic crustaceans (Nicol 2003) and in some ways function in a manner similar to fish. They school (Nicol 2003), they are relatively long-lived (Siegel 1987) and interannual abundances largely reflect variations in year-class strength (Loeb et al. 1997). This means that populations tend to be sustained by strong cohorts or year classes, so that when recruitment fails, senescence of the dominant age class can be important (Fraser & Hofmann 2003). Good recruitment is linked to early spawning and is correlated with sea-ice cover in the preceding year, reflecting the dependence of krill on ice algae as food during winter (Frazer et al. 2002). Loeb et al. (1997) suggested that krill may compete for food with salps (*Salpa thompsoni*), but other studies indicate spatial segregation (e.g. Nicol et al. 2000, Pakhomov 2004), as discussed previously, and the occurrence of krill and salp years around Elephant Island seems to be directly linked to sea ice conditions, that is, to the biotope that prevails rather than to competition. It is interesting that gelatinous salps have shown dramatic increases in the Southern Ocean (Atkinson et al. 2004), while gelatinous zooplankton in the form of medusae have shown similarly dramatic increases in abundance at the other end of the Earth in the Bering Sea (Brodeur et al. 1999).

Because krill and salps have quite different effects on carbon sequestration rates, long-term shifts towards a predominance of salps would imply improved carbon sequestration due to the large-scale vertical migration of salps and their formation of fast-sinking faeces (Wiebe et al. 1979). At the same time, a reduction in krill stocks has obvious negative implications for populations of krill eaters. Based on a global database, Cury et al. (2011) concluded that if the supply of 'fish' (including Southern Ocean krill) drops below approximately one-third of the maximum seen in long-term studies, seabird populations suffer reduced breeding success.

Rising sea temperatures

Clarke (1993) provides evidence that, despite its known effects on viscosity, gas solubility, carbonate solubility and so forth, temperature has rarely if ever caused the extinction of marine species through direct physiological effects. We do, however, need to bear in mind that many Southern Ocean species, including various molluscs and nototheniid fish, exhibit extreme stenothermy (e.g. Peck et al. 2004,

Beers & Jayasundara 2015). Instead, Clarke argues, extinctions are likely to be caused by ecological effects, sometimes quite subtle, influenced by temperature and other factors. These could include effects such as changes in breeding phenology (e.g. Barbraud et al. 2012), which could uncouple predator/prey relationships. One example of indirect temperature effects comes from Hill et al. (2013), who modelled the effects of rising sea surface temperature on the summer habitat of krill. The effects were mainly negative but could be reduced by localised increases in primary production. In fact, the authors point out that very high abundances and growth rates of krill occur around South Georgia, which is towards the species' northern limit. Physiological stress is expected to be high there but is compensated for by high food availability. Nevertheless, this represents a cost/benefit balance. Tagliorolo and McQuaid (2016) found that the intertidal mussel *Perna perna* can survive where temperatures are low, but food availability is very high, but it grows slowly there and is unable to breed, forming a sink population. Increased stress from rising temperatures could have similar sublethal consequences for krill around South Georgia. Critically, the modelled reductions in krill habitat would affect food availability within the foraging ranges of top predators that breed on South Georgia at a time when their energy demands are particularly high.

Certainly, rising sea temperature is likely to have profound indirect effects on the biota of the Southern Ocean. An overall warming is likely to result in a southward drift in the average positions of the circumpolar frontal systems. Southward shifts in frontal positions will move biogeographic boundaries, compressing cold-water species while allowing warmer-water species to expand their distributions (see previous section regarding the polar frontal zone). However, this effect also has implications for the efficiency of the biological pump because the dominant primary producers are determined largely by conditions of turbulence and nutrient availability and differ regionally (Longhurst 1998). This in turn has implications for which taxa of grazers can utilise primary production and so for food chain length and structure, as well as whether phytoplankton populations are likely to be grazer limited. Shifts between long and short food chains, based on the size of the organisms dominating the phytoplankton, occur both spatially and between seasons. Such shifts are a major characteristic of the Southern Ocean. As we have seen, long food chains based on small phytoplankton and involving a strong microbial loop with considerable recycling of material in surface waters are characteristic of the permanently open ocean and winter conditions, with short, diatom-based food chains dominating under more neritic and frontal conditions in summer (e.g. Pilskaln et al. 2004). Since the early twentieth century, it has been recognised that special communities of marine phytoplankton characterise particular water masses (Smayda & Reynolds 2001), and there is a strong link between environmental conditions of turbulence and nutrient availability and the taxa that dominate the phytoplankton (e.g. Margalef 1978). A model produced by Irwin et al. (2006) indicates that the link between environmental conditions and phytoplankton community structure can be explained by scaling up from individual cell nutrient requirements. The model does not depend on competitive interactions or grazing and successfully predicts the dominance of small taxa under oligotrophic conditions, with an increase in larger taxa under eutrophic conditions. This suggests that these communities are shaped primarily by resource availability, rather than biological interactions, although we know that local cell removal by grazing can exceed production (e.g. Dubischar & Bathmann 1997). A shift between short food chains with grazing of large cells dominated by the mesozooplankton and longer food chains based on picoplankton primarily grazed by small zooplankton has implications for patterns of energy transfer and the efficiency of the biological pump. On more local scales, episodic upwelling of nutrient-rich upper circumpolar deep water around the Antarctic Peninsula appears to be driven by non-seasonal physical forces and results in site-specific domination of the phytoplankton community by diatoms (Prézelin et al. 2004). This is consistent with the hypothesis that climatically induced changes in oceanic mixing have altered nutrient availability in the euphotic zone and driven macroevolutionary shifts in the size of marine pelagic diatoms through the Cenozoic (Finkel et al. 2005). In other words, changes to the mixing regime could ultimately alter size composition of phytoplankton not only through short-term ecological effects but arguably in the long term through the evolution of body size.

There are also more direct consequences of shifts in the taxa dominating primary production. Grazing pressure on phytoplankton decreases with size (Smetacek et al. 2004), and Smith and Lancelot (2004) suggest that the smaller phytoplankton of the Southern Ocean are controlled by combinations of light, iron and microzooplankton grazing, while the larger phytoplankton of the high nitrogen low chlorophyll (HNLC) regions are bottom-up controlled. A survey of the literature on environmental control of phytoplankton in the Southern Ocean by Boyd (2002) indicates that the key determinants across the various regions (other than grazing) are light, iron and silicic acid but that their interactions vary in time and space and among taxa. Among these factors, the one most likely to be directly affected by climate change is light availability, through water column stratification as rising temperatures are likely to result in shallowing of mixed layer depths globally (Boyd et al. 2015). This will also reduce upward nutrient flux. The modelling exercises of Sarmiento et al. (1998) imply the possibility of dramatic changes in phytoplankton community structure because of increased stratification. Arrigo et al. (1999) found that, in the Ross Sea, phytoplankton community structure was closely linked to mixed layer depth. Diatoms dominate areas with strong, shallow stratification while the prymnesiophyte *Phaeocystis antarctica* dominates deeply mixed areas. There is, however, a dramatic shift in community composition from *Phaeocystis* to diatoms (mainly *Nitzschia recurvata*) as the water column stabilises and the mixed layer becomes shallower over the course of about two weeks. In the context of predictions of increased stability, this has several important implications. Firstly, *Phaeocystis* and diatoms dominate the phytoplankton of both austral and boreal polar seas (El-Sayed & Fryxell, 1993, El-Sayed 2005), so the effects are likely to involve huge areas. Secondly, sedimentation of material from *Phaeocystis* blooms renders the biological pump particularly efficient (Hamm 2000). Arrigo et al. (1999) found that CO_2 drawdown and new production were both much lower for diatoms than for *Phaeocystis*, so that a phytoplankton community dominated by diatoms will be less efficient at atmospheric CO_2 drawdown. This would presumably result in a positive feedback between atmospheric CO_2 accumulation, water column stabilisation and decreased CO_2 drawdown. Thirdly, DiTullio et al. (2000) found that carbon export to deep waters is particularly early and rapid for blooms of *P. antarctica*. Sequestration is episodic but not necessarily linked to senescence, as previously supposed (Wassmann et al. 1990), and studies from the North Sea indicate that organic matter derived from *Phaeocystis* blooms can degrade while floating or in suspension so that it does not actually reach the sediment (Hamm & Rousseau 2003). Nevertheless, as with a change in the preponderance of krill and salps, a shift from a community dominated by *Phaeocystis* to one dominated by diatoms has enormous implications for the efficiency of the biological pump as well as for grazers.

Changes in wind regimes

Wind patterns are likely to change as meridional temperature gradients diminish (IPCC 1995), and Pierrehumbert (2000) has speculated that an extreme consequence of climate change could be a breakdown of the trade wind system. Others, however, predicted a strengthening and southward shift of wind stress in the Southern Ocean, and there are now clear indications of this (Marshall 2003, Fyfe & Saenko 2006). Some possible effects of changes in wind regime are quite specific, such as impairment of the foraging ability of albatrosses, which have evolved flight strategies dependant on predictably high wind stress (Weimerskirch et al. 2000). Other effects are much more fundamental and of global significance. For example, major current systems such as the Antarctic Circumpolar Current would be affected (see below), while the formation of Antarctic bottom water is believed to depend strongly on the winds that blow over the Southern Ocean (Rahmstorf & England 1997). Alterations to wind systems have obvious implications for mixing of the water column and levels of primary production. In fact, they have implications for the entire structure of the planet's oceanic biomes, as Longhurst's four primary biomes include westerlies and trades. We have seen that Southern Ocean primary production is generally limited by the availability of light, iron and silicic acid (Boyd 2002). Primary productivity in perhaps half the world's oceans may be limited by iron availability (Moore et al. 2001), and we can consider the ecosystems of Southern Ocean habitats as falling into two broad categories: those in which iron is not limiting to primary production and those in which it is (Smetacek et al. 2004).

Where (or when) iron is in ample supply, primary production is driven by diatoms or *Phaeocystis*, which can form the large phytoplankton blooms that support relatively short food chains with krill and large copepods as primary consumers and large species as apex consumers. This differs from those parts of the open ocean that are not supplied with iron and phytoplankton communities are dominated by pico and nanophytoplankton. Grazing of these by salps and small zooplankton, such as copepods and pteropods, is high, and the food chains they support tend to be longer, with considerable recycling of material through the microbial loop. Smetacek et al. (2004) suggest that these food webs may support unidentified higher-level organisms, perhaps squid, with smaller populations than those of species supported by iron-replete systems.

It follows that changes to the supply of iron could be among the most profound changes in Southern Ocean ecosystems (e.g. Olson et al. 2000). Global warming will not change the availability of iron but may affect its supply if wind stress decreases. Iron is supplied either from adjacent land masses, through upwelling or as wind-blown dust, with the latter forming the main input of iron to the ocean (Jickells et al. 2005), implying links among glacial/interglacial cycles, dust supply and the efficiency of the biological pump (Kumar et al. 1995). Both supply of dust and upwelling are likely to be subject to long-term change under a regime of global warming as upwelling in the Southern Ocean girdling Antarctica is wind-driven, so that small changes in wind stress around the continent could have marked effects (Pierrehumbert 2000). Reductions in wind stress should lead not only to increased stabilisation of the water column but also to a reduction in iron input and a broad shift towards food chains based on small phytoplankton. This implies increased importance of the microbial loop, less efficient carbon sequestration and loss of biomass among the largest consumers.

Alterations to major current systems

Comparatively small-scale changes in coastal currents following thinning of sea ice have the potential for erosion of the ice shelf in the Weddell Sea (Hellmer et al. 2012), but even open-ocean current patterns are likely to change as density gradients alter in response to warming, increased freshwater input through ice melt and changes in wind. In fact, Ducklow et al. (2007) suggest that ocean circulation is the mechanism that will link climate change with ecosystem changes. As discussed, Fyfe and Saenko (2005) and Gille (2008) suggest that recent decades have seen a poleward shift in the mean position of the Antarctic Circumpolar Current (ACC) in response to changes in the wind regime driven largely by the actions of man. Advection by current systems is of central importance in Southern Ocean ecosystems, and the ACC plays a major role partly because it is responsible for large-scale transport of zooplankton, including krill (Hofmann & Murphy, 2004). Transport by the ACC is largely driven by direct or indirect wind effects (Olbers et al. 2004), and several ecological effects in western Antarctica have been linked to meridional shifts in the ACC. For example, Trathan et al. (2003) found changes in krill abundance at South Georgia to be consistent with meso-scale or large-scale shifts in the southern ACC front. Meridional shifts in the position of the ACC would affect land–sea interactions; there are at least two examples of this. First, upwelling of upper circumpolar deep water around the Antarctic Peninsula drives the phytoplankton community to diatom domination, as described, and this upwelling has been linked to the position of the southern boundary of the ACC relative to the shelf edge (Prézelin et al. 2004). Second, the formation of both warm and cold core surface anomalies west of the Prince Edward Islands is caused by the interaction of the ACC with the Southwest Indian Ridge (Ansorge & Lutjeharms

2003, 2005). These eddies drift in an easterly direction towards the islands and carry high biomass of macrozooplankton (mainly euphausiids and amphipods) and micronekton (Pakhomov et al. 2003). Essentially, they originate from a frontal system with high biomass and carry this frontal signature with them in their biology as well as in their temperature and salinity (Pakhomov & Froneman 1999a, Bernard & Froneman 2003). High-prey biomass makes these features, especially their edges, important foraging grounds for some of the top predators found at the islands, such as the grey-headed albatross, *Thalassarche chrysostoma* (Nel et al. 2001). The importance of eddies as feeding grounds has likewise been shown for frigatebirds in the Moçambique Channel (Weimerskirch et al. 2004). A shift in the average position of the ACC will affect its interaction with bottom topography, with downstream consequences for island-based top predators.

Similarly, changes in the position of frontal systems will have important consequences for animals constrained to a fixed geographic location. Many predators forage at frontal systems and poleward shifts in their average positions will have critical implications for those that breed on land. Péron et al. (2012) estimate that by 2100 breeding and incubating king penguins at Crozet Island will have to travel twice the present distance to forage at the Antarctic polar front, with obvious implications for breeding success. We have already seen the effects of frontal shifts on seabird population at the Prince Edward Islands (Allan et al. 2013), and this is an excellent example of how species may be affected quite differently by climate change (e.g. Barbraud et al. 2011).

Ozone depletion and ultraviolet

Depletion of ozone in the lower stratosphere has been strongest over Antarctica, and although the ozone layer is now expected to recover (Perlwitz et al. 2008), this seems to be associated with changes to the climate of the southern hemisphere, including Antarctica, through its effects on the circulation of the Southern Ocean super-gyre (Cai 2006) and the Southern Annular Mode (Thompson et al. 2011, Manatsa et al. 2013). Thompson et al. (2011) suggest that trends in the Southern Annular Mode linked to the ozone hole may have contributed to observed changes such as shifts in the positions of the major fronts but are not clearly related to sea ice conditions. Ozone depletion allows increased ultraviolet (UV) radiation to reach the planet's surface, but other climate-driven changes, such as changes in albedo or water column quality and light penetration, may be even more important in determining UV-B levels and the balance between UV-B and photosynthetically active radiation (Häder et al. 2003). Nevertheless, depletion of atmospheric ozone is a major concern and has continued for over 20 years (Karentz & Bosch, 2001); the ozone hole over Antarctica during September 2006 was both the deepest and largest on record, covering an area of approximately 27.5×10^6 km^2 (NASA 2017). UV has direct negative effects on zooplankton, including euphausiids (Damkaer & Dey 1983), but concern with the indirect effects of ozone depletion on marine ecosystems has centred on the primary producers, and despite methodological problems discussed by Prézelin et al. (1994a,b), the negative effects of UV irradiation on the phytoplankton community have been well studied (Malloy et al. 1997). Smith et al. (1992) showed that the effects of increased UV-B flux were species-specific; *Phaeocystis* spp. showed much greater inhibition of growth than the diatom *Chaetoceros socialis*. Consequently, increased UV-B may not simply lead to increases in photoinhibition and reductions in photosynthesis, it may also cause shifts in species composition of the phytoplankton. Similarly, Malloy et al. (1997) provide evidence of DNA damage due to increased UV-B flux in fish larvae and krill and suggest that this may reduce recruitment in species with early life-history stages in the water column during spring when ozone depletion is greatest. Despite the obvious effects at the scales studied, and the possibility of subtle long-term effects on the ecosystem, there has been no evidence of large-scale catastrophic effects (Karentz & Bosch, 2001). Arrigo et al. (2003) modelled the hemisphere-scale seasonal effects of a deep ozone hole on primary production in the Southern Ocean and concluded that, although UV-induced loss of surface production was severe, when integrated to the 0.1% light depth, the effects were minimal. This was because of attenuation by the water column and sea ice and because, happily, maximum ozone depletion occurs in the months when sea-ice cover is greatest. Presumably longer-term decreases in sea-ice cover would reduce this mitigating effect on the consequences of high-UV irradiation so that the interaction of reduced ice cover and reduced ozone would have stronger negative effects on the phytoplankton.

A more indirect effect of ozone depletion concerns dimethyl sulphide (DMS). DMS is critical in cloud formation over the sea (Andreae & Crutzen 1997) and is produced by phytoplankton in response to various forms of stress, including attack by bacteria or viruses (Malin et al. 1998, Burkill et al. 2002). UV affects several processes controlling DMS concentration in seawater, including DNA damage to phytoplankton and bacteria, and a study by Kniveton et al. (2003) has shown a negative correlation between UV intensity and DMS production around Amsterdam Island in the Southern Ocean. The effect is independent of sea temperature, wind and photosynthetically active radiation (PAR), and the authors tentatively suggest that increased UV flux may result in decreased atmospheric DMS levels. As with most attempts to anticipate the consequences of long-term change, the effects of increased UV irradiation may be made more complex and difficult to predict because there is evidence that UV can cause DNA damage in marine viruses (Wilhelm et al. 2003) and affect protein synthesis in bacteria (Hernández et al. 2006), so that there be interactions among UV, viruses, bacteria, phytoplankton and DMS production.

Ocean acidification

Gutt et al. (2015) consider ocean acidification to be one of the major threats to the Southern Ocean because, according to the models of Orr et al. (2005), virtually the entire ocean will be affected. CO_2 oscillations have been implicated in the Permian Triassic mass extinction (Berner 2002), and it is estimated that changes to ocean pH over the next few centuries could exceed any seen over the last 300 million years (Caldeira & Wickett 2003). Increasing concentrations of CO_2 in surface waters will have effects not only on rates of primary production but also on water-breathing marine animals, especially pelagic species that are adapted to low CO_2 levels (Pörtner et al. 2004).

The aragonite and calcite horizons represent the depths below which organic precipitation of calcium carbonate becomes difficult or impossible, and these horizons are rapidly becoming shallower, in some places at rates of 1–2 m per annum (Guinotte & Fabry 2008), with parts of the Southern Ocean being particularly problematic (Feely et al. 2004). Orr et al. (2005) focussed on changes in surface pH in polar and subpolar regions and predicted under-saturation of aragonite throughout the entire Southern Ocean by 2100, a much shorter timescale than generally accepted (e.g. Caldeira & Wickett 2003). Subsequently, McNeil and Matear (2008) included seasonality to calculate that, given the IPCC IS92a scenario used by Orr et al. (2005), under-saturation of aragonite in the Southern Ocean in winter can be expected as early as 2030–2038, though with a substantial delay north of the Antarctic polar front. This clearly has huge implications for key planktonic species such as the pteropod *Limacina helicina*, which exhibits veliger development during winter.

In fact, ocean acidification has implications for all calcareous organisms, and increasing atmospheric CO_2 concentrations can have direct effects on the phytoplankton. Riebesell et al. (2000) conducted laboratory and field experiments that demonstrated reduced calcite production and increased frequency of malformed coccoliths in Subarctic coccolithophorids under conditions of raised CO_2 concentration. Foraminiferans make a major contribution to global carbonate flux (Schiebel 2002), and Southern Ocean sediment cores encapsulating the last 50,000 years reveal a negative relationship between atmospheric CO_2 levels and the shell weight of *Globigerina bulloides* (Moy et al. 2009). This is supported by similar work at modern volcanic vents indicating that foraminiferans are likely to be directly affected by changing pH levels (Dias et al. 2010, Uthicke et al. 2013).

For larger organisms, potentially lethal effects are likely to be preceded by negative long-term consequences at the population and species levels through effects on growth and reproduction such as the reduction of copepod egg production at raised CO_2 levels (Kurihara et al. 2004). Similarly, Kawaguchi et al. (2011, 2013) found that hatching success and larval development in *Euphausia superba* decreased dramatically as CO_2 was raised. Pteropods are another key element of Southern Ocean ecosystems, responsible for considerable vertical flux of organic material and carbonate (e.g. Collier et al. 2000). Reduced phytoplankton stocks in the Ross Sea led to the absence of the herbivorous pteropod *Limacina helicina* in the following year and marked reductions in the metabolism of one of its chief predators, another pteropod, *Clione antarctica*, which exhibited a 50% reduction in metabolic rates (Seibel & Dierssen 2003). In sediment traps, the shells of dead pteropods dissolve below the aragonite saturation layer (Honjo et al. 2000). Orr et al. (2005) exposed live Subarctic pteropods to low pH in shipboard experiments and found rapid (hours) dissolution of their aragonite shells. Subsequently, Bednaršek et al. (2012) provided striking evidence of shell dissolution of live *Limacina helicina antarctica* collected from the water column with no experimental manipulation. The implication is that habitable space for pteropods is already threatened and in the future will shrink not just meridionally but also vertically within the water column, with potential upward cascading effects through the food web as pteropods are eaten by fish, seabirds and even baleen whales.

Given the focus on threats to calcification, it is important to remember that populations can be limited at many different ontogenetic stages, and decreased pH can be important in other ways. While the gametes of invertebrates seem to be broadly robust to temperature and acidification, larvae are more vulnerable to hypercapnia and could form a population bottleneck, even if the embryo can tolerate ambient temperatures (Byrne 2011).

Alterations to the microbial plankton

The microbial loop involves a highly complex set of organisms and interactions that has a profound influence on pelagic dynamics, and there has been little speculation on how this will respond to long-term climate change. Both the structure and the functioning of microbial communities are likely to change in response to alterations to physico-chemical conditions in the water column, and this will have critical consequences for the microbial loop and the biological pump. Evans et al. (2011) examined microbial communities in the Subantarctic and subtropical zones south of Tasmania, where there are observations of southerly intrusions of subtropical waters. They found that variability in the microbial community was well explained by temperature, salinity and levels of inorganic nitrogen and concluded that climate change is likely to result in an increase in small phytoplankton and cyanobacteria in the Subantarctic zone. This would be coupled with similar increases in the abundances and activity of bacteria and viruses and result in enhanced importance of the microbial loop and a less efficient carbon pump.

The role of bacteria as a major component of the microbial loop is well recognised but may nevertheless have been underestimated. Microzooplankton can form a major component of the *Euphausia superba* diet (Perissinotto et al. 2000), and Sailley et al. (2013) used a model that included heterotrophic nanoflagellates and microzooplaknton to compare pelagic ecosystems in the northern and southern Western Antarctic Peninsula (WAP). They found that, on average, carbon flow through bacteria was 70% of that consumed by krill and concluded that the WAP ecosystem does not conform to the classic understanding of grazing by krill/zooplankton forming the key link between diatom production and top predators. Rather, they estimated that, except during salp blooms (e.g. Bernard et al. 2012), only up to 10% of primary production passed through zooplankton grazers. They concluded that the system is either in a state of transition or that it exists in different states between the north, where microbial grazing increasingly dominates, and the south, where grazing may be dominated by either microzooplankton or larger zooplankton.

Bacterial population dynamics reflect interactions among the effects of phytoplankton production and physical factors on resources (dissolved organic matter and nutrients) (Ducklow et al. 2007), with the top-down biological effects of grazing and viral attack. A classic example of the complexity of these communities comes from work in McMurdo Sound. Most eukaryotic phytoplankton and many bacteria require cobalamin or vitamin B12, which is produced only by other bacteria and archaea. In fact, Croft et al. (2005) found that at least some members of all algal phyla, including examples of open-ocean and coastal diatoms, require cobalamin and suggested that it is obtained through a symbiotic relationship with bacteria. Bertrand et al. (2007) showed that the growth of phytoplankton (mostly diatoms) in McMurdo Sound can be simultaneously and independently limited by both iron and cobalamin. Bertrand et al. (2015) later suggested that cobalamin is produced by populations of bacteria, but that these bacteria in turn depend on organic matter produced by phytoplankton in a mutualistic relationship. As other bacteria consume cobalamin, and its residence time in surface waters is short, this leads to a complex set of interactions involving positive and negative feedback loops. For example, Bertrand et al. (2015) propose that increased production of organic material by phytoplankton in response to aeolian-driven pulses of iron will promote bacterial growth and thus the release of cobalamin. This in turn would again enhance the growth of phytoplankton by releasing them from cobalamin limitation, but increased photosynthesis would also enhance cobalamin uptake by other bacteria. As the primary source of cobalamin, *Oceanospirillaceae* ASP10-02a, is ubiquitous throughout Southern Ocean surface waters, Bertrand et al. (2015) suggest that it is a key provider of cobalamin during summer.

In so far as the role of bacteria in degradation and nutrient recycling is driven by the supply of material, the microbial loop is likely to respond to change rather than driving it. Nevertheless, it is easy to imagine that complex feedback loops could be established that would modify ocean-climate interactions. Bacteria can also cause disease. Indeed, most planktonic algae carry viruses or bacteria (Park et al. 2004), and there have been reports of algicidal bacteria that may be able to affect the dynamics of algal blooms, prompting Mayali and Azam (2003) to suggest that we should include the actions of algicidal bacteria in our understanding of how algal communities are structured. Such interactions could shape and be shaped by changes in the marine environment (see below).

Importantly, the microbial plankton also includes viruses. Although their biomass is low, viruses are the most abundant particles in the ocean (Suttle 2005), with profoundly important and often underappreciated effects in their role as pathogens. Their role in polar waters is still poorly understood, but they can be a major cause of mortality in marine bacteria (Suttle 2007). Clearly, viruses play a critical role in the microbial loop. Conceptually, by causing lysis of phytoplankton and bacteria, viruses divert the carbon that these hosts contain away from their consumers and into the pool of dissolved organic matter (Wommack & Colwell 2000), decreasing the efficiency of the biological pump. Their role is accentuated both because they are enormously abundant and because they show extremely rapid turnover times (Fuhrman 1999).

Like bacteria (Gundersen et al. 1972), viral densities tend to be greatest near the surface (Culley & Welschmeyer 2002) despite their sensitivity to sunlight, and their abundances decline rapidly below the euphotic zone (Wommack & Colwell 2000). However, both bacteria and viruses show enormous spatial variability in abundances on extremely small scales, and Long and Azam (2001) suggest that strong antagonistic interactions among bacteria may contribute to the structuring of their communities at microscales. Seymour et al. (2006) showed 4–5-fold changes in abundances of viruses on scales of centimetres in the Southern Ocean. While virus and bacteria numbers are often correlated at larger scales (Wommack & Colwell 2000), Seymour et al. (2006) found no correlation at these very small scales, implying that the exposure of bacteria to viral infection is less frequent and less uniform than previously believed.

Part of the explanation for this may lie in the nature of polar viruses. Viral diversity appears to be lower in the Southern Ocean than at lower latitudes (Brum et al. 2015) and is dominated by so-called temperate or latent viruses (Brum et al. 2016), which exhibit a lysogenic stage. This involves lysogenic infection during which the viral genome is incorporated into the host genome and is reproduced along with the host genetic material without causing host mortality. This can later be triggered into lytic infection with reproduction of the virus by the host and bursting of the host cell (Fuhrman 1999). Lysogenic infection seems to be more common where and when prokaryotic productivity and abundance are low (Evans & Brussaard 2012). Brum et al. (2016) suggest that lysogenic infection can affect the ratio of bacterial to primary production by introducing a delay in bacterial responses at the onset of phytoplankton blooms. This is due to a reduction in host metabolism caused by lysogeny that is hypothesised to improve host survival when resources are scarce (Paul 2008). During full bloom conditions, lytic infection prevails, and bacterial and phytoplankton productivity become coupled.

Viruses are assumed to infect all living things in the ocean but are also presumed to have high strain specificity (Suttle 2005). Prokaryotes are the major hosts of viruses (Culley & Welschmeyer 2002), but they also attack eukaryotic phytoplankton (significantly this includes diatoms [Munn 2006]), and because of their high specificity, viruses may strongly influence the community composition of both bacteria (Hewson & Fuhrman 2006) and phytoplankton (Fuhrman 1999). There are also strong indications that both bacteria (Mayali & Azam 2004) and viruses may be implicated in the control of algal blooms (Munn 2006), including blooms of *Phaeocystis* once the colony skin has been breached following the active growth phase (Jacobsen et al. 1996, Hamm 2000). Consequently, viruses, although operating at the smallest biological scales, can have enormous community- and ecosystem-level effects (Wilhelm & Suttle 1999).

Viruses can directly affect the populations of larger organisms through their effects as pathogens. An anticipated global ecological response to climate change is the immigration or increase in frequency of epidemic diseases (Walther et al. 2002), and there is strong evidence that rates of epidemics have increased in the sea in the past few decades (Harvell et al. 1999, 2004). Disease outbreaks may reflect either a weakening of host resistance driven by the extra physiological stress imposed on organisms by climate change or viral host shifts (Harvell et al. 1999). The disease outbreaks recorded by Harvell et al. are restricted to large multicellular, mostly neritic organisms, but obviously there is a strong sampling bias in terms of both where outbreaks are likely to be noticed and how visible the hosts are. Such diseases may be caused by a wide variety of organisms including fungi, bacteria and so forth. In the case of viruses, a study examining a marine copepod showed no significant effects of elevated viral loads in laboratory cultures on mortality or sublethal levels such as egg production (Drake & Dobbs 2005), leading the authors to suggest that viral effects may be stronger for smaller organisms, such as protists. Certainly, there is a perception that parasitism (involving viruses as well as protozoans and fungi) should be included in our understanding of plankton dynamics (Park et al. 2004). There is no doubt that viruses can affect the population dynamics of larger marine organisms profoundly, and several of the disease outbreaks affecting vertebrates that are recorded by Harvell et al. (1999) were attributed to viruses.

It appears that some viruses move between the land and the sea (Suttle 2005), and this may provide one mechanism for the increasing rates of disease outbreaks in the sea. Terrestrial dust has been described as forming a biogeochemical link between the land, air and sea, with the vast quantities of terrestrial dust deposited annually in the sea transporting not only iron but also viable microorganisms (Ridgwell 2002, Garrison et al. 2003, Griffin & Kellogg 2004). Hayes et al. (2001) link climate change and increases in marine diseases by proposing that increased iron availability increases the pathogenicity of microbes. The argument is based on the fact that many organisms in the sea may be limited by iron availability, including Southern Ocean bacteria (Church et al. 2000), and that at least in mammals, host immune responses to pathogens involve withholding iron to inhibit pathogen growth through iron deprivation (Weinberg 1974). The conclusions of Hayes et al. (2001) refer primarily to macronutrient rich coastal regions, but a similar scenario could apply to the Southern Ocean.

Given the significant role of viruses in the Southern Ocean, we can recognise two ways in which they may be directly affected by long-term climate change, with indirect effects on the ecosystem as a whole. First, the ease of viral infection will react to temperature changes. Second, changes in wind regimes such as strengthening of the westerlies (Smith JLK et al. 2014, Smith

WO et al. 2014) will influence the dispersal of terrestrial dust into the ocean. This will affect the supply of iron and the supply of microbes to the marine realm (e.g. Bowman & Deming 2017). If or how climate shifts will interact with the ecological effects of viruses is unclear, but it is obvious that changes in the composition and abundance of virus communities are likely to have powerful consequences for ecosystem functioning in the open ocean. One possibility would be pathogen outbreaks linked to climate change altering phytoplankton populations, while an increase in viral attacks on phytoplankton could also lead to an increase in the production of dimethyl sulphide (DMS), at least in the short term. There are many other possibilities.

Conclusions

Like much of the world's oceans, the Southern Ocean has been subject to two vast perturbations linked to man's activities: the removal of the great whales and many of the finfish, followed by the effects of climate change. The first was a disruption of top-down effects through the removal of enormous levels of consumer biomass, leading to competitive release for many predators. The second will alter physical conditions within the system in complex, interacting ways. This will almost certainly lead to changes to primary producer community structure with bottom-up consequences operating through the quantity and nature of primary production. Four aspects should be borne in mind when attempting to understand the effects of climate change on Southern Ocean ecosystems.

First, the Southern Ocean is vast, with different patterns of physical and biological change among different regions, both meridionally and zonally. Conditions of warming and loss of sea ice differ between east and west Antarctica, while at this stage warming of the upper water column appears to be limited to the northern limits of the Southern Ocean.

Second, the ecosystems of the Southern Ocean are strongly physically forced at the base of the food web. Primary production is extremely pulsed and depends on different taxa in the open ocean, sea ice and neritic waters, often with short-term succession within these communities. Which taxa are responsible for primary production is fundamental to the efficiency of energy transfer to top predators, determines the efficiency of the biological pump and depends fundamentally on light availability (e.g. water column stability) and levels of macro and micronutrients. All of these vary in time and space but will almost certainly show longer-term climate-driven trends.

Third, these systems are particularly dependent on small numbers of key species. This is true for both the animals and the primary producers. The vulnerability of the animals to sublethal effects on their demography will have important consequences before physical conditions reach lethal levels. Similarly, the direct, physiological effects of increasing temperatures are likely to be relatively unimportant compared to the indirect effects of loss of the sea-ice habitat and changes to water column stability. These will affect krill abundances and primary producer community structure.

Fourth, most of our datasets are too short to differentiate noise from signal reliably. We can use ice and sediment cores to gain insight into past temperature fluctuations and changes to primary producer communities, and we can use archaeological evidence to elucidate penguin populations. At this stage, however, there is no way to describe historical population fluctuations in critical components such as krill, salps or myctophids as background against which we can understand ongoing fluctuations.

Superimposed on the 'first order' effects of climate change will be the disruption of existing species interactions and the establishment of new relationships as individual species respond in different ways and at different rates to changing conditions. In the shorter term, phenological changes will be important, while in the longer term, developmental mode and rates of evolution will come into play by determining the ability of species to respond phenotypically and genotypically. The levels of uncertainty that we face are captured by contrasting predictions of the effects of loss of sea ice on populations of krill, one of the best studied organisms in the Southern Ocean.

It is also important to recognise that in dealing with the Southern Ocean as a whole we are considering organisms with very different senses of place. For purely pelagic organisms such as phytoplankton, invertebrates, fish and cephalopods, biogeographic change means expansion, contraction or shifting of their biome. For organisms that are in some sense fixed in space, such as top predators that can only breed on land, the situation is different. Advection plays a key role in many Southern Ocean systems, allowing parts of the ocean to support predator populations in excess of those that local production alone could support. Disruption of existing patterns of advection and the shifting of favourable feeding grounds towards or away from breeding sites is already having dramatic consequences unconnected to direct physiological effects.

Because of the strong physical forcing of the ecosystems of the Southern Ocean, many of the initial consequences of climate change will operate through direct physiological effects on the primary producers and indirect effects on the larger organisms. All of this will lead to rebalancing of the system into a configuration that may not be one that we consider desirable. Of course, that is an anthropocentric value judgement. From the perspective of the planet, these ecosystems will persist in some configuration or other.

Lastly, it is important to recognise that alterations to the Southern Ocean will inevitably feed back into global climate so that the response of the Southern Ocean to climate change will have global consequences.

Acknowledgements

This work is based on research supported by the South African Research Chairs Initiative of the Department of Science and Technology and the National Research Foundation. I am extremely grateful to the Oppenheimer Memorial Trust Fund for the generous financial support that made the completion of this manuscript possible. It is a pleasure to thank the members of the Paterson lab at the University of St Andrews for their good-humoured patience and forbearing while hosting me during the writing. Lastly, I thank all the past and present members of the Southern Ocean Group at Rhodes University who taught me so much.

References

Abrams, R.W. 1985. Environmental determinants of pelagic seabird distribution in the African sector of the Southern Ocean. *Journal of Biogeography* **12**, 473–492.

Ackley, S., Wadhams, P., Comiso, J.C. & Worby, A.P. 2003. Decadal decrease of Antarctic sea ice extent inferred from whaling records revisited on the basis of historical and modern sea ice records. *Polar Research* **22**, 19–25.

Adams, N. & Brown, C. 1989. Dietary differentiation and trophic relationships in the sub-Antarctic penguin community at Marion Island. *Marine Ecology Progress Series. Oldendorf* **57**, 249–258.

Adams, N. & Klages, N. 1987. Seasonal variation in the diet of the king penguin (*Aptenodytes patagonicus*) at sub-Antarctic Marion Island. *Journal of Zoology* **212**, 303–324.

Ainley, D.G. 2002. The Ross Sea, Antarctica, where all ecosystem processes still remain for study, but maybe not for long. *Marine Ornithology* **30**, 55–62.

Ainley, D.G., Ballard, G. & Dugger, K.M. 2006. Competition among penguins and cetaceans reveals trophic cascades in the western Ross Sea, Antarctica. *Ecology* **87**, 2080–2093.

Ainley, D.G., Ballard, G., Ackley, S., Blight, L.K., Eastman, J.T., Emslie, S.D., Lescroël, A., Olmastroni, S., Townsend, S.E. & Tynan, C.T. 2007. Paradigm lost, or is top-down forcing no longer significant in the Antarctic marine ecosystem? *Antarctic Science* **19**, 283–290.

Ainley, D.G., Fraser, W.R., Smith, Jr., W.O., Hopkins, T.L. & Torres, J.J. 1991. The structure of upper level pelagic food webs in the Antarctic: Effect of phytoplankton distribution. *Journal of Marine Systems* **2**, 111–122.

Allan, L.E., William Froneman, P., Durgadoo, J.V., McQuaid, C.D., Ansorge, I.J. & Richoux, N.B. 2013. Critical indirect effects of climate change on sub-Antarctic ecosystem functioning. *Ecology and Evolution* **3**, 2994–3004.

Allcock, A.A. & Strugnell J.M. 2012. Southern Ocean diversity: New paradigms from molecular ecology. *Trends in Ecology and Evolution* **27**, 520–528.

Allison, I., Brandt, R.E. & Warren, S.G. 1993. East Antarctic sea ice: Albedo, thickness distribution, and snow cover. *Journal of Geophysical Research: Oceans* **98**, 12417–12429.

Andreae, M.O. & Crutzen, P.J. 1997. Atmospheric aerosols: biogeochemical sources and role in atmospheric chemistry. *Science* **276**, 1052–1058.

Ansorge, I. & Lutjeharms, J. 2003. Eddies originating at the South-West Indian ridge. *Journal of Marine Systems* **39**, 1–18.

Ansorge, I.J. & Lutjeharms, J.R. 2005. Direct observations of eddy turbulence at a ridge in the Southern Ocean. *Geophysical Research Letters* **32**, L14603, doi:10.1029/2005GL022588.

Armand, L.K., Cornet-Barthaux, V., Mosseri, J. & Quéguiner, B. 2008. Late summer diatom biomass and community structure on and around the naturally iron-fertilised kerguelen plateau in the Southern Ocean. *Deep Sea Research Part II: Topical Studies in Oceanography* **55**, 653–676.

Armand, L.K., Crosta, X., Romero, O. & Pichon, J.-J. 2005. The biogeography of major diatom taxa in Southern Ocean sediments: 1. Sea ice related species. *Palaeogeography, Palaeoclimatology, Palaeoecology* **223**, 93–126.

Armour, K.C., Marshall, J., Scott, J.R., Donohoe, A. & Newsom, E.R. 2016. Southern Ocean warming delayed by circumpolar upwelling and equatorward transport. *Nature Geoscience* **9**, 549–554.

Arndt, C.E. & Swadling, K.M. 2006. Crustacea in Arctic and Antarctic sea ice: distribution, diet and life history strategies. *Advances in Marine Biology* **51**, 197–315.

Arrigo, K.R., Lubin, D., Van Dijken, G.L., Holm-Hansen, O. & Morrow, E. 2003. Impact of a deep ozone hole on Southern Ocean primary production. *Journal of Geophysical Research: Oceans* **108**, 3154, doi:10.1029/2001JC001226, C5

Arrigo, K.R., Perovich, D.K., Pickart, R.S., Brown, Z.W., Van Dijken, G.L., Lowry, K.E., Mills, M.M., Palmer, M.A., Balch, W.M. & Bahr, F. 2012. Massive phytoplankton blooms under Arctic sea ice. *Science* **336**, 1408–1408.

Arrigo, K.R., van Dijken, G.L. & Long, M. 2008b. Coastal Southern Ocean a strong anthropogenic CO_2 sink. *Geophysical Research Letters* **35**, doi: 10.1029/2008GL035624.

Arrigo, K.R., Robinson, D.H., Worthen, D.L., Dunbar, R.B., DiTullio, G.R., VanWoert, M. & Lizotte, M.P. 1999. Phytoplankton community structure and the drawdown of nutrients and CO2 in the Southern Ocean. *Science* **283**, 365–367.

Arrigo, K.R. & Thomas, D.N. 2004. Large scale importance of sea ice biology in the Southern Ocean. *Antarctic Science* **16**, 471–486.

Arrigo, K.R. & Van Dijken, G.L. 2003. Phytoplankton dynamics within 37 Antarctic coastal polynya systems. *Journal of Geophysical Research: Oceans* **108**, 3271, doi:10.1029/2002JC001739, C8.

Arrigo, K.R., Van Dijken, G.L. & Bushinsky, S. 2008a. Primary production in the Southern Ocean, 1997–2006. *Journal of Geophysical Research: Oceans* **113**, C08004, doi:10.1029/2007JC004551.

Arthur, B., Hindell, M., Bester, M.N., Oosthuizen, W.C., Wege, M. & Lea, M. 2016. South for the winter? Within-dive foraging effort reveals the trade-offs between divergent foraging strategies in a free-ranging predator. *Functional Ecology* **30**, 1623–1637.

Atkinson, A., Siegel, V., Pakhomov, E. & Rothery, P. 2004. Long-term decline in krill stock and increase in salps within the Southern Ocean. *Nature* **432**, 100–103.

Atkinson, A., Siegel, V., Pakhomov, E., Rothery, P., Loeb, V., Ross, R., Quetin, L., Schmidt, K., Fretwell, P. & Murphy, E. 2008. Oceanic circumpolar habitats of Antarctic krill. *Marine Ecology Progress Series* **362**, 1–23.

Atkinson, A., Whitehouse, M., Priddle, J., Cripps, G., Ward, P. & Brandon, M. 2001. South Georgia, Antarctica: a productive, cold water, pelagic ecosystem. *Marine Ecology Progress Series* **216**, 279–308.

Azam, F., Fenchel, T., Field, J., Gray, J., Meyer-Reil, L. & Thingstad, F. 1983. The ecological role of water-column microbes in the sea. *Marine Ecology Progress Series. Oldendorf* **10**, 257–263.

Ballard, G., Jongsomjit, D., Veloz, S.D. & Ainley, D.G. 2012. Coexistence of mesopredators in an intact polar ocean ecosystem: the basis for defining a Ross sea marine protected area. *Biological Conservation* **156**, 72–82.

Banks, H. & Wood, R. 2002. Where to look for anthropogenic climate change in the ocean. *Journal of Climate* **15**, 879–891.

Barange, M., Pakhomov, E., Perissinotto, R., Froneman, P., Verheye, H., Taunton-Clark, J. & Lucas, M. 1998. Pelagic community structure of the subtropical convergence region south of Africa and in the mid-Atlantic Ocean. *Deep Sea Research Part I: Oceanographic Research Papers* **45**, 1663–1687.

Barbraud, C., Rivalan, P., Inchausti, P., Nevoux, M., Rolland, V. & Weimerskirch, H. 2011. Contrasted demographic responses facing future climate change in Southern Ocean seabirds. *Journal of Animal Ecology* **80**, 89–100.

Barbraud, C., Rolland, V., Jenouvrier, S., Nevoux, M., Delord, K. & Weimerskirch, H. 2012. Effects of climate change and fisheries bycatch on Southern Ocean seabirds: a review. *Marine Ecology Progress Series* **454**, 285–307.

Barbraud, C. & Weimerskirch, H. 2001a. Contrasting effects of the extent of sea-ice on the breeding performance of an Antarctic top predator, the snow petrel *Pagodroma Nivea*. *Journal of Avian Biology* **32**, 297–302.

Barbraud, C. & Weimerskirch, H. 2001b. Emperor penguins and climate change. *Nature* **411**, 183–186.

Barbraud, C., Weimerskirch, H., Guinet, C. & Jouventin, P. 2000. Effect of sea-ice extent on adult survival of an Antarctic top predator: the snow petrel *Pagodroma Nivea*. *Oecologia* **125**, 483–488.

Bargelloni, L., Zane, L., Derome, N., Lecointre, G. & Patarnello, T. 2000. Molecular zoogeography of Antarctic *Euphausiids and Notothenioids*: from species phylogenies to intraspecific patterns of genetic variation. *Antarctic Science* **12**, 259–268.

Barkai, A. & McQuaid, C. 1988. Predator-prey role reversal in a marine benthic ecosystem. *Science* **242**, 62–64.

Barrera-Oro, E. 2002. The role of fish in the Antarctic marine food web: differences between inshore and offshore waters in the Southern Scotia Arc and West Antarctic Peninsula. *Antarctic Science* **14**, 293–309.

Barrera-Oro, E. & Casaux, R. 2008. General ecology of coastal fish from the south Shetland Island and west Antarctic Peninsula areas. *Ber Polar Meeresforsch* **571**, 95–110.

Bathmann, U. 1996. Abiotic and biotic forcing on vertical particle flux in the Southern Ocean, V. Ittekkot et al. (eds). *Particle Flux in the Ocean*. SCOPE Report 57, London: Wiley & Sons, 243–250.

Bathmann, U., Scharek, R., Klaas, C., Dubischar, C. & Smetacek, V. 1997. Spring development of phytoplankton biomass and composition in major water masses of the Atlantic sector of the Southern Ocean. *Deep Sea Research Part II: Topical Studies in Oceanography* **44**, 51–67.

Batta-Lona, P.G., Bucklin, A., Wiebe, P.H., Patarnello, T. & Copley, N.J. 2011. Population genetic variation of the Southern Ocean krill, *Euphausia superba*, in the Western Antarctic Peninsula region based on mitochondrial single nucleotide polymorphisms (SNPs). *Deep Sea Research Part II: Topical Studies in Oceanography* **58**, 1652–1661.

Bearhop, S., Phillips, R.A., McGill, R., Cherel, Y., Dawson, D.A. & Croxall, J.P. 2006. Stable isotopes indicate sex-specific and long-term individual foraging specialisation in diving seabirds. *Marine Ecology Progress Series* **311**, 157–164.

Bednaršek, N., Tarling, G., Bakker, D., Fielding, S., Jones, E., Venables, H., Ward, P., Kuzirian, A., Lézé, B. & Feely, R. 2012. Extensive dissolution of live pteropods in the Southern Ocean. *Nature Geoscience* **5**, 881–885.

Beers, J.M. & Jayasundara, N. 2015. Antarctic notothenioid fish: what are the future consequences of 'losses' and 'gains' acquired during long-term evolution at cold and stable temperatures? *Journal of Experimental Biology* **218**, 1834–1845.

Belkin, I.M. 2009. Rapid warming of large marine ecosystems. *Progress in Oceanography* **81**, 207–213.

Bentley, M.J., Hodgson, D., Smith, J., Cofaigh, C., Domack, E., Larter, R., Roberts, S., Brachfeld, S., Leventer, A. & Hjort, C. 2009. Mechanisms of holocene palaeoenvironmental change in the Antarctic Peninsula region. *The Holocene* **19**, 51–69.

Bergstrom, D.M. & Chown, S.L. 1999. Life at the front: history, ecology and change on Southern Ocean islands. *Trends in Ecology & Evolution* **14**, 472–477.

Bernard, K.S. & Froneman, P.W. 2003. Mesozooplankton community structure and grazing impact in the Polar Frontal Zone of the Southern Ocean during austral summer 2002. *Polar Biology* **26**, 268–275.

Bernard, K.S., Steinberg, D.K. & Schofield, O.M. 2012. Summertime grazing impact of the dominant macrozooplankton off the western Antarctic Peninsula. *Deep Sea Research Part I: Oceanographic Research Papers* **62**, 111–122.

Berner, R.A. 2002. Examination of hypotheses for the permo–triassic boundary extinction by carbon cycle modeling. *Proceedings of the National Academy of Sciences* **99**, 4172–4177.

Bertrand, E.M., McCrow, J.P., Moustafa, A., Zheng, H., McQuaid, J.B., Delmont, T.O., Post, A.F., Sipler, R.E., Spackeen, J.L. & Xu, K. 2015. Phytoplankton–bacterial interactions mediate micronutrient colimitation at the coastal Antarctic sea ice edge. *Proceedings of the National Academy of Sciences* **112**, 9938–9943.

Bertrand, E.M., Saito, M.A., Rose, J.M., Riesselman, C.R., Lohan, M.C., Noble, A.E., Lee, P.A. & DiTullio, G.R. 2007. Vitamin B12 and iron colimitation of phytoplankton growth in the Ross Sea. *Limnology and Oceanography* **52**, 1079–1093.

Blanchet, M.-A., Biuw, M., Hofmeyr, G.G., de Bruyn, P.N., Lydersen, C. & Kovacs, K.M. 2013. At-sea behaviour of three krill predators breeding at bouvetøya—Antarctic fur seals, macaroni penguins and chinstrap penguins. *Marine Ecology Progress Series* **477**, 285–302.

Boden, B., Rae, C.D. & Lutjeharms, J. 1988. The distribution of the diatoms of the south-west Indian Ocean surface water between Cape Town and the Prince Edward Islands Archipelago. *South African Journal of Science* **84**, 811–818.

Boden, B.P. & Reid, F.M.H. 1989. Marine plankton diatoms between Cape Town and the Prince Edward Islands (SW Indian Ocean). *South African Journal of Antarctic Research* **19**, 2–49.

Boeckel, B., Baumann, K.-H., Henrich, R. & Kinkel, H. 2006. Coccolith distribution patterns in south Atlantic and Southern Ocean surface sediments in relation to environmental gradients. *Deep Sea Research Part I: Oceanographic Research Papers* **53**, 1073–1099.

Bortolotto, E., Bucklin, A., Mezzavilla, M., Zane, L. & Patarnello, T. 2011. Gone with the currents: lack of genetic differentiation at the circum-continental scale in the Antarctic krill *Euphausia superba*. *BMC genetics* **12**, 32.

Bost, C.-A., Cotté, C., Bailleul, F., Cherel, Y., Charrassin, J.-B., Guinet, C., Ainley, D.G. & Weimerskirch, H. 2009. The importance of oceanographic fronts to marine birds and mammals of the Southern Oceans. *Journal of Marine Systems* **78**, 363–376.

Bowman, J.S. & Deming, J.W. 2017. Wind-driven distribution of bacteria in coastal Antarctica: evidence from the Ross Sea region. *Polar Biology* **40**, 25–35.

Boyce, D.G., Lewis, M.R. & Worm, B. 2010. Global phytoplankton decline over the past century. *Nature* **466**, 591–596.

Boyd, P.W. 2002. Environmental factors controlling phytoplankton processes in the Southern Ocean. *Journal of Phycology* **38**, 844–861.

Boyd, P.W., Arrigo, K., Strzepek, R. & Dijken, G. 2012. Mapping phytoplankton iron utilization: insights into Southern Ocean supply mechanisms. *Journal of Geophysical Research: Oceans* **117**, C06009, doi:10.1029/2011JC007726.

Boyd, P.W., Lennartz, S.T., Glover, D.M. & Doney, S.C. 2015. Biological ramifications of climate-change-mediated oceanic multi-stressors. *Nature Climate Change* **5**, 71–79.

Boyd, P.W., Watson, A.J., Law, C.S., Abraham, E.R., Trull, T., Murdoch, R., Bakker, D.C., Bowie, A.R., Buesseler, K. & Chang, H. 2000. A mesoscale phytoplankton bloom in the polar Southern Ocean stimulated by iron fertilization. *Nature* **407**, 695–702.

Boysen-Ennen, E. & Piatkowski, U. 1988. Meso-and macrozooplankton communities in the Weddell Sea, Antarctica. *Polar Biology* **9**, 17–35.

Bracegirdle, T.J., Connolley, W.M. & Turner, J. 2008. Antarctic climate change over the twenty-first century. *Journal of Geophysical Research: Atmospheres* **113**, D03103, doi:10.1029/2007JD008933.

Bradshaw, C.J., Hindell, M.A., Best, N.J., Phillips, K.L., Wilson, G. & Nichols, P.D. 2003. You are what you eat: describing the foraging ecology of southern elephant seals (*Mirounga leonina*) using blubber fatty acids. *Proceedings of the Royal Society of London B: Biological Sciences* **270**, 1283–1292.

Braithwaite, J.E., Meeuwig, J.J., Letessier, T.B., Jenner, K.C.S. & Brierley, A.S. 2015. From sea ice to blubber: linking whale condition to krill abundance using historical whaling records. *Polar Biology* **38**, 1195–1202.

Branch, G.M., Attwood, C.G., Gianakouras, D. & Branch, M.L. 1993. Patterns in the benthic communities on the shelf of the sub-Antarctic Prince Edward Islands. *Polar Biology* **13**, 23–34.

Branch, T.A. & Williams, T.M. 2006. Legacy of industrial whaling. In *Whales, Whaling and Ocean Ecosystems*, J.A. Estes, D.P. Demaster, D.F. Doak, T.M. Williams & R.L. Brownell (eds). Berkeley, CA: University of California Press, 262–278.

Brandão, S.N., Sauer, J. & Schön, I. 2010. Circumantarctic distribution in Southern Ocean benthos? A genetic test using the genus macroscapha (Crustacea, Ostracoda) as a model. *Molecular Phylogenetics and Evolution* **55**, 1055–1069.

Brandt, R.E., Warren, S.G., Worby, A.P. & Grenfell, T.C. 2005. Surface albedo of the Antarctic sea ice zone. *Journal of Climate* **18**, 3606–3622.

Brewer, P.G. & Hester, K. 2009. Ocean acidification and the increasing transparency of the ocean to low-frequency sound. *Oceanography* **22**, 86–93.

Brierley, A., Demer, D., Watkins, J. & Hewitt, R. 1999a. Concordance of interannual fluctuations in acoustically estimated densities of Antarctic krill around South Georgia and Elephant Island: biological evidence of same-year teleconnections across the Scotia Sea. *Marine Biology* **134**, 675–681.

Brierley, A.S., Fernandes, P.G., Brandon, M.A., Armstrong, F., Millard, N.W., McPhail, S.D., Stevenson, P., Pebody, M., Perrett, J. & Squires, M. 2002. Antarctic krill under sea ice: elevated abundance in a narrow band just south of ice edge. *Science* **295**, 1890–1892.

Brierley, A.S. & Thomas, D.N. 2002. Ecology of Southern Ocean pack ice. *Advances in Marine Biology* **43**, 171–IN4.

Brierley, A., Watkins, J., Goss, C., Wilkinson, M. & Everson, I. 1999b. Acoustic estimates of krill density at South Georgia, 1981 to 1998. *CCAMLR Science* **6**, 47–57.

Brierley, A.S., Watkins, J.L. & Murray, A.W. 1997. Interannual variability in krill abundance at South Georgia. *Marine Ecology Progress Series* **150**, 87–98.

Briggs, J.C. & Bowen, B.W. 2012. A realignment of marine biogeographic provinces with particular reference to fish distributions. *Journal of Biogeography* **39**, 12–30.

Brodeur, R.D., Mills, C.E., Overland, J.E., Walters, G.E. & Schumacher, J.D. 1999. Evidence for a substantial increase in gelatinous zooplankton in the Bering Sea, with possible links to climate change. *Fisheries Oceanography* **8**, 296–306.

Bromwich, D.H., Nicolas, J.P., Monaghan, A.J., Lazzara, M.A., Keller, L.M., Weidner, G.A. & Wilson, A.B. 2013. Central west Antarctica among the most rapidly warming regions on earth. *Nature Geoscience* **6**, 139–145.

Brown, C. & Klages, N. 1987. Seasonal and annual variation in diets of macaroni (*Eudyptes chrysolophus chrysolophus*) and southern rockhopper *(E. chrysocome chrysocome)* Penguins at sub-Antarctic Marion Island. *Journal of Zoology* **212**, 7–28.

Brum, J.R., Hurwitz, B.L., Schofield, O., Ducklow, H.W. & Sullivan, M.B. 2016. Seasonal time bombs: dominant temperate viruses affect Southern Ocean microbial dynamics. *The ISME Journal* **10**, 437–449.

Brum, J.R., Ignacio-Espinoza, J.C., Roux, S., Doulcier, G., Acinas, S.G., Alberti, A., Chaffron, S., Cruaud, C., De Vargas, C. & Gasol, J.M. 2015. Patterns and ecological drivers of ocean viral communities. *Science* **348**, 1261498.

Buchholz, F. & Saborowski, R. 2000. Metabolic and enzymatic adaptations in northern krill, *Meganyctiphanes norvegica*, and Antarctic krill, *Euphausia superba*. *Canadian Journal of Fisheries and Aquatic Sciences* **57**, 115–129.

Budd, G. 2000. Changes in Heard Island glaciers, king penguins and fur seals since 1947. *Papers and Proceedings of the Royal Society of Tasmania* **133**, 47–60.

Burkill, P.H., Archer, S.D., Robinson, C., Nightingale, P.D., Groom, S.B., Tarran, G.A. & Zubkov, M.V. 2002. Dimethyl sulphide biogeochemistry within a coccolithophore bloom (DISCO): an Overview. *Deep Sea Research Part II: Topical Studies in Oceanography* **49**, 2863–2885.

Burkill, P.H., Edwards, E. & Sleight, M. 1995. Microzooplankton and their role in controlling phytoplankton growth in the marginal ice zone of the Bellingshausen Sea. *Deep Sea Research Part II: Topical Studies in Oceanography* **42**, 1277–1290.

Bye, J., May, J. & Simmonds, I. 2006. Control of the Antarctic ice sheet by ocean–ice interaction. *Global and Planetary Change* **50**, 99–111.

Byrne, M. 2011. Impact of ocean warming and ocean acidification on marine invertebrate life history stages: Vulnerabilities and potential for persistence in a changing ocean. *Oceanography and Marine Biology: An Annual Review* **49**, 1–42.

Caldeira, K. & Wickett, M.E. 2003. Oceanography: anthropogenic carbon and ocean pH. *Nature* **425**, 365–365.

Cardona, L., De Quevedo, I.Á., Borrell, A. & Aguilar, A. 2012. Massive consumption of gelatinous plankton by Mediterranean apex predators. *PloS One* **7**, e31329.

Chavtur, V. & Kruk, N. 2003. Latitudinal distribution of pelagic ostracods (Ostracoda, Halocyprinidae) in the Australian–New Zealand sector of the Southern Ocean. *Russian Journal of Marine Biology* **29**, 137–143.

Cherel, Y. & Duhamel, G. 2004. Antarctic jaws: cephalopod prey of sharks in Kerguelen waters. *Deep Sea Research Part I: Oceanographic Research Papers* **51**, 17–31.

Cherel, Y., Fontaine, C., Richard, P. & Labatc, J.-P. 2010. Isotopic niches and trophic levels of myctophid fishes and their predators in the Southern Ocean. *Limnology and Oceanography* **55**, 324–332.

Cherel, Y. & Hobson, K.A. 2007. Geographical variation in carbon stable isotope signatures of marine predators: a tool to investigate their foraging areas in the Southern Ocean. *Marine Ecology Progress Series* **329**, 281–287.

Cherel, Y., Hobson, K.A., Guinet, C. & Vanpe, C. 2007. Stable isotopes document seasonal changes in trophic niches and winter foraging individual specialization in diving predators from the Southern Ocean. *Journal of Animal Ecology* **76**, 826–836.

Cherel, Y., Ridoux, V. & Rodhouse, P. 1996. Fish and squid in the diet of king penguin chicks, *Aptenodytes patagonicus*, during winter at sub-Antarctic Crozet Islands. *Marine Biology* **126**, 559–570.

Cheung, W.W., Watson, R. & Pauly, D. 2013. Signature of ocean warming in global fisheries catch. *Nature* **497**, 365–368.

Chown, S., Gremmen, N. & Gaston, K. 1998. Ecological biogeography of Southern Ocean Islands: species-area relationships, human impacts, and conservation. *The American Naturalist* **152**, 562–575.

Chown, S.L. & Smith, V.R. 1993. Climate change and the short-term impact of feral house mice at the sub-Antarctic Prince Edward Islands. *Oecologia* **96**, 508–516.

Church, M.J., Hutchins, D.A. & Ducklow, H.W. 2000. Limitation of bacterial growth by dissolved organic matter and iron in the Southern Ocean. *Applied and Environmental Microbiology* **66**, 455–466.

Clarke, A. 1993. Temperature and extinction in the sea: a physiologist's view. *Paleobiology* **19**, 499–518.

Clarke, A., Barnes, D.K. & Hodgson, D.A. 2005. How isolated is Antarctica? *Trends in Ecology & Evolution* **20**, 1–3.

Clarke, A., Murphy, E.J., Meredith, M.P., King, J.C., Peck, L.S., Barnes, D.K. & Smith, R.C. 2007. Climate change and the marine ecosystem of the Western Antarctic Peninsula. *Philosophical Transactions of the Royal Society of London B: Biological Sciences* **362**, 149–166.

Clarke, M.R. 1966. A review of the systematics and ecology of oceanic squids. *Advances in Marine Biology* **4**, 91–300.

Coll, M., Navarro, J., Olson, R.J. & Christensen, V. 2013. Assessing the trophic position and ecological role of squids in marine ecosystems by means of food-web models. *Deep Sea Research Part II: Topical Studies in Oceanography* **95**, 21–36.

Collier, R., Dymond, J., Honjo, S., Manganini, S., Francois, R. & Dunbar, R. 2000. The vertical flux of biogenic and lithogenic material in the Ross Sea: moored sediment trap observations 1996–1998. *Deep Sea Research Part II: Topical Studies in Oceanography* **47**, 3491–3520.

Collins, M.A., Allcock, A.L. & Belchier, M. 2004. Cephalopods of the South Georgia slope. *Journal of the Marine Biological Association of the UK* **84**, 415–419.

Collins, M.A. & Rodhouse, P.G. 2006. Southern Ocean cephalopods. *Advances in Marine Biology* **50**, 191–265.

Comiso, J., McClain, C., Sullivan, C., Ryan, J. & Leonard, C. 1993. Coastal zone color scanner pigment concentrations in the Southern Ocean and relationships to geophysical surface features. *Journal of Geophysical Research: Oceans* **98**, 2419–2451.

Connan, M., Cherel, Y. & Mayzaud, P. 2007. Lipids from stomach oil of procellariiform seabirds document the importance of myctophid fish in the Southern Ocean. *Limnology and Oceanography* **52**, 2445–2455.

Connan, M., McQuaid, C.D., Bonnevie, B.T., Smale, M.J. & Cherel, Y. 2014. Combined stomach content, lipid and stable isotope analyses reveal spatial and trophic partitioning among three sympatric albatrosses from the Southern Ocean. *Marine Ecology Progress Series* **497**, 259–272.

Connan, M., Teske, P.R., Tree, A.J., Whittington, P.A. & McQuaid, C.D. 2015. The subspecies of Antarctic terns (*Sterna vittata*) wintering on the South African coast: evidence from morphology, genetics and stable isotopes. *Emu* **115**, 223–236.

Constable, A.J., Nicol, S. & Strutton, P.G. 2003. Southern Ocean productivity in relation to spatial and temporal variation in the physical environment. *Journal of Geophysical Research* **108**, 8079, doi: 10.1029/2001JC001270.

Convey, P., Chown, S.L., Clarke, A., Barnes, D.K., Bokhorst, S., Cummings, V., Ducklow, H.W., Frati, F., Green, T. & Gordon, S. 2014. The spatial structure of Antarctic biodiversity. *Ecological Monographs* **84**, 203–244.

Crampton, J.S., Cody, R.D., Levy, R., Harwood, D., McKay, R. & Naish, T.R. 2016. Southern Ocean phytoplankton turnover in response to stepwise Antarctic cooling over the past 15 million years. *Proceedings of the National Academy of Sciences* **113**, 6868–6873.

Croft, M.T., Lawrence, A.D., Raux-Deery, E., Warren, M.J. & Smith, A.G. 2005. Algae acquire vitamin B12 through a symbiotic relationship with bacteria. *Nature* **438**, 90–93.

Croll, D.A., Tershy, B.R., Hewitt, R.P., Demer, D.A., Fiedler, P.C., Smith, S.E., Armstrong, Popp, J.M., Kiekhefer, T., Lopez, V.R., Urban, J. & Gendron, D. 1998. An integrated approach to the foraging ecology of marine birds and mammals. *Deep Sea Research Part II* **45**, 1353–1371.

Crosta, X., Romero, O., Armand, L.K. & Pichon, J.-J. 2005. The biogeography of major diatom taxa in Southern Ocean sediments: 2. Open ocean related species. *Palaeogeography, Palaeoclimatology, Palaeoecology* **223**, 66–92.

Croxall, J.P. 1992. Southern Ocean environmental changes: Effects on seabird, seal and whale population. *Philosophical Transactions of the Royal Society of London B* **338**, 319–328.

Croxall, J.P. 2004. The potential effects of marine habitat change on Antarctic seabirds. *Ibis* **146**, 90–91.

Croxall, J.P., Reid, K. & Prince, P. 1999. Diet, provisioning and productivity responses of marine predators to differences in availability of Antarctic krill. *Marine Ecology Progress Series* **177**, 115–131.

Croxall, J.P., Trathan, P. & Murphy, E. 2002. Environmental change and Antarctic seabird populations. *Science* **297**, 1510–1514.

Culley, A.I. & Welschmeyer, N.A. 2002. The abundance, distribution, and correlation of viruses, phytoplankton, and prokaryotes along a Pacific Ocean transect. *Limnology and Oceanography* **47**, 1508–1513.

Cunningham, D.M. & Moors, P.J. 1994. The decline of rockhopper penguins *Eudyptes chrysocome* at Campbell Island, Southern Ocean and the influence of rising sea temperatures. *Emu* **94**, 27–36.

Curran, M.A., van Ommen, T.D., Morgan, V.I., Phillips, K.L. & Palmer, A.S. 2003. Ice core evidence for Antarctic sea ice decline since the 1950s. *Science* **302**, 1203–1206.

Cury, P.M., Boyd, I.L., Bonhommeau, S., Anker-Nilssen, T., Crawford, R.J., Furness, R.W., Mills, J.A., Murphy, E.J., Österblom, H. & Paleczny, M. 2011. Global seabird response to forage fish depletion—one-third for the birds. *Science* **334**, 1703–1706.

Daly, K.L. 2004. Overwintering growth and development of larval Euphausia superba: an interannual comparison under varying environmental conditions west of the Antarctic Peninsula. *Deep Sea Research Part II: Topical Studies in Oceanography* **51**, 2139–2168.

Daly, K.L. & Zimmerman, J.J. 2004. Comparisons of morphology and neritic distributions of *Euphausia Crystallorophias* and *Euphausia superba furcilia* during autumn and winter west of the Antarctic Peninsula. *Polar Biology* **28**, 72–81.

Damkaer, D.M. & Dey, D.B. 1983. UV damage and photoreactivation potentials of larval shrimp, *Pandalus platyceros*, and adult *Euphausiids, Thysanoessa raschii*. *Oecologia* **60**, 169–175.

De Broyer, C. & Danis, B. 2011. How many species in the Southern Ocean? Towards a dynamic inventory of the Antarctic marine species. *Deep Sea Research Part II: Topical Studies in Oceanography* **58**, 5–17.

De Lavergne, C., Palter, J.B., Galbraith, E.D., Bernardello, R. & Marinov, I. 2014. Cessation of deep convection in the open Southern Ocean under anthropogenic climate change. *Nature Climate Change* **4**, 278–282.

Dehnhard, N., Eens, M., Demongin, L., Quillfeldt, P., Suri, D. & Poisbleau, M. 2015. Limited individual phenotypic plasticity in the timing of and investment into egg laying in southern rockhopper penguins under climate change. *Marine Ecology Progress Series* **524**, 269–281.

de Jong, J., Schoemann, V., Lannuzel, D., Croot, P., Baar, H. & Tison, J. 2012. Natural iron fertilization of the Atlantic sector of the Southern Ocean by continental shelf sources of the Antarctic Peninsula. *Journal of Geophysical Research: Biogeosciences* **117**, G01029, doi:10.1029/2011JG001679.

Deppeler, S.L. & Davidson, A.T. 2017. Southern Ocean phytoplankton in a changing climate. *Frontiers in Marine Science* **4**, 40.

Desbruyères, D.G., Purkey, S.G., McDonagh, E.L., Johnson, G.C. & King, B.A. 2016. Deep and abyssal ocean warming from 35 years of repeat hydrography. *Geophysical Research Letters* **43**, 10356–10365, doi:10.1002/2016GL070413.

Detmer, A. & Bathmann, U. 1997. Distribution patterns of autotrophic pico- and nanoplankton and their relative contribution to algal biomass during spring in the Atlantic sector of the Southern Ocean. *Deep Sea Research Part II: Topical Studies in Oceanography* **44**, 299–320.

Dias, B., Hart, M., Smart, C. & Hall-Spencer, J. 2010. Modern seawater acidification: the response of foraminifera to high-CO_2 conditions in the Mediterranean Sea. *Journal of the Geological Society* **167**, 843–846.

Dickson, J., Morley, S. & Mulvey, T. 2004. New data on *Martialia hyadesi* feeding in the Scotia Sea during winter; with emphasis on seasonal and annual variability. *Journal of the Marine Biological Association of the UK* **84**, 785–788.

Dieckmann, G.S. & Hellmer, H.H. 2010. The importance of sea ice: an overview. *Sea Ice* **2**, 1–22.

DiTullio, G., Grebmeier, J., Arrigo, K., Lizotte, M., Robinson, D., Leventer, A., Barry, J., VanWoert, M. & Dunbar, R. 2000. Rapid and early export of *Phaeocystis antarctica* blooms in the Ross Sea, Antarctica. *Nature* **404**, 595–598.

Doake, C. & Vaughan, D. 1991. Rapid disintegration of the wordie ice shelf in response to atmospheric warming. *Nature* **350**, 328–330.

Dobricic, S., Vignati, E. & Russo, S. 2016. Large-scale atmospheric warming in winter and the Arctic Sea ice retreat. *Journal of Climate* **29**, 2869–2888.

Domack, E., Duran, D., Leventer, A., Ishman, S., Doane, S., McCallum, S., Amblas, D., Ring, J., Gilbert, R. & Prentice, M. 2005a. Stability of the Larsen b ice shelf on the Antarctic Peninsula during the Holocene epoch. *Nature* **436**, 681–685.

Domack, E., Ishman, S., Leventer, A., Sylva, S., Willmott, V. & Huber, B. 2005b. A chemotrophic ecosystem found beneath Antarctic ice shelf. *Eos, Transactions, American Geophysical Union* **86**, 269–271.

Drake, L.A. & Dobbs, F.C. 2005. Do viruses affect fecundity and survival of the copepod *Acartia tonsa dana*? *Journal of Plankton Research* **27**, 167–174.

Dubischar, C.D. & Bathmann, U.V. 1997. Grazing Impact of copepods and salps on phytoplankton in the Atlantic sector of the Southern Ocean. *Deep Sea Research Part II: Topical Studies in Oceanography* **44**, 415–433.

Dubischar, C., Pakhomov, E., von Harbou, L., Hunt, B. & Bathmann, U. 2012. Salps in the Lazarev Sea, Southern Ocean: II. Biochemical composition and potential prey value. *Marine Biology* **159**, 15–24.

Ducklow, H.W., Baker, K., Martinson, D.G., Quetin, L.B., Ross, R.M., Smith, R.C., Stammerjohn, S.E., Vernet, M. & Fraser, W. 2007. Marine pelagic ecosystems: the west Antarctic Peninsula. *Philosophical Transactions of the Royal Society of London B: Biological Sciences* **362**, 67–94.

Eastman, J.T. 2005. The nature of the diversity of Antarctic fishes. *Polar Biology* **28**, 93–107.

El-Sayed, S.Z. 2005. History and evolution of primary productivity studies of the Southern Ocean. *Polar Biology* **28**, 423–438.

El-Sayed, S.Z. & Fryxell, G.A. 1993. Phytoplankton. In *Antarctic Microbiology* ed. Friedmann, E.I. 65–122. New York: Wiley-Liss.

Emslie, S.D. & Patterson, W.P. 2007. Abrupt recent shift in $\delta^{13}C$ and $\delta^{15}N$ values in Adélie penguin eggshell in Antarctica. *Proceedings of the National Academy of Sciences of the United States of America* **104**, 11666–11669.

Emslie, S.D., Coats, L. & Licht, K. 2007. A 45,000 yr record of Adélie penguins and climate change in the Ross Sea, Antarctica. *Geological Society of America* **35**, 61–64, doi: 10.1130/G23011A.

Evans, C. & Brussaard, C.P. 2012. Regional variation in lytic and lysogenic viral infection in the Southern Ocean and its contribution to biogeochemical cycling. *Applied and Environmental Microbiology* **78**, 6741–6748.

Evans, C., Thomson, P.G., Davidson, A.T., Bowie, A.R., van den Enden, R., Witte, H. & Brussaard, C.P. 2011. Potential climate change impacts on microbial distribution and carbon cycling in the Australian Southern Ocean. *Deep Sea Research Part II: Topical Studies in Oceanography* **58**, 2150–2161.

Evans, K. & Hindell, M.A. 2004. The diet of sperm whales (*Physeter macrocephalus*) in Southern Australian waters. *ICES Journal of Marine Science: Journal du Conseil* **61**, 1313–1329.

Falkowski, P.G., Barber, R.T. & Smetacek, V. 1998. Biogeochemical controls and feedbacks on ocean primary production. *Science* **281**, 200–206.

Fauchald, P., Erikstad, K.E. & Skarsfjord, H. 2000. Scale-dependent predator–prey interactions: the hierarchical spatial distribution of seabirds and prey. *Ecology* **81**, 773–783.

Feely, R.A., Sabine, C.L., Lee, K., Berelson, W., Kleypas, J., Fabry, V.J. & Millero, F.J. 2004. Impact of anthropogenic CO2 on the CaCO3 system in the oceans. *Science* **305**, 362–366.

Fenchel, T. 1988. Marine plankton food chains. *Annual Review of Ecology and Systematics* **19**, 19–38.

Finkel, Z.V., Katz, M.E., Wright, J.D., Schofield, O.M. & Falkowski, P.G. 2005. Climatically driven macroevolutionary patterns in the size of marine diatoms over the cenozoic. *Proceedings of the National Academy of Sciences of the United States of America* **102**, 8927–8932.

Fitzwater, S., Johnson, K., Gordon, R., Coale, K. & Smith, W. 2000. Trace metal concentrations in the Ross Sea and their relationship with nutrients and phytoplankton growth. *Deep Sea Research Part II: Topical Studies in Oceanography* **47**, 3159–3179.

Flores, H., Van de Putte, A.P., Siegel, V., Pakhomov, E.A., Van Franeker, J.A., Meesters, E.H. & Volckaert, F.A. 2008. Distribution, abundance and ecological relevance of pelagic fishes in the Lazarev Sea, Southern Ocean. *Marine Ecology Progress Series* **367**, 271–282.

Flores, H., Van Franeker, J.-A., Cisewski, B., Leach, H., Van de Putte, A.P., Meesters, E.H., Bathmann, U. & Wolff, W.J. 2011. Macrofauna under sea ice and in the open surface layer of the Lazarev Sea, Southern Ocean. *Deep Sea Research Part II: Topical Studies in Oceanography* **58**, 1948–1961.

Flores, H., Van Franeker, J.A., Siegel, V., Haraldsson, M., Strass, V., Meesters, E.H., Bathmann, U. & Wolff, W.J. 2012. The association of antarctic krill *Euphausia superba* with the under-ice habitat. *PloS One* **7**, e31775.

Forcada, J. & Trathan, P.N. 2009. Penguin responses to climate change in the Southern Ocean. *Global Change Biology* **15**, 1618–1630.

Francois, R., Altabet, M.A., Goericke, R., McCorkle, D.C., Brunet, C. & Poisson, A. 1993. Changes in the δ13C of surface water particulate organic matter across the subtropical convergence in the SW Indian Ocean. *Global Biogeochemical Cycles* **7**, 627–644.

Fraser, C.I., Kay, G.M., du Plessis, M. & Ryan, P.G. 2016. Breaking down the barrier: dispersal across the Antarctic polar front. *Ecography* **39**, 001–003.

Fraser, W.R. & Hofmann, E.E. 2003. A predator's perspective on causal links between climate change, physical forcing and ecosystem response. *Marine Ecology Progress Series* **265**, 1–15.

Frazer, T.K., Quetin, L.B. & Ross, R.M. 2002. Abundance, sizes and developmental stages of larval krill, *Euphausia superba*, during winter in ice-covered seas west of the Antarctic Peninsula. *Journal of Plankton Research* **24**, 1067–1077.

Frenger, I., Gruber, N., Knutti, R. & Münnich, M. 2013. Imprint of Southern Ocean eddies on winds, clouds and rainfall. *Nature Geoscience* **6**, 608–612.

Frölicher, T.L., Sarmiento, J.L., Paynter, D.J., Dunne, J.P., Krasting, J.P. & Winton, M. 2015. Dominance of the Southern Ocean in anthropogenic carbon and heat uptake in CMIP5 Models. *Journal of Climate* **28**, 862–886.

Froneman, P.W. & Bernard, K.S. 2004. Trophic cascading in the Polar Frontal Zone of the Southern Ocean during austral autumn 2002. *Polar Biology* **27**, 112–118.

Froneman, P.W. & Pakhomov, E.A. 1998. Biogeographic study of the planktonic communities of the Prince Edward Islands. *Journal of Plankton Research* **20**, 653–669.

Froneman, P., McQuaid, C. & Perissinotto, R. 1995a. Biogeographic structure of the microphytoplankton assemblages of the south Atlantic and Southern Ocean during austral summer. *Journal of Plankton Research* **17**, 1791–1802.

Froneman, P., Pakhomov, E. & Meaton, V. 1998. Surface distribution of microphytoplankton of the south-west Indian Ocean along a repeat transect between Cape Town and the Prince Edward Islands. *South African Journal of Science* **94**, 124–129.

Froneman, P. & Perissinotto, R. 1996a. Microzooplankton grazing and protozooplankton community structure in the South Atlantic and in the Atlantic sector of the Southern Ocean. *Deep Sea Research Part I: Oceanographic Research Papers* **43**, 703–721.

Froneman, P. & Perissinotto, R. 1996b. Microzooplankton grazing in the Southern Ocean: implications for the carbon cycle. *Marine Ecology* **17**, 99–115.

Froneman, P., Perissinotto, R., McQuaid, C. & Laubscher, R. 1995b. Summer distribution of netphytoplankton in the Atlantic sector of the Southern Ocean. *Polar Biology* **15**, 77–84.

Fuhrman, J.A. 1999. Marine viruses and their biogeochemical and ecological effects. *Nature* **399**, 541–548.

Fyfe, J.C. & Saenko, O.A. 2005. Human-induced change in the Antarctic circumpolar current. *Journal of Climate* **18**, 3068–3073.

Fyfe, J.C. & Saenko, O.A. 2006. Simulated changes in the extratropical southern hemisphere winds and currents. *Geophysical Research Letters* **33**, L06701, doi:10.1029/2005GL025332

Garibotti, I.A., Vernet, M., Ferrario, M.E., Smith, R.C., Ross, R.M. & Quetin, L.B. 2003. Phytoplankton spatial distribution patterns along the western Antarctic Peninsula (Southern Ocean). *Marine Ecology Progress Series* **261**, 21–39.

García de la Rosa, S.B., Sanchez, F. & Figueroa, D. 1997. Comparative feeding ecology of patagonian toothfish (*Dissostichus eleginoides*) in the Southwestern Atlantic. *CCAMLR Science* **4**, 105–124.

Garrison, V.H., Shinn, E.A., Foreman, W.T., Griffin, D.W., Holmes, C.W., Kellogg, C.A., Majewski, M.S., Richardson, L.L., Ritchie, K.B. & Smith, G.W. 2003. African and Asian dust: From desert soils to coral reefs. *Bioscience* **53**, 469–480.

Garzio, L.M. & Steinberg, D.K. 2013. Microzooplankton community composition along the western Antarctic Peninsula. *Deep Sea Research Part I: Oceanographic Research Papers* **77**, 36–49.

Gent, P.R. 2016. Effects of southern hemisphere wind changes on the meridional overturning circulation in ocean models. *Annual Review of Marine Science* **8**, 79–94.

Gervais, F., Riebesell, U. & Gorbunov, M.Y. 2002. Changes in primary productivity and chlorophyll a in response to iron fertilization in the southern polar frontal zone. *Limnology and Oceanography* **47**, 1324–1335.

Gibbons, M. 1997. Pelagic biogeography of the south Atlantic Ocean. *Marine Biology* **129**, 757–768.

Gille, S.T. 2002. Warming of the Southern Ocean since the 1950s. *Science* **295**, 1275–1277.

Gille, S.T. 2008. Decadal-scale temperature trends in the southern hemisphere ocean. *Journal of Climate* **21**, 4749–4765.

Gleitz, M., Grossmann, S., Scharekm, R. & Smetacek, V. 1996. Ecology of diatom and bacterial assemblages in water associated with melting summer sea ice in the Weddell Sea, Antarctica. *Antarctic Science* **8**, 135–146.

Gloersen, P., Campbell, W.J., Cavalieri, D.J., Comiso, J.C., Parkinson, C.L. & Jay Zwally, H. 1993. Satellite passive microwave observations and analysis of Arctic and Antarctic Sea ice, 1978–1987. *Annals of Glaciology* **17**, 149–154.

González, A.F. & Rodhouse, P.G. 1998. Fishery biology of the seven star flying squid *Martialia hyadesi* at South Georgia during winter. *Polar Biology* **19**, 231.

Gordon, A.L., Zambianchi, E., Orsi, A., Visbeck, M., Giulivi, C.F., Whitworth, T. & Spezie, G. 2004. Energetic plumes over the western Ross Sea continental slope. *Geophysical Research Letters* **31**, L21302, doi:10.1029/2004GL020785

Grant, S., Constable, A., Raymond, B. & Doust, S. 2006. *Bioregionalisation of the Southern Ocean: report of experts workshop.* Experts Workshop on Bioregionalisation of the Southern Ocean, held in Hobart, Australia, 4–8 September 2006, hosted by WWF-Australia and the Antarctic Climate and Ecosystems Cooperative Research Centre.

Gravalosa, J.M., Flores, J.-A., Sierro, F.J. & Gersonde, R. 2008. Sea surface distribution of coccolithophores in the eastern Pacific sector of the Southern Ocean (Bellingshausen and Amundsen Seas) during the late austral summer of 2001. *Marine Micropaleontology* **69**, 16–25.

Greve, M., Mathakutha, R., Steyn, C. & Chown, S.L. 2017. Terrestrial invasions on sub-Antarctic Marion and Prince Edward Islands. *Bothalia-African Biodiversity & Conservation* **47**, 1–21.

Griffin, D.W. & Kellogg, C.A. 2004. Dust storms and their impact on ocean and human health: dust in Earth's atmosphere. *EcoHealth* **1**, 284–295.

Griffiths, H.J., Danis, B. & Clarke, A. 2011. Quantifying Antarctic marine biodiversity: the SCAR-MarBIN data portal. *Deep Sea Research Part II: Topical Studies in Oceanography* **58**, 18–29.

Guinet, C., Cherel, Y., Ridoux, V. & Jouventin, P. 1996. Consumption of marine resources by seabirds and seals in Crozet and Kerguelen waters: changes in relation to consumer biomass 1962–85. *Antarctic Science* **8**, 23–30.

Guinotte, J.M. & Fabry, V. 2008. Ocean acidification and its potential effects on marine ecosystems. *New York Academy of Sciences* **1134**, 320–342, doi: 10.1196/annals.1439.013.

Gundersen, K., Mountain, C., Taylor, D., Ohye, R. & Shen, J. 1972. Some chemical and microbiological observations in the Pacific Ocean off the Hawaiian Islands. *Limnology and Oceanography* **17**, 524–532.

Gutt, J., Bertler, N., Bracegirdle, T.J., Buschmann, A., Comiso, J., Hosie, G., Isla, E., Schloss, I.R., Smith, C.R., Tournadre, J. & Xavier, J.C. 2015. The Southern Ocean ecosystem under multiple climate change stresses-an integrated circumpolar assessment. *Global Change Biology* **21**, 1434–1453.

Haberman, K.L., Ross, R.M. & Quetin, L.B. 2003. Diet of the Antarctic krill (*Euphausia superba* Dana): II. Selective grazing in mixed phytoplankton assemblages. *Journal of Experimental Marine Biology and Ecology* **283**, 97–113.

Häder, D.-P., Kumar, H., Smith, R.C. & Worrest, R.C. 2003. Aquatic ecosystems: effects of solar ultraviolet radiation and interactions with other climatic change factors. *Photochemical & Photobiological Sciences* **2**, 39–50.

Hairston, N.G., Smith, F.E. & Slobodkin, L.B. 1960. Community structure, population control, and competition. *The American Naturalist* **94**, 421–425.

Hall, A. & Visbeck, M. 2002. Synchronous variability in the southern hemisphere atmosphere, sea ice, and ocean resulting from the annular mode. *Journal of Climate* **15**, 3043–3057.

Hamm, C.E. 2000. Architecture, ecology and biogeochemistry of *Phaeocystis* colonies. *Journal of Sea Research* **43**, 307–315.

Hamm, C.E. & Rousseau, V. 2003. Composition, assimilation and degradation of *Phaeocystis globosa*-derived fatty acids in the North Sea. *Journal of Sea Research* **50**, 271–283.

Hamner, W.M., Hamner, P.P., Strand, S.W. & Gilmer, R.W. 1983. Behavior of Antarctic krill, *Euphausia superba*: chemoreception, feeding, schooling, and molting. *Science* **220**, 433–435.

Hansen, F. & Van Boekel, W. 1991. Grazing pressure of the calanoid copepod Temora iongicorm's on a Phaeocystis dominated spring. *Marine Ecology Progress Series* **78**, 123–129.

Hansen, J., Sato, M., Hearty, P., Ruedy, R., Kelley, M., Masson-Delmotte, V., Russell, G., Tselioudis, G., Cao, J. & Rignot, E. 2016. Ice melt, sea level rise and superstorms: evidence from paleoclimate data, climate modeling, and modern observations that 2 C global warming could be dangerous. *Atmospheric Chemistry and Physics* **16**, 3761–3812.

Harvell, C.D., Kim, K., Burkholder, J.M., Colwell, R.R., Epstein, P.R., Grimes, D.J., Hofmann, E.E., Lipp, E.K., Osterhaus, A.D.M.E., Overstreet, R.M., Porter, J.W., Smith, G.W. & Vasta G.R. 1999. Emerging marine diseases--climate links and anthropogenic factors. *Science* **285**, 1505–1510.

Harvell, D., Aronson, R., Baron, N., Connell, J., Dobson, A., Ellner, S., Gerber, L., Kim, K., Kuris, A., McCallum, H., Lafferty, K., McKay, B., Porter, J., Pascual, M., Smith, G., Sutherland, K. & Ward, J. 2004. The rising tide of ocean diseases: Unsolved problems and research priorities. *Frontiers in Ecology* **2**, 375–382.

Hawkins, E. & Sutton, R. 2012. Time of emergence of climate signals. *Geophysical Research Letters* **39**, L01702, doi:10.1029/2011GL050087.

Hayes, M.L., Bonaventura, J., Mitchell, T.P., Prospero, J.M., Shinn, E.A., Van Dolah, F. & Barber, R.T. 2001. How are climate and marine biological outbreaks functionally linked? *Hydrobiologia* **460**, 213–220.

Hellmer, H.H., Kauker, F., Timmermann, R., Determann, J. & Rae, J. 2012. Twenty-first-century warming of a large Antarctic ice-shelf cavity by a redirected coastal current. *Nature* **485**, 225–228.

Hernández, K.L., Quiñones, R.A., Daneri, G. & Helbling, E.W. 2006. Effects of solar radiation on bacterioplankton production in the upwelling system off central-southern Chile. *Marine Ecology Progress Series* **315**, 19–31.

Hewitt, R.P., Demer, D.A. & Emery, J.H. 2003. An 8-year cycle in krill biomass density inferred from acoustic surveys conducted in the vicinity of the South Shetland Islands during the austral summers of 1991–1992 through 2001–2002. *Aquatic Living Resources* **16**, 205–213.

Hewson, I. & Fuhrman, J.A. 2006. Viral impacts upon marine bacterioplankton assemblage structure. *Journal of the Marine Biological Association of the United Kingdom* **86**, 577–589.

Hill, S.L., Phillips, T. & Atkinson, A. 2013. Potential climate change effects on the habitat of Antarctic krill in the Weddell quadrant of the Southern Ocean. *PloS One* **8**, e72246.

Hinke, J.T., Salwicka, K., Trivelpiece, S.G., Watters, G.M. & Trivelpiece, W.Z. 2007. Divergent responses of Pygoscelis penguins reveal a common environmental driver. *Oecologia* **153**, 845–855, doi: 10.1007/s00442-007-0781-4.

Hoffmann, L., Peeken, I., Lochte, K., Assmy, P. & Veldhuis, M. 2006. Different reactions of Southern Ocean phytoplankton size classes to iron fertilization. *Limnology and Oceanography* **51**, 1217–1229.

Hofmann, E.E. 1985. The large-scale horizontal structure of the Antarctic circumpolar current from FGGE drifters. *Journal of Geophysical Research: Oceans* **90**, 7087–7097.

Hofmann, E.E. & Murphy, E.J. 2004. Advection, krill, and Antarctic marine ecosystems. *Antarctic Science* **16**, 487–499.

Hogg, A.M.C., Meredith, M.P., Blundell, J.R. & Wilson, C. 2008. Eddy heat flux in the Southern Ocean: response to variable wind forcing. *Journal of Climate* **21**, 608–620.

Holm-Hansen, O. & Sakshaug, E. 2002. Polar marine phytoplankton. In *Encyclopedia of Environmental Microbiology*, G. Bitton (ed). New York: Wiley, https://doi.org/10.1002/0471263397.env257.

Honjo, S., Francois, R., Manganini, S., Dymond, J. & Collier, R. 2000. Particle fluxes to the interior of the Southern Ocean in the western Pacific sector along 170 W. *Deep Sea Research Part II: Topical Studies in Oceanography* **47**, 3521–3548.

Hopkins, T. 1985. Food web of an Antarctic midwater ecosystem. *Marine Biology* **89**, 197–212.

Hopkins, T. 1987. Midwater food web in McMurdo Sound, Ross Sea, Antarctica. *Marine Biology* **96**, 93–106.

Hunt, B., Pakhomov, E. & McQuaid, C. 2001. Short-term variation and long-term changes in the oceanographic environment and zooplankton community in the vicinity of a sub-Antarctic archipelago. *Marine Biology* **138**, 369–381.

Hunt, B., Pakhomov, E. & McQuaid, C. 2002a. Community structure of mesozooplankton in the Antarctic polar frontal zone in the vicinity of the Prince Edward Islands (Southern Ocean): small-scale distribution patterns in relation to physical parameters. *Deep Sea Research Part II: Topical Studies in Oceanography* **49**, 3307–3325.

Hunt, B., Pakhomov, E., Siegel, V., Strass, V., Cisewski, B. & Bathmann, U. 2011. The seasonal cycle of the Lazarev Sea macrozooplankton community and a potential shift to top-down trophic control in winter. *Deep Sea Research Part II: Topical Studies in Oceanography* **58**, 1662–1676.

Hunt, B.P. & Hosie, G.W. 2003. The continuous plankton recorder in the Southern Ocean: a comparative analysis of zooplankton communities sampled by the CPR and vertical net hauls along 140 E. *Journal of Plankton Research* **25**, 1561–1579.

Hunt Jr, G.L., Stabeno, P., Walters, G., Sinclair, E., Brodeur, R.D., Napp, J.M. & Bond, N.A. 2002b. Climate change and control of the south-eastern Bering Sea pelagic ecosystem. *Deep Sea Research Part II: Topical Studies in Oceanography* **49**, 5821–5853.

Huntley, M.E., Sykes, P.F. & Marin, V. 1989. Biometry and trophodynamics of *Salpa thompsoni* Foxton (Tunicata: Thaliacea) near the Antarctic Peninsula in austral summer, 1983–1984. *Polar Biology* **10**, 59–70.

Ichii, T., Bengston, J., Boveng, P., Takao, Y., Jansen, J., Hiruki-Raring, L., Cameron, M., Okamura, H., Hayashi, T. & Naganobu, M. 2007. Provisioning strategies of Antarctic fur seals and chinstrap penguins produce different responses to distribution of common prey and habitat. *Marine Ecology Progress Series* **344**, 277–297.

Ichii, T., Shinohara, N., Fujise, Y., Nishiwaki, S. & Matsuoka, K. 1998. Interannual changes in body fat condition index of minke whales in the Antarctic. *Marine Ecology Progress Series* **175**, 1–12.

Imber, M.J. 1978. The squid families *cranchiidae and gonatidae* (Cephalopoda: Teuthoidea) in the New Zealand region. *New Zealand Journal of Zoology* **5**, 445–484.

Inchausti, P., Guinet, C., Koudil, M., Durbec, J., Barbraud, C., Weimerskirch, H., Cherel, Y. & Jouventin, P. 2003. Inter-annual variability in the breeding performance of seabirds in relation to oceanographic anomalies that affect the Crozet and the Kerguelen sectors of the Southern Ocean. *Journal of Avian Biology* **34**, 170–176.

IPCC (Intergovernmental Panel on Climate Change). 1995. The science of climate change. Houghton, J.T., Meira Filho, L.G., Callander, B.A., Harris, N., Kattenberg, A. & Maskell, K. (eds) Cambridge, UK: Cambridge University Press.

Irwin, A.J., Finkel, Z.V., Schofield, O.M. & Falkowski, P.G. 2006. Scaling-up from nutrient physiology to the size-structure of phytoplankton communities. *Journal of Plankton Research* **28**, 459–471.

Jaccard, S., Hayes, C.T., Martínez-García, A., Hodell, D., Anderson, R.F., Sigman, D. & Haug, G. 2013. Two modes of change in Southern Ocean productivity over the past million years. *Science* **339**, 1419–1423.

Jacobsen, A., Bratbak, G. & Heldal, M. 1996. Isolation and characterization of a virus infecting phaeocystis pouchetii (Prymnesiophyceae) 1. *Journal of Phycology* **32**, 923–927.

Jaeger, A., Lecomte, V.J., Weimerskirch, H., Richard, P. & Cherel, Y. 2010. Seabird satellite tracking validates the use of latitudinal isoscapes to depict predators' foraging areas in the Southern Ocean. *Rapid Communications in Mass Spectrometry* **24**, 3456–3460.

Jarman, S., Elliott, N., Nicol, S. & McMinn, A. 2002. Genetic differentiation in the Antarctic coastal krill *Euphausia crystallorophias*. *Heredity* **88**, 280–287.

Jenouvrier, S., Barbraud, C., Cazelles, B. & Weimerskirch, H. 2005b. Modelling population dynamics of seabirds: importance of the effects of climate fluctuations on breeding proportions. *Oikos* **108**, 511–522.

Jenouvrier, S., Weimerskirch, H., Barbraud, C., Park, Y.-H. & Cazelles, B. 2005a. Evidence of a shift in the cyclicity of Antarctic seabird dynamics linked to climate. *Proceedings of the Royal Society of London B: Biological Sciences* **272**, 887–895.

Jickells, T., An, Z., Andersen, K.K., Baker, A., Bergametti, G., Brooks, N., Cao, J., Boyd, P., Duce, R. & Hunter, K. 2005. Global iron connections between desert dust, ocean biogeochemistry, and climate. *Science* **308**, 67–71.

Jue, N.K., Batta-Lona, P.G., Trusiak, S., Obergfell, C., Bucklin, A., O'Neill, M.J. & O'Neill, R.J. 2016. Rapid evolutionary rates and unique genomic signatures discovered in the first reference genome for the Southern Ocean salp, *Salpa thompsoni* (Urochordata, Thaliacea). *Genome Biology and Evolution* **8**, 3171–3186.

Kaehler, S., Pakhomov, E. & McQuaid, C. 2000. Trophic structure of the marine food web at the Prince Edward Islands (Southern Ocean) determined by δ13C and δ15N analysis. *Marine Ecology Progress Series* **208**, 13–20.

Kahru, M., Mitchell, B., Gille, S., Hewes, C. & Holm-Hansen, O. 2007. Eddies enhance biological production in the Weddell-Scotia confluence of the Southern Ocean. *Geophysical Research Letters* **34**, L14603, doi:10.1029/2007GL030430

Kaiser, S., Brandão, S.N., Brix, S., Barnes, D.K., Bowden, D.A., Ingels, J., Leese, F., Schiaparelli, S., Arango, C.P. & Badhe, R. 2013. Patterns, processes and vulnerability of Southern Ocean benthos: a decadal leap in knowledge and understanding. *Marine Biology* **160**, 2295–2317.

Kang, S.-H. & Lee, S. 1995. Antarctic phytoplankton assemblage in the western Bransfield Strait region, February 1993: composition, biomass, and mesoscale distributions. *Marine Ecology Progress Series* **129**, 253–267.

Karentz, D. & Bosch, I. 2001. Influence of ozone-related increases in ultraviolet radiation on Antarctic marine organisms. *Integrative and Comparative Biology* **41**, 3–16.

Kawaguchi, S., Ishida, A., King, R., Raymond, B., Waller, N., Constable, A., Nicol, S., Wakita, M. & Ishimatsu, A. 2013. Risk maps for Antarctic krill under projected Southern Ocean acidification. *Nature Climate Change* **3**, 843–847.

Kawaguchi, S., Kurihara, H., King, R., Hale, L., Berli, T., Robinson, J.P., Ishida, A., Wakita, M., Virtue, P. & Nicol, S. 2011. Will krill fare well under Southern Ocean acidification? *Biology Letters* **7**, 288–291.

Kear, A.J. 1992. The diet of Antarctic squid: comparison of conventional and serological gut contents analyses. *Journal of Experimental Marine Biology and Ecology* **156**, 161–178.

Kilada, R., Reiss, C.S., Kawaguchi, S., King, R.A., Matsuda, T. & Ichii, T. 2017. Validation of band counts in eyestalks for the determination of age of Antarctic krill, *Euphausia superba*. *PloS One* **12**, e0171773.

Kniveton, D., Todd, M., Sciare, J. & Mihalopoulos, N. 2003. Variability of atmospheric dimethylsulphide over the southern Indian Ocean due to changes in ultraviolet radiation. *Global Biogeochemical Cycles* **17**, 1096, doi:10.1029/2003GB002033, 4

Kock, K.-H., Barrera-Oro, E., Belchier, M., Collins, M., Duhamel, G., Hanchet, S., Pshenichnov, L., Welsford, D. & Williams, R. 2012. The role of fish as predators of krill (*Euphausia superba*) and other pelagic resources in the Southern Ocean. *CCAMLR Science* **19**, 115–169.

Kohfeld, K.E., Le Quéré, C., Harrison, S.P. & Anderson, R.F. 2005. Role of marine biology in glacial-interglacial CO2 cycles. *Science* **308**, 74–78.

Komar, P.D., Morse, A.P., Small, L.F. & Fowler, S.W. 1981. An analysis of sinking rates of natural copepod and euphausiid fecal pellets. *Limnology and Oceanography* **26**, 172–180.

Koubbi, P., Moteki, M., Duhamel, G., Goarant, A., Hulley, P.-A., O'Driscoll, R., Ishimaru, T., Pruvost, P., Tavernier, E. & Hosie, G. 2011. Ecoregionalization of myctophid fish in the Indian sector of the Southern Ocean: results from generalized dissimilarity models. *Deep Sea Research Part II: Topical Studies in Oceanography* **58**, 170–180.

Kozlov, A. 1995. A review of the trophic role of mesopelagic fish of the family Myctophidae in the Southern Ocean ecosystem. *CCAMLR Science* **2**, 71–77.

Kropuenske, L.R., Mills, M.M., van Dijken, G.L., Bailey, S., Robinson, D.H., Welschmeyer, N.A. & Arrigoa, K.R. 2009. Photophysiology in two major Southern Ocean phytoplankton taxa: photoprotection in *Phaeocystis Antarctica* and *Fragilariopsis cylindrus*. *Limnology and Oceanography* **54**, 1176–1196.

Kumar, N., Anderson, R., Mortlock, R. & Froelich, P. 1995. Increased biological productivity and export production in the glacial Southern Ocean. *Nature* **378**, 675.

Kurihara, H., Shimode, S. & Shirayama, Y. 2004. Sub-lethal effects of elevated concentration of CO2 on planktonic copepods and sea urchins. *Journal of Oceanography* **60**, 743–750.

Kwok, R. & Comiso, J.C. 2002. Spatial patterns of variability in Antarctic surface temperature: connections to the southern hemisphere annular mode and the southern oscillation. *Geophysical Research Letters* **29**, 1705, doi:10.1029/2002GL015415.

La Mesa, M., Eastman, J. & Vacchi, M. 2004. The role of notothenioid fish in the food web of the Ross Sea shelf waters: a review. *Polar Biology* **27**, 321–338.

Lampitt, R. & Antia, A. 1997. Particle flux in deep seas: regional characteristics and temporal variability. *Deep Sea Research Part I: Oceanographic Research Papers* **44**, 1377–1403.

Lancelot, C., Mathot, S., Veth, C. & de Baar, H. 1993. Factors controlling phytoplankton ice-edge blooms in the marginal ice-zone of the northwestern Weddell Sea during sea ice retreat 1988: field observations and mathematical modelling. *Polar Biology* **13**, 377–387.

Landschützer, P., Gruber, N., Haumann, F.A., Rödenbeck, C., Bakker, D.C., Van Heuven, S., Hoppema, M., Metzl, N., Sweeney, C. & Takahashi, T. 2015. The reinvigoration of the Southern Ocean carbon sink. *Science* **349**, 1221–1224.

Latif, M., Martin, T. & Park, W. 2013. Southern Ocean sector centennial climate variability and recent decadal trends. *Journal of Climate* **26**, 7767–7782.

Laubscher, R., Perissinotto, R. & McQuaid, C. 1993. Phytoplankton production and biomass at frontal zones in the Atlantic sector of the Southern Ocean. *Polar Biology* **13**, 471–481.

Laws, R. 1977. Seals and whales of the Southern Ocean. *Philosophical Transactions of the Royal Society of London B: Biological Sciences* **279**, 81–96.

Laws, R.M. 1985. The ecology of the Southern Ocean. *American Scientist* **73**, 26–40.

Le Fèvre, J., Legendre, L. & Rivkin, R.B. 1998. Fluxes of biogenic carbon in the Southern Ocean: roles of large microphagous zooplankton. *Journal of Marine Systems* **17**, 325–345.

Legendre, L. & Le Fèvre, J. 1995. Microbial food webs and the export of biogenic carbon in oceans. *Aquatic Microbial Ecology* **9**, 69–77.

Li, X., Holland, D.M., Gerber, E.P. & Yoo, C. 2014. Impacts of the north and tropical Atlantic Ocean on the Antarctic Peninsula and sea ice. *Nature* **505**, 538–542.

Lipinski, M. & Jackson, S. 1989. Surface-feeding on cephalopods by procellariiform seabirds in the southern Benguela region, South Africa. *Journal of Zoology* **218**, 549–563.

Liu, J., Curry, J.A. & Martinson, D.G. 2004. Interpretation of recent Antarctic sea ice variability. *Geophysical Research Letters* **31**, L02205, doi:10.1029/2003GL018732

Lizotte, M.P. 2001. The contributions of sea ice algae to Antarctic marine primary production 1. *American Zoologist* **41**, 57–73.

Llovel, W. & Terray, L. 2016. Observed southern upper-ocean warming over 2005–2014 and associated mechanisms. *Environmental Research Letters* **11**, 124023.

Loeb, V.J., Hofmann, E.E., Klinck, J.M., Holm-Hansen, O. & White, W.B. 2009. ENSO and variability of the Antarctic Peninsula pelagic marine ecosystem. *Antarctic Science* **21**, 135–148.

Loeb, V.J., Siegel, V., Holm-Hansen, O. & Hewitt, R. 1997. Effects of sea-ice extent and krill or salp dominance on the Antarctic food web. *Nature* **387**, 897.

Lohrer, A.M., Cummings, V.J. & Thrush, S.F. 2013. Altered sea ice thickness and permanence affects benthic ecosystem functioning in coastal Antarctica. *Ecosystems* **16**, 224–236.

Long, R.A. & Azam, F. 2001. Antagonistic interactions among marine pelagic bacteria. *Applied and Environmental Microbiology* **67**, 4975–4983.

Longhurst, A.L. 1998. *Ecological Geography of the Sea*. San Diego: Academic Press, 398 pp.

Longhurst, A.R. & Harrison, W.G. 1989. The biological pump: profiles of plankton production and consumption in the upper ocean. *Progress in Oceanography* **22**, 47–123.

Lowther, A.D., Lydersen, C., Biuw, M., de Bruyn, P.N., Hofmeyr, G.J. & Kovacs, K.M. 2014. Post-breeding at-sea movements of three central-place foragers in relation to submesoscale fronts in the Southern Ocean around Bouvetøya. *Antarctic Science* **26**, 533–544.

Lutjeharms, J. 1985. Location of frontal systems between Africa and Antarctica: some preliminary results. *Deep Sea Research Part A. Oceanographic Research Papers* **32**, 1499–1509.

Lutjeharms, J. 1990. The oceanography and fish distribution of the Southern Ocean. In *Fishes of the Southern Ocean*, O. Gon (ed). Grahamstown, South Africa: JLB Smith Institute of Ichthyology, 6–27.

Lutz, M.J., Caldeira, K., Dunbar, R.B. & Behrenfeld, M.J. 2007. Seasonal rhythms of net primary production and particulate organic carbon flux to depth describe the efficiency of biological pump in the global ocean. *Journal of Geophysical Research: Oceans* **112**, C10011, doi:10.1029/2006JC003706

Lyman, J.M., Good, S.A., Gouretski, V.V., Ishii, M., Johnson, G.C., Palmer, M.D., Smith, D.M. & Willis, J.K. 2010. Robust warming of the global upper ocean. *Nature* **465**, 334–337.

Malin, G., Wilson, W.H., Bratbak, G., Liss, P.S. & Mann, N.H. 1998. Elevated production of dimethylsulfide resulting from viral infection of cultures of *Phaeocystis pouchetii*. *Limnology and Oceanography* **43**, 1389–1393.

Malinverno, E., Maffioli, P. & Gariboldi, K. 2016. Latitudinal distribution of extant fossilizable phytoplankton in the Southern Ocean: planktonic provinces, hydrographic fronts and palaeoecological perspectives. *Marine Micropaleontology* **123**, 41–58.

Malloy, K.D., Holman, M.A., Mitchell, D. & Detrich, H.W. 1997. Solar UVB-induced DNA damage and photoenzymatic DNA repair in Antarctic zooplankton. *Proceedings of the National Academy of Sciences* **94**, 1258–1263.

Manatsa, D., Morioka, Y., Behera, S.K., Yamagata, T. & Matarira, C.H. 2013. Link between Antarctic ozone depletion and summer warming over southern Africa. *Nature Geoscience* **6**, 934–939.

Margalef, R. 1978. Life-forms of phytoplankton as survival alternatives in an unstable environment. *Oceanologica Acta* **1**, 493–509.

Marinov, I., Gnanadesikan, A., Toggweiler, J.R. & Sarmiento, J.L. 2006. The Southern Ocean biogeochemical divide. *Nature* **441**, 964–967.

Marshall, G.J. 2003. Trends in the southern annular mode from observations and reanalyses. *Journal of Climate* **16**, 4134–4143.

Marshall, J. & Speer, K. 2012. Closure of the meridional overturning circulation through Southern Ocean upwelling. *Nature Geoscience* **5**, 171–180.

Massom, R.A. & Stammerjohn, S.E. 2010. Antarctic Sea ice change and variability–physical and ecological implications. *Polar Science* **4**, 149–186.

Mayali, X. & Azam, F. 2003. Algicidal bacteria in the sea and their impact on algal blooms. *Journal of Eukaryotic Microbiology* **51**, 139–144.

Mayali, X. & Azam, F. 2004. Algicidal bacteria in the sea and their impact on algal blooms. *Journal of Eukaryotic Microbiology* **51**, 139–144.

Mayewski, P.A., Meredith, M.P., Summerhayes, C.P., Turner, J., Worby, A., Barrett, P.J., Casassa, G., Bertler, N.A.N., Bracegirdle, T., Naveira Garabato, A.C., Bromwich, D., Campbell, H., Hamilton, G.S., Lyons, W.B., Maasch, K.A., Aoki, S., Xiao, C. & van Omen, T. 2009. State of Antarctic and Southern Ocean climate system. *Reviews of Geophysics* **47**, 1–38.

McGlone, M.S., Turney, C.S., Wilmshurst, J.M., Renwick, J. & Pahnke, K. 2010. Divergent trends in land and ocean temperature in the Southern Ocean over the past 18,000 years. *Nature Geoscience* **3**, 622–626.

McIntyre, T., De Bruyn, P., Ansorge, I., Bester, M., Bornemann, H., Plötz, J. & Tosh, C. 2010. A lifetime at depth: vertical distribution of southern elephant seals in the water column. *Polar Biology* **33**, 1037–1048.

McLeod, D.J., Hallegraeff, G.M., Hosie, G.W. & Richardson, A.J. 2012. Climate-driven range expansion of the red-tide dinoflagellate *Noctiluca scintillans* into the Southern Ocean. *Journal of Plankton Research* **34**, 332–337.

McLeod, D.J., Hosie, G.W., Kitchener, J.A., Takahashi, K.T. & Hunt, B.P. 2010. Zooplankton atlas of the Southern Ocean: the SCAR SO-CPR Survey 1991–2008. *Polar Science* **4**, 353–385.

McMinn, A. & Hodgson, D. 1993. Summer phytoplankton succession in Ellis Fjord, eastern Antarctica. *Journal of Plankton Research* **15**, 925–938.

McNeil, B.I. & Matear, R.J. 2008. Southern Ocean acidification: a tipping point at 450-ppm atmospheric CO2. *Proceedings of the National Academy of Sciences* **105**, 18860–18864.

McQuaid, C.D. 2010. Marine connectivity: Timing is everything. *Current Biology* **20**, R938–R940, doi: 10.1016/j.cub.2010.09.049.

McQuaid, C.D. & Froneman, P.W. 2008. Biology in the oceanographic environment, in S.L. Chown & P.W. Froneman (eds). *The Prince Edward Islands: Land–Sea Interactions in a Changing Ecosystem*. Stellenbosch, South Africa: Sun Press, 97–120.

Melbourne-Thomas, J., Constable, A., Wotherspoon, S. & Raymond, B. 2013. Testing paradigms of ecosystem change under climate warming in Antarctica. *PLoS One* **8**, e55093.

Melbourne-Thomas, J., Corney, S., Trebilco, R., Meiners, K., Stevens, R., Kawaguchi, S., Sumner, M. & Constable, A. 2016. Under ice habitats for Antarctic krill larvae: could less mean more under climate warming? *Geophysical Research Letters* **43**, 10322–10327, doi:10.1002/2016GL070846

Menge, B.A. 2000. Top-down and bottom-up community regulation in marine rocky intertidal habitats. *Journal of Experimental Marine Biology and Ecology* **250**, 257–289.

Mercer, J.H. 1978. West Antarctic ice sheet and CO2 greenhouse effect- a threat of disaster. *Nature* **271**, 321–325.

Meredith, M.P. & King, J.C. 2005. Rapid climate change in the ocean west of the Antarctic Peninsula during the second half of the 20th century. *Geophysical Research Letters* **32**, L19604, doi:10.1029/2005GL024042

Meyer, B. 2012. The overwintering of Antarctic krill, *Euphausia superba*, from an ecophysiological perspective. *Polar Biology* **35**, 15–37.

Michels, J., Dieckmann, G.S., Thomas, D.N., Schnack-Schiel, S.B., Krell, A., Assmy, P., Kennedy, H., Papadimitriou, S. & Cisewski, B. 2008. Short-term biogenic particle flux under late spring sea ice in the western Weddell Sea. *Deep Sea Research Part II* **55**, 1024–1039.

Micol, T. & Jouventin, P. 2001. Long-term population trends in seven Antarctic seabirds at Pointe Géologie (Terre Adélie) human impact compared with environmental change. *Polar Biology* **24**, 175–185.

Mock, T., Otillar, R.P., Strauss, J., McMullan, M., Paajanen, P., Schmutz, J., Salamov, A., Sanges, R., Toseland, A. & Ward, B.J. 2017. Evolutionary genomics of the cold-adapted diatom *Fragilariopsis cylindrus*. *Nature* **541**, 536–540.

Moline, M.A., Claustre, H., Frazer, T.K., Schofield, O. & Vernet, M. 2004. Alteration of the food web along the Antarctic Peninsula in response to a regional warming trend. *Global Change Biology* **10**, 1973–1980.

Moline, M.A., Karnovsky, N.J., Brown, Z., Divoky, G.J., Frazer, T.K., Jacoby, C.A., Torres, J.J. & Fraser, W.R. 2008. High latitude changes in ice dynamics and their impact on polar marine ecosystems. *Annals of the New York Academy of Sciences* **1134**, 267–319.

Montes-Hugo, M., Doney, S.C., Ducklow, H.W., Fraser, W., Martinson, D., Stammerjohn, S.E. & Schofield, O. 2009. Recent changes in phytoplankton communities associated with rapid regional climate change along the Western Antarctic Peninsula. *Science* **323**, 1470–1473.

Moore, J.K. & Abbott, M.R. 2000. Phytoplankton chlorophyll distributions and primary production in the Southern Ocean. *Journal of Geophysical Research* **105**(C12), 28709–28722, doi:10.1029/1999JC000043

Moore, J.K. & Abbott, M.R. 2002. Surface chlorophyll concentrations in relation to the Antarctic polar front: seasonal and spatial patterns from satellite observations. *Journal of Marine Systems* **37**, 69–86.

Moore, J.K., Abbott, M.R., Richman, J.G. & Nelson, D.M. 2000. The Southern Ocean at the last glacial maximum: a strong sink for atmospheric carbon dioxide. *Global Biogeochemical Cycles* **14**, 455–475.

Moore, J.K., Doney, S.C., Glover, D.M. & Fung, I.Y. 2001. Iron cycling and nutrient-limitation patterns in surface waters of the world Ocean. *Deep Sea Research Part II: Topical Studies in Oceanography* **49**, 463–507.

Morgenthaler, A., Frere, E., Rey, A.R., Torlaschi, C., Cedrola, P., Tiberi, E., Lopez, R., Mendieta, E., Carranza, M.L., Acardi, S., Collm, N., Gandini, P. & Millones, A. 2018. Unusual number of Southern Rockhopper Penguins, Eudyptes chrysocome, molting and dying along the Southern Patagonian coast of Argentina: Pre-molting dispersion event related to adverse oceanographic conditions? *Polar Biology* **41**, 1041–1047.

Mori, M. & Butterworth, D.S. 2004. Consideration of multispecies interactions in the Antarctic: a preliminary model of the minke whale–blue whale–krill interaction. *African Journal of Marine Science* **26**, 245–259.

Morrison, A.K., Griffies, S.M., Winton, M., Anderson, W.G. & Sarmiento, J.L. 2016. Mechanisms of Southern Ocean heat uptake and transport in a global eddying climate model. *Journal of Climate* **29**, 2059–2075.

Moy, A.D., Howard, W.R., Bray, S.G. & Trull, T.W. 2009. Reduced calcification in modern Southern Ocean planktonic foraminifera. *Nature Geoscience* **2**, 276–280.

Munn, C.B. 2006. Viruses as pathogens of marine organisms—From bacteria to whales. *Journal of the Marine Biological Association of the United Kingdom* **86**, 453–467.

Murphy, E., Cavanagh, R., Hofmann, E., Hill, S., Constable, A., Costa, D., Pinkerton, M., Johnston, N., Trathan, P. & Klinck, J. 2012. Developing integrated models of Southern Ocean food webs: including ecological complexity, accounting for uncertainty and the importance of scale. *Progress in Oceanography* **102**, 74–92.

Murphy, E.J., Hofmann, E.E., Watkins, J.L., Johnston, N.M., Piñones, A., Ballerini, T., Hill, S.L., Trathan, P.N., Tarling, G.A., Cavanagh, G.A., Young, E.F., Thorpe, S.E. & Fretwell, P. 2013. Comparison of the structure and function of Southern Ocean regional ecosystems: The Antarctic Peninsula and South Georgia. *Journal of Marine Systems* **109–110**, 22–42.

Murphy, E.J., Thorpe, S.E., Watkins, J.L. & Hewitt, R. 2004. Modeling the krill transport pathways in the Scotia Sea: Spatial and environmental connections generating the seasonal distribution of krill. *Deep Sea Research Part II* **51**, 1435–1456.

Murphy, E.J., Tratham, P.N., Watkins, J.L., Reid, K., Meredith, M.P., Forcada, J., Thorpe, S.E., Johnston, N.M. & Rothery, P. 2007b. Climatically driven fluctuations in Southern Ocean ecosystems. *Proceedings of the Royal Society B* **274**, 3057–3067.

Murphy, E., Watkins, J., Reid, K., Trathan, P., Everson, I., Croxall, J., Priddle, J., Brandon, M., Brierley, A. & Hofmann, E. 1998. Interannual variability of the South Georgia marine ecosystem: biological and physical sources of variation in the abundance of krill. *Fisheries Oceanography* **7**, 381–390.

Murphy, E., Watkins, J., Trathan, P., Reid, K., Meredith, M., Thorpe, S., Johnston, N., Clarke, A., Tarling, G. & Collins, M. 2007a. Spatial and temporal operation of the Scotia Sea ecosystem: a review of large-scale links in a krill centred food web. *Philosophical Transactions of the Royal Society of London B: Biological Sciences* **362**, 113–148.

NASA. 2017. https://ozonewatch.gsfc.nasa.gov/statistics/annual_data.html. Accessed 24 April 2017.

Nel, D.C., Lutjeharms, J.R.E., Pakhomov, E.A., Ansorge, I.J., Ryan, P.G. & Klages, N.T.W. 2001. Exploitation of mesoscale oceanographic features by grey-headed albatross *Thalassarche chrysostoma* in the southern Indian Ocean. *Marine Ecology Progress Series* **217**, 15–26.

Nemoto, T., Okiyama, M., Iwasaki, N. & Kikuchi, T. 1988. Squid as predators on krill (*Euphausia superba*) and prey for sperm whales in the Southern Ocean, in D. Sahrhage (ed.). *Antarctic Ocean and Resources Variability*. Berlin, Heidelberg: Springer, 292–296.

Nemoto, T., Okiyama, M. & Takahashi, M. 1985. Aspects of the roles of squid in food chains of marine Antarctic ecosystems, in W.R. Siegfried et al. (eds). *Antarctic Nutrient Cycles and Food Webs*. Berlin, Heidelberg: Springer, 415–420.

Newman, P.A., Nash, E.R., Kawa, S.R., Montzka, S.A. & Schauffler, S.M. 2006. When will the Antarctic ozone hole recover? *Geophysical Research Letters* **33**, L12814, doi:10.1029/2005GL025232

Nicol, S. 1990. The age-old problem of krill longevity. *BioScience* **40**, 833–836.

Nicol, S. 2003. Living krill, zooplankton and experimental investigations: a discourse on the role of krill and their experimental study in marine ecology. *Marine and Freshwater Behaviour and Physiology* **36**, 191–205.

Nicol, S. 2006. Krill, currents, and sea ice: *Euphausia superba* and its changing environment. *Bioscience* **56**, 111–120.

Nicol, S., Clarke, J., Romaine, S., Kawaguchi, S., Williams, G. & Hosie, G. 2008. Krill (*Euphausia superba*) abundance and adélie penguin (*Pygoscelis adeliae*) breeding performance in the waters off the Béchervaise Island colony, East Antarctica in 2 years with contrasting ecological conditions. *Deep Sea Research Part II: Topical Studies in Oceanography* **55**, 540–557.

Nicol, S., Pauly, T., Bindoff, N.L., Wright, S., Thiele, D., Hosie, G.W., Strutton, P.G. & Woehler, E. 2000. Ocean circulation off East Antarctica affects ecosystem structure and sea-ice extent. *Nature* **406**, 504–507.

Nishikawa, J., Naganobu, M., Ichii, T., Ishii, H., Terazaki, M. & Kawaguchi, K. 1995. Distribution of salps near the South Shetland Islands during austral summer, 1990–1991 with special reference to krill distribution. *Polar Biology* **15**, 31–39.

Nixon, M. 1987. Cephalopod diets. *Cephalopod Life Cycles* **2**, 201–219.

Olbers, D., Borowski, D., Völker, C. & Wolff, J.-O. 2004. The dynamical balance, transport and circulation of the Antarctic circumpolar current. *Antarctic Science* **16**, 439–470.

Olson, R., Sosik, H., Chekalyuk, A. & Shalapyonok, A. 2000. Effects of iron enrichment on phytoplankton in the Southern Ocean during late summer: active fluorescence and flow cytometric analyses. *Deep Sea Research Part II: Topical Studies in Oceanography* **47**, 3181–3200.

Orr, J.C., Fabry, V.J., Aumont, O., Bopp, L., Doney, S.C., Feely, R.A., Gnanadesikan, A., Gruber, N., Ishida, A. & Joos, F. 2005. Anthropogenic Ocean acidification over the twenty-first century and its impact on calcifying organisms. *Nature* **437**, 681–686.

Orsi, A., Johnson, G. & Bullister, J. 1999. Circulation, mixing, and production of Antarctic bottom water. *Progress in Oceanography* **43**, 55–109.

Pace, M.L., Cole, J.J., Carpenter, S.R. & Kitchell, J.F. 1999. Trophic cascades revealed in diverse ecosystems. *Trends in Ecology & Evolution* **14**, 483–488.

Page, T.J. & Linse, K. 2002. More evidence of speciation and dispersal across the Antarctic polar front through molecular systematics of Southern Ocean limatula (Bivalvia: Limidae). *Polar Biology* **25**, 818–826.

Pakhomov, E. 2004. Salp/krill interactions in the eastern Atlantic sector of the Southern Ocean. *Deep Sea Research Part II: Topical Studies in Oceanography* **51**, 2645–2660.

Pakhomov, E.A. & Froneman, P.W. 1999a. The Prince Edward Islands pelagic ecosystem, south Indian Ocean: A review of achievements, 1976–1990. *Journal of Marine Systems* **18**, 355–367.

Pakhomov, E. & Froneman, P. 1999b. Macroplankton/micronekton dynamics in the vicinity of the Prince Edward Islands (Southern Ocean). *Marine Biology* **134**, 501–515.

Pakhomov, E. & Froneman, P. 2000. Composition and spatial variability of macroplankton and micronekton within the Antarctic polar frontal zone of the Indian Ocean during austral autumn 1997. *Polar Biology* **23**, 410–419.

Pakhomov, E.A., Ansorge, I.J., Kaehler S., Vumazonke, L.U., Gulekana, K., Bushula, T., Balt, C., Paul, D., Hargey, N., Stewart, H., Chang. N., Furno, L., Mkatshwa, S., Visser, C., Lutjeharms, J.R.E. & Hayes-Foley, P. 2003. Studying the impact of ocean eddies on the ecosystem of the Prince Edward Islands: DEIMEC II. *South African Journal of Science* **99**, 187–190.

Pakhomov, E.A., Froneman, P.W., Ansorge, I.J. & Lutjeharms, J.R.E. 2000b. Temporal variability in the physico-biological environment of the Prince Edward Islands (Southern Ocean). *Journal of Marine Systems* **26**, 75–95.

Pakhomov, E. & Froneman, P. 2004. Zooplankton dynamics in the eastern Atlantic sector of the Southern Ocean during the austral summer 1997/1998—Part 2: grazing impact. *Deep Sea Research Part II: Topical Studies in Oceanography* **51**, 2617–2631.

Pakhomov, E., Froneman, P. & Perissinotto, R. 2002. Salp/krill interactions in the Southern Ocean: spatial segregation and implications for the carbon flux. *Deep Sea Research Part II: Topical Studies in Oceanography* **49**, 1881–1907.

Pakhomov, E., McClelland, J., Bernard, K., Kaehler, S. & Montoya, J. 2004. Spatial and temporal shifts in stable isotope values of the bottom-dwelling shrimp *Nauticaris marionis* at the sub-Antarctic archipelago. *Marine Biology* **144**, 317–325.

Pakhomov, E. & McQuaid, C. 1996. Distribution of surface zooplankton and seabirds across the Southern Ocean. *Polar Biology* **16**, 271–286.

Pakhomov, E. & Perissinotto, R. 1996. Antarctic neritic krill *Euphausia crystallorophias*: spatio-temporal distribution, growth and grazing rates. *Deep Sea Research Part I: Oceanographic Research Papers* **43**, 59–87.

Pakhomov, E.A. & Perissinotto, R. 1997. Spawning success and grazing impact of *Euphausia crystallorophias* in the Antarctic shelf region, in B. Battaglia & J. Valencia (eds). *Antarctic Communities: Species, Structure and Survival*. Cambridge, UK: Cambridge University Press, 187–192.

Pakhomov, E., Perissinotto, R. & McQuaid, C. 1996. Prey composition and daily rations of myctophid fishes in the Southern Ocean. *Marine Ecology Progress Series* **134**, 1–14.

Pakhomov, E., Perissinotto, R., McQuaid, C. & Froneman, P. 2000a. Zooplankton structure and grazing in the Atlantic sector of the Southern Ocean in late austral summer 1993: Part 1. *Ecological zonation. Deep Sea Research Part I: Oceanographic Research Papers* **47**, 1663–1686.

Pakhomov, E., Perissinotto, R., McQuaid, C. & Froneman, P. 2000b. Zooplankton structure and grazing in the Atlantic sector of the Southern Ocean in late austral summer 1993: part 1. Ecological zonation. *Deep Sea Research Part I: Oceanographic Research Papers* **47**, 1663–1686.

Park, M.G., Yih, W. & Coats, D.W. 2004. Parasites and phytoplankton, with special emphasis on dinoflagellate infections. *Journal of Eukaryotic Microbiology* **51**, 145–155.

Parkinson, C.L. 1990. Search for the little ice age in Southern Ocean sea-ice records. *Annals of Glaciology* **14**, 221–225.

Parkinson, C.L. 2002. Trends in the length of the Southern Ocean Sea-ice season, 1979–99. *Annals of Glaciology* **34**, 435–440.

Parkinson, C.L. 2004. Southern Ocean sea ice and its wider linkages: insights revealed from models and observations. *Antarctic Science* **16**, 387–400.

Paul, J.H. 2008. Prophages in marine bacteria: dangerous molecular time bombs or the key to survival in the seas? *The ISME Journal* **2**, 579–589.

Pauly, D., Christensen, V., Dalsgaard, J., Froese, R. & Torres, F. 1998. Fishing down marine food webs. *Science* **279**, 860–863.

Pauly, D., Watson, R. & Alder, J. 2005. Global trends in world fisheries: impacts on marine ecosystems and food security. *Philosophical Transactions of the Royal Society B: Biological Sciences* **360**, 5–12.

Peck, L.S., Webb, K.E. & Bailey, D.M. 2004. Extreme sensitivity of biological function to temperature in Antarctic marine species. *Functional Ecology* **18**, 625–630.

Pedro, J., Martin, T., Steig, E.J., Jochum, M., Park, W. & Rasmussen, S.O. 2016. Southern Ocean deep convection as a driver of Antarctic warming events. *Geophysical Research Letters* **43**, 2192–2199.

Perissinotto, R. 1992. Mesozooplankton size-selectivity and grazing impact on the phytoplankton community of the Prince Edward archipelago (Southern Ocean). *Marine Ecology Progress Series. Oldendorf* **79**, 243–258.

Perissinotto, R., Gurney, L. & Pakhomov, E. 2000. Contribution of heterotrophic material to diet and energy budget of Antarctic krill, *Euphausia superba*. *Marine Biology* **136**, 129–135.

Perissinotto, R. & McQuaid, C.D. 1990. Role of the sub-Antarctic shrimp, *Nauticaris marionis* in coupling benthic and pelagic food webs. *Marine Ecology Progress Series* **64**, 81–87.

Perissinotto, R. & McQuaid, C.D. 1992. Land-based predator impact on vertically migrating zooplankton and micronekton advected to a Southern Ocean archipelago. *Marine Ecology Progress Series* **80**, 15–27.

Perissinotto, R. & Pakhomov, E. 1998a. Contribution of salps to carbon flux of marginal ice zone of the Lazarev Sea, Southern Ocean. *Marine Biology* **131**, 25–32.

Perissinotto, R. & Pakhomov, E.A. 1998b. The trophic role of the tunicate *Salpa thompsoni* in the Antarctic marine ecosystem. *Journal of Marine Systems* **17**, 361–374.

Perissinotto, R., Pakhomov, E., McQuaid, C. & Froneman, P. 1997. In situ grazing rates and daily ration of Antarctic krill *Euphausia superba* feeding on phytoplankton at the Antarctic polar front and the marginal ice zone. *Marine Ecology Progress Series* **160**, 77–91.

Perissinotto, R. & Rae, C.D. 1990. Occurrence of anticyclonic eddies on the Prince Edward plateau (Southern Ocean): effects on phytoplankton biomass and production. *Deep Sea Research Part A. Oceanographic Research Papers* **37**, 777–793.

Perissinotto, R., Rae, C.D., Boden, B. & Allanson, B. 1990. Vertical stability as a controlling factor of the marine phytoplankton production at the Prince Edward archipelago (Southern Ocean). *Marine Ecology Progress Series* **60**, 205–209.

Péron, C., Weimerskirch, H. & Bost, C.-A. 2012. Projected poleward shift of king penguins' (Aptenodytes patagonicus) foraging range at the Crozet Islands, southern Indian Ocean. *Proceedings of the Royal Society of London B: Biological Sciences* rspb20112705 **279**, 2515–2523.

Perlwitz, J., Pawson, S., Fogt, R.L. & Nielsen, J.E. 2008. Impact of stratospheric ozone hole recovery on Antarctic climate. *Geophysical Research Letters* **35**, L08714, doi: 10.1029/2008GL033317.

Phillips, B., Kremer, P. & Madin, L.P. 2009. Defecation by Salpa thompsoni and its contribution to vertical flux in the Southern Ocean. *Marine Biology* **156**, 455–467.

Phillips, K.L., Jackson, G.D. & Nichols, P.D. 2001. Predation on myctophids by the squid Moroteuthis ingens around Macquarie and Heard Islands: stomach contents and fatty acid analyses. *Marine Ecology Progress Series* **215**, 179–189.

Phillips, K.L., Nichols, P.D. & Jackson, G.D. 2003. Size-related dietary changes observed in the squid *Moroteuthis ingens* at the Falkland Islands: Stomach contents and fatty-acid analyses. *Polar Biology* **26**, 474–85.

Piatkowski, U. & Pütz, K. 1994. Squid diet of emperor penguins (*Aptenodytes forsteri*) in the eastern Weddell Sea, Antarctica during late summer. *Antarctic Science* **6**, 241–247.

Pierrehumbert, R. 2000. Climate change and the tropical Pacific: the sleeping dragon wakes. *Proceedings of the National Academy of Sciences* **97**, 1355–1358.

Pilskaln, C., Manganini, S., Trull, T., Armand, L., Howard, W., Asper, V. & Massom, R. 2004. Geochemical particle fluxes in the southern Indian Ocean seasonal ice zone: Prydz Bay region, east Antarctica. *Deep Sea Research Part I: Oceanographic Research Papers* **51**, 307–332.

Pinaud, D. & Weimerskirch, H. 2005. Scale-dependent habitat use in a long-ranging central place predator. *Journal of Animal Ecology* **74**, 852–863.

Piñones, A., Hofmann, E.E., Dinniman, M.S. & Klinck, J.M. 2011. Lagrangian simulation of transport pathways and residence times along the Western Antarctic Peninsula. *Deep Sea Research Part II: Topical Studies in Oceanography* **58**, 1524–1539.

Pollard, R.T., Salter, I., Sanders, R.J., Lucas, M.I., Moore, C.M., Mills, R.A., Statham, P.J., Allen, J.T., Baker, A.R. & Bakker, D.C. 2009. Southern Ocean deep-water carbon export enhanced by natural iron fertilization. *Nature* **457**, 577–580.

Pörtner, H.O., Langenbuch, M. & Reipschläger, A. 2004. Biological impact of elevated ocean CO_2 concentrations: lessons from animal physiology and Earth history. *Journal of Oceanography* **60**, 705–718.

Prézelin, B.B., Boucher, N.P. & Schofield, O. 1994a. Evaluation of field studies of UVB radiation effects on Antarctic marine primary productivity. In *Stratospheric Ozone Depletion/UV-B Radiation in the Biosphere*, C.S. Weiler & P.A. Penhale (eds). Berlin, Heidelberg: Springer, 181–194.

Prézelin, B.B., Boucher, N.P. & Smith, R.C. 1994b. *Marine Primary Production under the Influence of the Antarctic Ozone Hole: Icecolors' 90*. Washington, DC: American Geophysical Union.

Prézelin, B.B., Hofmann, E.E., Moline, M. & Klinck, J.M. 2004. Physical forcing of phytoplankton community structure and primary production in continental shelf waters of the Western Antarctic Peninsula. *Journal of Marine Research* **62**, 419–460.

Purkey, S.G. & Johnson, G.C. 2010. Warming of global abyssal and deep Southern Ocean waters between the 1990s and 2000s: contributions to global heat and sea level rise budgets. *Journal of Climate* **23**, 6336–6351.

Purkey, S.G. & Johnson, G.C. 2012. Global contraction of Antarctic bottom water between the 1980s and 2000s. *Journal of Climate* **25**, 5830–5844.

Purkey, S.G. & Johnson, G.C. 2013. Antarctic bottom water warming and freshening: contributions to sea level rise, ocean freshwater budgets, and global heat gain. *Journal of Climate* **26**, 6105–6122.

Quéguiner, B. 2013. Iron fertilization and the structure of planktonic communities in high nutrient regions of the Southern Ocean. *Deep Sea Research Part II: Topical Studies in Oceanography* **90**, 43–54.

Quetin, L. & Ross, R. 1985. Feeding by Antarctic krill, *Euphausia superba*: does size matter? In *Antarctic Nutrient Cycles and Food Webs*, W.R. Siegfried, P.R Condy, R.M. Laws (eds). Berlin: Springer, 372–377.

Raclot, T., Groscolas, R. & Cherel, Y. 1998. Fatty acid evidence for the importance of myctophid fishes in the diet of king penguins, *Aptenodytes patagonicus*. *Marine Biology* **132**, 523–533.

Rahmstorf, S. & England, M.H. 1997. Influence of southern hemisphere winds on North Atlantic deep water flow. *Journal of Physical Oceanography* **27**, 2040–2054.

Raymond, B., Shaffer, S.A., Sokolov, S., Woehler, E.J., Costa, D.P., Einoder, L., Hindell, M., Hosie, G., Pinkerton, M. & Sagar, P.M. 2010. Shearwater foraging in the Southern Ocean: the roles of prey availability and winds. *PloS One* **5**, e10960.

Razouls, S., Razouls, C. & De Bovée, F. 2000. Biodiversity and biogeography of Antarctic copepods. *Antarctic Science* **12**, 343–362.

Reid, K. & Croxall, J.P. 2001. Environmental response of upper trophic-level predators reveals a system change in an Antarctic marine ecosystem. *Proceedings of the Royal Society of London B: Biological Sciences* **268**, 377–384.

Reid, K., Croxall, J.P., Briggs, D.R. & Murphy, E.J. 2005. Antarctic ecosystem monitoring: quantifying the response of ecosystem indicators to variability in Antarctic krill. *ICES Journal of Marine Science: Journal du Conseil* **62**, 366–373.

Ribbe, J. 2004. Oceanography: the southern supplier. *Nature* **427**, 23–24.

Richoux, N.B., Jaquemet, S., Bonnevie, B.T., Cherel, Y. & McQuaid, C.D. 2010. Trophic ecology of grey-headed albatrosses from Marion Island, Southern Ocean: insights from stomach contents and diet tracers. *Marine Biology* **157**, 1755–1766.

Ridgwell, A.J. 2002. Dust in the Earth system: the biogeochemical linking of land, air and sea. *Philosophical Transactions of the Royal Society of London A: Mathematical, Physical and Engineering Sciences* **360**, 2905–2924.

Riebesell, U., Zondervan, I., Rost, B., Tortell, P.D., Zeebe, R.E. & Morel, F.M. 2000. Reduced calcification of marine plankton in response to increased atmospheric CO_2. *Nature* **407**, 364–367.

Riisgård, H.U. & Larsen, P.S. 2010. Particle capture mechanisms in suspension-feeding invertebrates. *Marine Ecology Progress Series* **418**, 255–293.

Rind, D., Chandler, M., Lerner, J., Martinson, D. & Yuan, X. 2001. Climate response to basin-specific changes in latitudinal temperature gradients and implications for sea ice variability. *Journal of Geophysical Research* **106**, 161–20.

Rind, D., Healy, R., Parkinson, C. & Martinson, D. 1995. The role of sea ice in $2\times$ CO_2 climate model sensitivity. Part I: the total influence of sea ice thickness and extent. *Journal of Climate* **8**, 449–463.

Roberts, S.J., Monien, P., Foster, L., Loftfield, J., Hocking, E., Schnetger, B., Pearson, E., Juggins, S., Fretwell, P. & Ireland, L. 2017. Past penguin colony responses to explosive volcanism on the Antarctic Peninsula. *Nature Communications* **8**, 14914.

Robinson, R.S., Sigman, D.M., DiFiore, P.J., Rohde, M.M., Mashiotta, T.A. & Lea, D.W. 2005. Diatom-bound 15N/14N: new support for enhanced nutrient consumption in the ice age sub-Antarctic. *Paleoceanography* **20**, PA3003, doi:10.1029/2004PA001114.

Rodhouse, P., Arnbom, T., Fedak, M., Yeatman, J. & Murray, A. 1992a. Cephalopod prey of the southern elephant seal, *Mirounga leonina* L. *Canadian Journal of Zoology* **70**, 1007–1015.

Rodhouse, P. & Nigmatullin, C.M. 1996. Role as consumers. *Philosophical Transactions of the Royal Society of London B: Biological Sciences* **351**, 1003–1022.

Rodhouse, P., White, M. & Jones, M. 1992b. Trophic relations of the cephalopod *Martialia hyadesi* (Teuthoidea: Ommastrephidae) at the Antarctic polar front, Scotia Sea. *Marine Biology* **114**, 415–421.

Roemmich, D., Church, J., Gilson, J., Monselesan, D., Sutton, P. & Wijffels, S. 2015. Unabated planetary warming and its ocean structure since 2006. *Nature Climate Change* **5**, 240–245.

Roemmich, D., Gould, W.J. & Gilson, J. 2012. 135 years of global ocean warming between the challenger expedition and the Argo programme. *Nature Climate Change* **2**, 425–428.

Roman, M.R., Dam, H.G., Gauzens, A.L. & Napp, J.M. 1993. Zooplankton biomass and grazing at the JGOFS Sargasso Sea time series station. *Deep Sea Research Part I: Oceanographic Research Papers* **40**, 883–901.

Rose, J.M., Feng, Y., DiTullio, G.R., Dunbar, R.B., Hare, C.E., Lee, P.A., Lohan, M., Long, M., Smith Jr, W.O., Sohst, B., Tozzi, S., Zhang, Y. & Hutchins, D.A. 2009. Synergistic effects of iron and temperature on Antarctic phytoplankton and microzooplankton assemblages. *Biogeosciences* **6**, 3131–3147.

Rott, H., Skvarca, P. & Nagler, T. 1996. Rapid collapse of northern Larsen ice shelf, Antarctica. *Science* **271**, 788.

Ruck, K.E., Steinberg, D.K. & Canuel, E.A. 2014. Regional differences in quality of krill and fish as prey along the Western Antarctic Peninsula. *Marine Ecology Progress Series* **509**, 39–55, doi: 10.3354/meps10868.

Sailley, S.F., Ducklow, H.W., Moeller, H.V., Fraser, W.R., Schofield, O.M., Steinberg, D.K., Garzio, L.M. & Doney, S.C. 2013. Carbon fluxes and pelagic ecosystem dynamics near two western Antarctic Peninsula Adélie penguin colonies: an inverse model approach. *Marine Ecology Progress Series* **492**, 253–272.

Sala, A., Azzali, M. & Russo, A. 2002. Krill of the Ross Sea: distribution, abundance and demography of *Euphausia superba* and *Euphausia crystallorophias* during the Italian Antarctic expedition (January-February 2000). *Scientia Marina* **66**, 123–133.

Sallée, J.B., Speer, K. & Morrow, R. 2008. Response of the Antarctic circumpolar current to atmospheric variability. *Journal of Climate* **21**, 3020–3039.

Salter, I., Kemp, A.E., Moore, C.M., Lampitt, R.S., Wolff, G.A. & Holtvoeth, J. 2012. Diatom resting spore ecology drives enhanced carbon export from a naturally iron-fertilized bloom in the Southern Ocean. *Global Biogeochemical Cycles* **26**, GB1014, doi:10.1029/2010GB003977

Sarmiento, J.L., Gruber, N., Brzezinski, M. & Dunne, J. 2004a. High-latitude controls of thermocline nutrients and low latitude biological productivity. *Nature* **427**, 56–60.

Sarmiento, J.L., Hughes, T.M., Stouffer, R.J. & Manabe, S. 1998. Simulated response of the ocean carbon cycle to anthropogenic climate warming. *Nature* **393**, 245–249.

Sarmiento, J.L., Slater, R., Barber, R., Bopp, L., Doney, S.C., Hirst, A., Kleypas, J., Matear, R., Mikolajewicz, U. & Monfray, P. 2004b. Response of ocean ecosystems to climate warming. *Global Biogeochemical Cycles* **18**, GB3003, doi:10.1029/2003GB002134

Saunders, R., Brierley, A., Watkins, J.L., Reid, K., Murphy, E.J., Enderlein, P. & Bone, D. 2007. Intra-annual variability in the density of Antarctic krill *(Euphausia superba)* at South Georgia, 2002–2005: within-year variation provides a new framework for interpreting previous 'annual'estimates of krill density. *CCAMLR Science* **14**, 27–41.

Schiebel, R. 2002. Planktic foraminiferal sedimentation and the marine calcite budget. *Global Biogeochemical Cycles* **16**, 1065, doi:10.1029/2001GB001459

Schmidt, K., Atkinson, A., Steigenberger, S., Fielding, S., Lindsay, M., Pond, D.W., Tarling, G.A., Klevjer, T.A., Allen, C.S. & Nicol, S. 2011. Seabed foraging by Antarctic krill: implications for stock assessment, bentho-pelagic coupling, and the vertical transfer of iron. *Limnology and Oceanography* **56**, 1411–1428.

Schmitz, O.J., Hambäck, P.A. & Beckerman, A.P. 2000. Trophic cascades in terrestrial systems: A review of the effects of carnivore removals on plants. *The American Naturalist* **155**, 141–153.

Schnack-Schiel, S.B. & Isla, E. 2005. The role of zooplankton in the pelagic-benthic coupling of the Southern Ocean. *Scientia Marina* **69**, 39–55.

Schneppenheim, R. & MacDonald, C. 1984. Genetic variation and population structure of krill (*Euphausia superba*) in the Atlantic sector of Antarctic waters and off the Antarctic Peninsula. *Polar Biology* **3**, 19–28.

Sedwick, P.N., DiTullio, G.R. & Mackey, D.J. 2000. Iron and manganese in the Ross Sea, Antarctica: seasonal iron limitation in Antarctic shelf waters. *Journal of Geophysical Research: Oceans* **105**, 11321–11336.

Seibel, B.A. & Dierssen, H.M. 2003. Cascading trophic impacts of reduced biomass in the Ross Sea, Antarctica: just the tip of the iceberg? *The Biological Bulletin* **205**, 93–97.

Seymour, J.R., Seuront, L., Doubell, M., Waters, R.L. & Mitchell, J.G. 2006. Microscale patchiness of virioplankton. *Journal of the Marine Biological Association* **86**, 551–561.

Siegel, V. 1987. Age and growth of Antarctic Euphausiacea (Crustacea) under natural conditions. *Marine Biology* **96**, 483–495.

Siegel, V. 2005. Distribution and population dynamics of *Euphausia superba*: summary of recent findings. *Polar Biology* **29**, 1–22.

Siegel, V. & Harm, U. 1996. The composition, abundance, biomass and diversity of the epipelagic zooplankton communities of the southern Bellingshausen Sea (Antarctic) with special references to krill and salps. *Archive of Fishery and Marine Research* **44**, 115–139.

Siegel, V. & Loeb, V. 1995. Recruitment of Antarctic krill *Euphausia superba* and possible causes for its variability. *Marine Ecology Progress Series* **123**, 45–56.

Siegel, V. & Piatkowski, U. 1990. Variability in the macrozooplankton community off the Antarctic Peninsula. *Polar Biology* **10**, 373–386.

Siegfried, W.R. 1985. Birds and mammals – oceanic birds of the Antarctic. In *Key Environments – Antarctica* ed. Bonner, W.N. & Walton, D.W.H. Pergamon Press Oxford, 242–263.

Sigman, D.M. & Boyle, E.A. 2000. Glacial/interglacial variations in atmospheric carbon dioxide. *Nature* **407**, 859–869.

Sigman, D.M., Hain, M.P. & Haug, G.H. 2010. The polar ocean and glacial cycles in atmospheric CO2 concentration. *Nature* **466**, 47–55.

Small, L., Fowler, S. & Ünlü, M. 1979. Sinking rates of natural copepod fecal pellets. *Marine Biology* **51**, 233–241.

Smayda, T.J. & Reynolds, C.S. 2001. Community assembly in marine phytoplankton: Application of recent models to harmful dinoflagellate blooms. *Journal of Plankton Research* **23**, 447–461.

Smetacek, V., Assmy, P. & Henjes, J. 2004. The role of grazing in structuring Southern Ocean pelagic ecosystems and biogeochemical cycles. *Antarctic Science* **16**, 541–558.

Smetacek, V., De Baar, H., Bathmann, U., Lochte, K. & Van Der Loeff, M.R. 1997. Ecology and biogeochemistry of the Antarctic circumpolar current during austral spring: a summary of Southern Ocean JGOFS cruise ANT X/6 of RV polarstern. *Deep Sea Research Part II: Topical Studies in Oceanography* **44**, 1–21.

Smetacek, V. & Nicol, S. 2005. Polar ocean ecosystems in a changing world. *Nature* **437**, 362–368.

Smetacek, V., Scharek, R. & Nöthig, E.-M. 1990. Seasonal and regional variation in the pelagial and its relationship to the life history cycle of krill. In *Antarctic Ecosystems*, K.R. Kerry & G. Hempel (eds). Berlin, Heidelberg: Springer, 103–114.

Smith, C.R., Grange, L.J., Honig, D.L., Naudts, L., Huber, B., Guidi, L. & Domack, E. 2011. A large population of king crabs in palmer deep on the west Antarctic Peninsula shelf and potential invasive impacts. *Proceedings of the Royal Society of London B: Biological Sciences* rspb20111496 **279**, 1017–1026.

Smith, J.L.K., Sherman, A., Huffard, C., McGill, P., Henthorn, R., Von Thun, S., Ruhl, H., Kahru, M. & Ohman, M. 2014. Large salp bloom export from the upper ocean and benthic community response in the abyssal northeast Pacific: day to week resolution. *Limnology and Oceanography* **59**, 745–757.

Smith, V.R. & Steenkamp, M. 1990. Climatic change and its ecological implications at a subantarctic island. *Oecologia* **85**, 14–24.

Smith, W.O. & Asper, V.L. 2001. The influence of phytoplankton assemblage composition on biogeochemical characteristics and cycles in the southern Ross Sea, Antarctica. *Deep Sea Research Part I* **48**, 137–161.

Smith, R.C., Ainley, D., Baker, K., Domack, E., Emslie, S., Fraser, B., Kennett, J., Leventer, A., Mosley-Thompson, E. & Stammerjohn, S. 1999. Marine ecosystem sensitivity to climate change historical observations and paleoecological records reveal ecological transitions in the Antarctic Peninsula region. *BioScience* **49**, 393–404.

Smith, R.C., Prezelin, B.B., Baker, K.S., Bidigare, R.R., Boucher, N.P., Coley, T., Karentz, D., MacIntyre, S., Matlick, H.A. & Menzies, D. 1992. Ozone depletion: ultraviolet radiation and phytoplankton biology in Antarctic waters. *Science* **255**, 952.

Smith, V. 1991. Climate change and its ecological consequences at Marion and Prince Edward Islands. *South African Journal of Antarctic Research* **21**, 223–224.

Smith, V. 2002. Climate change in the sub-Antarctic: an illustration from Marion Island. *Climatic Change* **52**, 345–357.

Smith, W.O., Ainley, D.G., Arrigo, K.R. & Dinniman, M.S. 2014. The oceanography and ecology of the Ross Sea. *Annual Review of Marine Science* **6**, 469–487.

Smith, W.O., Ainley, D.G. & Cattaneo-Vietti, R. 2007. Trophic interactions within the Ross Sea continental shelf ecosystem. *Philosophical Transactions of the Royal Society of London B: Biological Sciences* **362**, 95–111.

Smith, W.O. & Gordon, L.I. 1997. Hyperproductivity of the Ross Sea (Antarctica) polynya during austral spring. *Geophysical Research Letters* **24**, 233–236.

Smith, W.O. & Lancelot, C. 2004. Bottom-up versus top-down control in phytoplankton of the Southern Ocean. *Antarctic Science* **16**, 531–539.

Smith, W.O. & Nelson, D.M. 1985. Phytoplankton bloom produced by a receding ice edge in the Ross Sea: spatial coherence with the density field. *Science* **227**, 163–167.

Smith, W.O. & Nelson, D.M. 1990. Phytoplankton growth and new production in the Weddell Sea marginal ice zone in the austral spring and autumn. *Limnology and Oceanography* **35**, 809–821.

Somero, G.N. 2010. The physiology of climate change: How potentials for acclimatization and genetic adaptation will determine 'winners' and 'losers'. *Journal of Experimental Biology* **213**, 912–920, doi: 10.1242/jeb.037473.

Stammerjohn, S., Massom, R., Rind, D. & Martinson, D. 2012. Regions of rapid sea ice change: an inter-hemispheric seasonal comparison. *Geophysical Research Letters* **39**, L06501, doi:10.1029/2012GL050874

Stammerjohn, S. & Smith, R. 1997. Opposing Southern Ocean climate patterns as revealed by trends in regional sea ice coverage. *Climatic Change* **37**, 617–639.

Stankovic, A., Spalik, K., Kamler, E., Borsuk, P. & Weglenski, P. 2002. Recent origin of sub-Antarctic notothenioids. *Polar Biology* **25**, 203–205.

Stephens, B.B. & Keeling, R.F. 2000. The influence of Antarctic Sea ice on glacial–interglacial CO2 variations. *Nature* **404**, 171–174.

Stroeve, J.C., Serreze, M.C., Holland, M.M., Kay, J.E., Malanik, J. & Barrett, A.P. 2012. The Arctic's rapidly shrinking sea ice cover: a research synthesis. *Climatic Change* **110**, 1005–1027.

Sullivan, C., Arrigo, K., McClain, C., Comiso, J. & Firestone, J. 1993. Distributions of phytoplankton blooms in the Southern Ocean. *Science* **262**, 1832–1837.

Sutherland, K.R., Madin, L.P. & Stocker, R. 2010. Filtration of submicrometer particles by pelagic tunicates. *Proceedings of the National Academy of Sciences* **107**, 15129–15134.

Suttle, C.A. 2005. Viruses in the sea. *Nature* **437**, 356–361.

Suttle, C.A. 2007. Marine viruses—major players in the global ecosystem. *Nature Reviews Microbiology* **5**, 801–812.

Tagliarolo, M. & McQuaid, C.D. 2016. Field measurements indicate unexpected, serious underestimation of mussel heart rates and thermal tolerance by laboratory studies. *PLoS ONE* **11**, e0146341.

Takahashi, K. & Okada, H. 2000. Environmental control on the biogeography of modern coccolithophores in the southeastern Indian Ocean offshore of Western Australia. *Marine Micropaleontology* **39**, 73–86.

Tarling, G.A., Ward, P., Sheader, M., Williams, J.A. & Symon, C. 1995. Distribution patterns of macrozooplankton assemblages in the southwest Atlantic. *Marine Ecology Progress Series* **120**, 29–40.

Testa, J., Siniff, D., Ross, M. & Winter, J. 1985. Weddell Sea – Antarctic cod interactions in McMurdo Sound, Antarctica. In *Antarctic Nutrient Cycles and Food Webs*, W.R. Siegfried, P.R. Condy & R.M. Laws (eds). Berlin, Heidelberg: Springer, 561–565.

Tewksbury, J.J., Huey, R.B. & Deutsch, C.A. 2008. Putting the heat on tropical animals. *Science* **320**, 1296.

Thomalla, S., Fauchereau, N., Swart, S. & Monteiro, P. 2011. Regional scale characteristics of the seasonal cycle of chlorophyll in the Southern Ocean. *Biogeosciences* **8**, 2849.

Thomas, E.R., Marshall, G.J. & McConnell, J.R. 2008. A doubling in snow accumulation in the western Antarctic Peninsula since 1850. *Geophysical Research Letters* **35**, L01706, doi:10.1029/2007GL032529.

Thomas, P. & Green, K. 1988. Distribution of *Euphausia crystallorophias* within Prydz Bay and its importance to the inshore marine ecosystem. *Polar Biology* **8**, 327–331.

Thompson, D.W., Solomon, S., Kushner, P.J., England, M.H., Grise, K.M. & Karoly, D.J. 2011. Signatures of the Antarctic ozone hole in southern hemisphere surface climate change. *Nature Geoscience* **4**, 741–749.

Thompson, L.G., Peel, D., Mosley-Thompson, E., Mulvaney, R., Dal, J., Lin, P., Davis, M. & Raymond, C. 1994. Climate since AD 1510 on dyer plateau, Antarctic Peninsula: evidence for recent climate change. *Annals of Glaciology* **20**, 420–426.

Thorpe, S.E., Murphy, E.J. & Watkins, J.L. 2007. Circumpolar connections between Antarctic krill (*Euphausia superba*) populations: investigating the roles of ocean and sea ice transport. *Deep Sea Research Part I: Oceanographic Research Papers* **54**, 792–810.

Tomczak, M. & Godfrey, J.S. 2013. *Regional Oceanography: An Introduction*. Oxford: Elsevier.

Torres, J.J., Aarset, A.V., Donnelly, J., Hopkins, T.L., Lancraft, T.M. & Ainley, D.G. 1994. Metabolism of Antarctic micronektonic Crustacea as a function of depth of occurrence and season. *Oceanographic Literature Review* **5**, 385.

Trathan, P.N., Brierley, A.S., Brandon, M.A., Bone, D.G., Goss, C., Grant, S.A., Murphy, E.J. & Watkins, J.L. 2003. Oceanographic variability and changes in Antarctic krill (*Euphausia superba*) abundance at South Georgia. *Fisheries Oceanography* **12**, 569–583.

Trathan, P.N., Croxall, J.P. & Murphy, E.J. 1996. Dynamics of Antarctic penguin populations in relation to inter-annual variability in sea ice distribution. *Polar Biology* **16**, 321–330.

Trivelpiece, W.Z., Hinke, J.T., Millera, A.K., Reissa, C.S., Trivelpiece, S.G. & Watters, G.M. 2011. Variability in krill biomass links harvesting and climate warming to penguin population changes in Antarctica. *Proceedings of the National Academy of Sciences* **108**, 7625–7628.

Turner, J., Colwell, S.R., Marshall, G.J., Lachlan-Cope, T.A., Carleton, A.M., Jones, P.D., Lagun, V., Reid, P.A. & Iagovkina, S. 2005. Antarctic climate change during the last 50 years. *International Journal of Climatology* **25**, 279–294.

Turner, J., Lu, H., White, I., King, J.C., Phillips, T., Hosking, J.S., Bracegirdle, T.J., Marshall, G.J., Mulvaney, R. & Deb, P. 2016. Absence of 21st century warming on Antarctic Peninsula consistent with natural variability. *Nature* **535**, 411–415.

Turner, J.T. 2002. Zooplankton fecal pellets, marine snow and sinking phytoplankton blooms. *Aquatic Microbial Ecology* **27**, 57–102.

Turner, J.T. 2015. Zooplankton fecal pellets, marine snow, phytodetritus and the ocean's biological pump. *Progress in Oceanography* **130**, 205–248.

Tynan, C.T. 1998. Ecological importance of the southern boundary of the Antarctic circumpolar current. *Nature* **392**, 708–710.

Uthicke, S., Momigliano, P. & Fabricius, K.E. 2013. High risk of extinction of benthic foraminifera in this century due to ocean acidification. *Scientific Reports* **3**, 1769.

Vacchi, M., Mesa, M.L., Dalu, M. & Macdonald, J. 2004. Early life stages in the life cycle of Antarctic silverfish, *Pleuragramma antarcticum* in Terra Nova Bay, Ross Sea. *Antarctic Science* **16**, 299–305.

Vancoppenolle, M., Meiners, K.M., Michel, C., Bopp, L., Brabant, F., Carnat, G., Delille, B., Lannuzel, D., Madec, G., Moreau, S. & Tison, J.L. 2013. Role of sea ice in global biogeochemical cycles: emerging views and challenges. *Quaternary Science Reviews* **79**, 207–230.

Vaughan, D.G. & Doake, C.S.M. 1996. Recent atmospheric warming and retreat of ice shelves on the Antarctic Peninsula. *Nature* **379**, 328–331.

Vaughan, D.G., Marshall, G.J., Connolley, W.M., King, J.C. & Mulvaney, R. 2001. Devil in the detail. *Science* **293**, 1777–1779.

Vaughan, D.G., Marshall, G.J., Connolley, W.M., Parkinson, C., Mulvaney, R., Hodgson, D.A., King, J.C., Pudsey, C.J. & Turner, J. 2003. Recent rapid regional climate warming on the Antarctic Peninsula. *Climatic Change* **60**, 243–274.

Verity, P. & Smetacek, V. 1996. Organism life cycles, predation, and the structure of marine pelagic ecosystems. *Marine Ecology Progress Series* **130**, 277–293.

Vinje, T. 2001. Anomalies and trends of sea-ice extent and atmospheric circulation in the Nordic Seas during the period 1864–1998. *Journal of Climate* **14**, 255–267.

Visser, M.E., Both, C. & Lambrechts, M.M. 2004. Global climate change leads to mistimed avian reproduction. *Advances in Ecological Research* **35**, 89–110.

von Bröckel, K. 1985. Primary production data from the south-eastern Weddell Sea. *Polar Biology* **4**, 75–80.

Voronina, N.M. 1968. The distribution of zooplankton in the southern ocean and its dependence on the circulation of water. *Sarsia* **34**, 277–284.

Vumazonke, L.U., Pakhomov, E.A., Froneman, P.W. & McQuaid, C.D. 2003. Diet and daily ration of male and female caridean shrimp *Nauticaris marionis* at the Prince Edward archipelago. *Polar Biology* **26**, 420–422.

Wadhams, P. & Munk, W. 2004. Ocean freshening, sea level rising, sea ice melting. *Geophysical Research Letters* **31**, L11311.

Walsh, J.J., Dieterle, D.A. & Lenes, J. 2001. A numerical analysis of carbon dynamics of the Southern Ocean phytoplankton community: the roles of light and grazing in effecting both sequestration of atmospheric CO_2 and food availability to larval krill. *Deep Sea Research Part I: Oceanographic Research Papers* **48**, 1–48.

Walther, G.-R., Post, E., Convey, P., Menzel, A., Parmesan, C., Beebee, T.J.C., Fromentin, J.-M., Hoegh-Guldberg, O. & Bairlein, F. 2002. Ecological responses to recent climate change. *Nature* **416**, 389–395.

Waluda, C.M., Hill, S.L., Peat, H.J. & Trathan, P.N. 2017. Long-term variability in the diet and reproductive performance of penguins at Bird Island, South Georgia. *Marine Biology* **164**, 39.

Ward, P. & Shreeve, R.S. 2001. The deep-sea copepod fauna of the Southern Ocean: patterns and processes. *Hydrobiologia* **453–454**, 37–54.

Ward, P., Whitehouse, M., Shreeve, R., Thorpe, S., Atkinson, A., Korb, R., Pond, D. & Young, E. 2007. Plankton community structure south and west of South Georgia (Southern Ocean): links with production and physical forcing. *Deep Sea Research Part I: Oceanographic Research Papers* **54**, 1871–1889.

Wassmann, P., Vernet, M., Mitchell, B.G. & Rey, F. 1990. Mass sedimentation of *Phaeocystis pouchetii* in the Barents Sea. *Marine Ecology Progress Series* **66**, 183–195.

Weimerskirch, H. 2007. Are seabirds foraging for unpredictable resources? *Deep Sea Research Part II: Topical Studies in Oceanography* **54**, 211–223.

Weimerskirch, H., Corre, M.L., Jaquemet, S., Potier, M. & Marsac, F. 2004. Foraging strategy of a top predator in tropical waters: great frigatebirds in the Mozambique Channel. *Marine Ecology Progress Series* **275**, 297–308.

Weimerskirch, H., Guionnet, T., Martin, J., Shaffer, S.A. & Costa, D.P. 2000. Fast and fuel efficient? Optimal use of wind by flying albatrosses. *Proceedings of the Royal Society of London B: Biological Sciences* **267**, 1869–1874.

Weimerskirch, H., Inchausti, P., Guinet, C. & Barbraud, C. 2003. Trends in bird and seal populations as indicators of a system shift in the Southern Ocean. *Antarctic Science* **15**, 249–256.

Weinberg, E.D. 1974. Iron and susceptibility to infectious disease. *Science* **184**, 952–956.

White, W.B. & Peterson, R.G. 1996. An Antarctic circumpolar wave in surface pressure, wind, temperature and sea-ice extent. *Nature* **380**, 699–702.

Wiebe, P.H., Madin, L.P., Haury, L.R., Harbison, G.R. & Philbin, L.M. 1979. Diel vertical migration by *Salpa aspera* and its potential for large-scale particulate organic matter transport to the deep-sea. *Marine Biology* **53**, 249–255.

Wiedenmann, J., Cresswell, K.A. & Mangel, M. 2009. Connecting recruitment of Antarctic krill and sea ice. *Limnology and Oceanography* **54**, 799–811.

Wilhelm, S.W., Jeffrey, W.H., Dean, A.L., Meador, J., Pakulski, J.D. & Mitchell, D.L. 2003. UV radiation induced DNA damage in marine viruses along a latitudinal gradient in the southeastern Pacific Ocean. *Aquatic Microbial Ecology* **31**, 1–8.

Wilhelm, S.W. & Suttle, C.A. 1999. Viruses and nutrient cycles in the sea: Viruses play critical roles in the structure and function of aquatic food webs. *Bioscience* **49**, 781–788.

Wilkins, D., Lauro, F.M., Williams, T.J., Demaere, M.Z., Brown, M.V., Hoffman, J.M., Andrews-Pfannkoch, C., Mcquaid, J.B., Riddle, M.J. & Rintoul, S.R. 2013. Biogeographic partitioning of Southern Ocean microorganisms revealed by metagenomics. *Environmental Microbiology* **15**, 1318–1333.

Wilson, P.R., Ainley, D.G., Nur, N., Jacobs, S.S., Barton, K.J., Ballard, G. & Comiso, J.C. 2001. Adélie penguin population change in the pacific sector of Antarctica: relation to sea-ice extent and the Antarctic circumpolar current. *Marine Ecology Progress Series* **213**, 301–309.

Winder, M. & Schindler, D.E. 2004. Climate change uncouples trophic interactions in an aquatic ecosystem. *Ecology* **85**, 2100–2106.

Wommack, K.E. & Colwell, R.R. 2000. Virioplankton: Viruses in aquatic ecosystems. *Microbiolgy and Molecular Biology Reviews* **64**, 69–114.

Worby, A.P., Geiger, C.A., Paget, M.J., Van Woert, M.L., Ackley, S.F. & DeLiberty, T.L. 2008. Thickness distribution of Antarctic sea ice. *Journal of Geophysical Research: Oceans* **113**, C05S92, doi:10.1029/2007JC004254

Xavier, J.C., Croxall, J.P. & Reid, K. 2003. Interannual variation in the diets of two albatross species breeding at South Georgia: implications for breeding performance. *Ibis* **145**, 593–610.

Xavier, J.C., Rodhouse, P., Purves, M., Daw, T., Arata, J. & Pilling, G. 2002. Distribution of cephalopods recorded in the diet of the patagonian toothfish (*Dissostichus eleginoides*) around South Georgia. *Polar Biology* **25**, 323–330.

Yin, J., Overpeck, J.T., Griffies, S.M., Hu, A., Russell, J.L. & Stouffer, R.J. 2011. Different magnitudes of projected subsurface ocean warming around Greenland and Antarctica. *Nature Geoscience* **4**, 524–528.

Yoon, W., Kim, S. & Han, K. 2001. Morphology and sinking velocities of fecal pellets of copepod, molluscan, euphausiid, and salp taxa in the northeastern tropical Atlantic. *Marine Biology* **139**, 923–928.

Young, J.W., Hunt, B.P., Cook, T.R., Llopiz, J.K., Hazen, E.L., Pethybridge, H.R., Ceccarelli, D., Lorrain, A., Olson, R.J. & Allain, V. 2015. The trophodynamics of marine top predators: current knowledge, recent advances and challenges. *Deep Sea Research Part II: Topical Studies in Oceanography* **113**, 170–187.

Yuan, X. & Martinson, D.G. 2001. The Antarctic dipole and its predictability. *Geophysical Research Letters* **28**, 3609–3612.

Zane, L., Ostellari, L., Maccatrozzo, L., Bargelloni, L., Battaglia, B. & Patarnello, T. 1998. Molecular evidence for genetic subdivision of Antarctic krill (*Euphausia superba* Dana) populations. *Proceedings of the Royal Society of London B: Biological Sciences* **265**, 2387–2391.

Zwally, H.J., Comiso, J.C., Parkinson, C.L., Cavalieri, D.J. & Gloersen, P. 2002. Variability of Antarctic sea ice 1979–1998. *Journal of Geophysical Research: Oceans* **107**, 9–1.

Oceanography and Marine Biology: An Annual Review, 2018, **56**, 73-104
© S. J. Hawkins, A. J. Evans, A. C. Dale, L. B. Firth, and I. P. Smith, Editors
Taylor & Francis

PROTECTED AREAS: THE FALSE HOPE FOR CETACEAN CONSERVATION?

EUNICE H. PINN[1,2]*

*[1]Joint Nature Conservation Committee, Inverdee House, Baxter
Street, Aberdeen, AB11 9QA, United Kingdom
[2]School of Law, University of Aberdeen, Taylor Building, Aberdeen, AB24 3UB
*Corresponding author: Eunice H. Pinn
e-mail: eunice.pinn@abdn.ac.uk*

Abstract

In recent decades, marine protected areas (MPAs) have become the management measure of choice for cetacean conservation. They are a conspicuous conservation approach, the principle of which is readily understood by the public. Despite the drive to further develop the network of protected areas, the efficacy for cetacean conservation has rarely been considered. Policymakers need to make informed decisions on the best approaches for conservation of these highly charismatic species in circumstances where there are increasing pressures for the development of maritime industries combined with declining resources for conservation. As such, there is now a vital need for a synthesis of the available evidence for the value of MPAs for cetacean conservation. The results of the systematic review undertaken here indicate that in over 80 years of use and over 1000 designations, there is a distinct lack of rigorous published evidence for direct conservation benefits of MPAs for cetaceans. Two studies demonstrated improved survival as a result of MPA designation, although this was insufficient to reverse population declines. One study noted the continued use of an MPA because of management intervention, and another recorded a stable population within an MPA but did not demonstrate that this was attributable to the MPA. In contrast, there were considerably more examples of the failure of protected areas to attain their conservation goals for cetaceans. These were the result of changes in the spatial distribution of the populations for which the protected area had been designated, the lack of enforcement of measures within the site, or the need for additional measures beyond MPA boundaries. This review has highlighted that, for cetaceans at least, the main role of MPAs is in awareness raising and education rather than for direct ecological and/or population conservation purposes. Given the mobility of cetaceans and the widespread nature of many of the anthropogenic pressures to which they are exposed, pragmatic and effective solutions for conservation are required. An adaptive and dynamic approach to the management of human activities presents one possible solution.

Introduction

The International Union for Conservation of Nature and Natural Resources (IUCN) defines a marine protected area (MPA) as "Any area of intertidal or subtidal terrain, together with its overlying water and associated flora, fauna, historical and cultural features, which has been reserved by law or other effective means to protect part or all of the enclosed environment" (Dudley 2008, p. 56, Day et al. 2012, p. 12). In recent years, there has been much debate about the value of MPAs for conservation (Agardy et al.

2011, Caveen et al. 2013, Bennett & Dearden 2014, Edgar et al. 2014, Wilson 2016, Hilborn 2016, 2017, Pendleton et al. 2017). The public tend to perceive MPAs in the same way they see national parks and other such protected areas on land (Kearney et al. 2013). Terrestrial protected areas are generally thought to have a distinct lack of industrial activity and only relatively limited human presence compared to the surrounding land. In this context, farming and forestry are rarely considered to represent industrial activities and instead are thought to contribute to the maintenance of these 'unspoilt' areas. The public perception of MPAs is similarly that they are areas of sea where all industrial activities are excluded. The reality is, however, that management within these sites can often be little different from that outside them. Consequently, the rush to implement MPAs has led to strongly held differences of opinions within the marine conservation community and an ideological divide between those who believe that MPAs are the best available management measure and those who think either that MPAs cannot be effective, or at least that there is insufficient evidence for their effectiveness. These differences of opinion may be both between and within different professional groups, such as scientists, resource managers and policymakers (Agardy et al. 2011, Wilson 2016, Hilborn 2017).

From the biological perspective, if protected areas are to be effective, their design must take account of both the ecology (e.g. preferred habitats, movement corridors and migration routes) and life history (e.g. mode of reproduction, social system and ontogenetic habitat changes) of the species being protected, as well as the inherent variability of their habitats (e.g. oceanographic regimes, anthropogenic disturbance and climatic change) (Carleton & McCormick 1993, Boersma & Parrish 1999, Hyrenbach et al. 2000). Arguably, the simplest scenario involves protecting sedentary species or clearly delineated, geographically fixed habitats within defined areas; protected areas can be extremely effective in these circumstances (Agardy 1994, Boersma & Parrish 1999, Probert 1999). However, even in terrestrial systems, protected areas have often failed to meet their conservation objectives in cases where the species to be protected are highly mobile or migratory, particularly when impacts beyond the protected areas were not adequately understood or controlled (Robbins et al. 1989, Trouwborst 2009, Borgström 2012). In marine systems, where mobile species regularly range over thousands of kilometres, and where movement patterns are often dynamic and poorly understood, conservation remains a considerable challenge. With the increasing promotion of MPAs as the conservation measure of choice, combined with the potential socio-economic and political challenges associated with establishing protected areas, it has become imperative that the scientific evidence for the effectiveness of MPAs for cetaceans is evaluated and placed in the context of other conservation and management options.

The apparent societal preference for MPAs, largely advocated by the environmental non-governmental organizations (eNGOs), appears to be intimately linked to the historical development of nature conservation law. From circa 700 BC when formal reserves were established by Assyrian nobles right through until the 1800s, wildlife protection was perceived primarily in terms of the preservation of game and quarry species, with the formation of protected areas in which to hunt them (Prato & Fagre 2005). In the United Kingdom, the Royal Parks in London and the New Forest in south-central England provide legacy examples of this, with large areas of land remaining largely undeveloped, even today (Reid 2002, Bell et al. 2013, Corkindale 2016). Created in 1765 at the insistence of Soame Jenyns, member of Parliament for Cambridge, the Tobago Main Ridge Forest Reserve is the oldest legally protected forest (Armstrong 2005). This protection came about as the understanding of the links between the trees and the atmosphere grew, and the need to preserve the watershed to benefit the sugar and cocoa plantations was recognized (Arfin 2004). It was almost 100 years later that the United States designated the world's first national park, Yellowstone, in 1872, although there was no formal definition of a national park until the National Parks Service Organic Act 1916. It was during the latter part of the 1800s through to the early 1900s that many of today's eNGOs were founded, such as the Commons Preservation Society (1865), the Royal Society for the Protection of Birds (1889), the Sierra Club (1892), the Wildlife Conservation Society (1895), the Audubon Society (1905) and the UK Wildlife Trusts (1912). Possibly because legislation with the aim of wildlife protection was generally fragmented in nature, with no overarching design or official bodies to ensure enforcement outside of the United States, many of these early eNGOs developed a strategy focussed on protecting areas through ordinary property rights, that is, land ownership, which enabled control over access and the activities that could take place (Bell et al. 2013).

International law, through the Geneva Convention on Fishing and Conservation of the Living Resources of the High Seas and the United Nations Convention on the Law of the Sea 1958 (UNCLOS), clearly established the duty of all States to conserve living resources of the sea, if only to ensure the resource is available for exploitation by future generations. Of key importance for cetaceans in this respect is the International Convention for the Regulation of Whaling 1946 (ICRW), which enabled the establishment of the International Whaling Commission (IWC). It was not until the advent of the more modern regional and global instruments associated with biodiversity that cetacean conservation really came of age. Such conventions include the United Nations Convention on the Conservation of European Wildlife and Natural Habitats 1979 (the Bern Convention), which is implemented throughout the European Union in the 1992 Directive on the Conservation of Natural Habitats and of Wild Fauna and Flora (92/43/EEC), otherwise known as the Habitats Directive. For members of the European Union, this directive represents one of the most important statutes ever introduced for nature conservation (Reid 2002, Bell et al. 2013, Born et al. 2015). Other important conventions include the Convention on Migratory Species of Wild Animals 1979 (CMS or Bonn Convention) and the Convention on Biological Diversity 1992 (CBD). Associated with CMS are two cetacean specific agreements: the Agreement on the Conservation of Small Cetaceans in the Baltic, North East Atlantic, Irish and North Seas 1994 (ASCOBANS) and the Agreement on the Conservation of Cetaceans in the Black Sea, Mediterranean Sea and Contiguous Atlantic Area 2001 (ACCOBAMS). On a global basis, today most States also participate in the work of international organizations which have a general interest in conservation, such as the United Nations Environment Programme (UNEP) and/or the IUCN. These organizations encourage the development of general principles or guidelines for the management and conservation of species, with the latter publishing the 'Red List' – the most authoritative global assessment of the conservation status of species.

The use of MPAs has fast become a mainstream approach for achieving marine conservation, although the area encompassed within MPAs is still considerably less than the total extent of protected areas on land. A UNEP review of protected areas noted that 15.4% of the world's land area and 3.4% ocean area had been designated as protected (Juffe-Bignoli et al. 2014). As lobbying by advocates of MPAs has become more successful, so this conservation measure has become more politicized; for example, all main parties in the 2015 UK general election had manifesto commitments to create new MPAs and to extend the current network of sites. In contrast, the key conventions and agreements all recognize that the best instrument should be used to tackle conservation issues. Taken together, these instruments enable the development of strict protection measures, that is, the control of anthropogenic activities, as well as the use of protected areas, but all of them fundamentally require consideration of species conservation status.

Designation of MPAs is not a new approach to cetacean conservation. Glacier Bay National Park and Preserve, originally designated in 1925 as Glacier Bay National Monument, has the distinction of being the oldest protected area that includes cetaceans amongst its features (Hoyt 2011). The world's first MPA specifically for cetaceans, known as 'The Sanctuary', was established in the Southern Ocean in 1938. It was thought to protect one-quarter of all Antarctic whales from commercial whaling (Zacharias et al. 2006). However, in 1955, under pressure from the whaling industry, it was reopened to commercial whaling and within a year was supplying 25% of the total Antarctic whale catch (NZ DOC 2004). This area was reinstated as an MPA in 1994 (see section 'Mismatch of MPA scale to issue and context'). There are now over 1000 MPAs designated or proposed for cetaceans (Hoyt 2011, Notarbartolo di Sciara et al. 2016, Directory of Cetacean Protected Areas around the World [see http://www.cetcaeanhabitat.org]). Many of these are recent designations, with approximately half having been identified in the last decade. It should be noted that the definition of MPA used by Hoyt (2011) and Notarbartolo di Sciara et al. (2016) is much broader than that of the

IUCN, for example, incorporating exclusive economic zone (EEZ) designations which would not fall within the IUCN definition.

Numbers of protected areas, however, do not provide an indication of effectiveness. The pre-eminent societal assumption seems to be that the declaration of an MPA results in protection (Kearney et al. 2012, de Santo 2013), which is just not necessarily the case. There is a huge volume of literature related to MPAs and cetaceans, and also considerable debate about the effectiveness of protected areas as a conservation measure (e.g. Agardy 1997, Agardy et al. 2003, 2011, 2016, Caveen et al. 2013, Geijer & Jones 2015, Hilborn 2016, 2017, Wilson 2016, Claudet 2017, Obura 2017). As such, there is an urgent need for a synthesis of the available evidence on the value of protected areas for cetacean conservation to enable policymakers to make informed decisions on the best approaches for conserving these highly charismatic species in circumstances where there are increasing pressures for the development of maritime industries combined with declining resources for conservation.

Methodology

A synthesis of the available evidence that collates primary research on the effectiveness of MPAs can provide crucial connections between research, policy and practice (Woodcock et al. 2016). The systematic review approach (Pullin & Stewart 2006, Higgins & Green 2008) enables a robust and objective assessment of the available evidence to provide a comprehensive synthesis suitable for evidence-based decision-making (Roberts et al. 2006). Such an approach enables a transparent and critical appraisal of the available evidence, whilst minimizing the chance of bias (Stewart et al. 2005, Sciberras et al. 2015). To determine the evidence for or against the effectiveness of MPAs for cetacean conservation, three different approaches were adopted. First, a systematic literature review; second, scrutiny of species-specific conservation, management and/or recovery plans; and third, analysis of case studies focussed on the key indicators of MPA effectiveness that have been identified by Agardy et al. (2011).

In the context of the present review, a positive effect of an MPA was defined as an increase in survival or abundance of the relevant population and a negative effect as the opposite. It should be borne in mind that whether there are apparent positive, negative or no effects, there is never a proper control area for comparison to know what would have happened without the designation of an MPA. This review can only look at before/after comparisons, in which 'no change' to the protected population could result from 'no effect of the MPA' or a positive effect if the population would otherwise have declined. In the latter case, however, it is assumed that a decline would normally be evident before MPA designation.

Systematic review

The systematic review was undertaken using two bibliographic databases of peer-reviewed literature: Science Direct (cataloguing articles published since 1823) and ISI Web of Science (articles published since 1941). In addition, consideration was also given to the scientific 'grey literature' (i.e. non-peer–reviewed articles). However, there is a considerable volume of grey literature available on the Internet concerning cetacean conservation and the use of MPAs, much of it generated by eNGOs and/or presenting opinion rather than evidence. As such, the proceedings of the periodic conferences organized by the International Committee on Marine Mammal Protected Areas (ICMMPA [http://icmmpa.org/conference/]) were utilized as an adequate representation of the factual grey literature.

Peer-reviewed literature

Science Direct and ISI Web of Science were chosen because together they catalogue the highest level of reliable cited journal articles, provide easy third-party access, and allow for repeatability of searches. A Boolean search of each database was undertaken using a variety of primary terms

Table 1 Search terms used for the systematic review in Science Direct and ISI Web of Science. An advanced search was used where all fields were searched for the terms.

Primary terms	Secondary terms	Tertiary terms
cetace*	'Protected area'	evidence
whale	Reserve	effec*
dolphin	Sanctuary	efficacy
porpoise	Refugia	benefi*
	Refuge	succes*
	Park	failure

* Denotes the 'wildcard' representing zero or more possible characters.

associated with MPA terminology, secondary terms focussing on cetaceans and tertiary terms relating to evidence and effectiveness (Table 1). The literature searches were conducted on 27 March 2017 and encompassed the entire datasets available at that time. This process identified 222 articles in Science Direct and 76 in Web of Science. These were further refined by reviewing the article title and abstract to exclude irrelevant articles. Articles that were removed from the search results included those that dealt with changes in cetacean distribution within MPAs or changes in anthropogenic activities within MPAs over time but did not have a link to specific management practices (e.g. Gannier 1998, 2002, Panigada et al. 2011, Cagnazzi et al. 2013, Fossi et al. 2013, Singkran 2013, Tepsich et al. 2014, Tobena et al. 2014, Perez-Jorge et al. 2015, Coomber et al. 2016); articles that linked cetacean distribution with the need to change, extend or zone MPAs (e.g. Lusseau & Higham 2004, Slooten et al. 2006, Oviedo & Solis 2008); and those that considered anthropogenic pressures and potential management practices beyond MPA boundaries (e.g. Williams et al. 2006, Wiley et al. 2008, Slooten & Dawson 2010, Gende et al. 2011, Bombosch et al. 2014, Merchant et al. 2014, Aniceto et al. 2016, Heenehan et al. 2017). This reduced the total number of articles to 24. The full text of these articles was examined, and those that did not explicitly consider evidence in relation to the effectiveness of MPAs for cetacean conservation were removed. Of those not taken forward, some modelled options for management (e.g. Pedersen et al. 2009, Parrot et al. 2011), whilst others examined the need to create an MPA or provide additional legal protection within or beyond MPAs (e.g. Liu & Hills 1997, Ashe et al. 2010, Bearzi et al. 2011, Hinch & De Santo 2011, Lu et al. 2014, Embling et al. 2015). This left 14 articles in total for the systematic review, which included one freshwater example. Owing to the paucity of papers identified for the systematic review, this example was retained despite not being marine.

ICMMPA proceedings

ICMMPA was formed in 2006 to promote the use of MPAs for marine mammal conservation based on best practice and to ensure that marine mammals and their habitats are represented in relevant ocean conservation actions. ICMMPA collaborates with the IUCN's World Commission on Protected Areas (WCPA) and IUCN Species Survival Commission (SSC) to enhance the profile of MPAs for marine mammal conservation. The ICMMPA conferences bring together approximately 200 researchers, managers, government representatives and eNGOs from over 42 countries on a biennial basis. The proceedings of the conferences held in 2009, 2011 and 2014 had been published at the time of the present review was conducted. The proceedings of the 2016 conference were not available.

Species-specific conservation, management and/or recovery plans

A review of species-specific conservation, management and recovery plans or strategies was also undertaken. This search was not comprehensive but instead provided a representative selection restricted to those in the English language and those adopted as a matter of policy rather than being proposed by eNGOs. This entailed a search of the websites of the following organizations:

ACCOBAMS
ASCOBANS
Department of Conservation, New Zealand
Department of Environmental Affairs, South Africa
Department of Environment and Energy, Australia
Department of the Environment, Heritage and Local Government, Ireland
IUCN
International Whaling Commission (IWC)
National Oceanic and Atmospheric Administration (NOAA), United States
Species at Risk Public Registry, Canada
The Government of the Netherlands

For each plan, the key threats and pressures were identified, and the proposed conservation measures were checked for the explicit use of MPAs.

Case studies

Agardy et al. (2011) identified five shortcomings of MPAs or key considerations for the debate on MPA effectiveness. These are:

- Mismatch of MPA scale to issue and context;
- Inappropriate planning or management processes;
- Failure of MPAs due to degradation of the environment in the surrounding unprotected area;
- Damaging displacement of human activities and other unintended consequences of MPAs; and
- Illusions of protection created by MPAs.

Cetacean case studies demonstrating these issues were identified through a literature review, often with more than one of the issues applicable to an individual example. Similar to the systematic review, one freshwater example was identified and considered appropriate to retain.

Results

Systematic review

Peer-reviewed literature

Despite there being over 1000 MPAs designated or proposed for cetaceans, there is more evidence demonstrating this is not an effective approach than for it being beneficial. Of the 14 papers identified, 11 presented evidence indicating the general failure of the MPAs under consideration. These conclusions were the result of either changes in the distribution of the populations for which the MPA had been designated (e.g. Cheney et al. 2014, Hartel et al. 2014, La Manna et al. 2014) or the lack of enforcement and need for additional measures (e.g. Cameron et al. 1999, Rayment et al. 2010, Gerrodette & Rojas-Bracho 2011, Steckenreuter et al. 2012a,b, Mullen et al. 2013, Rojas-Bracho & Reeves 2013, Mosnier et al. 2015). For example, Hartel et al. (2014) identified changes in bottlenose dolphin (*Tursiops truncatus*) distribution and habitat use that meant the tourism exclusion zones designated in the Bay of Islands (New Zealand) to protect important resting locations were no longer being utilized by *T. truncatus*. In the Mediterranean,

La Manna et al. (2014) noted a decrease in use of the Isole Pelagie Marine Protected Area (Italy) by *T. truncatus* during daylight hours, which was attributed to the greater presence of tourism vessels. Cameron et al. (1999) found no evidence of a change in survival rate of Hector's dolphin (*Cephalorhynchus hectori*) following the establishment of the Banks Peninsula Marine Mammal Sanctuary (New Zealand). In the Gulf of California, the lack of enforcement of fisheries measures and also illegal fishing within MPAs designated to protect the vaquita (*Phocoena sinus*) led to a continuing decline in population estimates (Gerrodette & Rojas-Bracho 2011, Rojas-Bracho & Reeves 2013). Steckenreuter et al. (2012a,b) noted an alteration of behaviour and activity of Indo-Pacific bottlenose dolphins (*Tursiops aduncus*) in the Port Stephens–Great Lakes Marine Park (Australia), finding that speed-restriction zones introduced to reduce disturbance of *T. aduncus* by vessels were not effective because these areas were utilized more intensively by dolphin-watching operators following their introduction.

The remaining three papers provided some evidence of positive effects of MPAs, although in each case, these were insufficient to change the population trend of the species in question. Just over 10 years after the work of Cameron et al. (1999), Gormley et al. (2012) recorded an increase in the survival rate of Hector's dolphin because of enforcement of by-catch regulations within the Banks Peninsula Marine Mammal Sanctuary. However, the resulting level of population growth observed was considered insufficient to protect the population adequately in the longer term, that is, the increase in survival rate resulting from the sanctuary designation was too low to enable population recovery. Mintzer et al. (2015) investigated the interactions between fishermen and botos (Amazon river dolphin, *Inia geoffrensis*) both within and outside the Mamirauá Sustainable Development Reserve (Brazil). Incidents of boto entanglement as well as the cetaceans taking or damaging the catch in fishing nets were frequent, and botos were illegally harvested for bait in most of the study communities. Involvement of fishermen in research and ecotourism activities within the reserve, however, led to improved attitudes toward botos and reduced mortality of botos. The scope of influence was, however, restricted to the study communities, and it was concluded that this was insufficient to prevent the decline of the boto population. Lastly, O'Brien and Whitehead (2013) concluded that the management of the Gully MPA (United States) was effective because the population of northern bottlenose whales (*Hyperoodon ampullatus*) remained stable over a 23-year period. However, the effectiveness of management was not actually tested. Consequently, in the present review, the conclusion of O'Brien and Whitehead (2013) was considered to be neutral because it is impossible to know whether the population would have remained stable if the MPA had not been designated.

ICMMPA proceedings

The theme of the first ICMMPA in 2009 was 'Networks: making connections' and explored the design, management and networks of MPAs. Seventeen key conclusions emerged that can be grouped as follows:

- There is a need for worldwide effort towards the robust identification of protected areas, including a classification system for cetacean MPAs similar to that of the IUCN (seven recommendations);
- MPAs may not always provide the solution to conservation problems (one recommendation);
- Additional special protection measures will likely be required beyond MPAs (three recommendations);
- Better and more informed management of MPAs is required (four recommendations); and
- MPAs should become centres of research and innovation (two recommendations).

Very few examples demonstrating the positive effects of MPAs for cetacean conservation were outlined in the proceedings. One example, however, was the effective management of tourism at Samadai

Reef, Egypt, introduced in 2004 to protect spinner dolphins (*Stenella longirostris longirostris*). This led to the population continuing to use the area whilst restricting tourism (further detail provided by Notarbartolo di Sciara et al. 2009). In contrast, there were many examples indicating that threats such as by-catch, ship strike and tourism within or near MPAs were preventing the effective conservation of cetaceans, including the vaquita, Hector's dolphin, humpback whale (*Megaptera novaeangliae*) and fin whale (*Balaenoptera physalus*). There were two notable conclusions from ICMMPA (2009):

"Protected area designations should bring added conservation value beyond that derived from other tools … If, on one hand, a marine mammal population is already adequately protected by one or several conventional tools or mechanisms, there may be little or no management justification for also establishing a protected area on its behalf (although there may be other types of justification related, for example, to public education and research)" (ICMMPA 2009, p. 57) and "It is important for management to be pursued at scales consistent with the spatial and nutritional requirements of a species. More specifically, ecologically designated MPA networks have to accommodate the life history of the species of concern and the dynamics of their oceanic habitats if such tools are going to contribute meaningfully to conservation" (ICMMPA 2009, p. 69).

The second ICMMPA held in 2011 was subtitled 'Endangered Spaces, Endangered Species' with the goal of seeking solutions to shared problems related to marine mammal conservation. The need to tackle anthropogenic impacts to ensure effective conservation was clearly identified, with examples including the vaquita, the Indo-Pacific humpback dolphin (*Sousa chinensis*) and Hector's dolphin. In such situations, the MPA role may largely be one of awareness raising. Again there were two notable conclusions: "It is important not to lose sight of the big picture: we should strive ultimately to 'manage' the whole oceans wisely, not only a designated subset" (ICMMPA 2011, p. 12) and "MPAs are good at protecting benthos and resident species, whilst MPA networks can capture more of the ecological requirements of migratory or highly mobile species. MSP [Marine Spatial Planning] is more powerful still, since it can consider multiple threats to ecosystems over wide areas and find ways to accommodate different uses in a way that keeps resource use sustainable" (ICMMPA 2011, p. 19).

In 2014, the third ICMMPA 'Important Marine Mammal Areas – A Sense of Place, A Question of Size' was held. The main goals were to highlight the importance of place and size for MPAs with marine mammal features and to introduce and further develop the proposed criteria for IUCN Important Marine Mammal Area (IMMA) designation. It was recognized that there are many instances where populations have continued to decline, indicating that MPAs have not provided sufficient protection, for example, for the vaquita in Mexico; the Maui dolphin (*Cephalorhynchus hectori maui*, a subspecies of Hector's dolphin) in New Zealand; the Indo-Pacific humpback dolphin in Hong Kong; spinner dolphins (*Stenella longirostris longirostris*) in Hawaii, Brazil and the Egyptian Red Sea; dwarf spinner dolphins (*S. longirostris roseiventris*) in Bali; the Indo-Pacific bottlenose dolphin in Australia; and fin and sperm (*Physeter macrocephalus*) whales in the Mediterranean Sea. Notable conclusions include: "Most MMPAs are small, that is, <4000 km^2, and represent political compromises that give only token attention to marine mammal habitat" (ICMMPA 2014, p. 13) and "An analysis of protected areas, primarily on land, shows that to date our protected area designs have been based more on unsubstantiated belief systems than on empirical evidence" (ICMMPA 2014, p. 19). The introduction and development of criteria for IMMAs and their similarities with and differences from other large-scale designations, such as Key Biodiversity Areas (KBA) of the IUCN, Ecologically or Biologically Significant Areas (EBSA) of the Convention on Biological Diversity (CBD), Specially Protected Areas of Mediterranean Importance (SPAMI) of the Barcelona Convention, and Biologically Important Areas (BIA) in the United States and Australia, were also alluded to.

Reviews of conservation, recovery and management plans

Forty conservation and recovery plans covering 32 species or subspecies from across the globe were identified in the present review. Tables 2 to 4 summarize the key pressures and threats highlighted

Table 2   Conservation and recovery plans for cetaceans in the North Atlantic

Jurisdiction and species	Key threats and pressures	Are MPAs used for mitigation?
North Sea (ASCOBANS) Harbour porpoise (*Phocoena phocoena*)	By-catch, noise disturbance	No.
Western Baltic, the Belt Sea and the Kattegat (ASCOBANS) Harbour porpoise (*P. phocoena*)	By-catch, noise disturbance, chemical pollution	No.
Baltic Sea (ASCOBANS) Harbour porpoise (*P. phocoena*)	By-catch, noise disturbance, prey resource	Yes. Areas identified as important for the reproduction and survival of the Baltic Harbour porpoise should be designated.
United Kingdom Harbour porpoise (*P. phocoena*)	By-catch, noise disturbance, chemical pollution, prey resource	Not specifically, but potential to identify MPAs is noted.
Netherlands Harbour porpoise (*P. phocoena*)	By-catch, noise disturbance, prey resource	No. Explicitly notes that MPAs are not appropriate.
Ireland Harbour porpoise (*P. phocoena*) Atlantic white-sided dolphin (*Lagenorhynchus acutus*) White-beaked dolphin (*Lagenorhynchus albirostris*) Bottlenose dolphin (*Tursiops truncatus*) Common dolphin (*Delphinus delphis*) Risso's dolphin (*Grampus griseus*) Killer whale (*Orcinus orca*) Northern bottlenose whale (*Hyperoodon ampullatus*) Long-finned pilot whale (*Globicephala melas*) Sperm whale (*Physeter macrocephalus*) Cuvier's beaked whale (*Ziphius cavirostris*) Sowerby's beaked whale (*Mesoplodon bidens*) Minke whale (*Balaenoptera acutorostrata*) Blue whale (*B. musculus*) Fin whale (*B. physalus*)	By-catch, prey resource, chemical pollution, noise disturbance, vessel collisions, wildlife watching	No, although entire EEZ designated as a sanctuary.

Continued

Table 2 (Continued)   Conservation and recovery plans for cetaceans in the North Atlantic

Jurisdiction and species	Key threats and pressures	Are MPAs used for mitigation?

Sei whale
  (*B. borealis*)
Humpback whale
  (*Megaptera novaeangliae*)
Striped dolphin
  (*Stenella coeruleoalba*)		

Mediterranean (ACCOBAMS)
Common dolphin
  (*Delphinus delphis*)
	By-catch, prey resource, climate
	  change
		No. The need for areas of conserva-
		  tion importance (ACIs) identified.
		  These are not legally defined as
		  MPAs in national legislation.

Black Sea (ACCOBAMS)
Harbour porpoise
  (*Phocoena phocoena*)
Common dolphin
  (*Delphinus delphis*)
Bottlenose dolphin
  (*Tursiops truncatus*)
	By-catch, live capture, noise
	  disturbance
		No. The need for areas of conserva-
		  tion importance (ACIs) identified.
		  These are not legally defined as
		  MPAs in national legislation.

Canada: north-west Atlantic
Blue whale
  (*Balaenoptera musculus*)
	Whaling (historic), noise disturbance,
	  prey resources, climate change,
	  chemical pollution, vessel collisions
		Yes. Saguenay–St. Lawrence Marine
		  Park (SSLMP), the Gully MPA, St.
		  Lawrence Estuary MPA and the
		  Manicouagan MPA. Aim of MPAs
		  is to raise awareness and control
		  whale watching.

USA, North Atlantic
Blue whale
  (*B. musculus*)
	Vessel collisions, entrapment, habitat
	  degradation, military noise, hunting
		Not specifically, but action to
		  identify and protect areas of
		  importance is noted.

USA, North Atlantic
Fin whale
  (*B. physalus*)
	Vessel collisions, prey resource,
	  climate change, whaling (historic)
		Not specifically, but action to
		  identify and protect areas of
		  importance is noted.

USA, North Atlantic
Sei whale
  (*B. borealis*)
	Vessel collisions, entanglement, prey
	  resource, climate change, whaling
	  (historic)
		No.

Canada, North Atlantic
North Atlantic right whale
  (*E. glacialis*)
	Whaling (historic), vessel collisions,
	  entanglement, noise disturbance,
	  habitat degradation
		No. Critical habitat identified in the
		  Grand Manan Basin and the Bay of
		  Fundy Right Whale Conservation
		  Area. These are not legally defined
		  as MPAs in national legislation.

USA, North Atlantic
North Atlantic right whale
  (*E. glacialis*)
	Vessel collisions, entanglement	No. Critical habitat has been
		  identified. These are not legally
		  defined as MPAs.

USA, West Atlantic Ocean including
the Caribbean Sea and Gulf of
Mexico
Sperm whale
  (*Physeter macrocephalus*)
	Directed hunts, vessel collisions,
	  entanglement, prey resource,
	  climate change
		No.

Mexico Vaquita (*Phocoena sinus*)	By-catch	Yes. For control of fishing activities.

Continued

Table 2 (Continued)   Conservation and recovery plans for cetaceans in the North Atlantic

Jurisdiction and species	Key threats and pressures	Are MPAs used for mitigation?
Canada, Scotian shelf Northern bottlenose whale (*Hyperoodon ampullatus*)	Whaling (historic), entanglement, oil and gas activities, noise disturbance, contaminants, vessel collisions	Yes. The Gully Marine Protected Area. Aims to reduce ship collisions and limit noise disturbance. Critical habitat has been identified in the Shortland and Haldimand Canyons of the Scotian Shelf. These are not legally defined as MPAs in national legislation.
Canada, North Atlantic Sowerby's beaked whale (*Mesoplodon bidens*)	Noise disturbance, entanglement, vessel collisions, chemical pollution	Yes. Gully Marine Protected Area. The protection measures for northern bottlenose whales also protect Sowerby's beaked whale. Critical habitat: Shortland and Haldimand Canyons of Scotian shelf. These are not legally defined as MPAs in national legislation.

Table 3   Conservation and recovery plans for cetaceans in the Pacific Ocean

Jurisdiction and species	Key threats and pressures	Are MPAs used for mitigation?
Canada, North Pacific Humpback whale (*Megaptera novaeangliae*)	Vessel collisions, entanglement	No.
USA, North Pacific Humpback whale (*M. novaeangliae*)	Entanglement, vessel collisions, whale watch harassment, habitat degradation	Yes. Hawaiian Islands Humpback Whale National Marine Sanctuary, which aims to raise awareness and control wildlife watching.
Western North Pacific (IWC) Gray whale (*Eschrichtius robustus*)	Entanglement, vessel collisions, noise disturbance, chemical pollution	No.
Canada, Eastern Pacific Gray whale (*E. robustus*)	Climate change, prey resource, noise disturbance, oil spill	Yes. Pacific Rim National Park Reserve, Pacific North Coast Integrated Management Area, Gwaii Haanas National Marine Conservation Area and Haida Heritage Site. MPAs aims to maintain integrity and protect migration corridors.
Canada, Eastern Pacific North Pacific right whale (*E. japonica*)	Whaling (historic), vessel collisions, entanglement, noise disturbance, chemical pollution	No.
USA, North Pacific North Pacific right whale (*E. japonica*)	Noise disturbance, vessel collisions, prey resource, climate change	No. Critical habitat has been identified. These are not legally defined as MPAs in national legislation.
Canada, North Pacific Blue whale (*Balaenoptera musculus*) Fin whale (*B. physalus*) Sei whale (*B. borealis*)	Whaling (historic), vessel collisions, noise disturbance, chemical pollution, climate change	No. Identification of critical habitat required through Species At Risk Act, but this is not a legally defined MPA in national legislation.

Continued

Table 3 (Continued)   Conservation and recovery plans for cetaceans in the Pacific Ocean

Jurisdiction and species	Key threats and pressures	Are MPAs used for mitigation?
USA, North Pacific Fin whale (*B. physalus*)	Vessel collisions, prey resource, climate change, whaling	Not specifically, but action to identify and protect areas of importance is noted.
USA, North Pacific Sei whale (*B. borealis*)	Directed hunts, vessel collisions, entanglement, prey resource, climate change	No.
USA, North Pacific Blue whale (*B. musculus*)	Vessel collisions, entrapment, habitat degradation, military noise, hunting	Not specifically, but action to identify and protect areas of importance is noted.
USA, North Pacific Sperm whale (*Physeter macrocephalus*)	Directed hunts, vessel collisions, entanglement, prey resource, climate change	No.
Canada, North Pacific Transient killer whale (*O. orca*)	Chemical pollution, noise disturbance, physical disturbance	No.
Canada, North Pacific Northern and Southern resident killer whales (*O. orca*)	Chemical pollution, prey resource, noise disturbance	Yes. Rubbing beaches protected by the Robson Bight–Michael Bigg Ecological Reserve for northern residents. No vessels (including non-motorized) or consumptive activities are permitted. Critical habitat also identified includes Johnstone Strait, south-eastern Queen Charlotte Strait and the channels connecting these straits. These are not legally defined as MPAs in national legislation. Southern residents critical habitat: Haro Strait and Boundary Pass and adjoining areas in the Strait of Georgia and the Strait of Juan de Fuca. These are not legally defined as MPAs in national legislation.
Pacific Northwest, USA Southern resident killer whales (*O. orca*)	Habitat degradation, prey resource, chemical pollution, physical disturbance, live captures (historic)	Critical habitat: Haro Strait and waters around the San Juan Islands, Puget Sound and the Strait of Juan de Fuca. These are not legally defined as MPAs in national legislation.
St Lawrence Estuary, Canada Beluga whale (*Delphinapterus leucas*)	Industrialization and chemical pollution, habitat degradation, noise disturbance, whale watching activities, prey resource	Yes. Saguenay-St. Lawrence Marine Park, Manicouagan MPA which aims to reduce vessel collisions and limit noise disturbance, awareness raising.
Cook Inlet, Alaska, USA Beluga whale (*D. leucas*)	Unauthorized harvest, noise disturbance, habitat degradation, chemical pollution	No. Two critical habitat areas identified which are separated by an exclusion area. These are not legally defined as MPAs in national legislation.

in the species plans and note whether MPAs were specifically identified for conservation. In only 10 of the 40 plans were MPAs specifically identified as a conservation measure. Of these, only a single MPA – Robson Bight (Michael Bigg) Ecological Preserve, Canada – was a 'strictly protected' site. This small site (16.2 km²), was established in 1982 because the resident killer whales (*Orcinus orca*) use two gravel beaches for 'rubbing', a behaviour considered unique to this population.

Table 4 Conservation and recovery plans for cetaceans in the Southern Hemisphere

Jurisdiction and species	Key threats and pressures	Are MPAs used for mitigation?

Queensland, Australia
  Australian snubfin dolphin
  (*Orcaella heinsohni*)
	By-catch, habitat degradation, coastal development, noise disturbance, vessel collisions including jet skis	No.

Queensland, Australia
  Indo-Pacific humpback dolphin
  (*Sousa chinensis*)
	Habitat degradation, noise disturbance, by-catch, incidental capture in shark control programme, marine debris	No.

New South Wales, Australia
  Humpback whale
  (*Megaptera novaeangliae*)
	Vessel collisions, noise disturbance, entanglement, marine debris	No.

New South Wales, Australia
  Sperm whale
  (*Physeter macrocephalus*)
	Vessel collisions, noise disturbance, entanglement, marine debris	No.

Australian waters including Overseas Territories
  Southern right whales
  (*Eubalaena australis*)
	Whaling, climate change, noise disturbance, entanglement, vessel collisions, noise disturbance, chemical pollution, marine debris, prey resource	No.

Australian waters including Overseas Territories
  Blue whale
  (*Balaenoptera musculus*)
  Pygmy blue whale
  (*B. musculus brevicauda*)
  Antarctic blue whale
  (*B. musculus intermedia*)
	Whaling, climate change, noise disturbance, vessel collisions, chemical pollution, marine debris, prey resource	No. Requires identification of biologically important areas (BIAs). These are not legally defined as MPAs in national legislation.

Great Barrier Reef Marine Park Authority, Australia
  Indo-Pacific bottlenose dolphin
  (*Tursiops aduncus*)
	Habitat degradation, incidental capture in shark nets, by-catch, vessel activity, noise disturbance	Measures proposed for reducing anthropogenic pressure on the species within the World Heritage site.

New Zealand
  Hector's dolphin
  (*Cephalorhynchus hectori hectori*)
  Maui's dolphin
  (*C. hectori maui*)
	By-catch, vessel activity, coastal development including mining, chemical pollution	Yes. For control of fishing (commercial and recreational) activities.

USA, Southern Ocean
  Sei whale
  (*Balaenoptera borealis*)
	Directed hunts, vessel collisions, entanglement, prey resource, climate change	No.

USA, Southern Ocean
  Fin whale
  (*B. physalus*)
	Vessel collisions, prey resource, climate change, whaling	Not specifically, but action to identify and protect areas of importance is noted.

Southwest Atlantic (IWC – Argentina, Brazil, Chile and Uruguay)
  Southern right whales
  (*Eubalaena australis*)
	Entanglement, vessel collisions, coastal development and industrialization	No, although potential for identification of critical habitat is noted.

Eastern south Pacific (IWC – Chile and Peru)
  Southern right whales
  (*E. australis*)
	Entanglement, vessel collisions, noise disturbance, habitat degradation, chemical pollution, climate change, prey resource	No, but if breeding areas are identified, then establishment of MPAs should be considered.

Argentina, Brazil and Uruguay (IWC)
  Franciscana
  (*Pontoporia blainvillei*)
	By-catch, habitat degradation, prey resource, chemical pollution, noise disturbance	Development and implementation of MPAs identified with the aim of awareness raising.

No consumptive activities or vessels, including non-motorized ones, are permitted within the site. The other nine plans utilizing MPAs largely focussed on regulations to reduce anthropogenic pressures within the site. A key component of almost all the plans utilizing MPAs was the need for awareness raising. In addition to the 10 plans advocating the use of MPAs, a further 13 plans included the use of 'place-based' conservation, that is, the identification of critical habitat or biologically important areas for the species if appropriate. These do not legally constitute MPAs under the national legislation but instead recognize the importance of particular locations, potentially very large in scale, where additional management of human activities might be required at certain times.

Case studies illustrating key considerations for MPA effectiveness

Mismatch of MPA scale to issue and context

The optimal protected area for a specific population would encompass that population's entire distribution (Reeves 2000). Whilst the two International Whaling Commission (IWC) whale sanctuaries provide, to some extent, examples at this scale, it is generally not practical or feasible to adopt such an approach for most cetaceans. The majority (approximately 62%) of MPAs designated or proposed for cetaceans are national designations less than 1000 km² in size (Figure 1), with approximately 15% of all sites being <20 km² in area and another 15% being 20–100 km². For example, MPA designations for bottlenose dolphins range from numerous sites of <5 km² in size (e.g. the Gulf of Trieste Miramare MPA, Italy; Gökçeada Marine Reserve, Turkey; 12 Special Areas of Conservation in the Azores, Portugal; Tupinambás Ecological Station and Tupiniquins Ecological Station, Brazil), to a large international site over 85,000 km² (i.e. the Pelagos Sanctuary for Mediterranean Marine Mammals). Where sites are small, short-term movement patterns could result in a deterioration of site effectiveness (Tezanos-Pinto et al. 2013, Hartel et al. 2014, La Manna et al. 2014). Even where site boundaries have been identified based on the core area of usage (e.g. the Moray Firth Special Area of Conservation, UK, encompassing 1513 km²), changes in species distribution that result in use of the wider coastal area can mean that use of the MPA declines (Wilson et al. 2004, Stockin et al. 2006, Culloch & Robinson 2008, Cheney et al. 2014).

At the other end of the MPA size range, the international IWC whale sanctuaries are the largest globally. The Indian Ocean Sanctuary (IOS), designated in 1979, covers the whole of the Indian Ocean south to 55°S, an area of 104 million km², whilst the Southern Ocean Sanctuary (SOS),

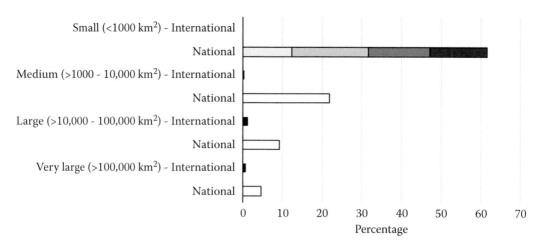

Figure 1 Variation in size of marine protected areas with cetaceans as a feature. The MPAs for sites <1000 km² were further subdivided: 1000–500 km² (pale grey), <500–100 km² (light grey), <100–20 km² (mid grey) and <20 km² (dark grey).

designated in 1994, covers the waters around Antarctica, an area of some 50 million km². Together these sanctuaries incorporate approximately 43% of the world's oceans by area and were designated with the primary aim of protecting cetaceans from hunting. However, by 1986, concern about the threat of commercial whaling to the remaining populations of blue, fin, humpback and right whales was so high that a global commercial whaling moratorium was established, that is, the allowable commercial catch was set to zero. Consequently, the IWC sanctuaries are considered by many to represent 'paper parks'. Nor does the IUCN recognize these sanctuaries as MPAs because they constitute areas where the species are protected by law across the entire region and not just within the sanctuary (Day et al. 2012). However, the sanctuaries are widely recognized as part of a portfolio of other treaties and measures to protect the seas more generally, bolstering biodiversity conservation efforts. For example, the Convention for the Protection, Management and Development of the Marine and Coastal Environment in the Eastern African Region (the Nairobi Convention) has declared its support for IOS, and the SOS is recognized by the Antarctic Treaty System, comprising the Convention for the Conservation of Antarctic Seals 1972 (Seals Convention), the Convention on the Conservation of Antarctic Marine Living Resources 1982 (CCAMLR), and the Protocol on Environmental Protection to the Antarctic Treaty 1991 (Madrid Protocol). Gerber et al. (2005) concluded that the IWC sanctuaries could be effective tools for conservation, but currently they were little more than 'paper parks' and would remain so without significant changes to the sanctuary programme by IWC members. Over a decade later, there has been no change to the sanctuary programme.

The political context for MPA development is clearly important. The Pelagos Sanctuary (originally called the Ligurian Sea Sanctuary) provides a good example of how sovereignty issues may drive boundary decisions that are not in the best interest of the species being protected. The Pelagos Sanctuary is situated in the north-western Mediterranean Sea and was designated to encompass the permanent Ligurian oceanographic front, which encourages high productivity resulting in good feeding opportunities for top predators, including 13 species of cetacean, although few are year-round residents (Notarbartolo di Sciara et al. 2008, Fossi et al. 2013). The main threats to the cetaceans in the area are fisheries, maritime traffic, military activities, the transport of hazardous materials through the area and whale watching. Following a decade of negotiations, Pelagos was established in 1999 through a trilateral agreement between France, Italy and the Principality of Monaco, coming into force in 2002. It encompasses approximately 87,000 km², 47% of which lies within the territorial waters of the three nations involved in the treaty, with the remaining 53% being in international waters (i.e. beyond 12 nautical miles of the coast). Political considerations (e.g. ensuring that the territorial waters included within the sanctuary were equitably subdivided) prevailed over ecological considerations (i.e. ensuring protection of the most important cetacean areas) in the boundary designation (Agardy et al. 2011). Consequently, a large zone of low cetacean density lying between Corsica and the Italian mainland was included within Pelagos boundaries, whilst a large portion of important pelagic cetacean habitat to the west of Corsica was left outside. In effect, Pelagos provides protection to some relatively low-value cetacean areas, while higher-value areas (from a cetacean perspective) have been left out. Notarbartolo di Sciara (2009) concluded that in the 10 years following the creation of Pelagos, it had failed to fulfil its main goal of significantly improving the conservation status of the area's marine mammal populations.

Governance of marine waters has evolved significantly since the sanctuary's designation. Following the declaration of an exclusive economic zone and an environmental protection zone by France and Italy, respectively, most of the Pelagos Sanctuary is now within European Union (EU) marine waters, meaning that the requirements of the Habitats Directive and other EU legal frameworks apply. Pelagos has also been entered into the list of Special Protected Areas of Mediterranean Interest (SPAMI) within the Protocol for Special Protected Areas and Biological Diversity of the Barcelona Convention. Under the auspices of the Barcelona Convention's Protocol, high-seas areas (i.e. those beyond national jurisdiction) identified as a SPAMI gain legal protection as the Protocol requires all parties adhering to the Convention to respect the protection measures established within each individual SPAMI. In 2014, European Commissioner Maria Damanaki stated that the Pelagos Sanctuary "holds great potential as a best practice example of how sustainable blue growth can work in busy and densely populated areas, while protecting marine life, including fish. We need to discuss how the EU's Common Fisheries Policy and Integrated Maritime Policy can ensure the Pelagos Sanctuary is operational and to support blue growth in the region" (Anonymous 2014). In 2016, a permanent secretariat for the sanctuary was put in place with the aim of consolidating the governance of the site, supported through cooperation with ACCOBAMS, the RAMOGE Agreement (under the Barcelona Convention) and the Mediterranean Science Commission (CIESM) (Permanent Secretariat of the Pelagos Agreement 2016).

Whilst there are examples of protected areas that match the scale likely to be required by many cetacean species, the political context within which such sites have been proposed or designated has seriously constrained their effectiveness. Where sites are much smaller in size, their effectiveness can be significantly reduced by changes in the distribution of the species being protected.

Inappropriate planning or management processes and failure
due to degradation of the unprotected surrounding ecosystem

Insufficient planning in the developmental stages of MPAs can lead to failure due to lack of engagement with stakeholders (Christie 2004), or once designated, a site may fail to meet the objectives for which it was established due to inadequate compliance and enforcement (Guidetti et al. 2008). Additionally, if functioning correctly, MPAs are, by their very nature, islands of protection. When the surrounding sea becomes degraded or threats are not managed outside MPAs, then the MPAs themselves may no longer be able to function properly to meet their conservation objectives (Murray et al. 2000, Notarbartolo di Sciara & Birkun 2010, Fossi et al. 2013). Although rather extreme, these issues are best exemplified by the baiji and the vaquita.

The baiji (*Lipotes vexillifer*, also known as the Chinese river dolphin or Yangtze river dolphin) was endemic to the Yangtze River, China. The species was protected through the Law of the People's Republic of China on the Protection of Wildlife as a First Category of National Key Protected Wildlife Species and, thereby should have received full legal protection throughout its range. There are no historical population estimates for the species, although it is thought to have declined rapidly from the 1950s to an estimated population of 400 by the 1980s (Zhou 1982). Key threats identified were habitat loss and degradation due to development initiatives such as the Three Gorges Dam (Zhou & Li 1989, Liu et al. 2000), vessel strikes (Zhou & Li 1989), by-catch in gill and fyke nets (Zhou & Wang 1994) and entanglement in rolling hook longlines and electro-fishing techniques (Zhou et al. 1998, Zhang et al. 2003). Conservation efforts for the species included the designation of a series of national and provincial reserves. As concerns for the species increased, more reserves were introduced along the river to help improve conservation efforts. However, despite the bans in place in relation to the use of rolling hook longlines, fyke nets and electro-fishing, these activities all continued (Turvey 2010). Eventually, the primary strategy introduced to prevent the baiji's extinction was *ex situ* conservation of a breeding population in semi-natural reserves to provide animals for the replenishment or re-establishment of the wild population. This approach, as one component of a broad-based conservation strategy, endorsed by international scientists (Perrin & Brownell 1989, Ellis et al. 1993, Perrin 1999) was unsuccessful (Liu et al. 1998). By 1999, the population had further declined to an estimated 13 animals (Zhang et al. 2003), and in 2006, a survey of the entire historical range of the baiji found no evidence of surviving animals (Turvey et al. 2007). Whilst protection from deliberate killing or injury appears to have been effective, prohibitions on certain fishing methods both within and outside/beyond the protected sites were ineffective and are considered to have led to the demise of this species (Turvey et al. 2007, Smith et al. 2008). This was the first functional extinction of any cetacean species caused by human activity.

The vaquita or Gulf of California porpoise (*Phocoena sinus*) has been considered one of the most endangered small cetacean species for several decades (Rojas-Bracho et al. 2008). It is endemic to the northern Gulf of California, Mexico, where the 1996 estimate of the population was 567 (95% confidence interval, CI: 177–1073) living in a core range of 2500 km^2 (Jaramillo-Legorreta et al. 1999). By 2008, the population had been reduced to 245 (95% CI: 68–884) (Jefferson et al. 2010), and extinction was a very real possibility. Rojas-Bracho et al. (2006, p. 179) noted that "The vaquita's survival does not depend on more or better science, but on improved management." By 2015, the population was estimated to be 59 (95% CI: 22–145) (CIRVA 2016a) and by 2016 was as low as 30 (95% credible interval 8–96) (CIRVA 2016b, Thomas et al. 2017). In spite of the creation of the Upper Gulf of California and Colorado River Delta Biosphere Reserve and the Vaquita Refuge encompassing the core range of the species, the continuation of gillnet fishing outside the reserve and illegal totoaba (*Totoaba macdonaldi*) fishing throughout the region, that is, also within the MPAs, has resulted in a continued decline (Gerrodette & Rojas-Bracho 2011). Between 1997 and 2015, the population had declined by 92% (CI: 80%–97%) (CIRVA 2016a).

The functional extinction of the baiji in China has helped provide additional impetus for the Mexican government to ensure better fisheries management. Efforts have focussed on reducing fishery interactions via vessel buyouts, improved surveillance and enforcement, providing socio-economic alternatives to fishermen, and testing alternative fishing gear. In April 2015, the Mexican president announced a plan for increased enforcement, a ban on gillnets for two years within the vaquita range and compensation for fishermen (Taylor 2015). The high levels of ongoing illegal gillnet fishing for totoaba led to a further 50% decline in abundance between 2015 and 2016 (CIRVA 2016b). In 2016, the IUCN recommended that a complete and permanent gillnet ban was put in place throughout the vaquita's range and called on all governments and competent international organizations, including the Convention on International Trade in Endangered Species (CITES) to assist in combating the illegal international trade in totoaba products. The IWC unanimously adopted a resolution calling for similar action (IWC 2016). The Mexican government introduced a complete ban on commercial fishing with gillnets in May 2017. There was, however, concern raised that it is still not illegal to sell or possess gillnets on land or at sea (CIRVA 2017). Although controversial, steps were also taken to progress *ex situ* conservation (Goldfarb 2016). The risks of capture and captive maintenance of vaquita were considered to be high but that these were greatly outweighed by the risk of entanglement in illegal gillnets and extinction in the wild (CIRVA 2017). During 2017, two female vaquitas were captured but released after showing signs of stress. One of these animals subsequently died, whilst the fate of the other is unknown (CIRVA 2018). As a result, further efforts to save the vaquita from extinction using *ex situ* conservation have been suspended. Instead, conservation efforts are now focussed on enhanced efforts to enforce the fisheries measures combined with the continued removal of active gillnets.

Despite protected areas encompassing the core range of the vaquita and baiji, the lack of enforcement of fisheries measures has meant these sites are failing, or have failed, the species for which they were designated. Only time will tell whether the most recent vaquita conservation efforts are sufficient to ensure that this species avoids the fate of the baiji.

MPAs that cause damaging displacement and other unintended consequences

Displacing an activity from within an MPA can have unintended consequences with a negative impact on the species for which the site has been designated. Hector's dolphin (*Cephalorhynchus hectori*) is endemic to New Zealand, with four distinct populations and limited gene flow between them. Of these, the North Island population, known as the Maui dolphin, has been identified as a separate subspecies (*C. hectori maui*) (Pichler et al. 1998, Baker et al. 2002) and listed as critically endangered (Slooten 2013). Such genetic differences over such a small geographic scale have not been observed in any other marine mammal (Dawson et al. 2001). There are over 7000 Hector's dolphins but only

50–70 Maui dolphins (Dawson et al. 2004, Gormley et al. 2005, Slooten et al. 2006, Currey et al. 2012). New Zealand's dolphins are exposed to a number of anthropogenic pressures, of which fishing with gillnets is considered to be the main threat, and population recovery is considered unlikely under the current protection measures (Slooten 2007, Slooten & Dawson 2010, Slooten & Davies 2012).

The Banks Peninsula Marine Mammal Sanctuary, South Island, was created in 1988 for Hector's dolphin. The management included fisheries regulations that prohibited the use of gillnets inside most of the sanctuary, although there were exemptions. To further reduce the impact on fisheries, the boundary of the sanctuary was also altered (Slooten 2013). The West Coast North Island Marine Mammal Sanctuary for Maui dolphin was created in 2003 and, again, fisheries exemptions were included. There was no protection for the dolphins outside the two protected areas, and nationwide, rapid population declines continued because the key threat, by-catch, was not controlled. In 2008, a more extensive package of protected areas and fisheries measures was implemented extending protection to most areas where the dolphins were found. However, in 2011 and again in 2012, fisheries exemptions were re-introduced to inshore areas (MPI & DOC 2012). The history of managing by-catch for New Zealand dolphins has demonstrated that managing fisheries in specific, relatively small protected areas is likely to displace the fisheries to other areas outside the protected areas that are still important to the species. In general, small protected areas are unlikely to provide an effective solution to a widespread problem, and the probability of population recovery is very low even if all gillnets were removed from the protected areas (Currey et al. 2012, Slooten 2013).

As noted earlier, Gormley et al. (2012) showed an increase in survival rates of Hector's dolphin and indicated that the previously rapid population decline can be slowed substantially where adequate fisheries measures are put in place. However, the resulting level of population growth observed is still considered insufficient to protect the population adequately from possible extinction. This study provides empirical evidence that MPAs can be effective for cetacean conservation but only if the threats to the population beyond the reserve are also tackled. The IUCN has recommended that both Hector's dolphin and the Maui dolphin should be protected throughout their entire range, including the removal of the compromises and exemptions to current fishing regulations (IUCN 2012). As a result, the West Coast North Island Marine Mammal Sanctuary for Maui dolphins was extended, and an interim ban on the use of set nets was implemented in the Taranaki region of the sanctuary. No additional measures were put in place for Hector's dolphin. Similar recommendations to those of the IUCN were subsequently made by IWC (IWC 2015), but no further action has been taken.

MPAs that create an illusion of protection

The public perception of MPAs is that they are areas of sea where all industrial activities are excluded. A consequence of this is that there is a risk that the public can be misled into thinking MPAs are doing their job when, in fact, they are not. Even when MPAs offer genuine protection, if they are designated to protect a site facing little threat, resources are diverted from the efforts required to address real threats (e.g. by-catch, noise disturbance, habitat degradation).

In 2000, the IWC received a proposal for the development of the South Pacific Whale Sanctuary (SPWS) from the governments of Australia and New Zealand. The proposal did not achieve the three-quarters majority of votes needed for establishment, with the main reason given by those voting against being that this would be a paper park providing no additional protection beyond the global moratorium on commercial whaling. The SPWS proposal was endorsed by the Pacific Island Forum (an intergovernmental group whose members include all 16 Pacific Island countries and territories within the proposed sanctuary area), the Convention on the Conservation of Nature in the South Pacific (Apia Convention) and the Convention for the Protection of Natural Resources and Environment of the South Pacific Region (SPREP Convention). Following the failure to achieve the three-quarters majority of votes to establish the sanctuary, many of the South Pacific countries whose waters were to be incorporated within SPWS have subsequently declared their exclusive economic zones (EEZs) as whale sanctuaries, encompassing an area of over 30 million km^2.

In the North Atlantic, the Irish government designated the Irish Whale and Dolphin Sanctuary encompassing their entire EEZ. This was based on an extension of the legal framework already in place in Irish law that bans hunting of cetaceans (Rogan & Berrow 1995). The sanctuary does not have the legal status of a reserve or refuge but protects cetaceans through the Whale Fisheries Act 1937 and the Wildlife Act 1976. The sanctuary has led to increased awareness of the 24 species of cetacean recorded in Irish waters but has been criticized for the lack of a management plan (Hoyt 2011). The designation of national sanctuaries has been seen as a token move by governments wishing to appear sympathetic to the environment either for internal (environmental perception) or external reasons (e.g. trade or tourism) (Hoyt 2011). Many other nations in the North Atlantic and Mediterranean regions have regulations in place that provide the same level of protection for cetaceans as the Irish sanctuary but have, however, chosen not to declare their waters as sanctuaries. As signatories to either ASCOBANS or ACCOBAMS, similar protection exists in the waters of Albania, Algeria, Belgium, Bulgaria, Croatia, Cyprus, Denmark, Egypt, Finland, France, Georgia, Germany, Greece, Italy, Lebanon, Lithuania, Malta, Morocco, Monaco, the Netherlands, Poland, Portugal, Romania, Slovenia, Spain, Sweden, Syria, Tunisia, Turkey, the UK and Ukraine.

Interestingly, there appears to be a difference in approach and attitude to such designations in different marine regions. In the North Atlantic and Mediterranean regions, nations tend to be criticized for the designation of so called 'paper parks' (De Santo et al. 2011, Leenhardt et al. 2013), whilst in the South Pacific region, New Zealand is criticized for not designating its entire EEZ as a whale sanctuary when the majority of its island neighbours have done so (Anonymous 2010). This is despite the same level of protection already being in place through New Zealand's Marine Mammals Protection Act.

Summary of the reviews on MPA effectiveness for cetaceans

There is considerably more evidence for the failure of MPAs for cetacean conservation than there is for success. Whilst recognizing that approximately half the >1000 MPAs designated or proposed for cetaceans are less than 10 years old and cetaceans often have long generation times, this conservation approach has been in use for over 80 years. As such, the lack of evidence for direct conservation benefits is concerning. Two studies have demonstrated potential benefits, although these were insufficient to reverse population declines (Gormley et al. 2012, Mintzer et al. 2015): one neutral example indicating a stable population over the longer term (O'Brien & Whitehead 2013) and one example of the positive effects of management resulting in continued use of an area (ICMMPA 2009). For cetaceans at least, the main role of MPAs appears to be in awareness raising and education rather than in direct ecological improvement for conservation purposes. As Slooten (2013) noted, area-based management has the potential to work, if the protected area is large enough, in the right place, effectively manages the key threats, ensures that the impacts are removed rather than displaced to other areas and no new threats are added. The present review identified no MPAs for cetaceans that meet these criteria, and the large majority appear to provide the illusion of protection whilst potentially distracting attention from more appropriate conservation measures.

More protected areas or integrated management with better spatial planning?

A very common theme running through almost all the examples identified in this review is the need to manage human activities regardless of where they occur (i.e. both within and outside MPAs), otherwise the conservation effort expended on MPA designation is wasted. Where pressures exist over large areas (e.g. throughout a species' range), conservation requires large-scale solutions. For mobile, transnational species, a wider-seas approach to management, that is, management focussed on reducing the impact of our activities, is likely to be more effective. This management, however, needs to be holistic: that is, it needs to be integrated across sectors and to take cumulative impacts into account. Whilst most legal requirements and agreement obligations for nature conservation advocate the use of the most appropriate management mechanisms, implementation has largely focussed on MPAs. This has led to conservation often appearing to clash with economic development and other legitimate interests in order to achieve its goals (Law Commission 2011, Morris 2011).

A pragmatic solution for the coexistence of nature conservation, particularly for cetaceans, and the sustainable development of the marine environment requires a much larger and more targeted vision than has been exhibited to date. We need the development of strategic, comprehensive, coordinated planning efforts for marine regions rather than a continued focus on small, localized areas and/or single sector management. The need for such approaches is not new and has been recognized both scientifically and legally (e.g. Juda 2007, Markus et al. 2011, ICMMPA 2011, Notarbartolo di Scara et al. 2016, Obura 2017, Chapter 17 of Agenda 21 and Section IV of the Johannesburg Plan from the United Nations World Summit on Sustainable Development by the Conference on Environment and Development [UNCED], Article 10 of the CBD). As noted by Redclift (1987), practitioners in the fields of ecology and economics both need to take a broader view, as well as there being a political commitment, for development to be truly sustainable.

Management in the marine environment is extremely complex both legally and practically (Potts et al. 2012, Boyes & Elliott 2014, 2015). With the current focus on further development of maritime industries across the globe, marine spatial planning with integrated surveillance seems to offer one of the better solutions to ensuring efficient and sustainable management of activities and to give authorities a clear understanding of environmental impacts. Cetacean conservation needs to move away from a focus on MPAs and become more flexible and proactive. Warren (1992) noted that environmental protection has been fundamentally misunderstood and that nature conservation does not mean an end to activities but instead requires integration into policy at a higher level to ensure coordination between different government departments. Unfortunately, 25 years on, the situation is little changed. Adaptive, dynamic and holistic management of human activities is required to achieve effective cetacean conservation hand-in-hand with sustainable development. Whilst this has been recognized in many international agreements and supranational legislation, there has been a failure to implement it.

Given the politicization of MPAs, the differences of opinion on their value and that the evidence is more indicative of their failure as a direct measure for cetacean conservation, does 'place-based' conservation have a role to play? The answer is 'yes' because all management is fundamentally area-based, for example, fishery regulations have geographic boundaries, be that in relation to fisheries closure areas, regulations for national waters or international agreements at the scale of ocean basins. Within the framework of marine spatial planning, cetacean conservation could be pursued through the identification of important areas without the MPA label and combined with sectoral management. The identification of such areas, however, needs to be undertaken robustly and with sufficient precision (Fernandez et al. 2017, Mannocci et al. 2017). This type of approach has been identified in over a third of the conservation and recovery plans reviewed. These areas can then be managed such that human activities are limited only when the animals are in the area and likely to be harmed.

The term 'critical habitat' is now frequently being used to denote important areas but often without its meaning being explained or defined. The United States' Endangered Species Act (ESA) legally defines critical habitat as specific areas: "within the geographical area occupied by the species at the time of listing, if they contain physical or biological features essential to conservation, and those features may require special management considerations or protection; and outside the geographical area occupied by the species if the agency determines that the area itself is essential for conservation" (ESA Section 3, paragraph 5A). A critical habitat designation is not a protected area. Instead, under Section 7 of the ESA, all federal agencies must ensure that any actions they authorize are not likely to jeopardize the continued existence of a listed species or destroy or adversely modify the designated critical habitat. Canada's Species at Risk Act (SARA) also legally defines critical habitat – "the habitat that is necessary for the survival or recovery of a listed wildlife species" – and, similarly, it has no regulatory status as a protected area. Instead these definitions recognize the importance of an area and the potential need for greater care to be taken. In complete contrast, Clark et al. (2010) equates 'critical habitat' not only with the general presence of the species of interest but also encompassing its prey resource. This deviation from the legal definition is not helpful, particularly when used as a basis for advocating MPAs.

Many nations have obligations through the Bern Convention, CBD, CMS and ASCOBANS to ensure that the favourable conservation status of listed species is maintained or improved utilizing a risk-based approach. These same international conventions and agreements provide an opportunity to further the idea of critical habitat for cetaceans where an enhanced risk-based approach to the management of human activities can be used (i.e. additional management measures are applied when evidence indicates the necessity). For example, such areas could be designated as Important Marine Mammal Areas (IMMAs) through CMS, Ecologically or Biologically Significant Marine Areas (EBSAs) through CBD, Specially Protected Areas of Mediterranean Importance (SPAMI) under the Barcelona Convention, Key Biodiversity Areas through IUCN, and/or Particularly Sensitive Sea Areas (PSSAs) or Areas to Be Avoided (ATBA) through the International Maritime Organization. With such listings, extremely large spatial areas can be incorporated and the appropriate sectoral management provided. EBSAs have been designated for humpback whale (*Megaptera novaeangliae*), sperm whale (*Physeter macrocephalus*), blue whale (*Balaenoptera musculus*), bottlenose dolphin (*Tursiops truncatus*), common dolphin (*Delphinus delphis*) and spinner dolphin (*Stenella longirostris*) (Kot et al. 2014). The Canary Islands PSSA incorporates important breeding areas for bottlenose dolphin, short-finned pilot whale (*Globicephala macrorhynchus*), rough-toothed dolphin (*Steno bredanensis*), Atlantic spotted dolphin (*Stenella frontalis*), common dolphin, striped dolphin (*S. coeruleoalba*), Risso's dolphin (*Grampus griseus*), sperm whale and Bryde's whale (*Balaenoptera edeni*). The IMO have moved shipping lanes and imposed speed restrictions at certain times to aid recovery of North Atlantic right whale (*Eubalaena glacialis*) (Conn & Silber 2013, Fisheries and Oceans Canada 2014, Waring et al. 2014) and to protect sperm whales (Silber et al. 2012). Whilst there are no examples of such management for cetacean by-catch, examples do exist in relation to by-catch of other species, including that of bluefin tuna (*Thunnus maccoyii*) (Collette et al. 2011) and yellowtail flounder (*Pleuronectes ferruginea*) (O'Keefe 2013, MarEx 2014). Such approaches highlight the importance of an area without the negative associations of MPA designation for economic development. However, it should be noted that some proponents consider IMMAs and EBSAs to represent an early stage for future MPA designation (e.g. CBD Secretariat 2011, Aquatic Mammals Working Group 2016). This should be strongly resisted unless progressed through a biosphere-type zoning approach within a marine spatial planning framework, where the most important areas become highly protected MPAs with strictly enforced measures. Such an approach would require a significant political commitment that has not been forthcoming to date. A more realistic alternative is the recognition that areas identified as being important for conservation purposes can overlap with areas for economic development. This is the reality of what is already happening. As Obura (2017) noted, all protection and management is, in fact, area-based, and it is time to move the debate onto integrated and effective management. We need to manage our activities such that the pressures we exert are reduced wherever this is necessary because whilst "humans are good at drawing lines ... the natural world and the ocean especially is not good at abiding by them" (Andrews 2015).

Sustainable development requires an ecosystem approach with an understanding of the ecosystem services provided by the environment. However, whilst the objectives of sustainable development and the ecosystem approach have been identified, there has been very little clarity on their pragmatic application, which can often seem contradictory (Kirk 2015). Many interpretations of sustainable development emphasize the importance of development and meeting socio-economic needs (e.g. French 2005, Ross 2012), whilst other interpretations constrain such development within the limits of ecosystem function, with a need to preserve that function and, therefore, provide future natural capital (e.g. the Marine Strategy Framework Directive 2008, Nicholson et al. 2009,

Schneiders et al. 2012, Bennett et al. 2015). Cetaceans have a significant role here. For example, the decline in whale numbers due to commercial whaling has been estimated to be at least 66% and perhaps as high as 90% and has likely altered the structure and function of the oceans (Roman et al. 2014). As highly mobile and/or migratory species, cetaceans support ecosystem services through nutrient cycling (Lavery et al. 2010, 2012, 2014; Roman & McCarthy 2010; Doughty et al. 2016). They also deliver important cultural services, providing both spiritual and recreational benefits across the globe (Ruiz-Frau et al. 2013). Conservation management decisions can and do influence development, so we must find sensible and pragmatic solutions for all. "The environmental movement has been criticized for being too negative – for being against progress, against economic growth … conservation is about more than leaving things alone: it is about hard choices with economic consequences" (Helm 2015, p. 3). An adaptive and dynamic approach to the management of human activities is required. For this to be effective, we need to recognize that continued monitoring and re-evaluation of the management approach is required and that, as circumstances change over time, so will the management measures.

Conclusions

Without effective management of the pressures on cetaceans, both within and outside MPAs, such sites are no better than paper parks. From the systematic literature review and evaluation of species-specific conservation plans and strategies, it can be concluded that the societal drive for the development of protected areas for large charismatic megafauna has potentially led to ineffective conservation, the proliferation of paper parks and, in the worst-case scenario, the first human-caused extinction of a cetacean when protected-area measures were not enforced. When areas are designated to protect megafauna under the impetus of public affection, there is rarely a solid theoretical foundation to the designation (Hooker & Gerber 2004, Leenhardt et al. 2013). Such an approach can result in the diversion of resources and conservation effort from more appropriate measures, which will ultimately be detrimental for the very species that are the focus of the conservation effort (Pinn 2016).

There is a substantial lack of evidence for the direct benefits of MPAs claimed by those advocating their use (Caveen et al. 2013, Wilson 2016). This is not a new or recent issue. Well over a decade ago, it was stated that "Scientists and managers need to become less accepting of having areas designated as sanctuaries without tangible protection" (Hooker & Gerber 2004, p. 38), and this is still true today. It also needs to be recognized that conservation goals and ambition are more often than not a question of politics and culture rather than science (Gerber et al. 2000, 2007, Heazle 2004, Freeman 2008). As concluded by Wilson (2016, p. 7) "The MPA concept has certainly come of age but we must keep in mind that it will not be appropriate in all instances and make sure that hard won tailored conservation efforts for mobile megafauna are not thoughtlessly steamrollered by the popular MPA bandwagon."

What is needed is clarity of definition, systematic testing of assumptions and adaptive application so that the appropriate mix of marine resource management can be elucidated and used depending upon the conditions that warrant them. Adaptive and dynamic management of human activities is required because we cannot manage cetaceans, only ourselves. In some cases, MPAs could be worse than the 'do nothing' option. The case of the baiji and the current plight of the vaquita provide very sobering examples of such failure.

Acknowledgements

It should be noted that the views presented here are the author's own and do not represent those of the organizations with which she is affiliated. This work was undertaken during an LLM Environmental Law studentship and was partially funded by the Clark Foundation. Thanks are due to my supervisor Anne-Michelle Slater and to two anonymous reviewers who helped to improve the manuscript.

References

Agardy, T. 1994. Advances in marine conservation: the role of marine protected areas. *Trends in Ecology and Evolution* **9**, 267–270.

Agardy, T. 1997. *Marine Protected Areas and Ocean Conservation*. Washington, DC: Academic Press Inc.

Agardy, T., Bridgewater, P., Crosby, M.P., Day, J., Dayton, P.K., Kenchington, R., Laffoley, D., McConney, P., Murray, P.A., Parks, J.E. & Peau, L. 2003. Dangerous targets? Unresolved issues and ideological clashes around marine protected areas. *Aquatic Conservation: Marine and Freshwater Ecosystems* **13**, 353–367.

Agardy, T., Claudet, J. & Day, J.C. 2016. 'Dangerous targets' revisited: old dangers in new contexts plague marine protected areas. *Aquatic Conservation: Marine and Freshwater Ecosystems* **26**, 7–23.

Agardy, T., Notarbartolo di Sciara, G. & Christie, P. 2011. Mind the gap: addressing the shortcomings of marine protected areas through large scale marine spatial planning. *Marine Policy* **35**, 226–232.

Andrews, S. 2015. Drawing lines. *The Marine Professional*, July 2015, 32–33.

Aniceto, A.S., Carroll, J., Tetley, M.J. & van Oosterhout, C. 2016. Position, swimming direction and group size of fin whales (*Balaenoptera physalus*) in the presence of a fast-ferry in the Bay of Biscay. *Oceanologia* **58**, 235–240.

Anonymous 2010. Tokelau declares whale sanctuary in its waters. *The People's Daily* 14 April 2010. Online. http://en.people.cn/90001/90777/90851/6950875.html (accessed 5 April 2017).

Anonymous 2014. Initiative to reenergise the Pelagos Sanctuary. In *World Fishing & Aquaculture* 31 July 2014. Fareham, Hampshire: Mercator Media Ltd. Online. http://www.worldfishing.net/news101/industry-news/initiative-to-reenergise-the-pelagos-sanctuary (accessed 24 February 2018).

Aquatic Mammals Working Group. 2016. Important marine mammal areas (IMMAs). In *1st Meeting of the Sessional Committee of the CMS Scientific Council (ScC-SC1) UNEP/CMS/ScC-SC1/Doc.10.4.2.1.* Bonn: UNEP/CMS Secretariat. Online. http://www.cms.int/en/document/important-marine-mammal-areas-immas (accessed 5 April 2017).

Arfin, F. 2004. Tobago: only the wet complain. *The Telegraph* 5 Apr 2004 Online. http://www.telegraph.co.uk/travel/destinations/centralamericaandcaribbean/trinidadandtobago/730078/Tobago-Only-the-wet-complain.html (accessed 8 February 2018).

Armstrong, H.G. 2005. Environmental education in Tobago's primary schools: a case study of coral reef education. *Revista de Biología Tropical* **53**(Suppl. 1), 229–238.

Ashe, E., Noren, D.P. & Williams, R. 2010. Animal behaviour and marine protected areas: incorporating behavioural data into the selection of marine protected areas for an endangered killer whale population. *Animal Conservation* **13**, 196–203.

Baker, A.N., Smith, A.N.H. & Pichler, F.B. 2002. Geographical variation in Hector's dolphin: recognition of new subspecies of *Cephalorhynchus hector*. *Journal of the Royal Society of New Zealand* **32**, 713–727.

Bearzi, G., Bonizzoni, S., Agazzi, S., Gonzalvo, J. & Currey, R.J.C. 2011. Striped dolphins and short-beaked common dolphins in the Gulf of Corinth, Greece: abundance estimates from dorsal fin photographs. *Marine Mammal Science* **27**(3), E165–E184.

Bell, S., McGillivray, D. & Pedersen, O. 2013. *Environmental Law*. Oxford: Oxford University Press, 8th edition.

Bennett, E.M., Cramer, W., Begossi, A., Cundill, G., Dıaz, S., Egoh, B.N., Geijzendorffer, I.R., Krug, C.B., Lavorel, S., Lazos, E., Lebel, L., Martın-Lopez, B., Meyfroidt, P., Mooney, H.A., Nel, J.L., Pascual, U., Payet, K., Harguindeguy, N.P., Peterson, G.D., Prieur-Richard, A.H., Reyers, B., Roebeling, P., Seppelt, R., Solan, M., Tschakert, P., Tscharntke, T., Turner II, B.L., Verburg, P.H., Viglizzo, E.F., White, P.C.L. & Woodward, G. 2015. Linking biodiversity, ecosystem services, and human well-being: three challenges for designing research for sustainability. *Current Opinion in Environmental Sustainability* **14**, 76–85.

Bennett, N.J. & Dearden, P. 2014. Why local people do not support conservation: community perceptions of marine protected area livelihood impacts, governance and management in Thailand. *Marine Policy* **44**, 107–116.

Boersma, P.D. & Parrish, J.K. 1999. Limiting abuse: marine protected areas, a limited solution. *Ecological Economics* **31**, 287–304.

Bombosch, A., Zitterbart, D.P., Van Opzeeland, I., Frickenhaus, S., Burkhardt, E., Wisz, M.S. & Boebel, O. 2014. Predictive habitat modelling of humpback (*Megaptera novaeangliae*) and Antarctic minke (*Balaenoptera bonaerensis*) whales in the Southern Ocean as a planning tool for seismic surveys. *Deep Sea Research Part I: Oceanographic Research Papers* **91**, 101–114.

Borgström, S. 2012. Legitimacy issues in Finnish wolf conservation. *Journal of Environmental Law* **24**, 451–476.

Born, C.-H., Cliquet, A., Schoukens, H., Misonne, D. & Van Hoorick, G. (eds) 2015. *The Habitats Directive in its EU Environmental Law Context: European Nature's Best Hope?* London: Routledge.

Boyes, S.J. & Elliott, M. 2014. Marine legislation – the ultimate 'horrendogram': international law, European directives & national implementation. *Marine Pollution Bulletin* **86**, 39–47.

Boyes, S.J. & Elliott, M. 2015. The excessive complexity of national marine governance systems – has this decreased in England since the introduction of the Marine and Coastal Access Act 2009? *Marine Policy* **51**, 57–65.

Cagnazzi, D., Fossi, M.C., Parra, G.J., Harrison, P.L., Maltese, S., Coppola, D., Soccodato, A., Bent, M. & Marsili, L. 2013. Anthropogenic contaminants in Indo-Pacific humpback and Australian snubfin dolphins from the central and southern Great Barrier Reef. *Environmental Pollution* **182**, 490–494.

Cameron, C., Barker, R., Fletcher, D., Slooten, E. & Dawson, S. 1999. Modelling survival of Hector's dolphins around Banks Peninsula, New Zealand. *Journal of Agricultural Biological and Environmental Statistics* **4**(2), 126–135.

Carleton, R.G. & McCormick, R.M.G. 1993. Challenges for biosphere reserves in coastal marine realms: representing ecological scales. In *Application of the Biosphere Reserve Concept to Coastal Marine Areas: Papers Presented at the UNESCO/IUCN San Francisco Workshop of 14–20 August 1989*, A. Price & S. Humphrey (eds). Gland, Switzerland: IUCN, 13–18.

Caveen, A.J., Gray, T.S., Stead, S.M. & Polunin, N.V.C. 2013. MPA policy: what lies behind the science? *Marine Policy* **37**, 3–10.

CBD Secretariat 2011. COP 11 Decision XI/17. Marine and coastal biodiversity: ecologically or biologically significant marine areas. Online. https://www.cbd.int/decision/cop/?id=13178 (accessed 5 April 2017).

Cheney, B., Corkrey, R., Durban, J.W., Grellier, K., Hammond, P.S., Islas-Villanueva, V., Janik, V.M., Lusseau, S.M., Parsons, K.M., Quick, N.J., Wilson, B. & Thompson, P.M. 2014. Long-term trends in the use of a protected area by small cetaceans in relation to changes in population status. *Global Ecology and Conservation* **2**, 118–128.

Christie, P. 2004. Marine protected areas as biological successes and social failures in Southeast Asia. *American Fisheries Society Symposium* **42**, 155–164.

CIRVA 2016a. *Seventh Meeting of the Comité Internacional para la Recuperación de la Vaquita Caracol Museo de Ciencias y Acuario*, May 10–13, 2016, Ensenada, BC, Mexico. Online. http://www.iucn-csg.org/wp-content/uploads/2010/03/cirva-7-final-report.pdf (accessed 5 April 2017).

CIRVA 2016b. *Eighth Meeting of the Comité Internacional para la Recuperación de la Vaquita (CIRVA-8)*. Southwest Fisheries Science Center, November 29–30th, 2016, La Jolla, CA. Online. http://www.iucn-csg.org/wp-content/uploads/2010/03/CIRVA-8-Report-Final.pdf (accessed 5 April 2017).

CIRVA 2017. *Ninth Meeting of the Comité Internacional para la Recuperación de la Vaquita (CIRVA-9)*. Southwest Fisheries Science Center. April 25–26, 2017, La Jolla, CA. Online. http://www.iucn-csg.org/wp-content/uploads/2010/03/CIRVA-9-Final-Report-May-11-2017.pdf (accessed 1 December 2017).

CIRVA 2018. *Tenth Meeting of the Comité Internacional para la Recuperación de la Vaquita (CIRVA-10)*. Southwest Fisheries Science Center. December 11–12, 2017, La Jolla, CA. Online. http://www.iucn-csg.org/wp-content/uploads/2018/01/CIRVA-10_final-report-2018.pdf (accessed 7 February 2018).

Clark, J., Dolman, S.J. & Hoyt, E. 2010. *Towards marine protected areas for cetaceans in Scotland, England and Wales: a scientific review identifying critical habitat with key recommendations*. Chippenham, UK: Whale and Dolphin Conservation Society.

Claudet, J. 2017. Six conditions under which MPAs might not appear effective (when they are). *ICES Journal of Marine Science*, doi: 10.1093/icesjms/fsx074

Collette, B., Chang, S.-K., Di Natale, A., Fox, W., Juan Jorda, M., Miyabe, N., Nelson, R., Uozumi, Y. & Wang, S. 2011. *Thunnus maccoyii*. The IUCN Red List of Threatened Species 2011. e.T21858A9328286. doi: 10.2305/IUCN.UK.2011-2.RLTS.T21858A9328286.en

Conn, P.B. & Silber, G.K. 2013. Vessel speed restrictions reduce risk of collision-related mortality for North Atlantic right whales. *Ecosphere* **4**(3), 1–16.

Coomber, F.G., D'Inca, M., Rosso, M., Tepsich, P., Notarbartolo di Sciara, G. & Moulins, A. 2016. Description of the vessel traffic within the north Pelagos Sanctuary: inputs for marine spatial planning and management implications within an existing international marine protected area. *Marine Policy* **69**, 102–113.

Corkindale, J. 2016. The environment as natural capital: towards a conservation strategy for Richmond Park. *Environmental Law and Management* **27**, 208–215.

Culloch, R.M. & Robinson, K.P. 2008. Bottlenose dolphins using coastal regions adjacent to a special area of conservation in the north-east Scotland. *Journal of the Marine Biological Association of the United Kingdom* **88**, 1237–1243.

Currey, R.J.C., Boren, L.J., Sharp, B.R. & Peterson, D. 2012. *A Risk Assessment of Threats to Maui's Dolphins.* Wellington, New Zealand: Ministry for Primary Industries and Department of Conservation. Online. http://www.doc.govt.nz/Documents/conservation/native-animals/marine-mammals/maui-tmp/mauis-dolphin-risk-assessment.pdf (accessed 5 April 2017).

Dawson, S., Pichler, F., Slooten, E., Russell, K. & Baker, C.S. 2001. The North Island Hector's dolphin is vulnerable to extinction. *Marine Mammal Science* **17**, 366–371.

Dawson, S., Slooten, E., Dufresne, E., Wade, P. & Clement, D. 2004. Small-boat surveys for coastal dolphins: line-transect surveys for Hector's dolphins (*Cephalorhynchus hectori*). *Fishery Bulletin* **201**, 441–451.

Day, J., Dudley, N., Hockings, M., Holmes, G., Laffoley, D., Stolton, S. & Wells, S. 2012. *Guidelines for Applying the IUCN Protected Area Management Categories to Marine Protected Areas.* Gland, Switzerland: IUCN.

De Santo, E.M. 2013. Missing marine protected area (MPA) targets: how the push for quantity over quality undermines sustainability and social justice. *Journal of Environmental Management* **124**, 137–146.

De Santo, E.M., Jones, P.J.S. and Miller, A.M.M. 2011. Fortress conservation at sea: a commentary on the Chagos marine protected area. *Marine Policy* **35**, 258–260.

Doughty, C.E., Roman, J., Faurby, S., Wolf, A., Haque, A., Bakker, E.S., Malhi, Y., Dunning, J.B., Jr. & Svenning, J.-C. 2016. Global nutrient transport in a world of giants. *Proceedings of the National Academy of Sciences of the United States of America* **113**, 868–873.

Dudley, N. (ed.) 2008. *Guidelines for Applying Protected Area Management Categories.* Gland, Switzerland: IUCN.

Edgar, G.J., Stuart-Smith, R.D., Willis, T.J., Kininmonth, S., Baker, S.C., Banks, S., Barrett, N.S., Becerro, M.A., Bernard, A.T.F., Berkhout, J., Buxton, C.D., Campbell, S.J., Cooper, A.T., Davey, M., Edgar, S.C., Försterra, G., Galván, D.E., Irigoyen, A.J., Kushner, D.J., Moura, R., Parnell, P.E., Shears, N.T., Soler, G., Strain, E.M.A. & Thomson, R.J. 2014. Global conservation outcomes depend on marine protected areas with five key features. *Nature* **506**, 216–220.

Ellis, S., Leatherwood, S., Bruford, M., Zhou, K. & Seal, U. 1993. Baiji (*Lipotes vexillifer*) population and habitat viability assessment preliminary report. *Species* **20**, 25–29.

Embling, C.B., Walters, A.E.M. & Dolman, S.J. 2015. How much effort is enough? The power of citizen science to monitor trends in coastal cetacean species. *Global Ecology and Conservation* **3**, 867–877.

Fernandez, M., Yesson, C., Gannier, A., Miller, P.I. & Azevedo, J.M.N. 2017. The importance of temporal resolution for niche modelling in dynamic marine environments. *Journal of Biogeography*, doi: 10.1111/jbi.13080

Fisheries & Oceans Canada 2014. *Recovery Strategy for the North Atlantic Right Whale (*Eubalaena glacialis*) in Atlantic Canadian Waters [Final]* (Species at Risk Act Recovery Strategy Series). Ottawa, Canada: Fisheries and Oceans. Online. https://www.sararegistry.gc.ca/virtual_sara/files/plans/rs_north_atl_right_whale_0609_e.pdf (accessed 5 April 2017).

Fossi, M.C., Panti, C., Marsili, L., Maltese, S., Spinsanti, G., Casini, S., Caliani, I., Gaspari, S., Muñoz-Arnanz, J., Jimenez, B. & Finoia, M.F. 2013. The Pelagos Sanctuary for Mediterranean marine mammals: marine protected area (MPA) or marine polluted area? The case study of the striped dolphin (*Stenella coeruleoalba*). *Marine Pollution Bulletin* **70**, 64–72.

Freeman, M.M.R. 2008. Challenges of assessing cetacean population recovery and conservation status. *Endangered Species Research* **6**, 173–184.

French, D. 2005. *International Law and Policy of Sustainable Development.* Manchester: Manchester University Press.

Gannier, A. 1998. Estimation of summer abundance of the striped dolphin *Stenella coeruleoalba* (Meyen, 1833) in the future northwestern Mediterranean international marine sanctuary. *Revue d Écologie – La Terre et la Vie* **53**, 255–272.

Gannier, A. 2002. Summer distribution of fin whales (*Balaenoptera physalus*) in the northwestern Mediterranean Marine Mammals Sanctuary. *Revue d Écologie – La Terre et la Vie* **57**, 135–150.

Geijer, C.K.A. & Jones, P.J.S. 2015. A network approach to migratory whale conservation: are MPAs the way forward or do all roads lead to the IMO? *Marine Policy* **51**, 1–12.

Gende, S.M., Hendrix, A.N., Harris, K.R., Eichenlaub, B., Nielsen, J. & Pyare, S. 2011. A Bayesian approach for understanding the role of ship speed in whale-ship encounters. *Ecological Applications* **21**, 2232–2240.

Gerber, L.R., DeMaster, D. & Roberts, S. 2000. Measuring success in conservation. Assessing efforts to restore populations of marine mammals is partly a matter of epistemology: how do you know when enough is enough? *American Science* **88**, 316–324.

Gerber, L.R., Hyrenbach, K.D. & Zacharias, M.A. 2005. Do the largest protected areas conserve whales or whalers? *Science* **307**, 525–526.

Gerber, L.R., Keller, A.C. & DeMaster, D. 2007. Ten thousand and increasing: is the western Arctic population of bowhead whale endangered? *Biological Conservation* **137**, 577–583.

Gerrodette, T. & Rojas-Bracho, L. 2011. Estimating the success of protected areas for the vaquita, *Phocoena sinus*. *Marine Mammal Science* **27**, E101–E125.

Goldfarb, B. 2016. Can captive breeding save Mexico's vaquita? *Science* **353**, 633–634.

Gormley, A., Dawson, S.M., Slooten, E. & Bräger, S. 2005. Mark-recapture estimates of Hector's dolphin abundance at Banks Peninsula, New Zealand. *Marine Mammal Science* **21**, 204–216.

Gormley, A.M., Slooten, E., Dawson, S.M., Barker, R.J., Rayment, W., du Fresne, S. & Bräger, S. 2012. First evidence that marine protected areas can work for marine mammals. *Journal of Applied Ecology* **49**, 474–480.

Guidetti, P., Milazzo, M., Bussotti, S., Molinari, A., Murenu, M., Pais, A., Spanò, N., Balzano, R., Agardy, T., Boero, F., Carrada, G., Cattaneo-Vietti, R., Cau, A., Chemello, R., Greco, S., Manganaro, A., Notarbartolo di Sciara, G., Russo, G.F. & Tunesi, L. 2008. Italian marine protected area effectiveness: does enforcement matter? *Biological Conservation* **141**, 699–709.

Hartel, E.F., Constantine, R. & Torres, L.G. 2014. Changes in habitat use patterns by bottlenose dolphins over a 10-year period render static management boundaries ineffective. *Aquatic Conservation: Marine and Freshwater Ecosystems* **25**, 701–711.

Heazle, M. 2004. Scientific uncertainty and the International Whaling Commission: an alternative perspective on the use of science in policy making. *Marine Policy* **28**, 361–374.

Heenehan, H.L., Van Parijs, S.M., Bejder, L., Tyne, J.A. & Johnston, D.W. 2017. Differential effects of human activity on Hawaiian spinner dolphins in their resting bays. *Global Ecology and Conservation* **10**, 60–69.

Helm, D. 2015. *Natural Capital: Valuing the Planet*. London: Yale University Press.

Higgins, J.P.T. & Green, S. (eds). 2008. *Cochrane Handbook for Systematic Reviews of Interventions Version 5.10*. London: The Cochrane Collaboration. Online. http://www.cochrane-handbook.org (accessed 25 February 2018).

Hilborn, R. 2016. Marine biodiversity needs more than protection. *Nature* **535**, 224–226.

Hilborn, R. 2017. Are MPAs effective? *ICES Journal of Marine Science* doi: 10.1093/icesjms/fsx068

Hinch, P.R. & De Santo, E.M. 2011. Factors to consider in evaluating the management and conservation effectiveness of a whale sanctuary to protect and conserve the North Atlantic right whale (*Eubalaena glacialis*). *Marine Policy* **35**, 163–180.

Hooker, S. & Gerber, L. 2004. Marine reserves as a tool for ecosystem-based management: the potential importance of megafauna. *Bioscience* **54**, 27–39.

Hoyt, E. 2011. *Marine Protected Areas for Whales, Dolphins and Porpoises*. London: Earthscan, 2nd edition.

Hyrenbach, K.D., Forney, K.A. & Dayton, P.K. 2000. Marine protected areas and ocean basin management. *Aquatic Conservation: Marine and Freshwater Ecosystems* **10**, 437–458.

ICMMPA 2009. *Proceedings of the First International Conference on Marine Mammal Protected Areas*. March 30 – April 3, 2009, Maui, Hawaii, R.R. Reeves (ed.). *Honolulu, Hawaii: International Committee on Marine Mammal Protected Areas*. Online. http://icmmpa.org/conference/the-1st-conference/the-1st-conference-documents/ (accessed 5 April 2017).

ICMMPA 2011. *Proceeding of the second International Conference on Marine Mammal Protected Areas. Endangered Spaces, Endangered Species*, November 7–11, 2011, Fort-de-France, Martinique, E. Hoyt (ed.). Online. http://second.icmmpa.org/files/2012/11/ICMMPA2-Final-Color-lo.pdf (accessed 5 April 2017).

ICMMPA 2014. *Proceeding of the third International Conference on Marine Mammal Protected Areas. Important Marine Mammal Areas–A Sense of Place, A Question of Size*, November 9–11, 2014, Adelaide, Australia, E. Hoyt (ed.). Online. http://icmmpa.org/wp-content/uploads/2016/02/REV01_ICMMPA3-2.pdf (accessed 5 April 2017).

IUCN 2012. Motion 35, IUCN actions to avert the extinctions of rare dolphins: Maui's dolphins, Hector's dolphins, vaquita and South Asian river dolphins (adapted at the IUCN World Conservation Congress 2012). Online. http://portals.iucn.org/2012motions/ (accessed 5 August 2016).

IWC 2015. Report of the Scientific Committee. *Journal of Cetacean Research and Management* **16**(suppl.), 1–87.

IWC 2016. Draft Resolution on the Critically Endangered Vaquita. IWC/66/20 Rev Agenda Item 6.7 Online. http://www.boycottmexicanshrimp.com/IWC-Resolution-on-Vaquita.pdf (accessed 5 April 2017).

Jaramillo-Legorreta, A.M., Rojas-Bracho, L. & Gerrodette, T. 1999. A new abundance estimate for vaquitas: first step for recovery. *Marine Mammal Science* **15**, 957–973.

Jefferson, T.A., Olson, P.A., Kieckhefer, T.R. & Rojas-Bracho, L. 2010. Photo-identification of the vaquita (*Phocoena sinus*): the world's most endangered cetacean. *Latin American Journal of Aquatic Mammals* **6**, 53–56.

Juda, L. 2007. The European Union and ocean use management: the marine strategy and the maritime policy. *Ocean Development and International Law* **38**, 259–282.

Juffe-Bignoli, D., Burgess, N.D., Bingham, H., Belle, E.M.S., de Lima, M.G., Deguignet, M., Bertzky, B., Milam, A.N., Martinez-Lopez, J., Lewis, E., Eassom, A., Wicander, S., Geldmann, J., van Soesbergen, A., Arnell, A.P., O'Connor, B., Park, S., Shi, Y.N., Danks, F.S., MacSharry, B. & Kingston, N. 2014. *Protected Planet Report 2014: Tracking Progress Towards Global Targets for Protected Species.* Cambridge, UK: UNEP-WCMC. Online. https://www.unep-wcmc.org/resources-and-data/protected-planet-report-2014 (accessed 25 February 2018).

Kearney, R., Buxton, C.D. & Farebrother, G. 2012. Australia's no-take marine protected areas: appropriate conservation or inappropriate management of fishing? *Marine Policy* **36**, 1064–1071.

Kearney, R., Farebrother, G., Buxton, C.D. & Goodsell, P. 2013. How terrestrial management concepts have led to unrealistic expectations of marine protected areas. *Marine Policy* **38**, 304–311.

Kirk, E.A. 2015. The ecosystem approach and the search for an objective and content for the concept of holistic ocean governance. *Ocean Development & International Law* **46**, 33–49.

Kot, C.Y., Halpin, P., Cleary, J. & Dunn, D. 2014. *A Review of Marine Migratory Species and the Information Used to Describe Ecologically or Biologically Significant Areas (EBSAs).* Assessment conducted by Marine Geospatial Ecology Lab, Duke University. Information document prepared by Global Ocean Biodiversity Imitative (GOBI) for the Convention on Migratory Species. Bonn: Convention on Migratory Species. Online. http://www.cms.int/sites/default/files/document/COP11_Inf_23_EBSA_final_report.pdf (accessed 25 February 2018).

La Manna, G., Manghi, M. & Sara, G. 2014. Monitoring the habitat use of common bottlenose dolphins (*Tursiops truncatus*) using passive acoustics in a Mediterranean marine protected area. *Mediterranean Marine Science* **15**, 327–337.

Lavery, T.J., Roudnew, B., Gill, P., Seymour, J., Seuront, L., Johnson, G., Mitchell, J.G. & Smetacek, V. 2010. Iron defecation by sperm whales stimulates carbon export in the Southern Ocean. *Proceedings of the Royal Society of London B* **277**, 3527–3531.

Lavery, T.J., Roudnew, B., Seuront, L., Mitchell, J.G. & Middleton, J. 2012. Can whales mix the oceans? *Biogeosciences Discussions* **9**, 8387–8403.

Lavery, T.J., Roudnew, B., Seymour, J., Mitchell, J.G., Smetacek, J. & Nicol, S. 2014. Whales sustain fisheries: blue whales stimulate primary production in the Southern Ocean. *Marine Mammal Science* **30**, 888–904.

Law Commission 2011. *Eleventh Programme of Law Reform. LAW COM No 330.* London: The Law Commission. Online. https://www.gov.uk/government/uploads/system/uploads/attachment_data/file/247223/1407.pdf (accessed 5 April 2017).

Leenhardt, P., Cazalet, B., Salvat, B., Claudet, J. and Feral, F. 2013. The rise of large-scale marine protected areas: conservation or geopolitics? *Ocean and Coastal Management* **85**, 112–118.

Liu, J.H. & Hills, P. 1997. Environmental planning, biodiversity and the development process: the case of Hong Kong's Chinese white dolphins. *Journal of Environmental Management* **50**, 351–367.

Liu, R., Wang, D. & Zhou, K. 2000. Effects of water development on river cetaceans in China. In *Biology and Conservation of Freshwater Cetaceans in Asia*, R.R. Reeves et al. (eds). Occasional Paper of the IUCN Species Survival Commission. Gland, Switzerland: IUCN, pp. 40–42.

Liu, R., Yang, J., Wang, D., Zhao, Q., Wei, Z. & Wang, X. 1998. Analysis on the capture, behaviour monitoring and death of the baiji (*Lipotes vexillifer*) in the Shishou semi-natural reserve at the Yangtze River, China. *IBI Reports* **8**, 11–22.

Lu, S.Y., Shen, C.H. & Chiau, W.Y. 2014. Zoning strategies for marine protected areas in Taiwan: case study of Gueishan Island in Yilan County, Taiwan. *Marine Policy* **48**, 21–29.

Lusseau, D. & Higham, J.E.S. 2004. Managing the impacts of dolphin-based tourism through the definition of critical habitats: the case of bottlenose dolphins (*Tursiops* spp.) in Doubtful Sound, New Zealand. *Tourism Management* **25**, 657–667.

Mannocci, L., Boustany, A.M., Roberts, J.J., Palacios, D.M., Dunn, D.C., Halpin, P.N., Viehman, S., Moxley, J., Cleary, J., Bailey, H., Bograd, S.J., Becker, E.A., Gardner, B., Hartog, J.R., Hazen, E.L., Ferguson, M.C., Forney, K.A., Kinlan, B.P., Oliver, M.J., Perretti, C.T., Ridoux, V., Teo, S.L.H. & Winship, A.J. 2017. Temporal resolutions in species distribution models of highly mobile marine animals: recommendations for ecologists and managers. *Diversity and Distributions* **23**, 1472–4642.

MAREX 2014. US fishing industry responds to bycatch criticism. *The Maritime Executive* 28 May 2014. Online. https://www.maritime-executive.com/article/US-Fishing-Industry-Responds-to-Bycatch-Criticism-2014-05-28#gs.rv955As (accessed 25 February 2018).

Markus, T., Schlacke, S. & Maier, N. 2011. Legal implications of integrated ocean policies: The EU's Marine Strategy Framework Directive. *The International Journal of Marine and Coastal Law* **26**, 59–90.

Merchant, N.D., Pirotta, E., Barton, T.R. & Thompson, P.M. 2014. Monitoring ship noise to assess the impact of coastal developments on marine mammals. *Marine Pollution Bulletin* **78**(2), 85–95.

Mintzer, V.J., Schmink, M., Lorenzen, K., Frazer, T.K., Martin, A.R. & Da Silva, V.M.F. 2015. Attitudes and behaviors toward Amazon River dolphins (*Inia geoffrensis*) in a sustainable use protected area. *Biodiversity and Conservation* **24**, 247–269.

Morris, R.K.A. 2011. The application of the Habitats Directive in the UK: compliance or gold plating? *Land Use Policy* **28**, 361–369.

Mosnier, A., Doniol-Valcroze, T., Gosselin, J.-F., Lesage, V., Measures, L.N. & Hammill, M.O. 2015. Insights into processes of population decline using an integrated population model: the case of the St. Lawrence Estuary beluga (*Delphinapterus leucas*). *Ecological Modelling* **314**, 15–31.

MPI & DOC 2012. Review of the Maui's dolphin threat management plan: consultation paper (Joint discussion paper No: 2012/18, 2012). Wellington, New Zealand: Ministry for Primary Industries (MPI) and Department of Conservation (DOC). Online. http://www.doc.govt.nz/Documents/conservation/native-animals/marine-mammals/maui-tmp/mauis-tmp-discussion-document-full.pdf (accessed 5 April 2017).

Mullen, K.A., Peterson, M.L. & Todd, S.K. 2013. Has designating and protecting critical habitat had an impact on endangered North Atlantic right whale ship strike mortality? *Marine Policy* **42**, 293–304.

Murray, K.T., Read, A.J. & Solow, A.R. 2000. The use of time/area closures to reduce bycatches of harbour porpoises: lessons from the Gulf of Maine sink gillnet fishery. *Journal of Cetacean Research and Management* **2**, 135–141.

Nicholson, E., Mace, G.M., Armsworth, P.R., Atkinson, G., Buckle, S., Clements, T., Ewers, R.M., Fa, J.E., Gardner, T.A., Gibbons, J., Grenyer, R., Metcalfe, R., Mourato, S., Muûls, M., Osborn, D., Reuman, D.C., Watson, C. & Milner-Gulland, E.J. 2009. Priority research areas for ecosystem services in a changing world. *Journal of Applied Ecology* **46**, 1139–1144.

Notarbartolo di Sciara, G. 2009. The Pelagos Sanctuary for the conservation of Mediterranean marine mammals: an iconic high seas MPA in dire straits. In *2nd International Conference on Progress in Marine Conservation in Europe*, Stralsund, Germany, 2–6 November 2009, H. von Nordheim et al. (eds). Bonn, Germany: Bundesamt für Naturschutz (BfN), 55–58. Online. https://www.bfn.de/fileadmin/MDB/documents/themen/meeresundkuestenschutz/downloads/Fachtagungen/PMCE-2009/PMCE_2009.pdf (accessed 5 April 2017).

Notarbartolo di Sciara, G. & Birkun, A. 2010. *Conserving Whales, Dolphins and Porpoises in the Mediterranean and Black Seas: An ACCOBAMS Status Report, 2010*. Monaco: ACCOBAMS. Online. http://www.eurobis.org/imis?module=ref&refid=210127&printversion=1&dropIMIStitle=1 (accessed 5 April 2017).

Notarbartolo di Sciara, G., Agardy, T., Hyrenbach, D., Scovazzi, T. & Van Klaveren, P. 2008. The Pelagos Sanctuary for Mediterranean marine mammals. *Aquatic Conservation: Marine and Freshwater Ecosystems* **18**, 367–91.

Notarbartolo di Sciara, G., Hanafy, M.H., Fouda, M.M., Afifi, A. & Costa, M. 2009. Spinner dolphin (*Stenella longirostris*) resting habitat in Samadai Reef (Egypt, Red Sea) protected through tourism management. *Journal of the Marine Biological Association of the United Kingdom* **89**, 211–216.

Notarbartolo di Sciara, G., Hoyt, E., Reeves, R., Ardron, J., Marsh, H., Vongraven, D. & Barr, B. 2016. Place-based approaches to marine mammal conservation. *Aquatic Conservation: Marine and Freshwater Ecosystems* **26**(Suppl. 2), 85–100.

NZ DOC [New Zealand Department of Conservation] 2004. *The Conservation of Whales in the 21st century. A Summary of the New Zealand Government's Policy on Whales and Whaling.* Wellington, New Zealand: New Zealand Department of Conservation. Online. http://www.doc.govt.nz/documents/conservation/native-animals/marine-mammals/conservation-whales-c21.pdf (accessed 5 April 2017).

O'Brien, K. & Whitehead, H. 2013. Population analysis of endangered northern bottlenose whales on the Scotian Shelf seven years after the establishment of a marine protected area. *Endangered Species Research* **21**, 273–284.

Obura, D.O. 2017. On being effective, and the other 90%. *ICES Journal of Marine Science* doi: 10.1093/icesjms/fsx096

O'Keefe, C. 2013. *SMAST Yellowtail Flounder Bycatch Avoidance Program.* Mid-Atlantic Fisheries Management Council, 4 February 2013. Dover, DE: Mid-Atlantic Fisheries Management Council. Online. http://www.mafmc.org/newsfeed/2013/2/21/yellowtail-bycatch-avoidance-program (accessed 25 February 2018).

Oviedo, L. & Solis, M. 2008. Underwater topography determines critical breeding habitat for humpback whales near Osa Peninsula, Costa Rica: implications for marine protected areas. *Revista De Biologia Tropical* **56**, 591–602.

Panigada, S., Lauriano, G., Burt, L., Pierantonio, N. & Donovan, G. 2011. Monitoring winter and summer abundance of cetaceans in the Pelagos Sanctuary (northwestern Mediterranean Sea) through aerial surveys. *PLoS One* **6**(7), e22878.

Parrott, L., Chion, C., Martins, C.C.A., Lamontagne, P., Turgeon, S., Landry, J.A., Zhens, B., Marceau, D., Michaud, R., Cantin, G., Menard, N. & Dionne, S. 2011. A decision support system to assist management of human activities in the St. Lawrence River Estuary, Canada. *Environmental Modelling and Software* **26**, 1403–1418.

Pedersen, S.A., Fock, H., Krause, J., Pusch, C., Sell, A.L., Bottcher, U., Rogers, S.I., Skold, M., Skov, H., Podolska, M., Piet, G.J. & Rice, J.C. 2009. Natura 2000 sites and fisheries in German offshore waters. *ICES Journal of Marine Science* **66**, 155–169.

Pendleton, L.H., Ahmadia, G.N., Browman, H.I., Thurstan, R.H., Kaplan, D.M. & Bartolino, V. 2017. Debating the effectiveness of marine protected areas. *ICES Journal of Marine Science*, doi: 10.1093/icesjms/fsx154

Perez-Jorge, S., Pereira, T., Corne, C., Wijtten, Z., Omar, M., Katello, J., Kinyua, M., Oro, D. & Louzao, M. 2015. Can static habitat protection encompass critical areas for highly mobile marine top predators? Insights from coastal East Africa. *PLoS One* **10**(7), e0133265.

Permanent Secretariat of the Pelagos Agreement 2016. *Permanent Secretariat of Pelagos Agreement has Reopened.* Genova, Italy: Pelagos Sanctuary. Online. http://www.sanctuaire-pelagos.org/en/news/all-news/511-the-pelagos-agreement-permanent-secretariat-has-reopened (accessed 24 February 2018).

Perrin, W.F. 1999. Selected examples of small cetaceans at risk. In *Conservation and Management of Marine Mammals*, J.R. Twiss & R.R. Reeves (eds). Washington, DC: Smithsonian Institution Press, 296–310.

Perrin, W.F. & Brownell Jr., R.L. 1989. Report of the workshop on biology and conservation of the platanistoid dolphins. In *Biology and Conservation of the River Dolphins*, W.F. Perrin et al. (eds). IUCN SSC Occasional Paper No. 3. Gland, Switzerland: IUCN, 1–22.

Pichler, F.B., Dawson, S.M., Slooten, E. & Baker, C.S. 1998. Geographic isolation of Hector's dolphin populations described by mitochondrial DNA sequences. *Conservation Biology* **12**, 676–682.

Pinn, E. 2016. Protected areas for harbour porpoise, but at what cost to their conservation? *Environmental Law Review* **18**, 97–103.

Potts, T., O'Higgins, T. & Hastings, E. 2012. Oceans of opportunity or rough seas? What does the future hold for developments in European marine policy? *Philosophical Transactions of the Royal Society A* **370**, 5682–5700.

Prato, T. & Fagre, D. 2005. *National Parks and Protected Areas: Approaches for Balancing Social, Economic and Ecological Values.* Oxford, UK: Blackwell Publishing.

Probert, P.K. 1999. Seamounts, sanctuaries and sustainability: moving towards deep-sea conservation. *Aquatic Conservation: Marine and Freshwater Ecosystems* **9**, 601–605.

Pullin, A.S. & Stewart, G.B. 2006. Guidelines for systematic review in conservation and environmental management. *Conservation Biology* **20**, 1647–1656.

Rayment, W., Dawson, S. & Slooten, E. 2010. Seasonal changes in distribution of Hector's dolphin at Banks Peninsula, New Zealand: implications for protected area design. *Aquatic Conservation: Marine and Freshwater Ecosystems* **20**, 106–116.

Redclift, M. 1987. *Sustainable Development: Exploring the Contradictions.* London: Routledge.

Reeves, R.R. 2000. *The Value of Sanctuaries, Parks and Reserves (Protected Areas) as Tools for Conserving Marine Mammals.* Final Report to the Marine Mammal Commission, contract number T74465385. Bethesda, MD: Marine Mammal Commission.

Reid, C.T. 2002. *Nature Conservation Law.* Edinburgh, UK: W. Green/Sweet & Maxwell.

Robbins, C.S., Sauer, J.R., Greenberg, R.S. & Droege, S. 1989. Population declines in North American birds that migrate to the neotropics. *Proceedings of the National Academy of Sciences* **86**, 7658–7862.

Roberts, P.D., Stewart, G.B. & Pullin, A.S. 2006. Are review articles a reliable source of evidence to support conservation and environmental management? A comparison with medicine. *Biological Conservation* **132**, 409–423.

Rogan, E. & Berrow, S.D. 1995. The management of Irish waters as a whale and dolphin sanctuary. In *Whales, Seals, Fish and Man*, A.S. Blix et al. (eds). Amsterdam, The Netherlands: Elsevier Science.

Rojas-Bracho, L. & Reeves, R.R. 2013. Vaquitas and gillnets: Mexico's ultimate cetacean conservation challenge. *Endangered Species Research* **21**, 77–87.

Rojas-Bracho, L., Reeves, R.R. & Jaramillo-Legorreta, A. 2006. Conservation of the vaquita *Phocoena sinus.* *Mammal Review* **36**, 179–216.

Rojas-Bracho, L., Reeves, R.R., Jaramillo-Legorreta, A. & Taylor, B.L. 2008. *Phocoena sinus.* The IUCN Red List of Threatened Species e.T17028A6735464. Gland, Switzerland: IUCN, doi: 10.2305/IUCN. UK.2017-2.RLTS.T17028A50370296.en

Roman, J., Estes, J.A., Morissette, L., Smith, C., Costa, D., McCarthy, J., Nation, J.B., Nicol, S., Pershing, A. & Smetacek, V. 2014. Whales as marine ecosystem engineers. *Frontiers in Ecology and the Environment* **12**, 377–385.

Roman, J. & McCarthy, J.J. 2010. The whale pump: marine mammals enhance primary productivity in a coastal basin. *PLoS One* **5**, e13255.

Ross, A. 2012. *Sustainable Development Law in the UK: From Rhetoric to Reality?* London: Earthscan.

Ruiz-Frau, A., Hinz, H., Edwards-Jones, G. & Kaiser, M.J. 2013. Spatially explicit economic assessment of cultural ecosystem services: non-extractive recreational uses of the coastal environment related to marine biodiversity. *Marine Policy* **38**, 90–98.

Schneiders, A., Van Daele, T., Van Landuyt, W. & Van Reeth, W. 2012. Biodiversity and ecosystem services: complementary approaches for ecosystem management? *Ecological Indicators* **21**, 123–133.

Sciberras, M., Jenkins, S.R., Mant, R., Kaiser, M.J., Hawkins, S.J. & Pullin, A.S. 2015. Evaluating the relative conservation value of fully and partially protected marine areas. *Fish and Fisheries* **16**, 58–77.

Silber, G.K., Vanderlaan, A.S.M., Tejedor Arceredillo, A., Johnson, L., Taggart, C.T., Brown, M.W., Bettridge, S. & Sagarminaga, R. 2012. The role of the international maritime organisation in reducing vessel threat to whales: process, options, action and effectiveness. *Marine Policy* **36**, 1221–1233.

Singkran, N. 2013. Classifying risk zones by the impacts of oil spills in the coastal waters of Thailand. *Marine Pollution Bulletin* **70**, 34–43.

Slooten, E. 2007. Conservation management in the face of uncertainty: effectiveness of four options for managing Hector's dolphin bycatch. *Endangered Species Research* **3**, 169–179.

Slooten, E. 2013. Effectiveness of area-based management in reducing bycatch of the New Zealand dolphin. *Endangered Species Research* **20**, 121–130.

Slooten, E. & Davies, N. 2012. Hector's dolphin risk assessments: old and new analyses show consistent results. *Journal of the Royal Society of New Zealand* **42**, 49–60.

Slooten, E. & Dawson, S.M. 2010. Assessing the effectiveness of conservation management decisions: likely effects of new protection measures for Hector's dolphin. *Aquatic Conservation: Marine and Freshwater Ecosystems* **20**, 334–347.

Slooten, E., Dawson, S.M., Rayment, W. & Childerhouse, S. 2006. A new abundance estimate for Maui's dolphin: what does it mean for managing this critically endangered species? *Biological Conservation* **128**, 576–581.

Smith, B.D., Zhou, K., Wang, D., Reeves, R.R., Barlow, J., Taylor, B.L. & Pitman, R. 2008. *Lipotes vexillifer.* The IUCN Red List of Threatened Species 2008:e.T12119A3322533. doi: 10.2305/IUCN.UK.2017-3. RLTS.T12119A50362206.en

Steckenreuter, A., Harcourt, R. & Moller, L. 2012b. Are speed restriction zones an effective management tool for minimising impacts of boats on dolphins in an Australian marine park? *Marine Policy* **36**, 258–264.

Steckenreuter, A., Moller, L. & Harcourt, R. 2012a. How does Australia's largest dolphin-watching industry affect the behaviour of a small and resident population of Indo-Pacific bottlenose dolphins? *Journal of Environmental Management* **97**, 14–21.

Stewart, G.B., Coles, C.F. & Pullin, A.S. 2005. Applying evidence-based practice in conservation management: lessons from the first systematic review and dissemination projects. *Biological Conservation* **126**, 270–278.

Stockin, K.A., Weir, C.R. & Pierce, G.J. 2006. Examining the importance of Aberdeenshire (UK) coastal waters for North Sea bottlenose dolphins (*Tursiops truncatus*). *Journal of the Marine Biological Association of the United Kingdom* **86**, 201–207.

Taylor, B. 2015. *Vaquita Gillnet Ban Begins April 29, 2015*. Anacortes, WA: The Society for Marine Mammalogy. Online. https://www.marinemammalscience.org/smm-news/vaquita-gillnet-ban-begins-april-29-2015/ (accessed 8 February 2018).

Tepsich, P., Rosso, M., Halpin, P.N. & Moulins, A. 2014. Habitat preferences of two deep-diving cetacean species in the northern Ligurian Sea. *Marine Ecology Progress Series* **508**, 247–260.

Tezanos-Pinto, G., Constantine, R., Brooks, L., Jackson, J., Mourão, F., Wells, S. & Baker, S. 2013. Decline in local abundance of bottlenose dolphins (*Tursiops truncatus*) in the Bay of Islands, New Zealand. *Marine Mammal Science* **29**(4), 390–410.

Thomas, L., Jaramillo-Legorreta, J., Cardenas-Hinojosa, G., Nieto-Garcia, E., Rojas-Bracho, L., Ver Hoef, J.M., Moore, J., Taylor, B., Barlow, J. & Tregenza, N. 2017. Last call: passive acoustic monitoring shows continued rapid decline of critically endangered vaquita. *The Journal of the Acoustical Society of America* **142**, EL512.

Tobena, M., Escanez, A., Rodriguez, Y., Lopez, C., Ritter, F. & Aguilar, N. 2014. Inter-island movements of common bottlenose dolphins *Tursiops truncatus* among the Canary Islands: online catalogues and implications for conservation and management. *African Journal of Marine Science* **36**, 137–141.

Trouwborst, A. 2009. International nature conservation law and the adaptation of biodiversity to climate change: a mismatch? Adaptation of biodiversity to climate change. *Journal of Environmental Law* **21**, 419–442.

Turvey, S. 2010. *Witness to Extinction: How We Failed to Save the Yangtze River Dolphin*. Oxford: Oxford University Press.

Turvey, S.T., Pitman, R.L., Taylor, B.L., Barlow, J., Akamatsu, T., Barrett, L.A., Zhao, X., Reeves, R.R., Stewart, B.S., Wang, K., Wei, Z., Zhang, X., Pusser, M., Richlen, L.T., Brandon, J.R. & Wang, D. 2007. First human-caused extinction of a cetacean species? *Biology Letters* **3**, 537–540.

Waring, G.T., Josephson, E., Maze-Foley, K. & Rosel, P.E. (eds). 2014. *U.S. Atlantic and Gulf of Mexico Marine Mammal Stock Assessments – 2014*. NOAA Technical Memorandum NMFS-NE-231. Silver Springs, Maryland: National Marine & Fisheries Service. doi: 10.7289/V5TQ5ZH0

Warren, L. 1992. Conservation: a secondary environmental consideration. *Journal of Law and Society* **18**, 64–80.

Wiley, D.N., Moller, J.C., Pace, R.M. & Carlson, C. 2008. Effectiveness of voluntary conservation agreements: case study of endangered whales and commercial whale watching. *Conservation Biology* **22**, 450–457.

Williams, R., Lusseau, D. & Hammond, P.S. 2006. Estimating relative energetic costs of human disturbance to killer whales (*Orcinus orca*). *Biological Conservation* **133**, 301–311.

Wilson, B. 2016. Might marine protected areas for mobile megafauna suit their proponents more than the animals? *Aquatic Conservation: Marine and Freshwater Ecosystems* **26**, 3–8.

Wilson, B., Reid, R.J., Grellier, K., Thompson, P.M. & Hammond, P.S. 2004. Considering the temporal when managing the spatial: a population range expansion impacts protected areas-based management for bottlenose dolphins. *Animal Conservation* **7**, 331–338.

Woodcock, P., O'Leary, B.C., Kaiser, M.J. & Pullin, A.S. 2016. Your evidence or mine? Systematic evaluation of reviews of marine protected area effectiveness. *Fish and Fisheries* **18**, 668–681, doi: 10.1111/faf.12196

Zacharias, M.A., Gerber, L.R. & Hyrenbach, K.D. 2006. Review of the Southern Ocean sanctuary: marine protected areas in the context of the International Whaling Commission sanctuary programme. *Journal of Cetacean Research and Management* **8**, 1–12.

Zhang, X., Wang, D., Liu, R., Wei, Z., Hua, Y., Wang, Y., Chen, Z. & Wang, L. 2003. The Yangtze River dolphin or baiji (*Lipotes vexillifer*): population status and conservation issues in the Yangtze River, China. *Aquatic Conservation: Marine and Freshwater Ecosystems* **13**, 51–64.

Zhou, K. 1982. On the conservation of the baiji. *Journal of Nanjing Normal College* **4**, 71–74.

Zhou, K. & Li, Y. 1989. Status and aspects of the ecology and behavior of the baiji, *Lipotes vexillifer*, in the lower Yangtze River. In *Biology and Conservation of the River Dolphins*, W.F. Perrin et al. (eds). IUCN SSC Occasional Paper No. 3. Gland, Switzerland: IUCN, 86–91.

Zhou, K., Sun, J., Gao, A. & Würsig, B. 1998. Baiji (*Lipotes vexillifer*) in the lower Yangtze River: movements, numbers, threats and conservation needs. *Aquatic Mammals* **24**, 123–132.

Oceanography and Marine Biology: An Annual Review, 2018, **56**, 105-236
© S. J. Hawkins, A. J. Evans, A. C. Dale, L. B. Firth, and I. P. Smith, Editors
Taylor & Francis

ANTARCTIC MARINE BIODIVERSITY: ADAPTATIONS, ENVIRONMENTS AND RESPONSES TO CHANGE

LLOYD S. PECK*

British Antarctic Survey, High Cross, Madingley Rd, Cambridge, CB3 0ET
*Corresponding author: Lloyd S. Peck
e-mail: l.peck@bas.ac.uk

Abstract

Animals living in the Southern Ocean have evolved in a singular environment. It shares many of its attributes with the high Arctic, namely low, stable temperatures, the pervading effect of ice in its many forms and extreme seasonality of light and phytobiont productivity. Antarctica is, however, the most isolated continent on Earth and is the only one that lacks a continental shelf connection with another continent. This isolation, along with the many millions of years that these conditions have existed, has produced a fauna that is both diverse, with around 17,000 marine invertebrate species living there, and has the highest proportions of endemic species of any continent. The reasons for this are discussed. The isolation, history and unusual environmental conditions have resulted in the fauna producing a range and scale of adaptations to low temperature and seasonality that are unique. The best known such adaptations include channichthyid icefish that lack haemoglobin and transport oxygen around their bodies only in solution, or the absence, in some species, of what was only 20 years ago termed the universal heat shock response. Other adaptations include large size in some groups, a tendency to produce larger eggs than species at lower latitudes and very long gametogenic cycles, with egg development (vitellogenesis) taking 18–24 months in some species. The rates at which some cellular and physiological processes are conducted appear adapted to, or at least partially compensated for, low temperature such as microtubule assembly in cells, whereas other processes such as locomotion and metabolic rate are not compensated, and whole-animal growth, embryonic development, and limb regeneration in echinoderms proceed at rates even slower than would be predicted by the normal rules governing the effect of temperature on biological processes. This review describes the current state of knowledge on the biodiversity of the Southern Ocean fauna and on the majority of known ecophysiological adaptations of cold-blooded marine species to Antarctic conditions. It further evaluates the impacts these adaptations have on capacities to resist, or respond to change in the environment, where resistance to raised temperatures seems poor, whereas exposure to acidified conditions to end-century levels has comparatively little impact.

Introduction

Antarctica and the Southern Ocean have many unique characteristics. Several are well known, including having the lowest temperatures and the greatest mass of ice on Earth. Others are less well known but have marked consequences for the organisms living there. These include isolation, where the combination of the separation of Antarctica's continental shelf from that of adjacent continental shelves is unique amongst the continents, and the circulation of the circumpolar current with the polar frontal zone produces a singular degree of biological isolation at the continental level. All of these factors have, over long evolutionary time, produced a range of attributes and adaptations in the Southern Ocean fauna that are either extreme because they are at one end of the relevant biological continuum, or only exist in Antarctica.

One of the earliest overviews of Antarctic marine biodiversity was given by Dell (1972). Since then there have been several good reviews on various aspects of Antarctic marine life, including Clarke (1983), who gave a seminal critique of adaptations to polar marine environments and followed this with more specific reviews of cold adaptation in 1991 and temperature and energetics in 1998. Arntz et al. (1994) reviewed Southern Ocean life from a biodiversity and life-history perspective, and Arntz et al. (1997) reviewed biodiversity with emphasis on community structure, biomass and spatial heterogenenity. Peck (2005a,b) concentrated on physiological performance, Peck et al. (2006b) evaluated the effects of environmental variability and predictability on organismal biology, Pörtner et al. (2007, 2012) evaluated thermal limits and temperature adaptation, and Clark and Peck (2009a) reviewed heat shock proteins and the stress response in Antarctic marine species. Adaptations in fish, including the recent discovery of neuroglobins (Cheng et al. 2009a) have been reviewed by Cheng and Detrich (2007), Patarnello et al. (2011) and Giordano et al. (2012b). Biodiversity patterns were evaluated by Convey et al. (2012) and physical gradients and their effects on biodiversity patterns by Convey et al. (2014). In recent years, a combination of increased access and scientific effort with, in some areas, dramatic improvements in technology have provided significant enhancements of understanding. This has been especially so in the 'omics technologies' where high-throughput next generation sequencing is enabling access to the fine-scale cellular and molecular functioning of Antarctic species (Huth & Place 2016a,b, Clark et al. 2017). Significant improvements in understanding have also come from technological breakthroughs in remote sensing and the ability to manipulate very large amounts of data. Many of these factors were highlighted in a recent Horizon Scan conducted by the Scientific Committee for Antarctic Research (SCAR) (Kennicutt et al. 2014, 2015).

Further to the above is the rapidly increasing need and impetus for research in the polar regions because they have become recognised as the fastest changing regions due to climate change impacts, and they contain faunas that are possibly the least capable of resisting change globally (e.g. Peck 2005a,b, Barnes & Peck 2008, Turner et al. 2009, 2013, Peck et al. 2014). Research here is of further importance because the coldest regions will be the first to disappear as the Earth warms.

This review aims to provide a comprehensive view of the current understanding of biodiversity patterns of benthic life in the Southern Ocean, including addressing questions about why diversity is higher than would be expected even a few decades ago, and also how the history of the continent has moulded the biodiversity. It also aims to describe and explain the wide range of life-history characteristics and physiological, cellular and molecular adaptations so far discovered in the fauna. These range from attributes that have been known for 50 years or more such as the presence of antifreeze in fish and the very slow growth rates exhibited by nearly all species, to attributes only recently identified including the lack of a heat shock response in many marine species and the discovery of new families of globin molecules, several of which remain to have their functions fully described. Finally, it aims to evaluate both the research that has been done on abilities to respond to altered environments and the impacts of the adaptations to extreme environments on capacities to respond. This is done predominantly in respect to temperature where most marine benthos appear limited in capacities to respond to temperature, and ocean acidification where conflicting results have been published, but many species appear little impacted by end-century conditions. The review focusses on benthic cold-blooded species, primarily animals because, although history will never accept difficulties as an excuse, to include a review of research on the microbes, plants, mammals and pelagic species that live in the Southern Ocean would have taken this work beyond the scope of a review of this type.

The physical environment

The Southern Ocean encircles the Antarctic continent (Figure 1). It ranges from the coastline to the polar frontal zone (PFZ), an area that fluctuates over time but represents a total area of around 22×10^6 km², around 6.1% of the world's oceans (www.NOAA.gov). The PFZ is the area where Southern Ocean waters abut those of the Pacific, Atlantic and Indian Oceans, and there is a sharp 3–5.5°C drop in temperature over a distance of 30–50 km when moving south into the Southern Ocean (www.eoearth.org). Ice is one of the dominant environmental factors in the region, both from the effects of scour from icebergs and from sea ice. At the winter maximum, sea ice extends over an area of around 20×10^6 km², and this is reduced by $10–15 \times 10^6$ km² at the summer minimum (Comiso 2010). Antarctica has 45,317 km of coastline, but over 80% of this is covered by ice shelves and glaciers in summer (Figure 1, Table 1). The world's total coastline length is 1.634×10^6 km (Burke et al. 2001). Antarctica thus accounts for 2.7% of the world's coastline but only 0.33% of the world's ice-free coastline in summer, and much less than this in winter. Significant portions of the ice-free coastline are in the South Orkney, South Shetland and Kerguelen Islands. However, in winter all the continental coastline is covered in ice. In terms of seabed, the area of continental

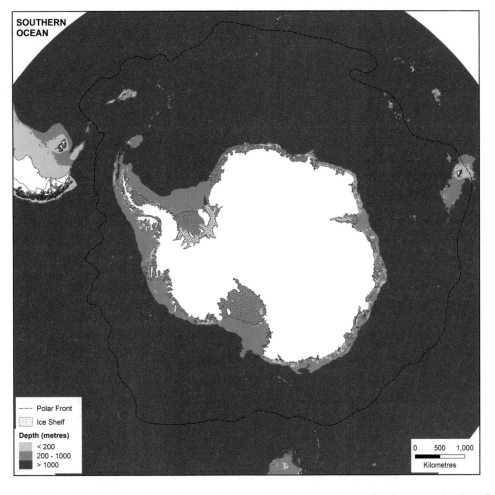

Figure 1 Map of the Southern Ocean (area south of the polar frontal zone) showing ice-covered and ice-free areas shallower than 200 m, 200–1000 m depth and deeper than 1000 m. (Image provided by P. Fretwell, British Antarctic Survey.)

Table 1 The areas in each of the categories shown in Figure 1 and the coastline length, both under, and free of glacier and ice shelf cover

Zone	Description	Area (km^2)	Length (km)
Southern Ocean	Total area (km^2)	36,415,505	
Southern Ocean	Not covered by ice	34,995,119	
Continental Shelf	Total area	4,623,202	
Continental Shelf	Not covered by ice	3,202,816	
Continental Shelf	Shallower than 200 m	593,547	
Continental Shelf	Shallower than 200 m (ice free)	390,071	
Antarctica coastline	Total length		45,317
Antarctica coastline	Ice-free length		5468

Sources: General Bathymetric Chart of the Oceans (GEBCO) and the SCAR Antarctic Digital Database (ADD) using a South Pole Azimuthal Equal Area projection for all areas quoted, and https://www.bas.ac.uk/science/science-and-society/education/antarctic-factsheet-geographical-statistics/

shelf available for biological colonisation, free from permanent ice, is around 4.62×10^6 km^2, but only 3.20×10^6 km^2 of this is not covered by ice shelves. This is approximately 7.3–10.6% of the world's continental shelf area (www.NOAA.gov, Barnes & Peck 2008). The continent and its shelf are depressed by the mass of ice on the land and previous ice scour at depth during glacial maxima (Huybrechts 2002), and the continental shelf break occurs at 800–1000 m depth as opposed to approximately 200 m depth elsewhere. Areas shallower than 200 m cover only 593,547 km^2, which is 2.1% of the world's ocean area shallower than 200 m.

The main physical factors affecting living organisms in polar marine environments are salinity, temperature, ice in its many forms, seabed topography and depth. Many physical factors change markedly with depth, especially between 0 and 100 m, but also to greater depths of 1000 m and beyond. These factors have large impacts upon the habitats in which marine organisms live. Phytoplankton productivity predominantly occurs in the top 100 m of the water column and mostly in the top 30–50 m (Clarke et al. 2008). Energy supply to primary consumers in habitats less than 50 m is usually dominated by phytoplankton. At progressively deeper depths, sedimenting biomass is modified during its passage through the water column, often aggregating, but significant fractions are consumed by pelagic species or broken down by microbial activity during the descent, and the material reaching the seabed is less available to suspension-feeders and more to deposit grazers (Kiørboe 2001). Light penetration is variable, depending on many factors including turbidity, phytoplankton productivity and, in polar regions, sea ice cover. Thus, a deeper continental shelf provides a markedly different environment for life in Antarctica than continental shelves elsewhere.

Ice has several major effects on habitats in the Southern Ocean. In open water, sea ice formation in winter reduces wind-induced mixing, causing stratification of the water column. It also markedly reduces light penetration to the water column beneath. The break-up of sea ice in summer allows mixing, bringing nutrients to photosynthetic depths, and increases light levels. However, the relationship between ice cover and water column productivity is complicated. Reductions in ice cover in a given system can increase or decrease productivity. If permanent ice cover is removed, increased light penetration can result in higher productivity (e.g. Arrigo et al. 2008, Montes-Hugo et al. 2009), and this can lead to increased productivity and biomass that is significant on a global scale (Peck et al. 2010a). Ice cover early in the growth season can increase stratification, which can reduce productivity (Vernet et al. 2008, Montes-Hugo et al. 2009, Venables et al. 2013, Meredith & Brandon 2017). However, more recent studies on the Antarctic Peninsula have shown that overall reductions in sea ice cover combined with low levels of water column stratification are associated with years of low phytoplankton biomass (Rozema et al. 2017). Salinity variation, from melting or freezing sea ice or from glacial melt, can also affect phytoplankton blooms (Moline et al. 2008). In nearshore Antarctic systems, removal of winter sea ice cover in spring can result in very intense blooms, probably because of the relatively high availability of key nutrients such as nitrate, phosphate and silicate, but also including iron from glacial meltwater run-off (e.g. Gerringa et al. 2012, Venables et al. 2013).

Sea ice itself has positive aspects for some biota, as productivity on or just within the basal surface of ice (epontic) can result in large standing biomass on the undersides of the ice that many species exploit, including juvenile krill in winter (Marschall 1988, Nicol 2006, Wiedenmann et al. 2009). A reduction in sea ice on the Antarctic Peninsula has been identified as related to reductions in krill numbers (Atkinson et al. 2004). High levels of productivity are also associated with the sea ice edge (Smith & Nelson 1985). There are specific epontic communities living in close association with sea ice (McMinn et al. 2010, Thomas & Diekmann 2010). These communities can include fish predators such as *Trematomus bernacchii, Pleuragramma antarctica* and *T. hansoni* (DeVries & Wohlschlag 1969).

On the seabed, ice disturbance is a major structuring ecological factor in both shallow and deep habitats down to 550 m (Barnes & Conlan 2007). Iceberg impacts have been shown to remove over 99.5% of all macrofauna and over 90% of meiofauna (Peck et al. 1999). Numbers of iceberg impacts on a regularly monitored grid at Rothera Point on the Antarctic Peninsula were shown to have increased between 2002 and 2010 by around 4-fold, and this increased the annual mortality of a common bryozoan, *Fenestrulina rugula*, from 89.5% to over 93% (Barnes & Souster 2011). Downstream impacts might be expected because *Fenestrulina rugula* is a pioneer species, often one of the first to settle and colonise newly cleared seabed following events such as iceberg impacts, and it plays a strong role in dictating later community composition (Barnes et al. 2014a).

Temperatures of water masses vary regionally and with depth. In the Ross Sea, possibly the coldest inhabited large water mass on Earth, winter temperatures are close to the freezing point of seawater ($-1.86°C$) in the water column. Temperatures can be even lower than this where salinity is raised during freezing events either associated with sea ice or ice growing on the seabed close to land. In summer, temperatures in shallow water only rise to around $-1.5°C$ (Orsi & Wiederwohl 2009). At depths below around 500 m in the Ross Sea, water temperatures are higher, up to around $+1.5°C$, as this is the depth that circumpolar deep water (CDW) intrudes to in this region. Circumpolar deep water is a large relatively warm saline water mass that occupies mid-water depths of the Antarctic circumpolar current. It is characteristically 2–4°C warmer than surface waters and is split into upper circumpolar deep water (UCDW) and lower circumpolar deep water (LCDW) by oceanographers.

Other regions of the Antarctic have warmer surface regimes. Along the Antarctic Peninsula and South Shetland and South Orkney Islands, near-surface winter temperatures are similar to those in the Ross Sea, but in the summer significantly positive values are achieved. At Rothera Station on Adelaide Island, summer peak temperatures are often around $+1.5°C$ (Clarke et al. 2007, 2008), as they are further north along the Antarctic Peninsula, for example, at Palmer Station on Anvers Island where temperatures can reach 2°C (Schram et al. 2015a), and on into the South Shetland Islands (Martinson et al. 2008) and South Orkney Islands (Clarke & Leakey 1996). Along the Antarctic Peninsula, UCDW penetrates onto the continental shelf to depths shallower than 300 m in most years (Martinson et al. 2008, Martinson & McKee 2012), and this can have strong influences on sea ice, coastal glaciers, ice shelves and marine productivity (Ducklow et al. 2013).

Salinity varies over both small and large spatial scales in the Southern Ocean. This variation comes from two main sources, the formation and melt of sea ice and run-off and melt from land-based glaciers and coastal ice shelves. The lowering of salinity associated with coastal run-off declines with distance from shore, but a reduced salinity signal is still detectable in surface waters hundreds of kilometres out across the continental shelf and beyond (Meredith et al. 2013). In nearshore environments, and over small spatial scales and shallow depths, salinities can be close to freshwater, which is an important factor for species inhabiting intertidal and intertidal fringe localities near melting glaciers. Very high salinities are produced locally when sea ice forms, which can result in brine channels in the ice and ice plumes under sea ice when cold brine sinks (Thomas & Diekmann 2010). Species inhabiting shallow water, especially those close to shore, experience very wide salinity fluctuations over both short periods when inundated with glacier run-off and seasonally when the environment freezes in winter.

The Antarctic marine environment is physically heterogeneous and patchy, both spatially and temporally, and this significantly impacts biodiversity patterns. This heterogeneity is evident not only on the seabed but also in the water column and associated ice cover. In the water column, patchiness exists over a range of scales due to variation in nutrient dynamics, length of the summer period of high light availability (which varies with latitude and factors such as ice cover), salinity changes and run-off from glaciers. Sea ice habitats are patchy over small spatial scales because of vertical gradients and strong salinity variation over small spatial scales (e.g. Petrich & Eiken 2010). They are patchy over larger scales because of interactions with differing levels of light input with latitude, because of the effects of land causing bottlenecks, and because of currents, heat transfer and wind that can cause open water polynyas all year round in the midst of otherwise continuous sea ice. Benthic and demersal organisms have even been shown to exist under ice shelves, sometimes many kilometres from the nearest open water (Littlepage & Pearse 1962, Lipps et al. 1979, Hain & Melles 1994, Domack et al. 2005). This includes a unique fauna that lives attached to the ice on the undersides of ice shelves. A recently described member of this fauna is an anemone, *Edwardsiella andrillae* that was discovered on the underside of the Ross Ice Shelf (Daly et al. 2013). Populations living under ice shelves must depend on particulate organic material advected from open water by currents or zooplankton moved under the ice by water currents during their daily vertical migrations. Many of the populations living under ice shelves must recruit by larval transport from distant locations.

Seabed patchiness in shallow water is primarily caused by variations in ice impacts (Barnes & Conlan 2012). Average annual wind speeds are highest globally in the latitudes between 50°S and 70°S, reaching values around 10 m s^{-1}, and average oceanic wave heights are also the largest in these latitudes at around 4 m (Barnes & Conlan 2007). These factors combine with the presence of ice in its many forms to make the shallow Southern Ocean seabed massively disturbed and only the most human-impacted seabeds due to trawling approach these levels of disturbance (Barnes & Conlan 2012). At depths of around 10–15 m, a site in North Cove, Rothera Point on Adelaide Island, Antarctic Peninsula was monitored for iceberg impacts, and over 90% of the site was impacted within one year, with several areas in the study experiencing multiple impacts. In another nearby bay, however, only around 40% of the area monitored was impacted per year (Brown et al. 2004). Shallower exposed sites are impacted more often. Other forms of ice also have strong effects on shallow benthos, with anchor ice growing from cold seabed extending down to 30 m depth at the highest latitudes. An ice foot often forms in the shallowest 2–3 m depth that can be several metres thick and persists for much of the year in some sites (Barnes & Conlan 2007).

In areas protected from scour, dense and diverse biological communities often develop. The interplay between levels of exposure to scour, wave action, protection and topography has led to the understanding that seabed communities in sites shallower than 100 m depth are usually held in various stages of early development because of the disturbance (Dayton et al. 1974), and this has been described as a dynamic mosaic (Barnes & Conlan 2012).

Biodiversity

Historical patterns

The biodiversity patterns seen in the Southern Ocean today are the product of both the prevailing environmental conditions and the history of the environment. Historically Antarctica was part of the Gondwana supercontinent bordered by the land masses to become South America and Australia, and 500 million years ago (mya), Gondwana was in the northern hemisphere (Crame 1994). It progressively moved across the globe and south until by 100 mya it was over the South Pole, still connected to South America and Australia, but its climate was warmer because of the transfer of heat from lower latitudes via large ocean currents (Crame 1994). The origins of some of the present fauna stem from this period and the subsequent break up of Gondwana. This is especially so for an element of the fauna that appeared through vicariance in the Weddellian Province (Clarke & Crame 1989, 2010). The notothenioid fishes may be one of the best examples of this and are one of the few examples of a marine species flock (Eastman & McCune 2000). They are currently the dominant group of Antarctic fishes. Their early diversification was along the Gondwana coast. There are now three non-Antarctic notothenioid groups (Eleginopsidae, Bovichtidae and Pseudaphritidae) living in South America, southern Australia and on Tasmania's coasts, and their distributions result from vicariance that occurred during the Gondwanan fragmentation 100–35 mya (Near 2004). Other groups with good evidence of origins during the Gondwanan break-up include the buccinid snails (Beu 2009).

Antarctica finally separated from South America, its last remaining Gondwanan neighbour, around 34 mya, when the Drake Passage opened. This initiated deep-water flow in the Antarctic circumpolar current, and, on land, large-scale glaciations began (Maldonado et al. 2003, 2014, Livermore et al. 2004). The link between the two continents became progressively weaker from around 40 mya, or even earlier (Livermore et al. 2005), but full separation did not occur for at least another 5 million years. After the separation, biodiversity patterns were mainly set by evolution *in situ* but with some movement of species to and from the deep sea and also by the same process along the Scotia Arc (e.g. Lipps & Hickman 1982). Clarke & Johnston (2003) argued that the relative representation of many of the major taxa may result from accidents of history rather than the nature of the Antarctic marine environment, with climatic change and glaciation causing extinction in some groups, which provided opportunities for others to expand.

Since the opening of the Drake Passage, temperatures in the Southern Ocean have generally cooled. There has been a steady decline in Southern Ocean temperature over the last 15 million years. Forty mya, sea temperatures around Antarctica were around 10–12°C warmer than at present (Zachos et al. 2008). There was a sharp fall in sea temperature around the separation of the continents approximately 35 mya of $\sim$3–4°C, followed by a rise back to preseparation temperatures during the early Miocene period (23–17 mya). From 15 to 17 mya, Southern Ocean temperatures declined gradually by around 10–12°C until current values were reached within the last 1 million years (Zachos et al. 2008). There were further small periods of warming in the late Miocene and early Pliocene (4.8–3.6 mya). These were, however, overlaid on the gradual cooling. Since Antarctica's isolation and cooling to current temperatures, there have been cooling and warming cycles in the environment caused by three main factors, the Milankovitch cycles: variations in the elliptical orbit of the Earth around the sun (eccentricity, 400,000- and 100,000-year cycles); changes in the tilt of the Earth's axis (obliquity, 41,000-year cycles); and wobbles in the rotational axis (precession, 23,000-year cycles) (Zachos et al. 2001). These have combined to give the glacial cycles observed and have been described from ice core records.

The cooling of the Southern Ocean has been accompanied by changes in sea ice, which have had profound effects on not only the present biodiversity but also how it was shaped during glacial cycles and interglacial periods. The extension of ice across the continental shelf reduced the area available for colonisation, made shallow sites unavailable and fragmented previously continuous ranges. This effect was increased by lowered sea level as the ice sheets increased in volume. The regular progression and retreat of ice restricted distributions to the outer sections of the continental shelf and to small areas on the continental shelf (refuges), only to return during interglacial periods. This process was identified as one that likely produced new species by isolation during periods of large ice extent, followed by range expansions in interglacials. Regular sequences of this around Antarctica during glacial cycles was called the 'biodiversity pump' by Clarke & Crame (2010).

In the early part of the first decade of the twenty-first century, data showing ice scour marks extending over the whole of the continental shelf suggested that ice sheets covered all available seabed, effectively reducing areas for biodiversity colonisation to zero (e.g. O'Cofaigh et al. 2002, Thatje et al. 2005). The idea of complete exclusion of life from the Antarctic continent was first challenged on land where continuous mountain and coastal refugia were identified as having been present throughout all previous glacial cycles from genetic analyses of living terrestrial groups (Stevens et al. 2006, Convey & Stevens 2007). This was followed by studies progressively showing marine life returned to the continental shelf and shallow sea from refugia on the continental shelf and possibly in shallow areas around Antarctica that persisted throughout previous glacial cycles (Graham et al. 2008, Thatje et al. 2008, Barnes & Kuklinski 2010). The fact that groups of organisms exist in the Southern Ocean which have persisted in isolation from other regions of the world since the Mesozoic confirms that sufficiently large refugia must have occurred during all previous glacial maxima. The important current questions lie around identifying where these refugia existed.

Current biodiversity

The extreme conditions and the long history of extremes in the Southern Ocean have produced a current fauna that is unique, with poor representation in some groups and with notable absences of representatives in many groups of fish, including scombrids and salmonids and some decapod crustaceans, including brachyuran crabs. Key papers developing many of the current ideas on the patterns and characteristics of current Antarctic marine diversity possibly began with Dearborn (1965), who worked in McMurdo Sound and was one of the first to assess benthic diversity. Hedgepeth (1971), Dell (1972) and White (1984) produced some of the first analyses of biodiversity patterns and the identification of provinces or regions. These were developed and clarified by Dayton (1990). Many of the more recent ideas on biodiversity patterns were erected by Arntz et al. (1994), Clarke (1996), Clarke & Crame (1992) and Clarke & Johnston (2003). More recently, the rapid development in genomic technologies has allowed an explosion in diversity research in Antarctica, but most of the patterns identified by the earlier researchers still remain as overarching paradigms. Overall diversity (in terms of species richness) is much higher than the non-specialist would expect from a polar ocean, with over 8000 species of marine invertebrate described to date (De Broyer et al. 2014) and an estimated at least 17,000 species living on the continental shelves (Gutt et al. 2004). Low levels of sampling and poor sampling in some areas (Clarke & Johnston 2003), combined with molecular taxonomic techniques identifying increasing numbers of cryptic species, means that such numbers, or more, are feasible, although there are examples where insufficient account has been taken of morphological variation when setting up taxonomic keys (e.g. Peck et al. 2018). However, such cases are rare. The fish fauna on the continental shelf is dominated by one group, the notothenioids that account for over 70% of the species present and over 90% of the biomass. They form a species flock that is similar in biodiversity terms to the cichlid fishes of the great lakes in Africa (Eastman & McCune 2000). In deeper water, another group, the snail fish (Liparidae), forms a large part of the fauna. Decapod crustaceans have representatives among the stone crabs (superfamily Lithodidae), but there is only a handful of species of caridean shrimps and no lobsters or true crabs (infraorder Brachyura), even though they are the most speciose crustacean group with over 6500 species worldwide. Other groups with poor representation include gastropod molluscs and balanomorph barnacles. The absence of some of these groups, such as the balanomorph barnacles, may be because elsewhere they predominantly inhabit shallow or intertidal sites, and the lack of suitable coastline may preclude their colonisation. There are also geographic areas or features where understanding, or even evaluations, of Antarctic marine biodiversity are poor. These include parts of the Bellingshausen and Amundsen Seas and also hydrothermal vents and seeps, which have only recently begun to be evaluated, but several of which have been located, for example along the Scotia Arc, including the Kemp Caldera (Rogers et al. 2012).

For some groups, on the other hand, a significant proportion of the world's total species is found in Antarctica. The sea spiders are especially well represented at around 21% of the world's total, but over

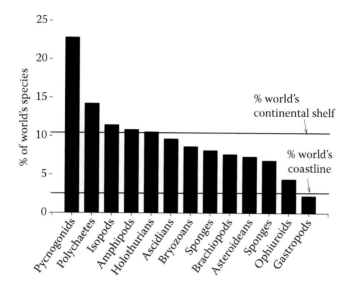

Figure 2 Proportion of total number of global species present in the Southern Ocean for several major animal groups. Data primarily from Clarke and Johnston (2003), updated using data by Linse et al. (2006), Barnes and Griffiths (2008), Munilla and Soler Membrives (2009), O'Loughlin et al. (2011), Emig (2017), Barnes (pers. comm.) (for bryozoans), and the World Register of Marine Species. Horizontal lines: proportions of global coastline and shelf area for the Southern Ocean. (Updated from Barnes, D.K.A. & Peck, L.S. 2008. *Climate Research* **37**, 149–163.)

13% of global polychaete species and over 10% of bryozoans, isopods and amphipods are also Southern Ocean species (Figure 2). These species all occur in an area of ocean that accounts for 7.3–10.6% of the world's continental shelf, 2.1% of marine seabed shallower than 200 m and 0.33% of global ice-free coastline. Thus, several taxa are represented at higher levels than would be expected on average for the areas of seabed or coastline available for colonisation and at higher levels than global averages. Several other groups, including brachiopods, echinoids, asteroids, holothuroids and bivalve molluscs, are all represented at levels near to 10% of global averages. The idea that the Antarctic benthic marine fauna is poorly represented is true for some of the most studied groups globally, such as molluscs and sponges, but it is not the case for several of the major taxa inhabiting the world's oceans.

A major issue when making biodiversity comparisons is making like-with-like analyses. The distributions of depths in the Southern Ocean are different from most others, with around two-thirds of them being deep sea beyond 3000 m, and the continental shelf being depressed by the mass of the ice sheet and previous scouring during glacial maxima to an average depth of around 450 m (Clarke & Crame 2010). The edge of the continental shelf is often taken as the 1000-m contour (e.g. Convey et al. 2012). However, depth makes a difference to the environment, especially in shallower depths as light, water movement, variations in food quality and salinity all change more on average in shallower depths. The issues then are making comparisons at set depth ranges, for instance, making comparisons at the usual continental shelf depths of 200–300 m, or to compare faunas on continental shelves that differ markedly in depth, and possibly average distance from the coastline and the enhanced nutrient inputs from that source. None of these are fully appropriate comparisons and should be made with caution and appropriate caveats. In many lower latitude regions, numbers of species living in the first 5–10 m below the intertidal can be very high. Numbers of scleractinian corals alone have been reported to be over 200 in some tropical intertidal habitats, and the associated faunas make this an exceptionally diverse, very shallow system (e.g. Richards et al. 2015). Further to this, both tropical and temperate intertidal and shallow wetlands can be highly diverse, which is especially so for mangroves (e.g. Dangan-Galon et al. 2016). Shallow-water diversity is particularly high in tropical coral reefs, formed by coral species with zooxanthellae, which predominantly exist in the top 100 m, and even within this are mainly within the top 50 m (Briggs 1996, Spalding et al. 2001). These are depths that experience high levels of ice disturbance in Antarctica. Future studies should attempt to account for these differences and limit comparisons to similar but restricted depth ranges. Investigations should also include assessments of the relative areas available for colonisation to those sampled, to identify where the biggest biodiversity differences occur between Antarctica and other regions. Further issues with factors such as depth are the lack of information for Antarctica and often for the polar regions in general. Thus, for sediments, some authors have suggested that diversity decreases with increasing latitude (e.g. Snelgrove et al. 2016), but Arctic sediment diversity has been demonstrated to be as high or higher than that in the Canadian Pacific and Atlantic (Archambault et al. 2010), and very recent research at a site on the Antarctic Peninsula has found higher levels of sediment species richness and animal abundance than at most lower latitude sites (Vause pers. comm.). Data are limited, but there may not be a latitudinal trend in diversity in soft sediment environments similar to that seen on hard substrata or on land.

The intertidal and sea ice

Two areas in Antarctica where understanding of biodiversity levels has advanced markedly in recent decades are for the intertidal and epontic sea ice communities. For many years, the intertidal of the Southern Ocean was thought to be only very sparsely colonised and for macro and meiofauna to be largely absent (e.g. Fogg 1998). In open intertidal areas, ice scour removes all obvious organisms with the possible exception of the mobile limpet *Nacella concinna* (Zacher et al. 2007). In areas protected from ice scour, however, such as in cobble boulder fields, a recognisable community can develop. Waller et al. (2006a) and Waller (2013) showed that over 40 species of marine invertebrate are present, 17 of which were sessile and attached, and some of those were at least four years old, demonstrating continuous multiyear survival. Zonation was also identified for the first time in the Antarctic intertidal, as communities changed with depth within a boulder field (Waller 2013).

Sea ice is far from uniform, and it provides a variety of habitats that supports a surprising level of biodiversity. It is colonised by a wide range of organisms that includes viruses, bacteria, meiofauna, invertebrates and fish that higher predators including birds and mammals can exploit (Thomas & Diekmann 2010). Only around a dozen species are categorised as truly ice-inhabiting and sympagic from both polar regions, but these species have to survive temperatures down to $-10°C$ and salinities varying from close to freshwater to up to three or four times more concentrated than seawater (Schnack-Schiel 2008). A much larger fauna has been identified as associated with sea ice as a habitat. This consists mainly of nematodes and rotifers in the Arctic, although copepods and turbellarians dominate in Antarctica. The distribution of organisms in the sea ice varies with the age of the ice and whether it is annual or multiyear ice. Most species occur in the lower layers of the sea ice, which are often flushed with seawater and physically closest to the subsea ice environment. This is where harpacticoid copepods are found, and even here they avoid narrow channels in the ice (Krembs et al. 2000). Some groups penetrate into the higher levels, and amongst these, some rotifers and turbellarians have been shown to pass through channels less than two-thirds of their body diameters by stretching and flexing their bodies (Schnack-Schiel 2008).

The isolation of the Antarctic marine environment, combined with environmental heterogeneity and historical environmental cycles that produced the biodiversity pump (Clarke & Crame 2010), has resulted in many new species evolving *in situ*. The result is that several Southern Ocean groups exhibit very high levels of endemism, often around 50%, including ascidians (Primo & Vasquez 2007), anemones (Rodriguez et al. 2007), bryozoans, bivalve molluscs and pycnogonids (Griffiths et al. 2009). Some other taxa exhibit even higher levels of endemism, including gastropod molluscs (around 75%, Griffiths et al. 2009), gammaridean amphipods (around 80%, De Broyer et al. 2007), and octopods (around 80%, Collins & Rodhouse 2006). It should be noted, however, that all of these figures have changed in recent years because of the significant efforts put into identifying new species through large-scale initiatives such as the Census of Antarctic Marine Life (CAML, www. caml.aq) and SCAR MarBIN (www.scarmarbin.be).

The overall outcome is a diverse and cold-adapted Southern Ocean fauna which has a range of general, and some unique, adaptations and life-history characteristics. These attributes have been shaped not only by physical factors such as isolation, cold, ice and seasonality, but also by biological factors such as predation.

Predation

From the earliest ecological studies of predation in Antarctic marine environments, it has been recognised that the types of predation, especially on the seabed, and numbers of specific types of predators are lower than elsewhere. However, there are trends that run contrary to this, and one is the high level of spongivory in Antarctic asteroids, where several species of starfish are primarily spongivorous (McClintock 1994, McClintock et al. 2005). Dayton (1990) and Arntz et al. (1994) highlighted differences with the Arctic in the absence or markedly reduced numbers of many bottom-feeding fish groups (e.g. sharks, rays, gadoid cods and flatfishes). The notothenioids are predominantly a demersal group, but there are no species so far identified in Antarctica that, for instance, regularly eat brittle star arms or the tips of infaunal bivalve mollusc siphons as do some lower latitude and Arctic flatfishes. There is also nothing analogous to the Arctic benthic feeding whales and walruses.

It can be argued that predation must be important for many species in Antarctica because chemical defences, or noxious, or repellent substances have been identified as present in the tissues of a wide range of species in the Southern Ocean. These include the brachiopod *Liothyrella uva* (McClintock et al. 1993, Mahon et al. 2003); the nudibranchs *Bathydoris hodgsoni* (Avila et al. 2000) and *Doris kerguelenensis* (Iken et al. 2002); the sponge *Latrunculia apicalis* (Furrow et al. 2003) and several other sponges (Peters et al. 2009); the ascidians *Distaplia cylindrica* (McClintock et al. 2004), *Cnemidocarpa verrucosa* (McClintock et al. 1991) and a range of other ascidians (Koplovitz et al. 2009); three soft corals (Slattery & McClintock 1995); and eggs, embryos and larvae of a range of invertebrate species (McClintock & Baker 1997). In a multispecies study, Núñez-Pons & Avila (2014) found that 17 of 31 Antarctic marine species contained lipophilic fractions that repelled the starfish *Odontaster validus*. Repellent substances have also been isolated from a range of Antarctic macroalgae (Amsler et al. 2005, Aumack et al. 2010, Bucolo et al. 2011). Large populations of herbivorous and omnivorous amphipods have been found associated with Antarctic macroalgae, which may explain the high incidence of macroalgal chemical defences (Amsler et al. 2014). It should be noted here that chemical defences are not the only factors that bear on likelihood of attack and success by predators, as the return compared to the effort needed to gain access has been identified as important (e.g. Peck 1993a, 2001a,b), and large size can provide a refuge from predation in prey species (Harper et al. 2009).

It is well recognised in the literature that predation pressure is a major structuring factor in the composition of biological communities, both in the present and over evolutionary time (Vermeij 1987). Several studies have described latitudinal gradients in predation pressure, although global-scale data are sparse (e.g. MacArthur 1972, Bertness et al. 1981), and few have produced data from the tropics to the poles. Recently Harper and Peck (2016) analysed frequencies of damage and repair in the shells of rhynchonelliform brachiopods to demonstrate a decrease in durophagous predation pressure across latitudes and with depth in the oceans. In the Southern Ocean, the main hypothesis is that the lack of durophagous-crushing predators over evolutionary time has resulted in a fauna that is archaic compared to lower latitude and Arctic marine benthic faunas, with similarities to Palaeozoic marine faunas (Aronson & Blake 2001, Arntz et al. 2005, Aronson et al. 2007). The top invertebrate predators are generally slow-moving asteroids and nemerteans and sessile anemones. Seabed protected from ice scour is often dominated by dense populations of epifaunal suspension- or deposit-feeders characterised by ophiuroids, crinoids, bryozoans, brachiopods, urchins and polychaetes that can extend over hundreds of square kilometres in deep water (Clarke et al. 2004a, Gili et al. 2006, Aronson et al. 2007, Convey et al. 2012). It should be noted that the fauna is often described as ancient, and the species composition is often thought to be reminiscent of past assemblages (e.g. Aronson et al. 2007). Some species, however, have evolved *in situ*, and there are recent colonisers, and in this sense, some are highly derived (Clarke & Crame 1989, Aronson et al. 1997). In this debate, however, it should be noted that the measured predation pressure is from durophagous predation. Levels of predation from engulfing predators such as nemerteans and anemones and from grazing predators such as urchins and limpets may be very intense, especially on early life stages, but these predation pressures have generally not been evaluated in the Southern Ocean benthos. Predation pressure is one of the main factors shaping the characteristics of the life histories of species living in a given environment. Other factors impacting life histories include competition and physical constraints, some of which can be clearer in extreme environments such as the Southern Ocean than elsewhere.

Life histories

The study of life histories is based around the concept that the phenotype is the product of a range of demographic traits including growth rate, overall size, age and size at maturity, number and size of offspring, reproductive investment and longevity. The interaction of these characteristics is key to setting the fitness of the individual in any particular environment (Stearns 1992). Life-history analyses of Antarctic marine species are rare, even though many of the relevant characteristics have been gathered for several decades (e.g. Arntz et al. 1994), and general physiological adaptations have been investigated for several groups (see reviews by Clarke 1983, 1998, Peck 2002a, Peck et al. 2006b, Pörtner et al. 2007, 2012).

Many life-history characteristics in studies of Antarctic benthic species in the early decades of marine biological research were identified as being K-selected attributes (Clarke 1979). These were detailed by Arntz et al. (1994) and included slow growth rates; seasonal growth; prolonged gametogenesis; seasonal reproduction; slow embryonic development; large, yolky eggs; low fecundity; delayed maturation; high incidences of brooding and protected development; extended longevity; large adult size; and low mortality.

Research in all of these areas has progressed markedly in the last 20 years, and understanding of the underlying mechanisms has advanced, but many of the attributes of polar marine ectotherms are still described as largely conforming to expectations of a K-selected fauna (e.g. Węslawski & Legezynska 2002). The next sections will deal with most of these characteristics in detail and show where understanding has improved to give greater insight into the adaptations of animals and limitations posed by living in a cold, highly seasonal polar ocean.

Growth

From the earliest studies of growth in Antarctic ectotherms, a pattern of slow growth has emerged for invertebrates and fish (Pearse 1965, Bregazzi 1972, Rakusa-Suszczewski 1972, Dayton et al. 1974, Everson 1977). These were followed by further studies showing generally slow or very slow growth across a wide range of benthic taxa, including fish (Eastman 1993), gastropods (Seager 1978, Picken 1979, 1980, Wägele 1988) and bivalve molluscs (Brey & Hain 1992, Nolan & Clarke 1993, Peck & Bullough 1993, Heilmayer et al. 2005, Higgs et al. 2009), decapods shrimps (Clarke & Lakhani 1979, Gorny et al. 1993), isopods (Luxmoore 1985), brachiopods (Brey et al. 1995b, Peck & Brey 1996, Peck et al. 1997b, Peck 2008), echinoids (Brey et al. 1995b), bryozoans (Barnes 1995, Brey et al. 1998, Bowden et al. 2006, Barnes et al. 2006b), octocorals (Peck & Brockington 2013) and polychaetes (Desbruyeres 1977). Some studies have reported relatively rapid growth in some ascidians

(Rauschert 1991, Kowalke et al. 2001) and bryozoans (Barnes 1995). Rapid growth was also reported for sponges by Dayton et al. (1974) and more recently for the giant sponge *Anoxycalyx joubini*, where a near 50-year study showed around 22 years of little growth and no recruitment followed by episodic recruitment and rapid growth at some time between 1989 and 2004 (Dayton et al. 2016). However, the rapid growth rates reported in these studies were still at least five times slower than the fastest growth rates reported for temperate species of the same groups. It should be noted here that most of the growth rates reported previously are for annual growth, and growth can be restricted to relatively small parts of the year. To identify correctly temperature effects on growth, comparisons should assess relative maximum rates of growth in species from different habitats. Data on maximum growth rates are rare globally but very rare in Antarctica. However, one way to address this issue may be to compare daily growth rings in skeletons of marine invertebrates living, or held, in different temperatures.

The question arises as to how to make reasonable comparisons of growth between Antarctic and lower latitude marine species. Comparisons using limited numbers or comparing the fastest rates in Antarctica with average or slow rates elsewhere are clearly flawed. Recently Peck (2016) collated data for von Bertalanffy or Richards growth functions for 37 species of echinoid sea urchins across latitudes from the tropics to the poles. These data showed a consistent decline in growth rate with latitude, even though the fastest Antarctic rates were faster than the slowest at temperate latitudes (Figure 3A). The most appropriate analysis of the effects of temperature on a biological rate is through an Arrhenius plot (Clarke 2017). This plots the logarithm of the rate against the inverse of temperature on the Kelvin scale as $1000/T$, and such plots are usually straight lines for biological functions, and the slope can be used to calculate the Arrhenius activation energy (E_a) for the reaction or process plotted (Hochachka & Somero 2002, Clarke 2017). When the K growth coefficients for echinoids in Figure 3A are replotted as an Arrhenius plot, a consistent relationship is obtained for temperate and tropical species living between 5°C and 30°C (Figure 3B). The K values for growth rates for the four Antarctic species investigated are all below the extension of the line for temperate and tropical species, and the difference between the Antarctic values and those predicted from this line is significant ($t = -2.83$, 4 d.f., P = 0.047). This shows that the growth rates for Antarctic urchins are slowed beyond the normal effects of temperature on biological systems. Peck (2016) termed this the cold marine physiological transition (CMPT).

An alternative mechanism for evaluating the effect of temperature on biological functions is by using the Q_{10} value of van't Hoff (Hochachka & Somero 2002, Clarke 2017). This metric expresses a change in a biological rate and converts it to that for a 10°C alteration in temperature using the equation:

$$Q_{10} = \left(\frac{R_2}{R_1}\right)^{10/(t_2 - t_1)}$$

where R_2 and R_1 are biological process rates at temperatures t_2 and t_1 respectively (Schmidt-Nielsen 1997). Most biological systems commonly follow Arrhenius relationships and there is a 2-fold to 3-fold increase for every 10°C temperature rise (a Q_{10} of 2–3), and a range of 1–4 covers all normal effects of temperature on enzyme mediated biological processes (Hochachka and Somero 2002). This 'rule' has been a cornerstone of temperature biology for around 100 years and is still quoted widely from research on molecular biology through biochemistry to physiology and ecology, and in all major texts (Hochachka & Somero 2002, Schmidt-Nielsen 1997, Clarke 2017) and reviews (e.g. Clarke 2004) on the thermal biology of biological systems.

Some studies of growth in Antarctic marine species have reported very large increases in growth for a small temperature rise in experiments. For example, in the scallop *Adamussium colbecki*, an increase from 0°C to 3°C produced a rise in growth with a Q_{10} of 71 (Heilmayer et al. 2005), and a 1°C rise in temperature gave a Q_{10} of 1000 for *in situ* growth in some Antarctic bryozoans

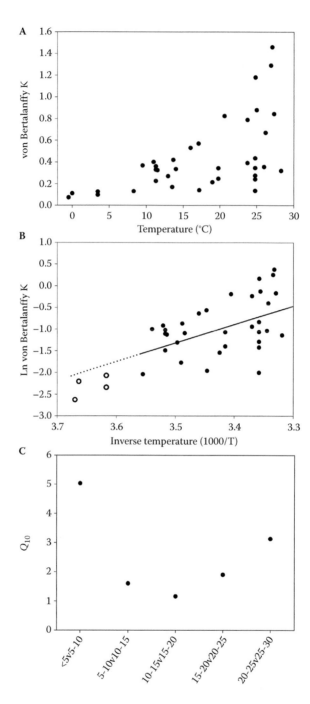

Figure 3 (A) Richards or von Bertalanffy K coefficients for echinoids from tropical to polar latitudes plotted against habitat temperature. (B) Arrhenius plot of K for echinoid growth rate. Solid line is the relationship for temperate and tropical species (Ln K = 13.72 – 4.297 1000/T; r² = 0.27, F = 25.1, 66 d.f., P < 0.001); dotted line is an extension of this relationship to polar temperatures. Open symbols are used to denote Antarctic species and emphasise they are below the Arrhenius line for growth in temperate and tropical echinoids. (C) Q_{10} coefficients for comparisons of mean K coefficient values for each 5°C block compared to its neighbour. (Modified from Peck, L.S. 2016. *Trends in Ecology and Evolution* **31**, 13–26.) Data for figures (A) and (B) are for 39 species with a minimum of 30 individuals per species.

and spirorbid worms (Ashton et al. 2017a,b). An analysis of Q_{10} values comparing mean echinoid von Bertalanffy K coefficients for 5°C blocks in Figure 3A, across the full temperature range and each block compared with adjacent neighbours, shows values are in the expected 1–4 range for temperatures between 5°C and 30°C (Figure 3C). The value for the comparison between species living below 5°C and in the 5–10°C range was, however, 5.2, suggesting that some factor other than the normal impact of temperature on enzyme mediated systems is having an effect. Combining these data with analyses of embryonic development rate and the duration of the postprandial rise in metabolism (the specific dynamic action [SDA] of feeding) showing similar slowing outside the normal effects of temperature led Peck (2016) to suggest that the likely cause is due to problems associated with protein synthesis and folding.

Growth and seasonality versus temperature

Two of the main reasons put forward to explain the observed slow annual growth in polar marine ectotherms are the extreme seasonality of the environment and the low temperature (Dehnel 1955, Dunbar 1968, 1970, Dayton et al. 1974, Clarke 1980, 1983, 1988, 1991, Clarke & Peck 1991). Most marine ectotherms grow when they feed (e.g. Peck et al. 1996), and this has been identified as part of the SDA for many years (Peck 1998, Secor 2009), and growth has also been demonstrated to be highly seasonal in several species (Bregazzi 1972, Seager 1978, Picken 1979, Richardson 1979, Sagar 1980, Clarke 1988, Berkman 1990, Urban & Mercuri 1998, Fraser et al. 2002, Ahn et al. 2003, Bowden et al. 2006). The relatively fast growth reported for some species (although still markedly slower than fast-growing species at lower latitudes) has been used in support of the idea that seasonality is the reason for slow growth because it shows that at least in some taxa the biochemical and physiological machinery to allow fast growth exists. It has also been argued that there should exist a capacity for fast growth at low temperatures because metabolic rates, and therefore metabolic costs, are greatly reduced at low temperature (Clarke 1983, Peck 2016). Metabolic costs are the measure of the instantaneous requirement for energy and are essentially a loss to the animal. They often include a large cost for homeostasis. Growth, reproduction and metabolism usually form the largest three fractions of an animal's energy budget (Kleiber 1961, Bayne 1976, Peck et al. 1987b), although in marine snails, mucus can be a large fraction, from between 5% and 10% of ingested energy (Paine 1965) to values between 25% and 30% (Peck et al. 1987b), and even up to 70% (Horn 1986). The reduction in metabolic losses at low temperature should allow more energy for growth, increasing growth efficiency as the proportion of energy consumed devoted to growth, and also potentially overall growth rate (Clarke 1983, 1987a, 1991, 1998). As seen earlier in this section, however, growth rates are generally low in Antarctic marine species and often slower than would be expected from the normal effects of temperature on biological processes (Peck 2016), which suggests other factors than energy availability are affecting growth.

Not all growth is seasonal in Antarctic marine species, even in some suspension-feeding species dependent on the short phytoplankton bloom. Barnes (1995) found that growth was consistent through the year in some bryozoans but was highly seasonal in others, and the differences were attributed to differences in the size range of phytoplankton consumed. Peck et al. (2000) found that shell growth in juveniles of the sediment dwelling bivalve mollusc *Yoldia* (now *Aequiyoldia*) *eightsii*, continued at the same rate in winter as summer, whereas tissue mass increased in summer but decreased in winter. In a more extreme case, in a mark-recapture tagging field experiment, winter shell growth was 12 times faster than summer increments in the brachiopod *Liothyrella uva* (Peck et al. 1997b). Like the bivalve *Aequiyoldia eightsii*, the brachiopod *Liothyrella uva* had a tissue mass cycle in phase with summer phytoplankton productivity. The conclusion drawn in both cases was that either tissue and shell growth are decoupled, or all growth is decoupled from the period of summer productivity, and growth is fuelled from stored reserves. These studies also emphasised that measuring growth only as change in mass or in length can lead to misleading conclusions as to the seasonality of the various growth processes.

The currently accepted major reason for slow annual growth in Antarctic marine species is that it is due to the effects of low temperature. Low temperature has been widely proposed as the cause of slow overall growth in polar latitudes from some of the earliest studies (Dunbar 1970, Kinne 1970, Arnaud 1974, 1977, Arntz et al. 1994). The alternative argument that adaptation to polar temperatures should allow fast growth, and that seasonality and resource limitation are the proximate causes of slow growth has also been a dominant idea since the 1970s (Dunbar 1970, Clarke 1980, 1983, 1988, 1991, White 1984). Seasonality clearly does restrict the biology of some species, but the argument that low temperature is the major factor has gained more traction in the last two decades. This change has come mainly from two directions, the demonstration that the slowing of development is most likely a temperature-related phenomenon (e.g. Hoegh-Guldberg & Pearse 1995, and this review), and also the finding that the synthesis of fully functional proteins is much more difficult at temperatures around and below 0°C than in warmer habitats (Fraser et al. 2007, Peck 2016). This difficulty at low temperature results in a larger proportion of the proteins being made on ribosomes being recycled immediately, and a smaller proportion of the protein made being deposited or retained for growth. This adds a large extra unseen cost to growth and reduces the efficiency of growth markedly in terms of the manufacture of structural protein and the increase in body size.

Protein synthesis, retention and folding

All organisms grow. Growth of cells is mainly via, and ultimately dependent on, the synthesis of proteins both for structural and functional purposes and the retention of those proteins post synthesis. The total protein content of an organism is known as its protein pool, and this pool is dynamic with newly synthesised proteins adding to the pool, and degradation removing them. Changes in the total protein content of an organism is called protein growth. The combination of synthesis, degradation and growth is called protein metabolism (Fraser & Rogers 2007). Studies of protein metabolism and its components at low temperature have only been conducted for the last 20 years.

Most research on the effects of temperature on proteins in the last 30 years has focussed on enzyme activities and questions such as are the function rates of enzymes adapted to low temperature or are concentrations of enzymes changed to compensate for reduced temperature (Hochachka & Somero 2002). Some cellular processes such as microtubule assembly (Detrich et al. 2000), and a few enzyme function rates (Fields et al. 2001, Kawall et al. 2002), have been demonstrated to be cold-compensated and to proceed at rates similar to warmer water orthologues. The enzyme function rate is measured as the rate that a substrate is converted to product per unit time per active site in an enzyme catalysed reaction and is called k_{cat}. From the very few studies conducted, it seems the overall outcome of adaptation to different temperatures has resulted in higher k_{cat} values in polar species than those from lower latitudes and, hence, more active enzymes (Fields & Somero 1998), although it seems that concentrations of enzymes are not modified to a great extent (Hochachka & Somero 2002). It should be noted here that comparisons of the effect of temperature on cellular or physiological process rates require clear interpretation of what is being measured. In many studies, process rates are measured at the studied organism's ambient temperature (and in ectotherms body temperature is predominantly very close to ambient). In these cases, polar investigations are conducted at lower temperatures than studies on temperate or tropical species. In such comparisons, the rate in the polar species is lower than that of the warmer analogues, and this is predominantly true for measures of cellular processes, for example, k_{cat} through to whole-animal physiologies. In the case of cellular processes, the requirement for complete or perfect temperature compensation of function would be to increase the amounts of enzymes in the cell to considerably higher levels. This seems unlikely from two aspects: energetic costs and physical space available. A considerable increase in the levels of thousands of proteins within a cell might not be possible due to limitations in cell size (but this would be a driver towards the observed larger cell size at lower temperatures – see later section 'Egg size, fecundity, reproductive effort and life histories'). However, even if this were possible, it seems unlikely that this level of protein metabolism could be maintained at low temperatures, especially because there is evidence that proteins are less stable at low temperature than in warmer regions (Fields et al. 2015, Peck 2016).

For the majority of comparisons, for example, whole-animal respiration rate, differences between warm-water and polar taxa are in line with predicted differences from Arrhenius relationships. However, where studies are conducted at overlapping temperatures, warmed colder-water species usually have faster rates than cooled species from warmer environments. This indicates that either there is some compensation for the lower temperature but that compensation is incomplete, or that the slope of the relationship between the biological function rate and temperature is different for a single species or population than for between species and population comparisons.

Protein synthesis rate

The main cellular energetic costs during normal function are protein synthesis, RNA/DNA synthesis, proton leak, Na^+/K^+-ATPase and $Ca2^+$-ATPase. Protein synthesis is a major component of cellular costs accounting for between 11% and 42% of the cellular budget in temperate species (Hawkins et al. 1989, Houlihan et al. 1995, Podrabsky & Hand 2000). The requirements for functioning proteins remain high in Antarctic marine species at low temperature, even though metabolic rates have been demonstrated to be low (Fraser et al. 2002). Studies of whole-animal protein synthesis in Antarctic marine ectotherms started in the 1990s with measures of the fraction of whole-animal protein content that is synthesised each day (k_s). Values were reported for the isopod *Glyptonotus antarcticus* of 0.24% day^{-1} in fed individuals at 0°C (Whiteley et al. 1996), and k_s ranged from 0.16% day^{-1} and 0.38% day^{-1} in starved and fed animals, respectively, also at 0°C (Robertson et al. 2001). In the holothurian *Heterocucumis steineni*, k_s values ranged from 0.23% day^{-1} in winter to 0.35% day^{-1} for body wall in summer. In the limpet *Nacella concinna*, whole-animal winter values were reported at 0.29% day^{-1} and 0.35% day^{-1}, whereas in summer, these were 0.40% day^{-1} and 0.56% day^{-1} (Fraser et al. 2002), but higher values (0.56% day^{-1} to 0.84% day^{-1}) were recorded for the same species in a later study (Fraser et al. 2007). The interannual differences in the studies on *N. concinna* were attributed to different food availability in different years (Fraser et al. 2007). These data indicate a reduction in protein synthesis of between 25% and 50% in winter over summer in Antarctic species. The cause for the reduction is probably a mixture of lower temperature and seasonal resource reduction in winter.

The few studies of protein synthesis in Antarctic species have all produced values much lower than those for warmer-water species (Fraser & Rogers 2007), with k_s averaging around 0.48% (±0.05 s.e.) day^{-1}, compared to 3.7% (±0.72 s.e.) day^{-1} for temperate marine species. Fractional rates of protein synthesis are thus around eight times lower than those for temperate species, and there is evidence that k_s declines markedly at temperatures below 5°C, which parallels the trend for growth.

Very few studies have investigated the effect of temperature on protein synthesis rates in Antarctic species, but Fraser et al. (2007) found that both fractional and absolute protein synthesis rates were fastest at approximately 1°C in *N. concinna*, declining both above and below this temperature. They argued that predicted warming in Antarctic waters could result in reduced, rather than increased, rates of protein synthesis and, in turn, possibly growth if changes happened faster than the animals' abilities to adapt (Fraser et al. 2007). The logical conclusion to be drawn from these data is that some factor associated with protein synthesis is very sensitive to temperatures near 0°C. Peck (2016) has suggested this is likely protein folding as folding is sensitive to viscosity, and seawater viscosity changes markedly more at low polar temperatures than at temperate or tropical latitudes. Protein stability decreases and unfolding increases at low temperature, which appears to be due to changes in the free energy relationships between nonpolar groups and water. This results in the penetration of water molecules into the edges of the protein, which weakens the hydrophobic bonding strength and makes the protein core less stable (Lopez et al. 2008, Dias et al. 2010).

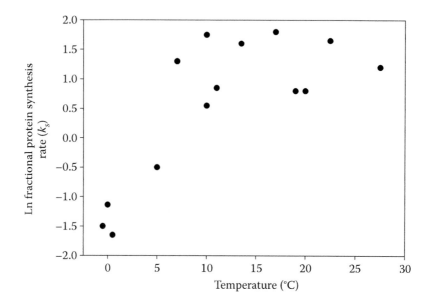

Figure 4 Mass standardised whole-animal fractional protein synthesis (k_s) rates plotted for ectotherms (fish, molluscs and crustaceans) at natural habitat temperatures for polar to tropical species. Note k_s values are Ln transformed. (Figure modified from Fraser, K.P.P. & Rogers, A.D. 2007. *Advances in Marine Biology* **52**, 267–362.)

Making good comparisons of protein synthesis rates between species living at different temperatures requires the same measures to be made and also comparable techniques used. Fraser & Rogers (2007) compiled data for a range of ectotherms across latitudes and a temperature range of more than 25°C for whole-animal rates of protein synthesis. Temperate and tropical animals living at temperatures between 5°C and 30°C had k_s values in the range 1.65%–6.3%, whereas the one species studied living at 5°C, the isopod *Saduria entomon*, had a k_s of 0.6% (Robertson et al. 2001), and the two Antarctic species where data were available had values of 0.24% day^{-1} (*Glyptonotus antarcticus*, Whiteley et al. 1996) and 0.40% (*Nacella concinna*, Fraser & Rogers 2007). These data all fitted a consistent pattern when the k_s values were logarithmically transformed, showing similar values for species living at temperate and tropical temperatures, but a marked decline occurs as temperatures fall below 5°C (Figure 4). Antarctic whole-animal fractional protein synthesis rates are thus around an order of magnitude lower than temperate and tropical species.

Much higher rates of protein synthesis have been reported in larvae, with values similar to those for temperate species in the Antarctic starfish *Acodontaster hodgsoni, Glabraster* (previously *Porania) antarctica* and *Odontaster meridionalis* (Ginsburg & Manahan 2009), and the echinoid *Sterechinus neumayeri* (Marsh et al. 2001). The high rates were associated with very high efficiencies of production (Pace & Manahan 2006, 2007). These were later correlated with high rates of biosynthesis of protein at the ribosome in embryos of *S. neumayeri* (Pace & Manahan 2010). There is some uncertainty over the efficiencies of protein synthesis reported for Antarctic urchin embryos, as Fraser and Rogers (2007) noted that the efficiencies quoted were beyond the theoretical thermodynamic limits, in terms of the number of adenosine triphosphate (ATP) equivalents needed per peptide bond during protein synthesis (usually four).

Protein synthesis retention efficiency

Protein synthesis is only the first part of the process leading to the production of well-conformed functional proteins. The second aspect is how well made the proteins are, and this is measured as the protein synthesis retention efficiency (PSRE). When protein synthesis results in a poorly conformed or non-functional protein, a mechanism is entrained to identify these proteins and break them down for recycling of the amino acids. The identification of badly formed protein that is then degraded is usually achieved via tagging with ubiquitin. PSRE is a measure of the proportion of synthesised proteins that are well formed (the percent retained as opposed to recycled) and functional (Hochachka & Somero 2002). It is usually calculated as:

$$PSRE = \frac{k_g}{k_s} \times 100(\%)$$

where k_g is the change in total body protein content (protein growth, % day^{-1}), and k_s is fractional protein synthesis (% day^{-1}). Protein synthesis retention efficiency has only been measured in one Antarctic species to date, the limpet *Nacella concinna*, where values of 15.7% in winter and 20.9% in summer were reported (Fraser et al. 2007). If the various studies on tissues and not whole animals are included, then Antarctic species exhibit low PSRE values, and there is a general decline in PSRE at lower temperatures that follows an Arrhenius relationship (Fraser et al. 2007, Figure 5). If *N. concinna* is representative of Antarctic ectotherms in general, then PSRE is two to six times lower in polar marine species than those living in temperate and tropical latitudes (Figure 5).

There is further strong support for low PSRE levels in Antarctic marine species from measures of ubiquitination, where very high levels have been reported in fish (Todgham et al. 2007), and high levels of expression of genes associated with ubiquitination have been reported in fish tissues (Shin et al. 2012). Low PSRE in Antarctic species has important consequences because it suggests that growth in polar ectotherms is less efficient than in lower latitude species because of the greater losses. This probably results from biochemical constraints on protein synthesis. This result contrasts with a few studies that have reported higher growth efficiencies at lower temperatures (e.g. for scallops, Heilmayer et al. 2004), but reduced efficiency at low temperature appears the more common outcome. Why a higher proportion of body proteins are degraded at low temperatures is currently unclear, although there is good evidence that the proportion of an animal's proteome that is denatured

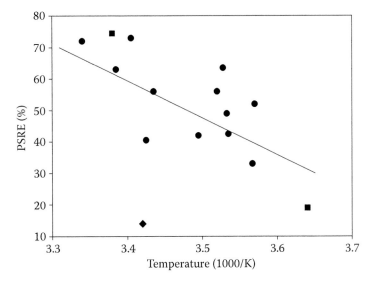

Figure 5 Mass standardised protein synthesis retention efficiency (PSRE) plotted against habitat temperature for a range of tropical to polar ectotherms. Data fit an Arrhenius relationship (*PSRE* = 450–115 T, r^2 = 0.29, 14 d.f., *P* < 0.05). Circles denote data for fish species, squares for molluscs and the diamond for a crustacean. (Redrawn from Fraser, K.P.P. et al. 2007. *Journal of Experimental Biology* **210**, 2691–2699.)

is higher in species living at cold polar water temperatures than elsewhere (Buckley et al. 2004, Place et al. 2004, Hofmann et al. 2005, Place & Hofmann 2005, Peck 2016).

Earlier it was shown that larvae of Antarctic marine invertebrates exhibit high rates of protein synthesis. These high rates of synthesis appear to be accompanied by low or very low rates of protein retention or deposition. Thus, Ginsburg and Manahan (2009) measured protein retention in larvae of the starfish *Odontaster meridionalis* and the urchin *Sterechinus neumayeri*, and found values of 5.1% and 3.8%, respectively. These compared with protein retention of $\geq 28\%$ for larvae of the temperate starfish *Patiria* (previously *Asterina*) *miniata* (Pace & Manahan 2007) and 21% for the temperate urchin *Lytechinus pictus* (Pace & Manahan 2006). Fast growth is often seen as one of the most important attributes of embryonic and larval stages of marine invertebrates, to minimise exposure to predators in what is seen as the most vulnerable part of the life cycle (Pechenik 1991, Pechenik & Levine 2007). It thus seems likely that the poor rate of protein deposition reported for Antarctic marine larvae constrains their growth rates, and one adaptation to minimise this problem is to increase, as far as possible, rates of protein synthesis. The compensation is, however, clearly incomplete because larval development rates in Antarctic marine ectotherms are much slower than for temperate species and slowed beyond the normal effects of temperature on biological systems (see section on 'Embryonic and larval development').

Costs of protein synthesis

Studies reporting the cost of synthesising proteins in temperate and tropical species vary somewhat, but most values are in the range 5–15 mmol O_2 g^{-1}, when energetic cost is expressed as a metabolic oxygen requirement (Fraser & Rogers 2007). Reports of the costs of protein synthesis in Antarctic ectotherms vary very widely, from values of 0.92 mmol O_2 g^{-1} protein in the urchin *Sterechinus neumayeri* (Marsh et al. 2001), which is below the theoretical minimum thermodynamic costs for synthesising protein, to 4.0 mmol O_2 g^{-1} protein for the starfish *Odontaster validus*, to 7 mmol O_2 g^{-1} protein for the scallop *Adamussium colbecki* (Storch et al. 2003), and up to 147 mmol O_2 g^{-1} protein for the isopod *Glyptonotus antarcticus* (Whiteley et al. 1996). The value for *Adamussium colbecki* was measured in a comparative study with the temperate scallop *Aequipecten opercularis*, which had a cost of 9 mmol O_2 g^{-1} protein. The Antarctic and temperate scallops thus had very similar protein synthesis costs when measured at the same time using the same techniques in the same laboratory. It is not clear why the various studies measuring costs of protein synthesis in Antarctic marine species have produced such massively variable data covering two orders of magnitude. It is also not clear why the cost *per se* should vary markedly between tropical, polar and temperate species because the effect of temperature should be measured on a Kelvin scale, and hence differences should be around 10% when comparing species living in Antarctica at $-2°C$ with tropical species at 25°C. From a pure physics standpoint, the rate that a biochemical reaction changes with temperature over the normal physiological range (273–313 K) is governed by the change in mean molecular speed as this dictates the energy of collisions between molecules (Clarke 2017). The Q_{10} for mean molecular speed over this range is 1.07. That most biological systems have Q_{10} values between one and four shows that temperature generally has a much larger effect than would be dictated by changes in molecular speed and that other, possibly many other factors affect changes in the rate of biological functions with temperature. Further studies of the rates, efficiencies and cost of protein synthesis are needed on a wide range of Antarctic marine species before several of the above issues can be resolved.

RNA to protein ratios

The signal to make proteins from deoxyribonucleic acid (DNA) in the nucleus is ribonucleic acid (RNA). The amount of RNA produced is a measure of the strength of the signal. In studies of protein metabolism, the ratio of the amount of RNA to the protein content of the organism is an indication

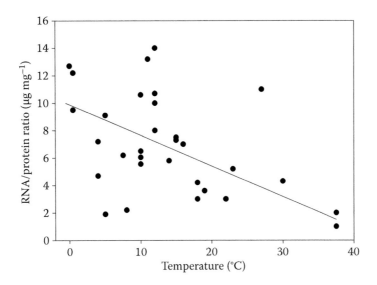

Figure 6 Whole-animal RNA-to-protein ratios mass standardised to a body mass of 129 g plotted against temperature. Masses were standardised using a scaling coefficient of 0.24. The ratio declines with temperature following the relationship shown (RNA: protein = 9.34–0.175 temperature, $r^2 = 0.22$, 30 d.f., $F = 8.18$, $P = 0.008$). (Figure modified from Fraser, K.P.P. et al. 2002. *Marine Ecology Progress Series* **242**, 169–177, adding data from, and quoted in Rastrick, S.P.S. & Whiteley, N.M. 2013. *PLoS ONE* **8**, e60050.) Where data are available for a species at more than one temperature, values shown are means for RNA-to-protein ratio and temperature.

of the difficulty involved in making proteins. The lower the signal to final protein product, the more efficient the processes between, and the less error involved, and the larger the signal to product ratio, the more difficulty and the greater the error. In the few studies of RNA-to-protein ratios in Antarctic species, there were clear seasonal variations in both RNA concentrations and RNA-to-protein ratios. These followed similar patterns to the variation in k_s. Seasonal changes have been documented in RNA concentrations and RNA-to-DNA ratios in tissues of temperate species (Bulow et al. 1981, Robbins et al. 1990, Kent et al. 1992, Melzner et al. 2005). Food intake and quality affect RNA concentrations in tissues, which also has an impact on k_s (Fraser et al. 2007).

RNA-to-protein ratios for Antarctic marine species are high when compared with marine ectotherms from temperate and tropical latitudes. When ratios are plotted against habitat temperature, a decline from polar to tropical species is evident, where on average values decline by 2.3 for each 10°C rise in temperature (Figure 6). An increase in RNA signal at low temperature is one way that thermal compensation of protein synthesis could be achieved. As seen previously, however, rates of protein synthesis are lower in Antarctic species and any compensation is thus not complete, and problems making fully conformed functional proteins only partially overcome. A further reason why this compensation may be only partial is that RNA activity appears to decline at lower temperatures (Fraser et al. 2002), compounding problems at the ribosome and during the folding phases of protein synthesis.

Gamete development and gametogenic cycles

Gametogenic cycles have been described for a range of Antarctic marine invertebrates. These include: in the molluscs, the small brooding bivalve *Kidderia subquadratum* (Shabica 1974), the infaunal clam *Laternula elliptica* and limpet *Nacella concinna* (Powell 2001); and in the echinoderms, the starfish *Odontaster validus* (Pearse 1965, Chiantore et al. 2002, Grange et al. 2007), the urchin *Sterechinus neumayeri* (Brockington 2001a, Chiantore et al. 2002, Brockington et al. 2007) and the brittle star *Ophionotus victoriae* (Grange et al. 2004, Grange 2005). Other taxa studied include the actinarian anemone *Epiactis georgiana* (Rodriguez et al. 2013). These species all take 18–24 months from the initiation of gametogenesis to spawning (e.g. Figure 7) except *Kidderia subquadratum*, which takes 15–19 months. These were the first species identified to take longer than 12 months for gametogenesis, where most temperate and tropical species complete the process within six months. Studies of Antarctic fish have also identified extended periods of gametogenesis, with predominantly two-year periods required to complete oogenesis (Everson 1970a, Sil'yanova 1982, Everson 1984, Kock & Kellerman 1991, Parker & Grimes 2010, Hanchet et al. 2015). Similar extended periods required to complete gametogenesis have also been reported for high Arctic species living permanently at temperatures around or below 0°C (Falk-Petersen & Lønning 1983, Pearse & Cameron 1991, Junquera et al. 2003).

A small number of Antarctic species complete gametogenesis faster than 18–24 months; specifically, the nemertean worm *Parborlasia corrugatus* requires 15 months from initiation of oocyte development to spawning (Grange et al. 2011a), and the scallop *Adamussium colbecki* needs only 12 months (Chiantore et al. 2002, Tyler et al. 2003). Both of these, however, like the molluscs and echinoderms, take significantly longer for gametogenesis than related or ecologically similar temperate species, although Tyler et al. (2003) concluded gametogenesis in *A. colbecki* is 'more scallop than Antarctic' and Lau et al. (2018) concluded that reproduction in the Antarctic nuculanid bivalve *Aequiyoldia eightsii* fitted neither the expected Antarctic patterns nor the characteristic nuculanid pattern. Overall, the general slowing in Antarctic marine invertebrates compared to temperate species is around five times (Peck 2016).

An unusual gametogenic cycle is demonstrated by the rhynchonelliform brachiopod *Liothyrella uva*. In this species, histological analyses of the gonad over a two-year period failed to reveal the double cohorts of developing eggs as seen in nearly all other Antarctic species investigated (Meidlinger et al. 1998). There was considerable interannual variation in reproductive output and also in numbers of the smallest size class of oocytes. There was, however, no seasonal signal present, and the conclusion drawn was that the absence of seasonal trends for all oocyte size classes showed that oocyte maturation in the population was continuous but asynchronous (Meidlinger et al. 1998). *Liothyrella uva* is a brooding species, but broods were very variable, as seen in other rhynchonelliforms (Hoverd 1985, Chuang 1994). Different females sampled at the same time held broods at markedly different developmental stages, and some females even contained broods with several developmental stages at the same time (Meidlinger et al. 1998). Single females of this species have also been noted to release larvae of markedly different developmental stages from swimming gastrulas to fully competent 3-lobed larvae (Peck et al. 2001). The marked variability exhibited by *L. uva* in its gonad and embryonic/larval development suggests there is extreme plasticity in the reproductive cycle of this species.

It might be expected that with markedly extended gametogenesis Antarctic marine ectotherms would not reproduce annually. Every species investigated to date does, however, spawn on an annual basis. This is achieved in most cases by having two cohorts of eggs developing simultaneously in the female gonad, one to be spawned in the current season and a second to be spawned in the following year (Figure 7). This double egg cohort in female gonads was first identified in the starfish *Odontaster validus* at McMurdo Sound by Pearse (1965) but has since been recognised as the most common gametogenic developmental cycle exhibited by Antarctic marine species (Pearse & Cameron 1991, Pearse et al. 1991, Gutt et al. 1992, Chiantore et al. 2002, Grange et al. 2004, 2007, 2011a, Servetto & Sahade 2016). Although, as noted previously, a few species do not follow this pattern, including the scallop *Adamussium colbecki* (Tyler et al. 2003), the nuculanid *Aequiyoldia eightsii* (Lau et al. 2018), and the brachiopod *Liothyrella uva* (Meidlinger et al. 1998).

Most multiyear studies of reproduction in Antarctic marine ectotherms have reported large, or very large levels of interannual variation. In the brittle star *Ophionotus victoriae*, reproductive effort was assessed as the proportional decrease in gonad size on spawning over a four-year period between 1997 and 2000 on Adelaide Island, Antarctic Peninsula. Decreases were: 12.5% (1997), 90% (1998),

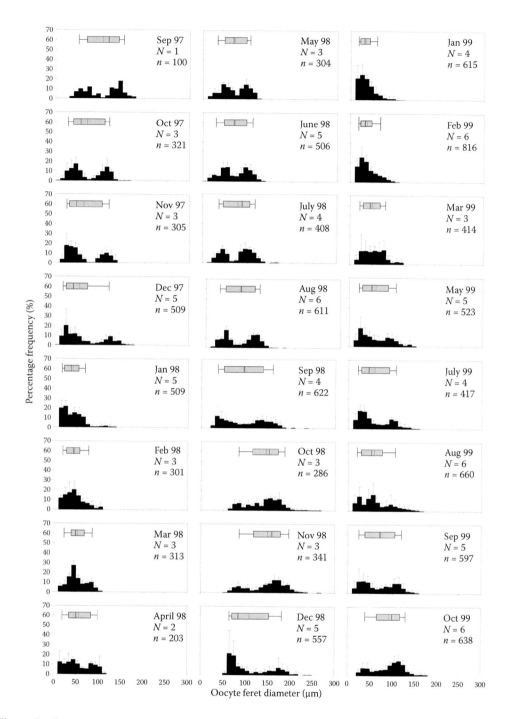

Figure 7 Oocyte diameters for eggs of the brittle star *Ophionotus victoriae* sampled from Rothera Point, Adelaide Island, Antarctic Peninsula. Data shown are for monthly samples collected between September 1997 and October 1999. Measures are feret diameters (±SD), measured on a compound microscope. N = number of females assessed each month, n = number of oocytes measured. Box plot margins indicate the 25th and 75th percentiles for oocyte size, whiskers on boxes indicate 10th and 90th percentiles. Note in most months there are two peaks in the distribution denoting cohorts of eggs, and oogenesis requires 18–24 months to complete. (Figure adapted from Grange, L.J. et al. 2004. *Marine Ecology Progress Series* **278**, 141–155.)

96% (1999) and 88% (2000), indicating nearly an 8-fold change in reproductive effort between smallest and largest years (Grange et al. 2004). In the starfish *Odontaster validus*, interannual variation in proportion of gonad spawned between 1997 and 2000 at the same site as the *Ophionotus victoriae* study showed less variation than the brittle star, but values still ranged from 34% to 62% for females and 52% to 78% for males (Grange et al. 2007). The smaller interannual variation in *Odontaster validus* was attributed to the highly catholic (broad) diet of the starfish.

Very few studies have examined variation in reproductive characters between populations of the same species in Antarctica. In such an investigation of the sea urchin *Sterechinus neumayeri*, gonad mass varied markedly between three sites, less than 10 km apart (Brockington et al. 2007). At one site, gonad dry mass, expressed as the value for a standard 30 mm diameter urchin, varied between 0.06 g dry mass and 0.17 g dry mass between April 1997 and October 1998 and then increased to 0.33 g dry mass by January 1999, and there was no evidence of spawning, as a decrease in gonad mass, over the whole period. At a second site, gonad mass ranged from 0.62 to 0.74 g dry mass between March and November 1997. It then declined to 0.16 g dry mass over the next two months, a loss of 77% on spawning, and values gradually increased to 0.45 g dry mass by January 1999. At the third site, gonad mass ranged between 0.15 and 0.4 g dry mass in the first half of the study and between 0.34 and 0.57 g dry mass in the second year of the investigation, again with no evidence of spawning. The site with the largest gonad mass thus had values around six to seven times higher than the lowest, and two sites showed little evidence of spawning, while the third, which was between the other two and less than 5 km from each, exhibited a very large spawning event over the same period (Brockington et al. 2007). There can thus be very large variations in gonad status and reproductive effort over both small spatial and temporal scales. The main explanation of this is variation in food supply through environmental factors such as seasonality, ice cover affecting productivity and advection, and also through resource limitation due to biotic factors such as competition (Clarke 1987b, 1988, 1991, Clarke & Peck 1991, Brockington et al. 2007, Grange et al. 2011a).

Embryonic and larval development

Development in marine invertebrates and fishes is affected by a wide range of factors (Pechenik 1986, 1999). Development in Antarctic species, especially echinoderms, has been reviewed by Pearse (1994), Hoegh-Guldberg and Pearse (1995), Pearse et al. (1991) and Peck (2002a). Temperature is generally accepted as the major factor controlling differences in the rate of development between species and across latitudes (Hoegh-Guldberg & Pearse 1995). One of the key pieces of evidence in this argument is that feeding (planktotrophic) and non-feeding (lecithotrophic) larval development rates are equally slowed at low temperature compared to temperate and tropical species, and resource limitation should impact planktotrophy more than lecithotrophy. Furthermore, development rates in brooding species appear to be slowed as much, if not more than for broadcast reproducing species (e.g. Peck et al. 2006b, Peck 2016), which again would not be predicted as an outcome of seasonal resource limitation effects on embryonic and larval development rate.

From the first investigation of embryonic and larval development in Antarctic marine ectotherms in the 1960s (Pearse 1969), it has been clear that the times required to reach a given developmental stage are greatly extended in comparison with temperate and tropical species. Since then, investigations of development in broadcast spawning species, including the starfish *Odontaster validus* (Pearse 1969, 1994, Stanwell-Smith & Peck 1998), the sea urchin *Sterechinus neumayeri* (Bosch et al. 1987, Pearse et al. 1991, Shilling & Manahan 1994, Stanwell-Smith & Peck 1998), the bivalve molluscs *Laternula elliptica* and *Adamussium colbecki* (Peck et al. 2007b), the limpet *Nacella concinna* (Peck et al. 2016a), the ascidian *Cnemidocarpa verrucosa* (Strathmann et al. 2006), and the nemertean *Parborlasia corrugatus* (Peck 1993b), have all exhibited development rates slowed by around or more than an order of magnitude compared to lower latitude species. Investigations of brooding species, including the isopods *Ceratoserolis trilobitoides* (Wägele 1987), *Aega antarctica*

(Wägele 1990) and *Glyptonotus antarcticus* (White 1970), the amphipod *Eusirus perdentatus* (Klages 1993), the caridean shrimps *Chorismus antarcticus* (Clarke 1985), *Notocrangon antarcticus* and *Nematocarcinus lanceopes* (Arntz et al. 1992, Gorny et al. 1992, 1993), the actinarian anemone *Epiactis georgiana* (Rodriguez et al. 2013), the gastropods *Trophonella* (previously *Trophon*) *scotiana, Neobuccinum eatoni, Doris* (previously *Austrodoris*) *kerguelenensis, Antarctophiline alata, Nuttallochiton mirandus* (Hain 1991), *Torellia mirabilis* and *Marseniopsis mollis* (Peck et al. 2006b), and the bivalve *Lissarca notorcadensis* (Brey & Hain 1992) have all demonstrated that brooding periods are markedly extended in Antarctica compared to lower latitudes.

In most cases, the comparisons of development rates between Antarctic and lower latitude species involve a limited number of species, and evaluations of changes of rate either with temperature or between regions do not contain enough data for useful comparisons to be made. In a few cases, however, sufficient data do exist to allow such analyses. The first comprehensive analysis across latitudes was done for echinoids by Bosch et al. (1987), and this showed a marked slowing in development rate at temperatures around 0°C and below, compared to values at 5°C and above. Similar results were published for echinoids by Stanwell-Smith and Peck (1998) and for bivalve molluscs by Peck et al. (2007b). One of the best examples of this pattern is in brooding period for marine gastropods, where sufficient numbers of published evaluations for tropical, temperate and polar species exist to allow a good analysis of changes in rate across latitudes (Peck 2016, Peck et al. 2006b, Figure 8A). Brooding period ranges from a few days to one to two weeks for tropical species and up to 15 weeks for temperate species living at 15–20°C. At temperatures around 0°C and below brooding period ranges from 17 to 102 weeks. When the values for brooding period are replotted as an Arrhenius plot, a clear linear relationship between temperature and Ln development rate is apparent for temperate and tropical species (Figure 8B). Values for Antarctic brooding gastropods are, however, all below an extrapolation of the relationship for temperate and tropical species when

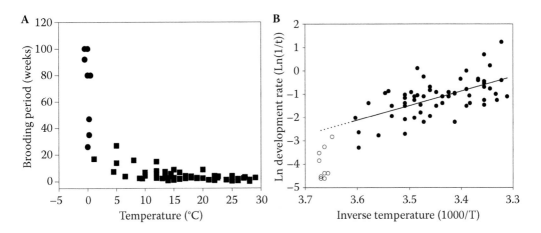

Figure 8 Brooding period and development rates of brooding marine gastropod snails at ambient temperatures for tropical to polar species. (A) Time from brood initiation to release (1/development rate). Circles denote Antarctic species whereas squares denote Arctic, temperate and tropical species. In most cases, release is of crawling juveniles, but for two Antarctic species, *Torellia mirabilis* and *Marseniopsis mollis* release is of veliger larvae, and development time to juvenile is approximately double that of brooding, *per se* (Peck et al. 2006a). Data for 68 gastropod species, nine of which live at temperatures around 0°C, and are the full development period to juvenile. (B) Arrhenius plot of Ln developmental rate to juvenile stage for brooding gastropod molluscs. Open circles denote species living at temperatures below 3°C and closed circles show data for species living above 3°C. Fitted line is for temperate and tropical species (brooding rate (1/weeks) = 20.37 – 6.25 1000/T; $r^2 = 0.36$, $F = 32.4$, 58 d.f., $P < 0.001$). (Figure from Peck, L.S. 2016. *Trends in Ecology and Evolution* **31**, 13–26.)

it is extended to lower temperatures (Figure 8B), and the difference is highly significant ($t = -6.73$, 8 d.f., $P < 0.0001$). Peck (2016) further showed that the Q_{10} for the slowing of development in Antarctic brooding gastropods is over 12. This means that development rates in Antarctic species are slowed more than the normally expected effect of temperature where Q_{10} values are in the range one to four (Clarke 2004) and beyond the Arrhenius relationship for warmer species. Peck (2016) argued that, as for growth, this showed a factor other than the normal impact of temperature on biological systems is having an effect in Antarctica, and that problems with synthesising proteins, especially protein folding, is the likely cause.

Ecological implications of slow development rates

The very slow development rates of Antarctic marine ectotherms have significant consequences for life-history traits, and these differ markedly between species with broadcast and brooding modes of reproduction. The earliest life-history stages, embryos and larvae, usually have much higher rates of mortality than later stages due to factors such as predation, advection to areas unsuitable for settlement, and starvation (Thorson 1950, Morgan 1995, Pechenik 1999, Marshall & Morgan 2011). Marshall et al. (2012) also showed that planktonic larvae are more common in regions where food levels and temperature are high. Recent work has, however, demonstrated that larval mortality is significantly lower in Antarctica than previously expected, and life-history and population dynamics models need to be modified in this respect (White et al. 2014). This suggests that mortality in the later settlement, metamorphosis and very early juvenile stages may be higher than previously thought, and that projections and possibly conservation measures should focus on models that include evaluations across life-history stages, as advocated by Marshall et al. (2012).

Periods that larvae spend in the water column increase as development rate decreases and the time spent at the various development stages increases. This factor has been recognised as potentially important for Antarctic broadcast spawning species for several decades (e.g. Pearse & Bosch 1986, Bosch et al. 1987, Stanwell-Smith & Peck 1998, Peck & Prothero-Thomas 2002, Peck et al. 2007b, Ginsburg & Manahan 2009). Indeed, the time required for embryos of the starfish *Odontaster validus* to hatch is around nine days at $-1.8°C$ (Stanwell-Smith & Peck, 1998), and for larvae to develop to settlement, around 180 days are likely needed (Pearse & Bosch 1986). Antarctic coastal current speeds range from 2.5 to 25 cm sec^{-1} (Fahrbach et al. 1992), and average speeds are in the range 5–10 cm sec^{-1} (Le & Shi 1997). At a conservative 5 cm sec^{-1} and a pelagic period of 170 days, a larva could travel 734 km, and at 10 cm sec^{-1}, over 1400 km would be possible.

However, dispersal in Antarctic broadcast spawning species may be less than previously expected in some taxa because of larval behaviours in relation to water currents. There are few data on larval behaviours in Antarctic marine species beyond those often reported during culture studies, such as swimming towards the surface or towards the bottom of culture vessels. In wild populations, behaviours resulting in larvae spending much of their developmental period in benthic habitats or avoiding moving water have been reported for several temperate marine benthic species (Cowen & Sponaugle 2009, Pringle et al. 2014). Such behaviours often result in settlement that is more local to parent populations than realised previously. Although such mechanisms are yet to be documented for Antarctic species, it is highly likely that some will fit this pattern.

Protected development reduces mortality during development. It also reduces dispersal during the development phase to zero. It has been recognised for over three decades that Antarctica has significantly higher levels of protected development than species living at lower latitudes in many taxa (Picken 1979, 1980, White 1984, Arntz et al. 1994). Marshall et al. (2012), in a meta-analysis, showed that around 80% of Antarctic marine species have a planktonic development, most of which exhibit protected development systems. Different taxa, however, exhibit markedly different levels of protected development, with molluscs and echinoderms showing strong decreases in the prevalence of planktonic larvae with latitude in Antarctica, but annelids do not (Marshall et al.

2012). Interestingly, their study also demonstrated that trends in reductions of planktonic forms with latitude are much stronger in the southern than in the northern hemisphere.

It has been suggested that protected development at high latitude may be an evolutionary response to the increased mortality at the larval stage that is a consequence of the greatly slowed development rates at temperatures around and below 0°C, possibly to avoid larvae being carried away from suitable habitat during development. These data have been used regularly in the past as evidence to support Thorson's rule, that there is a strong trend towards non-pelagic development and to brood protection in polar seas because of the problems associated with extended larval periods and very short, highly seasonal periods of phytoplankton productivity (Thorson 1950). Thorson's rule was mainly based on observations comparing Arctic sampling expeditions with data from temperate sites, especially Denmark (Thorson 1936, 1946). Although it was first called Thorson's rule by Mileikovsky (1971), Thorson was more circumspect on the universality of the paradigm than later authors. In Antarctica, there were several studies aimed at testing this rule in the 1980s and 1990s. Clarke (1992) showed a trend of markedly reduced pelagic development in gastropod molluscs at high latitude. Bosch and Pearse (1990) and Pearse (1994) showed that in echinoderms the proportion of species that exhibit protected development is very similar in Antarctica and California. They found that Thorson's rule did not apply to echinoderms in the proportions of brooding species, but that there was a shift in proportions of planktotrophy (feeding larvae) versus lecithotrophy (non-feeding larvae that depend on stored reserves), with higher levels of lecithotrophy in Antarctica. This trend to non-feeding larval development is often confused with Thorson's rule, which concerns the prevalence of pelagic versus protected development in marine faunas in different regions. Several studies showed that proportions of different larval types are similar in South America and Antarctica in echinoderms and molluscs (Pearse et al. 1991, Hain & Arnaud 1992, Gallardo & Penchaszadch 2001). Later studies showed that numbers of pelagic larval taxa in Signy Island, South Orkneys (Stanwell-Smith et al. 1999) and at Rothera Point, Adelaide Island, Antarctic Peninsula (Bowden et al. 2009) were similar to those from temperate latitudes, and similar to those reported by Thorson (1946) for Denmark, but an order of magnitude higher than values for the Arctic. Fetzer and Arntz (2008) and Sewell (2005) also noted high numbers of planktotrophic larvae at high latitudes.

Studies of larval development and physiology have also demonstrated there is no low temperature barrier to completing development (e.g. Bosch et al. 1987, Peck et al. 2006b, Pace & Manahan 2007). Thus, on three counts, data appear contrary to Thorson's rule, and the conclusions drawn by Pearse (1994) and Pearse and Lockhart (2004) that the rule does not apply are generally accepted. More recent investigations conducting meta-analyses of developmental mode from tropical to polar sites and using very large datasets have confirmed that there is no trend away from pelagic larval phases in the development of marine invertebrates, but there is an increase in proportions of non-feeding pelagic larvae with latitude (Marshall et al. 2012, Marshall & Burgess 2015).

Indeed, many of the most common and abundant species living in shallow Antarctic marine habitats are broadcast reproducers, for example, the limpet *Nacella concinna*, the sea urchin *Sterechinus neumayeri*, the starfish *Odontaster validus*, the bivalve *Aequiyoldia eightsii* and the brittle star *Ophionotus victoriae*. A similar situation, where the most common species living in shallow habitats are predominantly broadcast spawners, while proportions are lower in rarer taxa and deeper habitats was noted for the Arctic over 60 years ago (Thorson 1950). It may be that in shallow environments, where physical disturbance from ice regularly both destroys the existing local fauna and also clears new areas for colonisation, the benefit from larval dispersal to newly available sites outweighs the negative impact from increased larval mortality. Antarctic benthic communities often contain high proportions of deposit-feeding and suspension-feeding taxa, with values higher than proportions at most lower latitude sites (Arntz et al. 1994, Gutt & Starmans 1998, Starmans et al. 1999, Clarke & Johnston 2003, Clarke et al. 2004a, Gili et al. 2006, Aronson et al. 2007, Barnes & Conlan 2012). These suspension-feeders are major predators of marine invertebrate larvae (Pechenik 1987, 1999, Morgan 1995).

The Antarctic benthos contains possibly the largest differences in dispersal capacity of benthic faunas globally. This is because the very slow development rates mean the planktonic phases for many of the broadcast breeding species are much longer than those at lower latitudes, and the direct developing species alongside them have very poor dispersal capacities. Brooding species do disperse, but genetic studies have shown dispersal is limited to a few kilometres or less in, for example, the brooding Antarctic gastropod *Margarella antarctica* (Hoffman et al. 2011). Similar limited dispersal has also been shown for lower latitude species with protected development (e.g. Sherman et al. 2008, Keever et al. 2009).

Beyond larval dispersal, marine animals can drift as adults as an active process such as in the anemone *Dactylanthus antarcticus*, which inflates to large size when it has completed feeding on its soft coral prey. It then releases from the substratum and drifts until it contacts a new prey item (Peck & Brockington 2013). There have also been observations of neutrally buoyant adults of some species drifing above the seabed when they have been disturbed by ice, and this includes the colonial tunicate *Distaplia cylindrica* (J. McClintock personal communication).

One mechanism for dispersal that is possible in the polar regions that is not available elsewhere is rafting on icebergs. Seafloor debris, rocks, rubble, sediment and even boulders are often present on upturned icebergs. Living marine invertebrates have been seen amongst this debris on some icebergs, including the gastropods *Margarella antarctica* and *Nacella concinna* and the urchin *Sterechinus neumayeri* (L. Peck, pers. obs.). The frequency of this occurrence, distances travelled by rafted invertebrates and variation between species in contribution to dispersal remains to be quantified but is likely to be one of the more important mechanisms in slow-moving species with protected development to juvenile stages.

Dispersal can be achieved by other means than motile or drifting early development stages or by being carried by icebergs. Postmetamorphic stages of over 1200 marine species have been identified as being capable of dispersal attached to a range of both natural and man-made substrata (Thiel & Gutow 2005). Natural floating substrata that have aided dispersal of attached species for millions of years include macroalgae, pumice and wood, the most common materials, but large animals including whales and turtles have also been used. Marine litter, especially plastics, has dramatically changed the opportunities for rafting species to disperse over the last 50–100 years (Barnes 2002, Barnes et al. 2009, Eriksen et al. 2014, Bergmann et al. 2015).

In Antarctica, marine debris is much rarer than at lower latitudes for several reasons, including the circulation of the Southern Ocean around Antarctica, isolating it from lower latitude oceans, the absence of coastal forested areas that are the source of floating debris elsewhere, and the absence of large human populations and their associated debris. Marine debris has, however, been regularly identified and collected at several sites. A survey in 2012 found unexpectedly high incidences of debris in the Southern Ocean, with plastics and plastic fragments present at around 50,000 pieces per km^2 (http://www.theguardian.com/environment/2012/sep/27/plastic-debris-southern-ocean-pristine). Marine debris deposited annually on selected Antarctic shorelines is also monitored, as is marine debris in selected bird colonies and occurrences of debris entangled on marine mammals under a CCAMLR programme (www.ccamlr.org/en/science/marine-debris). Several species have been documented attached to floating plastics in the Southern Ocean, especially bryozoans, and this includes reproductively active species (Barnes & Fraser 2003). None of the organisms identified as rafting or attached to floating debris in Antarctica are from outside the Southern Ocean, and hence, there is no evidence that alien species have used or are using this mechanism as a route to enter Antarctic waters. The reason for this has often been assumed that the sea surface freezes all the way around Antarctica in winter, and there is no unfrozen coastline for aliens to colonise and establish. With climate change, however, it has been suggested that open coastline is likely to occur before 2100 on the Antarctic Peninsula, which would change the likelihood of establishment of alien species, and this has been identified as a future risk to biodiversity (Kennicutt et al. 2014, 2015, Sutherland et al. 2015).

Egg size, fecundity, reproductive effort and life histories

The slow development rates in Antarctic marine species described above have often been associated with, where reported, large egg size and hence reduced numbers of eggs released per spawning event. The production of larger eggs at higher latitudes and colder environments has been recognised since at least the 1930s (Thorson 1936, 1950, Clarke 1985, 1993a, Pearse 1994, Collin 2003, Moran & McAlister 2009) and was part of the paradigm of reduced pelagic development and increased protected development at high latitude, called Thorson's rule that was rejected when phylogenetic considerations were taken into account (see 'Ecological implications of slow development rates' section previously). In Antarctica, large egg size, with diameters generally two to five times greater than related temperate species, was reported in the 1970s for amphipod crustaceans (Bone 1972, Bregazzi 1972) and in the 1980s and 1990s for shrimps (Clarke 1985, 1992, 1993a,b). There are many reports of large egg size for individual species in Antarctica, again generally two to five times larger than comparison with lower latitude species (Table 2).

Because of the requirement for data for fisheries purposes, studies of fish reproduction are more common than for invertebrates, and the data collected often include egg size. A trend from smaller, pelagic eggs producing smaller juveniles at low latitude to larger demersal eggs with larger juveniles at higher latitudes is well documented (Leis et al. 2013). This is despite phylogeny having a large impact on egg size, with, for example, salmonids having very large eggs in the range 2–9 mm diameter but mostly in the 2–6 mm range. Most other temperate and tropical fish have eggs in the range 0.5–4 mm diameter (Bagenal 1971, Robertson & Collin 2015). The Antarctic fish fauna is dominated by the suborder Notothenioidea, most of which produce eggs with diameters between 3 and 5 mm, similar in size to the largest produced in cool temperate latitudes and well above the mean size for all species in lower latitude regions. The most common explanation used to explain large egg size at high latitude is temperature, but at smaller, regional scales, resource availability has been demonstrated to be important with several studies showing smaller eggs are more common in regions of high food availability and larger eggs where food supplies are limited (Lessios 1990, Marko & Moran 2002, Robertson & Collin 2015).

The numbers of studies made within species are smaller than for comparisons between species, but these comparisons have also produced data showing increases in egg size with latitude. In the isopod *Ceratoserolis trilobitoides*, eggs almost doubled in size with a less than 15° increase in latitude, where egg dry mass ranged from 3.3 to 3.9 mg dry mass in sub-Antarctic South Georgia to 6.5 mg dry mass in the high Weddell Sea (Wägele 1987, Clarke & Gore 1992, Gorny et al. 1992). Within species, trends to larger egg size at higher latitude in Antarctica have also been demonstrated in the philobryid bivalves *Lissarca miliaris* (Reed et al. 2014) and *Lissarca notorcadensis* (Brey & Hain 1992), the caridean decapod *Notocrangon antarcticus* (Lovrich et al. 2005) and in some nototheniid fish (Kock & Kellerman 1991).

In a detailed study of animal size in amphipod crustaceans, Chapelle and Peck (2004) showed that not only the largest species of amphipod at any given site increased with latitude and seawater oxygen content, but the minimum size of a species also increased towards the poles, although at a smaller rate than for larger size classes. It was further demonstrated by Chapelle and Peck (2004) that minimum amphipod size across latitude correlated with the size of egg produced. It has long been suggested in amphipod crustaceans that minimum size for a species is set by egg size and the capacity to brood embryos (Mills 1967). Strong evidence to support this came from correlations between amphipod size and the numbers of eggs produced and embryos in broods, where the smallest species are the only ones producing broods of single eggs (Sainte-Marie 1991), and that in very small species, males are smaller than females (Chapelle & Peck 2004).

Futhermore, a link between reproductive mode and egg size has been recognised for nearly 100 years (Mortensen 1921, 1936, Thorson 1936). The move to more protected development and fewer species with pelagic developmental phases (Thorson's rule) seemed to be part of a consistent

Table 2 Maximum egg diameter measurements for a range of Southern Ocean marine invertebrates

Species	Maximum egg diameter (mm)	References
Polychaete worms		
Leodamas marginatus	0.58	Hardy (1977)
Laetmonice producta	0.32	Micaletto et al. (2002)
Molluscs		
Pareledone charcoti (cephalopod)	11	Kühl (1988)
Pareledone turqueti (cephalopod)	19	Kühl (1988)
Adelieledone polymorpha (cephalopod)	10	Kühl (1988)
Nuttallochiton mirandus (polyplacophoran)	0.94	Hain & Arnaud (1992)
Neomeniomorpha (solenogastre)	0.78	Hain & Arnaud (1992)
Prodoris clavigera (nudibranch)	2.1	Wägele (1988)
Tritonia challengeriana (nudibranch)	0.34 ± 0.007 (s.e.)	Woods & Moran (2008)
Tritonia tetraquetra (nudibranch)	0.102 ± 0.001 (s.e.)	Woods & Moran (2008)
Laevilitorina caliginosa (littorinid)	0.20	Simpson (1977)
Adacnarca nitens (bivalve)	0.040	Higgs et al. (2009)
Gaimardia trapesina (bivalve)	0.40	Simpson (1977)
Plaxiphora aurata (polyplacophoran)[a]	0.27	Simpson (1977)
Hemiarthrum setulosum (polyplacophoran)[d]	0.80	Simpson (1977)
Crustaceans		
Antarctomysis maxima (mysid shrimp)	1.75 (stage 1 embryo)	Siegel & Mühlenhardt-Siegel (1988)
Mysidetes (previously *Antarctomysis*) *posthon* (mysid shrimp)	0.77 (stage 1 embryo)	Siegel & Mühlenhardt-Siegel (1988)
Eusirus perdentatus (amphipod)	2.75	Klages (1991, 1993)
Ampelisca richardsoni (amphipod)	1.1	Klages (1991)
Paraceradocus gibber (amphipod)	1.7	Klages (1991)
Echinoderms		
Heterocucumis steineni (holothurian)	1.0	Gutt et al. (1992)
Psolus dubiosus (holothurian)	1.3	Gutt et al. (1992)
Odontaster validus (starfish)[a]	0.17	Bosch & Pearse (1990)
Odontaster meridionalis (starfish)[a]	0.19	Bosch & Pearse (1990)
Glabraster (previously *Porania*) *antarctica* (starfish)[a]	0.55	Bosch & Pearse (1990)
Bathybiaster loripes (starfish)[b]	0.93	Bosch & Pearse (1990)
Psilaster charcoti (starfish)[b]	0.95	Bosch & Pearse (1990)
Acodontaster conspicuus (starfish)[b]	0.70	Bosch & Pearse (1990)
Acodontaster elongatus (starfish)[b]	0.54	Bosch & Pearse (1990)
Acodontaster hodgsoni (starfish)[b]	0.55	Bosch & Pearse (1990)
Lophaster gaini (starfish)[b]	1.28	Bosch & Pearse (1990)
Macroptychaster accrescens (starfish)[c]	1.28	Bosch & Pearse (1990)
Perknaster fuscus (starfish)[c]	1.20	Bosch & Pearse (1990)
Diplasterias brucei (starfish)[d]	2.80	Bosch & Pearse (1990)
Notasterias armata (starfish)[d]	3.50	Bosch & Pearse (1990)
Sterechinus neumayeri (sea urchin)	0.20–0.21	Moore & Manahan (2007)
Sterechinus neumayeri (sea urchin)	0.15–0.17	Suckling et al. (2015)
Ascidian		
Cnemidocarpa verrucosa	0.24	Strathmann et al. (2006)
Scleractinian corals		

Continued

Table 2 (Continued)   Maximum egg diameter measurements for a range of Southern Ocean marine invertebrates

Species	Maximum egg diameter (mm	References
Flabellum thouarsii	4.80	Waller et al. (2008)
Flabellum curvatum	5.12	Waller et al. (2008)
Flabellum impensum	5.20	Waller et al. (2008)

Data for echinoderms and polyplacophorans are split into the following notes:

[a] Broadcast spawning species with feeding larvae;

[b] Broadcast spawners with lecithotrophic (non-feeding) larvae;

[c] Unknown, but probably broadcast with lecithotrophic larvae; and

[d] Brooding species.

Similar data to the above categories are only available for the species shown.

global trend that was accepted until the 1990s, when detailed within taxon analyses were used to disprove the hypothesis (Bosch & Pearse 1990, Clarke 1992, 1996, Pearse 1994, Pearse & Bosch 1994). Large egg and embryo size in Antarctica was identified as part of a global trend of increase in these characteristics with latitude from tropics to polar regions in a large meta-analysis of marine invertebrate life histories (Marshall et al. 2012), and the increase in egg size with latitude and large eggs in Antarctica remains an accepted trend and requires explanation.

The explanations for larger egg size at higher latitude, lower temperature sites have been mainly ecological and developed over several decades (Thorson 1950, Arnaud 1977, Simpson 1977, Picken 1980, Jablonski & Lutz 1983, Rohde 1985, 2002, Clarke 1987b, 1993a,b, Arntz et al. 1994). These include: (1) slow development at polar temperatures means most species cannot complete development during the brief phytoplankton bloom, which pelagic planktotrophic larvae feed on, requiring a longer period for development before reaching critical stages; (2) larger size at metamorphosis might enhance survival post settlement and thus select for larger eggs and non-pelagic development; (3) slower development increases time in the water column and increases predation in this phase, suggesting longer intracapsular developmental phases without feeding, which would require larger reserves in the egg before release to the water column would be advantageous; (4) synchronising embryo hatching to the larval phase and metamorphosis to juveniles with the phytoplankton bloom is more difficult, and hence, development during the previous winter at very low food levels would be required; (5) it is easier for offspring that do not pass through a pelagic phase to settle close to parents, where habitats are likely to be favourable, and this may be more important in cold-water environments where factors like iceberg scour may have an effect; (6) pelagic larvae may suffer osmotic difficulties in summer due to dilution of Arctic and Antarctic waters by melting ice, again giving some advantage to development in winter when food supplies for larvae are low. Rohde (2002) excluded seasonality of food supply, timing of developmental events and increased predation on the basis that larger egg size and reduced incidence of pelagic development are attributes also of parasitic taxa living at high latitudes. Settling close to parents as a driver for these adaptations is also unlikely as the same attributes are shared by species where individuals live at large distances from parental stock as well as those that colonise in close proximity. It is further unlikely that osmotic stress plays a role as species living at depths beyond the effects of melting ice have similar levels of reduced pelagic development as shallow species. An ecological reason thus still needs to be agreed upon.

It is possible that there may be a physiological explanation for the increased egg size at low temperatures, and this might also explain why there is reduced pelagic development, as larger egg size is correlated with protected development. There are three factors that would lead to increased egg size at low temperature from physiological consideration. First, that development rates are slowed more at low temperature than routine or standard metabolic rates, where development rates in brooding gastropod molluscs were five to ten times slower than temperate species, but oxygen consumption in bivalve molluscs was only two to three times slower (Peck 2016, Figure 8). Because of this difference, the proportion of overall costs during development devoted to maintenance will be significantly higher. Extending the development period by five times but increasing maintenance metabolism 2-fold increases overall metabolic costs by 2.5 times. Slowing development by an order of magnitude and slowing maintenance metabolism by a factor of three increases overall metabolic costs by 3.3 times. These extra costs need to be met either by more food intake and/or stored reserves over the whole development period in planktotrophic developers, or from greater stored reserves in species with protected development. This analysis would lead to the prediction that, on average, increase in egg size at low temperature would be greater in species with protected development than broadcast spawners, as the option of increased overall food intake is not possible. An increased requirement for food during pelagic development may also be a driver towards protected development, especially in highly seasonal environments where food supplies are only available for restricted periods during the year.

The second physiological explanation for larger egg size is through the demonstration that it is more difficult to produce fully functional proteins at low polar temperatures because of problems with stability, or at the synthesis stage (Peck 2016, plus earlier section in this review on 'Protein synthesis, retention and folding'). The high levels of ubiquitination measured in Antarctic marine species (Todgham et al. 2007, Shin et al. 2012) suggest that during development a higher proportion of the full protein complement will be held in a cycle of ubiquitination, degradation and resynthesis than in temperate species. Under these circumstances, greater quantities of amino acids in the egg would allow protein synthesis to continue while significant amounts are cycling through degradation. Another way of overcoming some of these problems would be to change the balance of free amino acids and fully synthesised proteins in the egg, reducing the need to synthesise protein *de novo* in the early embryo stages. This hypothesis is yet to be tested, but in fish, a correlation between protein content, and reduced free amino acid content with increasing egg size has been reported (Rønnestadt & Fyhn 1993).

Third, Van der Have and de Jong (1996) evaluated changes in cell size and cell number during development of ectotherms. They concluded that growth, as increase in cell size and differentiation, as opposed to change in numbers of cells is affected differently by temperature. For a given stage of development, the number of cells present is roughly similar in different species, and the outcome of larger eggs is larger cells and hence larger size at metamorphosis in Antarctic marine species. Chapelle and Peck (1999, 2004) and Peck and Chapelle (2003) demonstrated that maximum size attained by amphipods across the globe was highly correlated with ambient oxygen levels in the environment, and Peck and Maddrell (2005) demonstrated that when oxygen levels are reduced, maximum size is lower, and cell sizes are smaller in *Drosophila melanogaster*. Levels of dissolved oxygen in seawater increase with latitude such that the concentration of dissolved oxygen at 0°C is 1.82 times higher than at 30°C. Diffusion distances for oxygen supply into developing embryos is described by the Fick equation (Dejours 1981), and a main driver of this is the concentration gradient between external environment and site of oxygen use. For a salinity of 35, the higher concentration of oxygen in polar oceans at 0°C (347.9 μmol kg^{-1}) compared to a tropical ocean at 30°C (190.7 μmol kg^{-1}) should allow eggs and embryos to be around 1.8 times larger in diameter and, hence, around six times heavier (Peck & Chapelle 1999). It thus seems that cell size relations with temperature and oxygen would suggest egg, embryo and adult size should increase at lower temperatures and, hence, with latitude. As oxygen and temperature covary in seawater, both relationships may be a result of a single underlying mechanistic explanation, that Chapelle and Peck (1999, 2004), Peck and Chapelle (2003) and Peck and Maddrell (2005) attributed to oxygen supply limitations.

The corollary of large egg size is smaller numbers of eggs released at each spawning event, and there are data that support this contention. Reports on this topic include studies on fish (La Mesa et al. 2008), mysid shrimps (Siegel & Mühlenhardt-Siegel 1988), gastropod and bivalve molluscs (Simpson 1977, Reed et al. 2014), nudibranch molluscs (Wägele 1988, Woods & Moran 2008), and caridean shrimps (Gorny et al. 1992, Clarke 1993a, Arntz et al. 1994). As larger eggs are generally correlated with fewer eggs produced in Antarctic species, a related question then emerges around how much energy and other resources are invested in reproduction and how high latitude species compare with those in temperate or tropical regions.

Reproductive effort

All organisms acquire energy and use it for a variety of purposes. The analysis of what organisms do with this energy often involves the construction of energy budgets based on the energy equation of Winberg (1956), elucidated well by Kleiber (1961). The equation states: consumption equals the sum of somatic and gonadal growth, metabolic costs and waste products. On this basis the proportion of an individual's available energy put into reproduction, the reproductive effort (RE) can be calculated (e.g. Clarke 1987b). RE is not easy to measure because it involves not only the investment in gametes and other reproductive tissues but also the metabolic costs of synthesis and energy used in activities associated with reproduction, some of which can be costly. Even this assessment does not include potential future losses that might be incurred from an investment in reproduction that increases the likelihood of mortality (e.g. Stearns 1992). Much more frequently reproductive output (RO) is measured as the change in gonad mass across spawning periods or the collection of released gametes combined with assessments of gonad and/or gamete energy contents. RO is very useful in that it allows both temporal and spatial comparisons to be made. Several of the studies listed previously have shown a decrease in not only the number of eggs produced per year by Antarctic species, but a reduction in the mass of body tissue devoted to reproduction, reduced RO.

As RE is measured in relation to the energy taken in by an organism (absorbed by an animal), and both somatic growth (see 'Growth' section previously) and metabolic rates are lower in Antarctica than in temperate and tropical regions, the comparison of RE with warmer-water species is less clear. Furthermore, Antarctic marine species generally live much longer than warmer-water relatives (Peck et al. 2006b) and are reproductively active over a longer period. Because of this, RE needs to be considered across the whole lifetime of an individual (Clarke 1987a). Lifetime RE is further complicated by the need to include not only the energy allocated to reproduction each year, but also the likelihood of surviving to that age, or including an estimate of the average age that individuals in a given population will survive to (e.g. Myers 2002). Very few studies of polar species have calculated lifetime RE, but Clarke (1987b) made this calculation for the Northern hemisphere deep-water shrimp *Pandalus borealis*. This species has a very wide geographic range spanning mean annual temperatures ranging from 0°C to around 10°C. When the calculation of RE was made assuming no change in metabolic costs due to temperature, RE increased from the colder to warmer sites, but when metabolic costs were remodelled to vary with temperature following Arrhenius predictions (see later section on Energy use, oxygen consumption and metabolic rate), there was no change in lifetime RE across the whole range inhabited by the shrimp (Figure 9).

Few studies have looked at RE differences between the sexes, but the assumption is usually that females invest more in reproduction than males because of the larger effort put into making an egg than sperm, and that in species with parental care, females are usually the primary carers. Females are also usually used in studies where RE is assessed because it is easier to collect eggs released during spawning and, hence, to quantify energy invested than it is to collect sperm released by males (e.g. Clarke 1979, Gremare & Olive 1986, Hess 1993, Brante et al. 2003, Béguer et al. 2010). In the few studies where direct comparisons have been made, however, results have suggested male RE is either similar to female RE or is greater. For example, Vahl and Sundet (1985) concluded that males have higher RE than females in the Iceland scallop *Chlamys islandica*, and Morricone (1999) showed that gonad index (measured as the mass of gonad compared to the foot) was higher in males than females prior to spawning and decreased more on spawning in males in the limpet *Nacella deaurata* in Tierra del Fuego. In Antarctica, Tyler et al. (2003) measured changes in gonad size in the scallop *Adamussium colbecki*, and a similar analysis on the starfish *Odontaster validus* also showed RE was higher in males than females (Grange et al. 2007).

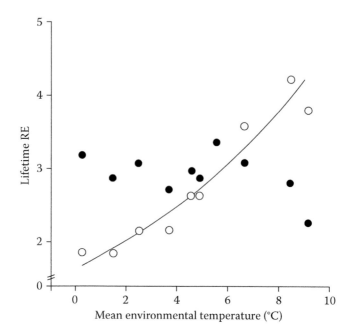

Figure 9 Lifetime reproductive effort in relation to mean environmental temperature for the caridean shrimp *Pandalus borealis*. Open symbols denote data calculated without allowing for variation in standard metabolic rate with temperature. Line fitted to Ln transformed data for both variables but shown as linear plot. Closed symbols denote data calculated assuming standard metabolic rate varies with temperature as in Clarke (1987b). There is no significant trend in the dark data. (Figure redrawn from Clarke, A. 1987b. *Marine Ecology Progress Series* **38**, 89–99.)

However, where studies of reproductive cycles have been conducted for several years, large interannual differences in RE have been identified. In a four-year analysis of reproduction in the brittle star *Ophionotus victoriae*, RO was measured as the decrease in gonad mass on spawning, and this changed by at least 5-fold between years, although the variation may be higher than this because in 1997 an estimate of 12% reduction in gonad mass was used, compared to over 80% in 1999, and the 12% value was not significantly different from zero (Grange et al. 2004). There were also very large differences, around an order of magnitude, between years in the relative size of the gonads, at peak just prior to spawning, in this species (Figure 10). Very large differences in gonad mass both between sites and between years for standard-sized animals were reported for the urchin *Sterechinus neumayeri* (Brockington et al. 2007). Large interannual differences in gonad size just prior to spawning were also reported for the Antarctic scallop *Adamussium colbecki* from Terra Nova Bay by Chiantore et al. (2002), but differences in maximum gonad size between years were smaller and not significant in the starfish *Odontaster validus* and the brittle star *Ophionotus victoriae*. This was, however, only a two-year study, and as can be seen from Figure 10, a longer investigation may be needed to see the large variations reported by Grange et al. (2004).

Variation in RO between years was much lower in the starfish *Odontaster validus*, where the changes in gonad index in the largest years were 2.25 times higher than the smallest years for females and 1.7 times for males (Grange et al. 2007). In a four-year study of the predatory nemertean *Parborlasia corrugatus*, there were no notable interannual differences in reproductive characters, and the authors reported that there was almost no variation in oocyte size between years in this species, which is very unusual in these types of investigation. It should, however, be noted that in *P. corrugatus*, as is typical for nemertean worms, it is not possible to separate gonads from other body tissues, and the gonad is not a discrete tissue, which precludes the measurement of a gonad index

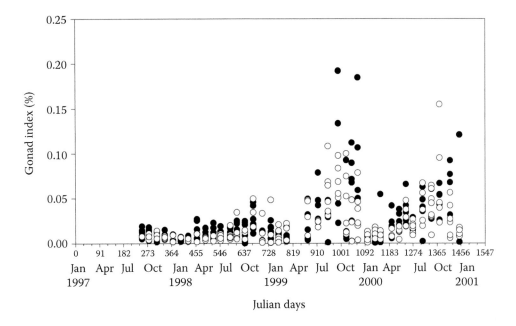

Figure 10 *Ophionotus victoriae* gonad index (the wet mass of the gonad relative to the volume of the disc for male (●) and female (○) brittle stars between August 1997 and December 2000. Note the marked difference (at least 5-fold) in change in gonad index, and hence mass of gametes released on spawning between years. (From Grange, L.J. et al. 2004. *Marine Ecology Progress Series* **278**, 141–155.)

or calculations of RO and RE. Few, if any, other studies have carried out interannual evaluations of reproductive characters over periods of more than three years.

Although the data are limited, it is notable that there were stronger interannual signals in the reproductive biology of the filter-feeding scallop than the omnivores studied by Chiantore et al. (2002). Furthermore, the brittle star *Ophionotus victoriae* and *Odontaster validus*, both of which are omnivores (McClintock 1994), had stronger interannual effects on their RO than the predatory nemertean *Parborlasia corrugatus* (Grange et al. 2004, 2007, 2011a). This difference across trophic groups would be expected from considerations of the effects of seasonal and interannual variation in resource availability and quality in different levels of the food web (Dunbar 1970, Clarke 1983, 1988, 1991). Similar differences in the impact of seasonal and interannual environmental factors on different trophic levels have been noted in other physiological characteristics such as feeding biology and oxygen consumption (e.g. Obermüller et al. 2010). Spatial variation in food supply has also been identified as a factor influencing biological characteristics, including growth and reproduction in populations of marine invertebrates on either side of McMurdo Sound (Dayton et al. 1969, 1970, Dayton 1990). These data have led authors to suggest that the primary mechanism for differences in interannual variation in reproductive characters between trophic groups is likely food availability.

Of interest here is that the timing of reproductive cycles appears to be markedly influenced by light regime. Pearse and Bosch (2002) conducted a series of elegant experiments on the common circumpolar omnivorous starfish *Odontaster validus* that demonstrated that gametogenic cycles could be shifted 180° out of phase by manipulating light regime. When specimens were held in a light regime that was six months out of phase, spawning occurred almost exactly six months later than in the wild, and this change was completed within a year.

There is a clear need to improve understanding of patterns of RE in Antarctic marine species, how it varies compared to species at lower latitudes, and also how it varies within the Southern Ocean and the factors determining the variation. Reproduction of many species is very likely to be affected by long-term cycles in the environment such as the Southern Annular Mode (SAM) and other factors such as El Niño. These are overlaid on long-term environmental change such as the Antarctic Peninsula warming that occurred over the last half of the twentieth century. Studies collecting data over periods in excess of 10 years, or repeat studies at decadal intervals, similar to the multiple year analyses reported by Chiantore et al. (2002) and Grange et al. (2004, 2007, 2011a), are needed to identify effects of these cycles and trends. There is also a need to improve understanding of differences between male and female RE, both in Antarctica and in other oceans around the world, and to identify if the balance of RE between the sexes varies with global region. RE is a major factor affecting life histories of animals that dictates success or failure of a species in a given environment.

Life-history theory

Understanding of life histories and the development of life-history theory during the 1950s and 1960s was based around the concept of trade-offs in the allocation of resources and the impact on fitness of diverting resources into one direction over another (Medawar 1952, Williams 1957, Hamilton 1966). Current life-history theory models were developed initially by Gadgil & Bossert (1970). The classic trade-off identified in life-history theory is that between investment in the individual's own survival and investment in offspring (Charnov & Krebs 1973, Stearns 1992, Kaplan & Gangestad 2005), and this trade-off varies with a range of factors including age and ecological context (Hoyle et al. 2008). Other important trade-offs occur as well, a prime example being the one between number of offspring and investment per propagule.

The trade-off between investment in individual offspring and numbers of offspring is a central part of life-history theory (Vance 1973, Smith & Fretwell 1974). David Lack first placed life histories in an evolutionary context (for animals with parental care) when working on birds. Lack (1947) hypothesised that natural selection will favour the clutch size that produces the most offspring that survive. It should be noted that this trade-off is very unlikely to be independent of other trade-offs, including reproductive investment versus parental survival and offspring size versus offspring survival. From this it is clear that there might be several interlinked 'optimum' solutions to the trade-off between offspring number and offspring size that also vary with external environmental conditions (Stearns 1989, 1992, Walker et al. 2007). Since then the field has developed to say that the number of offspring that survive and reproduce should be optimised, and that where adult resources are finite, which is generally true for animals, then number and size of embryo produced directly trade-off.

Many authors over several decades have stressed that early-stage survival and performance strongly influence population dynamics in marine and aquatic systems (Thorson 1946, Strathmann 1985, Roughgarden et al. 1988, Pechenik 1999, Moran & Emlet 2001, Underwood & Keough 2001, Marshall et al. 2006, 2012). As seen earlier, there is a very common if not universal trend towards larger egg sizes at higher latitudes, and it is clear that greater investment per egg, and hence individual offspring, is the optimum life-history solution in Antarctica, and this likely maximises embryonic and larval survival to juvenile stages. The reasons for this remain to be identified but could be due to problems making proteins at low temperature (Peck 2016), or for reasons given earlier around slowed development rates and overall energy requirements. The final outcome in terms of importance to life histories would be to enhance early-stage survival and hence overall fitness.

Longevity and age at first reproduction

Longevity and age at first reproduction are key life-history traits. They allow generation time and number of years over which an individual reproduces to be calculated. In marine environments, data for these two characteristics are often best quantified and understood for fish because of their value, for example, in setting catch limits in fisheries. In Antarctica, data are sparse for invertebrates but available for several fish species and good for some, for example, *Notothenia coriiceps* (Kock 1992), *Dissostichus eleginoides* (Kock et al. 1985).

Temperate fish usually mature at ages between one and four years. Sole has a 50% age at first reproduction of 1.7–2.5 years (Mollet et al. 2013); for haddock, this is three years, whiting two years and cod three to five years (Hunter et al. 2015). Arruda et al. (1993) reported seven goby species in the Ria de Aveiro lagoon, Portugal, all mature at less than one year of age, while O'Brien et al. (1993) showed that 14 of 19 species of finfish investigated on the Georges Bank reached maturity in less than 2.5 years, and only one required more than five years. A few species take longer, and the sea bass *Dicentrarchus labrax* requires three to six years to reach maturity in North Wales (Carroll 2014). Antarctic fish generally take much longer than temperate species to become mature. *Notothenia coriiceps, N. rossii, Chaenocephalus aceratus* and *Gobionotothen gibberifrons* all require six to eight years to reach maturity, and *Pleuragramma antarctica* and *Trematomus eulepidotus* need seven to nine years (Kock 1992). *Dissostichus eleginoides* does not mature before it is 8–10 years old (Kock et al. 1985), *Trematomus bernacchii* reaches maturity at around 10 years (Gon & Heemstra 1990), and *Dissostichus mawsoni* needs 13–17 years to achieve maturity (Parker & Grimes 2010). Most of the previous species live to between 15 and 30 years, with later-maturing species having greater longevity (Gon & Heemstra 1990). Smaller Antarctic species generally mature earlier and have less longevity. *Lepidonotothen* (formerly *Nototheniops*) *nudifrons* reaches maturity at four to five years (Radtke & Hourigan 1990), *Harpagifer antarcticus* lives to nine years of age (Daniels 1983) and matures at around five years, or 55% of maximum age (La Mesa & Eastman 2012), and Kock and Kellerman (1991) estimates that most small notothenioids in the pack-ice zone live for less than 10 years and begin spawning at three to four years.

Most fish species achieve maturity between 30%–40% of maximum age and 50%–80% of maximum size (Kock 1992, He & Stewart 2001). Antarctic species, therefore, reproduce over many years and for longer periods, on average, than warmer water fish. This increases lifetime RE compared to temperate and tropical species. Reduced numbers of gametes released decreases RE relatively, but large egg size may offset the decrease. Calculating lifetime RE is, therefore, not a trivial exercise.

Data are generally poorer for longevity and age at first reproduction for Antarctic marine invertebrates. The Antarctic scallop *Adamussium colbecki* lives for more than 10 years and becomes mature at between six and seven years of age (Cattaneo-Vietti et al. 1997). The large infaunal bivalve *Laternula elliptica* lives up to 36 years (Philipp et al. 2005), begins to produce gonads at 30–40 mm length, aged four to five years, but does not begin reproducing until around nine to ten years of age (Urban & Mercuri 1998, Clark et al. 2013). In brooding bivalves *Mysella* (previously *Kidderia*) *subquadrata* at Palmer station matures at 23–27 months post fertilisation, 18–22 months after juveniles are released from broods (Shabica 1974), whereas the smallest brooding female of *Adacnarca nitens* in the Weddell Sea is around 3.5–4 years old, and 50% brooding is thus not reached until individuals are at least five years of age (Higgs et al. 2009). The common limpet *Nacella concinna* begins to produce gonads with mature gametes at 18–20 mm length at Rothera Point, Adelaide Island (S. Morley, personal communication), which from growth rate data at the same site equates to four to five years of age (Clarke et al. 2004b). In the brachiopod *Liothyrella uva*, broods are first seen in individuals around 31 mm in length (Meidlinger et al. 1998), which equates to an age around 17–18 years (Peck et al. 1997b). Seasonal changes in body mass associated with reproduction in the sea urchin *Sterechinus neumayeri* at Rothera Point, Adelaide Island become apparent at a test diameter of around 15 mm (T. Souster, pers. comm.). From growth rate data from the same site, this equates to an age of eight to nine years (Brey et al. 1995a, Brockington 2001a). The reviews of the biology and reproduction of the amphipods have concluded that polar species are characterised by slow growth, great longevity and deferred maturity (Sainte-Marie 1991, Johnson et al. 2001). Evidence for this includes the Antarctic *Paramoera walkeri* which lives for up to four years and matures at 19 months of age (Rakusa-Suszczewski 1972, Brown et al. 2015), *Eusirus perdentatus* which lives for five to eight years and matures at three years (Klages 1993), and

Orchomenella franklini which lives for three years and begins reproducing after two years (Baird & Stark 2013). Most warmer-water marine invertebrate species mature in less than a year, and tropical species often have more than one generation per year.

One species where maturation has been stated to be early, and contrary to the usually identified situation of deferred maturity, is in the pennatulid sea pen *Malacobelemnon daytoni* (Servetto et al. 2013). In this species, which grows to over 120 mm rachis length, a measure of overall size (López-González et al. 2009), the smallest mature colony was 15 mm long. Octocorals from different regions have been reported to mature at between 2 and 13 years of age (Coma et al. 1995). One remaining issue with the observation of early maturation in *M. daytoni* is that there are no data on its growth rate, and although the size for maturity is small, it is possible that like other Antarctic octocorals growth is very slow (e.g. Peck & Brockington 2013), which would extend the age at first or 50% maturity.

Overall maturity appears to be delayed in Antarctic fish and invertebrates. Data are not strong enough to be definitive on how much it is slowed, but in most cases, the slowing appears to be in the range 2-fold to 10-fold.

Energy use, oxygen consumption and metabolic rate

A basic requirement for all organisms is to use energy to perform the biological functions necessary for life across all their life-history stages. This is true from the molecular level where gene expression requires energy to locomotion in whole animals. The ability to perform these functions and the rate at which they are carried out ultimately depend on an adequate energy supply. In this context, all organisms transform energy. Photosynthesising groups (plants, cyanobacteria, etc.) transform light energy into chemical energy. Animals predominantly consume energy in the form of biomass and use it in a variety of ways to maintain their body tissues, to grow, to reproduce, to produce defences, in activity, and so forth. These processes can be aggregated into an energy budget:

$$C - F = P_s + P_r + R + U + M \text{ (following Peck et al. 1987b)}$$

where C = food consumed, F = faeces produced, P_s = somatic growth, P_r = reproductive investment, R = respiration, U = excretion, and M = mucus production. This form of the energy budget was chosen to emphasise that in some groups non-standard components can be large parts of the budget. In the case shown, the budget was formulated for a gastropod mollusc, *Haliotis tuberculata*, where mucus accounted for up to nearly 30% of the energy consumed (Peck et al. 1987b). There are other cases where an unusual term has been included in the energy budget, including a third growth or productivity component in crustaceans to represent the energy lost as cast exoskeletons during moulting (e.g. Luxmoore 1985).

There have been very few energy budgets constructed for Antarctic marine species, and none of these have measured all of the parameters of the budget at the same time. They have combined published data for some parts of the budget with new measurements. There are clearly some problems with these approaches, as the studies that data were collated from were carried out at different times and sometimes different places. The only study to date that measured all of the major components of the energy budget was Luxmoore (1985) who studied the carnivorous isopod *Paraserolis* (previously *Serolis*) *polita* at Signy Island. Assimilation efficiency (=absorption efficiency) on a diet of the amphipod *Cheirimedon femoratus* was 80%, which is similar to other carnivorous isopods (Luxmoore 1985). The study covered more time than any other Antarctic energy budget investigation as measurements were made every month for two years. Data were size-corrected and expressed as energy values for the whole population in Borge Bay, Signy Island, South Orkney Islands. Energy consumption was estimated from measurements of absorption efficiency multiplied by the sum of all other components of the energy budget. Energy consumed was estimated at 289 J m^{-2} yr^{-1}, somatic growth accounted for

78 J m^{-2} yr^{-1}, reproductive growth 24 J m^{-2} yr^{-1}, losses through moulted skeleton 13 J m^{-2} yr^{-1}, oxygen consumption 108 J m^{-2} yr^{-1}, ammonia excretion 9 J m^{-2} yr^{-1}, and faecal production 58 J m^{-2} yr^{-1}. The largest components of the budget were somatic growth and oxygen consumption, and somatic growth cost over three times as much energy as reproductive growth.

Kowalke (1998) carried out a broad-scale analysis of energetics of two Antarctic sponges, four ascidians and the bivalve mollusc *Laternula elliptica,* all of which are sessile suspension-feeders. He estimated energy consumption from measurements of pumping rate combined with assessments of the energy content of suspended material in the water around his target species, which was dominated by phytoplankton in summer months and resuspended benthic particulate material in winter. He used published values for respiration and somatic and gonadal production to calculate energy budgets. One drawback of this approach is that there was a large discrepancy (25%–30% for the bivalve *L. elliptica* and 52%–83% for the ascidians) between estimated annual energy intake, which may have been due to an overestimation of the amount of food available in winter. Energy budget values were expressed as summations for the populations present in Potter Cove, King George Island and were corrected for the size ranges of animals present. Populations of the bivalve *L. elliptica* allocated similar amounts of energy to respiration (480 kJ m^{-2} yr^{-1}), faecal production (436 kJ m^{-2} yr^{-1}) and somatic production (467 kJ m^{-2} yr^{-1}), but less on reproductive production (324 kJ m^{-2} yr^{-1}). These proportions would, however, be expected to vary if population demography were different. In three of the four ascidian species, *Ascidia challengeri, Cnemidocarpa verrucosa* and *Molgula pedunculata*, faecal egestion accounted for over twice as much energy used as respiration and somatic growth, and reproductive growth accounted for only a small part of the budget. In the fourth species, *Corella eumyota*, respiration and somatic growth each accounted for around 40%–45% of the energy used, while faecal egestion and reproductive growth were smaller at 7.0%–7.5% each.

Aguera et al. (2015) calculated a dynamic energy budget (DEB) for the starfish *Odontaster validus* using predominantly published data from sites in the Ross Sea and Antarctic Peninsula and collected in different years, up to more than 30 years apart. The budget showed similar levels of reproductive investment as temperate seastars, the ability to mobilise reserves rapidly, very slow growth, only reaching maximum size after 35 years and beginning to reproduce after seven years. It did not, however, reflect the very large interannual reproductive output measured in this species by Grange et al. (2007). A similar approach was used to produce a DEB for the large infaunal bivalve *Laternula elliptica*, again based on literature, data collected at a range of sites and over multiple years by Aguera et al. (2017). They described *L. elliptica* as having a 'metabolism specifically adapted to low temperatures, with a low maintenance cost and a high capacity to uptake and mobilise energy, providing this organism with a level of energetic performance matching that of related species from temperate regions'. These are the first budgets of this type, and it will be interesting to see the variation in findings from future exercises similar to this on other Antarctic taxa, as *L. elliptica* is one of the largest and fastest growing bivalves from this region.

For fish, Everson (1970b) produced an energy budget for *Notothenia neglecta* (now *coriiceps*) in Borge Bay, Signy Island, South Orkney Islands. He determined age (and hence growth from von Bertalanffy size to age relationships) from scale and otolith ring assessments, somatic production from annual changes in body mass, and reproductive production from changes in gonad mass on spawning. Oxygen consumption was assessed at the same time on the same population but reported elsewhere (Ralph & Everson 1968). Amounts of energy consumed were estimated as proportions of energy used in respiration and growth ($1.25 \times (P_s + P_r + R)$), and faecal production and nitrogen excretion calculated as 20% of energy consumed. On this basis, growth was slow, and the main component of the energy budget was respiration, which accounted for more than 25 times the energy used in somatic and reproductive growth combined and nearly four times as much as the energy lost in faeces and nitrogen excretion.

Brodte et al. (2006) also investigated energetics in a fish, but they studied the non-notothenioid Antarctic eelpout *Pachycara brachycephalum*. Growth was measured at different temperatures in

143

laboratory trials along with respiration, nitrogen excretion and body condition. Faecal production was calculated from an equation relating food intake and temperature for brown trout by Elliott (1976). *Pachycara brachycephalum* inhabits deeper waters where the temperatures are a few degrees warmer than shallower Antarctic sites, so rates of growth (etc.) are not as slow as for several of the shallower notothenioids. The paper compared the Antarctic species with a temperate eelpout *Zoarces viviparus* from the North Sea where temperatures at collection were 17.5–19.5°C. Both species were fed *ad libitum* with cockle meat (*Cerastoderma edule*) on every second day but had very low absorption efficiencies (~7%–25% for *Pachycara brachycephalum* and ~2%–9% for *Zoarces viviparus*). Values for carnivorous fish are usually in the range 60%–90% and for herbivorous fish 40%–75% (Wooton 2012). Measurements were made at 0°C, 2°C, 4°C and 6°C, and energy used in growth, oxygen consumption, ammonia excretion and faecal production was ~15%–30% greater than energy consumed at 2°C and 4°C. In the temperate species, 50%–85% of absorbed energy was used in metabolism, whereas in the Antarctic species, this was 45%–75% (Brodte et al. 2006). The Antarctic species also grew faster than the temperate species in this study at these low temperatures.

Energy budgets may be more difficult to obtain for cold-water species because of their slow growth rates (Peck 2016), extreme seasonality of the environment where some species cease feeding for several months in winter, and other aspects of physiology can be highly seasonal (Gruzov 1977, Whitaker 1982, Clarke 1988, Clarke et al. 1988, Clarke & Peck 1991, Pearse et al. 1991, Arntz et al. 1994, Barnes & Clarke 1995, Brockington et al. 2001, Jazdzewski et al. 2001, Fraser et al. 2002, 2004, Barnes & Peck 2005, Morley et al. 2007, 2016c, Obermüller et al. 2010, 2013). The very large interannual variation in biology, presumably caused by long-term variations in food supply quality and quantity, and evidenced by very large interannual variation in reproduction (e.g. Grange et al. 2004) adds further to the complexity. This means that all energy budgets in polar regions need to be evaluated regularly during the year and over several years to cover all of these aspects. However, the lack of good data for energy budgets for Antarctic marine species requires attention, and good studies in this will likely advance understanding of differences compared to lower latitude species and also trade-offs during periods of stress or environmental change.

Energy budget theory has developed over the last 20 years into the field of dynamical energy budgeting where the static 'snapshot' of most previous energy budget studies that calculated the budget for a set moment in time was extended to follow the changes of the energy fluxes through an organism over time and in its full formulation over the full life cycle of that organism (Kooijman 2000). This approach has recently been applied to the Antarctic starfish *Odontaster validus* (Aguera et al. 2015) and infaunal bivalve mollusc *Laternula elliptica* (Aguera et al. 2017), as described earlier. The dynamic energy budget approach has distinct advantages in that it is mechanistically based and allows analyses of seasonal and ontogenetic changes in energy use, can identify critical times when energy may be limiting, and can interpolate from stage to stage and time to time using a set of well-developed equations. Its main drawback is the requirement for large amounts of data on several aspects of the biology of an organism at the same time, and for several parts of the life cycle, and these need to be repeated at different times of the year to allow for seasonal aspects to be covered, all of which require non-trivial levels of effort and expertise. A major element of all energy budgets is the cost of maintenance of the organism and the costs of activities. These are usually measured via oxygen consumption.

Oxygen consumption, metabolic cold adaptation and metabolism

All of the major processes in an energy budget (covering all of the energy used by the organism) have an associated metabolic cost accrued when ATP is used in that process. In fully aerobic conditions, this can be measured via oxygen consumption (MO_2), and oxygen consumption is a measure of the immediate energy requirement under these conditions. Following the rationale laid out in Clarke (1987a), the energy budget then becomes:

$$C = \boxed{F} + \boxed{\begin{array}{c} P_s \\ + \\ R_s \end{array}} + \boxed{\begin{array}{c} P_r \\ + \\ R_r \end{array}} + \boxed{R_m} + \boxed{R_a} + \boxed{U} + \boxed{M}$$

where R_s represents the respiratory costs associated with somatic production, R_r is the respiratory costs of reproductive production, R_m is maintenance metabolic cost and R_a is respiratory costs associated with activity. From the previous equation, it is clear that measured oxygen uptake in an organism is a complex entity made up of several costs ($R_s + R_r + R_m + R_a$) from a range of sources (Clarke 1987a).

There have been many investigations of whole-animal oxygen consumption in Antarctic marine invertebrates and fish. These include: amphipods (e.g. Opalinski & Jazdzewski 1978, Rakusa-Suszczewski 1982, Chapelle et al. 1994, Chapelle & Peck 1995, Doyle et al. 2012, Gomes et al. 2013, 2014); isopods (e.g. Belman 1975, Luxmoore 1984, Robertson et al. 2001); the nemertean *Parborlasia corrugatus* (Clarke & Prothero-Thomas 1997, Davison & Franklin 2002, Obermüller et al. 2010); bivalves (Ralph & Maxwell 1977, Davenport 1988, Ahn & Shim 1998, Kowalke 1998, Pörtner et al. 1999b, Peck & Conway 2000, Brockington 2001a, Peck et al. 2002, Heilmayer & Brey 2003, Heilmayer et al. 2004, Pörtner et al. 2006, Morley et al. 2007, Cummings et al. 2011); gastropods (Ralph & Maxwell, 1977, Houlihan & Allan 1982, Peck 1989, Peck & Veal 2001, Fraser et al. 2002, Harper & Peck 2003, Obermüller et al. 2010, Morley et al. 2012a, Watson et al. 2013, Peck et al. 2015b, Suda et al. 2015); cephalopods (Daly & Peck 2000, Oellermann et al. 2012); bryozoans (Peck & Barnes 2004, Barnes & Peck 2005); brachiopods (Peck et al. 1986a,b,c, 1987a, 1997a,b, Peck 1989, 1996); ascidians (Kowalke 1998, Torre et al. 2012); sponges (Kowalke 1998, Gatti et al. 2002, Morley et al. 2016b); cnidarians (Torre et al. 2012, Henry & Torres 2013); echinoids (Belman & Giese 1974, Brockington & Peck 2001, Watson et al. 2013); asteroids (Belman & Giese 1974, Peck et al. 2008, Obermüller et al. 2010); holothurians (Fraser et al. 2004); and ophiuroids (Obermüller et al. 2010).

Studies of resting or routine MO_2 date back to the early 1960s (e.g. Wohlschlag 1963, 1964, Hemmingsen et al. 1969, Everson & Ralph 1970, Holeton 1970, Wells 1978, 1987, White et al. 2012). MO_2 in Antarctic fish has also been measured in relation to several other factors, including feeding (e.g. Johnston & Battram 1993, Boyce & Clarke 1997, Sandblom et al. 2012); effects of temperature (e.g. Johnston et al. 1991); acclimation of metabolism to altered environmental conditions, especially elevated temperature (e.g. Wilson et al. 2002, Robinson & Davison 2008, Strobel et al. 2012, Enzor et al. 2013, Enzor & Place 2014, Peck et al. 2014, Morley et al. 2016c); field levels of activity and seasonality (Campbell et al. 2008, Obermüller et al. 2010); antioxidant and reactive oxygen species production (Abele & Puntarulo 2004, Heise et al. 2004, Benedetti et al. 2008, Mueller et al. 2011, Enzor & Place 2014, Almroth et al. 2015); and the functioning of reversible oxygen binding pigments in oxygen delivery (e.g. Davino et al. 1994, di Prisco et al. 2002, Cheng & Detrich 2007, Verde et al. 2008, 2011, Cheng et al. 2009a,b, Giordano et al. 2010, 2012a, 2015).

MO_2 increases across latitudes towards the tropics, and as temperature increases, in fish (Clarke & Johnston 1999) and invertebrates (Peck & Conway 2000, Peck et al. 2006b, Peck 2016). Early in the last century, Krogh (1916) recognised that polar cold-blooded animals are active at low temperatures, but temperate ectotherms are not when cooled to similar temperatures. This led him to propose the hypothesis that polar species should have elevated metabolic rates to overcome the effects of low temperature. The first studies on metabolic rates in Antarctic ectotherms supported this idea, and this led to the concept of 'metabolic cold adaptation' (MCA) (Scholander et al. 1953, Wohlschlag 1964). Studies of several molecular and cellular processes have also showed compensation for low temperature in high latitude species, for example, elevated mitochondrial ATP synthesis capacity (Sommer & Pörtner 2004), increased enzyme activities (Crockett & Sidell 1990), greater mitochondrial volume density in fish muscles (Johnston et al. 1998, Guderley 2004, Lurman *et al.* 2010a,b), and microtubule assembly (Pucciarelli et al. 1997, 2013, Detrich et al. 2000).

Note that for some of these adaptations some taxa do not follow the trend, for instance adaptation to low temperature allows Antarctic clams of the genus *Laternula* to bury at the same rate as temperate congeners, but this is achieved through having larger burying muscles and not through increases in mitochondrial content of the muscle, as seen in fish swimming muscles, which in the Antarctic species are the same, or lower than temperate and tropical clams (Morley et al. 2009b,c). More recent whole-animal research on MO_2, primarily investigating within species trends, evaluating populations living at different latitudes and temperatures, provided further support for MCA (e.g. Hodkinson 2003, Schaefer & Walters 2010, Gaitán-Espitia & Nespolo 2014). These findings mainly show that populations living at higher latitudes and cooler temperatures have higher metabolic rates (and usually higher cellular process rates such as enzyme activity) than populations or congeners living at warmer temperatures when both are measured at the same temperature. When measured at their normal habitat temperature, the colder populations and species have lower metabolic rates than their warmer counterparts. The same outcome of lower physiological rates, mainly MO_2 at lower temperatures, was the finding of studies comparing large numbers of species across tropical to polar latitudes in fish (Clarke & Johnston 1999) and bivalve molluscs (Peck & Conway 2000, Peck 2016), both of which showed the change in MO_2 with temperature matched Arrhenius predictions, with Q_{10} values predominantly in the range two to three. These latter studies argued strongly against the MCA hypothesis, and within species, MCA studies have recently concluded that while there is some compensation of metabolism, evolutionary adaptation and phenotypic thermal plasticity appear insufficient to fully compensate for the thermodynamic effects of reduced temperature (e.g. White et al. 2012). Thus, it seems that while some compensation is evident in Antarctic marine species in certain cellular processes, and in some aspects of whole-animal metabolism, it is insufficient to move large-scale comparisons of oxygen consumption with temperate and tropical species outside Arrhenius expectations. It should be noted here, however, that there is good evidence for MCA in terrestrial insects (Addo-Bediako et al. 2002) and the difference between terrestrial and marine animals in this respect remains to be explained.

Beyond comparisons of routine, basal or standard metabolism, valuable comparisons of abilities to raise metabolic rates to do work in marine animals are also possible. In Antarctic marine species, research in this area has mainly been on the rise in metabolism after feeding.

Specific dynamic action of feeding (SDA) and metabolic scope

Metabolic rates of animals rise after consuming food. They stay high for a period and then return to pre-feeding levels. The postprandial rise and fall of metabolism is called the specific dynamic action of feeding (SDA) or the heat increment of feeding (HIF). This phenomenon has been known since the first half of the twentieth century when it was identified in domesticated animals (Brody 1945). The rise in metabolism is usually assessed via oxygen consumption, which in fully aerobic conditions is a proxy for proximate energy use and comprises the total costs of processing food (handling and digestion) and a variety of postabsorptive processes. These include the breakdown and synthesis of proteins, transport of absorbed materials, storage and growth (Peck 1998, Secor 2009, Khan et al. 2015). Studies manipulating diets with materials that are not absorbed, such as cellulose, indicated that only 5%–30% of the SDA is used in handling food and digestion (Tandler & Beamish 1979, Carefoot 1990), and some studies showed food handling accounted for less than 3% of the metabolic rise (Cho & Slinger 1979). Other studies further showed protein synthesis can form a large part of the SDA (e.g. Houlihan et al. 1995). Furthermore, studies injecting amino acids into animals showed that an SDA was produced of similar size to a meal with the same amino acid content (Brown & Cameron 1991, Peck 1998). This also suggests the major components of the SDA are post absorptive. The conclusion from all of these studies is that in most cold-blooded species the largest portion of the rise in metabolism after feeding is in postabsorptive processes, mainly in protein turnover.

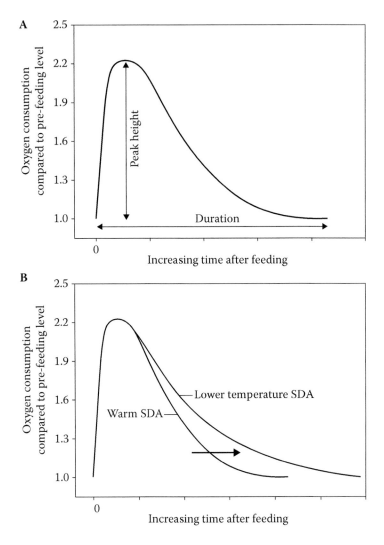

Figure 11 An idealised representation of the Specific Dynamic Action of feeding (SDA). (A) Oxygen consumption by an idealised animal following feeding. MO_2 rises to a peak, after which it declines back to pre-feeding levels. Peak height is measured as a factorial multiplier of pre-feeding MO_2. In the example given here, the peak is 2.2 times higher than the pre-feeding MO_2. When SDA studies are carried out, the usual statistics quoted are the factorial peak height and the duration, which is the time from the beginning of the rise above pre-feeding MO_2 to when oxygen consumption returns to that level). 0 shows the day when animals were fed. (B) Figure demonstrating that at lower temperatures factorial peak height is the same as in warmer SDAs, but the duration is extended. (Modified from Peck, L.S. 2016. *Trends in Ecology and Evolution* **31**, 13–26.)

There are two main aspects of the SDA that are usually reported. These are peak height and the duration of the rise in metabolism (Figure 11). Peak height is usually expressed as the factorial rise above the pre-feeding standard or routine metabolic rate where a doubling of metabolism after feeding produces an SDA peak height of x2 and a trebling produces one of x3. In most marine ectotherms, the SDA peak rise is in the range two to four (Peck 1998), but some species can have much higher values than this, with a value of 45 reported for Burmese pythons eating a Thomson's gazelle (Secor & Diamond 1995). This is similar to the maximum aerobic scopes for racehorses, which are the highest reported for any vertebrate (Birlenbach & Leith 1994). The duration of the SDA is the time from the initial rise in metabolism after feeding to the return to pre-feeding levels.

The area under the curve is called the SDA coefficient. It represents the total amount of energy used in digestive and postabsorptive processes associated with a feeding event.

There have been several investigations of SDA in a range of Antarctic marine species. These include fish (Johnston & Battram 1993, Boyce & Clarke 1997, Brodeur et al. 2002, Vanella et al. 2010, Sandblom et al. 2012), amphipods (Chapelle et al. 1994) and isopods (Robertson et al. 2001), rhynchonelliform brachiopods (Peck 1996), nemertean worms (Clarke & Prothero-Thomas 1997), limpets (Peck & Veal 2001) and starfish (Peck et al. 2008). These studies all demonstrated that the peak heights of the SDAs measured were similar to those of related or ecologically similar species from lower latitudes, but the SDA duration was markedly extended. Peck (2016) took this analysis further and compared SDA characteristics for marine invertebrates and fish across the globe from the tropics to the polar regions. He showed that peak heights did not vary with latitude or temperature (Figure 12A), but SDA duration increased as temperature decreased at higher latitudes (Figure 12B). He further demonstrated that the increase in SDA duration matched expected Arrhenius relationships for the effect of temperature on biological systems for temperate and tropical species, but the durations for extreme low temperatures, around or below 0°C, in the polar regions were markedly longer than they should be. They were significantly above predicted values for polar temperatures from the Arrhenius relationship for temperate and tropical species (Figure 12C).

This slowing of process rate for the SDA to levels well below that expected from Arrhenius temperature relationships for temperate and tropical species is very similar to that seen for growth (Figure 3) and embryonic development (Figure 8). It emphasises that marine species living at low polar temperatures around or below 0°C experience extra difficulties with physiological processes that require significant protein synthesis, and it highlights that problems with protein synthesis or folding are the likely cause of these dramatic slowing of rates (Peck 2016). The impact of adaptation to low temperature on performance of the above biological functions is clear and dramatically reduces their rates compared to warmer-water species. A major function not yet considered is whole-animal activity. This is important in a number of ecological contexts, including foraging, predator–prey interactions, reproductive behaviours and competition.

Activity

The rate at which activity is carried out is rarely measured in polar marine invertebrates and fish. The studies that have been conducted show activity is usually carried out much more slowly than in temperate and tropical species, and in general, Antarctic species are 2–10 times slower than temperate species living in 10–15°C warmer conditions (Figure 13). These have been previously reviewed in Peck (2002a), Pörtner (2002a) and Peck et al. (2006b). Some examples are: clap frequency in the Antarctic scallop *Adamussium colbecki* was measured to be performed at approximately 50% the rate of temperate scallops (Bailey et al. 2005); Antarctic bivalve molluscs and anemones require between 5 and 10 times longer than related or ecologically similar temperate species when burying (Ansell & Peck 2000, Peck et al. 2004a); average pumping rates, when feeding, were 15–50 times slower in Antarctic sponges, ascidians, and the bivalve mollusc *Laternula elliptica* than related or ecologically similar temperate species (Kowalke 1998); the Antarctic predatory snail *Trophonella longstaffi* takes 28 days to drill through the shell of its prey and complete a meal, compared to 10–12 days for temperate predatory snails (Harper & Peck 2003); the limpet *Nacella concinna* routinely walks at speeds of 0.13–0.25 mm s^{-1}, and 0.25–1.0 mm s^{-1} during escape responses from predators, rates that are on average just over half as fast as temperate limpets (Peck et al. 1993, 2006b, Peck, 2002a, Markowska & Kidawa 2007); the spatangoid urchin *Abatus ingens* moves through sediment at 0.3–1.95 cm h^{-1} on average, with maximum speeds of 1.1–3.3 cm h^{-1} (Thomson & Riddle 2005), which compares with the temperate *Echinocardium cordatum* that moves at 2.0–8.0 cm h^{-1} and lives at temperatures between 6°C and 15°C (Buchanan 1966); and locomotion speeds in the Antarctic starfish *Odontaster validus* were measured at 0.06–0.55 mm s^{-1} (McClintock et al. 2008), which

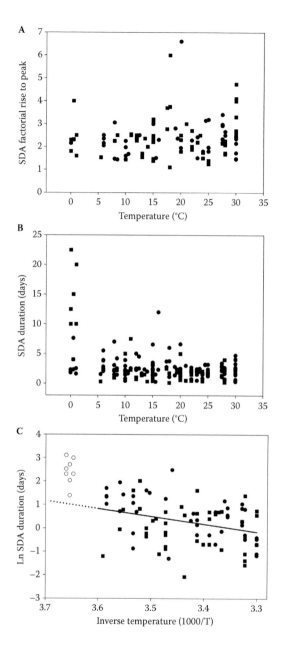

Figure 12 Specific Dynamic Action of feeding (SDA) data for marine invertebrates and fish at normal habitat temperatures from the tropics to the poles. (A) SDA peak height (factorial peak increase in MO_2 over pre-feeding level). There is no significant relationship with temperature (FP $= 0.232 + 0.009$ T, $r^2 = 0.00$, $F_{1,110} = 0.85$, P $= 0.358$, VIF $= 1.00$). (B) SDA duration (days: the time taken from the initial rise in MO_2 after feeding to the return to pre-feeding levels). (C) An Arrhenius plot of SDA duration. The solid line represents the relationship for temperate and tropical species (Ln SDA (days) $= -11.13 + 3.32(1000/K)$; $r^2 = 0.107$, F $= 10.4$, P < 0.0001, 88 d.f.); the dotted line shows the extension of the relationship for tropical and temperate species to polar temperatures. In all plots, each data point represents a single species, and where the literature contains more than one record, the value plotted is the mean for duration and temperature. Closed symbols are for temperate and tropical species living at mean temperatures above 5°C; open symbols are for polar species living permanently near or below 0°C; • = marine invertebrates; ■ = marine fish. (Figure from Peck, L.S. 2016. *Trends in Ecology and Evolution* **31**, 13–26.)

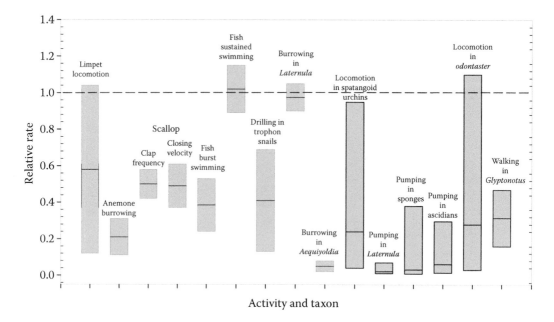

Figure 13 Rates of accomplishing various activities for a range of Antarctic marine invertebrates compared with related or ecologically similar temperate species. The hatched line indicates representative rates for temperate species and is set at a value of 1. Data sources for comparisons of limpet locomotion, anemone burrowing, scallop and fish swimming, and burrowing in bivalve molluscs are all taken from Peck (2002a) and Peck et al. (2004a, 2006b). Data for drilling in the predatory snail *Trophonella longstaffi* are from Harper and Peck (2003); for pumping of water in association with feeding currents from Kowalke (1998); for locomotion in spatangoid urchins from Thomson and Riddle (2005); for locomotion in the starfish *Odontaster validus* from McClintock et al. (2008); and walking in the isopod *Glyptonotus antarcticus* from Young et al. (2006a). Boxes show the full range of values, with the midline indicating the mean. (Figure based on Peck, L.S. et al. 2004a. *Polar Biology* **27**, 357–367 and Peck, L.S. et al. 2006b. *Biological Reviews* **81**, 75–109.)

compares with data for temperate species that usually fall in the range 0.5–2 mm s^{-1}, with some species over 20 mm s^{-1} (echinoblog.blogspot.co.uk, accessed 6 Feb 2016).

To date, only two whole-animal activities have been identified in Antarctic marine invertebrates and fish that appear fully compensated for temperature. One is in Notothenioid fish where sustained swimming is maintained at the same speed (as number of body lengths per second) as temperate species. This temperature compensation is supported by an increase in the numbers of mitochondria in the red muscles. The polar species have roughly twice as many mitochondria per gram of muscle tissue as temperate species, but this is only in red, not white muscle (Johnston et al. 1998). The second activity fully compensated for low temperature is burrowing in the infaunal Antarctic clam *Laternula elliptica*. In this species, the adaptation that allows the activity to be performed at similar rates as temperate and tropical congeneric species has been identified. It is not via an increase in mitochondria to improve energy generation but by an increase in size of the muscle that is used for burying, the foot, which is around two to three times larger than in warmer-water congeners (Morley et al. 2007, 2009b,c).

Although the data are limited, there are some patterns worth noting in Figure 13. Excluding the two activities where temperature compensation has been identified, and a mechanism explaining the compensation described (sustained swimming in fish and burrowing in *Laternula elliptica*), six of the seven remaining activities involving whole-animal movement (locomotion, burying, walking and swimming) are, on average, two to five times slower than rates for temperate species living at 10–15°C. Drilling by the predatory snail *Trophonella longstaffi* is also within this range. Slowing of this magnitude matches expected temperature effects on biological systems, which should be in the range of one to four times for a 10°C change in temperature (Peck 2016). The only activities outside this range are those for water pumping, which are all from one study (Kowalke 1998), and the comparison for burying in the small infaunal bivalve *Yoldia* (now *Aequiyoldia*) *eightsii* (Peck et al. 2004a). Both of these are much slower than expected, being on average more than 10 times slower than temperate species. These findings require further study to verify they are slowed by this amount. If they are slowed more than other activity comparisons, then it would be difficult to explain why burying in one bivalve species (*Laternula elliptica*) should be fully temperature-compensated while another species living in the same habitat is slowed more than expected, and other activities based around muscular contraction (walking, swimming) are slowed by the expected amount. The extra slowing for pumping in bivalve molluscs, ascidians and sponges (Kowalke 1998), is an activity based around ciliary pumping mechanisms, and if verified, these findings would suggest that there is an extra impact of low temperature on such mechanisms. Viscosity of seawater increases as temperatures fall to values around 0°C (Peck 2016), and this has been cited in the past as a major factor affecting swimming in organisms using cilia for propulsion, for example, ciliates and protists (Beveridge et al. 2010) and also in marine invertebrate larvae (Podolsky & Emlet 1993, Chan 2012). A similar effect on ciliary feeding pumps would explain the markedly lower pumping rates in Antarctic sponges, ascidians and bivalve molluscs observed by Kowalke (1998) and could be a low temperature constraint on this feeding mechanism.

Temperature affects the rate at which cellular biochemical and biological processes proceed (Hochachka & Somero 2002, Peck 2016). In whole animals, physiologies generally speed up to an optimum temperature and then decline above this before failure. There is also a hierarchical arrangement where complexity of organisation reduces the optimum and maximum temperature for function with cellular functions continuing to higher temperatures (Pörtner 2002a, Pörtner et al. 2007, 2012). There have been very few studies of activity of Antarctic marine invertebrates and fish when temperatures are reduced to values below 0°C. One such was an investigation of walking and righting in the isopod *Glyptonotus antarcticus* and the amphipod *Paraceradocus gibber* (Young et al. 2006a). In both species, walking speed increased consistently between −2°C and +5°C, when experiments were halted to avoid reaching the thermal limit. In contrast, there was no significant relationship between the time required to right when turned over in *Glyptonotus antarcticus* at temperatures between −2°C and +5°C. There have been many studies where data have been collected in positive temperatures. These have produced variable results, with some showing a decline in capability with temperature, but others showing an increase. Those that show an increase include the study by Young et al. (2006a) above on *G. antarcticus* and *Paraceradocus gibber*. The starfish *Odontaster validus* has also been reported to demonstrate an increase in activity rate with temperature. The time required to right when turned over decreases up to 7.5°C, but turning rate then slows markedly before the animal's upper thermal limit at around 10°C (Peck et al. 2008). However, Kidawa et al. (2010) found that increasing temperature from 0°C to 5°C reduced the number of individuals capable of righting and also slowed responses to food, food odour and rate of locomotion. Of the species showing declining activity rates with temperature, swimming in the scallop *Adamussium colbecki* and righting in the limpet *Nacella concinna* are very temperature sensitive, with a complete loss of capability at temperatures between 2°C and 5°C in both species (Peck et al. 2004b). Locomotion in *N. concinna*, in populations at its northern limit in South Georgia is, however, fastest at 2°C but continues until temperatures reach 14°C (Davenport 1997). Reburying when removed from sediment declines markedly between 0°C and 5°C in the infaunal clam *Laternula elliptica*, with large animals losing the ability to bury at lower temperatures than small ones, and all sizes incapable of reburying above 5°C (Peck et al. 2004b). The isopod *Paraserolis* (previously *Serolis*) *polita* exhibits declining rates of righting with increasing temperature between 0°C and 5°C (Janecki et al. 2010). Reduced salinity also slowed the rate of righting in *P. polita*, and the salinity effect was stronger at higher temperatures.

Investigations of the effect of temperature on the performance of activity and other physiological processes are important, because although there has been significant research on absolute temperature limits in several species, the relationship between absolute limits and limits for performing functions is poorly understood, especially in terms of diversity and differences between species. A few studies do exist, and research has demonstrated that upper temperature limits to successful fertilisation (>50%) of eggs in the starfish *Odontaster validus* and the bivalve *Laternula elliptica* were around 5°C and 6°C, respectively (Grange et al. 2011b). Normal early embryo development declined markedly to values below 50% at 2–3°C in the urchin and 6–8°C in the starfish, and for surviving larvae, development rate was not affected by elevated temperature up to 5.7°C (the maximum temperature tested) in the urchin and between 7°C and 9°C in the starfish. Similar, though less extensive data were obtained for the same species by Stanwell-Smith and Peck (1998), and these values are lower than upper temperature limits for acute warming experiments (1°C day^{-1}) for adults of these species (Peck et al. 2009a) but are similar to limits at slow rates of warming (1°C month^{-1}) of 3–4°C for *Sterechinus neumayeri* and 6°C for *Odontaster validus* (Peck et al. 2009a), and adult upper temperature limits for acclimation of similar or slightly lower values (Peck et al. 2014). At least in some Antarctic species, temperature limits for rates of development thus seem to be similar to adult limits at slow, ecologically more relevant rates of warming. Data for the limpet *Nacella concinna* show that duration tenacity, the time an individual can stay attached to a surface against an applied force, declines by 50% between 0°C and 5°C (Morley et al. 2012a), but radula rasping while eating continued to temperatures between 10°C and 12°C (Morley et al. 2014). Both of these experiments were carried out at a rapid, acute rate of warming of 0.2°C h^{-1}, and at the slower but still rapid rate of warming of 1°C day^{-1}, adult thermal maxima were between 11°C and 12°C. In an acute warming experiment involving the starfish *Odontaster validus*, the rate at which an individual righted itself when turned over increased from 0°C to 8°C but declined rapidly when warmed at around 10°C, and the proportion of animals feeding when offered food was 100% between 0°C and 6°C but declined to below 50% between 8°C and 10°C (Peck et al. 2008). In the same experiments, adult thermal limits (50% survival) were between 12°C and 15°C. In the infaunal clam *Laternula elliptica*, 50% of individuals lose the capacity to rebury when removed from sediment at 3°C, and 50% of individuals of the limpet *Nacella concinna* failed to right when turned over at the same temperature when warmed at 0.1°C h^{-1}. In the scallop *Adamussium colbecki*, there was a complete loss of capacity to swim, a high-energy activity, when temperatures were raised to between 1°C and 2°C (Peck et al. 2004b). At a slower rate of warming, which should lower thermal limits (Peck et al. 2009a, 2014), *Nacella concinna* had an upper lethal temperature (50% survival) of 11–12°C and *Laternula elliptica* of 14–15°C (Peck et al. 2009a).

Differences between whole-animal upper temperature limits and essential biological functions vary markedly amongst functions and between species. The differences between species in the effect of temperature on functions will, however, dictate many of the changes in community and ecosystem structure that will occur in the coming decades because they will be one of the main influences on critical ecological factors such as competition and predation. Small differences in relationships between performance and temperature will bear heavily on competitive and predator–prey outcomes in a warming world. Despite this, the relationships between loss of critical function and upper survival limit on warming remains very poorly understood in all but a very small number of species in Antarctica, and globally.

Temperature, power and crushing predators

As seen previously, temperature has a marked effect on the performance of several physiological attributes of marine ectotherms, including oxygen consumption and activity that appear to follow Arrhenius predictions in Antarctic species and development, growth and postprandial processes associated with feeding that are slowed beyond expected levels. The performance of animal muscles is also affected by temperature, which affects many aspects of behaviour. As noted previously, some of these have specific adaptations to overcome temperature limitations such as burying in the clam *Laternula elliptica* and sustained swimming in fish, whereas most, including burst swimming in fish, do not (Figure 13). It is generally accepted that the force generated by muscles does not vary with temperature because the force generated is dependent on the number of cross bridges involved in the contraction of the muscle. Observations, however, suggest that in ectotherms there may be a small increase in force generated by muscles at higher temperatures when measured at normal habitat temperatures, and the increase may be as much as 10% for each 10°C rise in temperature (Shadwick & Lauder 2006). Rate characteristics such as the time involved in force development, the cycle time of cross bridges in muscles and relaxation time vary to a much greater extent with temperature (Johnson & Johnston 1991, Shadwick & Lauder 2006). The outcome of changes in these characteristics is that power generation varies with temperature (Wakeling & Johnston 1998). In fish muscles, power generation increases by around 2.5 times for each 10°C rise in temperature from polar to tropical species (Figure 14), which is in line with expected Arrhenius temperature effects.

Muscle power increases with muscle diameter and, hence, muscle mass. To fully compensate for a reduction in power output as temperature decreases along the lines shown in Figure 14, a muscle would have to increase its mass by around 2.5 times for every 10°C drop in temperature, or to produce the same power, an Antarctic ectotherm would need a muscle over 15 times larger than a tropical species at 30°C.

Force may often be thought of as the important criterion limiting the ability of durophagous predators to complete a successful attack on prey items. Power is the rate of doing work, and work is force times distance moved. Force exerted by muscle does not vary with temperature, but power

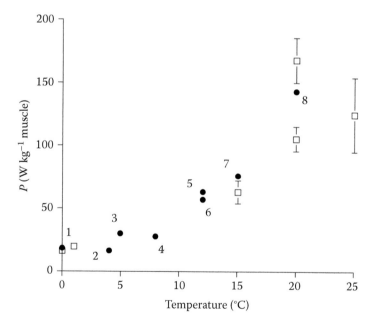

Figure 14 Fish muscle mass-specific power output with environmental ambient temperature. Open symbols denote data for *Notothenia coriiceps* and *Notothenia rossii* from Antarctica, *Myoxocephalus scorpius* from the North Sea, *Scorpaena notata* and *Serranus cabrilla* from the Mediterranean and *Paracirrhites forsteri* from the Indo-West-Pacific. Numbers with filled circles denote data for: 1 *Notothenia coriiceps* from Franklin and Johnston (1997); 2 and 4 *Gadus morhua* from Moon et al. (1991) and Anderson and Johnston (1990); 3, 6 and 7 *Myoxocephalus scorpius* from Altringham and Johnston (1990), James and Johnston (1998) and Wakeling and Johnston (1998); 5 *Pollachius virens* from Altringham et al. (1993); and 8 *Scorpaena notata*. (Figure from Wakeling, J.M. & Johnston, I.A. 1998. *Journal of Experimental Biology* **201**, 1505–1526.)

does (Wakeling & Johnston 1998). During the handling phase of an attack, crushing predators such as reptant decapods will often apply force several times to identify the best position for an attack before applying full force (Hughes & Seed 1995). Two main ways that durophagous crustaceans attack their prey are either by direct crushing or by chipping or cutting the shell edge and inserting chelae (Elner & Hughes 1978). The former is generally used for small prey and the latter for larger prey (Ameyaw-Akumfi & Hughes 1987). These predators have also been observed applying pressure multiple times to a prey bivalve, up to 200 times in a method termed a pressure pulse attack that is thought to weaken shells by producing microfractures (Boulding & Labarbera 1986). Larger prey have significantly higher rewards per unit effort for predators and should be taken preferentially (MacArthur & Pianka 1966). However, durophagous crustaceans often take less preferred prey on energy maximisation criteria because of other factors such as potential damage to claw crushing surfaces (Smallenge & Van der Meer 2003). Large size has also been identified as a refuge from predation because the prey organism grows beyond the size that the predator can successfully attack (e.g. Harper et al. 2009).

Reductions in muscle performance at low temperature, such as slower cross-bridge cycling time and relaxation, that result in reduced power output, are likely important factors in the absence of durophagous predators such as brachyuran crabs in Antarctica (Aronson et al. 2007). The 'picking' method of feeding by the lithodid crabs that are present in Antarctica requires less force and power to make successful attacks. The power versus temperature relationship shown in Figure 14 would suggest that power limitation is likely to preclude crushing predators from Antarctic waters for many decades, if not centuries to come. The forces and power that predators can exert are two of the factors affecting how robust skeletons of their prey need to be, and this has clear implications for life at low temperature in the Southern Ocean.

Shell thickness and energy budgets

Antarctic marine ectotherms have often been described as having thin or small shells (Arnaud 1977, Vermeij 1978, Arntz et al. 1994, Aronson et al. 2007, McClintock et al. 2009). Early studies noted that bivalve molluscs tend to be smaller in the polar regions (Nicol 1964, 1966) and have less ornate shells (Nicol 1967), and that gastropod molluscs also have smaller shells at high latitudes (Graus 1974). These findings have been more recently identified as part of a global trend of decreasing skeleton size from tropical to polar latitudes in groups with external shells (rhynchonelliform brachiopods, bivalve and gastropod molluscs) and also in echinoid urchins (Watson et al. 2012). Decreasing skeleton size with latitude, therefore, appears to be a general trend across a wide range of taxa and not limited to molluscs.

Two main hypotheses have been proposed to explain the decrease in skeleton size towards the poles. The first is that on physical environmental and energetic considerations the cost of shell production increases towards the poles. Most marine ectotherm skeletons are made from calcium carbonate ($CaCO_3$). As temperature decreases in the sea, the solubility of calcium carbonate increases (Revelle & Fairbridge 1957). As a consequence, the saturation state of $CaCO_3$ (Ω_{CaCO3}) in seawater decreases, where saturation state is defined as: the product of the concentrations of dissolved calcium and carbonate ions in seawater divided by their product at equilibrium. At lower temperatures, the ions used to make skeletons are more soluble, and more thermodynamic work is required to remove them from seawater. This has been proposed by several authors to be the possible reason why skeletons are smaller at high latitude because it increases the energy needed to complete shell production (Graus 1974, Vermeij 1978, Clarke 1983, Clarke 1990, Clarke 1993c, Vermeij 1993). This argument has been combined with the poleward decrease in metabolic rates in marine ectotherms described earlier, and elsewhere (e.g. Peck 2016), to show not only should the costs of building skeletons increase in real terms at lower temperatures, but this effect should be greater in relation to metabolic costs (Clarke & Johnston 1999, Peck & Conway 2000, Watson et al. 2017).

The second major hypothesis to explain the observed latitudinal variation in skeleton size is that it is caused by changes in predation pressure. Direct evidence for latitudinal gradients in predation pressure is limited (e.g. Vermeij 1978), although recently Harper and Peck (2016) completed the first global-scale analysis of predation pressure by evaluating repairs from shell damage caused by durophagous predators from tropical to polar latitudes. They demonstrated that the incidence of repaired shell damage, and hence durophagous predation pressure, declines towards the poles. There are some examples covering smaller geographic scales that appear to show levels of predation pressure are higher at lower latitudes (Paine 1966, MacArthur 1972). Furthermore, durophagous predators such as brachyuran crustaceans and fish are missing from, or are rare in, shallow nearshore habitats (<100 m depth) in Antarctica (Aronson & Blake 2001, Clarke et al. 2004a, Aronson et al. 2007), being restricted to areas with seawater temperatures above 0°C (Griffiths et al. 2013). One reason put forward for the absence of brachyurans in Antarctica, in addition to muscle power constraints cited earlier, is that reptant decapod crustaceans have a high haemolymph magnesium ion concentration compared to other crustaceans. Magnesium ions act as a narcotic for many marine invertebrates, and if, as has been shown for some species, extracellular magnesium increases with decreasing temperature (Morritt & Spicer 1993), the narcotising effect may remove them from sites permanently at 0°C and below (Frederich et al. 2000, 2001), making them unable to inhabit Antarctic shallow coastal waters.

Recently Watson et al. (2017) compared the energetic cost of making skeleton with those of respiration and growth, which are two of the major components of the energy budget. The costs of making skeletons in bivalve and gastropod molluscs were small as proportions of the energy budget, being 0.4%–7.4% of the total calculated budget in gastropod molluscs and 0.2%–3.4% in bivalve molluscs. Costs were greater at higher latitudes than in warmer sites, and predicted ocean acidification effects raised cost estimates to over 10% of the total energy budget by the end of the century in polar gastropods. The conclusion drawn by Watson et al. (2017) was that both cost and durophagous predation pressure might be major drivers of skeleton size in bivalve and gastropod molluscs. Data are insufficient to separate out the effects of predation and cost on skeleton size, but similar trends of reduced skeleton size towards the poles exist in other taxa. For example, in echinoid urchins the impact of durophagous predators is very different across latitudes from the effect of crushing predators on bivalve and gastropod molluscs. Overall, more work is needed to identify conclusively the factors controlling skeleton size variation with latitude and the reason for small skeletons in Antarctica. It is still not possible to exclude calcification costs or predation as the main factor in any taxon.

Seasonality

Seasonality is often overlooked in investigations of adaptation and effect of environment on organism biology. Seasonality is caused by variations in the incident energy arriving at the Earth's surface through the year, which is approximated by day length, but also varies in some regions with seasonal weather patterns and cloud cover. There is little variation in this factor in tropical latitudes but very great variation in the polar regions, where periods of 24 hours of sunlight in summer are counterbalanced with periods of 24 hours of dark in winter. This pattern is caused by the tilt of the Earth.

Variations in climate and small changes in seasonality over long time periods are caused by three main factors. These are: wobbles in the Earth's rotational axis, called axial precession, which produces a 23,000-year cycle in incident light energy arriving at the surface; changes in the tilt angle of the Earth, called obliquity, which gives a 41,000-year cycle in incident radiation; and changes in the shape of the Earth's orbit, called eccentricity, which results in 100,000-year and 400,000-year cycles in incident radiation (Zachos et al. 2001).

Over geological time, some aspects of current seasonality are as strong as or possibly stronger than previous periods (Clarke & Crame 2010). This is especially so in the polar regions because the poles are currently isolated. In the Arctic, a small ocean is nearly surrounded by land, and in the Antarctic, a large continent is isolated by a circulating ocean that has a strong temperature discontinuity with other oceans at the polar frontal zone (PFZ). Up until approximately 34 mya, the poles were less isolated because large amounts of energy were moved to the high latitudes by ocean currents, and the polar regions were significantly warmer than today (Florindo & Siegert 2009). This resulted in less ice at the poles, and coastal ice and sea ice are large drivers of the current seasonality in the polar regions, making conditions more extreme in these respects now than in most previous geological periods (De Conto et al. 2007). Most previous geological periods saw the distribution of continents in one or two large masses, including Pangea and Gondwana, where one or both poles had little terrestrial continental cover, even when there were polar land masses that were continuous with lower latitude continents.

The minimum and maximum extent of sea ice around Antarctica vary markedly between years. In the last 30–40 years, the satellite era, Antarctic sea ice in a typical year has varied from a minimum of around 3×10^6 km^2 to a maximum around 18×10^6 km^2 (http://earthobservatory.nasa.gov/Features/WorldOfChange/sea_ice_south.php). The largest maximum extent was 20.1×10^6 km^2, which was reached in September 2014. The growth and recession of around 15×10^6 km^2 of sea ice annually has a dramatic effect on the seasonality of the ocean beneath. Sea ice reduces the amount of light penetrating into the water column and reduces mixing due to wind.

This extreme variation in some aspects of the physical environment contrasts with the temperature stability that characterises the Southern Ocean. Sea temperatures are possibly the most stable of any ocean on Earth. At high Antarctic sites (e.g. McMurdo Sound), mean annual temperature variation is less than 0.5°C, and the most variable sites recorded to date (e.g. Signy Island, South Orkney Islands) have annual ranges around 4.5°C (Barnes et al. 2006a).

Seasonality is a major structuring characteristic for ecology, adaptation biology and biodiversity in Antarctica (Clarke 1988). One of the most obvious impacts on organisms of the intense seasonality in light and ice cover in Antarctica is the effect on primary productivity, and this strong signal of seasonality propagates through the food web. Phytoplankton blooms tend to be more intense in Antarctica in nearshore environments, where chlorophyll (Chl) standing stock levels average 10–15 mg m^{-3} for up to three months, and peak values can exceed 30 mg m^{-3} (Clarke & Leakey 1996, Clarke et al. 2008). In offshore areas, Chl levels in blooms typically reach values around an order of magnitude lower than this, but in some areas, such as the ice edge and in polynyas, productivity levels are often twice as high, or more than in open ocean blooms (Arrigo et al. 2015). In offshore blooms, productivity is limited by nutrient availability, often iron (Boyd et al. 2012), whereas in coastal polynyas, melting ice shelves increase iron availability, enhancing productivity. In a recent study of 46 polynyas, this factor explained more than twice as much variation in productivity as any other factor (Arrigo et al. 2015).

The main part of nearshore phytoplankton blooms is often dominated by large diatoms and colonial phytobionts of the microphytoplankton (Clarke et al. 2008), which peak between early December and mid-March. The main part of the nanophytoplankton bloom (5–20 μm diameter) follows a similar time course, but there is detectable productivity between September and June, and their Chl concentration during the summer averages around only 0.5–1.0 mg m^{-3}, more than an order of magnitude less than that in the larger diatom fraction. The small nanophytoplankton bloom (2–5 μm diameter) also lasts between September and June but achieves peak Chl levels of only 0.2–0.4 mg m^{-3} (Clarke et al. 2008).

The intense seasonality of ice and phytoplankton above 5-μm diameter, led Clarke (1988) to suggest there should be periods of low activity in terms of growth and other physiological attributes in winter, and biological activity in at least primary consumers should be linked to the short summer period of main phytoplankton biomass. This idea was part of a larger hypothesis on the impact of seasonality on organism biology, namely the seasonality hypothesis, which posits that seasonal signals should be stronger in primary consumers than predators or scavengers less directly dependent on algal productivity (Clarke 1988). It also argues that as seasonality in light regime becomes more extreme towards the poles then seasonal limitation of organism biology should be stronger at higher latitudes, which has been observed to be the case (Clarke & Peck 1991). There are also very strong seasonal signals in the Antarctic in marine benthic community development through a combination of intense seasonality of reproductive cycles and recruitment/establishment on surfaces (e.g. Bowden 2005, Bowden et al. 2006, 2009).

The combination of the idea of seasonal limitation of biological capacity with observations that superficially appear in conflict with the hypothesis, of high biomass and biodiversity (Gutt et al. 1992, Clarke & Johnston 2003, Gutt et al. 2013a), has led to the suggestion that there is an Antarctic paradox. Support for a paradox has also cited work that shows: some Antarctic taxa grow to very large size; some groups reproduce and grow at similar rates to temperate taxa (e.g. Teixidó et al. 2004); that there are sediment 'food banks' of large amounts of primary productivity that settle on the seabed, not consumed by pelagic organisms that are available for months over spring, summer and autumn on some parts of the continental shelf (Mincks et al. 2005); tidal and ice associated water movements that resuspend sediment material in winter making it available for benthic suspension-feeders (Peck et al. 2005, Smith et al. 2006); some Antarctic benthic suspension-feeders consume small particles whereas their temperate relatives consume zooplankton (Orejas et al. 2003); some suspension-feeders feed throughout the year (Barnes & Clarke 1994, 1995); and there have been observations of unusual feeding behaviours such as in the upright nephtheid soft coral *Gersemia antarctica*, that supplements suspension-feeding by bending its whole body to deposit feed on benthic organisms in surface sediments (Slattery et al. 1997).

Seasonality of feeding

Most of the observations used to support the idea of a seasonality paradox do not conflict with the idea that seasonality constrains the biology of many of the organisms living there. In fact, some of these observations could be used to support seasonal constraints. For instance, the finding that some suspension-feeding groups consume small particles instead of zooplankton is support for the idea that seasonal limitations of availability of suitable food sources have forced these taxa to move to alternative food supplies. Similarly, the suspension-feeders that feed all year round (Barnes & Clarke 1994, 1995) feed on the smaller size fractions, again possibly because of seasonal restrictions in the availability of larger size fractions that would be consumed in less seasonal environments. Furthermore, the dramatic sedimentation events that result in high levels of organic material in sediments could be evidence that the intensity of seasonality is so strong that the normal processes where much of the organic material produced in the water column is broken down before reaching the seabed are overwhelmed. Reproduction and growth reported at similar rates to temperate species are often small number comparisons, and where large numbers are considered, the fastest Antarctic rates may be similar to temperate species elsewhere, but they are significantly slower than the fastest temperate and tropical species, although care is needed when making comparisons of rates (Peck 2016). Most of the arguments about resuspension of food in winter due to currents do not apply to the very large number of organisms living on most rock surfaces, and this effect is primarily one of sediment dwelling species.

Very strong effects of seasonality on the biology of Antarctic marine invertebrates and fish are evident from a very large number of studies, some of which have been reviewed in Clarke (1988), Arntz et al. (1994) and Peck et al. (2006b). In one of the earliest studies of seasonality, Gruzov (1977) showed that while the biomass of detritivores and omnivores did not noticeably vary from summer to winter, large changes in numbers of species consuming plankton occurred (e.g. some hydroids), and some species entered a state of diapause in winter (e.g. the holothurian *Oswaldella antarctica*).

Part of the seasonality limitation hypothesis is that effects would be expected to be stronger in taxa directly dependent on primary productivity (herbivores) as opposed to detritivores, omnivores and carnivores. As noted earlier, a few benthic marine species have been demonstrated to feed throughout the year (Barnes & Clarke 1994, 1995, Obermüller et al. 2010). However, in the majority of species, feeding either ceases in winter or is markedly curtailed. This has been noted for the sea urchin *Sterechinus neumayeri* (Brockington 2001a, Brockington et al. 2001), for the sea cucumber *Heterocucumis steineni* (Fraser et al. 2004), and for the infaunal bivalve mollusc *Laternula elliptica* (Ahn et al. 2001, Morley et al. 2007). In an extensive year-round study of suspension-feeding taxa at Signy Island, Barnes and Clarke (1995) found that of 10 bryozoan species studied, one fed throughout the year, and the other nine ceased feeding for periods between one and five months in winter. They also showed that a hydroid and a suspension-feeding polychaete ceased feeding for one month each year, while the holothurian *Cucumaria georgiana* stopped for between three and five months. Fraser et al. (2002) showed that the limpet *Nacella concinna* feeds all year round, but there was still a very large decrease in feeding (as estimated from faecal egestion of freshly caught individuals) to a value only around 10%–15% that of summer feeding levels. All of the species mentioned previously are either directly dependent on primary productivity or are grazers or omnivores that consume large amounts of phytodetritus. The seasonal signal in feeding is very strong, such that over 95% of species that directly consume phytoplankton, benthic microalgae or macroalgae cease feeding in winter, and those that do continue feeding in winter consume only a small fraction of the amounts eaten in summer.

There have been few investigations of year-round feeding in higher trophic level groups in Antarctica. Obermüller et al. (2010), however, studied seasonal variation in feeding and metabolic characteristics in three carnivores and two omnivores near Adelaide Island, Antarctic Peninsula. Feeding rate varied significantly with season in the sponge-eating nudibranch *Doris kerguelenensis* and the predator/scavenger amphipod *Paraceradocus miersi*. There was no seasonal signal in feeding in the predatory fish *Harpagifer antarcticus* and nemertean *Parborlasia corrugatus*, and the scavenging brittle star *Ophionotus victoriae*. Other studies on the predatory fish *Harpagifer antarcticus* and *Notothenia coriiceps* found that acclimating them to winter light regimes resulted in lower food intake even when excess food was available (Targett 1990, Johnston & Battram 1993, Coggan 1996).

Seasonality of activity, metabolism and growth

The strength of seasonal signals in other aspects of the biology than feeding varies markedly. The Antarctic fish *Notothenia coriiceps* has been identified as entering a low activity state in winter similar to hibernation in some mammals (Campbell et al. 2008). Gruzov (1977) noted a marked decline in activity, growth and metabolism of several groups of marine invertebrates near Myrny Station, on the Zukov Islands in the Davis Sea, which he described as similar to diapause. Species noted to exhibit very large seasonal activity changes included the hydroid *Oswaldella antarctica*, the holothurian *Staurocucumis turqueti* (previously *Cucumaria spatha*) and the alcyonarian *Eunephthya* sp., although Gruzov (1977) did note that this phenomenon is widespread beyond the previous examples, especially in holothurians and sponges.

Many studies have shown strong seasonality of growth in Antarctic fish (e.g. Everson 1970b, Kawaguchi et al. 1989, Casaux et al. 1990, Targett 1990, Coggan 1996, 1997, La Mesa & Vacchi 2001) and invertebrates (e.g. Arntz et al. 1994, Brethes et al. 1994, Barnes 1995, Barnes & Clarke 1998, Peck et al. 2000, Ahn et al. 2003, Heilmayer et al. 2005, Bowden et al. 2006, Peck & Brockington 2013). One unusual finding was that the brachiopod *Lioythyrella uva* grew five times faster in winter than summer in shell growth but increased mass in summer, not winter, when reserves were laid down as protein in the outer mantle (Peck et al. 1997b). This was interpreted as minimising overall annual energy costs by separating growth from feeding periods. Some studies have shown only small differences in growth rates between summer and winter in Antarctic marine invertebrates, such as in juveniles of the infaunal deposit feeding bivalve mollusc *Yoldia* (now *Aequiyoldia*) *eightsii* at Signy Island, where seasonal differences in growth were not significant in specimens less than 10 mm in length (Peck et al. 2000). The explanation for this was year-round availability of food. This argument, however, should also apply to many scavengers and carnivores, for example, fish that do show strong seasonality in growth.

Growth can be difficult to assess in many species, especially soft-bodied animals such as sea anemones or jellyfish. One approach is to quantify changes in the total amount of protein in the animal with time. Many authors have argued that in the soft tissues of most animal species growth occurs mainly through the synthesis and retention of proteins, and that process exceeds degradation of proteins that are either malformed or have come to the end of their useful life (e.g. Fraser & Rogers 2007). The seasonality of protein synthesis has been measured across the year in some Antarctic marine species and has been shown to be highly seasonal, for example, in the holothurian *Heterocucumis steineni* (Fraser et al. 2004) and the limpet *Nacella concinna* (Fraser et al. 2007). These data are in line with studies showing growth to be seasonal in many marine species and they support the strong links between protein synthesis and growth, although it should also be noted that protein synthesis is essential for many other functions that are not growth related, such as protection from freezing or responding to environmental stress.

Seasonal studies of metabolic rates in Antarctic ectotherms have generally demonstrated strong seasonal signals. The data on seasonal changes in oxygen consumption in predators and primary consumers/grazers are summarised in Figure 15. Summer to winter changes in oxygen consumption have a smaller range in predators (1.1–2.2) than primary consumers/grazers (0.8–5.8). Predators generally feed throughout the year (Obermüller et al. 2010), and this is one of the differences between higher and lower trophic levels in the seasonality hypothesis (Clarke 1988). Some primary consumers/ grazers have also been shown to feed year-round on benthic productivity, for example, the limpet *N. concinna* (Obermüller et al. 2011, 2013) and in suspension-feeders on nano or picoplankton, for example, the bryozoan *Arachnopusia inchoata* (Barnes & Clarke 1995), which may explain some of the lower seasonal changes for primary consumers in Figure 15. The seasonal change shown for the scallop *Adamussium colbecki* may also be underestimated as in this study standard or starved oxygen consumption was measured (Heilmayer & Brey 2003), and feeding usually increases metabolic rates in marine invertebrates by 2 to 3 times (Peck 2016). The differences between Antarctic primary consumers/grazers and predators in the impact of season on metabolic rates shown in Figure 15 are not significant (data not normal after Ln, double Ln, arcsin or SQRT transforms, Kruskal Wallis H = 2.68, P = 0.10).

With increasing amounts of data gathered on seasonality and its impact on metabolic rates, it is now becoming clear that there is a signal in all trophic groups. It does not appear to be stronger in primary consumers than secondary consumers, as predicted by the seasonality hypothesis, but that other factors, such as year-round food availability for some species mean that comparisons of trophic levels may not be assessing groups where seasonal effects differ the most.

Seasonality of body mass and composition

Several studies have documented seasonal variation in body mass and proximate composition in Antarctic marine species. The brachiopod *Liothyrella uva* has strong seasonal cycles in tissue mass and also in total protein, carbohydrate and lipid composition (Peck et al. 1987a, Peck & Holmes 1989). The strongest cycle was in protein. Oxygen-to-nitrogen ratios from respiration and excretion studies were used to show this species fuels its metabolism predominantly from protein (Peck et al. 1986c, 1987b). Furthermore, seasonal protein stores have been identified in temperate brachiopods (James et al. 1992). The authors concluded that protein was being used as the seasonal storage material in *L. uva*. In the large infaunal bivalve *Laternula elliptica*, there were strong seasonal cycles in organic mass, protein and lipid, and these cycles differed between tissues (Ahn et al. 2003). Large mass changes occurred in muscle, gonads and digestive gland during spawning, and protein

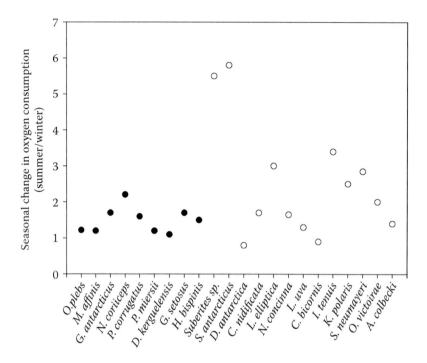

Figure 15 Seasonal changes in oxygen consumption for polar predators and primary consumers/grazers. Data shown are the factorial change from summer to winter. Data are for the primary consumers: rhynchonelliform brachiopod *Liothyrella uva* (Peck et al. 1986c, 1987a, 1997a,b), the infaunal bivalve *Laternula elliptica* (Brockington 2001b, Morley et al. 2007), the scallop *Adamussium colbecki* (Heilmayer & Brey 2003), the bryozoans *Camptoplites bicornis*, *Isosecuriflustra tenuis* and *Kymella polaris* (all Barnes & Peck 2005), the sponges *Suberites* sp., *Sphaerotylus antarcticus*, *Dendrilla antarctica* and *Clathria nidificata* (all Morley et al. 2016c), and the limpet *Nacella concinna* (Fraser et al. 2002). In the grazers, the sea urchin *Sterechinus neumayeri* (Brockington & Peck 2001) and the brittle star *Ophionotus victoriae* (Obermüller et al. 2010). Predatory species shown are the amphipods *Abyssorchomene* (previously *Orchomene*) *plebs* (Rakusa-Suszczewski 1982) *Gammarus setosus* (Weslawski & Opalinski 1997) and *Paraceradocus miersi* (Obermüller et al. 2010), the isopod *Glyptonotus antarcticus* (Janecki & Rakusa-Suszczewski 2006), the sponge-eating nudibranch *Doris kerguelenensis*, the nemertean *Parborlasia corrugatus* and the fish *Harpagifer bispinis* (all Obermüller et al. 2010). Data for the fish *Notothenia coriiceps* are from Campbell et al. (2008). Lehtonen (1996) studied seasonal metabolic rates in the Arctic benthic amphipod *Monoporeia affinis*, and these data are included to make the comparison one of polar species and not just Antarctic. Open symbols denote primary consumers and grazers, closed symbols denote predatory species. (Figure modified and redrawn from Morley, S.A. et al. 2016c. *Biodiversity* **17**, 34–45.)

was again identified as the main overwintering reserve, accounting for 60% of the energy used in winter compared to 20% each for lipid and carbohydrate. Pearse and Giese (1966a,b) also noted high-protein and low-carbohydrate contents in a survey of the biochemical composition of seven species of benthic marine invertebrates from McMurdo Sound. Cycles in organic content have been noted in the limpet *Nacella concinna,* with summer levels being around twice those of winter levels (Fraser et al. 2004), in the urchin *Sterechinus neumayeri* (Brockington & Peck 2001), and in three bryozoan species (Barnes & Peck 2005), although the seasonal changes in the bryozoans were small. Heilmayer and Brey (2003) found no seasonal change in soft tissue mass from summer to winter in the scallop *Adamussium colbecki*, which is surprising as this is a suspension-feeding bivalve, and the expectation is that species dependent for their nutrition on the highly seasonal short-duration phytoplankton bloom should exhibit strong seasonal signals in their body composition and mass.

It should be noted that much stronger seasonal signals in mass and proximate composition are seen in some pelagic species, especially some herbivorous crustaceans, and these signals are much stronger in high-latitude sites than in temperate or tropical regions (Clarke & Peck 1991). For example, calanoid copepods can be more than 50% lipid at the end of summer (Båmstedt 1986). Metabolic rates and activity levels in pelagic invertebrates are also generally higher than in benthic species, and it has been argued that these characteristics increase the impact of seasonality on their biology, resulting in stronger seasonal signals in mass and body composition (Clarke & Peck 1991). Whole-animal adaptations and responses to seasonal signals are underpinned by adaptations within the cells of the animals being studied.

Cellular level adaptations

When temperature changes, selection is expected to favour organisms maintaining their physiological functions at rates as close to those at the higher temperature as is feasible and consistent with maximising reproductive fitness (Clarke 1991, 2003, Hochachka & Somero 2002). They can do this by using three general mechanisms that affect the proteins involved in metabolomics processes: by modulating the intracelleular milieu, by using the cytosol to reduce the effects of the temperature alteration on protein reactivity; to use different protein variants or isoforms that have better thermal characteristics at the new temperature and/or by changing the concentrations of reactants, usually enzymes, in cells (Hochachka & Somero 2002). Most studies in this area have been on single factors in cells, or single proteins, especially enzymes (Clarke 2003). They have in some cases, however, shown strong links between protein function and whole-animal capacities or performance (e.g. Neargarder et al. 2003).

Although understanding in this field of general cellular adaptations is far from complete, patchy and still controversial in many areas, there are some well described adaptations and hypotheses. One of these is the imidazole alphastat hypothesis first proposed by Reeves (1972). This hypothesis is based around the premise that protein function is optimised by the modulation of intracellular and extracellular pH. This is achieved, at least in part, by using imidazole groups to buffer the temperature induced changes. Overall imidazole and protein ionisation are held at constant levels, which maintains cellular functions (Pörtner et al. 1998, Burton 2002). The alphastat hypothesis is generally well accepted, and there are far more studies supporting it than against, but some authors have argued strongly against its generality on several grounds (e.g. Heisler 1986), and some studies reported data that do not support alphastat (e.g. Taylor et al. 1999). Alphastat is portrayed as a universal mechanism for adaptation of cellular function as temperature varies. There are, however, many factors beyond this that assist with adaptation or acclimation to altered temperature.

Another general mechanism that has received significant levels of research and support is the modulation of the composition of cell membranes. Alterations in lipid composition occur when temperature changes the optimum balance between flexibility and rigidity in the liquid-crystalline membrane bilayer and also between lipid phases including gel and crystalline, sometimes called lamellar and non-lamellar phases. The process of increasing levels of unsaturated fatty acids in membranes at lower temperature has been documented in species across the globe and is often seen in acclimation to temperature change. The general principle is that the levels of unsaturated fatty acids increase in cell membranes at lower temperature, a process called 'homeoviscous adaptation' (Dey et al. 1993, Hazel 1995). Increased levels of unsaturated fatty acids in cellular membranes have been reported for Antarctic fish (reviewed in Verde et al. 2006). Logue et al. (2000) studied membrane fatty acid composition in 17 species of fish from across the globe, including four from Antarctica (*Pagothenia borchgrevinki*, *Trematomus bernacchii*, *Dissostichus mawsoni* and *Lycodichthys dearborni*) and one from the sub-Antarctic South Georgia (*Notothenia neglecta*, now *N. coriiceps*). By assessing the fluorescence anisotropy of a probe (1,6-diphenyl-1,3,5-hexatriene), they identified that there was a high level of temperature compensation in membrane static order (a measure of fluidity or viscosity) in the Antarctic species. The membranes in the synaptosomes (neuronal synapses, or nerve cell junctions, isolated for research purposes) of the Antarctic fish were, however, less fluid than predicted from first principles, and temperature compensation was only partial and not complete. It may be that perfect compensation of membrane fluidity is not possible, and on this basis, low-temperature adaptation of many membrane related processes are, at best, imperfect. Data are very limited on the relative functioning of membrane associated processes versus those within the cell milieu, but the compensation of cytochrome c oxidase in the Antarctic eelpout is less well compensated for temperature than citrate synthase, which is not membrane associated (Hardewig et al. 1999a), and a similar imperfect adaptation of membrane related processes was also found for Arctic cod (Lucassen et al. 2006).

Changing fatty acid composition in membranes has been shown to affect proton leak in mitochondria and, hence, the costs of maintaining these organelles (Porter et al. 1996, Brookes et al. 1998). It also affects the functioning of membrane-embedded proteins such as membrane pumps to a greater or lesser extent and other functions such as the electron transport chain. Imperfect compensation for temperature would thus be expected to have significant effects on the functioning of many cellular processes in Antarctic ectotherms.

It is interesting to note here that the fatty acid profiles of marine mammals, both pinnipeds and cetaceans, are consistent with the temperature-related adaptations seen in ectotherms. Thus, the superficial, colder layers of the skin in seals, walruses and whales are enriched in unsaturated fatty acids, whereas the deeper, warmer layers are enriched in saturated fatty acids and long-chain monounsaturated fatty acids (Fredheim et al. 1995, Best et al. 2003, Bagge et al. 2012). These changes in composition are usually interpreted as being related to dietary or storage functions, but at least one author has identified a potential functional relationship with temperature (Strandberg et al. 2008).

Several specific adaptations have been identified in Antarctic marine ectotherms. Some of them, such as antifreeze in fish, are general low-temperature adaptations seen in both poles, and some have only been identified in Antarctic species, but this is likely a function of the age of the available habitat and isolation from other faunas. Some of the main cellular and physiological adaptations are described below.

Antifreeze

The most well-known adaptation of marine species to polar waters is the production of antifreeze to overcome the problem of their tissues freezing. The temperature of the seawater in much of the Southern Ocean is below $-1°C$, for at least parts of the year and at the highest latitude marine sites for most of the year (Barnes et al. 2006a). This poses problems for organisms that have body fluids that freeze at temperatures close to $0°C$. Vulnerable species can respond to this type of challenge in one of two ways, allowing their extracellular body fluids to freeze (freeze tolerant), or to employ mechanisms to ensure body fluids do not freeze (freeze avoidance) (Lee & Denlinger 1991, Duman 2015). Antarctic marine species studied to date use a variety of methods to avoid freezing. Marine invertebrates have body fluid concentrations similar to seawater and so should not freeze until the water around them freezes. Antarctic marine invertebrates might, therefore, be expected to have few problems as long as they avoid solid ice. Many species, however, live in close contact with ice, with some even using ice as their main habitat, feeding off epontic productivity (e.g. Wiedenmann et al. 2009) and using ice as a refuge from predation (Thomas & Diekmann 2010), and this also includes some fish species (DeVries & Wohlschlag 1969). In these conditions, very small ice crystals are often present in seawater, and there is a need to inhibit their growth in tissues and cells.

Antarctic fish have a significant problem living in these conditions because water-based solutions with solute concentrations similar to their body fluids freeze at temperatures between $-0.5°C$ and $-1.0°C$ (Eastman & DeVries 1986, Peck 2015). Early studies of high latitude Southern Ocean fishes demonstrated that the freezing point of their circulating fluids was below $-2°C$, and the extra freezing resistance was due to the presence of antifreeze glycoproteins (AFGP) in their blood serum

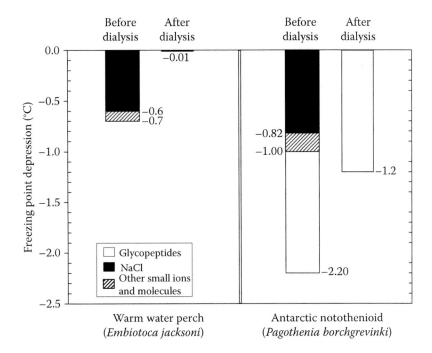

Figure 16 The effect of NaCl, other small ions and molecules, and glycopeptides on the freezing point of blood plasma from a warm-water fish (the perch *Embiotoca jacksoni*) and an Antarctic notothenioid (*Pagothenia borchgrevinki*). The higher salt content of the plasma of *P. borchgrevinki* produces a larger freezing point depression than in the warm-water *E. jacksoni*, but over half the freezing point depression in the Antarctic species is due to glycoproteins. The higher salt concentration in Antarctic fish plasma gives an extra 0.3°C freezing point depression over the temperate perch, but a further 1.2°C is added by the presence of AFGP molecules. These extra lowerings of the freezing point seem small, but they make Antarctic fish safe in an environment where temperatures rarely fall below −2°C. (Figure from Peck, L.S. 2015. *Journal of Experimental Biology* **218**, 2146–2147.)

(DeVries & Wohlschlag 1969, DeVries 1971). Antarctic fish also have higher solute concentrations in their blood (osmotic concentration of 550–625 mOsm kg^{-1}) than temperate and tropical fish (320–380 mOsm kg^{-1}). This higher concentration of low-molecular-weight solutes accounts for around half of the total freezing point depression (to around −1.1°C), with the rest (to just below −2°C) accounted for by the action of AFGPs (Figure 16) (Eastman & DeVries 1986).

A surprising number of Antarctic marine invertebrate species inhabit the intertidal (Waller et al. 2006a, Waller 2013) where they can be exposed to temperatures much lower than the freezing point of seawater. In a study of 11 intertidal species, Waller et al. (2006b) showed their 50% freezing points ranged from −5.5°C in the limpet *Nacella concinna* to −23.1°C in the marine mite *Rhombognathus gressitti*, and freezing point depression was inversely correlated with size. Further investigation of eight species of intertidal invertebrates showed only three had different freezing points between intertidal and subtidal populations, and one of these three, the copepod *Tigriopus angulatus*, had a lower freezing point in the subtidal population. One species studied by Waller et al. (2006b), the nemertean *Antarctonemertes valida*, showed evidence of the presence of antifreeze proteins through thermal hysteresis of the supercooling point of its haemolymph. Thermal hysteresis is the separation of the freezing point and the melting point of a liquid. It is characteristic of antifreeze proteins and glycoproteins and is not seen when the freezing point is lowered by the addition of solutes.

More recent research showed that the mucus secreted by the limpet *Nacella concinna* contributed to freezing resistance, possibly because its high viscosity inhibits ice crystal proliferation (Hawes et al. 2010). These studies show that Antarctic marine invertebrates have freezing points well below those expected from knowledge of their body fluid solute concentrations, but the evidence suggests few have antifreeze proteins because of the lack of thermal hysteresis. Identifying the mechanisms conferring this extra freezing resistance requires further research.

Over the last 20 years, the rapid development of molecular and 'omics' technologies have had a marked impact on understanding antifreeze mechanisms and the evolution of antifreeze molecules. This has aided in understanding the biodiversity and evolution of antifreeze mechanisms with AFGPs now known globally from insects, frogs, plants, bacteria, ciliates, diatoms and copepods as well as fish (Kiko 2010, Storey & Storey 2013, Duman 2015, Pucciarelli et al. 2015).

Many different-size isoforms of antifreeze glycoproteins have been identified in Antarctic notothenioid fishes. They are encoded in large families of polyprotein genes (Chen et al. 1997). These large gene families arose by gene duplication, enabling the synthesis of large quantities of protective antifreeze glycoproteins and antifreeze proteins. Fourteen antifreeze glycoprotein polyprotein genes have now been identified. They each encode multiple (up to 30–40) antifreeze glycoprotein molecules. This emphasises how important this adaptation was in the evolution of Antarctic fishes (Nicodemus-Johnson et al. 2011).

Antifreeze glycoproteins in notothenioid fish evolved from a pancreatic trypsinogen serine protease progenitor following duplications of a coding element between intron one and exon two of the trypsin gene (Chen et al. 1997). It is now accepted that this was a critical evolutionary step that gave a great advantage to notothenioid fish over other taxa as the Southern Ocean cooled to subzero temperatures. Structurally almost identical antifreeze glycoproteins have been found in some Arctic gadoids (DeVries & Cheng 2005), but they arose from a different evolutionary route. The precursor for the Arctic AFGP gene is not known, but it did not derive from a trypsin-like gene (Cheng & Chen 1999, Near et al. 2012).

In excess of the AFGPs of Antarctic notothenioids, four different types of antifreeze proteins (AFP) that work in a similar way to the AFGPs, inhibiting ice crystal growth, have been identified in fish around the globe (DeVries & Cheng 2005, Davies 2014). Type I AFPs were identified first in *Pseudopleuronectes americanus*, the American flounder, but have subsequently been found in other flounder species and some sculpins (Duman & DeVries 1976, Graham et al. 2013). The taxonomic diversity suggests type I AFPs have evolved independently on several occasions. Type II AFPs are present in smelt, herring and the sea raven. They are globular, and some need calcium ions to be active, whereas other forms do not. This variation in action has been suggested to indicate different routes of evolution are involved (e.g. Ewart et al. 1998). Type III AFPs have been identified in zooarcid fish from Antarctica and also from the northern hemisphere. There are two forms of this group of AFPs, one that does not appear to produce a thermal hysteresis effect but produces a larger hysteresis when both forms are combined (Nishimiya et al. 2005). Type III AFP seems to have evolved from a sialic acid precursor (Cheng & DeVries 1989, Baardsnes & Davies 2001). Type IV AFP was first identified in *Myoxocephalus octodecemspinosus*, the long-horned sculpin by Deng et al. (1997). Transcripts of type IV AFP have now been found in *Carassius auratus*, the carp and *Danio rerio*, the zebrafish, both of which are freshwater fish, which should not need to be protected against freezing, but the AFP appears important in developmental processes (Xiao et al. 2014).

In addition to the AFGPs that Antarctic notothenioids produce, many fish in this group also produce an AFP that lacks carbohydrate and is around 15 kDa in size (Jin 2003, DeVries 2004, Yang et al. 2013). This AFP has little thermal hysteresis on its own, but when it is present in combination with the AFGPs, the thermal hysteresis effect is doubled. Because of this enhancing effect, the protein was called an antifreeze-potentiating protein (AFPP). To achieve the full thermal hysteresis effect seen in Antarctic fish plasma, both the AFPP and the AFGPs need to be present because the AFGPs and the AFPP bind to different parts of the growing ice crystal.

All AFPs and AFGPs work in the way that DeVries (1971) proposed, which was later confirmed by Raymond and DeVries (1977) in what is now called the adsorption-inhibition mechanism. Ice crystals grow when water molecules attach to the surface plane of the crystal. AFPs and AFGPs slow or halt the growth by adsorbing on to the crystal surface at specific sites where growth is preferred, which is usually on a prism plane. Once attached, water molecules are precluded from attaching to the ice crystal at these sites, and growth can only occur in regions between the AF(G)P molecules in non-preferred sites (Raymond & DeVries 1977, Raymond et al. 1989, Knight et al. 1991, Duman 2015). The regions between attached AF(G)Ps have a high radius of curvature. The curvature of a surface affects many factors such as the stability of small droplets and the growth of small crystals in what is known as the Kelvin effect (Berg 2010), which contributes to inhibition of further growth of ice crystals. Different molecules often adsorb onto different surfaces, and the strength of the thermal hysteresis (and the freezing point depression) depends on the site of adsorption. The mechanism that binds the AFPs or AFGPs to the crystal surface was first thought to be via hydrogen bonding to oxygen in the ice lattice (DeVries 1971, DeVries & Cheng 1992, Sicheri & Yang 1995). This hypothesis is based on the presence of repeat structures in the AFP and AFGP molecules. More recent work has indicated that another mechanism may be important too, and this involves the organisation of water molecules into a crystal-like arrangement around more hydrophobic residues. The 'organised' water then freezes to the ice crystal surface, adhering the AFP or AFGP at the same time (Jia & Davies 2002, Garnham et al. 2011, Davies 2014, Sun et al. 2014). Some studies have suggested that both mechanisms may be used and their importance varies between AFPs and also between AFGPs (Ebbinghaus et al. 2012, Meister et al. 2013, 2014). An interesting outcome of antifreeze molecules binding to ice crystals to stop them growing is that this process seems to inhibit melting of ice crystals in positive temperatures such that they can become super-heated (Cziko et al. 2014).

There is significant developmental and ecological variation in antifreeze properties of Antarctic fishes. Thus, in a study of the larvae of three Antarctic fish species, one, *Pagothenia borchgrevinki* had larvae with high levels of AFPs, similar to those in adults, but the other two, *Gymnodraco acuticeps* and *Pleuragramma antarctica*, had larvae that did not contain any AFPs or AFGPs but still did not freeze in significantly subzero temperatures. It was argued that protection seems to be gained from a physical barrier around the larva from the integument and a reduction in susceptible external tissues such as gills (Cziko et al. 2006). Analyses of samples collected in recent ecological surveys have identified that blood serum antifreeze activity of Antarctic *Trematomus* fishes varies across habitat temperature and depth (Jin & DeVries 2006, Fields & DeVries 2015). The mechanisms for achieving this variability are still to be identified.

The evolution of antifreeze molecules was thus a step that allowed fish species to succeed in polar marine conditions. A very wide range of AFPs and AFGPs has been identified in fish across the globe, and several types can occur in a single species. They have evolved many times from different sources and via different routes. The remarkable similarity of AFGPs across fish species, especially Antarctic and Arctic species that are only distantly related, is one of the best known examples of convergent evolution.

Tubulins

One of the best described examples of a molecular-level adaptation to low temperature in polar seas is in tubulins and their polymerisation into microtubules. Microtubules are polymers that perform a cytoskeletal function in the cells of all eukaryotes. They are involved in a very wide range of physiological and structural processes including cell division, maintaining the shape of cells, transport within cells, secretion, cell motility, the mitotic spindle, centrioles, basal bodies, cilia and cell polarisation. They are especially important in neuronal tissues and for neuronal function. Microtubules are composed of α- and β-tubulin heterodimer subunits joined together into linear protofilaments. A single microtubule contains 10–15 protofilaments that are linked laterally in the construction of a hollow cylinder 24-nm diameter (Rusan et al. 2008). Microtubules are polar structures because of the way the $\alpha\beta$ heterodimers are arranged. They also have different rates of polymerisation at each end because of this arrangement. Protofilaments are structured with the β-tubulin monomer aligned towards the plus, faster growing end, and the α-tubulin monomer at the other slower-growing end. There is a third isoform γ-tubulin that is important as a template in microtubule assembly. Microtubules also have associated heterogeneous proteins (microtubule-associated proteins or MAPs) attached to their surfaces.

When microtubules polymerise, structured water is released from sites where the subunits make contact (Correia & Williams 1983). Polymerisation is also entropically driven. Because of these factors, microtubule assembly and polymerisation are temperature sensitive (Detrich 1998). In mammals, microtubule assembly functions best at temperatures near 37°C, and they depolymerise at temperatures around or below 15°C, whilst in temperate fish, microtubules also depolymerise at temperatures below 5°C (Detrich & Overton 1986). Microtubules of Antarctic fish are stable at all marine temperatures experienced in the Southern Ocean, down to below −2°C (Detrich et al. 1987). They also polymerise at subzero temperatures (Detrich 1991). The low-temperature stability and polymerisation of microtubules in Antarctic fish is achieved by changes in the sequence of the tubulin proteins that increases the flexibility of domains that are involved in contact between dimers, and these substitutions occur post translationally (Wallin & Stromberg 1995, Detrich et al. 2000). This is probably particularly important to the low-temperature polymerisation of tubulins (Shearwin & Timasheff 1992, Detrich 1998). Other potentially important factors are changes to the hydrophobic properties of the dimer surfaces (Detrich et al. 2000) and changes in the electrostatic properties of tubulin dimers to reduce repulsion (Detrich 1998).

With regard to other species, much recent work has focussed on identifying and characterising tubulins in the Antarctic ciliate *Euplotes focardii*. Genes for one α-tubulin and three β-tubulins were first identified in the 1990s (Miceli et al. 1994, 1996). Subsequently, a fourth β-tubulin was discovered in this species (Pucciarelli et al. 2009). *Euplotes focardii* has been described as being exceptionally rich in microtubules, and several genes encoding tubulins have been characterised. Three isotypes of β-tubulin have been identified in this species that have small sequence substitutions compared to the β-tubulin of the temperate *E. crassus*. These changes increase the flexibility of the tubulin protein (Chiappori et al. 2012). Each isotype exhibits different flexibility in regions involved in lateral and longitudinal contact phases when microtubules assemble.

Recent studies on tubulins in Antarctic species have focussed on the folding mechanism and the factors involved in obtaining successful folding to the functional microtubule. In *E. focardii*, this has included the analysis of the role of chaperonins and co-factors in the folding process (Pucciarelli et al. 2013). Cuellar et al. (2014) investigated the role of the cytosolic chaperonin (CCT) in tubulin folding in the Antarctic fish *Gobionotothen gibberifrons* and concluded that the folding cycle of CCT is 'partially compensated at their habitat temperature, probably by means of enhanced CP-binding affinity and increased flexibility of the CCT subunits'. Overall, it appears that there are many adaptations to facilitate the production of microtubules at low temperature in Antarctic marine species, that full compensation for temperature is probably not achieved, and that a significant part of the adaptation is achieved from amino acid substitutions that increase flexibility of the proteins involved.

Haemoglobin and oxygen transport

To biologists not familiar with Antarctic marine species, one of the most surprising and difficult adaptations to account for is the absence of red blood cells and haemoglobin in the channichthyid icefish. This trait was noted in the early part of the twentieth century by naturalists visiting Antarctica and was described scientifically by Ruud (1954) who, when he first heard of the fish that had clear blood on a visit to South Georgia over 30 years before, was initially sceptical of their existence. Haemoglobinless fish evolved in the evolutionary long-term, low-temperature habitats in the seas around Antarctica, where the oxygen concentration in seawater is 1.8–1.9 times higher than tropical seawater, and metabolic rates are around 10–25 times lower than species living at 30°C

(Clarke & Johnston 1999, Peck & Conway 2000). The condition has been described in 16 species, all members of the family Channichthyidae, or crocodile icefish. This phenotype is not possible at higher temperatures where minimum, basal, metabolic rates and costs are higher because of the Arrhenius effects of temperature on biological reaction rates.

The lack of haemoglobin in channichthyid blood means oxygen is carried only in solution, and the blood is only capable of carrying around 10% of the oxygen of red-blooded relatives (Holeton 1970, Cheng & Detrich 2012). The haemoglobinless condition has been described as maladaptive, disadaptive or disadvantageous (e.g. Montgomery & Clements 2000). Adaptation via natural selection would argue, however, that when functional haemoglobin was lost there must have been a strong selective advantage for this trait to have persisted, or that the trait arose under conditions where selection pressures were relaxed and has then persisted with little selection pressure acting against (Cheng & Detrich 2012). The argument that the loss of haemoglobin may have occurred under conditions of low selection pressure is supported by the fact that the channichthyids are sedentary and sluggish. They evolved from an ancestor with these characteristics that diverged from their nearest relatives the dragon fishes (Bathydraconidae) around 6–12 mya (Near et al. 2012, Verde et al. 2012a). On the basis of genetic data, the channichthyids diversified and lost myoglobin genes sometime during the last 2–5.5 mya when there were ice ages and Antarctica had periodic ice-sheet extensions and contractions (Bargelloni et al. 2000, Sidell & O'Brien 2006). This would have produced conditions where deep fjords were periodically available for colonisation by a relatively depauperate fish fauna, leading to conditions of low selection pressure (Sidell & O'Brien 2006, Cheng & Detrich 2012). This is further supported by phylogenetic analyses that suggest the loss of genes for haemoproteins occurred on four separate occasions in the channichthyid lineage (Near et al. 2006, Sidell & O'Brien 2006). The other alternative, that there may have been selective advantages with the loss of haemoglobin, has been relatively poorly investigated. Channichthyid blood viscosity has been identified as being significantly lower than that of red-blooded notothenioids (Egginton 1996, Egginton et al. 2002, 2006, Kock 2005, Garofalo et al. 2009). The viscosity of the icefish *Chaenocephalus aceratus* is around 3.3–3.5 centipoises (cP) at normal habitat temperatures compared to 5.5–6.0 cP for the red-blooded *Notothenia coriiceps* (Egginton 1996). Temperate fish blood viscosity increases more than 80% when temperatures are decreased from 25°C to 5°C (Sidell & Hazel 1987). Antarctic fish blood is made more viscous by the increase in osmolarity to around twice those of temperate fish, from raised solute concentrations, and by the presence of AFPs and AFGPs, which are essential for fish to live at significantly subzero temperatures (Egginton et al. 2006). Further increases in viscosity will occur in the presence of small ice crystals beginning to grow around nuclei when temperatures are near the freezing point of the fish blood. It may be that under these conditions channichthyid icefish with low-viscosity blood have an advantage over red-blooded species.

If the very strong reduction in the ability to carry oxygen around the body in channichthyids is disadaptive and persists because of a lack of selection pressure, we would expect to see impacts on a range of life-history traits from growth rate to reproductive investment. There are good data for the latter. Kock et al. (2000) measured gonad index (GI) in four species of haemoglobinless channichthyids and 10 species of red-blooded nototheniids and concluded that all species had GI values of 20%–25% at spawning with the exception of *Lepidonotothen squamifrons*, a species with haemoglobin, which has a lower value. Kock and Kellerman (1991) measured GI in three species of channichthyid and 12 species of nototheniid. The mean GI for the former was 19.28% (s.e. = 3.71) and for the latter was 19.33% (s.e. = 2.36). There is thus little support from these data that the lack of haemoglobin puts overall restrictions on the balance between energy intake and energetic costs in channichthyids compared to red-blooded nototheniids.

The evolution of this trait in channichthyids has produced a range of other adaptations to compensate for the reduced oxygen-carrying capacity of the blood (Verde et al. 2012b, 2011, di Prisco & Verde 2015). At the cellular level, these include increased densities of mitochondria in the muscle cells of the heart of channichthyids (O'Brien et al. 2000), where mitochondria account for 36% of the cell volume in *Chaenocephalus aceratus*, which does not express myoglobin in its muscle tissues, compared to 20% in *Chionodraco rastrospinosus*, a channichthyid that does express myoglobin, and only 16% in *Gobionotothen gibberifrons*, a red-blooded nototheniid. Of the 16 Antarctic channichthyids that are haemoglobinless, six completely lack myoglobin in their muscles (Grove et al. 2004, Sidell & O'Brien 2006, O'Brien 2016). Myoglobin is thought to have a critical role in the storage and diffusion of oxygen in cells (Wittenberg & Wittenberg 2003), and it was described as an 'essential hemoprotein in striated muscle' (Ordway & Garry 2004). The finding that all channichthyids lack myoglobin in their striated muscle (Grove et al. 2004), and that some lack it in heart muscle as well (Moylan & Sidell 2000, Sidell & O'Brien 2006), changed the appreciation in this field. As the lack of oxygen-binding proteins has been argued as a disaptation and negative in terms of animal fitness (e.g. Montgomery & Clements 2000), it is interesting that it has recently been suggested that the loss of these oxygen carriers reduces levels of oxygen damage in cells and tissues and may provide a benefit in terms of reduced costs for the repair of proteins following reactive oxygen species (ROS) damage (O'Brien 2016).

To compensate for the reduction in oxygen-carrying capacity of the blood of haemoglobinless fish, it is further thought that an increase of mitochondrial density may play a role. In this case, oxygen delivery would be enhanced via diffusion through the lipids in mitochondrial membranes. This would also increase ATP production because the folding of the internal surfaces, the cristae density, is lower in haemoglobinless species, which reduces the capacity for oxidative phosphorylation (O'Brien et al. 2000, 2003, O'Brien 2016). This would be offset to some degree by more efficient oxygen supply through the lipids in membranes.

At the tissue level, haemoglobinless species have large hearts that pump four to five times as much blood, and they have a total blood volume two to four times that of red-blooded species (Hemmingsen 1991). The heart itself is composed of spongy myocardium infused with a high density of capillaries, and it pumps at lower pressure than hearts of red-blooded fish, but pumps a large volume with each contraction. The output of the heart is generally accepted in these species to be regulated by heart rate, with stroke volume (volume of haemolymph pumped per heart contraction) varying little (Axelsson et al. 1992). Haemoglobinless fish have capillaries that have diameters around 1.5 times greater than red-blooded relatives, and Egginton et al. (2002) concluded that wider capillaries are essential for the maintenance of tissue oxygenation in the absence of respiratory pigments. They also have lower numbers of circulating blood cells and high lipid contents in the plasma compared to temperate species (Davison et al. 1997), which are also thought to be adaptations to low temperature and high environmental oxygen concentrations, especially increased blood viscosity and slowed biochemical reactions (Farrell & Steffensen 2005, Campbell et al. 2009).

Resistance to blood flow is governed by two main factors, the viscosity of the circulating fluid and the characteristics of the blood vessels (Dejours 1966, Schmidt-Nielsen 1997). The rate of liquid flow in tubes is described by Poiseille's equation:

$$Q = \Delta p \frac{\pi}{8} \frac{r^4}{l\eta} = \frac{p}{R}$$

where Q = rate of fluid flow; Δp = the pressure drop along the tube (p = pressure); r = tube radius; l = tube length; η = viscosity; and R = resistance). The important relationships in this equation are that resistance to flow is proportional to the length of the tube and to viscosity, and it is inversely proportional to the fourth power of the radius. Increasing capillary diameter in haemoglobinless fish by ×1.5 thus reduces resistance to flow more than 5-fold, that is, to less than 20% of the resistance of the narrower capillaries. Water viscosity increases at lower temperatures and more than doubles as temperature is reduced from 30°C to 0°C. Furthermore, it changes more at low temperatures, around and below 0°C, than at higher temperatures. Many of the attributes of haemoglobinless fish blood can thus be interpreted as adaptations to allow increased blood flow and overcome problems of increased viscosity and capillary resistance in a low-temperature world, where the likelihood of the presence of micro ice crystals, that also increase fluid viscosity, is high. Recent studies have further suggested that the loss of Hb and Mb, their associated NO-oxygenase activity and subsequent elevation of nitric oxide (NO) levels up to twice those observed in red-blooded notothenioids (Beers et al. 2010), may explain the unique cardiovascular and physiological traits evolved in icefish to assure higher blood volume and cardiac output (Sidell & O'Brien 2006), thus posing the question whether other globin family members (neuroglobin and citoglobin) might compensate such losses in icefish.

Neuroglobin and cytoglobin

Haemoglobin has been known as a component of animal blood, and associated with oxygen transport since the mid-nineteenth century (Hoppe-Seyler 1864, Fenn & Rahn 1964). Myoglobin has been recognised as important in oxygen relations in vertebrate muscles for over 50 years and was one of the first molecules to have its 3-dimensional structure elucidated using X-ray diffraction techniques (Kendrew et al. 1958). Two other members of the superfamily of globins have recently been identified in vertebrates: neuroglobin, which is present in the central and peripheral components of the nervous system; and cytoglobin, which is present in all major tissues. Both were discovered by the same team around the turn of the century (Burmester et al. 2000, 2002). The functions of these molecules remain to be fully elucidated, but neuroglobin probably has at least some oxygen delivery or scavenging related role, as it confers protection from hypoxic neuronal injury *in vitro* and ischaemic cerebral injury *in vivo* in mice (Sun et al. 2001, Greenberg et al. 2008).

Recently the genes for neuroglobin (Cheng et al. 2009a,b, Boron et al. 2011, Giordano et al. 2012a) and cytoglobin (Shin et al. 2012) have been discovered in Antarctic fish. Neuroglobin is a 17-kDa monomeric hexa-coordinated hemoprotein with the classical globin folding pattern. It has a high oxygen affinity (in the range of typical myoglobin values (half saturation pressure $P_{50} = 0.9$–2.2 Torr (0.12–0.29 kPa))). Cytoglobin is a 21-kDa hemoprotein with the same globin folding pattern and oxygen affinity in the myoglobin-like range of 1 Torr (0.13 kPa). Antarctic fish neuroglobin and cytoglobin have been shown to bind oxygen and carbon monoxide reversibly. They also have high oxygen affinity, which is similar to that of human cytoglobin, but the high oxygen affinity means they are unlikely to be involved in oxygen transport (Giordano et al. 2015, Verde et al. 2011). Other globins, including globin X and Y, found in teleosts, have not been found so far in Antarctic fish (Giordano et al. 2015). Neuroglobin has been found in the haemoglobinless channichthyids, most of which also lack myoglobin. It has been hypothesised that their retention in this group is likely to be associated with protection of nervous tissues from nitrosative effects and oxidative damage to tissues in the oxygen-rich waters of the Southern Ocean (Giordano et al. 2015). One other way that the function and importance of cytoglobins has been studied is through their genes and gene expression.

Cytoglobin (Cygb) genes are expressed in all vertebrate tissues, in a range of species. Cygb concentration is higher in the brain, eyes, skeletal muscles, heart and liver than other tissues (Burmester et al. 2002, Fordel et al. 2004). Cygb functions in cells and tissues are poorly described to date, but several possible roles have been suggested. These include: nitrite reduction and nitric oxide (NO) generation during anaerobiosis (Li et al. 2012); regulation of intracelluar NO concentrations (Liu et al. 2012); oxygen supply to the respiratory chain in mitochondria (Kawada et al. 2001, Hankeln et al. 2005), and during the synthesis of collagen (Schmidt et al. 2004), and to protect cells against oxidative stress (Li et al. 2007). There are many stresses that animals face, both in the Antarctic and elsewhere. There is a range of responses to stresses, but the most well known are possibly the heat shock response (HSR) and mechanisms to reduce or repair damage from reactive oxygen species. These are discussed below.

Stress responses: the HSR and reactive oxygen species

The heat shock response (HSR) is the only response to stress identified in organisms that has been claimed to be universal (Gross 2004). The HSR classically involves the production of heat shock proteins (HSP) in response to a thermal challenge (Gross 2004). They are also, however, produced in response to a wide range of other stressors, including dehydration in plants and insects (Feder & Hofmann 1999). The HSP family of proteins is large, and there are many forms that differ among species. They have a wide range of functions, which include helping misfolded proteins to reach their functional state or to return to that state if it is lost. They have also been recognised to assist in the identification of degraded proteins and to have a role in the regulation of their removal from the cell. This process is an important part of the prevention of the formation of cytotoxic aggregates in cells (Parsell & Lindquist 1993). The best studied of the HSP proteins are the HSP70 proteins, so called because they are the 70 kDa family.

At the start of this millennium, Hofmann et al. (2000) made a surprising discovery when investigating the HSR in Antarctic fishes. When they warmed *Trematomus bernacchii*, the expected classic HSR was not elicited; HSP70 production did not increase when they raised temperatures beyond its normal temperature range. This was the first time such an HSR had not been demonstrated on warming beyond normally experienced temperatures in any multicellular organism more complex than a hydroid (La Terza et al. 2001, 2004, 2007). Studies that followed showed that the majority of Antarctic fishes lack the ability to increase HSP70 production when challenged with any of the commonly used stressors (Tomanek 2010, Beers & Jayasundara 2014). The reason for this lack of HSR is thought to be due to a mutation in the *hsp70* gene, specifically in its promoter region that inhibits the binding and following transcription of HSF1 (Buckley et al. 2004). All Antarctic fish do, however, constitutively express at high levels the inducible form of the HSP70 protein that is usually subject to an increase in production in response to stress in all non-Antarctic invertebrates and fish (Place & Hofmann 2005, Clark et al. 2008a).

Antarctic marine invertebrates differ from the fish in that they exhibit a range of HSP70 responses following exposure to heat stress. A few studies have found an HSR in some species, while others have reported a lack of increase in HSP production in Antarctic invertebrate species when warmed. Recently the sea urchin *Sterechinus neumayeri* was shown to increase HSP70 in response to exposure to 3°C temperatures for 48 hours (González et al. 2016). Two molluscs, the clam *Laternula elliptica* and the limpet *Nacella concinna*, both exhibited an HSR, but an increase in production of a range of HSPs only occurred in experiments when temperatures were raised to 8–10°C and 15°C, respectively, with no HSR at lower temperatures (Clarke et al. 2008). These temperatures are well above any that *Laternula elliptica* and *Nacella concinna* have been exposed to for millions of years. The possibility that there might be a functional need for an HSP in these species was later shown in the limpet when the induction of several members of the HSP70 family was demonstrated to occur during the tidal cycle in individuals living in the intertidal (Clark et al. 2008c) and also in association with the spring thaw of sea ice and exposure over highly extended periods to low levels of warming (Clark & Peck 2009b). To date, only one species of invertebrate has not been shown to exhibit an HSP70 HSR, the starfish, *Odontaster validus* (Clark et al. 2008b). However, these heat shock experiments only evaluated the HSR using *hsp70* genes cloned by degenerate polymerase chain reaction (PCR). Thus, it is likely that not all members of the HSP70 family were identified. In the same series of experiments, no HSR was detected in the amphipod *Paraceradocus miersi*, (Clark et al. 2008b), but subsequent thermal challenges using discovery-led next-generation sequencing (NGS) did in fact reveal the possession of a thermally inducible form of HSP70 (Clark et al. 2016). In the Antarctic krill *Euphausia superba* and *E. crystallorophias*, gene duplication events were shown to have occurred that produced several forms of HSP70 but even so change has to have only resulted in a weak HSR (Cascella et al. 2015).

In a recent study using acute warming of 1°C h^{-1}, Clark et al. (2016) showed that the HSR varied markedly between six different species of Antarctic marine invertebrate, with each species having its own response to warming. They used a combined transcriptomic and metabolomics approach to evaluate the response to acute warming and demonstrated that the upregulation of the production of members of the HSP70 family occurred in three species, (the amphipod *Paraceradocus miersi*, the rhynchonelliform brachiopod *Liothyrella uva* and the bivalve mollusc *Laternula elliptica*), whereas in three others (the gastropod mollusc *Marseniopsis mollis*, the holothurian *Cucumaria georgiana* and the bivalve mollusc *Aequiyoldia eightsii*), there was no indication of any change in HSP70 production at this particular rate of thermal change. This work demonstrated that the response varies with the challenge applied.

Interestingly, numerous HSP family gene duplications have been identified in Antarctic marine species, including in krill, which indicates an ongoing requirement for their production. The hypothesis that living in polar marine conditions where temperatures are permanently around or below 0°C requires constitutively high expression of one or more *hsp70* genes is still the main paradigm in this area of science.

Antarctic marine species have been noted for nearly two decades to have elevated resting or constitutive levels of the proteins involved in pathways conferring resistance to, and providing protection from, damage from reactive oxygen species (ROS), primarily superoxide dismutase (SOD), catalase and the glutathione enzymes (Abele & Puntarulo, 2004, Chen et al. 2008, Clark et al. 2010, 2011). This is explained as an adaptation to living in polar marine environments, where ambient oxygen levels are the highest in the world's oceans, and the capacity for body fluids to carry dissolved oxygen is the highest on Earth, making the organisms living there more vulnerable to ROS damage than those from lower latitudes. In their study of the effects of acute warming on six Antarctic marine invertebrates, Clark et al. (2017) found that none of them produced the expected response to reactive oxygen damage, with no detectable upregulation of pathways conferring resistance as evidenced by expression levels of SOD, catalase or glutathione enzyme genes. Previous research on bivalves focussed on SOD and the pro-oxidative product malondialdehyde (MDA) in Antarctic bivalve molluscs (*Adamussium colbecki* and *Aequiyoldia eightsii*) had shown that SOD levels decreased, but MDA levels increased when the animals were warmed (Regoli et al. 1997, Abele et al. 2001). MDA is often used as an indicator of oxidative stress because it is part of the signalling system deciding cell survival or death (Ayala et al. 2014). Because of this, the increase in MDA combined with a decrease in SOD with warming was explained via an evolutionary maximisation of the antioxidant system to cope with the very high levels of oxygen in the ambient environment, but this came with a reduction in thermal stability of the antioxidant system. It has been suggested that these could be 'examples of species-specific enhanced sensitivity of critical enzymes which directly impact on organism physiology and survival' (Clark et al. 2017). Antarctic marine species are often thought to have evolved fine-scale adaptation to extreme low temperature conditions because of the long evolutionary period, in excess of 10 million years, they have evolved in isolation from lower latitude faunas (Clarke & Crame 1992). Despite this, recent work has questioned whether the proteins of Antarctic marine ectotherms are fully adapted to temperatures around and below 0°C (Peck 2016, Clark et al. 2017).

It is the specific adaptations to any environment that produce the cellular and physiological attributes of all species, and these adaptations enhance or constrain their capacities to respond to change in the environment. The fine-scale, and often unique, adaptations described for Antarctic marine species previously will dictate their future success or failure.

Impacts of environmental change

The discussion so far has concentrated on adaptations to the Antarctic environment at the whole-animal and cellular levels. The cryosphere is being increasingly affected by climate change, and it might be expected that cold-adapted endemic polar species may be particularly vulnerable (Somero & DeVries 1967, Peck 2002a,b, Peck et al. 2006a,b, 2009a,b, 2014, Pörtner et al. 2007, 2012).

Human-driven environmental change is a global phenomenon, with clear impacts on biodiversity in all regions of the Earth (UNEP 2012). The areas of most rapid change to date have been in the polar regions where the fastest rates of regional warming on Earth have been reported. This warming has been accompanied by dramatic loss of sea ice and recession of coastal glaciers and ice shelves (IPCC 2014). In Antarctica, many regions have not exhibited significant change over the last 50–100 years, but the Antarctic Peninsula has seen some of the fastest change over the last 50 years (Turner et al. 2009). The atmospheric warming on the peninsula may be less now than in the past (Turner et al. 2016), but coastal ice is still receding, and oceanic systems are still in flux (Cook et al. 2016). In parts of this region, air temperatures increased by over 3°C in the last half of the twentieth century, and sea temperatures to the west of the Peninsula increased by 1°C over the same period (Meredith & King 2005). The major challenges that marine species face from the current environmental change in the polar regions comes from three main sources: increased temperature, ocean acidification and altered levels of sea ice and iceberg scour in the benthos, although salinity (Clark & Peck 2009a,b) and hypoxia (Tremblay & Abele 2016) have also been noted as potentially important stressors.

Responding to change at the level of the organism

Organisms can respond to changes in their environment in a large number of ways that vary across process scales from cellular and molecular to community and ecosystem, and these responses vary with the scale of the change in spatial and temporal contexts (Figure 17). The various responses can be classified across scales. Within cells, biochemical buffering is at the smallest scale, and beyond this, gene expression and phenotypic plasticity through physiological flexibility buffer changes that occur over hours to weeks. At larger scales still, gene frequency alterations and selection in populations work alongside behavioural modifications. At slow rates of change, evolutionary gene modification and speciation are important responses. These are the more mechanistic types of response that organisms can have, and processes involving ecological interactions and migration have effects across these scales.

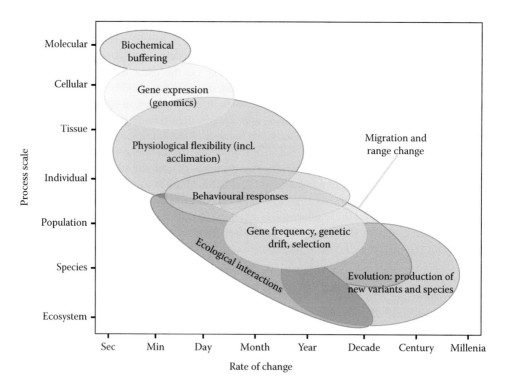

Figure 17 Schematic diagram of biological responses to changes in the environment in relation to the rate of change and the scale of process involved. In mechanistic processes, at the smallest scales biochemical buffering is entrained. This is followed by a cascade of processes through gene expression, phenotypic plasticity and population gene frequency modification through to evolutionary change. Phenotypic plasticity links mechanistic process understanding with ecological and evolutionary responses. (Figure redrawn from Peck, L.S. 2011. *Marine Genomics* **4**, 237–243.)

The type of response an organism produces varies depending on the rate of environmental change. Immediate responses required to survive rapid change often result in an increase in energy demand for muscle contraction or homeostasis. In marine invertebrates this is fuelled from phospho-L-arginine (PLA) stores in cells (e.g. Bailey et al. 2003, Morley et al. 2009a), whilst in vertebrates, the energy store is creatine phosphate (Nakayama & Clarke 2003). The size of PLA or creatine phosphate stores are one of the limitations of capacities to respond to rare rapid environmental change over short timescales of seconds to minutes, such as when seawater is heated by volcanic activity and escape responses are needed for survival. On the scale of seconds to weeks or months, several processes can have effects. Gene expression and its modulation through post translational processes such as epigenetic mechanisms are important in whole-organism responses, including those involved in stress such as the heat shock response (Gross 2004, Clark & Peck 2009a), in damage repair (Sleight et al. 2015), or in longer-term processes such as seasonal changes in the lower lethal temperature limit of the intertidal snail *Littorina brevicula* (Chiba et al. 2016) and thermal acclimation (e.g. Heinrich et al. 2012, Ravaux et al. 2012). Some of these mechanisms can persist for long periods when epigenetic factors can, at least temporarily, fix a gene expression level (Metzger & Schulte 2016, Putnam et al. 2016, Clark et al. 2018). Phenotypic plasticity is effective over a very large range of timescales from changes in heartbeat rate driven by adrenergic changes in fish (Farrell et al. 2009) to seasonal or life-history changes such as the modulation of activity (e.g. Morley et al. 2007, Aguzzi & Sarda 2008, Oystein 2012), metabolic or growth rates (e.g. Bayne 2004, Obermüller et al. 2010, 2011), or the production of antifreeze (e.g. Cheng & Detrich 2007).

Responses at the population level can be affected by modifications of the genepool from mutation or, more likely, gene flow within and between populations. Genome level responses are also dictated by life-history factors including generation time, population size, number of embryos produced per reproductive event and ocean currents (for connectivity between separate populations). In viruses and microbes, nucleotide substitution rates can be very rapid, and evolutionary change can occur in days to weeks (Peel & Wyndham 1999, Duffy et al. 2008). However, decades may be required in species with few offspring per reproductive event and long generation times such as elephants, some Antarctic marine invertebrates such as brachiopods (Peck 2008) and bivalve molluscs (Román-González et al. 2017), and whales (Jackson et al. 2009). The different responses shown in Figure 17 thus cover a range of process scales and rates of change that are much wider than the response of any single species or population, as different taxa can have very different timescales of response. Climate change-driven alterations to the environment are at the slower end of the range of rates of change seen in Figure 17, and it might be expected that only the processes relevant at decadal or longer timescales might be important when considering responses to climate change. Several authors, however, including Pörtner et al. (2007, 2012), Helmuth (2009) and Peck (2011) have emphasised that processes at shorter timescales produce a cascade of responses that play a large part in determining outcomes at larger scales. It is not only a question of population persistence but also one of maintaining reproductive fitness and being able to produce future generations. These factors are also of great importance when identifying sensitivities that are important to making future predictions.

Somero (2010, 2012, 2015), amongst other authors, argued that the most important responses organisms have to confer survival and maintain fitness in a climate change context are flexibility of the phenotype (often called phenotypic plasticity) via acclimatisation of physiological processes and via alteration of the genepool in populations, genetic adaptation. It is the former that is thought to be particularly important in long-lived species with long generation times. Organisms with very short generation times and rapid reproductive processes need less phenotypic plasticity to survive as their ability to modify their genetic complement is high, and their response is to produce new variants and not survive as individuals. Conversely, when generation times are long, including many Antarctic marine species, there will be a strong requirement for phenotypic plasticity via mechanisms such as acclimatisation to allow survival until adequate genetic adaptation can be achieved (Peck 2011).

173

Ecological change

In Antarctica the effects of reduced sea ice and increased levels of iceberg scour, mainly recorded in the Antarctic Peninsula region, have been predominantly evaluated as ecological impacts. They have primarily been assessed in terms of ecological outcomes and responses (e.g. Lipps et al. 1979, Gutt & Starmans 1998, 2011, 2015, Peck et al. 1999, 2010a, Riddle et al. 2007, Bertolin & Schloss 2009, Barnes & Souster 2011, Ducklow et al. 2012, 2013, Fillinger et al. 2013, Cape et al. 2014, Sahade et al. 2015, Barnes 2016, Hauquier et al. 2016).

Studies here have, for instance, demonstrated that climate-related warming has been accompanied by strengthening of winds on the Antarctic Peninsula, and this has reduced sea ice extent and duration (Spence et al. 2017). Some authors support the idea that the oceanic warming has been produced by the development of stronger southern hemisphere mid-latitude winds that assist the movement of warmer offshore waters onto the continental shelf (Martinson et al. 2008, Spence et al. 2014). An alternative explanation has, however, come from fine-scale models that suggest the heat transfer could be due to mesoscale eddies and tides (Stewart & Thompson 2013, Flexas et al. 2015). This heat transfer resulted in an increase in the ice-free period on the Bellingshausen side of the peninsula by over three months per annum between 1979–1980 and 2010–2011 (Stammerjohn et al. 2008, 2012). Amongst other effects, the loss of sea ice was attributed to be the cause of a reduction in the numbers of krill present in the region over a similar period (Atkinson et al. 2004), as sea ice is an important overwintering resource for this species (Flores et al. 2012). Recent work has also emphasised the importance of nearshore fjords for overwintering in krill (Cleary et al. 2016), and these fjords are experiencing rapid change, especially on the west Antarctic Peninsula where these environments have warmed, and the majority of glaciers have retreated significantly in the last 50 years. It is difficult to see how animals such as krill can adapt to reduced sea ice, even with altered behaviours, when they rely on it for food and shelter in the critical juvenile period over winter.

Ice shelf loss and coastal glacier retreat are, however, not only affecting those species that rely on them for overwintering, but they are markedly changing the productivity and biodiversity in these areas (Peck et al. 2010a, Barnes 2015). Ice shelf loss has increased areas of open water and seabed, where a major biotic response has been the development of new productivity and biological communities and ecosystems (Arrigo et al. 2008, Peck et al. 2010a, Gutt et al. 2013b, Constable et al. 2014, Barnes 2015). The effects of this enhanced productivity on new and surrounding communities has yet to be fully evaluated.

Sea ice loss and increased coastal glacier and ice shelf retreat have also resulted in an increase in the frequency of iceberg impacts on the seabed on the Antarctic Peninsula, which has removed biota in recent decades and is limiting the capacity for biological communities to grow and sequester carbon (Barnes DKA et al. 2014a, Barnes 2016). It is not possible for the animals inhabiting these sites to adapt to such catastrophic events, and with their slow development, Antarctic animals will recolonise these devastated areas more slowly than when similar events (e.g. bottom trawling, volcanic eruptions) take place in temperate or tropical regions of the globe. Beyond ecological analyses of the impacts of changes in sea ice and iceberg scour, the other two major problems identified with global change in the Southern Ocean, rising temperatures and ocean acidification, have received considerable attention from the science community. These have been evaluated from ecological, physiological and genetic approaches, and here adaptation is more likely. These two factors are analysed in more detail below.

Rising temperatures

There have been many studies of physiological capacities to respond to temperature stress in Antarctic marine species over recent decades on a very wide range of different taxa, including fish (e.g. Macdonald & Montgomery 1982, Hardewig et al. 1999b, Hofmann et al. 2000, 2005,

Wilson et al. 2001, 2002, Podrabsky & Somero 2006, Franklin et al. 2007, Robinson & Davison 2008, Bilyk & DeVries 2011, Strobel et al. 2012, Todgham et al. 2017), molluscs (Peck 1989, Urban & Silva 1998, Pörtner et al. 1999b, 2006, Peck et al. 2002, 2004a, Clark et al. 2008a,b, Morley et al. 2010, 2011, 2012a,b,c, Reed et al. 2012, Reed & Thatje 2015), echinoderms (Stanwell-Smith & Peck 1998, Clark et al. 2008b, Peck et al. 2008, 2009b, Morley et al. 2012b, 2016c), amphipods (Young et al. 2006a,b, Clark et al. 2008b, Doyle et al. 2012, Gomes et al. 2013, 2014, Clusella-Trullas et al. 2014, Faulkner et al. 2014, Schram et al. 2015b), isopods (Whiteley et al. 1996, 1997, Robertson et al. 2001, Young et al. 2006a,b, Janecki et al. 2010, Clusella-Trullas et al. 2014, Faulkner et al. 2014), brachiopods (Peck 1989, 2008, Peck et al. 1997a), sponges (Fillinger et al. 2013), and macroalgae or phytoplankton (Montes-Hugo et al. 2009, Schloss et al. 2012). There have also been many assessments of the effects of elevated temperature using a larger-scale approach, both experimentally and using field observations identifying multispecies response or evaluating community, ecosystem or overall biodiversity level responses (e.g. Fraser & Hofmann 2003, Peck et al. 2004b, 2010b, 2013, Aronson et al. 2007, Clarke et al. 2007, Barnes & Peck 2008, Schofield et al. 2010, Richard et al. 2012, Gutt et al. 2015, Morley et al. 2016a, Clark et al. 2017).

The vast majority of these studies have shown that Antarctic marine ectotherms have poor capacities to survive elevated temperatures in experiments, when compared with lower latitude species, which was first identified in the 1960s (Somero & DeVries 1967). Antarctic fish have generally higher capacities to tolerate warming in laboratory experiments than invertebrates (e.g. Podrabsky & Somero 2006, Franklin et al. 2007, Robinson & Davison 2008, Bilyk & DeVries 2011), and some invertebrates have possibly the poorest reported abilities to survive experimentally elevated temperatures of any marine species on Earth, including the brachiopod *Liothyrella uva* (Peck 1989), the brittle star *Ophionotus victoriae* (Peck et al. 2009b) and the bivalve mollusc *Limopsis marionensis* (Pörtner et al. 1999b), amongst several others.

Very recently some of the first *in situ* warming experiments have been deployed in Antarctica to assess the effects of warming on natural field communities of biofouling organisms (Ashton et al. 2017b). This study used an embedded heating system to raise the temperature of the surface and overlying boundary layer of water of settlement panels by 1°C or 2°C above ambient. The experiment ran for nine months, and, except for temperature, all other environmental variables were unchanged. All warming treatments produced large changes in the assemblage structure, with a pioneer species of bryozoan *Fenestrulina rugula* dominating in warmed conditions. Growth rates of the common species present nearly doubled for a 1°C warming, indicating that some transition had occurred in the organisms' physiology because this level of effect is well beyond any possible direct effect of temperature on enzyme mediated biological systems, as the Q_{10} for growth was ~1000. This unexpected effect of warming could possibly be related to the problems associated with protein synthesis at low temperature discussed in the earlier section on 'Growth'. Ashton et al. (2017b) found that a warming of 2°C produced variable results between species, and this may indicate that the more sensitive taxa were nearing their thermal limits.

Rates of warming

One early confounding factor in assessing whether Antarctic marine species were generally more sensitive to warming in experiments was that experimental protocols varied between laboratories. Such experiments also do not mimic real-time temperature changes. Thus, a major factor here is the rate of warming used in experiments. This factor has been recognised for a longer period in experiments on terrestrial species and is known as the rate hypothesis reviewed by Terblanche et al. (2011). In Antarctic marine species, Peck et al. (2009a,b) demonstrated a marked effect of rate of warming in experiments on thermal limits. In a study assessing upper temperature limits in 14 species from six phyla at warming rates from 1°C day^{-1} to 1°C month^{-1}, they showed that thermal maxima were 2.5–8 times lower at the slowest rate of warming than the fastest (Figure 18). However,

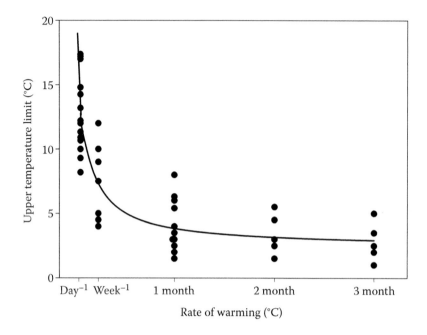

Figure 18 Upper temperature limit (CT$_{max}$) values for 14 species of Antarctic marine invertebrates at different rates of warming. Figure updated and adapted from Peck et al. (2014). Upper temperature limits quoted are values above the current average summer maximum temperature (1.0°C at the study site).

these results using several rates of warming also enabled, for the first time, extrapolation to long-term survival temperatures. This is particularly important in long-lived Antarctic species, where husbandry of species is often unknown or poorly understood, and it is not possible to keep many of them long-term in aquaria. In this case, more rapid thermal challenge experiments are essential, and upper thermal limits can be used as proxies for predicting the difference in resiliences or sensitivity between species and, therefore, enable modelling of future biodiversity change.

Demonstrating a relationship between the rate of warming in experiments and the upper temperature limits of species, as shown in Figure 18 and modelled by Richard et al. (2012), is of great importance when linking laboratory experiments with field observations or predictions. Sunday et al. (2011, 2012) showed there is a link between upper temperature limits and range boundaries in ectothermic animals, and the link is stronger in marine than terrestrial species. There is further a general consensus that including mechanistic biological capacities, through the use of physiological and phenotypical characteristics, into models predicting future consequences of climate change is an important step (e.g. Pennisi 2005, Gaston et al. 2009, Helmuth 2009, Peck et al. 2009a, Morley et al. 2016a). Variation in the assessment of upper temperature limits of the order seen in Figure 18 is a significant factor that can provide large errors when making predictions of climate impacts on biotas, including mechanistic traits, and one that needs to be included when making models to provide such predictions. Further important factors in the development of this macrophysiological approach that need to be quantified and allowed for include the various ways in which different aspects of phenotypic plasticity affect range shifts in response to environmental change (e.g. Chown et al. 2010, Sunday et al. 2012). Interactions between this and the effects of different rates of change are likely to be complex (Chown et al. 2009, Clusella-Trullas et al. 2011, 2014, Faulkner et al. 2014).

An additional way of analysing rate of warming data (as shown in Figure 18) and the relationship to environmental change is to use the data obtained from different rates of warming to estimate the difference between the long-term physiological temperature limit and the current maximum or

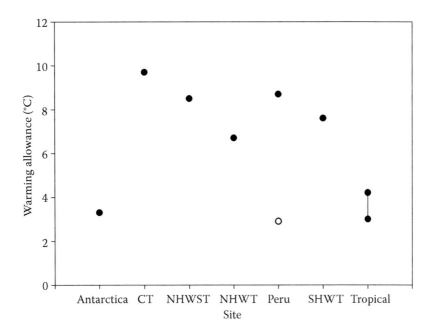

Figure 19 Warming allowances for multispecies assessments at seven sites from polar to tropical latitudes. Warming allowance here is the difference between long-term CT_{max} (see Figure 18) and the maximum environmental temperature at the site studied (°C). The sites range from Antarctica to Singapore (tropical) with abbreviations following Richard et al. (2012) and Peck et al. (2014): CT = cold temperate site (west coast of Scotland); NHWST = northern hemisphere warm shallow temperate site (depths less than 5 m); NHWT = northern hemisphere warm temperate site (deeper than 5 m; South of France and west coast of the United States); Peru (depths deeper than 5 m); and SHWT = southern hemisphere warm temperate (all depths deeper than 5 m). The line between the points for tropical data indicate the range of values, whilst the closed circle in the Peru data indicates the warming allowance outside an El Niño year, and the open symbol is for data in an El Niño. (Figure redrawn from Peck, L.S. et al. 2014. *Journal Experimental Biology* **217**, 16–22.)

mean temperatures experienced in the environment. This gives a measure of the buffer provided by the phenotypic plasticity of the population or species to warming and has been termed the warming tolerance by Deutsch et al. (2008), or the warming allowance by Richard et al. (2012). When this approach is used to compare Antarctic with temperate and tropical marine species, a pattern of small allowances in Antarctica and the tropics and larger allowances in temperate regions is apparent (Figure 19).

Figure 19 shows that Antarctic and tropical marine ectotherms have a warming allowance around 3–4°C above currently experienced maximum environmental temperatures, whereas for temperate species, this is around or more than double these values. Interestingly, in Peru, outside an El Niño, the warming allowance is around 9°C, but during an El Niño, where there are usually large mortalities in many species, it is around 3°C and similar to Antarctic and tropical species, suggesting both the very low and high-latitude species are living close to their thermal limits. However, the El Niño effect is slightly different from the tropics and poles, in that this, whilst a regular event, is infrequent and random enough that the animals are not able to adapt to the elevated temperatures it causes and as such can be viewed along similar lines to corrosive upwelling events that cause similar devastation along the US coast (Feely et al. 2008).

Having established that Antarctic marine cold-blooded species have restricted temperature limits compared to temperate species and poor capacities to respond to warming, it is now important to understand the mechanisms they can employ to resist, or respond to warming.

Mechanisms of resistance to warming

There have been many studies attempting to produce a synthetic, mechanistic understanding of the responses of marine species to warming environments or to provide an integrative evaluation of those responses (e.g. Pörtner 2001, 2002a,b, 2006, Peck 2002a,b, 2005a,b, 2011, Pörtner et al. 2005a,b, 2007, 2012, Peck et al. 2009a, 2014, Somero 2010, 2012, Ingels et al. 2012), and many of these have included understanding of Antarctic species or been built upon research carried out in Antarctica. Furthermore, the development of modern molecular methods has allowed analyses of resistance to warming at a finer scale than previously and allows a focus on specific mechanisms such as the heat shock response, the electron transport chain or anaerobic processes (e.g. Huth & Place 2016a,b, Clark et al. 2017, Enzor et al. 2017).

The most commonly cited and prevalent current theory on the mechanisms dictating resistance to warming is the oxygen and capacity limited thermal tolerance (OCLTT) hypothesis that was developed from the earlier oxygen limitation hypothesis initially proposed by Hans Pörtner (Pörtner et al. 2000, 2004, Pörtner 2001, 2002a,b, Pörtner & Farrell 2008). This hypothesis suggests that there is an optimal temperature window for an organism to function in and that at either side of this window a point is reached where performance declines because of aerobic (oxygen supply) constraints (Figure 20). The points where performance declines are called the pejus (=getting worse) thresholds (T_p in Figure 20), and these are identified by a decline in whole-animal aerobic scope (Pörtner et al. 2012). The loss of performance is attributed to the balance of oxygen demand and supply moving away from the optimal position. With continued warming or decrease in temperature, the aerobic scope eventually falls to zero when the upper or lower critical threshold temperatures (T_c in Figure 20) are reached. These points can be identified by the accumulation of anaerobic end products in cells and tissues, such as succinate in marine invertebrates and lactate in vertebrates. These effects are posited to occur before more severe physiological temperature effects such as membrane disruption or protein denaturation, which occur closer to the temperature of death (T_d in Figure 20).

The OCLTT paradigm was based around several pieces of information, many of which were obtained in studies of Antarctic marine species. These included the finding that upper critical temperatures are characterised by the accumulation of anaerobic metabolic end products such as succinate, for example, in the bivalves *Limopsis marionensis* (Pörtner et al. 1999b), *Laternula elliptica* (Peck et al. 2002) and *Adamussium colbecki* (Bailey et al. 2003), and in fish in the eelpout *Pachycara brachycephalum* (Van Dijk et al. 1999). These data were supported by findings that showed the capacity to perform activity declined before critical thresholds were achieved and well before upper lethal temperatures were reached. This was a demonstration that aerobic scope declines as temperature is progressively increased, and it was demonstrated in Antarctic species for, amongst others, burying in the clam *Laternula elliptica* and righting in the limpet *Nacella concinna* (Peck et al. 2004b), where in both species 50% of the population lost the ability to perform the activity being tested when warmed to between 2°C and 3°C, and none were capable of burying or righting, respectively, at 5°C. In the scallop *Adamussium colbecki*, increased temperature had an even stronger impact on its ability to swim, a higher energy activity requiring greater aerobic scope, with complete failure at 2°C (Bailey 2001, Bailey et al. 2005). Furthermore, the Antarctic fish *Pagothenia borchgrevinki* survives up to 11°C when warmed acutely. Its swimming performance is, however, more tightly limited, with its fastest swimming speed only observed between 1°C and 2°C, followed by a progressive decline in speed under further warming such that maximum speed at 7°C is only 50% that at 2°C (Wilson et al. 2002). Other Antarctic fish are likely to be similarly, or more, limited in their ability to maintain activity, as *P. borchgrevinki* appears to be less sensitive to elevated temperature than other species (Seebacher et al. 2005, Bilyk & DeVries 2011). A progressive loss of the capacity to perform aerobic activity with raised temperature is in line with predictions from the OCLTT hypothesis because it should reflect progressive reduction in aerobic scope or

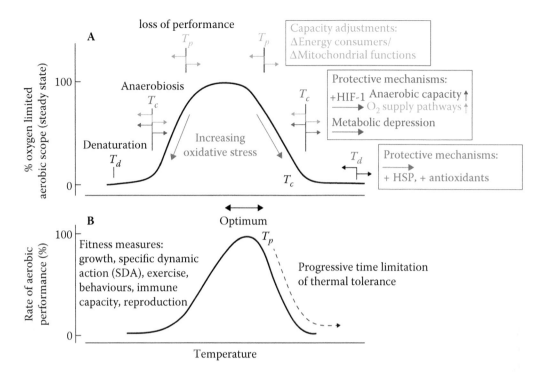

Figure 20 Schematic diagram of concept of oxygen limited thermal tolerance (following Pörtner et al. 2000, 2004, Pörtner 2001, 2002a,b). (A) An optimum temperature range is flanked by zones of progressive loss of aerobic scope. The start of these zones of progressive decline in capacity is denoted by T_p, the pejus threshold. Beyond the pejus zone is where aerobic scope declines to zero and anaerobic end products begin to accumulate. This zone is entered at the critical threshold, T_c and survival beyond T_c is time limited, depending on the rate of accumulation of anaerobic products and the species tolerance to these products. T_d denotes the zone where death occurs in the short term. All of these thresholds can shift when phenotypic plasticity mechanisms are entrained or from genetic adaptation. Boxes denote typical physiological and molecular mechanisms that dominate in each zone. (B) A typical aerobic performance curve, where whole-animal maximum scope and performance is usually asymmetric, with the peak of the curve nearer the upper T_p. (Figure modified from Figure 13.6 in Pörtner, H.O. et al. 2012. In *Antarctic Ecosystems: An Extreme Environment in a Changing World*, A. Rogers et al. (eds). Wiley Interscience, 379–416.)

capacity. Another strong support for this paradigm was obtained from studies that showed varying oxygen levels in seawater affected the upper temperature limits. Thus, when oxygen concentrations in seawater were increased, upper temperature limits and the capacity to burrow were increased, and both decreased when oxygen concentrations were lowered, in the clam *Laternula elliptica* (Peck et al. 2004b, 2007a, Pörtner et al. 2006). Several later studies provided further support for OCLTT. In Antarctica, these included that the concentration of haematocrit (the amount of haemoglobin) circulating in the blood was strongly correlated with temperature limits in five species of Antarctic fish (Beers & Sidell 2011).

Outside the Antarctic literature progressively more reports have been published that question the validity of OCLTT as a universal overarching concept explaining thermal limits (e.g. Ejbye-Ernst et al. 2016, Verberk et al. 2016). This has been especially the case in insects, where studies investigating low-temperature limits do not support the paradigm, and investigations of upper-temperature limits show stronger support in water-respiring species than air-breathing taxa (e.g. Klok et al. 2004, Stevens et al. 2010, Boardman & Terblanche 2015, Verberk et al. 2016, Shiehzadegan et al. 2017). Studies on other taxa have also questioned the universality of the hypothesis including in the snake *Python regius* (Fobian et al. 2014), the toad *Rhinella marina* (Overgaard et al. 2012), the freshwater shrimp *Macrobrachium rosenbergii* and several fish species (e.g. Clark T. et al. 2013, Norin et al. 2014, Wang et al. 2014, Lefevre et al. 2016).

In the Antarctic literature some studies have reported data not consistent with OCLTT. Clark et al. (2017) investigated the molecular responses to warming at 1°C h^{-1} by analysing complementary metabolomics and transcriptomic data, at temperatures close to their upper limits in six marine invertebrate species. Responses were diverse, with only two of the six species showing evidence of the accumulation of anaerobic end products, whilst only three increased expression of HSP70, the classical heat shock response. Each species appeared to have its own molecular response profile, and the conclusion drawn was there is no overall unifying mechanism determining temperature limits. Further to this, Devor et al. (2016) showed that increasing the oxygen content of seawater did not affect the upper thermal limits of either red-blooded or haemoglobinless Antarctic fishes.

As discussed earlier, the rate of warming in experiments markedly affects the upper temperatures to which marine species can survive. There is also mounting evidence that the mechanisms setting these upper temperature limits differ at different rates of warming. Several studies of Antarctic fish and crustaceans at rapid rates of warming (1°C h^{-1} or faster) have cast doubt on OCLTT. These have been based around evaluations of the functions of nervous systems showing failure, of at least some parts at similar temperatures to the whole-animal upper temperature limits. In fish, there is evidence that conduction at central nervous system (CNS) synapses fails in *Pagothenia borchgrevinki* near their upper limits (Hochachka & Somero 2002). In another study on the same species, specimens produced a sharp increase in the release of acetylcholine (a neurotransmitter) at neuromuscular junctions at temperatures between 12°C and 14°C, which is near this species' upper-temperature limit at the rate of warming used (Macdonald & Montgomery 1982). A similar finding of failure of neuromuscular junction function at temperatures near to whole-animal upper limits was reported for the Antarctic amphipod *Paraceradocus gibber* and the isopod *Glyptonotus antarcticus* (Young 2004, Young et al. 2006b). This was also shown in molecular evaluations of the potential mechanism behind failure for *Paraceradocus miersi* when warmed at 1°C h^{-1} (Clark et al. 2016).

As previously noted, many studies have provided support for OCLTT, and these have predominantly been conducted at rates of warming between 1°C h^{-1} and 1°C week^{-1}. However, since the most relevant rates of warming in the sea in relation to responses to climate change (those involving acclimatory responses) are much slower than this, and at slower rates of change, support for OCLTT has been limited. Several studies at medium and especially at slow rates of warming have suggested that other mechanisms rather than oxygen supply might be important in setting temperature limits. These include: (1) a limited capacity of the antioxidant system in Antarctic species to combat damage from reactive oxygen, resulting in the accumulation of toxic oxidised proteins such as protein carbonyls (Regoli et al. 1997, Abele et al. 2001, Powell et al. 2005, Heise et al. 2007, Clark et al. 2013); (2) high-temperature sensitivity of critical enzymes limits function of essential pathways at higher temperatures (Clark et al. 2016); (3) limitation of energy reserves, where increased maintenance costs exceed energy acquisition (e.g. Sørensen & Loeschcke 2007, Peck et al. 2010b, 2014) and (4) key processes are slowed at low temperature, which results in the observed very long times required for acclimation in Antarctica (Peck et al. 2014) and the poor acclimation of a range of cellular processes sets limits.

The first of the mechanisms, limitation of the antioxidant system, describes an imbalance between increased production of toxic metabolic end products from increased levels of oxidative damage at elevated temperatures and the ability of the cellular mechanisms involved to break these products down. As cells and tissues become energetically compromised during prolonged heat exposure, for example, protein turnover starts to shut down as an energy saving strategy (Hochachka et al. 1996). This produces a depression in metabolic rate, and under these conditions, cells may fail to remove efficiently cellular oxidative damage products, such as protein carbonyls (oxidised proteins) and malondialdehyde (MDA), which is a biomarker of oxidative stress and an initial product of lipid peroxidation. Accumulation of these products occurs as metabolic rates decline and autophagic, and proliferative activities become reduced (summarised by Philipp & Abele 2010). This results in a progressive increase in toxic end products in cells until a level is reached that the organisms can no longer tolerate, and apoptosis dominates the cellular processes (Powell et al. 2005, Zhang et al. 2008).

The second is a direct limitation of metabolic function by limited capacities of key enzymes. Experimental studies on the environmental stress response of *Aequiyoldia eightsii* and the scallop *Adamussium colbecki* indicated that whilst MDA levels increased in response to oxidative stress, levels of the antioxidant enzyme superoxide dismutase were decreasing (Regoli et al. 1997, Abele et al. 2001). At the time, it was suggested that whilst Antarctic species may have maximised the activity of their antioxidant system to work at near freezing temperatures to combat the associated high oxygenation levels, this came with a trade-off of reduced thermal stability (Abele et al. 2001). In similar studies, the thermal denaturation of enzymes involved in the Krebs cycle have been suggested as the underlying cause of the accumulation of succinate as a stress metabolite (Van Den Thillart & Smit 1984, Michaud et al. 2008). This field of investigation is expanding due to the rapid advances in sequencing technologies, which allow for the evaluation of thousands of genes in any one species when subjected to different environmental stressors. In the study of six different marine invertebrates warmed at 1°C h^{-1}, as described previously, different responses were seen in each species. Whilst it was impossible to narrow down organism failure to one specific enzyme or group of enzymes, it was noteworthy that in the bivalve *Aequiyoldia eightsii,* the metabolite *O*-propionyl carnitine was produced, which is involved in fatty acid and energy metabolism. This chemical is critical for the induction of antioxidant defence against lipid peroxidation in mammals, which is cytotoxic (Sayed-Ahmed et al. 2001). Hence, although limited in number, these studies show that there is species-specific enhanced sensitivity of critical enzymes, which may directly impact on organism physiology and survival.

The third mechanism for temperature limitation, via insufficiency of energy reserves, can work in two main ways: (1) As temperature increases metabolic rates in ectotherms rise, the minimum maintenance costs rise, usually increasing by a factor of two to three for every 10°C of warming (Peck 2016). Increased temperature increases the rate at which, for instance, digestive enzymes function and the rate at which food can be processed, but these changes will be small and only affect resource gained over part of the year. If overall resource availability does not change, or changes to a very small extent, and there is little or no ability to increase resources gained, then an increase in maintenance metabolic costs reduces energy available for other functions such as growth or reproduction, and such reductions can prejudice long-term survival of populations; (2) The interplay between resource availability, seasonality and increased costs could have specific impacts. Most Antarctic primary consumers feed on the summer phytoplankton bloom, which is only available for a short period each year (Clarke 1988, Clarke & Leakey 1996, Ducklow et al. 2013, Venables et al. 2013). In response to the low food availability in winter, many Antarctic species stop feeding and markedly reduce metabolic rates in winter, sometimes to a hypometabolic state (Barnes & Clarke 1994, 1995, Brockington & Peck 2001, Barnes & Peck 2005, McClintock et al. 2005, Morley et al. 2016c), and this winter hypometabolic state is not limited to primary consumers, having been observed in fish (Campbell et al. 2008), bivalve molluscs (Morley et al. 2012a), and brittle stars and nudibranch molluscs (Obermüller et al. 2010). Seasonality of primary production is primarily driven by light and nutrient availability, and these factors are not likely to change significantly the duration of food availability for primary consumers in the Southern Ocean over the coming decades of climate warming, although changes in ice cover are likely to have large effects (see sections on 'The intertidal and sea ice', and 'Ecological change'). The warming of the environment will, however, increase the metabolic energy costs of the overwintering period, and if stored reserves are insufficient, then capacity to exploit the following season's productivity could be reduced, or survival may be compromised during the winter.

The final potential mechanism for temperature limitation is the requirement for long periods for Antarctic marine species to acclimate their physiology to altered conditions. Studies of long-term exposure to raised temperature have been undertaken in Antarctic fish (Bilyk & DeVries 2011), and there is evidence that even after prolonged periods acclimation is incomplete in cardiorespiratory capacity (Egginton & Campbell 2016), which would reduce resistance to warming. In the invertebrates, very long periods, between three and nine months, have been reported as required to complete acclimation in Antarctic marine invertebrates (Peck et al. 2009b, 2010b, 2014, Morley et al. 2011, 2016c, Suckling et al. 2015). Such long periods required to complete acclimation mean that physiological mechanisms are not optimised for several months when seasons change. Although for the next 100 years winter sea temperatures in Antarctica's shallows will still be close to the freezing point of seawater, winters will be shorter and summer maxima higher. Hence, exploitation of altered conditions, especially as the annual temperature range will increase with a warmer sea, may not be possible as the animals are almost permanently in the process of acclimating their physiology.

One reason why acclimation might require longer to complete in Antarctic species is the reported lack of ability to modulate the fatty acid saturation of cellular membranes when animals are warmed. Gonzalez-Cabrera et al. (1995), held two notothenioid fish species, *Trematomus bernacchii* and *Trematomus newnesi*, for five weeks at 4°C and showed that, although some physiological characteristics such as serum osmolarity and Na^+/K^+-ATPase activity changed over time, there was no alteration of cell membrane fatty acid saturation. Further to this, Macdonald et al. (1988) found that the release of acetylcholine from synaptic vesicles increased markedly at temperatures above 6°C in the fish *Pagothenia borchgrevinki*. As noted previously by Young (2004) and Young et al. (2006b), failure of neuromuscular junctions also occurred at temperatures near upper thermal limits in two Antarctic marine crustaceans. A failure of mitochondrial function was reported at significantly lower temperatures in Antarctic fish than temperate species by Weinstein and Somero (1998) and Strobel et al. (2013), and mitochondrial failure at the lowest reported temperature for any marine species (9°C) was reported by Pörtner et al. (1999a) for the Antarctic clam, *Laternula elliptica*. These were both identified by a sharp change in slope of an Arrhenius regression of rate of oxygen consumption versus temperature for the mitochondria. Pörtner et al. (2007, 2012) interpreted these results as indicating that the absence of acclimation could be due to the evolutionary loss or malfunction of one or more of the enzyme systems needed to restructure lipids in membranes when animals are warmed. Many of the previous failures can be explained by an uncontrolled increase in membrane fluidity, and loss of control of membrane fluidity would affect a very wide range of cellular functions from pumping of ions for homeostasis to depolarisations for nervous conduction.

It is currently unclear which of the previous mechanisms is the most important in limiting any given species' capacity to resist or survive environmental warming, and as concluded by Clark et al. (2017), for resistance to rapid warming, different factors may be the most important in limiting different species, and there might not be a single overriding paradigm. The factor dictating survival might depend on both intrinsic factors such as levels of phenotypic plasticity or extrinsic factors such as rising temperature increasing costs with little increase in resource availability, and these factors will differ between species because of, for instance, different trophic requirements, activity levels, life histories, and so forth. It is likely that one or more of the mechanisms described will impact fitness in the majority of Antarctic marine species, and possibly all of the mechanisms will need to be assessed for a wide range of species before a reliable mechanistic understanding of the responses to environmental change can be obtained for Antarctic marine species.

Effects of age and life-history stage on resistance to warming

Thermal windows differ across life-history stages (Pechenik 1987, Pörtner & Farrell 2008, Philipp & Abele 2010, Peck 2011, Clark et al. 2013, Peck et al. 2013). They are often quoted to be narrowest in early developmental stages (Vernon 1900, Spicer & Gaston 1999), and it has been suggested that thermal constraints in early life stages might be a major factor limiting geographical distributions (Andronikov 1975, Pechenik 1999, Byrne 2011, 2012, Karelitz et al. 2017). Recent studies have especially emphasised the sensitivity of larval stages (Przeslawski et al. 2015, Karelitz et al. 2017, Clark pers. comm.).

In Antarctica, the effects of temperature on development of embryos and larvae have been studied for over four decades, with research primarily on echinoderms and molluscs (Pearse 1969, Bosch et al. 1987, Clarke 1992, Stanwell-Smith & Peck 1998, Peck et al. 2006a, 2007b, 2016a). The major finding is that temperature has a much more marked effect on development rates at temperatures around or below zero than in warmer areas, that temperature is the main criterion slowing development in Antarctic species (Hoegh-Guldberg & Pearse 1995), and that problems associated with protein synthesis are probably a major factor (Peck 2016). In the urchin *Sterechinus neumayeri*, for example, development rate of eggs and embryos increased monotonically between −2°C and +0.2°C but did not increase at temperatures above this (Stanwell-Smith & Peck 1998). Mortality, on the other hand, was low and stable between −2°C and +1.7°C, above which it increased rapidly. There was thus a window between +0.2°C and +1.7°C where development was at its fastest and mortality low. Two other species studied, the starfish *Odontaster validus* and *O. meridionalis*, had different patterns, with development rate in both increasing monotonically across the temperature range studied (−2°C to +3°C). Mortality was different between the starfish species studied, with *O. meridionalis* having a linear increase in mortality across the temperature range, from around 5% to 20%, whereas *O. validus* had constant mortality between 10% and 15% across the range. For all three species, levels of mortality of the eggs and embryos were well below the normally quoted 50% in thermal tolerance studies, so the upper temperature limit for eggs and embryonic development was well above 3°C. Upper temperature limits at slow rates of warming have been measured for adult *Sterechinus neumayeri* as between 4°C and 5°C (Peck et al. 2014), and at around 7°C for adult *Odontaster validus* (Peck et al. 2008). These results show that there are marked differences in the temperature relationships for development in Antarctic marine species, and there may not be a universal pattern or difference between early life stages and adult temperature limits, but that for some at least, temperature limits to early stage development are similar to those of adults in Antarctic marine species. So far, only a single study has been conducted on the effects of pressure on the urchin *Sterechinus neumayeri* that showed increasing pressure reduced the thermal windows of development for eggs and embryos (Tyler et al. 2000).

Most research on temperature limits in the past has compared early stages with adults either at similar rates of warming, or where adults have been warmed acutely. On these measures, early life stages are generally thought to be more sensitive to elevated temperature (e.g. Storch et al. 2011, Collin & Chan 2016). As seen previously, however, temperature limits for embryonic development are close to the long-term limits, or limits at slow rates of warming for at least some of the Antarctic species studied. There may thus be a perceived difference because relevant rates of warming have not been used for each life-history stage in the comparisons. Thus, if we take into account the numbers of cell divisions involved, where early stages are conducting cell division at rates much faster than later life stages, then a comparison with long-term limits at slow rates of warming for adults would be the most appropriate, and the differences between thermal limits of early and later life-history stages might not be as extreme as often perceived. This would argue for a limiting mechanism at the cellular level where problems accumulate with cell division or metabolic processes that run faster in early stages than adults and hence would argue in favour of mechanisms that explain temperature limitation from the accumulation of toxic end products or accumulated molecular damage.

At developmental stages between larvae and adults, there have been several studies of thermal tolerance and resistance to warming in juveniles or early juveniles of Antarctic marine invertebrates. The general conclusion in most of these studies is that juveniles are more resistant to warming than adults. Thus, in a study of four Antarctic species (*Laternula elliptica, Sterechinus neumayeri, Odontaster validus* and *Heterocucumis steineni*) warmed at rates between 1°C h⁻¹ and 1°C every 3 days, Peck et al. (2013) showed that early juveniles either had the same or higher upper temperature limits than large adults. Peck et al. (2007a) further showed that juveniles of the clam *Laternula elliptica* maintained the ability to rebury into sediment at higher temperatures than adults. In other studies on Antarctic marine species, juveniles have been found to be more resistant than adults to a range of environmental stressors. In *Laternula elliptica*, small individuals were less impacted than large adults by disturbance from increased sedimentation, and survival after injury (Philipp et al. 2011). In the same species, the immune system performed better in juveniles than adults in response to physical damage and starvation (Husmann et al. 2011), and again in *L. elliptica*, juveniles were more resistant to hypoxia than adults (Clark et al. 2013).

These whole-animal results were supported by evidence from several molecular analyses. For example, a study of haemocytes and siphon tissue that showed stronger upregulation in juveniles than adults of genes involved in antioxidant defence (*Le*-SOD and *Le*-catalase), wound repair (*Le*-TIMP and *Le*-chitinase), and stress and immune response (*Le*-HSP70, *Le*-actin, and *Le*-theromacin) (Husmann et al. 2014). Additionally, in a study of resistance to warming in *L. elliptica*, younger individuals had a more robust response, as demonstrated by a strong upregulation of transcripts of chaperones and antioxidants, not seen in adults (Clark et al. 2016). In that study, Clark et al. (2016) also showed that as individuals aged the proportion of their body allocated to muscle progressively declined, which may partly explain the lack of activity in older animals. These studies of responses to elevated temperature and other stresses all support the disposable-soma-theory of ageing: in long-lived species that reproduce many times, resources are diverted from tissue maintenance to reproduction progressively with age (Abele et al. 2009), and this leaves them less capable of responding to environmental insults.

In addition to the physical effects of warming on different species, there is a significant but underexplored consequence of the increased development rates that are a consequence of warming. That is, species with a pelagic phase, either as larvae or buoyant eggs, will experience a decrease in dispersal potential as the larval developmental phase is passed through more rapidly. Consequences are likely to differ between species and also depend on distances between habitable environmental patches and current regimes, but in *Sterechinus neumayeri*, raising the temperature from $-2°C$ to $+0.5°C$ reduces the time from egg fertilisation to settlement from over 120 days to under 90 days, or a 25%–30% reduction. Similar reductions in larval lifetime would likely remove some current dispersal routes for several or many species. This is clearly an area that needs more investigation in the future.

Resistance to altered temperature is the most studied environmental factor in relation to predicted future change, and it is often cited as the most important factor in this respect. However, the most recent investigations have involved combined evaluations of the impacts of altered temperature and carbonate saturation on larval development (e.g. Karelitz et al. 2017). The main conclusions of this type of work are that development rates increase with temperature, but that mortality and developmental abnormality increases markedly above a certain temperature. The two combined produce an optimal temperature window for development. This is discussed in more detail below.

Ocean acidification and organismal responses

Changing ocean chemistry

Human outputs of CO_2 since the industrial revolution have increased global temperatures by absorbing radiation from the Earth that would otherwise have passed out of the atmosphere. A second major consequence is that around 30% of the CO_2 released has been absorbed by the oceans, which has resulted in a decrease of ocean pH of around 0.1 units, and the major part of this change has occurred in the last 50 years (Caldeira & Wickett 2003, 2005, IPCC 2014). Carbonate solubility varies strongly with pH, temperature and pressure, with higher solubility at lower pH, higher pressure and lower temperature. Thus, for the same amounts dissolved per unit seawater and the same concentrations, the solubility state is lower in all three cases. This results in the deep oceans being undersaturated and the polar oceans having lower carbonate saturation at the surface than oceans at lower latitudes (Orr et al. 2005, Fabry et al. 2009). Future predictions are that acidification will continue to be strongest in the polar regions, especially in Antarctica such that surface waters will be undersaturated for aragonite, the more soluble of the common forms of carbonate, by the middle of this century (Figure 21, McNeil & Matear 2008, Feely et al. 2009a,b, 2012). This undersaturation will occur first in winter, which has

Carbonate levels predicted to drop as ocean acidifies

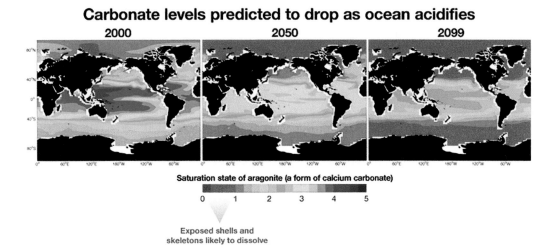

Figure 21 Model derived saturation states for aragonite in the world's oceans for 2000, 2050 and 2099. Where aragonite saturation is below 1, it is undersaturated, and materials made of this form of $CaCO_3$ should dissolve. (Figure produced by Woods Hole Oceanographic Institution based on Feely, R.A. et al. 2009a. *Oceanography* **22**, 36–47.)

colder conditions than summer, and there is already a seasonal signal in carbonate saturation in the Southern Ocean (McNeil et al. 2011, Hauri et al. 2016, Legge et al. 2017).

General predictions are that as carbonate solubility increases organisms will need to expend more energy to extract this essential material from seawater, and species heavily dependent on carbonates to make skeletons, such as molluscs, echinoderms and crustaceans, will be particularly at risk. Indeed, trends in skeletons across latitudes have also been observed that are consistent with higher solubility of carbonates at higher latitudes (e.g. Watson et al. 2012, 2017), and in an observational study, Sewell and Hoffmann (2011) using an analysis of depth ranges and calcification of Antarctic echinoderms concluded with the prediction that future acidification will produce a shallowing of the carbonate compensation depth (the depth at which the solubility (Ω) of $CaCO_3 = 1$ and below this it dissolves), which will remove heavily calcified forms from progressively shallower depths. In this respect, these groups in the polar regions can be viewed as likely being the first to suffer from acidification effects and as 'a canary in the mine' type of indicator of ecosystem state or health. Given the importance of Antarctica as the place where acidification is predicted to show its effects first, it is surprising that more research has not been carried out on the impact it is having, or will have, on Southern Ocean species. From a Web of Science search in September 2017, nearly 6000 papers have been published on ocean acidification in international scientific journals, but fewer than 250 of these include material on Antarctica.

The importance of Antarctic studies in a global context has been highlighted by several authors and a range of reasons quoted. Thus Whiteley (2011) stated 'The species most at risk [from ocean acidification] are exclusively marine and have limited physiological capacities to adjust to environmental change. They are poor iono- and osmoregulators and have limited abilities to compensate for acid–base disturbances. The problems are compounded in slow-moving, relatively inactive species because they have low circulating protein levels and low buffering capacities. Species living in low-energy environments, such as deep-sea and polar habitats, are particularly vulnerable, because they are metabolically limited with respect to environmental change'. This is the generally accepted position in ocean acidification research. However, despite the perceived importance of polar research and the seeming vulnerability of its fauna, there are relatively few publications and data are scarce.

Global responses to ocean acidification

Early investigations of the effects of altered pH generally found strongly negative effects on organisms of acidified conditions, especially in groups such as corals (e.g. Hoegh-Guldberg 1999, Raven et al. 2005, Kleypas et al. 2006, Hoegh-Guldberg et al. 2007, Kuffner et al. 2008, Kroeker et al. 2010, 2013). This area of research particularly expanded after the publishing of the seminal paper 'The other CO_2 problem' (Doney et al. 2009). Studies around this time on the effects of lowered pH on early life stages usually reported particularly large negative impacts (e.g. Kurihara & Shirayama 2004, Kurihara et al. 2007, Dupont et al. 2008, Watson et al. 2009). The majority of research in this field to date has concentrated on temperate species and studies range from field observations, for example in pteropods (Bednaršek et al. 2012), through laboratory manipulations to historical evaluations of skeletons in sediments (e.g. Mackas & Galbraith 2012) or from museum collections (Cross et al. 2018). The most frequent approach has primarily been based on laboratory manipulations and evaluating responses in terms of survival, integrity of skeletal structures or evaluations of physiological mechanisms such as metabolism or acid–base regulation (e.g. Gazeau et al. 2007, 2011, McDonald et al. 2009, Melzner et al. 2009a,b, 2011, Beniash et al. 2010, Talmage & Gobler 2010, Welladsen et al. 2010, Baumann et al. 2012, Comeau et al. 2012a, Ziveri et al. 2014).

Many of the earlier studies suffered from being short-term, acute exposures, and this problem began to be highlighted in the field around 2011 (e.g. Byrne 2011, Whiteley 2011). Whilst short-term studies may be useful in highlighting the differences in resilience between species, progressively more research has been conducted using longer-term exposures of weeks to months. Many of these have reported greatly reduced or no measureable effects of lowered pH on marine species in adult resistance (e.g. Coleman et al. 2014, Hazan et al. 2014, Cross et al. 2015, 2016), energetics (Morley et al. 2016c), reproductive characteristics (Havenhand & Schlegel 2009, Parker et al. 2012, Suckling et al. 2015, Munday et al. 2016, Runge et al. 2016), and embryonic and larval development (e.g. Suckling et al. 2014a,b, Bailey et al. 2016). These studies demonstrate that the impacts of acidification are much less when long-term exposures are used, and negative reproductive effects are lessened greatly when parental broodstock are conditioned for long periods before spawning (e.g. Suckling et al. 2015). One step further on from long-term studies is that of multigenerational effects. To date, very few studies have investigated the capacities of species to respond to acidified conditions across generations, and generally, significant adaptation is found in most of these studies (reviewed in Sunday et al. 2014, Stillman & Paganini 2015). In microbes, performance returns to normal levels within a few cell divisions (e.g. Collins 2012, Lohbeck et al. 2013). In multicellular animals, the requirement for such research has been highlighted, although data are very limited (e.g. Munday et al. 2013, Reusch 2014, Sunday et al. 2014). In the very few multigenerational studies conducted, performance of animals held in acidified conditions returns to levels very similar to that of individuals held in control conditions in only a few generations, and this has been demonstrated in the marine polychaete *Ophryotrocha labronica* (Rodríguez-Romero et al. 2016) and the sea urchin *Psammechinus miliaris* (Suckling & Clark pers. comm.), but there may also be lasting subtle effects such as impaired learning and altered neurotransmitter and retinal function (Nilsson et al. 2012, Chivers et al. 2014, Chung et al. 2014, Lai et al. 2015, Roggatz et al. 2016). An unusual very recent approach has been to investigate skeletons of marine calcifiers from samples collected over the last 120 years from museum specimens. In such a study, Cross et al. (2018) showed that oceanic changes since the early 1900s have had no discernible effect on shell structure or chemistry using a wide range of analytical techniques. Another unusual, very recent approach assessed changes in shell shape (Telesca et al. 2018) and composition (Telesca pers. comm.) in mussels of the genus *Mytilus* from sites across 3980 km of coastline stretching from the Arctic to the Mediterranean Sea and showed that seawater pH had a very small effect on these keystone bivalve and that temperature salinity and food supply were the major drivers. The only change detected in the study was a 3% increase in shell density. Furthermore, studies at the molecular level have shown significant amounts of genetic variability in response to ocean acidification (Pespeni et al. 2013a,b), which may promote resilient populations.

Another more recent development in the field is to investigate multiple species, or community-level responses to acidification. Here studies have generally shown that some species are negatively impacted by lowered seawater pH, others are affected little by conditions predicted up to 2100, while some species have improved performance in altered conditions (e.g. Dijkstra et al. 2011, Hale et al. 2011, Peck et al. 2015a, Schram et al. 2017). There has also been a move to study multifactorial effects, primarily combined with temperature but also with other factors such as salinity. These investigations have usually found that temperature has a larger impact on the biology of the studied species than acidification (e.g. Wood et al. 2010, Noisette et al. 2014, 2015, Collard et al. 2016, Zhang et al. 2016, Glandon & Miller 2017, Karelitz et al. 2017).

The impacts of acidification on marine calcifiers that reduce skeletal integrity are often discussed in the context of predator/prey interactions and poorer abilities to defend against, especially, crushing predators. Such discussions usually ignore likely impacts on the predators, where data are generally more limited, but impacts could be greater on durophagous predators because their skeletons are more soluble, and their skeletons dissolve faster in exposures to acidified conditions in general than their molluscan prey (E.M. Harper, personal communication). The crushing molars on the dactylus pivot point of crab and lobster chelae are the hardest and most carbonate-dense structures in either crustacean or molluscan skeletons. It is likely that they will, therefore, be more difficult to make in a future high CO_2 world than less calcium-dense structures. Further to this, recent research has demonstrated that when Antarctic macroalgae are held in altered conditions, their ability to deter amphipod herbivores remains unchanged, despite there being changes in their biochemical composition (Schram et al. 2017). More research is needed on crustacean predators and their likely future under climate change to better inform debate on predator/prey interactions in the coming decades.

Despite the increasing number of studies demonstrating that impacts of acidification are much less when long-term exposures are used and multiple generations examined, there are species differences. In some species, there are rapid adaptations to altered pH conditions across generations, but there are still species where acidification has significant negative impacts and areas, such as predation pressure, where there are very few data. The current challenge is to identify which species are being, and will be, negatively impacted in future and how that will affect ecosystem stability and the capacity of the oceans to provide services for human societies.

Ocean acidification in Antarctica

In a similar fashion to studies at lower latitudes, several Antarctic studies have reported negative impacts of ocean acidification. This was especially so for the early research where laboratory exposures to altered pH were rapid and acute. Again, as understanding improved and experiments have become more sophisticated, more recently laboratory-based research has used longer-term exposures and often reported smaller impacts, or none at all, for mid- or end-century predicted conditions. Recent studies involving more than one environmental stressor have also usually found acidification to have a smaller impact than other factors, especially temperature (e.g. Huth & Place 2016a,b, Enzor et al. 2017).

Enzor et al. (2017) working on Antarctic fish, using modern molecular methods, showed that temperature had a much larger effect on animal physiologies and capacity to acclimate than altered pH, but there was a small combined effect beyond that of temperature alone. In a 42- to 56-day acclimation experiment, they demonstrated that *Pagothenia borchgrevinki* was able to acclimate its oxygen consumption and aerobic capacity to temperatures around 4°C and a *pCO$_2$* of 1000 µatm, but *Trematomus newnesi and T. bernacchii* only achieved partial compensation, indicating there were energetic limitations or compromises. Temperature is usually identified as the major factor in experiments on Antarctic fish species when more than one variable is considered. There are, however, very large differences in the responses to multiple environmental stressors between Antarctic fish species that have been observed at molecular, cellular, tissue and whole-animal levels (e.g. Huth & Place 2016a,b). This echoes the work identifying species-specific responses and species-specific factors setting limits to warming in Antarctic marine invertebrates (Clark et al. 2017).

Most of the polar, and more specifically Antarctic, work on altered seawater chemistry has focussed on sea urchins, as these are very easy to maintain in aquaria and spawn. Recent laboratory investigations have focussed on multiple stressors and attempted to identify the relative impact of acidification and any synergistic effects. Ho et al. (2013) conducted fertilisation experiments on the urchin *Sterechinus neumayeri* at temperatures 2°C and 4°C above, and 0.2–0.4 pH units below, ambient. They found that temperature and sperm concentration affected fertilisation rates, but acidification did not. Ericson et al. (2012) studied fertilisation and early development in the same species and concluded that both attributes were highly resistant to altered pH and temperature conditions over the coming decades, but that predicted conditions for 2100 and beyond had measureable negative effects for both temperature and pH and their interactions. Similar conclusions were earlier drawn by Ericson et al. (2010) for both *S. neumayeri* and the nemertean *Parborlasia corrugatus*, where in both species fertilisation and early development were not noticeably affected by a pH decrease of 0.3–0.5 units. In a later study on *Sterechinus neumayeri*, Yu et al. (2013) also found very limited effects of near future acidification conditions (to 2100) on early development, and Clark et al. (2009) found that embryonic and larval development in *S. neumayeri* is more resilient to future predicted acidification conditions than in urchins from lower latitudes. In an investigation on the Subantarctic urchin *Arbacia dufresnii*, Catarino et al. (2012) concluded there was no measurable effect on embryo or larval morphology or behaviour of lowered pH (to 7.7 or 7.4), and the only identified impact was a slowing of development rate. There have been several investigations of combined impacts of predicted climate-driven alterations to temperature and acidification on Antarctic species. These have predominantly reported significant temperature but only mild or no effects of acidification. This has been demonstrated for early development in the urchin *Sterechinus neumayeri* (Byrne et al. 2013, Kapsenberg & Hofmann 2014). More recently, Karelitz et al. (2017) came to similar conclusions for the urchin *S. neumayeri* and the starfish *Odontaster validus*, that elevated temperature has a far greater impact on embryo and larval development then lowered pH, that there is little synergistic effect, and that temperature will have far more impact on future distributions than acidification of the oceans in these species. Similarly, in the infaunal bivalve mollusc *Laternula elliptica*, Bylenga et al. (2015) showed there was little effect of altered pH on fertilisation and early development, but elevated temperature (from −1.6°C to −0.5°C) had a positive effect across all pH treatments. In the single stressor experiment on the starfish *Odontaster validus*, Gonzalez-Bernat et al. (2013) reported negative effects of lowered pH on larval survival at end-century predicted conditions but not in near future conditions. They also reported small negative impacts on fertilisation success in end-century conditions but only at low-sperm concentrations. It thus seems that from the limited number of species studied, Antarctic urchins are resistant to predicted acidification conditions to the end of the century and possibly beyond.

In terms of the impacts on adults, Cummings et al. (2011) studied the effects of lowered pH on the bivalve *Laternula elliptica* and found significantly higher basal metabolic rates and more indication of stress as indicated by HSP70 expression after 21 days. They further found no significant differences between altered pH treatments and controls after 120 days in HSP70 expression or animal condition, but basal metabolism was elevated. They suggested that long-term increased costs could have large impacts on survival of the species. Further to this, McClintock et al. (2009) showed that shells of three dead mollusc and one dead brachiopod species suffered significant dissolution within 35 days of exposure to pH 7.4. Schram et al. (2014) found that altered pH and elevated temperatures to predicted end-century levels had no measurable effect on righting responses in the limpet *Nacella concinna* or the gastropod *Margarella antarctica*. Further to this, Schram et al. (2016a) found little or no effect of the same conditions on shell morphology, proximate body composition, growth or net calcification in the same species. One study at least, however, has reported stronger acidification than temperature effects at end-century conditions. Schram et al. (2016b), working on the macroalgal associated amphipods *Gondogeneia antarctica* and *Paradexamine fissicauda*, identified a range of sublethal effects caused by elevated temperature and lowered pH, but the latter produced a large increase in mortality in both species. As discussed in the previous section on 'Global responses to ocean acidification', our knowledge of predator–prey interactions in a high CO_2 world is poor. However, in Antarctica, this point might be different from lower latitudes because of the lack of durophagous predators (e.g. Aronson et al. 2007), but skeletal integrity and strength is likely to be important here in defence against drilling predators such as the snail *Trophonella shackletoni* and also against the abundant soft-bodied predators including starfish, anemones and nemertean worms.

A few studies in Antarctica have assessed the impacts of altered pH on macroalgae, which are important in their own right but also serve an extra key function as community habitats. Schoenrock et al. (2015) studied combined end-century temperature and acidification effects on *Desmarestia anceps* and *D. menziesii* and found no negative responses to either stressor in either species. The species did, however, respond differently, which led the authors to conclude that future change could alter algal community structure and, hence, communities associated with and dependent on macroalgae. After this, Schoenrock et al. (2016) showed that amongst the macroalgae crustose fleshy species responded better to altered conditions than calcified species, again suggesting significant change in community structure is likely in the coming decades.

Responses to altered environments can require very long periods in Antarctic marine invertebrates as acclimation to elevated temperature can take three to nine months (Peck et al. 2009b, 2010b, 2014, Bilyk and DeVries 2011), and acidified conditions can extend this period (Suckling et al. 2015). Thus, long-term studies are paramount in these species to ensure they are fully acclimated to the altered conditions, as stated in the previous section 'Global responses to ocean acidification' on temperature, and to allow evaluations of realistic responses to predicted environmental change. For example, *Laternula elliptica* requires very long times to complete shell repair after damage (Sleight et al. 2015), and hence, it is likely that some of the results obtained in the study by Cummings et al. (2011) may have differed if longer times had been allowed for full acclimation. In fact, long-exposure experiments on Antarctic brachiopods showed that they are able to repair shell damage and also produce new growth in end-century pH conditions at the same rates as animals in controls (Cross et al. 2015). Similarly, in a two-year exposure study of reproduction in the urchin *Sterechinus neumayeri*, Suckling et al. (2015) showed that up to eight months might be required for adults to acclimate fully their metabolism to altered pH and temperature, and embryo and larval development were significantly enhanced in altered conditions when the parents had been exposed to end-century conditions for 17 months as opposed to six months. In the same species, Morley et al. (2016c) found no effect of lowered pH (-0.3 and -0.5 units) or elevated temperature ($2°C$) on the growth of somatic or reproductive tissue, or scope for growth after 40 months of exposure. In general, even though the number of species studied to date is small, these very long-term exposures show at least some Antarctic marine species have the potential for significant phenotypic plasticity to respond to altered pH.

In contrast to shallow-water species, research on some pelagic taxa has consistently demonstrated negative impacts of acidification. Thus, Bednaršek et al. (2012) showed that Southern Ocean pteropods have significant dissolution of their shells that is consistent with increased carbonate solubility in the region. A similar result was also demonstrated for pteropods in low-pH upwelling water in California (Bednaršek et al. 2014). Other studies have demonstrated measurable impacts of acidification on pteropods in experiments (Lischka et al. 2011) and from observations of changes in latitude and time (Comeau et al. 2012b). However, more recently, Peck et al. (2016b,c) have cast some doubt on how important some of these observations are as they showed that in healthy animals the external protein covering on pteropod shells, the periostracum, and internal repair mechanisms can provide a powerful defence against dissolution at reduced pH. They further showed that individuals that have damaged periostracum, either from earlier failed predation attacks or from handling in experiments, suffer significant dissolution that can be confused with dissolution from altered conditions, *per se*. Other Antarctic pelagic species investigated include copepods (Bailey et al. 2016), where acidification had little or no effect on development of early life stages, and krill where

Kawaguchi et al. (2011) found no effect on embryos and larvae of elevating CO_2 to 1000 µatm but found negative impacts at 2000 µatm. In a microcosm-based community-level study, Tarling et al. (2016) showed that exposure to pCO_2 levels of 750 and 1000 µatm altered the community balance, and the interaction between copepods and their dinoflagellate prey was particularly affected. It should be noted that all of the pelagic experimental studies, because of restrictions to shipboard experiments, have been short-term, often with large changes to levels of pCO_2 in the system to values beyond those predicted for the year 2100, and as seen for benthic species, increasing the duration of exposure significantly alters the outcomes of experimental trials. The future challenge will be the development of longer-term trials in pelagic species.

In summary, there are very few long-term studies in Antarctic species. Where these have been performed, they generally show resilience. However, to date we have no knowledge of the underlying genetics of either skeletal production, what makes a more resilient animal or indeed and what proportion of the population may contain genes that confer resilience. These are clearly targets for future research.

Summary and conclusions

Biodiversity in the Southern Ocean is much higher than would be expected from standard texts on global biodiversity that highlight the trend of reducing diversity with latitude. This high Antarctic biodiversity is particularly evident in the benthos, where it is estimated that as many as 17,000–20,000 species of invertebrate are likely living on the Antarctic continental shelves. The fauna has evolved in isolation from other continents and continental shelves, and a range of unique adaptations has been produced, the most well known of which are antifreeze in fish, an absence of red blood cells in channichthyid fish, the absence of the standard heat shock response in several species and large size in some taxa. There are likely many more subtle changes that are opaque to the methods used to assess adaptations in the past, but that will be uncovered in the coming years because of the rapid improvement in the 'omics' technologies and the explosion of their use on Antarctic species in the last five years.

As well as many cellular and physiological adaptations being different in Antarctic marine species, life histories and physiological rates are generally slower than those at lower latitudes. Some traits such as growth and embryo/larval development are slowed well beyond the normal expected effects of temperature on biological systems. This has been attributed to problems associated with the manufacture of proteins at low temperatures, which is carried out at rates around eight times slower than in temperate species. This problem has also been suggested to play a role in the observation that Antarctic marine species also generally have larger eggs than those at lower latitude, and there seems to be a general trend of larger eggs in lower temperature habitats. One factor suggested to be potentially important in making proteins more difficult to synthesise in Antarctica is increased viscosity effects on protein folding at temperatures near or below 0°C. Increased viscosity has also been proposed in this review as possibly implicated in pumping rates slowed beyond expected temperature effects in Antarctic filter feeders. This contrains feeding abilities but also has strong implications for all ciliary-based activity at polar temperatures, including swimming in marine invertebrate larvae and sperm and protists. A further viscosity-related point that needs attention is the possibility that the absence of haemoglobin in channichthyid icefish might give advantages at low subzero temperatures when blood viscosity is raised further by the presence of small ice crystals and the attendant antifreezes.

Slower physiological rates appear to be linked to lower capacities for activity and to perform work. This has been suggested to be the possible cause for the lack of crushing predators in Antarctic marine habitats, as power production in muscles is related to environmental temperature. It is suggested here that predation patterns in Antarctica may also differ from lower latitudes as they are dominated by soft-bodied, engulfing predators such as starfish, nemertean worms and anemones. This means predation would differ in kind from tropical to polar regions and makes assessments of predation pressure across latitudes more complex than considered in previous studies.

Global environmental change is a major issue for biologists and ecologists across the planet. It is particularly important in polar regions because the most rapid change has occurred in these areas in recent decades. In Antarctica, rapid change has been confined mainly to areas on and around the Antarctic Peninsula, where air temperatures rose by around 3°C in the last half of the twentieth century, and to the west of the Antarctic Peninsula, where sea temperatures at the surface increased by 1°C in the same time. Other areas of Antarctica have not warmed significantly, and this has been explained by the tight wind patterns around the continent. The ozone hole has contributed to the tightening and strengthening of the circumpolar westerly winds around Antarctica in the last century, which may have helped to keep the east Antarctic from warming (Marshall 2003, Barnes EA et al. 2014b, https://legacy.bas.ac.uk/met/gjma/sam.html). The future strength of the circumpolar westerlies and hence their impact on temperature in east Antarctica is affected both by the filling of the ozone hole and global CO_2 concentrations. These opposing effects have different outcomes in different Intergovernmental Panel on Climate Change (IPCC) model predictions (Barnes EA et al. 2014b). Some predictions indicate there will be more rapid warming in east and parts of west Antarctica, which would expand considerably the potential area where ecosystems are under threat.

Two main challenges for organisms in the Southern Ocean from climate change have been identified: warming and ocean acidification. There are two main ways organisms can respond effectively to such change, by coping with the plasticity of their phenotype and adapting their genetic make-up to make them more fit for the new conditions. Antarctic marine species appear to have poor capacities to respond to environmental warming compared to temperate species and in this respect are similar to tropical species. Unlike both temperate and tropical species, however, they need very long periods to reset their physiology after a change in temperature, requiring up to as much as nine months to acclimate. This means their ability to cope, through phenotypic plasticity, with fluctuations in temperatures is less than species living elsewhere. Their long generation times and production of larger, but fewer eggs when they spawn are also both factors reducing their likelihood of either generating mutations that are beneficial in increasing survival or exchanging useful genes between populations.

In the context of ocean acidification, calcification, especially to produce skeletons, has a greater energetic cost for high latitude marine species than those living at lower latitudes because CO_2 and carbonate solubility is higher in colder waters, which reduces carbonate availability. This cost is even greater when assessed in relation to the animal's overall energy budget because metabolic and growth rates fall markedly at temperatures below 0°C, but costs of calcification do not. Evidence for this is seen in the trend for marine calcifiers to have smaller skeletons in colder waters. Furthermore, taking into account regional impacts of climate change, carbonate ion availability will reach critical levels of undersaturation in polar oceans before areas at lower latitudes. Data suggest, however, that Antarctic marine species have better than expected capacities to resist altered seawater pH, and that temperature is likely to be a greater challenge for many species than acidification, at least until the end of this century. In this context, there is a fundamental need for long-term studies of organismal responses to change over several months to years. There is a further need for research incorporating transgenerational effects. Such studies should incorporate temperature, altered pH, oxygen and salinity amongst other factors, as well as working at higher ecological scales incorporating several species or working at the community or assemblage level. There are, however, many unknowns, and each species has different responses to change. The data are too poor to be able to identify which species are most vulnerable to change and if any of these are crucial to continued ecosystem functioning. This is the challenge for the future.

Long-term studies over decadal scales are needed to identify changes in physiological processes, especially reproductive effort and success, in Antarctic marine species. Data to date show large interannual variation. However, identifying trends, the role of natural cycles such as Southern Oscillation Index (SOI) or El Niño-Southern Oscillation (ENSO), or the effects of extreme years is essential to predicting impacts of future change, and this is not possible without very long-term data.

In addition, future change will also bring the collapse of ice shelves and coastal glaciers, which will expose new areas of coastal ocean to sunlight for primary production, both in the water column or, in shallower areas, on the seabed itself. This will allow new communities to establish, and at least in the short-term, overall levels of biomass in the Southern Ocean are likely to increase due to these processes. It could also affect the dispersal capabilities of many species, as it has been suggested here that transport on icebergs is a likely important mechanism for gene flow in the sedentary or slow moving benthos that brood their offspring, and these species dominate assemblages on the Antarctic seabed. There is much uncertainty over future changes in the physical marine environment, especially in relation to frequency of iceberg scour, currents and circulation patterns, so detailed predictions of biodiversity responses are not possible. It is highly likely, however, that if and when there is ice-free coastline year-round in Antarctica, the prospects for the establishment of non-native species will rise dramatically.

Overall, the marine life in the Antarctic is characterised by a unique cold-adapted fauna. Much current research has concentrated on the effects of change on marine invertebrates which are integral to the Antarctic food web and Antarctic marine ecology. Changes to the numbers and the balance of these invertebrates will significantly impact commercial fisheries, the ecology of marine protected areas (MPAs) and the populations of the charismatic larger animals (seals, whales, albatrosses, etc.) on which much of the burgeoning tourist industry is based. However, we currently do not know if the ecology of the polar regions critically hinges on one or two or even a handful of species. There are insufficient data at present for any level of certainty. However, with the emergence of new technologies, we are now in a better position than ever before to unlock the secrets of Antarctic marine life. These developments are rapidly changing our abilities to obtain information, produce and interrogate models, and test and erect hypotheses. This is especially so in the 'omics' technologies, where a revolution in understanding seems to be happening now. Dramatic improvements in understanding are also likely from the use of unmanned vehicles such as drones and gliders and the increasing sophistication of satellite imaging and analysis.

In terms of the science to be done in Antarctica in the coming decades, there are more questions and challenges now than ever before. The future may be daunting in many respects for many marine species, with the prospects of large negative impacts from climate change, but it also holds very exciting times for science in Antarctica, especially in understanding adaptations and improving the knowledge base from the use of novel technologies.

Acknowledgements

The author thanks a wide range of collaborators, students and assistants who helped to gather data cited in this review. He especially thanks Andrew Clarke for decades of science discussion relevant to this work. Others who have played key roles in aiding the development of ideas in this article are Melody Clark, David Barnes, Simon Morley, Alistair Crame, Hans Pörtner, John Pearse, John Spicer, Paul Tyler, Laura Grange, Ian Johnston, Richard Aronson, Simon Maddrell and Bob Boutilier. Andrew Clarke, Jim McClintock, John Spicer, Cinzia Verde, and especially Melody Clark made significant helpful comment on the manuscript. There have been many others that have made contributions that I should thank, but the list is too long, and the errors are my own.

References

Abele, D., Brey, T. & Philipp, E.E.R. 2009. Bivalve models of aging and the determination of molluscan lifespans. *Experimental Gerontology* **44**, 307–315.

Abele, D. & Puntarulo, S. 2004. Formation of reactive species and induction of antioxidant defence systems in polar and temperate marine invertebrates and fish. *Comparative Biochemistry and Physiology A-Molecular and Integrative Physiology* **138A**, 408–415.

Abele, D., Tesch, C., Wencke, P. & Pörtner, H.O. 2001. How does oxidative stress relate to thermal tolerance in the Antarctic bivalve Yoldia eightsi? *Antarctic Science* **13**, 111–118.

Addo-Bediako, A., Chown, S.L. & Gaston, K.J. 2002. Metabolic cold adaptation in insects: a large-scale perspective. *Functional Ecology* **16**, 332–338.

Aguera, A., Ahn, I.-Y., Guillaumot, C. & Danis, B. 2017. A dynamic energy budget (DEB) model to describe Laternula elliptica (King, 1832) seasonal feeding and metabolism. *PLoS ONE* **12**, e0183848.

Aguera, A., Collard, M., Jossart, Q., Moreau, C. & Danis, B. 2015. Parameter estimations of dynamic energy budget (DEB) model over the life-history of a key Antarctic species: the Antarctic sea star Odontaster validus Koehler, 1906. *PLoS ONE* **10**, e0140078.

Aguzzi, J. & Sarda, F. 2008. A history of recent advancements on Nephrops norvegicus behavioural and physiological rhythms. *Reviews in Fish Biology and Fisheries* **18**, 235–248.

Ahn, I.-Y., Surh, J., Park, Y.-G., Kwon, H., Choi, K.-S., Kang, S.-H., Choi, J.H., Kim, K.-W. & Chung, H. 2003. Growth and seasonal energetic of the Antarctic bivalve Laternula elliptica from King George Island, Antarctica. *Marine Ecology Progress Series* **257**, 99–110.

Ahn, I.-Y., Chung, H. & Choi, K.S. 2001. Some ecological features of the Antarctic clam, Laternula elliptica (King & Broderip) in a nearshore habitat on King George Island. *Ocean and Polar Research* **23**, 419–424.

Ahn, I.-Y. & Shim, J.H. 1998. Summer metabolism of the Antarctic clam, Laternula elliptica (King & Broderip) in Maxwell Bay, King George Island and its implications. *Journal of Experimental Marine Biology and Ecology* **224**, 253–264.

Almroth, B.C., Asker, N., Wassmur, B., Rosengren, M., Jutfelt, F., Grans, A., Sundell, K., Axelsson, M. & Sturve, J. 2015. Warmer water temperature results in oxidative damage in an Antarctic fish, the bald nothoten. *Journal of Experimental Marine Biology and Ecology* **215**, 130–137.

Altringham, J.D. & Johnston, I.A. 1990. Modelling muscle power output in a swimming fish. *Journal of Experimental Biology* **148**, 395–402.

Altringham, J.D., Wardle, C.S. & Smith, C.I. 1993. Myotomal muscle function at different locations in the body of a swimming fish. *Journal of Experimental Biology* **182**, 191–206.

Ameyaw-Akumfi, C. & Hughes, R.N. 1987. Behaviour of Carcinus maenas feeding on large Mytilus edulis. How do they assess the optimal diet? *Marine Ecology Progress Series* **38**, 213–216.

Amsler, C.D., Iken, K., McClintock, J.B., Amsler, M.O., Peters, K.J., Hubbard, J.M., Furrow, F.B. & Baker, B.J. 2005. Comprehensive evaluation of the palatability and chemical defences of subtidal macroalgae from the Antarctic Peninsula. *Marine Ecology Progress Series* **294**, 141–159.

Amsler, C.D., McClintock, J.B. & Baker, B.J. 2014. Chemical mediation of mutualistic interactions between macroalgae and mesograzers structure unique coastal communities along the western Antarctic Peninsula. *Journal of Phycology* **50**, 1–10.

Anderson, M. & Johnston, I.A. 1990. Scaling of power output in fast muscle fibres of the Atlantic cod during cyclical contractions. *Journal of Experimental Biology* **170**, 143–154.

Andronikov, V. 1975. Heat resistance of gametes of marine invertebrates in relation to temperature conditions under which the species exist. *Marine Biology* **30**, 1–11.

Ansell, A.D. & Peck, L.S. 2000. Burrowing in the Antarctic anemone Halcampoides sp. from Signy Island, Antarctica. *Journal of Experimental Marine Biology and Ecology* **252**, 45–55.

Archambault, P., Snelgrove, P.V.R., Fisher, J.A.D., Gagnon, J.M., Garbary, D.J., Harvey, M., Kenchington, E.L., Lesage, V., Lévesque, M., Lovejoy, C., Mackas, D.L., McKindsey, C.W., Nelson, J.R., Pepin, P., Piché, L., Poulin, M. 2010. From sea to sea: Canada's three oceans of biodiversity. *PLoS ONE* **5**, e12182.

Arnaud, P.M. 1974. Contribution à la bionomie marine benthique des régions antarctiques et subantarctiques. *Téthys* **6**, 567–653.

Arnaud, P.M. 1977. Adaptations within the Antarctic marine benthic ecosystem. In *Adaptations within Antarctic Ecosystems*, G.A. Llano (ed.). Houston, TX: Gulf Publishing Co., 135–137.

Arntz, W.E., Brey, T. & Gallardo, V.A. 1994. Antarctic Zoobenthos. *Oceanography and Marine Biology: An Annual Review* **32**, 241–304.

Arntz, W.E., Brey, T., Gerdes, D., Gorny, M., Gutt, J., Hain, S. & Klages, M. 1992. Patterns of life-history and population dynamics of benthic invertebrates under the high Antarctic conditions of the Weddell Sea. In *Proceedings of the 25th European Marine Biology Symposium, Ferrara, Italy*, G. Colombo et al. (eds). Fredensborg, Denmark: Olsen & Olsen, 221–230.

Arntz, W.E., Gutt, J. & Klages, M. 1997. Antarctic marine biodiversity. In *Antarctic Communities, Species Structure and Survival*, B. Battaglia et al. (eds). Cambridge, UK: Cambridge University Press, 3–14.

Arntz, W.E., Thatje, S., Gerdes, D., Gili, J.-M., Gutt, J., Jacob, U., Montiel, A., Orejas, C. & Teixidó, N. 2005. The Antarctic-Magellan connection: Macrobenthos ecology on the shelf and upper slope, a progress report. *Scientia Marina* **69**(Suppl. 2), 237–69.

Aronson, R.B. & Blake, D.B. 2001. Global climate change and the origin of modern benthic communities in Antarctica. *American Zoologist* **41**, 27–39.

Aronson, R.B., Blake, D.B. & Oji, T. 1997. Retrograde community structure in the late Eocene of Antarctica. *Geology* **25**, 903–906.

Aronson, R.B., Thatje, S., Clarke, A., Peck, L.S., Blake, D.B., Wilga, C.D. & Seibel, B.A. 2007. Climate change and invisibility of the Antarctic benthos. *Annual Review of Ecology, Evolution, and Systematics* **38**, 129–154.

Arrigo, K.R., van Dijken, G.I. & Bushinsky, S. 2008. Primary production in the Southern Ocean 1997–2006. *Journal of Geophysical Research* **113**, C08004.

Arrigo, K.R., van Dijken, G.L. & Strong, A.L. 2015. Environmental controls of marine productivity hot spots around Antarctica. *Journal of Geophysics Research – Oceans* **120**, 5545–5565.

Arruda, L.M., Azevedo, J.N. & Neto, A.I. 1993. Abundance, age structure and growth, and reproduction of gobies (Pisces: Gobiidae) in the Ria de Aveiro lagoon (Portugal). *Estuarine, Coastal and Shelf Science* **37**, 509–523.

Ashton, G., Barnes, D.K.A., Morley, S.A. & Peck, L.S. 2017a. Response to 'Are Q_{10}s of more than 1000 realistic'? *Current Biology* **27**, R1303–R1304.

Ashton, G., Morley, S.A., Barnes, D.K.A., Clark, M.S. & Peck, L.S. 2017b. Warming by 1°C drives species and assemblage level responses in Antarctica's marine shallows. *Current Biology* **27**, 2698–2705.

Atkinson, A., Siegel, V., Pakhamov, E. & Rothery, P. 2004. Long-term decline in krill stock and increase in salps within the Southern Ocean. *Nature* **432**, 100–103.

Aumack, C.F., Amsler, C.D., McClintock, J.B. & Baker, B.J. 2010. Chemically mediated resistance to mesoherbivory in finely branched macroalgae along the western Antarctic Peninsula. *European Journal of Phycology* **45**, 19–26.

Avila, C., Iken, K., Fontana, A. & Gimino, G. 2000. Chemical ecology of the Antarctic nudibranch Bathydoris hodgsoni Eliot, 1907: Defensive role and origin of its natural products. *Journal of Experimental Biology and Ecology* **252**, 27–44.

Axelsson, M., Davison, W., Forster, M.E. & Farrell, A.P. 1992. Cardiovascular responses of the red-blooded Antarctic fishes Pagothenia bernacchii and P. borchgrevinki. *Journal of Experimental Biology* **167**, 179–201.

Ayala, A., Munoz, M.F. & Arguelles, S. 2014. Lipid peroxidation: Production, metabolism, and signalling mechanisms of malondialdehyde and 4-hydroxy-2-nonenal. *Oxidative Medicine and Cellular Longevity* **2014**, 360438.

Baardsnes, J. & Davies, P.L. 2001. Sialic acid synthase: The origin of fish type III antifreeze protein? *Trends in Biochemical Science* **26**, 468–469.

Bagenal, T.B. 1971. The interrelation of the size of fish eggs, the date of spawning and the production cycle. *Journal of Fish Biology* **3**, 207–219.

Bagge, L.E., Koopman, H.N., Rommel, S.A., McLellan, W.A. & Pabst D.A. 2012. Lipid class and depth-specific thermal properties in the blubber of the short-finned pilot whale and the pygmy sperm whale. *Journal of Experimental Biology* **215**, 4330–4339.

Bailey, A., Thor, P., Browman, H.I., Fields, D.M., Runge, J., Vermont, A., Bjelland, R., Thompson, C., Shema, S., Durif, C.M.F. & Hop, H. 2016. Early life stages of the Arctic copepod Calanus glacialis are unaffected by increased seawater pCO_2. *ICES Journal of Marine Science* **74**, 996–1004.

Bailey, D.M. 2001. *The thermal dependence of swimming and muscle physiology in temperate and Antarctic scallops*. PhD thesis, St Andrews University, UK, 140 pp.

Bailey, D.M., Johnston, I.A. & Peck, L.S. 2005. Invertebrate muscle performance at high latitude: swimming activity in the Antarctic scallop, Adamussium colbecki. *Polar Biology* **28**, 464–469.

Bailey, D.M., Peck, L.S., Pörtner, H.O. & Bock, C. 2003. High-energy phosphate metabolism during exercise and recovery in temperate and Antarctic scallops – an *in vivo* ^{31}P-NMR study. *Physiological and Biochemical Zoology* **76**, 622–633.

Baird, H.P. & Stark, J.S. 2013. Population dynamics of the ubiquitous Antarctic benthic amphipod Orchomenella franklini and its vulnerability to environmental change. *Polar Biology* **36**, 155–167.

Bargelloni, L., Marcatao, S., Lorenzo, Z. & Patarnello, T. 2000. Mitochondrial phylogeny of Notothenioids: a molecular approach to Antarctic fish evolution and biogeography. *Systematic Biology* **49**, 114–129.

Barnes, D.K.A. 1995. Seasonal and annual growth in erect species of Antarctic bryozoans. *Journal of Experimental Marine Biology and Ecology* **188**, 181–198.

Barnes, D.K.A. 2002. Invasions by marine life on plastic debris. *Nature* **416**, 808–809.

Barnes, D.K.A. 2015. Antarctic sea-ice losses drive gains in benthic carbon drawdown. *Current Biology* **25**, R789–R790.

Barnes, D.K.A. 2016. Iceberg killing fields limit huge potential for benthic blue carbon in Antarctic shallows. *Global Change Biology* **23**, 2649–2659.

Barnes, D.K.A. & Clarke, A. 1994. Seasonal variation in the feeding activity of four species of Antarctic bryozoa in relation to environmental factors. *Journal of Experimental Marine Biology and Ecology* **181**, 117–133.

Barnes, D.K.A. & Clarke, A. 1995. Seasonality of feeding activity in Antarctic suspension feeders. *Polar Biology* **15**, 335–340.

Barnes, D.K.A. & Clarke, A. 1998. Seasonality of polypide recycling and sexual reproduction in some erect Antarctic bryozoans. *Marine Biology* **131**, 647–658.

Barnes, D.K.A. & Conlan, K.E. 2007. Disturbance, colonization and development of Antarctic benthic communities. *Philosophical Transactions of the Royal Society of London B* **362**, 11–38.

Barnes, D.K.A. & Conlan, K.E. 2012. The dynamic mosaic. In *Antarctic Ecosystems: An Extreme Environment in a Changing World. Chapter 9*, A. Rogers et al. (eds). Wiley Interscience, 255–290.

Barnes, D.K.A., Fenton, M. & Cordingley, A. 2014a. Climate-linked iceberg activity massively reduces spatial competition in Antarctic shallow waters. *Current Biology* **24**, R553–R554.

Barnes, D.K.A. & Fraser, K.P.P. 2003. Rafting by five phyla on man-made flotsam in the Southern Ocean. *Marine Ecology Progress Series* **262**, 289–291.

Barnes, D.K.A., Fuentes, V., Clarke, A., Schloss, I.R. & Wallace, M.I. 2006a. Spatial and temporal variation in shallow seawater temperatures around Antarctica. *Deep-Sea Research II* **53**, 853–858.

Barnes, D.K.A., Galgani, F., Thompson, R.C. & Barlaz, M. 2009. Accumulation and fragmentation of plastic debris in global environments. *Philosophical Transactions of the Royal Society of London B* **364**, 1985–1998.

Barnes, D.K.A. & Griffiths, H.J. 2008. Biodiversity and biogeography of southern temperate and polar bryozoans. *Global Ecology and Biogeography* **17**, 84–99.

Barnes, D.K.A. & Kuklinski, P. 2010. Bryozoans of the Weddell Sea continental shelf, slope and abyss: did marine life colonize the Antarctic shelf from deep water, outlying islands or in situ refugia following glaciations? *Journal of Biogeography* **37**, 1648–1656.

Barnes, D.K.A. & Peck, L.S. 2005. Extremes of metabolic strategy in Antarctic Bryozoa. *Marine Biology* **147**, 979–988.

Barnes, D.K.A. & Peck, L.S. 2008. Vulnerability of Antarctic shelf biodiversity to predicted regional warming. *Climate Research* **37**, 149–163.

Barnes, D.K.A. & Souster, T. 2011. Reduced survival of Antarctic benthos linked to climate-induced iceberg scouring. *Nature Climate Change* **1**, 365–368.

Barnes, D.K.A., Webb, K. & Linse, K. 2006b. Slow growing Antarctic bryozoans show growth to increase over 20 years and be anomalously high in 2003. *Marine Ecology Progress Series* **314**, 187–195.

Barnes, E.A., Barnes, N.W. & Polvani, L.M. 2014a. Delayed Southern Hemisphere climate change induced by stratospheric ozone recovery, as projected by the CMIP5 models. *Journal of Climate* **27**, 857–892.

Baumann, H., Talmage, S.C. & Gobler, C.J. 2012. Reduced early life growth and survival in a fish in direct response to increased carbon dioxide. *Nature Climate Change* **2**, 38–41.

Bayne, B.L. 1976. *Marine Mussels: Their Ecology and Physiology*. Cambridge, UK: Cambridge University Press.

Bayne, B.L. 2004. Phenotypic flexibility and physiological tradeoffs in the feeding and growth of marine bivalve molluscs. *Integrative and Comparative Biology* **44**, 425–432.

Bednaršek, N., Feely, R.A., Reum, J.C.P., Peterson, B., Menkel, J., Alin, S.R. & Hales, B. 2014. Limacina helicina shell dissolution as an indicator of declining habitat suitability owing to ocean acidification in the California current ecosystem. *Proceedings of the Royal Society of London B* **281**, 2014.0123.

Bednaršek, N., Tarling, G.A., Bakker, D.C.E., Fielding, S., Jones, E.M., Venables, H.J. & Feely, R.A. 2012. Extensive dissolution of live pteropods in the Southern Ocean. *Nature Geoscience* **5**, 881–885.

Beers, J.M., Borley, K.A. & Sidell, B.D. 2010. Relationship among circulating hemoglobin, nitric oxide synthase activities and angiogenic poise in red- and white-blooded Antarctic notothenioid fishes. *Comparative Biochemistry and Physiology* **156A**, 422–429.

Beers, J.M. & Jayasundara, M. 2014. Antarctic notothenioid fish: what are the future consequences of 'losses' and 'gains' acquired during long-term evolution at cold and stable temperatures? *Journal of Experimental Biology* **218**, 1834–1845.

Beers, J.M. & Sidell, B.D. 2011. Thermal tolerance of Antarctic notothenioid fishes correlates with level of circulating hemoglobin. *Physiological and Biochemical Zoology* **84**, 353–362.

Belman, B.W. 1975. Oxygen-consumption and ventilation of Antarctic isopod Glyptonotus. *Comparative Biochemistry and Physiology* **50**, 149–151.

Belman, B.W. & Giese, A.C. 1974. Oxygen-consumption of an asteroid and an echinoid from the Antarctic. *Biological Bulletin* **146**, 157–164.

Benedetti, M., Martuccio, G., Nigro, M. & Regoli, F. 2008. Comparison of antioxidant efficiency in the Antarctic nototheniid species, Trematomus bernachii, Trematomus newnesi and Trematomus hansoni. *Marine Environmental Research* **66**, 98–99.

Beniash, E., Ivanina, A., Lieb, N.S., Kurochkin, I. & Sokolova, I.M. 2010. Elevated level of carbon dioxide affects metabolism and shell formation in oysters Crassostrea virginica. *Marine Ecology Progress Series* **419**, 95–108.

Berg, J. 2010. *An Introduction to Interfaces and Colloids, the Bridge to Nanoscience*. Singapore: World Scientific Publishing Co. Pte. Ltd., 804 pp.

Bergmann, M., Gutow, L. & Klages, M. 2015. *Marine Anthropogenic Litter*. Springer.com, ebook, 447 pp.

Berkman, P.A. 1990. The population biology of the Antarctic scallop, Adamussium colbecki (Smith, 1902) at New Harbor, Ross Sea. In *Antarctic Ecosystems. Ecological Change and Conservation*, K.R. Kerry & G. Hempel (eds). Berlin Heidelberg New York: Springer, 281–288.

Bertness, M.D., Garrity, S.D. & Levings, S.C. 1981. Predation-pressure and gastropod foraging: a tropical-temperate comparison. *Evolution* **35**, 995–1007.

Bertolin, M.L. & Schloss, I.R. 2009. Phytoplankton production after the collapse of the Larsen A ice shelf, Antarctica. *Polar Biology* **32**, 1435–1446.

Best, N.J., Bradshaw, C.J., Hindell, M.A. & Nichols, P.D. 2003. Vertical stratification of fatty acids in the blubber of southern elephant seals (Mirounga leonina): implications for diet analysis. *Comparative Biochemistry and Physiology* **134B**, 253–263.

Beu, A.G. 2009. Before the ice: biogeography of Antarctic Paleogene molluscan faunas. *Palaeogeography Palaeoclimatology, Palaeoecology* **284**, 191–226.

Beveridge, O.S., Petchey, O.L. & Humphries, S. 2010. Mechanisms of temperature-dependent swimming: the importance of physics, physiology and body size in determining protist swimming speed. *Journal of Experimental Biology* **213**, 4223–4231.

Bilyk, K.T. & DeVries, A.L. 2011. Heat tolerance and its plasticity in Antarctic fishes. *Comparative Biochemistry and Physiology* **158**, 382–390.

Birlenbach, U.S. & Leith, D.E. 1994. Intraspecific adaptations for different aerobic capacities: thoroughbred and draft horses. *FASEB Journal* **8**, A2.

Boardman, L. & Terblanche, J.S. 2015. Oxygen safety margins set thermal limits in an insect model system. *Journal of Experimental Biology* **218**, 1677–1685.

Bone, D.G. 1972. Aspects of the biology of the Antarctic amphipod Bovallia gigantean Pfeffer at Signy Island, South Orkney Islands. *British Antarctic Survey Bulletin* **27**, 105–122.

Boron, I., Russo, R., Boechi, L., Cheng, C.-H.C., di Prisco, G., Estrin, D.A., Verde, C. & Nadra, A.D. 2011. Structure and dynamics of Antarctic fish neuroglobin assessed by computer simulations. *IUBMB Life* **63**, 206–213.

Bosch, I., Beauchamp, K.A., Steele, M.E. & Pearse, J.S. 1987. Development, metamorphosis and seasonal abundance of embryos and larvae of the Antarctic sea urchin Sterechinus neumayeri. *Biological Bulletin* **173**, 126–135.

Bosch, I. & Pearse, J.S. 1990. Developmental types of shallow water asteroids of McMurdo Sound, Antarctica. *Marine Biology* **104**, 41–46.

Boulding, G.E. & Labarbera, M. 1986. Fatigue damage: repeated loading enables crabs to open larger bivalves. *Biological Bulletin* **171**, 538–547.

Bowden, D.A. 2005. Seasonality of recruitment in Antarctic sessile marine benthos. *Marine Ecology Progress Series* **297**, 101–118.

Bowden, D.A., Clarke, A. & Peck, L.S. 2009. Seasonal variation in the diversity and abundance of pelagic larvae of Antarctic marine invertebrates. *Marine Biology* **156**, 2033–2047.

Bowden, D.A., Clarke, A., Peck, L.S. & Barnes, D.K.A. 2006. Antarctic sessile marine benthos: colonization and growth on artificial substrata over 3 years. *Marine Ecology Progress Series* **316**, 1–16.

Boyce, S.J. & Clarke, A. 1997. Effect of body size and ration on specific dynamic action in the Antarctic plunderfish, Harpagifer antarcticus Nybelin 1947. *Physiological Zoology* **70**, 679–690.

Boyd, P.W., Arrigo, K.R., Strzepek, R. & van Dijken, G.L. 2012. Mapping phytoplankton iron utilization: insights into Southern Ocean supply mechanisms. *Journal of Geophysical Research* **117**, C06009.

Brante, A., Fernández, M., Eckerle, L., Mark, F., Pörtner, H.O. & Arntz, W.E. 2003. Reproductive investment in the crab Cancer setosus along a latitudinal cline: egg production, embryo losses and embryo ventilation. *Marine Ecology Progress Series* **251**, 221–232.

Bregazzi, P.K. 1972. Life cycles and seasonal movements of Cheirimedon femoratus (Pfeffer) and Tryphosella kergueleni (Miers) (Crustacea: Amphipoda). *British Antarctic Survey Bulletin* **30**, 1–34.

Brethes, J.C., Ferreyra, G. & Delavega, S. 1994. Distribution, growth and reproduction of the limpet Nacella-(Patinigera)-concinna (Strebel 1908) in relation to potential food availability, in Esperanza bay (Antarctic Peninsula). *Polar Biology* **14**, 161–170.

Brey, T., Gutt, J., Mackensen, A. & Starmans, A. 1998. Growth and productivity of the high Antarctic bryozoans Melicerita oblique. *Marine Biology* **132**, 327–333.

Brey, T. & Hain, S. 1992. Growth, reproduction and production of Lissarca notorcadensis (Bivalvia: Phylobryidae) in the Weddell Sea, Antarctica. *Marine Ecology Progress Series* **82**, 219–226.

Brey, T., Pearse, J.S., Basch, L., Mclintock, J.B. & Slattery, M. 1995a. Growth and production of Sterechinus neumayeri (Echinoidea: Echinodermata) in McMurdo Sound, Antarctica. *Marine Biology* **124**, 279–292.

Brey, T., Peck, L.S., Gutt, J., Hain, S. & Arntz, W. 1995b. Population dynamics of Magellania fragilis, a brachiopod dominating a mixed-bottom macrobenthic assemblage on the Antarctic shelf. *Journal of the Marine Biological Association of the U.K.* **75**, 857–870.

Briggs, J.C. 1996. Tropical diversity and conservation. *Conservation Biology* **10**, 713–718.

Brockington, S. 2001a. *The seasonal ecology and physiology of Sterechinus neumayeri (Echinodermata: Echinoidea) at Adelaide Island, Antarctica.* PhD thesis, British Antarctic Survey, Cambridge, 209 pp.

Brockington, S. 2001b. The seasonal energetics of the Antarctic bivalve Laternula elliptica (King & Broderip) at Rothera Point, Adelaide Island. *Polar Biology* **24**, 523–530.

Brockington, S., Clarke, A. & Chapman, A.L.G. 2001. Seasonality of feeding and nutritional status during the austral winter in the Antarctic sea urchin Sterechinus neumayeri. *Marine Biology* **139**, 127–138.

Brockington, S. & Peck, L.S. 2001. Seasonality of respiration and ammonia excretion in the Antarctic echinoid Sterechinus neumayeri. *Marine Ecology Progress Series* **259**, 159–168.

Brockington, S., Peck, L.S. & Tyler, P.A. 2007. Gametogenesis and gonad mass cycles in the common circumpolar Antarctic echinoid Sterechinus neumayeri. *Marine Ecology Progress Series* **330**, 139–147.

Brodeur, J., Peck, L.S. & Johnston, I.A. 2002. Feeding increases MyoD and PCNA expression in myogenic progenitor cells of Notothenia coriiceps. *Journal of Fish Biology* **60**, 1475–1485.

Brodte, E., Knust, R. & Pörtner, H.O. 2006. Temperature-dependent energy allocation to growth in Antarctic and boreal eelpout (Zoarcidae). *Polar Biology* **30**, 95–107.

Brody, S. 1945. *Bioenergetics and Growth.* New York: Reinhold Publ. Corp.

Brookes, P.S., Buckingham, J.A., Tenreiro, A.M., Hulbert, A.J. & Brand, M.D. 1998. The proton permeability of the inner membrane of liver mitochondria from ectothermic and endothermic vertebrates and from obese rats: correlations with standard metabolic rate and phospholipid fatty acid composition. *Comparative Biochemistry and Physiology B* **19**, 325–334.

Brown, C.R. & Cameron, J.N. 1991. The relationship between specific dynamic action (SDA) and protein synthesis rates in the channel catfish. *Physiological Zoology* **64**, 298–309.

Brown, K.E., King, C.K. & Harrison, P.L. 2015. Reproduction, growth and early life-history of the Antarctic gammarid amphipod Paramoera walkeri. *Polar Biology* **38**, 1583–1596.

Brown, K.M., Fraser, K.P.P., Barnes, D.K.A. & Peck, L.S. 2004. Links between the structure of an Antarctic shallow-water community and ice-scour frequency. *Oecologia* **141**, 121–129.

Buchanan, J.B. 1966. The biology of Echinocardium cordatum (Echinodermata: Spatangoida) from different habitats. *Journal of the Marine Biological Association of the U.K.* **46**, 97–1.

Buckley, B.A., Place, S.P. & Hofmann, G.E. 2004. Regulation of heat shock genes in isolated hepatocytes from an Antarctic fish, Trematomus bernacchii. *Journal of Experimental Biology* **207**, 3649–3656.

Bucolo, P., Amsler, C.D., McClintock, J.B. & Baker, B.J. 2011. Palatability of the Antarctic rhodophyte Palmaria decipiens (Reinsch) RW Ricker and its endo/epiphyte Elachista antarctica Skottsberg to sympatric amphipods. *Journal of Experimental Marine Biology and Ecology* **396**, 202–206.

Bulow, F.J., Zeaman, M.E., Winningham, J.R. & Hudson, W.F. 1981. Seasonal variations in RNA-DNA ratios and in indicators of feeding, reproduction, energy storage, and condition in a population of bluegill, Lepomis macrochirus Rafinesque. *Journal of Fish Biology* **18**, 237–244.

Burke, L., Kura, Y., Kassem, K., Revenga, C., Spalding, M. & McAllister, D. 2001. *Pilot Analysis of Global Ecosystems: Coastal Ecosystems.* Washington, DC: World Resources Institute.

Burmester, T., Ebner, B., Weich, B. & Hankeln, T. 2002. Cytoglobin: a novel globin type ubiquitously expressed in vertebrate tissues. *Molecular Biology and Evolution* **19**, 416–421.

Burmester, T., Weich, B., Reinhardt, S. & Hankeln, T. 2000. A vertebrate globin expressed in brain. *Nature* **407**, 520–523.

Burton, R.F. 2002. Temperature and acid—base balance in ectothermic vertebrates: the imidazole alphastat hypotheses and beyond. *Journal of Experimental Biology* **205**, 3587–3600.

Bylenga, C.H., Cummings, V.J. & Ryan, K.G. 2015. Fertilisation and larval development in an Antarctic bivalve, Laternula elliptica, under reduced pH and elevated temperatures. *Marine Ecology Progress Series* **536**, 187–201.

Byrne, M. 2011. Impact of ocean warming and ocean acidification on marine invertebrate life-history stages: vulnerabilities and potential for persistence in a changing ocean. *Oceanography and Marine Biology: an Annual Review* **49**, 1–42.

Byrne, M. 2012. Global change ecotoxicology: identification of early life-history bottlenecks in marine invertebrates, variable species responses and variable experimental approaches. *Marine Environmental Research* **76**, 3–15.

Byrne, M., Ho, M.A., Koleits, L., Price, C., King, C.K., Virtue, P., Tilbrook, B. & Lamare, M. 2013. Vulnerability of the calcifying larval stage of the Antarctic sea urchin Sterechinus neumayeri to near-future ocean acidification and warming. *Global Change Biology* **19**, 2264–2275.

Båmstedt, U. 1986. Chemical composition and energy content. In *The Biological Chemistry of Marine Copepods*, E.D.S. Corner & S.C.M. O'Hara (eds). Oxford, UK: Clarendon Press, 1–58.

Béguer, M., Bergé, J., Girardin, M. & Boët, P. 2010. Reproductive biology of Palaemon longirostris (Decapoda: Palaemonidae) from Gironde Estuary (France), with a comparison with other European populations. *Journal of Crustacean Biology* **30**, 175–185.

Caldeira, K. & Wickett, M.E. 2003. Anthropogenic carbon and ocean pH. *Nature* **425**, 365.

Caldeira, K. & Wickett, M.E. 2005. Ocean model predictions of chemistry changes from carbon dioxide emissions to the atmosphere and ocean. *Journal of Geophysics Research* **110**, C09S04.

Campbell, H.A., Davison, W., Fraser, K.P.P., Peck, L.S. & Egginton, S. 2009. Heart rate and ventilation in Antarctic fishes are largely determined by ecotype. *Journal of Fish Biology* **74**, 535–552.

Campbell, H.A., Fraser, K.P.P., Bishop, C.M., Peck, L.S. & Egginton, S. 2008. Hibernation in an Antarctic fish: on ice for winter. *PLoS ONE* **3**, e1743.

Cape, M.R., Vernet, M., Kahru, M. & Spreen, G. 2014. Polynya dynamics drive primary production in the Larsen A and B embayments following ice shelf collapse. *Journal of Geophysical Research-Oceans* **119**, 572–594.

Carefoot, T.H. 1990. Specific dynamic action (SDA) in the supralittoral isopod Ligia palasii: identification of components of apparent SDA and effects of amino acid quality and content on SDA. *Comparative Biochemistry and Physiology* **95A**, 309–316.

Carroll, A. 2014. *Population dynamics of the European sea bass (Dicentrarchus labrax) in Welsh waters.* MSc, Bangor University, 50 pp.

Casaux, R.J., Mazzotta, A.S. & Barrera-Oro, S. 1990. Seasonal aspects of the biology and diet of nearshore nototheniid fish at Potter Cove, South Shetland Islands, Antarctica. *Polar Biology* **11**, 63–72.

Cascella, K., Jollivet, D., Papot, C., Léger, N., Corre, E., Ravaux, J., Clark, M.S. & Toullec, J.-Y. 2015. Diversification, evolution and sub-functionalization of 70 kDa heat-shock proteins in two sister species of Antarctic krill: differences in thermal habitats, responses and implications under climate change. *PLoS ONE* **10**, e0121642.

Catarino, A., De Ridder, C., Gonzalez, M., Gallardo, P. & Dubois, P. 2012. Sea urchin Arbacia dufresnei (Blainville 1825) larvae response to ocean acidification. *Polar Biology* **35**, 455–461.

Cattaneo-Vietti, R., Chiantore, M. & Albertelli, G. 1997. The population structure and ecology of the Antarctic scallop Adamussium colbecki (Smith, 1902) at Terra Nova Bay (Ross Sea, Antarctica). *Scientia Marina* **61**, 15–24.

Chan, K.Y.K. 2012. Biomechanics of larval morphology affect swimming: insights from the sand dollars Dendraster excentricus. *Integrative and Comparative Biology* **52**, 458–469.

Chapelle, G. & Peck, L.S. 1995. Acclimation and substratum effects on metabolism of the Antarctic amphipods Waldeckia obesa (Chevreux, 1905) and Bovallia gigantea (Pfeffer, 1888). *Polar Biology* **15**, 225–232.

Chapelle, G. & Peck, L.S. 1999. Polar gigantism dictated by oxygen availability. *Nature* **399**, 114–115.

Chapelle, G. & Peck, L.S. 2004. Amphipod crustacean size spectra: new insights in the relationship between size and oxygen. *Oikos* **106**, 167–175.

Chapelle, G., Peck, L.S. & Clarke, A. 1994. Feeding and starvation effects on metabolism in the necrophagous Antarctic amphipod Waldeckia obesa (Chevreux, 1905). *Journal of Experimental Marine Biology and Ecology* **183**, 63–76.

Charnov, E.L. & Krebs, J.R. 1973. On clutch size and fitness. *Ibis* **116**, 217–219.

Chen, L., DeVries, A.L. & Cheng, C.-H.C. 1997. Evolution of antifreeze glycoprotein gene from a trypsinogen gene in Antarctic notothenioid fish. *Proceedings of the National Academy of Sciences USA* **94**, 3811–3816.

Chen, Z.Z., Cheng, C.-H.C., Zhang, J.F., Cao, L.X., Chen, L., Zhou, L.H., Jin, Y.D., Ye, H., Deng, C., Dai, Z.H., Xu, Q.H., Hu, P., Sun, S.H., Shen, Y. & Chen, L.B. 2008. Transcriptomic and genomic evolution under constant cold in Antarctic notothenioid fish. *Proceedings of the National Academy of Sciences USA* **105**, 12944–12949.

Cheng, C.-H.C. & Chen, L. 1999. Evolution of an antifreeze glycoprotein. *Nature* **401**, 443–444.

Cheng, C.-H.C. & Detrich III, H.W. 2007. Molecular ecophysiology of Antarctic notothenioid fishes. *Philosophical Transactions of the Royal Society B - Biological Sciences* **362**, 2215–2232.

Cheng, C.-H.C. & Detrich III, H.W. 2012. Molecular ecophysiology of Antarctic notothenioid fishes. In *Antarctic Ecosystems: An Extreme Environment in a Changing World. Chapter 12*, A. Rogers et al. (eds). Wiley Interscience, 357–378.

Cheng, C.-H.C. & DeVries, A.L. 1989. Structures of antifreeze peptides from the Antarctic eel pout, Austrolycicthys brachycephalus. *Biochimica et Biophysica Acta* **997**, 55–64.

Cheng, C.-H.C., di Prisco, G. & Verde, C. 2009a. Cold-adapted Antarctic fish: the discovery of neuroglobin in the dominant suborder Notothenioidei. *Gene* **433**, 100–101.

Cheng, C.-H.C., di Prisco, G. & Verde, C. 2009b. The "icefish paradox". Which is the task of neuroglobin in Antarctic hemoglobin-less icefish? *IUBMB Life* **61**, 184–188.

Chiantore, M., Cattaneo-Vietti, R., Elia, L., Guidetti, M. & Antonini, M. 2002. Reproduction and condition of the scallop Adamussium colbecki (Smith 1902), the sea-urchin Sterechinus neumayeri (Meissner 1900) and the sea-star Odontaster validus (Koehler 1911) at Terra Nova Bay (Ross Sea): different strategies related to inter-annual variations in food availability. *Polar Biology* **25**, 251–255.

Chiappori, F., Pucciarelli, S., Merelli, I., Ballarini, P., Miceli, C. & Milanesi, L. 2012. Structural thermal adaptation of b-tubulins from the Antarctic psychrophilic protozoan Euplotes focardii. *Proteins* **80**, 1154–1166.

Chiba, S., Iida, T., Tomioka, A., Azuma, N., Kurihara, T. & Tanaka, K. 2016. Population divergence in cold tolerance of the intertidal gastropod Littorina brevicula explained by habitat-specific lowest air temperature. *Journal of Experimental Marine Biology and Ecology* **481**, 49–56.

Chivers, D.P., McCormik, M.I., Nilsson, G.E., Munday, P.L., Watson, S.-A., Meekan, M., Mitcheel, M.D., Corkill, K.C. & Ferrari, M.C.O. 2014. Impaired learning of predators and lower prey survival under elevated CO_2: a consequence of neurotransmitter interference. *Global Change Biology* **20**, 515–522.

Cho, C.Y. & Slinger, S.J. 1979. Apparent digestibility measurements in feedstuffs for rainbow trout. *Proceedings of the World Symposium on Finfish Nutrition and Fishfeed Technology* **2**, 251–255.

Chown, S.L., Hoffmann, A.A., Kristensen, T.N., Angilletta, M.J. Jr, Stenseth, N.C. & Pertoldi, C. 2010. Adapting to climate change: a perspective from evolutionary physiology. *Climate Research* **43**, 3–15.

Chown, S.L., Jumbam, K.R., Sørensen, J.G. & Terblanche, J.S. 2009. Phenotypic variance, plasticity and heritability estimates of critical thermal limits depend on methodological context. *Functional Ecology* **23**, 133–140.

Chuang, S.H. 1994. Observations on the reproduction and development of Liothyrella neozelanica Thomson, 1918 (Terebratulacea, Articulata, Brachiopoda). *Journal of the Royal Society of New Zealand* **24**, 209–218.

Chung, W.S., Marshall, N.J., Watson, S.-A., Munday, P.L. & Nilsson, G.E. 2014. Ocean acidification slows retinal function in a damselfish through interference with GABA receptors. *Journal of Biology* **217**, 323–326.

Clark, D., Lamare, M. & Barker, M. 2009. Response of sea urchin pluteus larvae (Echinodermata: Echinoidea) to reduced seawater pH: a comparison among a tropical, temperate, and a polar species. *Marine Biology* **156**, 1125–1137.

Clark, M.S., Fraser, K.P.P. & Peck, L.S. 2008a. Antarctic marine molluscs do have an HSP70 heat shock response. *Cell Stress and Chaperones* **13**, 39–49.

Clark, M.S., Fraser, K.P.P. & Peck, L.S. 2008b. Lack of an HSP70 heat shock response in two Antarctic marine invertebrates. *Polar Biology* **31**, 1059–1065.

Clark, M.S., Geissler, P., Waller, C., Fraser, K.P.P., Barnes, D.K.A. & Peck, L.S. 2008c. Low heat shock thresholds in wild Antarctic inter-tidal limpets (Nacella concinna). *Cell Stress and Chaperones* **13**, 51–58.

Clark, M.S., Husmann, G., Thorne, M.A.S., Burns, G., Truebano, M., Peck, L.S., Abele, D. & Phillip, E.E.R. 2013. Hypoxia impacts large adults first: consequences in a warming world. *Global Change Biology* **19**, 2251–2263.

Clark, M.S. & Peck, L.S. 2009a. HSP70 Heat shock proteins and environmental stress in Antarctic marine organisms: a mini-review. *Marine Genomics* **2**, 11–18.

Clark, M.S. & Peck, L.S. 2009b. Triggers of the HSP70 stress response: environmental responses and laboratory manipulation in an Antarctic marine invertebrate (Nacella concinna). *Cell Stress and Chaperones* **14**, 649–660.

Clark, M.S., Sommer, U., Kaur, J., Thorne, M.A.S., Morley, S.A., King, M., Viant, M. & Peck, L.S. 2017. Biodiversity in marine invertebrate responses to acute warming revealed by a comparative multi-omics approach. *Global Change Biology* **23**, 318–330.

Clark, M.S., Thorne, M.A.S., Burns, G. & Peck, L.S. 2016. Age-related thermal response: the cellular resilience of juveniles. *Cell Stress and Chaperones* **21**, 75–85.

Clark, M.S., Thorne, M.A.S., King, M., Hipperson, H., Hoffman, J.I. & Peck, L.S. 2018. Life in the intertidal: cellular responses, methylation and epigenetics. *Functional Ecology*. https://doi.org/10.1111/1365-2435.13077.

Clark, M.S., Thorne, M.A.S., Toullec, J.Y., Meng, Y., Guan, L.L., Peck, L.S. & Moore, S. 2011. Antarctic krill 454 pyrosequencing reveals chaperone and stress transcriptome. *PLoS ONE* **6**, e15919.

Clark, M.S., Thorne, M.A.S., Vieira, F.A., Cardoso, J.C.R., Power, D.M. & Peck, L.S. 2010. Insights into shell deposition in the Antarctic bivalve Laternula elliptica: gene discovery in the mantle transcriptome using 454 pyrosequencing. *BMC Genomics* **11**, 362.

Clark, T.D., Sandblom, E. & Jutfelt, F. 2013. Aerobic scope measurements of fishes in an era of climate change: respirometry, relevance and recommendations. *Journal of Experimental Biology* **216**, 2771–2782.

Clarke, A. 1979. On living in cold water: K strategies in Arctic benthos. *Marine Biology* **55**, 111–119.

Clarke, A. 1980. A reappraisal of the concept of metabolic cold adaptation in polar marine invertebrates. *Biological Journal of the Linnean Society* **14**, 77–92.

Clarke, A. 1983. Life in cold water: the physiological ecology of polar marine ectotherms. *Oceanography and Marine Biology: An Annual Review* **21**, 341–453.

Clarke, A. 1985. The reproductive biology of the polar hippolytid shrimp Chorismus antarcticus at South Georgia. In *Marine Biology of Polar Regions and Effects of Stress on Marine Organisms*, J.S. Gray & M.E. Christiansen (eds). New York: John Wiley, 237–246.

Clarke, A. 1987a. The adaptation of aquatic animals to low temperatures. In *The Effects of Low Temperatures on Biological Systems*, B.W.W. Grout & G.J. Morris (eds). London: Edward Arnold, 315–348.

Clarke, A. 1987b. Temperature, latitude and reproductive effort. *Marine Ecology Progress Series* **38**, 89–99.

Clarke, A. 1988. Seasonality in the Antarctic marine environment. *Comparative Biochemistry and Physiology* **90**, 461–473.

Clarke, A. 1990. Temperature and evolution: Southern Ocean cooling and the Antarctic marine fauna. In *Antarctic Ecosystems: Ecological Change and Conservation*, K.R. Kerry & G. Hempel (eds). Berlin Heidelberg: Springer-Verlag, 9–22.

Clarke, A. 1991. What is cold adaptation and how should we measure it? *American Zoologist* **31**, 81–92.

Clarke, A. 1992. Reproduction in the cold: Thorson revisited. *Invertebrate Reproduction and Development* **22**, 175–184.

Clarke, A. 1993a. Reproductive trade-offs in caridean shrimps. *Functional Ecology* **7**, 411–419.

Clarke, A. 1993b. Egg size and egg composition in polar shrimps (Caridea: Decapoda). *Journal of Experimental Marine Biology and Ecology* **168**, 189–203.

Clarke, A. 1993c. Temperature and extinction in the sea: a physiologist's view. *Paleobiology* **19**, 499–518.

Clarke, A. 1996. Marine benthic populations in Antarctica: patterns and processes. In Foundations for Ecological Research West of the Antarctic Peninsula. *Antarctic Research Series* **70**, 373–388.

Clarke, A. 1998. Temperature and energetics: an introduction to cold ocean physiology. In *Cold Ocean Physiology*, H.O. Pörtner & R. Playle (eds). Cambridge, UK: Cambridge University Press, 3–30.

Clarke, A. 2003. Costs and consequences of evolutionary temperature adaptation. *Trends in Ecology and Evolution* **18**, 573–581.

Clarke, A. 2004. Is there a universal temperature dependence of metabolism? *Functional Ecology* **18**, 243–251.

Clarke, A. 2017. *Principles of Thermal Ecology: Temperature, Energy, and Life*. Oxford: Oxford University Press, 480 pp.

Clarke, A., Aronson, R.B., Crame, J.A., Gili, J.-M. & Blake, D.B. 2004a. Evolution and diversity of the benthic fauna of the Southern Ocean continental shelf. *Antarctic Science* **16**, 559–568.

Clarke, A. & Crame, J.A. 1989. The origin of the Southern Ocean marine fauna. In *Origins and Evolution of the Antarctic Biota*, vol. 47, J. A. Crame (ed.), London, UK: Special Publications of Geological Society of London, 253–268.

Clarke, A. & Crame, J.A. 1992. The Southern Ocean benthic fauna and climate change - a historical perspective. *Philosophical Transactions of the Royal Society of London Series B-Biological Sciences* **338**, 299–309.

Clarke, A. & Crame, J.A. 2010. Evolutionary dynamics at high latitudes: speciation and extinction in polar marine faunas. *Philosophical Transactions of the Royal Society B* **365**, 3655–3666.

Clarke, A. & Gore, D.J. 1992. Egg size and composition in Ceratoserolis (Crustacea: Isopoda) from the Weddell Sea. *Polar Biology* **12**, 129–134.

Clarke, A., Holmes, L.J. & White, M.G. 1988. The annual cycle of temperature, chlorophyll and major nutrients at Signy Island, South Orkney Islands, 1969–82. *British Antarctic Survey Bulletin* **80**, 65–86.

Clarke, A. & Johnston, N. 1999. Scaling of metabolic rate and temperature in teleost fish. *Journal of Animal Ecology* **68**, 893–905.

Clarke, A. & Johnston, N.M. 2003. Antarctic marine benthic diversity. *Oceanography and Marine Biology: An Annual Review* **41**, 47–114.

Clarke, A. & Lakhani, K.H. 1979. Measures of biomass, moulting behaviour and the pattern of early growth in Chorismus Antarctica (Pfeffer). *British Antarctic Survey Bulletin* **47**, 61–88.

Clarke, A. & Leakey, R.J. 1996. The seasonal cycle of phytoplankton, macronutrients and the microbial community in a nearshore Antarctic marine ecosystem. *Limnology and Oceanography* **41**, 1281–1294.

Clarke, A., Meredith, M.P., Wallace, M.I., Brandon, M.I. & Thomas, D.N. 2008. Seasonal and interannual variability in temperature, chlorophyll and macronutrients in northern Marguerite Bay, Antarctica. *Deep-Sea Research Part II* **55**, 1988–2006.

Clarke, A., Murphy, E.J.M., Meredith, M.P., King, J.C., Peck, L.S., Barnes, D.K.A. & Smith, R.C. 2007. Climate change and the marine ecosystem of the western Antarctic Peninsula. *Philosophical Transactions of the Royal Society of London B* **362**, 149–166.

Clarke, A. & Peck, L.S. 1991. The physiology of polar marine zooplankton. In *Proceedings Of the Pro Mare Symposium on Polar Ecology, Trondheim*, E. Sakshaug et al. (eds). Polar Research 10, 355–369.

Clarke, A. & Prothero-Thomas, E. 1997. The effect of feeding on oxygen consumption and nitrogen excretion in the Antarctic nemertean Parborlasia corrugatus. *Physiological Zoology* **70**, 639–649.

Clarke, A., Prothero-Thomas, E., Beaumont, J.C., Chapman, A.L. & Brey, T. 2004b. Growth in the limpet Nacella concinna from contrasting sites in Antarctica. *Polar Biology* **28**, 62–71.

Cleary, A.C., Durbin, E.G., Cassas, M.Z. & Zhou, M. 2016. Winter distribution and size structure of Antarctic krill Euphausia superba populations in-shore along the West Antarctic Peninsula. *Marine Ecology Progress Series* **552**, 115–129.

Clusella-Trullas, S., Blackburn, T.M. & Chown, S.L. 2011. Climatic predictors of temperature performance curve parameters in ectotherms imply complex responses to climate change. *The American Naturalist* **177**, 738–751.

Clusella-Trullas, S., Boardman, L., Faulkner, K.T., Peck, L.S. & Chown, S.L. 2014. Effects of temperature on heat-shock responses and survival of two species of marine invertebrates from sub-Antarctic Marion Island. *Antarctic Science* **26**, 145–152.

Coggan, R. 1996. Growth: Ration relationships in the Antarctic fish Notothenia coriiceps R. maintained under different conditions of temperature and photoperiod. *Journal of Experimental Marine Biology and Ecology* **210**, 23–35.

Coggan, R. 1997. Seasonal and annual growth rates in the Antarctic fish Notothenia coriiceps R. *Journal of Experimental Marine Biology and Ecology* **213**, 215–229.

Coleman, D.W., Byrne, M. & Davis, A.R. 2014. Molluscs on acid: gastropod shell repair and strength in acidifying oceans. *Marine Ecology Progress Series* **509**, 203–211.

Collard, M., Rastrick, S.P.S., Calosi, P., Demolder, Y., Dille, J., Findlay, H.S., Hall-Spencer, J.M., Milazzo, M., Moulin, L., Widdicombe, S., Dehairs, F. & Dubois, P. 2016. The impact of ocean acidification and warming on the skeletal mechanical properties of the sea urchin Paracentrotus lividus from laboratory and field observations. *ICES Journal of Marine Science* **73**, 727–738.

Collin, R. 2003. World-wide patterns of development in calyptraeid gastropods. *Marine Ecology Progress Series* **247**, 103–122.

Collin, R. & Chan, K.Y.K. 2016. The sea urchin Lytechinus variegatus lives close to the upper thermal limit for early development in a tropical lagoon. *Ecology and Evolution* **6**, 5623–5634.

Collins, M.A. & Rodhouse, P.G.K. 2006. Southern Ocean cephalopods. *Advances in Marine Biology* **50**, 191–265.

Collins, S. 2012. Marine microbiology: evolution on acid. *Nature Geoscience* **5**, 310–311.

Coma, R., Zabala, M. & Gili, J.-M. 1995. Reproduction and cycle of gonadal development in the Mediterranean gorgonian Paramuricea clavata. *Marine Ecology Progress Series* **117**, 173–183.

Comeau, S., Alliouane, S. & Gattuso, J.P. 2012a. Effects of ocean acidification on overwintering juvenile Arctic pteropods Limacina helicina. *Marine Ecology Progress Series* **456**, 279–284.

Comeau, S., Gattuso, J.P., Nisumaa, A. & Orr, J. 2012b. Impact of aragonite saturation state changes on migratory pteropods. *Proceedings of the Royal Society B* **279**, 732–738.

Comiso, J.C. 2010. Variability and trends of the global sea-ice cover. In *Sea-Ice*, 2nd edition D.N. Thomas & G.S. Diekmann (eds). Oxford: Wiley Blackwell, 205–246.

Constable, A.J., Melbourne-Thomas, J., Corney, S.P., Arrigo, K.R., Barbraud, C., Barnes, D.K.A., Bindoff, N.L., Boyd, P.W., Brandt, A., Costa, D.P., Davidson, A.T., Ducklow, H.W., Emmerson, L., Fukuchi, M., Gutt, J., Hindell, M.A., Hofmann, E.E., Hosie, G.W., Iida, T., Jacob, S., Johnston, N.M., Kawaguchi, S., Kokubun, N., Koubbi, P., Lea, M.-A., Makhado, A., Massom, R.A., Meiners, K., Meredith, M.P., Murphy, E.J., Nicol, S., Reid, K., Richerson, K., Riddle, M.J., Rintoul, S.R., Smith, W.O. Jr., Southwell, C., Stark, J.S., Sumner, M., Swadling, K.M., Takahashi, K.T., Trathan, P.N., Welsford, D.C., Weimerskirch, H., Westwood, K.J., Wienecke, B.C., Wolf-Gladrow, D., Wright, S.W., Xavier, J.C. & Ziegler, P. 2014. Climate change and Southern Ocean ecosystems I: how changes in physical habitats directly affect marine biota. *Global Change Biology* **20**, 3004–3025.

Convey, P., Barnes, D.K.A., Griffiths, H.J., Grant, S.M., Linse, K. & Thomas, D.N. 2012. Biogeography and regional classification of Antarctica. In *Antarctic Ecosystems. Chapter 15*, A. Rogers et al. (eds). Wiley Interscience, 471–491.

Convey, P., Chown, S.L., Clarke, A., Barnes, D.K.A., Bokhorst, S., Cummings, V., Ducklow, H.W., Frati, F., Green, T., Gordon, S., Griffiths, H.J., Howard-Williams, C., Huiskes, A.H.L., Laybourn-Parry, J., Lyons, W., Mcminn, A., Morley, S.A., Peck , L.S., Quesada, A., Robinson, S.A., Schiaparelli, S. & Wall, D.H. 2014. The spatial structure of Antarctic biodiversity. *Ecological Monographs* **84**, 203–244.

Convey, P. & Stevens, M.I. 2007. Antarctic biodiversity. *Science* **317**, 1877–1878.

Cook, A.J., Holland, P.R., Meredith, M.P., Murray, T., Luckman, A. & Vaughan, D.G. 2016. Ocean forcing of glacier retreat in the western Antarctic Peninsula. *Science* **353**, 283–286.

Correia, J.J. & Williams, R.C. Jr. 1983. Mechanisms of assembly and disassembly of microtubules. *Annual Review of Biophysics and Bioengineering* **12**, 211–235.

Cowen, R.K. & Sponaugle, S. 2009. Larval dispersal and marine population connectivity. *Annual Review of Marine Science* **1**, 443–66.

Crame, J.A. 1994. Evolutionary history of Antarctica. In *Antarctic Science, Global Concerns*, G. Hempel (ed.). Berlin: Springer-Verlag, 188–214.

Crockett, E.L. & Sidell, B.D. 1990. Some pathways of energy metabolism are cold adapted in Antarctic fishes. *Physiological Zoology* **63**, 472–488.

Cross, E.L., Harper, E.M. & Peck, L.S. 2018. A 120-year record of resilience to environmental change in brachiopods. *Global Change Biology* **24**, 2262–2271.

Cross, E.L., Peck, L.S. & Harper, E.M. 2015. Ocean acidification does not impact shell growth or repair of the Antarctic brachiopod Liothyrella uva (Broderip, 1833). *Journal of Experimetal Marine Biology and Ecology* **462**, 29–35.

Cross, E.L., Peck, L.S., Lamare, M.D. & Harper, E.M. 2016. No ocean acidification effects on shell growth and repair in the New Zealand brachiopod Calloria inconspicua (Sowerby, 1846). *ICES Journal of Marine Science* **73**, 920–926.

Cuellar, J., Yebenes, H., Parker, S.K., Carranza, G., Serna, M., Valpuesta, J.M., Zabala, J.C. & Detrich, H.W. III 2014. Partially compensated at their habitat temperature, probably by means of enhanced CP-binding affinity and increased flexibility of the CCT subunits. *Biology Open* **3**, 261–270.

Cummings, V., Hewitt, J., Van Rooyen, A., Currie, K., Beard, S., Thrush, S., Norkko, J., Barr, N., Heath, P., Halliday, N.J., Sedcole, R., Gomez, A., McGraw, C. & Metcalf, V. 2011. Ocean acidification at high latitudes: potential effects on functioning of the Antarctic bivalve Laternula elliptica. *PLoS ONE* **6**, e16069.

Cziko, P.A., DeVries, A.L., Evans, C.W. & Cheng, C.-H.C. 2014. Antifreeze protein-induced superheating of ice inside Antarctic notothenioid fishes inhibits melting during summer warming. *Proceedings of the National Academy of Sciences USA* **111**, 14583–14588.

Cziko, P.A., Evans, C.W., Cheng, C.-H.C. & DeVries, A.L. 2006. Freezing resistance of antifreeze-deficient larval Antarctic fish. *Journal of Experimental Biology* **209**, 407–420.

Daly, H. & Peck, L.S. 2000. Metabolism and temperature in the Antarctic octopus Pareledone charcoti (Joubin). *Journal of Experimental Marine Biology and Ecology* **245**, 197–214.

Daly, M., Rack, F. & Zook, R. 2013. Edwardsiella andrillae, a new species of sea anemone from Antarctic Ice. *PLoS ONE* **8**, e83476.

Dangan-Galon, F., Dolorosa, R.G., Sespeñe, J.S. & Mendoza, N.I. 2016. Diversity and structural complexity of mangrove forest along Puerto Princesa Bay, Palawan Island, Philippines. *Journal of Marine and Island Cultures* **5**, 118–125.

Daniels, R.A. 1983. Demographic characteristics of an Antarctic plunderfish, Harpagifer bispinis antarcticus. *Marine Ecology Progress Series* **13**, 181–187.

Davenport, J.A. 1988. Oxygen-consumption and ventilation rate at low-temperatures in the Antarctic protobranch bivalve mollusk Yoldia (=Aequiyoldia) eightsi (Courthouy). *Comparative Biochemistry and Physiology* **90A**, 511–513.

Davenport, J.A. 1997. Comparisons of the biology of the intertidal subantarctic limpets Nacella concinna and Kerguelenella lateralis. *Journal of Molluscan Studies* **63**, 39–48.

Davies, P.L. 2014. Ice-binding proteins: a remarkable diversity of structures for stopping and starting ice growth. *Trends in Biochemical Science* **39**, 548–555.

Davino, R., Caruso, C., Tamburrini, M., Romano, M., Rutigliano, B., Delaureto, P.P., Camardella, L., Carratore, V. & di Prisco, G. 1994. Molecular characterization of the functionally distinct hemoglobins of the Antarctic fish Trematomus newnesi. *Journal of Biological Chemistry* **269**, 9675–9681.

Davison W., Axelsson, M., Nilsson S. & Forster, M.E. 1997. Cardiovascular control in Antarctic notothenioid fishes. *Comparative Biochemistry and Physiology* **118A**, 1001–1008.

Davison, W. & Franklin, C.E. 2002. The Antarctic nemertean Parborlasia corrugatus: an example of an extreme oxyconformer. *Polar Biology* **25**, 238–240.

Dayton, P., Jarrell, S., Kim, S., Thrush, S., Hammerstrom, K., Slattery, M. & Parnell, E. 2016. Surprising episodic recruitment and growth of Antarctic sponges: implications for ecological resilience. *Journal of Experimental Marine Biology and Ecology* **482**, 38–55.

Dayton, P.K. 1990. Polar Benthos. In *Polar Oceanography, part B: Chemistry, Biology and Geology*, W.O. Smith (ed.). London: Academic Press, 631–685.

Dayton, P.K., Robilliard, G.A. & Paine, R.T. 1969. Anchor ice formation in McMurdo Sound, and its biological effects. *Science* **163**, 273–274.

Dayton, P.K., Robilliard, G.A. & Paine, R.T. 1970. Benthic faunal zonation as a result of anchor ice at McMurdo Sound, Antarctica. In *Antarctic Ecology*, Vol. 1, M.W. Holdgate (ed.). New York: Academic Press, 244–258.

Dayton, P.K., Robilliard, G.A., Paine, R.T. & Dayton, L.B. 1974. Biological accommodation in the benthic community at McMurdo Sound, Antarctica. *Ecological Monographs* **44**, 105–128.

De Broyer, C., Koubbi, P., Griffiths, H.J., Raymond, B., Udekem d'Acoz, C. d', Van de Putte, A.P., Danis, B., David, B., Grant, S., Gutt, J., Held, C., Hosie, G., Huettmann, F., Post, A. & Ropert-Coudert, Y. (eds). 2014. *Biogeographic Atlas of the Southern Ocean*. Scientific Committee on Antarctic Research, Cambridge, XII + 498 pp.

De Broyer, C., Lowry, J.K., Jazdzewski, K. & Robert, H. 2007. Catalogue of the Gammaridean and Corophiidean Amphipoda (Crustacea) of the Southern Ocean, with distribution and ecological data. *Bulletin, Institut Royal des Sciences Naturelles de Belgique* **77**, 1–325.

De Conto, R., Pollard, D. & Harwood, D. 2007. Sea-ice feedback and Cenozoic evolution of Antarctic climate and ice sheets. *Paleoceanography* **22**, PA3214.

Dearborn, J.H. 1965. *Ecological and faunistic investigations of the marine benthos at McMurdo Sound, Antarctica.* PhD dissertation, Stanford University, 238 pp.

Dehnel, P.A. 1955. Rates of growth of gastropods as a function of latitude. *Physiological Zoology* **28**, 115–144.

Dejours, P. 1966. *Respiration.* New York: Oxford University Press.

Dejours, P. 1981. *Principles of Comparative Respiratory Physiology.* New York: Elsevier.

Dell, R.K. 1972. Antarctic benthos. *Advances in Marine Biology* **10**, 1–216.

Deng, G., Andrews, D.W. & Laursen, R.A. 1997. Amino acid sequence of a new type of antifreeze protein, from the longhorn sculpin Myoxocephalus octodecimspinosis. *FEBS Letters* **402**, 17–20.

Desbruyeres, D. 1977. Evolution des populations de trois espèces d'annelides polychètes en milieu sub-Antarctique. *Comité National Francais des Recherches Antarctiques* **42**, 135–169.

Detrich, H.W. III 1991. Polymerization of microtubule proteins from Antarctic fish. In *Biology of Antarctic fish*, G. di Prisco et al. (eds). Berlin: Springer, 248–262.

Detrich, H.W. III 1998. Molecular adaptation of microtubules and microtubule motors from Antarctic fish. In *Fishes of Antarctica. A Biological Overview*, G. di Prisco et al. (eds). Milan: Springer-Verlag Italia, 139–149.

Detrich, H.W. III & Overton, S.A. 1986. Heterogeneity and structure of brain tubulins from cold-adapted Antarctic fishes: comparison to tubulins from a temperate fish and a mammal. *Journal of Biological Chemistry* **261**, 10922–10930.

Detrich, H.W. III, Parker, S.K., Williams, R.C., Nogales, E. & Downing, K.H. 2000. Cold adaptation of microtubule assembly and dynamics: structural interpretation of primary sequence changes present in the α- and β-tubulins of Antarctic fishes. *Journal of Biological Chemistry* **278**, 37038–37047.

Detrich, H.W. III, Prasad, V. & Ludueña, R.F. 1987. Cold-stable microtubules from Antarctic fishes contain unique α tubulins. *Journal of Biological Chemistry* **262**, 8360–8366.

Deutsch, C.A., Tewksbury, J.J., Huey, R.B., Sheldon, K.S., Ghalambor, C.K., Haak, D.C. & Martin, P.R. 2008. Impacts of climate warming on terrestrial ectotherms across latitude. *Proceedings of the National Acadamy of Science USA* **105**, 6668–6672.

Devor, D.P., Kuhn, D.E., O'Brien, K.M. & Crockett, E.L. 2016. Hyperoxia does not extend critical thermal maxima (CT_{max}) in white- or red-blooded Antarctic notothenioid fishes. *Physiological and Biochemical Zoology* **89**, 1–9.

DeVries, A.L. 1971. Glycoproteins as biological antifreeze agents in Antarctic fishes. *Science* **172**, 1152–1155.

DeVries, A.L. 2004. Ice antifreeze proteins and antifreeze genes in polar fishes. In *Life in the Cold*, B.M. Barnes & H.V. Carey (eds). Fairbanks, AK: Institute of Arctic Biology, University of Alaska, 275–282.

DeVries, A.L. & Cheng, C.-H.C. 1992. The role of antifreeze glycopeptides and peptides in the survival of cold water fishes. In *Water and Life; Comparative Analysis of Water Relationships at the Organismic, Cellular, and Molecular Levels*, G.N. Somero et al. (eds). Berlin, Heidelberg, Germany: Springer-Verlag, 303–315.

DeVries, A.L. & Cheng, C.-H.C. 2005. Antifreeze proteins and organismal freezing avoidance in polar fishes. In *Fish Physiology*, vol. 22, A. P. Farrell & J. F. Steffensen (eds). San Diego, CA: Academic Press, 155–201.

DeVries, A.L. & Wohlschlag, D.E. 1969. Freezing resistance in some Antarctic fishes. *Science* **163**, 1073–1075.

Dey, I., Buda, C., Wiik, T., Halver, J.E. & Farkas, T. 1993. Molecular and structural composition of phospholipid membranes in livers of marine and freshwater fish in relation to temperature. *Proceedings of the National Acadamy of Sciences* **90**, 7498–7502.

Dias, C.L., Ala-Nissila, T., Wong-ekkabut, J., Vattulainen, I., Grant, M. & Karttunen, M. 2010. The hydrophobic effect and its role in cold denaturation. *Cryobiology* **60**, 91–99.

Dijkstra, J.A., Westerman, E.L. & Harris, L.G. 2011. The effects of climate change on species composition, succession and phenology: a case study. *Global Change Biology* **17**, 2360–2369.

di Prisco, G., Cocca, E., Parker, S.K. & Detrich, H.W. 2002. Tracking the evolutionary loss of hemoglobin expression by the white-blooded Antarctic icefishes. *Gene* **295**, 185–191.

di Prisco, G. & Verde, C. 2015. The Ross Sea and its rich life: research on molecular adaptive evolution of stenothermal and eurythermal Antarctic organisms and the Italian contribution. *Hydrobiologia* **761**, 335–361.

Domack, E., Ishman, S., Leventer, A., Sylva, S., Willmott, V. & Huber, B. 2005. A chemotrophic ecosystem found beneath Antarctic ice shelf. *EOS Transactions of the American Geophysical Union* **86**, 269–276.

Doney, S.C., Fabry, V.J., Feely, R.A. & Kleypas, J.A. 2009. Ocean acidification: the other CO_2 problem. *Annual Reviews in Marine Science* **1**, 169–192.

Doyle, S.R., Momo, F.R., Brethes, J.C. & Ferrera, G.A. 2012. Metabolic rate and food availability of the Antarctic amphipod Gondogeneia antarctica (Chevreux 1906): seasonal variation in allometric scaling and temperature dependence. *Polar Biology* **25**, 413–424.

Ducklow, H., Clarke, A., Dickhut, R., Doney, S.C., Geisz, H.N., Huang, K., Martinson, D.G., Meredith, M.P., Moeller, H.V., Montes-Hugo, M., Schofield, O.M., Stammerjohn, S.E., Steinberg, D. & Frazer, W. 2012. The marine ecosystem of the West Antarctic Peninsula. In *Antarctica: An Extreme Environment in a Changing World*, A. Rogers et al. (eds). Oxford: Blackwell, 121–159.

Ducklow, H.W., Fraser, W.R., Meredith, M.P., Stammerjohn, S.E., Doney, S.C., Martinson, D.G., Sailley, S.F., Schofield, O.M., Steinberg, D.K., Venables, H.J. & Amsler, C.D. 2013. West Antarctic Peninsula: an ice-dependent coastal marine ecosystem in transition. *Oceanography* **26**, 190–203.

Duffy, S., Shackelton, L.A. & Holmes, E.C. 2008. Rates of evolutionary change in viruses: patterns and determinants. *Nature Reviews Genetics* **9**, 267–276.

Duman, J.G. 2015. Animal ice-binding (antifreeze) proteins and glycolipids: an overview with emphasis on physiological function. *Journal of Experimental Biology* **218**, 1846–1855.

Duman, J.G. & DeVries, A.L. 1976. The isolation, characterization and physical properties of antifreeze protein from the winter flounder, Pseudopleuronectes americanus. *Comparative Biochemistry and Physiology* **54**, 375–380.

Dunbar, M.J. 1968. *Ecological Development in the Polar-Regions*. Englewood Cliffs, NJ: Prentice-Hall.

Dunbar, M.J. 1970. Ecosystem adaptation in marine polar environments. In *Antarctic Ecology*, M.W. Holdgate (ed.). London: Academic Press, 105–114.

Dupont, S., Havenhand, J., Thorndyke, W., Peck, L. & Thorndyke M. 2008. CO_2-driven ocean acidification radically affects larval survival and development in the brittlestar Ophiothrix fragilis. *Marine Ecology Progress Series* **373**, 285–294.

Eastman, J.T. 1993. *Antarctic Fish Biology: Evolution in a Unique Environment*. London: Academic Press.

Eastman, J.T. & DeVries, A.L. 1986. Antarctic fishes. *Scientific American* **254**, 106–114.

Eastman, J.T. & McCune, A.R. 2000. Fishes on the Antarctic continental shelf: evolution of a marine species flock? *Journal of Fish Biology* **57**, 84–102.

Ebbinghaus, S., Meister, K., Prigozhin, M.B., DeVries, A.L., Havenith, M., Dzubiella, J. & Gruebele, M. 2012. Functional importance of short-range binding and long-range solvent interactions in helical antifreeze peptides. *Biophysics Journal* **103**, L20–L22.

Egginton, S. 1996. Blood rheology of Antarctic fishes: viscosity adaptations at very low temperatures. *Journal of Fish Biology* **48**, 513–521.

Egginton, S. & Campbell, H.A. 2016. Cardiorespiratory responses in an Antarctic fish suggest limited capacity for thermal acclimation. *Journal of Experimental Biology* **219**, 1283–1286.

Egginton, S., Campbell, H.A. & Davison, W. 2006. Cardiovascular control in Antarctic fish. *Deep Sea Research II Topical Studies in Oceanography* **53**, 1115–1130.

Egginton, S., Skilbeck, C., Hoofd, L., Calvo, J. & Johnston, I.A. 2002. Peripheral oxygen transport in skeletal muscle of Antarctic and sub-Antarctic notothenioid fish. *Journal of Experimental Biology* **205**, 769–779.

Ejbye-Ernst, R., Michaelsen, T.Y., Tirsgaard, B., Wilson, J.M., Jensen, L.F., Steffensen, J.F., Pertoldi, C., Aarestrup, K. & Svendsen, J.C. 2016. Partitioning the metabolic scope: the importance of anaerobic metabolism and implications for the oxygen-and capacity-limited thermal tolerance (OCLTT) hypothesis. *Conservation Physiology* **4**, cow019.

Elliott, J.M. 1976. Energy losses in the waste products of brown trout (Salmo trutta L.). *Journal of Animal Ecology* **45**, 561–580.

Elner, R.W. & Hughes, R.N. 1978. Energy maximization in the diet of the shore crab, Carcinus maenas. *Journal of Animal Ecology* **47**, 103–116.

Emig, C.C. 2017. *Atlas of Antarctic and Sub-Antarctic Brachiopoda*. Madrid: Carnets de Géologie, CG2017_B03.

Enzor, L.A., Hunter, E.M. & Place, S.P. 2017. The effects of elevated temperature and ocean acidification on the metabolic pathways of notothenioid fish. *Conservation Physiology* **5**, cox019.

Enzor, L.A. & Place, S.P. 2014. Is warmer better? Decreased oxidative damage in notothenioid fish after long-term acclimation to multiple stressors. *Journal of Experimental Biology* **217**, 3301–3310.

Enzor, L.A., Zippay, M.L. & Place, S.P. 2013. High latitude fish in a high CO_2 world: synergistic effects of elevated temperature and carbon dioxide on the metabolic rates of Antarctic notothenioids. *Comparative Biochemistry and Physiology* **164**, 154–161.

Ericson, J.A., Ho, M.A., Miskelly, A., King, C.K., Virtue, P., Tilbrook, B. & Byrne, M. 2012. Combined effects of two ocean change stressors, warming and acidification, on fertilization and early development of the Antarctic echinoid Sterechinus neumayeri. *Polar Biology* **35**, 1027–1034.

Ericson, J.A., Lamare, M.D., Morley, S.A. & Barker, M.F. 2010. The response of two ecologically important Antarctic invertebrates (Sterechinus neumayeri and Parborlasia corrugatus) to reduced seawater pH: effects on fertilisation and embryonic development. *Marine Biology* **157**, 2689–2702.

Eriksen, M., Lebreton, L.C.M., Carson, H.S., Thiel, M., Moore, C.J., Borerro, J.C., Galgani, F., Ryan, P.G. & Reisser, J. 2014. Plastic pollution in the world's oceans: more than 5 trillion plastic pieces weighing over 250,000 tons afloat at sea. *PLoS ONE* **9**, e111913.

Everson, I. 1970a. Reproduction in Notothenia neglecta Nybelin. *British Antarctic Survey Bulletin* **23**, 81–92.

Everson, I. 1970b. The population dynamics and energy budget of Notothenia neglecta Nybelin at Signy Island, South Orkney Islands. *British Antarctic Survey Bulletin* **23**, 25–50.

Everson, I. 1977. Antarctic marine secondary production and the phenomenon of cold adaptation. *Philosophical Transactions of the Royal Society of London B* **279**, 55–66.

Everson, I. 1984. Fish biology. In *Antarctic Ecology*, vol. 2, R.M. Laws (ed.). London: Academic Press, 491–532.

Everson, I. & Ralph, R. 1970. Respiratory metabolism of Chaenocephalus aceratus. In *Antarctic Ecology*, Vol. 1, M.W. Holdgate (ed.). New York: Academic Press Inc., 315–319.

Ewart, K.V., Li, Z., Yang, D.S.C., Fletcher, G.L. & Hew, C.L. 1998. The icebinding site of Atlantic herring antifreeze protein corresponds to the carbohydrate binding site of C-type lectins. *Biochemistry* **37**, 4080–4085.

Fabry, V.J., McClintock, J.B., Mathis, J.T. & Grebmeier, J.M. 2009. Ocean acidification at high latitudes: the bellweather. *Oceanography* **22**, 160–171.

Fahrbach, E., Rohardt, G. & Krause, G. 1992. The Antarctic coastal current in the southeastern Weddell Sea. *Polar Biology* **12**, 171–182.

Falk-Petersen, I.-B. & Lønning, S. 1983. Reproductive cycles of two closely related sea urchin species, Strongylocentrotus droebachiensis (O.F. Müller) and Strongylocentrotus pallidus (G.O. Sars). *Sarsia* **68**, 157–164.

Farrell, A.P. & Steffensen, J.F. (eds). 2005. *Fish Physiology*, vol. 22. London: Academic Press.

Farrell, A.P., Eliason, E.J., Sandblom, E. & Clark, T.D. 2009. Fish cardiorespiratory physiology in an era of climate change. *Canadian Journal of Zoology* **87**, 835–851.

Faulkner, K., Clusella-Trullas, S., Peck, L.S. & Chown, S. 2014. Lack of coherence in the warming responses of marine crustaceans. *Functional Ecology* **2**, 895–903.

Feder, M.E. & Hofmann, G.E. 1999. Heat shock proteins, molecular chaperones and their stress response: evolutionary and ecological physiology. *Annual Reviews in Physiology* **61**, 243–282.

Feely, R.A., Doney, S.C. & Cooley, S.R. 2009a. Ocean acidification: present conditions and future changes in a high-CO_2 world. *Oceanography* **22**, 36–47.

Feely, R.A., Orr, J., Fabry, V.J., Kleypas, J.A., Sabine, C.L. & Langdon, C. 2009b. Present and future changes in seawater chemistry due to ocean acidification. In *Carbon Sequestration and Its Role in the Global Carbon Cycle*. Geophysical Monograph Series, 183, B.J. McPherson & E.T. Sundquist (eds). Washington, DC: AGU, 175–188.

Feely, R.A., Sabine, C.L., Byrne, R.H., Millero, F.J., Dickson, A.G., Wanninkhof, R., Murata, A., Miller, L.A. & Greeley, D. 2012. Decadal changes in the aragonite and calcite saturation state of the Pacific Ocean. *Global Biogeochemical Cycles* **26**, 1–15.

Feely, R.A., Sabine, C.L., Hernandez-Ayon, M., Ianson, D. & Hales, B. 2008. Evidence for upwelling of corrosive "acidified" water onto the continental shelf. *Science* **320**, 1490–1492.

Fenn, W.O. & Rahn, W. 1964. *Handbook of Physiology, Section 3: Respiration*. Washington, DC: American Physiological Society,.

Fetzer, I. & Arntz, W.E. 2008. Reproductive strategies of benthic invertebrates in the Kara Sea (Russian Arctic): adaptation of reproduction modes to cold waters. *Marine Ecology Progress Series* **356**, 189–202.

Fields, L.G. & DeVries, A.L. 2015. Variation in blood serum antifreeze activity of Antarctic Trematomus fishes across habitat temperature and depth. *Comparative Biochemistry and Physiology* **185A**, 43–50.

Fields, P.A., Dong, Y., Meng, X. & Somero, G.N. 2015. Adaptations of protein structure and function to temperature: there is more than one way to 'skin a cat'. *Journal of Experimental Biology* **281**, 1801–1811.

Fields, P.A. & Somero, G.N. 1998. Hot spots in cold adaptation: localised increases in conformational flexibility in lactate dehydrogenase A_4 orthologues of Antarctic notothenioid fishes. *Proceedings of the National Academy of Sciences USA* **95**, 11476–11481.

Fields, P.A., Wahlstrand, B.D. & Somero, G.N. 2001. Intrinsic versus extrinsic stabilization of enzymes: the interaction of solutes and temperature on A4-lactate dehydrogenase orthologs from warm-adapted and cold-adapted marine fishes. *FEBS Journal* **268**, 4497–4505.

Fillinger, L., Janussen, D., Lundälv, T. & Richter, C. 2013. Rapid glass sponge expansion after climate-induced Antarctic ice shelf collapse. *Current Biology* **23**, 1330–1334.

Flexas, M., Schodlok, M.P., Padman, L., Menemenlis, D. & Orsi, A.H. 2015. Role of tides on the formation of the Antarctic slope front at the Weddell-Scotia confluence. *Journal of Geophysics Research* **120**, 3658–3680.

Flores, H., van Franeker, J.A., Siegel, V., Haraldsson, M., Strass, V., Meesters, E.H., Bathmann, U. & Wolff, W.J. 2012. The Association of Antarctic krill Euphausia superba with the under-ice habitat. *PLoS ONE* **7**, e31775.

Florindo, F. & Siegert, M. (eds). 2009. *Antarctic Climate Evolution. Developments in Earth and Environmental Sciences*, vol. 8. Amsterdam: Elsevier.

Fobian, D., Overgaard, J. & Wang, T. 2014. Oxygen transport is not compromised at high temperature in pythons. *Journal of Experimental Biology* **217**, 3958–3961.

Fogg, G.E. 1998. *The Biology of Polar Habitats*. Oxford: Oxford University Press.

Fordel, E., Geuens, E., Dewilde, S., De Coen, W. & Moens, L. 2004. Hypoxia/ischemia and the regulation of neuroglobin and cytoglobin expression. *IUBMB Life* **56**, 681–687.

Franklin, C.E., Davison, W. & Seebacher, F. 2007. Antarctic fish can compensate for rising temperatures: thermal acclimation of cardiac performance in Pagothenia borchgrevinki. *Journal of Experimental Biology* **210**, 3068–3074.

Franklin, C.E. & Johnston, I.A. 1997. Muscle power output during escape responses in an Antarctic fish. *Journal of Experimental Biology* **200**, 703–712.

Fraser, K.P.P., Clarke, A. & Peck, L.S. 2002. Feast and famine in Antarctica: seasonal physiology in the limpet, Nacella concinna (Strebel, 1908). *Marine Ecology Progress Series* **242**, 169–177.

Fraser, K.P.P., Clarke, A. & Peck, L.S. 2007. Growth in the slow lane: protein metabolism in the Antarctic limpet Nacella concinna (Strebel, 1908). *Journal of Experimental Biology* **210**, 2691–2699.

Fraser, K.P.P., Peck, L.S. & Clarke, A. 2004. Protein synthesis, RNA concentrations, nitrogen excretion, and metabolism vary seasonally in the Antarctic holothurian Heterocucumis steineni (Ludwig, 1898). *Physiological and Biochemical Zoology* **77**, 556–569.

Fraser, K.P.P. & Rogers, A.D. 2007. Protein metabolism in marine animals: the underlying mechanism of growth. *Advances in Marine Biology* **52**, 267–362.

Fraser, W.R. & Hofmann, E.E. 2003. A predator's perspective on causal links between climate change, physical forcing and ecosystem response. *Marine Ecology Progress Series* **265**, 1–15.

Frederich, M., Sartoris, F.J., Arntz, W.E. & Pörtner, H.O. 2000. Haemolymph Mg^{2+} regulation in decapod crustaceans: physiological correlates and ecological consequences in polar areas. *Journal of Experimental Biology* **203**, 1383–1393.

Frederich, M., Sartoris, F.J. & Pörtner, H.O. 2001. Distribution patterns of decapod crustaceans in polar areas: a result of magnesium regulation? *Polar Biology* **24**, 719–723.

Fredheim, B., Holen, S., Ugland, K.I. & Grahl-Nielsen, O. 1995. Fatty acids composition in blubber heart and brain from phocid seals. In *Whales, Seals, Fish and Man*, A.S. Bli et al. (eds). Amsterdam: Elsevier Science, 153–168.

Furrow, F.B., Amsler, C.D., McClintock, J.B. & Baker, B.J. 2003. Surface sequestration of chemical feeding deterrents in the Antarctic sponge Latrunculia apicalis as an optimal defense against sea star spongivory. *Marine Biology* **143**, 443–449.

Gadgil, M. & Bossert, W. 1970. Life-history consequences of natural selection. *American Naturalist* **104**, 1–24.

Gaitán-Espitia, J.D. & Nespolo, R. 2014. Is there metabolic cold adaptation in terrestrial ectotherms? Exploring latitudinal compensation in the invasive snail Cornu aspersum. *Journal of Experimental Biology* **217**, 2261–2267.

Gallardo, C.S. & Penchaszadch, P.E. 2001. Hatching mode and latitude in marine gastropods: revisiting Thorson's paradigm in the southern hemisphere. *Marine Biology* **138**, 547–552.

Garnham, C.P., Campbell, R.L. & Davies, P.L. 2011. Anchored clathrate waters bind antifreeze proteins to ice. *Proceedings of the National Academy of Sciences USA* **108**, 7363–7367.

Garofalo, F., Pellegrino, D., Amelio, D. & Tota, B. 2009. The Antarctic hemoglobinless icefish, fifty-five years later: a unique cardiocirculatory interplay of disaptation and phenotypic plasticity. *Comparative Biochemistry and Physiology* **154**, 10–28.

Gaston, K.J., Chown, S.L., Calosi, P., Bernardo, J., Bilton, D.T., Clarke, A., Clusella-Trullas, S., Ghalambor, C.K., Konarzewski, M., Peck, L.S., Porter, W.P., Pörtner, H.O., Rezende, E.L., Schulte, P.M., Spicer, J.I., Stillman, J.H. Terblanche, J.S. & van Kleunen, M. 2009. Macrophysiology: a conceptual re-unification. *American Naturalist* **174**, 595–612.

Gatti, S., Brey, T., Müller, W.E.G., Heilmayer, O. & Holst, G. 2002. Oxygen microoptodes: a new tool for oxygen measurements in aquatic animal ecology. *Marine Biology* **140**, 1075–1085.

Gazeau, F., Gattuso, J.P., Greaves, M., Elderfield, H., Peene, J., Heip, C.H.R. & Middelburg, J.J. 2011. Effect of carbonate chemistry alteration on the early embryonic development of the Pacific oyster (Crassostrea gigas). *PLoS ONE* **6**, e23010.

Gazeau, F., Quiblier, C., Jansen, J.M., Gattuso, J.–P., Middelburg, J.J. & Heip, C.H.R. 2007. Impact of elevated CO_2 on shellfish calcification. *Geophysics Research Letters* **34**, L07603.

Gerringa, L.J.A., Alderkamp, A.-C., Laan, P., Thuróczy, C.-E., de Baar, H.J.W., Mills, M.M., van Dijken, G.L., van Haren, H. & Arrigo, K.R. 2012. Iron from melting glaciers fuels the phytoplankton blooms in Amundsen Sea Southern Ocean: iron biogeochemistry. *Deep-Sea Research II* **71–76**, 16–31.

Gili, J.-M., Arntz, W.E., Palanques, A., Orejas, C., Clarke, A., Dayton, P.K., Isla, E., Teixidó, N., Rossi, S. & López-González, P.J. 2006. A unique assemblage of epibenthic sessile suspension feeders with archaic features in the high-Antarctic. *Deep-Sea Research II* **53**, 1029–1052.

Ginsburg, D.T. & Manahan, D.W. 2009. Developmental physiology of Antarctic asteroids with different life-history modes. *Marine Biology* **156**, 2391–2402.

Giordano, D., Boron, I., Abbruzzetti, S., Van Leuven, W., Nicoletti, F.P., Forti, F., Bruno, S., Cheng, C.-H.C., Moens, L., di Prisco, G., Nadra, A.D., Estrin, D., Smulevich, G., Dewilde, S., Viappiani, C. & Verde, C. 2012a. Biophysical characterisation of neuroglobin of the icefish, a natural knockout for hemoglobin and myoglobin. Comparison with human neuroglobin. *PLoS ONE* **7**, e44508.

Giordano, D., Russo, R., Coppola, D., Altomonte, G., di Prisco, G., Bruno, S. & Verde, C. 2015. 'Cool' adaptations to cold environments: globins in Notothenioidei (Actynopterygii, Perciformes). *Hydrobiologia* **761**, 293–312.

Giordano, D., Russo, R., Coppola, D., Di Prisco, G. & Verde, C. 2010. Molecular adaptations in haemoglobins of notothenioid fishes. *Journal of Fish Biology* **76**, 301–318.

Giordano, D., Russo, R., di Prisco, G. & Verde, C. 2012b. Molecular adaptations in Antarctic fish and marine microorgansisms. *Marine Genomics* **6**, 1–6.

Glandon, H.L. & Miller, T.J. 2017. No effect of high pCO_2 on juvenile blue crab, Callinectes sapidus, growth and consumption despite positive responses to concurrent warming. *ICES Journal of Marine Science* **17**, 1201–1209.

Gomes, V., Passos, M.J.A.C.R., Rocha, A.S., Santos, T.C.A., Hasue, F.M. & Ngan, P.V. 2014. Oxygen consumption and ammonia excretion of the Antarctic amphipod Bovallia gigantea Pfeffer, 1888, at different temperatures and salinities. *Brazilian Journal of Oceanography* **62**, 315–321.

Gomes, V., Passos, M.J.A.C.R., Rocha, A.S., Santos, T.C.A., Machado, A.S.D. & Phan, V.N. 2013. Metabolic rates of the Antarctic amphipod Gondogeneia antarctica at different temperatures and salinities. *Brazilian Journal of Oceanography* **61**, 243–249.

Gon, O. & Heemstra, P.C. (eds). 1990. *Fishes of the Southern Ocean*. J.L.B. Grahamstown: Smith Institute of Ichthyology.

Gonzalez-Bernat, M.J., Lamare, M. & Barker, M. 2013. Effects of reduced seawater pH on fertilisation, embryogenesis and larval development in the Antarctic seastar Odontaster validus. *Polar Biology* **36**, 235–247.

Gonzalez-Cabrera, P.J., Dowd, F., Pedibhotla, V.K., Rosario, R., Stanley-Samuelson, D. & Petzel, D. 1995. Enhanced hypo-osmoregulation induced by warm-acclimation in Antarctic fish is mediated by increased gill and kidney Na^+/K^+-ATPase activities. *Journal of Experimental Biology* **198**, 2279–2291.

González, K., Gaitán-Espitia, J., Font, A., Cárdenas, C.A. & González-Aravena, M. 2016. Expression pattern of heat shock proteins during acute thermal stress in the Antarctic sea urchin, Sterechinus neumayeri. *Revista Chilena de Historia Natural* **89**, 2 only.

Gorny, M., Arntz, W.E., Clarke, A. & Gore, D.J. 1992. Reproductive biology of caridean decapods from the Weddell Sea. *Polar Biology* **12**, 111–120.

Gorny, M., Brey, T., Arntz, W.E. & Bruns, T. 1993. Growth, development and productivity of Chorismus antarcticus (Pfeffer) (Crustacea: Decapoda: Natantia) in the eastern Weddell Sea, Antarctica. *Journal of Experimental Marine Biology and Ecology* **174**, 261–275.

Graham, A.G.C., Fretwell, P.T., Larter, R.D., Hodgson, D.A., Wilson, C.K., Tate, A.J. & Morris, P. 2008. A new bathymetric compilation highlighting extensive paleo-ice sheet drainage on the continental shelf, South Georgia, sub-Antarctica. *Geochemistry Geophysics and Geosystems* **9**, Q07011.

Graham, L.A., Hobbs, R.S., Fletcher, G.L. & Davies, P.L. 2013. Helical antifreeze proteins have independently evolved in fishes on four occasions. *PLoS ONE* **8**, e81285.

Grange, L.J. 2005. *Reproductive success in Antarctic marine invertebrates*. PhD thesis, Southampton University, Southampton, School of Ocean and Earth Sciences.

Grange, L.J., Peck, L.S. & Tyler, P.A. 2011a. Reproductive ecology of the circumpolar Antarctic nemertean Parborlasia corrugatus: No evidence for inter-annual variation. *Journal of Experimental Marine Biology and Ecology* **404**, 98–107.

Grange, L.J., Tyler, P.A. & Peck, L.S. 2007. Multi-year observations on the gametogenic ecology of the Antarctic seastar Odontaster validus. *Marine Biology* **153**, 15–23.

Grange, L.J., Tyler, P.A. & Peck, L.S. 2011b. Fertilization success of the circumpolar Antarctic seastar Odontaster validus (Koehler, 1906): a diver-collected study. In *Diving for Science 2011*, N.W. Pollock (ed.). *Proceedings of the American Academy of Underwater Sciences 30th Symposium*. Dauphin Island, AL: AAUS, 140–151.

Grange, L.J., Tyler, P.A., Peck, L.S. & Cornelius, N. 2004. Long-term interannual cycles of the gametogenic ecology of the Antarctic brittle star Ophionotus victoriae. *Marine Ecology Progress Series* **278**, 141–155.

Graus, R.R. 1974. Latitudinal trends in the shell characteristics of marine gastropods. *Lethaia* **7**, 303–314.

Greenberg, D.A., Jin, K. & Khan, A.A. 2008. Neuroglobin: an endogenous neuroprotectant. *Current Opinions in Pharmacology* **8**, 20–24.

Gremare, A. & Olive, P.J.W. 1986. A preliminary study of fecundity and reproductive effort in two polychaetous annelids with contrasting reproductive strategies. *International Journal of Invertebrate Reproduction and Development* **9**, 1–16.

Griffiths, H.J., Barnes, D.K.A. & Linse, K. 2009. Towards a generalised biogeography of the Southern Ocean benthos. *Journal of Biogeography* **36**, 162–177.

Griffiths, H.J., Whittle, R.J., Roberts, S.J., Belchier, M. & Linse, K. 2013. Antarctic crabs: nvasion or endurance? *PLoS ONE* **8**, e66981.

Gross, M. 2004. Emergency services: a bird's eye perspective on the many different functions of stress proteins. *Current Protein and Peptide Science* **5**, 213–223.

Grove, T.J., Hendrickson, J.W. & Sidell, B.D. 2004. Two species of Antarctic icefishes (Genus Champsocephalus) share a common genetic lesion leading to the loss of myoglobin expression. *Polar Biology* **27**, 579–585.

Gruzov, E.N. 1977. Seasonal alterations in coastal communities in the Davies Sea. In *Adaptations within Ecosystems*, G. Llano (ed.). Washington, DC: Smithsonian Press, 263–278.

Guderley, H. 2004. Metabolic responses to low temperature in fish muscle. *Biological Reviews* **79**, 409–427.

Gutt, J., Barnes, D.K.A., Lockhart, S.J. & van de Putte, A. 2013a. Antarctic macrobenthic communities: A compilation of circumpolar information. *Nature Conservation* **4**, 1–13.

Gutt, J., Barratt, I., Domack, E., d'Udekem d'Acoz, C., Dimmler, W., Grémare, A., Heilmayer, O., Isla, E., Janussen, D., Jorgensen, B., Kock, K.-H., Lehnert, L.S., López-Gonsáles, P., Langner, S., Linse, K., Manjón-Cabeza, M.E., Meißner, M., Montiel, A., Raes, M., Robert, H., Rose, A., Schepisi, E.S., Saucede,

T., Scheidat, M., Schenke, H.-W, Seiler, J. & Smith, S. 2011. Biodiversity change after climate-induced ice-shelf collapse in the Antarctic. *Deep-Sea Research Part II Topical Studies in Oceanography* **58**, 74–83.

Gutt, J., Bertler, N., Bracegirdle, T.J., Buschmann, A., Comiso, J., Hosie Isla, G., Schloss, E., Smith, I.E., Tournadre, J. & Xavier, J.C. 2015. The Southern Ocean ecosystem under multiple climate stresses - an integrated circumpolar assessment. *Global Change Biology* **21**, 1434–1453.

Gutt, J., Cape, M., Dimmler, W., Fillinger, L., Isla, E., Lieb, V., Lundälv, T. & Pulcher, C. 2013b. Shifts in Antarctic megabenthic structure after ice-shelf disintegration in the Larsen area east of the Antarctic Peninsula. *Polar Biology* **36**, 895–906.

Gutt, J., Gerdes, D. & Klages, M. 1992. Seasonality and spatial variability in the reproduction of two Antarctic holothurians (Echinodermata). *Polar Biology* **11**, 533–544.

Gutt, J., Sirenko, B.I., Smirnov, I.S. & Arntz, W.E. 2004. How many macrozoobenthic species might inhabit the Antarctic shelf? *Antarctic Science* **16**, 11–16.

Gutt, J. & Starmans, A. 1998. Structure and biodiversity of mega-benthos in the Weddell and Lazarev Seas (Antarctica): ecological role of physical parameters and biological interactions. *Polar Biology* **20**, 229–247.

Hain, S. 1991. Maintenance and culture of living benthic molluscs from high Antarctic shelf areas. *Aquaculture and Fisheries Management* **23**, 1–11.

Hain, S. & Arnaud, P.M. 1992. Notes on the reproduction of high-Antarctic mollusks from the Weddell Sea. *Polar Biology* **12**, 303–312.

Hain, S. & Melles, M. 1994. Evidence for a marine shallow water mollusc fauna beneath ice shelves in the Lazarev and Weddell Seas, Antarctica, from shells of Adamussium colbecki and Nacella (Patinigera) cf. concinna. *Antarctic Science* **6**, 29–36.

Hale, R., Calosi, P., McNeill, L., Mieszkowska, N. & Widdicombe, S. 2011. Predicted levels of future ocean acidification and temperature rise could alter community structure and biodiversity in marine benthic communities. *Oikos* **120**, 661–674.

Hamilton, W.D. 1966. The moulding of senescence by natural selection. *Journal of Theoretical Biology* **12**, 12–45.

Hanchet, S., Dunn, A., Parker, S., Horn, P., Stevens, D. & Mormede, S. 2015. The Antarctic toothfish (Dissostichus mawsoni): biology, ecology, and life-history in the Ross Sea region. *Hydrobiologia* **761**, 397–414.

Hankeln, T., Ebner, B., Fuchs, C., Gerlachm F., Haberkampm M., Laufs, T. L., Roesner, A., Schmidt, M., Weich, B., Wystub, S., Saaler-Reinhardt, S., Reuss, S., Bolognesi, M., De Sanctis, D., Marden, M.C., Kiger, L., Moens, L., Dewilde, S., Nevo, E., Avivi, A., Weber, R.E., Fago, A. & Burmester, T. 2005. Neuroglobin and cytoglobin in search of their role in the vertebrate globin family. *Journal of Inorganic Biochemistry* **299**, 110–119.

Hardewig, I., van Dijk, P.L.M., Moyes, C.D. & Pörtner, H.O. 1999a. Temperature-dependent expression of cytochrome c oxidase in fish: a comparison between temperate and Antarctic and eelpout. *American Journal of Physiology* **277**, R508–R516.

Hardewig, I., Peck, L.S. & Pörtner, H.O. 1999b. Thermal sensitivity of mitochondrial function in the Antarctic Notothenioid Lepidonotothen nudifrons. *Comparative Biochemistry and Physiology* **124A**, 179–189.

Hardy, P. 1977. Scoloplos marginatus mcleani lifecycle and adaptations to the Antarctic benthic environment. In *Adaptations within Antarctic Ecosystems*, G.A. Llano (ed.). Houston, TX: Gulf Publishing Co., 209–226.

Harper, E.M. & Peck, L.S. 2003. Predatory behaviour and metabolic costs in the Antarctic muricid gastropod Trophon longstaffi. *Polar Biology* **26**, 208–217.

Harper, E.M. & Peck, L.S. 2016. Latitudinal and depth gradients in predation pressure. *Global Ecology and Biogeography* **25**, 670–678.

Harper, E.M., Peck, L.S. & Hendry, K.R. 2009. Patterns of shell repair in articulate brachiopods indicate size constitutes a refuge from predation. *Marine Biology* **156**, 1993–2000.

Hauquier, F., Ballesteros-Redondo, L., Gutt, J. & Vanreusel, A. 2016. Community dynamics of nematodes after Larsen ice-shelf collapse in the eastern Antarctic Peninsula. *Ecology and Evolution* **6**, 305–317.

Hauri, C., Friedrich, T. & Timmermann, A. 2016. Abrupt onset and prolongation of aragonite undersaturation events in the Southern Ocean. *Nature Climate Change* **2**, 172–176.

Havenhand, J.N. & Schlegel, P. 2009. Near-future levels of ocean acidification do not affect sperm motility and fertilisation kinetics in the oyster Crassostrea gigas. *Biogeosciences* **6**, 3009–3015.

Hawes, T.C., Worland, M.R. & Bale, J.S. 2010. Freezing in the Antarctic limpet, Nacella concinna. *Cryobiology* 61, 128–132.

Hawkins, A.J.S., Widdows, J. & Bayne, B.L. 1989. The relevance of whole-body protein metabolism to measured costs of maintenance and growth in Mytilus edulis. *Physiological Zoology* 62, 745–763.

Hazan, Y., Wangensteen, O.S. & Fine, M. 2014. Tough as a rockboring urchin: adult Echinometra sp. EE from the Red Sea show high resistance to ocean acidification over long-term exposures. *Marine Biology* 161, 2531–2545.

Hazel, J.R. 1995. Thermal adaptation in biological membranes: is homeoviscous adaptation the explanation? *Annual Review of Physiology* 57, 19–42.

He, J.X. & Stewart, D.J. 2001. Age and size at first reproduction of fishes: predictive models based only on growth trajectories. *Ecology* 82, 784–792.

Hedgepeth, J.W. 1971. Perspectives of benthic ecology in Antarctica. In *Research in the Antarctic*, L.O. Quam (ed.). Washington, DC: American Association for the Advancement of Science, 93–136.

Heilmayer, O. & Brey, T. 2003. Saving by freezing? Metabolic rates of Adamussium colbecki in a latitudinal context. *Marine Biology* 143, 477–484.

Heilmayer, O., Brey, T. & Pörtner, H.O. 2004. Growth efficiency and temperature in scallops: a comparative analysis of species adapted to different temperatures. *Functional Ecology* 18, 641–647.

Heilmayer, O., Honnen, C., Jacob, U., Chiantore, C., Cattaneo-Vietti, R. & Brey, T. 2005. Temperature effects on summer growth rates in the Antarctic scallop Adamussium colbecki. *Polar Biology* 28, 523–527.

Heinrich, S., Valentin, K., Frickenhaus, S., John, U. & Wiencke, C. 2012. Transcriptomic analysis of acclimation to temperature and light stress in Saccharina latissima (Phaeophyceae). *PLoS ONE* 7, e44342.

Heise, K., Estevez, M.S., Puntarulo, S., Galleano, M., Nikinmaa, M., Pörtner, H.O. & Abele, D. 2007. Effects of seasonal and latitudinal cold on oxidative stress parameters and activation of hypoxia inducible factor (HIF-1) in zoarcid fish. *Journal of Comparative Physiology* 177B, 765–77.

Heise, K., Puntarulo, S., Nikinmaa, M., Pörtner, H.O. & Abele, D. 2004. Physiological stress response to warm and cold acclimation in polar and temperate fish. *Free Radical Biology and Medicine* 36, S141–S142.

Heisler, N. 1986. Comparative aspects of acid—base regulation. In *Acid-Base Regulation in Animals*, N. Heisler (ed.). Amsterdam: Elsevier, 397–450.

Helmuth, B. 2009. From cells to coastlines: how can we use physiology to forecast the impacts of climate change? *Journal of Experimental Biology* 212, 753–760.

Hemmingsen, E.A. 1991. Respiratory and cardiovascular adaptation in hemoglobin-free fish: resolved and unresolved problems. In *Biology of Antarctic Fish*, G. di Prisco et al. (eds). New York: Springer-Verlag, 191–203.

Hemmingsen, E.A., Douglas, E.L. & Grigg, G.C. 1969. Oxygen consumption in an Antarctic hemoglobin-free fish, Pagetopsis macropterus, and in three species of Notothenia. *Comparative Biochemistry and Physiology* 29, 467–470.

Henry, L.V. & Torres, J.J. 2013. Metabolism of an Antarctic solitary coral, Flabellum impensum. *Journal of Experimental Marine Biology and Ecology* 449, 17–21.

Hess, H.C. 1993. The evolution of parental care in brooding spirorbid polychaetes: the effect of scaling constraints. *American Naturalist* 141, 577–596.

Higgs, N.D., Reed, A.J., Hooke, R., Honey, D., Heilmayer, O. & Thatje, S. 2009. Growth and reproduction in the Antarctic brooding bivalve Adacnarca nitens (Philobryidae), from the Ross Sea. *Marine Biology* 156, 1073–1081.

Ho, M.A., Price, C., King, C.K., Virtue, P. & Byrne, M. 2013. Effects of ocean warming and acidification on fertilization in the Antarctic echinoid Sterechinus neumayeri across a range of sperm concentrations. *Marine Environmental Research* 90, 136–141.

Hochachka, P.W., Buck, L.T., Doll, C.J. & Land, S.C. 1996. Unifying theory of hypoxia tolerance: molecular/metabolic defense and rescue mechanisms for surviving oxygen lack. *Proceedings of the National Academy of Science USA* 93, 9493–9498.

Hochachka, P.W. & Somero, G.N. 2002. *Biochemical Adaptation*. Princeton, NJ: Princeton University Press.

Hodkinson, I.D. 2003. Metabolic cold adaptation in arthropods: a smaller-scale perspective. *Functional Ecology* 17, 562–567.

Hoegh-Guldberg, O. 1999. Climate change, coral bleaching and the future of the world's coral reefs. *Marine and Freshwater Research* 50, 839–866.

Hoegh-Guldberg, O., Mumby, P.J., Hooten, A.J., Steneck, R.S., Greenfield, P., Gomez, E., Harvell, C.D., Sale, P.F., Edwards, A.J., Caldeira, K., Knowlton, N., Eakin, C.M., Iglesias-Prieto, R., Muthiga, N., Bradbury, R.H., Dubi, A. & Hatziolos, M.E. 2007. Coral reefs under rapid climate change and ocean acidification. *Science* **318**, 1737–1742.

Hoegh-Guldberg, O. & Pearse, J.S. 1995. Temperature, food availability and the development of marine invertebrate larvae. *American Zoologist* **35**, 415–425.

Hoffman, J.I., Clarke, A., Linse, K. & Peck, L.S. 2011. Effects of brooding and broadcasting reproductive modes on the population genetic structure of two Antarctic gastropod molluscs. *Marine Biology* **158**, 287–296.

Hofmann, G.E., Buckley, B.A., Airaksinen, S., Keen, J.E. & Somero, G.N. 2000. Heat-shock protein expression is absent in the Antarctic fish Trematomus bernacchii (Family Nototheniidae). *Journal of Experimental Biology* **203**, 2331–2339.

Hofmann, G.E., Lund, S.G., Place, S.P. & Whitmer, A.C. 2005. Some like it hot, some like it cold: the heat shock response is found in New Zealand but not Antarctic notothenioid fishes. *Journal of Experimental Marine Biology and Ecology* **316**, 79–89.

Holeton, G.F. 1970. Oxygen uptake and circulation by a hemoglobinless Antarctic fish (Chaenocephalus aceratus Lonnberg) compared with 3 red-blooded Antarctic fish. *Comparative Biochemistry and Physiology* **34**, 457–471.

Hoppe-Seyler, F. 1864. Über die chemischen und optischaften des Blutfarbstoffes. *Virchows Archiv A Pathological Anatomy and Histopathology* **13**, 233–235.

Horn, P.L. 1986. Energetics of Chiron pelliserpenris (Quor & Gaimard, 1835) (Mollusca: Polyplacophora) and the importance of mucus in its energy budget. *Journal of Experimental Marine Biology and Ecology* **101**, 119–141.

Houlihan, D.F. & Allan, D. 1982. Oxygen consumption of some Antarctic and British gastropods: an evaluation of cold adaptation. *Comparative Biochemistry and Physiology* **73A**, 383–387.

Houlihan, D.F., Carter, C.G. & McCarthy, I.D. 1995. Protein turnover in animals. In *Nitrogen Metabolism and Excretion*, P.J. Walsh & P. Wright (eds). Boca Raton, FL: CRC Press, 1–32.

Hoverd, W.A. 1985. Histological and ultrastructural observations of the lophophore and larvae of the brachiopod, Notosaria nigricans (Sowerby, 1846). *Journal of Natural History* **19**, 831–850.

Hoyle, A., Bowers, R.G., White, A. & Boots, M. 2008. The influence of trade-off shape on evolutionary behaviour in classical ecological scenarios. *Journal of Theoretical Biology* **250**, 498–511.

Hughes, R.N. & Seed, R. 1995. Behavioural mechanisms of prey selection in crabs. *Journal of Experimental Marine Biology and Ecology* **193**, 225–238.

Hunter, A., Spiers, D.C. & Heath, M.R. 2015. Fishery-induced changes to age and length dependent maturation schedules of three demersal fish species in the Firth of Clyde. *Fisheries Research* **170**, 14–23.

Husmann, G., Abele, D., Rosenstiel, P., Clark, M.S., Kraemer, L. & Philipp, E.E.R. 2014. Age-dependent expression of stress and antimicrobial genes in the hemocytes and siphon tissue of the Antarctic bivalve, Laternula elliptica, exposed to injury and starvation. *Cell Stress and Chaperones* **19**, 15–32.

Husmann, G., Philipp, E.E.R., Rosenstiel, P., Vazquez, S. & Abele, D. 2011. Immune response of the Antarctic bivalve Laternula elliptica to physical stress and microbial exposure. *Journal of Experimental Marine Biology and Ecology* **398**, 83–90.

Huth, T.P. & Place, S.P. 2016a. RNA-seq reveals a diminished acclimation response to the combined effects of ocean acidification and elevated seawater temperature in Pagothenia borchgrevinki. *Marine Genomics* **28**, 87–97.

Huth, T.P. & Place, S.P. 2016b. Transcriptome wide analyses reveal a sustained cellular stress response in the gill tissue of Trematomus bernacchii after acclimation to multiple stressors. *BMC Genomics* 17, 127.

Huybrechts, P. 2002. Sea-level changes at the LGM from ice-dynamic reconstructions of the Greenland and Antarctic ice sheets during the glacial cycles. *Quaternary Science Reviews* **22**, 203–231.

Iken, K., Avila, C., Fontana, A. & Gavagnin, M. 2002. Chemical ecology and origin of defensive compounds in the Antarctic nudibranch Austrodoris kerguelenensis (Opisthobranchia: Gastropoda). *Marine Biology* **141**, 101–109.

Ingels, J., Vanreusel, A., Brandt, A., Catarino, A.I., David, B., De Ridder, C., Dubois, P., Gooday, A.J., Martin, P., Pasotti, F. & Robert, H. 2012. Possible effects of global environmental changes on Antarctic benthos: a synthesis across five major taxa. *Ecology and Evolution* **2**, 453–485.

IPCC 2014. Climate Change 2014: Synthesis Report. Contribution of Working Groups I, II and III to the Fifth Assessment Report of the Intergovernmental Panel on Climate Change (Core Writing Team, R.K. Pachauri & L.A. Meyer (eds)). IPCC, Geneva, Switzerland.

Jablonski, D. & Lutz, R.A. 1983. Larval ecology of marine benthic invertebrates: paleobiological implications. *Biological Reviews* **58**, 21–89.

Jackson, J.A., Baker, C.S., Vant, M., Steel, D.J., Medrano-Gonzalez, L. & Palumbi, S.R. 2009. Big and slow: phylogenetic estimates of molecular evolution in baleen whales (suborder Mysticeti). *Molecular Biology and Evolution* **26**, 2427–2440.

James, M.A., Ansell, A.D., Collins, M.J., Curry, G.B., Peck, L.S. & Rhodes, M.C. 1992. Recent advances in the study of living brachiopods. *Advances in Marine Biology* **28**, 175–387.

James, R.S. & Johnston, I.A. 1998. Scaling of muscle performance during escape responses in the short-horn sculpin (Myoxocephalus scorpius). *Journal of Experimental Biology* **201**, 913–923.

Janecki, T., Kidawa, A. & Potocka, M. 2010. The effects of temperature and salinity on vital biological functions of the Antarctic crustacean Serolis polita. *Polar Biology* **33**, 1013–1020.

Janecki, T. & Rakusa-Suszczewski, S. 2006. Biology and metabolism of Glyptonotus antarcticus (Eights) (Crustacea: Isopoda) from Admiralty Bay, King George Island, Antarctica. *Polar Bioscience* **19**, 29–42.

Jazdzewski, K., De Broyer, C., Pudlarz, M. & Zielinski, D. 2001. Seasonal fluctuations of vagile benthos in the uppermost sublittoral of a maritime Antarctic fiord. *Polar Biology* **24**, 910–917.

Jia, Z. & Davies, P.L. 2002. Antifreeze proteins: an unusual receptor-ligand interaction. *Trends in Biochemical Science* **27**, 101–106.

Jin, Y. 2003. *Freezing avoidance of Antarctic fishes: the role of a novel antifreeze potentiating protein and the antifreeze glycoproteins.* PhD dissertation, University of Illinois-Urbana, Champaign, IL.

Jin, Y.M. & DeVries, A.L. 2006. Antifreeze glycoprotein levels in Antarctic notothenioid fishes inhabiting different thermal environments and the effect of warm acclimation. *Comparative Biochemistry and Physiology* **144B**, 290–300.

Johnson, T.P. & Johnston, I.A. 1991. Temperature adaptation and the contractile properties of live muscle fibres from teleost fish. *Journal of Comparative Physiology* **161B**, 27–36.

Johnson, W.S., Stevens, M. & Watling, L. 2001. Reproduction and development of marine peracaridans. *Advances in Marine Biology* **39**, 105–260.

Johnston, I.A. & Battram, J. 1993. Feeding energetics and metabolism in demersal fish species from Antarctic, temperate and tropical environments. *Marine Biology* **115**, 7–14.

Johnston, I.A., Calvo, J., Guderley, H., Fernandez, D. & Palmer L. 1998. Latitudinal variation in the abundance and oxidative capacities of muscle mitochondria in perciform fishes. *Journal of Experimental Biology* **201**, 1–12.

Johnston, I.A., Clarke, A. & Ward, P. 1991. Temperature and metabolic-rate in sedentary fish from the Antarctic, North-Sea and Indo-West Pacific-Ocean. *Marine Biology* **109**, 191–195.

Junquera, S., Román, E., Morgan, J., Sainza, M. & Ramilo, G. 2003. Time scale of ovarian maturation in Greenland halibut (Reinhardtius hippoglossoides, Walbaum). *ICES Journal of Marine Science* **60**, 767–773.

Kaplan, H.S. & Gangestad, S.W. 2005. Life-history theory and evolutionary psychology. In *The Handbook of Evolutionary Psychology*, D.M. Buss (ed.). Hoboken, NJ: John Wiley and Sons Inc, 68–95.

Kapsenberg, L. & Hofmann, G.E. 2014. Signals of resilience to ocean change: high thermal tolerance of early stage Antarctic sea urchins (Sterechinus neumayeri) reared under present-day and future pCO_2 and temperature. *Polar Biology* **37**, 967–980.

Karelitz, S.E., Uthicke, S., Foo, S.A., Barker, M.F., Byrne, M., Pecorino, D. & Lamare, M. 2017. Ocean acidification has little effect on developmental thermal windows of echinoderms from Antarctica to the tropics. *Global Change Biology* **23**, 657–672.

Kawada, N., Kristensen, D.B., Asahina, K., Nakatani, K., Minamiyama, Y., Seki, S. & Yoshizato, K. 2001. Characterization of a stellate cell activation-associated protein (STAP) with peroxidase activity found in rat hepatic stellate cells. *Journal of Biological Chemistry* **276**, 25318–25323.

Kawaguchi, K., Ishikawa, S., Matsude, O. & Naito, Y. 1989. Tagging experiments of nototheniid fish, Trematomus bernachii B. under the coastal fast ice in Lutzow- Holm Bay Antarctica. *Polar Biology* **2**, 111–116.

Kawaguchi, S., Kurihara, H., King, R., Hale, L., Berli, T., Robinson, J. P., Ishida, A., Wakita, M., Virtue, P., Nicol, S. & Ishimatsu, A. 2011. Will krill fare well under Southern Ocean acidification? *Biology Letters* **7**, 288–291.

Kawall, H.G., Torres, J.J., Sidell, B.D. & Somero, G.N. 2002. Metabolic cold adaptation in Antarctic fishes: evidence from enzymatic activities of brain. *Marine Biology* **140**, 279–286.

Keever, C.C., Sunday, J., Puritz, J.B., Addison, J.A., Toonen, R.J., Grosberg, K. & Hart, M.W. 2009. Discordant distribution of populations and genetic variation in a sea star with high dispersal potential. *Evolution* **63**, 3214–3222.

Kendrew, J.C., Bodo, G., Dintzis, H.M., Parrish, R.G., Wyckoff, H. & Phillips, D.C. 1958. A three-dimensional model of the myoglobin molecule obtained by x-ray analysis. *Nature* **181**, 662–666.

Kennicut II, M.C., Chown, S.L., Cassano, J.J., Liggett, D., Massom, R., Peck, L.S., Rintoul, S. R., Storey, J.W.V., Vaughan, D.G., Wilson, T.J. & Sutherland, W.J. 2014. Six priorities for Antarctic Science. *Nature* **512**, 23–25.

Kennicutt, M.C., Chown, S.L., Cassano, J.J., Liggett, D., Peck, L.S., Massom, R., Rintoul, S.R., Storey, J., Vaughan, D.G., Wilson, T.J., Allison, I., Ayton, J., Badhe, R., Baeseman, J., Barrett, P.J., Bell, R.E., Bertler, N., Bo, S., Brandt, A., Bromwich, D., Cary, S.C., Clark, M.S., Convey, P., Costa, E.S., Cowan, D., Deconto, R., Dunbar, R., Elfring, C., Escutia, C., Francis, J., Fricker, H.A., Fukuchi, M., Gilbert, N., Gutt, J., Havermans, C., Hik, D., Hosie, G., Jones, C., Kim, Y.D., Le Maho, Y., Lee, S.H., Leppe, M., Leitchenkov, G., Li, X., Lipenkov, V., Lochte, K., López-Martínez, J., Lüdecke, C., Lyons, W., Marenssi, S., Miller, H., Morozova, P., Naish, T., Nayak, S., Ravindra, R., Retamales, J., Ricci, C.A., Rogan-Finnemore, M., Ropert-Coudert, Y., Samah, A.A., Sanson, L., Scambos, T., Schloss, I.R., Shiraishi, K., Siegert, M.J., Simões, J.C., Storey, B., Sparrow, M.D., Wall, D.H., Walsh, J.C., Wilson, G., Winther, J.G., Xavier, J.C., Yang, H. & Sutherland, W.J. 2015. A roadmap for Antarctic and Southern Ocean Science for the next two decades and beyond. *Antarctic Science* **27**, 3–18.

Kent, J., Prosser, C.L. & Graham, G. 1992. Alterations in liver composition of channel catfish (Ictalurus punctatus) during seasonal acclimatization. *Physiological Zoology* **65**, 867–884.

Khan, J.R., Pether, S., Bruce, M., Walker, S.P. & Herbert, N.A. 2015. The effect of temperature and ration size on specific dynamic action and production performance in juvenile hapuku (Polyprion oxygeneios). *Aquaculture* **437**, 67–74.

Kidawa, A., Potocka, M. & Janecki, T. 2010. The effects of temperature on the behaviour of the Antarctic sea star Odontaster validus. *Polish Polar Research* **31**, 273–284.

Kiko, R. 2010. Acquisition of freeze protection in a sea-ice crustacean through horizontal gene transfer? *Polar Biology* **33**, 543–556.

Kinne, O. 1970. Temperature: invertebrates. In *Marine Ecology*, O. Kinne (ed.). London: Wiley Intersciecne, 407–514.

Kiørboe, T. 2001. Formation and fate of marine snow: small-scale processes with large-scale implications. *Scientia Marina* **65**, 57–71.

Klages, M. 1991. *Biologische und populationsdynamische Untersuchungen an ausgewählten Gammariden (Crustacea: Amphipoda) des südöstlichen Weddellmeeres, Antarktis.* PhD thesis, University of Bremen.

Klages, M. 1993. Biology of the Antarctic gammaridean amphipod, Eusirus perdentatus Chevreaux 1912 (Crustacea: Amphipoda): distribution, reproduction and population dynamics. *Antarctic Science* **5**, 349–359.

Kleiber, M. 1961. *The Fire of Life: An Introduction to Animal Energetics.* New York: Wiley.

Kleypas, J.A., Feely, R.A., Fabry, V.J., Langdon, C., Sabine, C.L. & Robbins, L.L. 2006. Impact of ocean acidification on coral reefs and other marine calcifiers: a guide for future research, report of a workshop held 18–20 April 2005, St. Petersburg, FL, sponsored by NSF, NOAA, and the US Geological Survey.

Klok, C.J., Sinclair, B.J. & Chown, S.L. 2004. Upper thermal tolerance and oxygen limitation in terrestrial arthropods. *Journal of Experimental Biology* **207**, 2361–2370.

Knight, C.A., Cheng, C.-H.C. & DeVries, A.L. 1991. Adsorption of alphahelical antifreeze peptides on specific ice crystal surface planes. *Biophysics Journal* **59**, 409–418.

Kock, K.-H. 1992. *Antarctic Fish and Fisheries.* Cambridge, UK: Cambridge University Press.

Kock, K.-H. 2005. Antarctic icefishes (Channichthyidae): a unique family of fishes. *Polar Biology* **28**, 862–895.

Kock, K.-H., Duhamel, G. & Hureau, J.-C. 1985. Biology and status of exploited Antarctic fish stocks: a review. *BIOMASS Scientific Series No* **6**, 143 pp. SCAR.

Kock, K.-H., Jones, C. & Wilhelms, S. 2000. Biological characteristics of Antarctic fish stocks in the southern Scotia Arc region. *CCAMLR Science* **7**, 1–42.

Kock, K.-H. & Kellerman, A. 1991. Reproduction in Antarctic notothenioid fish. *Antarctic Science* **3**, 125–150.

Kooijman, S.A.L.M. 2000. *Dynamic Energy and Mass Budgets in Biological Systems*, 2nd edition. Cambridge, UK: Cambridge University Press.

Koplovitz, G., McClintock, J.B., Amsler, C.D. & Baker, B.J. 2009. Palatability and chemical anti-predatory defenses in common ascidians from the Antarctic Peninsula. *Aquatic Biology* **7**, 81–92.

Kowalke, J. 1998. Energy budgets of benthic suspension feeding animals of the Potter Cove (King George Island, Antarctica). *Berichte zur Polarfoschung* **286**, 1–147.

Kowalke, J., Tatián, M., Sahade, R. & Arntz, W.E. 2001. Production and respiration of Antarctic ascidians. *Polar Biology* **24**, 663–669.

Krembs, C.R., Gradinger, R. & Spindler, M. 2000. Implications of brine channel geometry and surface area for the interaction of sympagic organisms in Arctic sea-ice. *Journal of Experimental Marine Biology and Ecology* **243**, 55–80.

Kroeker, K.J., Kordas, R.L., Crim, R., Hendriks, I.E., Ramajo, L., Singh, G.S., Duarte, C.M. & Gattuso, J.-P. 2013. Impacts of ocean acidification on marine organisms: quantifying sensitivities and interaction with warming. *Global Change Biology* **19**, 1884–1896.

Kroeker, K.J., Kordas, R.L., Crim, R.N. & Singh, G.G. 2010. Meta-analysis reveals negative yet variable effects of ocean acidification on marine organisms. *Ecology Letters* **13**, 1419–1434.

Krogh, A. 1916. *Respiratory Exchange of Animals and Man.* London: Longmans, Green.

Kuffner, I.B., Andersson, A.J., Jokiel, P.L., Rodgers, K.S. & Mackenzie, F.T. 2008. Decreased abundance of crustose coralline algae due to ocean acidification. *Nature Geoscience* **1**, 114–117.

Kurihara, H., Kato, S. & Ishimatsu, A. 2007. Effects of increased seawater pCO_2 on early development of the oyster *Crassostrea gigas*. *Aquatic Biology* **1**, 91–98.

Kurihara, H. & Shirayama, Y. 2004. Effects of increased atmospheric CO_2 on sea urchin early development. *Marine Ecology Progress Series* **274**, 161–169.

Kühl, S. 1988. A contribution to the reproductive biology and geographical distribution of Antarctic Octopodidae (Cephalopoda). *Malacologia* **29**, 89–100.

La Mesa, M., Caputo, V. & Eastman, J.T. 2008. The reproductive biology of two epibenthic species of Antarctic nototheniid fish of the genus Trematomus. *Antarctic Science* **20**, 355–364.

La Mesa, M. & Eastman, J.T. 2012. First data on age and sexual maturity of the Tristan klipfish, Bovichtus diacanthus (Bovichtidae) from Tristan da Cunha, South Atlantic. *Antarctic Science* **24**, 115–120.

La Mesa, M. & Vacchi, M. 2001. Age and growth of high Antarctic notothenioid fish. *Antarctic Science* **13**, 227–235.

La Terza, A.L., Miceli, C. & Luporini, P. 2001. Divergence between two Antarctic species of the ciliate Euplotes, E. focardii and E. nobilii, in the expression of heat-shock protein 70 genes. *Molecular Ecology* **10**, 1061–1067.

La Terza, A.L., Miceli, C. & Luporini, P. 2004. The gene for the heat-shock protein 70 of Euplotes focardii, an Antarctic psychrophilic ciliate. *Antarctic Science* **16**, 23–28.

La Terza, A.L., Passini, V., Barchetta, S. & Luporini, P. 2007. Adaptive evolution of the heatshock response in the Antarctic psychrophillic ciliate, Euplotes focardii: hints from comparative determination of the hsp70 gene structure. *Antarctic Science* **19**, 239–244.

Lack, D. 1947. The significance of clutch size 1. Intraspecific variation. *Ibis* **89**, 302–352.

Lai, F., Jutfelt, F. & Nilsson, G.E. 2015. Altered neurotransmitter function in CO_2 -exposed stickleback (Gasterosteus aculeatus): a temperate model species for ocean acidification research. *Conservation Physiology* **3**, cov018.

Lau, C,Y.S., Grange, L.J., Peck, L.S. & Reed, A.J. 2018. The reproductive ecology of the Antarctic bivalve Aequiyoldia eightsi follows neither Antarctic nor taxonomic patterns. *Polar Biology*, online: https://doi.org/10.1007/s00300-018-2309-2

Le, K. & Shi, J. 1997. A study of circulation and mixing in the region of Prydz Bay, Antarctica. *Studia Marina Sinica* **38**, 39–52.

Lee, R.E. & Denlinger, D.L. 1991. *Insects at Low Temperature.* New York: Chapman and Hall.

Lefevre, S., Findorf, I., Bayley, M., Huong, D.T.T. & Wang, T. 2016. Increased temperature tolerance of the air-breathing Asian swamp eel Monopterus albus after high-temperature acclimation is not explained by improved cardio-respiratory performance. *Journal of Fish Biology* **88**, 418–432.

Legge, O.J., Bakker, D.C.E., Meredith, M.P., Venables, H.J., Brown, P.J., Jones, E.M. & Johnson, M.T. 2017. The seasonal cycle of carbonate system processes in Ryder Bay, West Antarctic Peninsula. *Deep Sea Research Part II: Topical Studies in Oceanography* **139**, 167–180.

Lehtonen, K.K. 1996. Ecophysiology of the benthic amphipod Monoporeia affinis in an open-sea area of the northern Baltic Sea: seasonal variations in oxygen consumption and ammonia excretion. *Marine Biology* **126**, 645–654.

Leis, J.M., Caselle, J.E., Bradbury, I.R., Kristiansen, T., Llopiz, J.K., Miller, M.J., O'Connor, M.I., Paris, C.B., Shanks, A.L., Sogard, S.M., Swearer, S.E., Treml, E.A., Vetter, R.D. & Warner, R.R. 2013. Does fish larval dispersal differ between high and low latitudes? *Proceedings of the Royal Society of London B* **280**, 20130327.

Lessios, H.A. 1990. Adaptation and phylogeny as determinants of egg size in echinoderms from the two sides of the Isthmus of Panama. *American Naturalist* **135**, 1–13.

Li, D., Chen, X. Q., Li, W.J., Yang, Y.H., Wang, J.Z. & Yu, A.C. 2007. Cytoglobin up-regulated by hydrogen peroxide plays a protective role in oxidative stress. *Neurochemical Research* **32**, 1375–1380.

Li, H., Hemann, C., Abdelghany, T.M., El-Mahdy, M.A. & Zweier, J.L. 2012. Characterization of the mechanism and magnitude of cytoglobin-mediated nitrite reduction and nitric oxide generation under anaerobic conditions. *Journal of Biological Chemistry* **287**, 36623–36633.

Linse, K., Griffiths, H.J., Barnes, D.K.A. & Clarke, A. 2006. Biodiversity and biogeography of Antarctic and sub-Antarctic Mollusca. *Deep-Sea Research II* 53, 985–1008.

Lipps, J.H. & Hickman, C.S. 1982. Origin, age and evolution of Antarctic and deep-sea faunas. In *Environment of the Deep Sea*, W.G. Ernst & J.G. Morris (eds). Englewood Cliffs, NJ: Prentice-Hall, 324–356.

Lipps, J.H., Ronan, T.E. & Delaca, T.E. 1979. Life below the Ross ice shelf, Antarctica. *Science* **203**, 447–449.

Lischka, S., Büdenbender, J., Boxhammer, T. & Riebesell, U. 2011. Impact of ocean acidification and elevated temperatures on early juveniles of the polar shelled pteropod Limacina helicina: mortality, shell degradation, and shell growth. *Biogeosciences* **8**, 919–932.

Littlepage, J.L. & Pearse, J.S. 1962. Biological and oceanographic observations under and Antarctic ice shelf. *Science* **137**, 679–681.

Liu, X., Follmer, D., Zweier, J.R., Huang, X., Hemann, C., Liu, K., Druhan, L.J. & Zweier, J.L. 2012. Characterization of the function of cytoglobin as an oxygen-dependent regulator of nitric oxide concentration. *Biochemistry* **51**, 5072–5082.

Livermore, R., Eagles, G., Morris, P. & Maldonado, A. 2004. Shackleton fracture zone: no barrier to early circumpolar ocean circulation. *Geology* **32**, 797–800.

Livermore, R., Nankivell, A., Eagles, G. & Morris, P. 2005. Paleogene opening of Drake Passage. *Earth and Planetary Science Letters* **236**, 459–470.

Logue, J.A., DeVries, A.L., Fodor, E. & Cossins, A.R. 2000. Lipid compositional correlates of temperature-adaptive interspecific differences in membrane physical structure. *Journal of Experimental Biology* **203**, 2105–2115.

Lohbeck, K.T., Riebesell, U., Collins, S. & Reusch, T.B.H. 2013. Functional genetic divergence in high CO_2 adapted Emiliania huxleyi populations. *Evolution* **67**, 1892–1900.

Lopez, C.F., Darst, R.K. & Rossky, P.J. 2008. Mechanistic elements of protein cold denaturation. *Journal of Physical Chemistry B* **112**, 5961–5967.

Lovrich, G.A., Romero, M.C., Tapella, F. & Thatje, S. 2005. Distribution, reproductive and energetic conditions of decapod crustaceans along the Scotia Arc (Southern Ocean). *Scientia Marina* **69**, 183–193.

Lucassen, M., Koschnick, N., Eckerle, L.G. & Pörtner, H.O. 2006. Mitochondrial mechanisms of cold adaptation in cod (Gadus morhua L.) populations from different climatic zones. *Journal of Experimental Biology* **209**, 2462–2471.

Lurman, G., Blaser, T., Lamare, M., Peck, L.S. & Morley, S.A. 2010a. Mitochondrial plasticity in brachiopod (Liothyrella spp.) smooth adductor muscle as a result of season and latitude. *Marine Biology* **157**, 907–913.

Lurman, G., Blaser, T., Lamare, M., Tan, K.-S., Pörtner, H.O., Peck, L.S. & Morley, S.A. 2010b. Pedal muscle ultrastructure as a function of thermal habitat in marine patellogastropod limpets (Cellana sp. and Nacella concinna). *Marine Biology* **157**, 1705–1712.

Luxmoore, R.A. 1984. A comparison of the respiration rate of some Antarctic isopods with species from lower latitudes. *British Antarctic Survey Bulletin* **62**, 53–65.

Luxmoore, R.A. 1985. The energy budget of a population of the Antarctic isopod Serolis polita. In *Antarctic Nutrient Cycles and Food Webs*, W.R. Seigfried et al. (eds). Berlin: Springer, 389–396.

López-González, P.J., Gili, J.-M. & Fuentes, V. 2009. A new species of shallow-water sea pen (Octocorallia: Pennatulacea: Kophobelemnidae) from Antarctica. *Polar Biology* **32**, 907–914.

MacArthur, R.H. 1972. *Geographical Ecology: Patterns in the Distribution of Species*. New York: Harper and Row.

MacArthur, R.H. & Pianka, E.R. 1966. On the use of a patchy environment. *American Naturalist* **100**, 603–610.

Macdonald, J.A. & Montgomery, J.C. 1982. Thermal limits of neuromuscular function in an Antarctic fish. *Journal of Comparative Physiology* **147**, 237–250.

Macdonald, J.A., Montgomery, J.C. & Wells, R.M.G. 1988. The physiology of McMurdo Sound fishes: current New Zealand research. *Comparative Biochemistry and Physiology* **90B**, 567–578.

Mackas, D.L. & Galbraith, M.D. 2012. Pteropod time-series from the NE Pacific. *ICES Journal of Marine Science* **69**, 448–459.

Mahon, A.R., Amsler, C.D., McClintock, J.B., Amsler, M.O. & Baker, B.J. 2003. Tissue-specific palatability and chemical defences against macropredators and pathogens in the common articulate brachiopod Liothyrella uva from the Antarctic Peninsula. *Journal of Experimental Marine Biology and Ecology* **290**, 197–210.

Maldonado, A., Barnolas, A., Bohoyo, F., Galindo-Zaldívar, J., Hernández-Molina, J., Lobo, F., Rodríguez-Fernández, X., Somoza, L. & Váquez, J.T. 2003. Contourite deposits in the central Scotia Sea: the importance of the Antarctic circumpolar current and the Weddell Gyre flows. *Palaeogeography, Palaeoclimatology, Palaeoecology* **198**, 187–221.

Maldonado, A., Bohoyo, F., Galindo-Zaldívar, J., Hernández-Molina, J., Lobo, F., Lodolo, E., Martos, Y.M. & Pérez, L.F. 2014. A model of oceanic development by ridge jumping: opening of the Scotia Sea. *Global and Planetary Change* **123**, 152–173.

Marko, P.B. & Moran, A.L. 2002. Correlated evolutionary divergence of egg size and a mitochondrial protein across the Isthmus of Panama. *Evolution* **56**, 1303–1309.

Markowska, M. & Kidawa, A. 2007. Encounters between Antarctic limpets, Nacella concinna, and predatory sea stars, Lysasterias sp., in laboratory and field experiments. *Marine Biology* **151**, 1959–1966.

Marschall, H.-P. 1988. The overwintering strategy of Antarctic krill under the pack-ice of the Weddell Sea. *Polar Biology* **9**, 129–135.

Marsh, A.G., Maxson, R.E. & Manahan, D.T. 2001. High macromolecular synthesis with low metabolic cost in Antarctic sea urchin embryos. *Science* **291**, 1950–1952.

Marshall, D.J., Bolton, T.F. & Keough, M.J. 2006. Offspring size affects the post-metamorphic performance of a colonial marine invertebrate. *Ecology* **84**, 3131–3137.

Marshall, D.J. & Burgess, S.C. 2015. Deconstructing environmental predictability: seasonality, environmental colour and the biogeography of marine life histories. *Ecology Letters* **18**, 174–181.

Marshall, D.J., Krug, P.J., Kupriyanova, E.K., Byrne, M. & Emlet, R.B. 2012. The biogeography of marine invertebrate life histories. *Annual Review of Ecology, Evolution, and Systematics* **43**, 97–114.

Marshall, D.J. & Morgan, S.G. 2011. Ecological and evolutionary consequences of linked life-history stages in the sea. *Current Biology* **21**, R718–725.

Marshall, G.J. 2003. Trends in the Southern Annular Mode from observations and reanalyses. *Journal of Climatology* **16**, 4134–4143.

Martinson, D.G. & McKee, D.C. 2012. Transport of warm upper circumpolar deep water onto the western Antarctic Peninsula continental shelf. *Ocean Science* **8**, 433–442.

Martinson, D.G., Stammerjohn, S.E., Iannuzzi, R.A., Smith, R.C. & Vernet, M. 2008. Western Antarctic Peninsula physical oceanography and spatio-temporal variability. *Deep Sea Research Part II* **55**, 1964–1987.

McClintock, J.B. 1994. Trophic biology of Antarctic shallow-water echinoderms. *Marine Ecology Progress Series* **111**, 91–202.

McClintock, J.B., Amsler, M.O., Amsler, C.D., Southworth, K.J., Petrie, C. & Baker, B.J. 2004. Biochemical composition, energy content and chemical antifeedant and antifoulant defenses of the colonial Antarctic ascidian Distaplia cylindrica. *Marine Biology* **145**, 885–894.

McClintock, J.B., Amsler, C.D., Baker, B.J. & VanSoest, R.W.M. 2005. Ecology of Antarctic marine sponges: an overview. *Integrative and Comparative Biology* **45**, 359–368.

McClintock, J.B., Angus, R.A., Ho, C., Amsler, C.D. & Baker, B.J. 2008. A laboratory study of behavioral interactions of the Antarctic keystone sea star Odontaster validus with three sympatric predatory sea stars. *Marine Biology* **154**, 1077–1084.

McClintock, J.B., Angus, R.A., McDonald, M.R., Amsler, C.D., Catledge, S.A. & Vohra, Y.K. 2009. Rapid dissolution of shells of weakly calcified Antarctic benthic macroorganisms indicates high vulnerability to ocean acidification. *Antarctic Science* **21**, 449–456.

McClintock, J.B. & Baker, B.J. 1997. Palatability and chemical defense of eggs, embryos and larvae of shallow-water antarctic marine invertebrates. *Marine Ecology Progress Series* **154**, 121–131.

McClintock, J.B., Heine, J., Slattery, M. & Weston, J. 1991. Biochemical and energetic composition, population biology, and chemical defense of the Antarctic ascidian Cnemidocarpa verrucosa lesson. *Journal of Experimental Marine Biology and Ecology* **147**, 163–175.

McClintock, J.B., Slattery, M. & Thayer, C.W. 1993. Energy content and chemical defense of the articulate Brachiopod Liothyrella uva (Jackson, 1912) from the Antarctic Peninsula. *Journal of Experimental Marine Biology and Ecology* **169**, 103–116.

McDonald, M.R., McClintock, J.B., Amsler, C.D., Rittschof, D., Angus, R.A., Orihuela, B. & Lutostanki, K. 2009. Effects of ocean acidification over the life-history of the barnacle Amphibalanus amphitrite. *Marine Ecology Progress Series* **385**, 179–187.

McMinn, A., Pankowskii, A., Ashworth, C., Bhagooli, R., Ralph, P. & Ryan, K. 2010. In situ net primary productivity and photosynthesis of Antarctic sea-ice algal, phytoplankton and benthic algal communities. *Marine Biology* **157**, 1345–1356.

McNeil, B.I. & Matear, R.J. 2008. Southern Ocean acidification: a tipping point at 450-ppm atmospheric CO_2. *Proceedings of the National Academy of Science USA* **105**, 18860–18864.

McNeil, B.I., Sweeney, C. & Gibson, J.A.E. 2011. Natural seasonal variability of aragonite saturation state within two Antarctic coastal ocean sites. *Antarctic Science* **23**, 411–412.

Medawar, P.D. 1952. *An Unsolved Problem in Biology.* London: H.K. Lewis & Co.

Meidlinger, K., Tyler, P.A. & Peck, L.S. 1998. Reproductive patterns in the Antarctic brachiopod Liothyrella uva. *Marine Biology* **132**, 153–162.

Meister, K., Duman, J.G., Yu, Y., DeVries, A.L., Leitner, D.M. & Havenith, M. 2014. The role of sulfates on antifreeze protein activity. *Journal of Physical Chemistry B* **118**, 7920–7924.

Meister, K., Ebbinghaus, S., Xu, Y., Duman, J.G., DeVries, A.L., Gruebelle, S., Leitner, D.M. & Havenith, M. 2013. Long-range protein-water dynamics in hyperactive insect antifreeze proteins. *Proceedings of the National Academy of Sciences USA* **110**, 1617–1622.

Melzner, F., Forsythe, J.W., Lee, J.B., Wood, P.G., Piatkowski, U. & Clemmesen, C. 2005. Estimating recent growth in the cuttlefish Sepia officinalis: are nucleic acid-based indicators for growth and condition the method of choice? *Journal of Experimental Marine Biology and Ecology* **317**, 35–51.

Melzner, F., Gutowska, M.A., Hu, M. & Stumpp, M. 2009a. Acid-base regulatory capacity and associated proton extrusion mechanisms in marine invertebrates: an overview. *Comparative Biochemistry and Physiology* **153A**, S80.

Melzner, F., Gutowska, M.A., Langenbuch, M., Dupont, S., Lucassen, M., Thorndyke, M.C., Bleich, M. & Pörtner, H.O. 2009b. Physiological basis for high CO_2 tolerance in marine ectothermic animals: pre-adaptation through lifestyle and ontogeny? *Biogeosciences* **6**, 2313–2331.

Melzner, F., Stange, P., Trubenbach, K., Thomsen, J., Casties, I., Panknin, U., Gorb, S.N. & Gutowska, M.A. 2011. Food supply and seawater pCO2 impact calcification and internal shell dissolution in the blue mussel Mytilus edulis. *PLoS ONE* **6**, e24223.

Meredith, M.P. & Brandon, M.A. 2017. Oceanography and sea ice in the Southern Ocean. In *Sea Ice*, D.N. Thomas (ed.). Chichester, UK: John Wiley & Sons, Ltd, 216–238.

Meredith, M.P. & King, J.C. 2005. Climate change in the ocean to the west of the Antarctic Peninsula during the second half of the 20th century. *Geophysics Research Letters* **32**, L19604.

Meredith, M.P., Venables, H.J., Clarke, A., Ducklow, H.W., Erickson, M., Leng, M.J., Lenaerts, J.T.M. & van den Broeke, M.R. 2013. The freshwater system west of the Antarctic Peninsula: spatial and temporal changes. *Journal of Climate* **26**, 1669–1684.

Metzger, D.C.H. & Schulte, P.M. 2016. Epigenomics in marine fishes. *Marine Genomics* **30**, 43–54.

Micaletto, G., Gambi, M.C. & Piraino, S. 2002. Observations on population structure and reproductive features of Laetmonice producta Grube (Polychaeta, Aphroditidae) in Antarctic waters. *Polar Biology* **26**, 327–333.

Miceli, C., Ballarini, P., Di Giuseppe, G., Valbonesi, A. & Luporini, P. 1994. Identification of the tubulin gene family and sequence determination of one b-tubulin gene in a cold-poikilotherm protozoan, the Antarctic ciliate Euplotes focardii. *Journal of Eukaryotic Microbiology* **41**, 420–427.

Miceli, C., Pucciarelli, S., Ballarini, P., Valbonesi, A. & Luporini, P. 1996. The b-tubulin gene family of the Antarctic ciliate Euplotes focardii: determination of the complete sequence of the b-T1 gene. In *Antarctic Communities*, B. Battaglia & J. Walton (eds). London: Cambridge University Press, 300–306.

Michaud, M., Benoit, J.B., Lopez-Martinez, G., Elnitsky, M.A., Lee, R.E. & Denlinger, D.L. 2008. Metabolomics reveals unique and shared metabolic changes in response to heat shock, freezing and desiccation in the Antarctic midge, Belgica antarctica. *Journal of Insect Physiology* **54**, 645–655.

Mileikovsky, S.A. 1971. Types of larval development in marine bottom invertebrates, their distribution and ecological significance: a re-evaluation. *Marine Biology* **10**, 193–213.

Mills, E.L. 1967. The biology of an ampeliscid amphipod crustacean sibling species pair. *Journal of the Fisheries Research Board of Canada* **24**, 305–355.

Mincks, S.L., Smith, C.R. & Demaster, D.J. 2005. Persistence of labile organic matter and microbial biomass in Antarctic shelf sediments: evidence of a sediment 'food bank'. *Marine Ecology Progress Series* **300**, 3–19.

Moline, M.A., Karnovsky, N.J., Brown, Z., Divoky, G.J., Frazer, T.K., Jacoby, C.A., Torres, J.J. & Fraser, W.R. 2008. High latitude changes in ice dynamics and their impact on polar marine ecosystems. *Annals of the New York Academy of Sciences* **1134**, 267–319.

Mollet, F.M., Engelhard, G.H., Vainikka, A., Laugen, A.T., Rijnsdorp, A.D. & Ernande, B. 2013. Spatial variation in growth, maturation schedules and reproductive investment of female sole Solea solea in the Northeast Atlantic. *Journal of Sea Research* **84**, 109–121.

Montes-Hugo, M., Doney, S.C., Ducklow, H.W., Fraser, W., Martinson, D., Stammerjohn, S.E. & Schofield, O.M. 2009. Recent changes in phytoplankton communities associated with rapid regional climate change along the Western Antarctic Peninsula. *Science* **323**, 1470–1473.

Montgomery, J. & Clements, K. 2000. Disaptation and recovery in the evolution of Antarctic fishes. *Trends in Ecology and Evolution* **15**, 267–271.

Moon, T.W., Altringham, J.D. & Johnston, I.A. 1991. Energetics and power output of isolated fish fast muscle fibres performing oscillatory work. *Journal of Experimental Biology* **158**, 261–273.

Moore, M. & Manahan, D.T. 2007. Variation among females in egg lipid content and developmental success of echinoderms from McMurdo Sound, Antarctica. *Polar Biology* **30**, 1245–1252.

Moran, A.L. & Emlet, R.B. 2001. Offspring size and performance in variable environments: field studies on a marine snail. *Ecology* **82**, 1597–1612.

Moran, A.L. & McAlister, J.S. 2009. Egg size as a life-history characteristics of marine invertebrates: is it all it's cracked up to be. *Biological Bulletin* **216**, 226–242.

Morgan, S.G. 1995. Life and death in the plankton: larval mortality and adaptation. In *Ecology of Marine Invertebrate Larvae*, L. McEdward (ed.). Boca Raton, FL: CRC, 279–322.

Morley, S.A., Bates, A., Lamare, M., Richard, J., Nguyen, K., Brown, J. & Peck, L.S. 2016a. Rates of warming and the global sensitivity of shallow marine invertebrates to elevated temperature. *Journal of the Marine Biological Association of the United Kingdom* **96**, 159–165.

Morley, S.A., Berman, J., Barnes, D.K.A., Carbonell, D.D.J., Downey, R.V. & Peck, L.S. 2016b. Extreme phenotypic plasticity in metabolic physiology of Antarctic demosponges. *Frontiers in Ecology and Evolution* **3**, 157.

Morley, S.A., Chien-Hsian, L., Clarke, A., Tan, K.S., Thorne, M.A.S., Peck, L.S. 2014. Limpet feeding rate and the consistency of physiological response to temperature. *Journal of Comparative Physiology* **184B**, 563–570.

Morley, S.A., Clark, M.S. & Peck, L.S. 2010. Depth gradients in shell morphology correlate with thermal limits for activity and ice disturbance in Antarctic limpets. *Journal of Experimental Marine Biology and Ecology* **390**, 1–5.

Morley, S.A., Hirse, T., Pörtner, H.O. & Peck, L.S. 2009a. Geographical variation in thermal tolerance within Southern Ocean marine ectotherms. *Comparative Biochemistry and Physiology Part* **153A**, 154–161.

Morley, S.A., Hirse, T., Thorne, M.A.S., Pörtner, H.O. & Peck, L.S. 2012a. Physiological plasticity, long term resistance or acclimation to temperature, in the Antarctic bivalve, Laternula elliptica. *Comparative Biochemistry and Physiology Part* **162A**, 16–21.

Morley, S.A., Lemmon, V., Obermüller, B.E., Spicer, J.I., Clark, M.S. & Peck, L.S. 2011. Duration tenacity: a method for assessing acclimatory capacity of the Antarctic limpet, Nacella concinna. *Journal of Experimental Marine Biology and Ecology* **399**, 39–42.

Morley, S.A., Lurman, G.L., Skepper, G.N., Pörtner, H.O. & Peck, L.S. 2009b. Thermal plasticity of mitochondria: a latitudinal comparison between Southern Ocean molluscs. *Comparative Biochemistry and Physiology* **152A**, 423–430.

Morley, S.A., Martin, S.M., Bates, A.E., Clark, M.S., Ericson, J., Lamare, M. & Peck, L.S. 2012b. Spatial and temporal variation in the heat tolerance limits of two abundant Southern Ocean invertebrates. *Marine Ecology Progress Series* **450**, 81–92.

Morley, S.A., Martin, S.M., Day, R.W., Ericson, J., Lai, C.-H., Lamare, M., Tan, K.-S., Thorne, M.A.S. & Peck, L.S. 2012c. Thermal reaction norms and the scale of temperature variation: latitudinal vulnerability of intertidal nacellid limpets to climate change. *PLoS ONE* **7**, e52818.

Morley, S.A., Peck, L.S., Miller, A. & Pörtner, H.O. 2007. Hypoxia tolerance associated with activity reduction is a key adaptation for Laternula elliptica seasonal energetics. *Oecologia* **153**, 29–36.

Morley, S.A., Suckling, C.S., Clark, M.S., Cross, E.L. & Peck, L.S. 2016c. Long-term effects of altered pH and temperature on the feeding energetics of the Antarctic sea urchin, Sterechinus neumayeri. *Biodiversity* **17**, 34–45.

Morley, S.A., Tan, K.S., Day, R.W., Martin, S.M., Pörtner, H.O. & Peck, L.S. 2009c. Thermal dependency of burrowing in three species within the bivalve genus Laternula: a latitudinal comparison. *Marine Biology* **156**, 1977–1984.

Morricone, E. 1999. Reproductive biology of the limpet Nacella (P.) deaurata (Gmelin, 1791) in Bahía Lapataia (Beagle Channel). *Scientia Marina* **63**, 417–426.

Morritt, D. & Spicer, J.I. 1993. A brief re-examination of the function and regulation of extracellular magnesium and its relationship to activity in crustacean arthropods. *Comparative Biochemistry and Physiology* **106**, 19–23.

Mortensen, T. 1921. *Studies of the Development and Larval Forms of Echinoderms.* G.E.D. Grad: Copenhagen.

Mortensen, T. 1936. Echinoidea and Ophiuroidea. *Discovery Reports* **12**, 199–348.

Moylan, T.J. & Sidell, B.D. 2000. Concentrations of myoglobin and myoglobin mRNA in heart ventricles from Antarctic fishes. *Journal of Experimental Biology* **203**, 1277–1286.

Mueller, I.A., Grim, J.M., Beers, J.M., Crockett, E.L. & O'Brien, K.M. 2011. Inter-relationship between mitochondrial function and susceptibility to oxidative stress in red- and white-blooded Antarctic notothenioid fishes. *Journal of Experimental Biology* **214**, 3732–3741.

Munday, P.L., Warner, R.R., Monro, K., Pandolfi, J.M. & Marshall, D.J. 2013. Predicting evolutionary responses to climate change in the sea. *Ecology Letters* **16**, 1488–1500.

Munday, P.L., Watson, S.-A., Parsons, D.M., King, A., Barr, N.G., Mcleod, I.M., Allan, B.J. M. & Pether, S.M.J. 2016. Effects of elevated CO_2 on early life-history development of the yellowtail kingfish, Seriola lalandi, a large pelagic fish. *ICES Journal of Marine Science* **73**, 641–649.

Munilla, T. & Soler Membrives, A. 2009. Check-list of the pycnogonids from Antarctic and sub-Antarctic waters: zoogeographic implications. *Antarctic Science* **21**, 99–111.

Myers, R.A. 2002. Recruitment: understanding density-dependence in fish populations. In *Handbook of Fish Biology and Fisheries, Volume 1: Fish Biology. Chapter 6*, P.J.B. Hart & J.D. Reynolds (eds). Chichester, UK: Blackwell, 123–148.

Nakayama, S. & Clarke, J.F. 2003. Smooth muscle and NMR review: an overview of smooth muscle metabolism. *Molecular and Cellular Biochemistry* **244**, 17–30.

Near, T.J. 2004. Estimating divergence times of notothenioid fishes using a fossil-calibrated molecular clock. *Antarctic Science* **16**, 37–44.

Near, T.J., Dornburg, A., Kuhn, K.L., Eastman, J.T., Pennington, J.N., Patarnello, T., Zane, F., Fernández, D.A. & Jones, C.D. 2012. Ancient climate change, antifreeze, and the evolutionary diversification of Antarctic fishes. *Proceedings of the National Academy of Science USA* **109**, 3434–3439.

Near, T.J., Parker, S.K. & Detrich III, H.W. 2006. A genomic fossil reveals key steps in haemoglobin loss by the Antarctic icefishes. *Molecular Biology and Evolution* **23**, 2008–2016.

Neargarder, G., Dahlhoff, E.P. & Rank, N.E. 2003. Variation in thermal tolerance is linked to phosphoglucose isomerase genotype in a montane leaf beetle. *Functional Ecology* **17**, 213–221.

Nicodemus-Johnson, J., Silic, S., Ghigliotti, L., Pisano, E. & Cheng, C.H. 2011. Assembly of the antifreeze glycoprotein/trypsinogen-like protease genomic locus in the Antarctic toothfish Dissostichus mawsoni (Norman). *Genomics* **98**, 194–201.

Nicol, D. 1964. An essay on size of marine pelecypods. *Journal of Paleontology* **38**, 968–974.

Nicol, D. 1966. Size of pelecypods in Recent marine faunae. *Nautilus* **79**, 109–113.

Nicol, D. 1967. Some characteristics of cold-water marine pelecypods. *Journal of Paleontology* **41**, 1330–1340.

Nicol, S. 2006. Krill, currents, and sea-ice: Euphausia superba and its changing environment. *Bioscience* **56**, 111–120.

Nilsson, G.E., Dixson, D.L., Domenici, P., McCormick, M.I., Sørensen, C., Watson, S.-A. & Munday, P.L. 2012a. Near-future carbon dioxide levels alter fish behaviour by interfering with neurotransmitter function. *Nature Climate Change* **2**, 201–204.

Nishimiya, Y., Sato, R., Takamichi, M., Miura, A. & Tsuda, S. 2005. Co-operative effect of the isoforms of type III antifreeze protein expressed in notched-fin eelpout, Zoarces elongatus Kner. *FEBS Journal* **272**, 482–492.

Noisette, F., Bordeyne, F., Davoult, D. & Martin, S. 2015. Assessing the physiological responses of the gastropod Crepidula fornicata to predicted ocean acidification and warming. *Limnology and Oceanography* **61**, 430–444.

Noisette, F., Richard, J., Le Fur, I., Peck, L.S., Davoult, D. & Martin, S. 2014. Metabolic responses to temperature stress under elevated pCO_2 in Crepidula fornicata. *Journal of Molluscan Studies* **81**, 238–246.

Nolan, C.P. & Clarke, A. 1993. Growth in the bivalve Yoldia eightsi at Signy Island, Antarctica determined from internal shell increments and Calcium-45 incorporation. *Marine Biology* **117**, 243–250.

Norin, T., Malte, H. & Clark, T.D. 2014. Aerobic scope does not predict the performance of a tropical eurythermal fish at elevated temperatures. *Journal of Experimental Biology* **217**, 244–251.

Núñez-Pons, L. & Avila, C. 2014. Deterrent activities in the crude lipophilic fractions of Antarctic benthic organisms: chemical defences against keystone predators. *Polar Research* **33**, 21624.

Obermüller, B.E., Morley, S.A., Clark, M.S., Barnes, D.K.A. & Peck, L.S. 2011. Antarctic intertidal limpet ecophysiology: a winter-summer comparison. *Journal of Experimental Marine Biology and Ecology* **403**, 39–45.

Obermüller, B.E., Peck, L.S., Barnes, D.K.A. & Morley, S.A. 2010. Seasonal physiology of Antarctic marine benthic predators and scavengers. *Marine Ecology Progress Series* **415**, 109–126.

Obermüller, B.E., Truebano, M., Peck, L.S., Eastman, J.T. & Morley, S.A. 2013. Reduced seasonality in elemental CHN composition of Antarctic marine benthic predators and scavengers. *Journal of Experimental Marine Biology and Ecology* **446**, 328–333.

O'Brien, K.M. 2016. New Lessons from an old fish: what Antarctic icefishes may reveal about the functions of oxygen-binding proteins. *Integrative and Comparative Biology* **56**, 531–541.

O'Brien, K.M., Skilbeck, C., Sidell, B.D. & Egginton, S. 2003. Muscle fine structure may maintain the function of oxidative fibres in haemoglobinless Antarctic fishes. *Journal of Experimental Biology* **206**, 411–421.

O'Brien, K.M., Xue, H. & Sidell, B.D. 2000. Quantification of diffusion distance within the spongy myocardium of hearts from Antarctic fishes. *Respiratory Physiology* **122**, 71–80.

O'Brien, L., Burnett, J. & Mayo, R.K. 1993. Maturation of Nineteen Species of Finfish off the Northeast Coast of the United States, 1985–1990. NOAA Technical Reports NMFS 113.

O'Cofaigh, C., Pudsey, C.J., Dowdeswell, J.A. & Morris, P. 2002. Evolution of subglacial bedforms along a paleo-ice stream, Antarctic Peninsula continental shelf. *Geophysics Research Letters* **29**, 1199.

Oellermann, M., Lieb, B., Poertner, H.O., Semmens, J.M. & Mark, F.C. 2012. Blue blood on ice: modulated blood oxygen transport facilitates cold compensation and eurythermy in an Antarctic octopod. *Frontiers in Zoology* **12**, 6.

O'Loughlin, P.M., Paulay, G., Davey, N. & Michonneau, F. 2011. The Antarctic region as a marine biodiversity hotspot for echinoderms: diversity and diversification of sea cucumbers. *Deep Sea Research Part II Topical Studies in Oceanography* **58**, 264–275.

Opalinski, K.W. & Jazdzewski, K. 1978. Respiration of some Antarctic amphipods. *Polish Archives of Hydrobiology* **25**, 643–655.

Ordway, G.A. & Garry, D.J. 2004. Myoglobin: an essential hemoprotein in striated muscle. *Journal of Experimental Biology* **207**, 3441–3446.

Orejas, C., Gili, J.-M. & Arntz, W.E. 2003. Role of small-plankton communities in the diet of two Antarctic octocorals (Primnoisis antarctica and Primnoella sp.). *Marine Ecology Progress Series* **250**, 105–116.

Orr, J.C., Fabry, V.J., Aumont, O., Bopp, L., Doney, S.C., Feely, R.A., Gnanadesikan, A., Gruber, N., Ishida, A., Joos, F., Key, R.M., Lindsay, K., Maier-Reimer, E., Matear, R., Monfray, P., Mouchet, A., Najjar, R.G., Plattner, G.K., Rodgers, K.B., Sabine, C.L., Sarmiento, J.L., Schlitzer, R., Slater, R.D., Totterdell, I.J., Weirig, M.F., Yamanaka, Y. & Yool, A. 2005. Anthropogenic ocean acidification over the twenty-first century and its impact on calcifying organisms. *Nature* **437**, 681–686.

Orsi, A.H. & Wiederwohl, C.L. 2009. A recount of Ross Sea waters. *Deep Sea Research II* **56**, 778–795.

Overgaard, J., Andersen, J.L., Findsen, A., Pedersen, P.B.M., Hansen, K., Ozolina, K. & Wang, T. 2012. Aerobic scope and cardiovascular oxygen transport is not compromised at high temperatures in the toad Rhinella marina. *Journal of Experimental Biology* **215**, 3519–3526.

Oystein, V. 2012. Fitness and phenology: annual routines and zooplankton adaptations to seasonal cycles. *Journal of Plankton Research* **34**, 267–276.

Pace, D.A. & Manahan, D.T. 2006. Fixed metabolic costs for highly variable rates of protein synthesis in sea urchin embryos and larvae. *Journal of Experimental Biology* **209**, 158–170.

Pace, D.A. & Manahan, D.T. 2007. Cost of protein synthesis and energy allocation during development of Antarctic sea urchin embryos and larvae. *Biological Bulletin* **212**, 115–129.

Pace, D.A. & Manahan, D.T. 2010. Ribosomal analysis of rapid rates of protein synthesis in the Antarctic sea urchin Sterechinus neumayeri. *Biological Bulletin* **218**, 48–60.

Paine, R.T. 1965. Natural history, limiting factors and energetics of the opisthobranch Navanax inermis. *Ecology* **46**, 603–619.

Paine, R.T. 1966. Food web complexity and species diversity. *American Naturalist* **100**, 65–75.

Parker, L.M., Ross, P.M., O'Connor, W.A., Borysko, L., Raftos, D.A. & Pörtner, H.-O. 2012. Adult exposure influences offspring response to ocean acidification in oysters. *Global Change Biology* **18**, 82–92.

Parker, S.J. & Grimes, P J. 2010. Length and age at spawning of Antarctic toothfish (Dissostichus mawsoni) in the Ross Sea. *CCAMLR Science* **17**, 53–73.

Parsell, D.A. & Lindquist, S. 1993. The function of heat-shock proteins in stress tolerance - degradation and reactivation of damaged proteins. *Annual Reviews of Genetics* **27**, 437–496.

Patarnello, T., Verde, C., di Prisco, G., Bargelloni, L. & Zane, L. 2011. How will fish that evolved at constant sub-zero temperatures cope with global warming? Notothenioids as a case study. *Bioessays* **33**, 260–268.

Pearse, J.S. 1965. Reproductive periodicities in several contrasting populations of Odontaster validus Koehler, a common Antarctic asteroid. *Antarctic Research Series (American Geophysical Union)* **5**, 39–85.

Pearse, J.S. 1969. Slow developing demersal embryos and larvae of the Antarctic sea star Odontaster validus. *Marine Biology* **3**, 110–116.

Pearse, J.S. 1994. Cold-water echinoderms break Thorson's rule. In *Reproduction, Larval Biology and Recruitment in Deep-Sea Benthos*, C.M. Young & K.J. Eckelbarger (eds). New York: Columbia University, 26–43.

Pearse, J.S. & Bosch, I. 1986. Are the feeding larvae of the commonest Antarctic asteroid really demersal. *Bulletin of Marine Science* **39**, 177–484.

Pearse, J.S. & Bosch, I. 1994. Brooding in the Antarctic: Ostergren had it nearly right. In *Echinoderms through Time (Proceedings of the 8th International Echinoderm Conference)*, B. David et al. (eds). Rotterdam: Balkema, 111–120.

Pearse, J.S. & Bosch, I. 2002. Photoperiodic regulation of gametogenesis in the Antarctic sea star Odontaster validus Koehler: evidence for a circannual rhythm modulated by light. *Invertebrate Reproduction and Development* **41**, 73–81.

Pearse, J.S. & Cameron, R.A. 1991. Echinodermata: Echinoidea. In *Reproduction of Marine Invertebrates VI Echinoderms and Lophophorates*, A.C. Giese et al. (eds). Pacific Grove, CA: The Boxwood Press, 514–662.

Pearse, J.S. & Giese, A. 1966a. Food, reproduction and organic constitution of the common Antarctic echinoid Sterechinus neumayeri (Meissner). *Biological Bulletin* **130**, 387–401.

Pearse, J.S. & Giese, A.C. 1966b. The organic constitution of several benthonic invertebrates from McMurdo Sound, Antarctica. *Comparative Biochemistry and Physiology* **18**, 47–5.

Pearse, J.S. & Lockhart, S.J. 2004. Reproduction in cold water: paradigm changes in the 20th century and a role for cidaroid sea urchins. *Deep-Sea Research, Part II* **51**, 1533–1549.

Pearse, J.S., McClintock, J.B. & Bosch, I. 1991. Reproduction of Antarctic marine invertebrates: tempos, modes and timing. *American Zoologist* **31**, 65–80.

Pechenik, J.A. 1986. Field evidence for delayed metamorphosis of larval gastropods: Crepidula plana Say, C. fornicata (L.), and Bittium alternatum. *Journal of Experimental Marine Biology and Ecology* **97**, 313–319.

Pechenik, J.A. 1987. Environmental influences on larval survival and development. In *Reproduction of Marine Invertebrates*, Vol. IX, A.C. Giese & J.S. Pearse (eds). Pacific Grove, CA: Boxwood Press, 551–608.

Pechenik, J.A. 1991. *Biology of the Invertebrates*, 2nd edition. Wm. C. Brown: Dubuque, Iowa.

Pechenik, J.A. 1999. On the advantages and disadvantages of larval stages in benthic marine invertebrate life cycles. *Marine Ecology Progress Series* **177**, 269–297.

Pechenik, J.A. & Levine, S.H. 2007. A new approach to estimating the magnitude of planktonic larval mortality using the marine gastropods Crepidula fornicata and C. plana. *Marine Ecology Progress Series* **344**, 107–118.

Peck, L.S. 1989. Temperature and basal metabolism in two Antarctic marine herbivores. *Journal of Experimental Marine Biology and Ecology* **127**, 1–12.

Peck, L.S. 1993a. The tissues of articulate brachiopods and their value to predators. *Philosophical Transactions of the Royal Society of London B* **339**, 17–32.

Peck, L.S. 1993b. Larval development in the Antarctic nemertean Parborlasia corrugatus (Heteronemertea, Lineidae). *Marine Biology* **116**, 301–310.

Peck, L.S. 1996. Feeding and metabolism in the Antarctic brachiopod Liothyrella uva: a low energy lifestyle species with restricted metabolic scope. *Proceedings of the Royal Society of London B* **263**, 223–228.

Peck, L.S. 1998. Feeding, metabolism and metabolic scope in Antarctic marine ectotherms. In *Cold Ocean Physiology*. Society for Experimental Biology Seminar Series 66, H.O. Pörtner & R. Playle (eds). Cambridge, UK: Cambridge University Press, 365–390.

Peck, L.S. 2001a. Physiology. In *Short Course on Brachiopods, Chapter 7*, S. Carlson & M. Sandy (eds). Kansas: Geological Society of the USA & University of Kansas, 89–104.

Peck, L.S. 2001b. Ecology of articulates. In Short Course on Brachiopods, Chapter 11, S. Carlson & M. Sandy (eds). Kansas: Geological Society of the USA & University of Kansas, 171–184.

Peck, L.S. 2002a. Ecophysiology of Antarctic marine ectotherms: limits to life. *Polar Biology* **25**, 31–40.

Peck, L.S. 2002b. Coping with Change: stenothermy, physiological flexibility and environmental change in Antarctic seas. In *Climate Changes: Effects on Plants, Animals and Humans*, C.L. Bolis et al. (eds). *14th ICCP*, September 2000, Troina, Italy, 1–13.

Peck, L.S. 2005a. Prospects for surviving climate change in Antarctic aquatic species. *Frontiers in Zoology* **2**, 9.

Peck, L.S. 2005b. Prospects for survival in the Southern Ocean: extreme temperature sensitivity of benthic species. *Royal Society Special Issue of Antarctic Science* **17**, 497–507.

Peck, L.S. 2008. Brachiopods and climate change. *Philosophical Transactions of the Royal Society of Edinburgh Earth and Environmental Sciences* **98**, 451–456.

Peck, L.S. 2011. Organisms and responses to environmental change. *Marine Genomics* **4**, 237–243.

Peck, L.S. 2015. DeVries: the Art of not freezing fish. Classics series. *Journal of Experimental Biology* **218**, 2146–2147.

Peck, L.S. 2016. A cold limit to adaptation in the sea. *Trends in Ecology and Evolution* **31**, 13–26.

Peck, L.S., Ansell, A.D., Curry, G. & Rhodes, M. 1997a. Physiology and metabolism. In *Treatise on Invertebrate Palaeontology Part (H) Brachiopoda, Chapter III*, R. Kaesler (ed). United States Kansas: Geological Society and the University of Kansas, 213–242.

Peck, L.S., Ansell, A.D., Webb, K.E. Hepburn, L. & Burrows, M. 2004a. Burrowing in Antarctic bivalve molluscs. *Polar Biology* **27**, 357–367.

Peck, L.S., Baker, A. & Conway, L.Z. 1996. Strontium labelling of the shell of the Antarctic limpet Nacella concinna. *Journal of Molluscan Studies* **62**, 315–325.

Peck, L.S. & Barnes, D.K.A. 2004. Metabolic flexibility: the key to long-term evolutionary success in bryozoa? *Proceedings of the Royal Society of London B* **271**, S18–S21.

Peck, L.S., Barnes, D.K.A., Cook, A.J., Fleming, A.H. & Clarke, A. 2010a. Negative feedback in the cold: ice retreat produces new carbon sinks in Antarctica. *Global Change Biology* **16**, 2614–2623.

Peck, L.S., Barnes, D.K.A. & Wilmott, J. 2005. Responses to extreme seasonality in food supply: diet plasticity in Antarctic brachiopods. *Marine Biology* **147**, 453–464.

Peck, L.S. & Brey, T. 1996. Bomb signals in old Antarctic brachiopods. *Nature* **380**, 207–208.

Peck, L.S. & Brockington, S.A. 2013. Growth in the Antarctic octocoral Primnoella scotiae and predation by the anemone Dactylanthus antarcticus. *Deep Sea Research II* **92**, 73–78.

Peck, L.S., Brockington, S. & Brey, T. 1997b. Growth and metabolism in the Antarctic brachiopod Liothyrella uva. *Philosophical Transactions of the Royal Society of London B* **352**, 851–858.

Peck, L.S., Brockington, S., VanHove, S. & Beghyn, M. 1999. Community recovery following catastrophic iceberg impacts in Antarctica. *Marine Ecology Progress Series* **186**, 1–8.

Peck, L.S. & Bullough, L.W. 1993. Growth and population structure in the infaunal bivalve Yoldia eightsi in relation to iceberg activity at Signy Island, Antarctica. *Marine Biology* **117**, 235–241.

Peck, L.S. & Chapelle, G. 1999. Amphipod gigantism dictated by oxygen availability? *Ecology Letters* **2**, 401–403.

Peck, L.S. & Chapelle, G. 2003. Reduced oxygen at high altitude limits maximum size. *Proceedings of the Royal Society of London B* **270**, S166–S167.

Peck, L.S., Clark, M.S., Morley, S.A., Massey, A. & Rossetti, H. 2009a. Animal temperature limits and ecological relevance: effects of size, activity and rates of change. *Functional Ecology* **23**, 248–253.

Peck, L.S., Clark, M.S., Power, D., Reis, J., Batista, F.M. & Harper, E.M. 2015a. Acidification effects on biofouling communities: winners and losers. *Global Change Biology* **21**, 1907–1913.

Peck, L.S., Clarke, A. & Chapman, A.L. 2006a. Metabolism and development of pelagic larvae of Antarctic gastropods with mixed reproductive strategies. *Marine Ecology Progress Series* **318**, 213–220.

Peck, L.S., Clark, M.S. & Dunn, N.I. 2018. Morphological variation in taxonomic characters of the Antarctic starfish *Odontaster validus*. *Polar Biology on Line*. doi.org/10.1007/s00300-018-2344-z.

Peck, L.S., Clarke, A. & Holmes, L. 1987a. Summer metabolism and seasonal biochemical changes in the brachiopod Liothyrella uva (Jackson, 1912). *Journal of Experimental Marine Biology and Ecology* **114**, 85–97.

Peck, L.S., Colman, J.G. & Murray, A.W.A. 2000. Growth and tissue mass cycles in the infaunal bivalve Yoldia eightsi at Signy Island, Antarctica. *Polar Biology* **23**, 420–428.

Peck, L.S., Convey, P. & Barnes, D.K.A. 2006b. Environmental constraints on life histories in Antarctic ecosystems: tempos, timings and predictability. *Biological Reviews* **81**, 75–109.

Peck, L.S. & Conway, L.Z. 2000. The myth of metabolic cold adaptation: oxygen consumption in stenothermal Antarctic bivalves. In *The Evolutionary Biology of the Bivalvia*, E.M. Harper et al. (eds). London: Geological Society, Special Publications, 177, 441–445.

Peck, L.S., Culley, M.B. & Helm, M.M. 1987b. A laboratory energy budget for the ormer Haliotis tuberculata L. *Journal of Experimental Marine Biology and Ecology* **106**, 103–123.

Peck, L.S., Heiser, S. & Clark, M.S. 2016a. Very slow embryonic and larval development in the Antarctic limpet Nacella polaris. *Polar Biology* **39**, 2273–2280.

Peck, L.S. & Holmes, L.J. 1989. Seasonal and ontogenetic changes in tissue size in the Antarctic brachiopod Liothyrella uva (Broderip, 1983). *Journal of Experimental Marine Biology and Ecology* 134, 25–36.

Peck, L.S. & Maddrell, S.H.P. 2005. The limitation of size by oxygen in the fruit fly Drosophila melanogaster. *Journal of Experimental Zoology* **303A**, 968–975.

Peck, L.S., Massey, A., Thorne, M.A.S. & Clark, M.S. 2009b. Lack of acclimation in Ophionotus victoriae: brittle stars are not fish. *Polar Biology* **32**, 399–402.

Peck, L.S., Meidlinger, K. & Tyler, P.A. 2001. Developmental and settlement characteristics of the Antarctic brachiopod Liothyrella uva (Broderip 1833). In *Brachiopods Past and Present, Proc. 4th International Congress on Brachiopods*, C.H.C. Brunton et al. (eds). London: The Systematics Association, July 1999, 80–90.

Peck, L.S., Morley, S.A. & Clark, M.S. 2010b. Poor acclimation capacities in Antarctic marine ectotherms. *Marine Biology* **157**, 2051–2059.

Peck, L.S., Morley, S.A., Pörtner, H.O. & Clark, M.S. 2007a. Small increases in temperature and reductions in oxygen availability limit burrowing capacity in the Antarctic clam Laternula elliptica. *Oecologia* **154**, 479–484.

Peck, L.S., Morley, S.A., Richard, J. & Clark, M.S. 2014. Acclimation and thermal tolerance in Antarctic marine ectotherms. *Journal Experimental Biology* **217**, 16–22.

Peck, L.S., Morris, D.J. & Clarke, A. 1986a. Oxygen consumption and the role of caeca in the recent Antarctic brachiopod Liothyrella uva notorcadensis (Jackson, 1912). *Biostratigraphie du Paleozoique* **5**, 349–356.

Peck, L.S., Morris, D.J. & Clarke, A. 1986b. The caeca of punctate brachiopods: a respiring tissue not a respiratory organ. *Lethaia* **19**, 232.

Peck, L.S., Morris, D.J., Clarke, A. & Holmes, L.J. 1986c. Oxygen consumption and nitrogen excretion in the Antarctic brachiopod Liothyrella uva (Jackson, 1912) under simulated winter conditions. *Journal of Experimental Marine Biology and Ecology* **104**, 203–213.

Peck, L.S., Powell, D.K. & Tyler, P.A. 2007b. Very slow development in two Antarctic bivalve molluscs, the infaunal clam, Laternula elliptica and the scallop Adamussium colbecki. *Marine Biology* **150**, 1191–1197.

Peck, L.S. & Prothero-Thomas, E. 2002. Temperature effects on the metabolism of larvae of the Antarctic starfish Odontaster validus, using a novel micro-respirometry method. *Marine Biology* **141**, 271–276.

Peck, L.S., Prothero-Thomas, E. & Hough, N. 1993. Pedal mucus production by the Antarctic limpet Nacella concinna (Strebel 1908). *Journal of Experimental Marine Biology and Ecology* **174**, 177–192.

Peck, L.S., Pörtner, H.O. & Hardewig, I. 2002. Metabolic demand, oxygen supply and critical temperatures in the Antarctic bivalve Laternula elliptica. *Physiological and Biochemical Zoology* **75**, 123–133.

Peck, L.S., Souster, T. & Clark, M.S. 2013. Juveniles are more resistant to warming than adults in 4 species of Antarctic marine invertebrates. *PLoS ONE* **8**, e66033.

Peck, L.S., Thorne, M.A.S., Hoffman, J.I., Morley, S.A. & Clark, M.S. 2015b. Variability among individuals is generated at the gene expression level. *Ecology* **96**, 2004–2014.

Peck, L.S. & Veal, R. 2001. Feeding, metabolism and growth in the Antarctic limpet Nacella concinna (Strebel 1908). *Marine Biology* **138**, 553–560.

Peck, L.S., Webb, K.E. & Bailey, D. 2004b. Extreme sensitivity of biological function to temperature in Antarctic marine species. *Functional Ecology* **18**, 625–630.

Peck, L.S., Webb, K.E., Clark, M.S., Miller, A. & Hill, T. 2008. Temperature limits to activity, feeding and metabolism in the Antarctic starfish Odontaster validus. *Marine Ecology Progress Series* **381**, 181–189.

Peck, V.L., Tarling, G.A., Manno, C., Harper, E.M. & Tynan, E. 2016b. Outer organic layer and internal repair mechanism protects pteropod Limacina helicina from ocean acidification. *Deep-Sea Research II* **127**, 41–52.

Peck, V.L., Tarling, G.A., Manno, C., Harper, E.M. & Tynan, E. 2016c. Response to comment "Vulnerability of pteropod (Limacina helicina) to ocean acidification: Shell dissolution occurs despite an intact organic layer" by Bednarsek et al. *Deep Sea Research II* **127**, 57–59.

Peel, M.C. & Wyndham, R.C. 1999. Selection of clc, cba, and fcb chlorobenzoate-catabolic genotypes from groundwater and surface waters adjacent to the Hyde Park, Niagara Falls, chemical landfill. *Applied and Environmental Microbiology* **65**, 1627–1635.

Pennisi, E. 2005. What determines species diversity? *Science* **309**, 90.

Pespeni, M.H., Barney, B.T. & Palumbi, S.R. 2013a. Differences in the regulation of growth and biomineralization genes revealed through long-term common-garden acclimation and experimental genomics in the purple sea urchin. *Evolution* **67**, 1901–1914.

Pespeni, M.H., Sanford, E., Gaylord, B., Hill, T.M., Hosfelt, J.D., Jaris, H.K., LaVigne, M., Lenz, E.A., Russell, A.D., Young, M.K. & Palumbi, S.R. 2013b. Evolutionary change during experimental ocean acidification. *Proceedings of the National Acadamy of Science USA* **110**, 6937–6941.

Peters, K.J., Amsler, C.D., McClintock, J.B., van Soest, R.W.M. & Baker, B.J. 2009. Palatability and chemical defenses of sponges from the western Antarctic Peninsula. *Marine Ecology Progress Series* **385**, 77–85.

Petrich, C. & Eiken, H. 2010. Growth, structure and properties of sea-ice. In *Sea-Ice*, 2nd edition, D.N. Thomas & G.S. Diekmann (eds). Oxford: Wiley Blackwell, 23–78.

Philipp, E., Brey, T., Pörtner, H.O. & Abele, D. 2005. Chronological and physiological ageing in a polar and a temperate mud clam. *Mechanisms of Ageing and Development* **126**, 598–609.

Philipp, E.E.R. & Abele, D. 2010. Masters of longevity: lessons from long-lived bivalves - a mini review. *Gerontology* **56**, 55–65.

Philipp, E.E.R., Husmann, G. & Abele, D. 2011. The impact of sediment deposition and iceberg scour on the Antarctic soft shell clam Laternula elliptica at King George Island, Antarctica. *Antarctic Science* **23**, 127–138.

Picken, G.B. 1979. Growth, production and biomass of the Antarctic gastropod Laevilacunaria antarctica Marten 1885. *Journal of Experimental Marine Biology and Ecology* **40**, 71–79.

Picken, G.B. 1980. The distribution, growth, and reproduction of the Antarctic limpet Nacella (Patinigera) concinna (Strebel 1908). *Journal of Experimental Marine Biology and Ecology* **42**, 71–85.

Place, S.P. & Hofmann, G.E. 2005. Constitutive expression of a stress inducible heat shock protein gene, hsp70, in phylogenetically distant Antarctic fish. *Polar Biology* **28**, 261–267.

Place, S.P., Zippay, M.L. & Hofmann, G.E. 2004. Constitutive roles for inducible genes: evidence for the alteration in expression of the inducible hsp70 gene in Antarctic nototheioid fishes. *American Journal of Physiology* **287**, R429–R436.

Podolsky, R. & Emlet, R. 1993. Separating the effects of temperature and viscosity on swimming and water movement by sand dollar larvae (Dendraster excentricus). *Journal of Experimental Biology* **176**, 207–222.

Podrabsky, J.E. & Hand, S.C. 2000. Depression of protein synthesis during diapause in embryos of the annual killifish, Astrofundulus limnaeus. *Physiological Biochemistry and Zoology* **73**, 799–800.

Podrabsky, J.E. & Somero, G.N. 2006. Inducible heat tolerance in Antarctic nototheniod fishes. *Polar Biology* **30**, 39–43.

Porter, R.K., Hulbert, A. & Brand, M.D. 1996. Allometry of mitochondrial proton leak: influence of membrane surface area and fatty acid composition. *American Journal of Physiology – Regulatory, Integrative and Comparative Physiology* **271**, R1550–R1560.

Powell, D.K. 2001. *The reproductive ecology of antarctic free spawning molluscs.* PhD thesis, Southampton University, Southampton, School of Ocean and Earth Science.

Powell, S.R., Wang, P., Divald, A., Teichberg, S., Haridas, V., McCloskey, T.W., Davies, K.A.J. & Katzeff, H. 2005. Aggregates of oxidized proteins (lipofuscin) induce apoptosis through proteasome inhibition and dysregulation of proapoptotic proteins. *Free Radical Biology and Medicine* **38**, 1093–1101.

Primo, C. & Vasquez, E. 2007. Zoogeography of the Antarctic ascidian fauna in relation to the sub-Antarctic and South America. *Antarctic Science* **19**, 321–336.

Pringle, J.M., Byers, J. E., Pappalardo, P., Wares, J.P. & Marshall, D. 2014. Circulation constrains the evolution of larval development modes and life histories in the coastal ocean. *Ecology* **95**, 1022–1032.

Przeslawski, R., Byrne, M. & Mellin, C. 2015. A review and meta-analysis of the effects of multiple abiotic stressors on marine embryos and larvae. *Global Change Biology* **21**, 2122–2140.

Pucciarelli, S., Ballarini, P. & Miceli, C. 1997. Cold-adapted microtubules: characterization of tubulin posttranslational modifications in the Antarctic ciliate Euplotes focardii. *Cell Motility and the Cytoskeleton* **38**, 329–40.

Pucciarelli, S., Chiappori, F., Sparvoli, D., Milanesi, L., Miceli, C. & Melki, R. 2013. Tubulin folding: the special case of a beta-tubulin isotype from the Antarctic psychrophilic ciliate Euplotes focardii. *Polar Biology* **36**, 1833–1838.

Pucciarelli, S., Devaraj, R.R., Mancini, A., Ballarini, P., Castelli, M., Schrallhammer, M., Petroni, G. & Miceli, C. 2015. Microbial consortium associated with the Antarctic marine ciliate Euplotes focardii: an investigation from genomic sequences. *Microbial Ecology* **70**, 484–497.

Pucciarelli, S., La, T.A., Ballarini, P., Barchetta, S., Yu, T., Marziale, F., Passini, V., Methé, B., Detrich, H.W., III & Miceli, C. 2009. Molecular cold-adaptation of protein function and gene regulation: the case for comparative genomic analyses in marine ciliated protozoa. *Marine Genomics* **2**, 57–66.

Putnam, H.M., Davidson, J.M. & Gates, R.D. 2016. Ocean acidification influences host DNA methylation and phenotypic plasticity in environmentally susceptible corals. *Evolutionary Applications* **9**, 1165–1178.

Pörtner, H.O. 2001. Climate change and temperature-dependent biogeography: oxygen limitation of thermal tolerance in animals. *Naturwissenschaften* **88**, 137–146.

Pörtner, H.O. 2002a. Physiological basis of temperature dependent biogeography: trade-offs in muscle design and performance in polar ectotherms. *Journal of Experimental Biology* **205**, 2217–2230.

Pörtner, H.O. 2002b. Climate variations and the physiological basis of temperature dependent biogeography: systemic to molecular hierarchy of thermal tolerance in animals. *Comparative Biochemistry and Physiology A* **132**, 739–761.

Pörtner, H.O. 2006. Climate dependent evolution of Antarctic ectotherms: an integrative analysis. *Deep Sea Research II* **53**, 1071–1104.

Pörtner, H.O. & Farrell, A.M. 2008. Physiology and climate change. *Science* **322**, 690–692.

Pörtner, H.O., Hardewig, I. & Peck, L.S. 1999a. Mitochondrial function and critical temperature in the Antarctic bivalve, Laternula elliptica. *Comparative Biochemistry and Physiology* **124A**, 179–189.

Pörtner, H.O., Hardewig, I., Sartoris, F.J. & van Dijk, P.L.M. 1998. Energetic aspects of cold adaptation: Critical temperatures in metabolic, ionic and acid—base regulation? In *Cold Ocean Physiology.* Society for Experimental Biology Seminar Series, 66, H.O. Pörtner & R.C. Playle (eds). Cambridge, UK: Cambridge University Press, 88–120.

Pörtner, H.O., Lucassen, M. & Storch, D. 2005a. Metabolic biochemistry: its role in thermal tolerance and in the capacities of physiological and ecological function. In *The Physiology of Polar Fishes*, A.P. Farrell & J.F. Steffensen (eds). Fish Physiology 22, (W.S. Hoar, D.R. Randall & A.P. Farrell, series eds). San Diego, CA: Elsevier Academic Press, 79–154.

Pörtner, H.O., Mark, F.C. & Bock, C. 2004. Oxygen limited thermal tolerance in fish? Answers obtained by nuclear magnetic resonance techniques. *Respiratory Physiology and Neurobiology* **141**, 243–260.

Pörtner, H.O., Peck, L.S. & Hirse, T. 2006. Hyperoxia alleviates thermal stress in the Antarctic bivalve, Laternula elliptica: evidence for oxygen limited thermal tolerance? *Polar Biology* **29**, 688–693.

Pörtner, H.O., Peck, L.S. & Somero, G.N. 2007. Thermal limits and adaptation in marine Antarctic ectotherms: an integrative view. *Philosophical Transactions of the Royal Society of London* **362**, 2233–2258.

Pörtner, H.O., Peck, L.S. Zielinski, S. & Conway, L.Z. 1999b. Temperature and metabolism in the highly stenothermal bivalve mollusc Limopsis marionensis from the Weddell Sea, Antarctica. *Polar Biology* **22**, 17–30.

Pörtner, H.O., Somero, G. N. & Peck, L.S. 2012. Thermal limits and adaptation in marine Antarctic ectotherms: an integrative view. In *Antarctic Ecosystems: An Extreme Environment in a Changing World*, A. Rogers et al. (eds). Wiley Interscience, 379–416.

Pörtner, H.O., Storch, D. & Heilmayer, O. 2005b. Constraints and trade-offs in climate dependent adaptation: energy budgets and growth in a latitudinal cline. *Scientia Marina* **69**, 271–285.

Pörtner, H.O., Van Dijk, P.L.M., Hardewig, I. & Sommer, A. 2000. Levels of metabolic cold adaptation: tradeoffs in eurythermal and stenothermal ectotherms. In *Antarctic Ecosystems: Models for a Wider Understanding*, W. Davison & C.W. Williams (eds). Christchurch, New Zealand: Caxton Press, 109–122.

Radtke, R.L. & Hourigan, T.F. 1990. Age and growth of the Antarctic fish Nototheniops nudifrons. *Fishery Bulletin of the U.S.* **88**, 557–581.

Rakusa-Suszczewski, S. 1972. The biology of Paramoera walkeri Stebbing (Amphipoda) and the Antarctic sub-fast ice community. *Polskie Archivum Hydrobiologii* **19**, 11–36.

Rakusa-Suszczewski, S. 1982. The biology and metabolism of Orchomene plebs (Hurley 1965) (Amphipoda: Gammaridea) from McMurdo Sound, Ross Sea, Antarctica. *Polar Biology* **1**, 47–54.

Ralph, R. & Everson, I. 1968. The respiratory metabolism of some Antarctic fish. *Comparative Biochemistry and Physiology* **27**, 299–307.

Ralph, R. & Maxwell, J.G.H. 1977. The oxygen consumption of the Antarctic limpet Nacella concinna. *British Antarctic Survey Bulletin* **45**, 19–24.

Rastrick, S.P.S. & Whiteley, N.M. 2013. Influence of natural thermal gradients on whole animal rates of protein synthesis in marine gammarid amphipods. *PLoS ONE* **8**, e60050.

Rauschert, M. 1991. Ergebnisse der faunistischen arbeiten im benthal von King George Island (Südshetlandinseln, Antarktis). *Berichte fur Polarforschung* **76**, 1–75.

Ravaux, J., Léger, N., Rabet, N., Morini, M., Zbinden, M., Thatje, S. & Shillito, B. 2012. Adaptation to thermally variable environments: capacity for acclimation of thermal limit and heat shock response in the shrimp Palaemonetes varians. *Journal of Comparative Physiology* **182**, 899–907.

Raven, J.A., Caldeira, K., Elderfield, H., Hoegh-Guldberg, O., Liss, P., Riebsell, U., Shepherd, J., Turley, C. & Watson, A. 2005. *Ocean acidification due to increasing atmospheric carbon dioxide*. Report to the Royal Society.

Raymond, J.A. & DeVries, A.L. 1977. Adsorption inhibition as a mechanism of freezing resistance in polar fishes. *Proceedings of the National Academy of Sciences USA* **74**, 2589–2593.

Raymond, J.A., Wilson, P. & Devries, A.L. 1989. Inhibition of growth of nonbasal planes in ice by fish antifreezes. *Proceedings of the National Academy of Sciences USA* **86**, 881–885.

Reed, A.J., Linse, K. & Thatje, S. 2014. Differential adaptations between cold-stenothermal environments in the bivalve Lissarca cf. miliaris (Philobryidae) from the Scotia Sea islands and Antarctic Peninsula. *Journal of Sea Research* **88**, 11–20.

Reed, A.J. & Thatje, S. 2015. Long-term acclimation and potential scope for thermal resilience in Southern Ocean bivalves. *Marine Biology* **162**, 2217–2224.

Reed, A.J., Thatje, S. & Linse, K. 2012. Shifting baselines in Antarctic ecosystems; ecophysiological response to warming in Lissarca miliaris at Signy Island, Antarctica. *PLoS ONE* **7**, e53477.

Reeves, R.B. 1972. An imidazole alphastat hypothesis for vertebrate acid–base regulation: tissue carbon dioxide content and body temperature in bullfrogs. *Respiratory Physiology* **14**, 219–236.

Regoli, F., Principato, G.B., Bertoli, E., Nigro, M. & Orlando, E. 1997. Biochemical characterization of the antioxidant system in the scallop Adamussium colbecki, a sentinel organism for monitoring the Antarctic environment. *Polar Biology* **17**, 251–258.

Reusch, T.B.H. 2014. Climate change in the oceans: evolutionary versus phenotypically plastic responses of marine animals and plants. *Evolutionary Applications* **7**, 104–122.

Revelle, R. & Fairbridge, R. 1957. Carbonates and carbon dioxide. *Memoirs of the Geological Society of America* **67**, 239–296.

Richard, J., Morley, S.A. & Peck, L.S. 2012. Estimating long-term survival temperatures at the assemblage level in the marine environment: Towards macrophysiology. *PLoS ONE* **7**, e34655.

Richards, Z.T., Garcia, R.A., Wallace, C.C., Rosser, N.L. & Muir, P.R. 2015. A diverse assemblage of reef corals thriving in a dynamic intertidal reef setting (Bonaparte Archipelago, Kimberley, Australia). *PLoS ONE* **10**, e0117791. Online. http://doi.org/10.1371/journal.pone.0117791

Richardson, M.G. 1979. The ecology and reproduction of the brooding Antarctic bivalve Lissarca miliaris. *British Antarctic Survey Bulletin* **49**, 91–115.

Riddle, M.J., Craven, M., Goldsworthy, P.M. & Carsey, F. 2007. A diverse benthic assemblage 100 km from open water under the Amery Ice Shelf, Antarctica. *Paleoceanography* **22**, PA1204.

Robbins, I., Lubet, P. & Besnard, J. 1990. Seasonal variations in the nucleic acid content and RNA: DNA ratio of the gonad of the scallop Pecten maximus. *Marine Biology* **105**, 191–195.

Robertson, D.R. & Collin, R. 2015. Inter- and Intra-specific variation in egg size among reef fishes across the Isthmus of Panama. *Frontiers in Ecology and Evolution* **2**, article 84.

Robertson, R.F., El-Haj, A.J., Clarke, A., Peck, L.S. & Taylor, E.W. 2001. The effects of temperature on metabolic rate and protein synthesis following a meal in the isopod Glyptonotus antarcticus eights (1852). *Polar Biology* **24**, 677–686.

Robinson, E. & Davison, W. 2008. The Antarctic notothenioid fish Pagothenia borchgrevinki is thermally flexible: acclimation changes oxygen consumption. *Polar Biology* **31**, 317–326.

Rodriguez, E., López-González, P.J. & Gili, J.-M. 2007. Biogeography of Antarctic sea anemones (Anthozoa, Actinaria): what do they tell us about the origin of the Antarctic benthic fauna? *Deep Sea Research II* **54**, 1876–1904.

Rodriguez, E., Orejas, C., López-González, P.J. & Gili, J.-M. 2013. Reproduction in the externally brooding sea anemone Epiactis georgiana in the Antarctic Peninsula and the Weddell Sea. *Marine Biology* **160**, 67–80.

Rodríguez-Romero, A., Jarrold, M.D., Massamba-N'Siala, G., Spicer, J.I. & Calosi, P. 2016. Multi-generational responses of a marine polychaete to a rapid change in seawater p CO_2. *Evolutionary Applications* **9**, 1082–1095.

Rogers, A.D., Tyler, P.A., Connelly, D.P., Copley, J.T., James, R., Larter, R.D., Linse, K., Mills, R.A., Garabato, A.N., Pancost, R.D., Pearce, D.A., Polunin, N.V.C., German, C.R., Shank, T., Boersch-Supan, P.H., Alker, B.J., Aquilina, A., Bennett, S.A., Clarke, A., Dinley, R.J.J., Graham, A.G.C., Green, D.R.H., Hawkes, J.A., Hepburn, L., Hilario, A., Huvenne, V.A.I., Marsh, L., Ramirez-Llodra E., Reid, W.D.K., Roterman, C.N., Sweeting, C.J., Thatje, S. & Zwirglmaier, K. 2012. The discovery of new deep-sea hydrothermal vent communities in the Southern Ocean and implications for biogeography. *PLoS Biology* **10**, e1001234.

Roggatz, C.C., Lorch, M., Hardege, J.D. & Benoit, D.M. 2016. Ocean acidification affects marine chemical communication by changing structure and function of peptide signalling molecules. *Global Change Biology* **22**, 3914–3926.

Rohde, A. 1985. Increased viviparity of marine parasites at high latitudes. *Hydrobiologia* **127**, 197–201.

Rohde, K. 2002. Ecology and biogeography of marine parasites. *Advances in Marine Biology* **43**, 1–86.

Román-González, A., Scourse, J.D., Butler, P.G., Reynolds, D.J., Richardson, C.J., Peck, L.S., Brey, T. & Hall, I.A. 2017. Analysis of ontogenetic growth trends in two marine Antarctic bivalves Yoldia eightsi and Laternula elliptica: implications for sclerochronology. *Palaeogeography, Palaeoclimatology, Palaeoecology* **465**, 300–306.

Roughgarden, J., Gaines, S. & Possingham, H. 1988. Recruitment dynamics in complex life cycles. *Science* **241**, 1460–1466.

Rozema, P.D., Venables, H.J., van de Poll, W.H., Clarke, A., Meredith, M.P. & Buma, A.G.J. 2017. Interannual variability in phytoplankton biomass and species composition in northern Marguerite Bay (West Antarctic Peninsula) is governed by both winter sea ice cover and summer stratification. *Limnology and Oceanography* **62**, 235–252.

Runge, J.A., Fields, D.M., Thompson, C.R.S., Shema, S.D., Bjelland, R.M., Durif, C.M.F., Skiftesvik, A.B. & Browman, H.I. 2016. End of the century CO_2 concentrations do not have a negative effect on vital rates of Calanus finmarchicus, an ecologically critical planktonic species in North Atlantic ecosystems. *ICES Journal of Marine Science* **73**, 937–950.

Rusan, N.M., Akong, K. & Peifer, M. 2008. Putting the model to the test: are APC proteins essential for neuronal polarity, axon outgrowth, and axon targeting? *Journal of Cell Biology* **183**, 203–212.

Ruud, J.T. 1954. Vertebrates without erythrocytes and blood pigment. *Nature* **173**, 848–850.

Rønnestadt, I. & Fyhn, H.J. 1993. Metabolic aspects of free amino acids in developing marine fish eggs and larvae. *Reviews in Fisheries Science* **1**, 239–259.

Sagar, P.M. 1980. Life cycle and growth of the Antarctic gammarid amphipod Paramoera walker (Stebbing 1906). *Journal of the Royal Society of New Zealand* **10**, 259–270.

Sahade, R., Lagger, C., Torre, L., Momo, F., Monien, P., Schloss, I., Barnes, D.K.A., Servetto, N., Tarantelli, S., Tatián, M., Zamboni, N. & Abele, D. 2015. Climate change and glacier retreat drive shifts in an Antarctic benthic ecosystem. *Scientific Advances* **1**, e1500050.

Sainte-Marie, B. 1991. Review of the reproductive bionomics of aquatic gammaridean amphipods: variation of life-history traits with latitude, depth, salinity and superfamily. *Hydrobiologia* **223**, 189–227.

Sandblom, E., Davison, W. & Axelsson, M. 2012. Cold physiology: postprandial blood flow dynamics and metabolism in the Antarctic fish Pagothenia borchgrevinki. *PLoS ONE* **7**, e33487.

Sayed-Ahmed, M.M., Khattab, M.M., Gad, M.Z. & Mostafa, N. 2001. L-carnitine prevents the progression of atherosclerotic lesions in hypercholesterolaemic rabbits. *Pharmacological Research* **44**, 235–242.

Schaefer, J. & Walters, A. 2010. Metabolic cold adaptation and developmental plasticity in metabolic rates among species in the Fundulus notatus species complex. *Functional Ecology* **24**, 1087–1094.

Schloss, I.R., Abele, D., Moreau, S., Demers, S., Bers, A.V., González, O. & Ferreyra, G.A. 2012. Response of phytoplankton dynamics to 19-year (1991–2009) climate trends in Potter Cove (Antarctica). *Journal of Marine Systems* **92**, 53–66.

Schmidt, M., Gerlach, F., Avivi, A., Laufs, T., Wystub, S., Simpson, J.C., Nevo, E., Saaler-Reinhardt, S., Reuss, S., Hankeln, T. & Burmester, T. 2004. Cytoglobin is a respiratory protein in connective tissue and neurons, which is upregulated by hypoxia. *Journal of Biological Chemistry* **279**, 8063–8069.

Schmidt-Nielsen, K. 1997. *Animal Physiology: Adaptation and Environment*, 4th edtion. Cambridge, UK: Cambridge University Press.

Schnack-Schiel, S.B. 2008. The macrobiology of sea-ice. In *Sea-Ice: An Introduction to its Physics, Chemistry, Biology and Geology, Chapter 7*, D.N. Thomas & G.S. Diekmann (eds). Oxford: Blackwell, 211–239.

Schoenrock, K.M., Schram, J.B., Amsler, C.D., McClintock, J.B. & Angus, R.A. 2015. Climate change impacts on overstory Desmarestia spp. from the western Antarctic Peninsula. *Marine Biology* **162**, 377–389.

Schoenrock, K.M., Schram, J.B., Amsler, C.D., McClintock, J.B., Angus, R.A. & Vohra, Y.K. 2016. Climate change confers a potential advantage to fleshy Antarctic crustose macroalgae over calcified species. *Journal of Experimental Marine Biology and Ecology* **474**, 58–66.

Schofield, O., Ducklow, H.W., Martinson, D.G., Meredith, M.P., Moline, M.A. & Fraser, W.R. 2010. How do polar marine ecosystems respond to rapid climate change? *Science* **328**, 1520–1523.

Scholander, P.F., Flagg, W., Walters, V. & Irving, L. 1953. Climatic adaptation in Arctic and tropical poikilotherms. *Physiological Zoology* **26**, 67–92.

Schram, J.B., McClintock, J.B., Amsler, C.D. & Baker, B.J. 2015a. Impacts of acute elevated seawater temperature on the feeding preferences of an Antarctic amphipod toward chemically deterrent macroalgae. *Marine Biology* **162**, 425–433.

Schram, J.B., Schoenrock, K.M., McClintock, J.B., Amsler, C.D. & Angus, R.A. 2014. Multiple stressor effects of near-future elevated seawater temperature and decreased pH on righting and escape behaviors of two common Antarctic gastropods. *Journal of Experimental Marine Biology and Ecology* **457**, 90–96.

Schram, J.B., Schoenrock, K.M., McClintock, J.B., Amsler C.D. & Angus, R.A. 2015b. Multi-frequency observations of seawater carbonate chemistry on the central coast of the western Antarctic Peninsula. *Polar Research* **34**, 25582.

Schram, J.B., Schoenrock, K.M., McClintock, J.B., Amsler, C.D. & Angus, R.A. 2016a. Testing Antarctic resilience: the effects of elevated seawater temperature and decreased pH on two gastropod species. *ICES Journal of Marine Science: Journal du Conseil* **73**, 739–752.

Schram, J.B., Schoenrock, K.M., McClintock, J.B., Amsler, C.D. & Angus, R.A. 2016b. Seawater acidification more than warming presents a challenge for two Antarctic macroalgal-associated amphipods. *Marine Ecology Progress Series* **554**, 81–97.

Schram, J.B., Schoenrock, K.M., McClintock, J.B., Amsler, C.D. & Angus, R.A. 2017. Ocean warming and acidification alter Antarctic macroalgal biochemical composition but not amphipod grazer feeding preferences. *Marine Ecology Progress Series* **581**, 41–56.

Seager, J.S. 1978. *The ecology of an Antarctic opisthobranch mollusc: Philline gibber strebel*. PhD thesis, University College, Cardiff, UK.

Secor, S.M. 2009. Specific dynamic action: a review of the postprandial metabolic response. *Journal of Comparative Physiology* **179**, 1–56.

Secor, S.M. & Diamond, J. 1995. Determinants of post-feeding metabolic response in Burmese pythons (Python molurus). *Physiological Zoology* **70**, 202–212.

Seebacher, F., Davison, W., Lowe, C.J. & Franklin, C.E. 2005. A falsification of the thermal specialization paradigm: compensation for elevated temperatures in Antarctic fish. *Biology Letters* **2**, 151–154.

Servetto, N. & Sahade, R. 2016. Reproductive seasonality of the Antarctic sea pen Malacobelemnon daytoni (Octocorallia, Pennatulacea, Kophobelemnidae). *PLoS ONE* **11**, e0163152.

Servetto, N., Torre, L. & Sahade, R. 2013. Reproductive biology of the Antarctic 'sea pen' Malacobelemnon daytoni (Octocorallia, Pennatulacea, Kophobelemnidae). *Polar Research* **32**, 20–40.

Sewell, M. 2005. Examination of the meroplankton community in the south-western Ross Sea, Antarctica, using a collapsible plankton net. *Polar Biology* **28**, 119–131.

Sewell, M.A. & Hofmann, G.E. 2011. Antarctic echinoids and climate change: a major impact on the brooding forms. *Global Change Biology* **17**, 734–744.

Shabica, S.V. 1974. *Reproductive biology of the brooding Antarctic Lamellibranch,* Kidderia subquadratum *(Pelseneer).* PhD thesis, University of Oregon.

Shadwick, R.E. & Lauder, G.V. 2006. *Fish Biomechanics.* San Diego, CA: Elsevier Academic Press.

Shearwin, K.E. & Timasheff, S.N. 1992. Linkage between ligand binding and control of tubulin conformation. *Biochemistry* **31**, 8080–8089.

Sherman, C.D.H., Hunt, A. & Ayre, D.J. 2008. Is life-history a barrier to dispersal? Contrasting patterns of genetic differentiation along an oceanographically complex coast. *Biological Journal of the Linnean Society* **95**, 106–116.

Shiehzadegan, S., Le Vinh Thuy, J., Szabla, N., Angilletta, M.J. & VandenBrooks, J.M. 2017. More oxygen during development enhanced flight performance but not thermal tolerance of Drosophila melanogaster. *PLoS ONE* **12**(5), e0177827.

Shilling, F.M. & Manahan, D.T. 1994. Energy metabolism and amino acid transport during early development of Antarctic and temperate echinoderms. *Biological Bulletin* **187**, 398–407.

Shin, S.C., Kim, S.J., Lee, J.K., Ahn, D.H., Kim, M.G., Lee, H., Lee, J., Kim, B.-K. & Park, H. 2012. Transcriptomics and comparative analysis of three antarctic notothenioid fishes. *PLoS ONE* **16**, e43762.

Sicheri, F. & Yang, D.S.C. 1995. Ice-binding structure and mechanism of an antifreeze protein from winter flounder. *Nature* **375**, 427–431.

Sidell, B.D. & Hazel, J.R. 1987. Temperature affects the diffusion of small molecules through cytosol of fish muscle. *Journal of Experimental Biology* **129**, 191–203.

Sidell, B.D. & O'Brien, K.M. 2006. When bad things happen to good fish: the loss of hemoglobin and myoglobin expression in Antarctic icefishes. *Journal of Experimental Biology* **209**, 1791–1802.

Siegel, V. & Mühlenhardt-Siegel, U. 1988. On the occurrence and biology of some Antarctic Mysidacea (Crustacea). *Polar Biology* **8**, 181–190.

Sil'yanova, Z.S. 1982. Oogenesis and stages of maturity of fishes of the family Nototheniidae. *Voprosy Ichthiologii*, **21**(4), 687–694. Translated as *Journal of Ichthyology* **21**, 81–89.

Simpson, R.D. 1va977. The reproduction of some littoral molluscs from Macquarie Island (sub-Antarctic). *Marine Biology* **44**, 125–142.

Slattery, M. & McClintock, J.B. 1995. Population structure and feeding deterrence in three shallow-water Antarctic soft corals. *Marine Biology* **122**, 461–470.

Slattery, M., McClintock, J.B. & Bowser, S.S. 1997. Deposit feeding: a novel mode of nutrition in the Antarctic colonial soft coral Gersemia antarctica. *Marine Ecology Progress Series* **149**, 299–304.

Sleight, V.A., Thorne, M.A.S., Peck, L.S. & Clark, M.S. 2015. Transcriptomic response to shell damage in the Antarctic clam, Laternula elliptica: time scales and spatial localization. *Marine Genomics* **20**, 45–55.

Smallenge, I.M. & Van der Meer, J. 2003. Why do shore crabs not prefer the most profitable mussels? *Journal of Animal Ecology* **72**, 599–607.

Smith, C.C. & Fretwell, S.D. 1974. Optimal balance between size and number of offspring. *American Naturalist* **108**, 499–506.

Smith, C.R., Minks, S. & Demaster, D.J. 2006. A síntesis of bentho-pelagic coupling on the Antarctic shelf: Food banks, ecosystem inertia and global climate change. *Deep-Sea Research II* **53**, 875–894.

Smith, W.O. & Nelson, D.M. 1985. Phytoplankton bloom produced by a receding ice edge in the Ross Sea: spatial coherence with the density field. *Science* **227**, 163–166.

Snelgrove, P., Vanden Berghe, E., Miloslavich, P., Archambault, P., Bailly, N., Brandt, A., Bucklin, A., Clark, M., Dahdouh-Guebas, F., Halpin, P., Hopcroft, R., Kaschner, K., Lascelles, B., Levin, L.A., Menden-Deuer, S., Obura, D., Reeves, R.R., Rynearson, T., Stocks, K., Tarzia, M., Tittensor, D., Tunnicliffe, V., Wallace, B., Wanless, R., Webb, T., Bernal, P., Rice, J. & Rosenberg, A. 2016. Global patterns in marine biodiversity. Chapter 34, In UN Assessment of the Ocean 2016. Online. http://www.un.org/Depts/los/global_reporting/WOA_RegProcess.htm

Somero, G.N. 2010. The physiology of climate change: how potentials for acclimatization and genetic adaptation will determine 'winners' and 'losers'. *Journal of Experimental Biology* **213**, 912–920.

Somero, G.N. 2012. The physiology of global change: linking patterns to mechanisms. *Annual Reviews in Marine Science* **4**, 39–61.

Somero, G.N. 2015. Temporal patterning of thermal acclimation: from behavior to membrane biophysics. *Journal of Experimental Biology* **218**, 167–169.

Somero, G.N. & DeVries, A.L. 1967. Temperature tolerance of some Antarctic fishes. *Science* **156**, 257–258.

Sommer, A.M. & Pörtner, H.O. 2004. Mitochondrial function in seasonal acclimatisation versus latitudinal adaptation to cold, in the lugworm Arenicola marina (L.). *Physiological and Biochemical Zoology* **77**, 174–186.

Spalding, M., Ravilious, C. & Green, E. 2001. *World Atlas of Coral Reefs*. Berkeley, CA: University of California Press and UNEP/WCMC, ISBN 0520232550.

Spence, P., Griffies, S.M., England, M.H., Hogg, A.M.C., Saenko, O.A. & Jourdain, N.C. 2014. Rapid subsurface warming and circulation changes of Antarctic coastal waters by poleward shifting winds. *Geophysics Research Letters* **41**, 4601–4610.

Spence, P., Holmes, R.M., Hog, A.McC, Griffies, S.M., Stewart, K.D. & England, M.W. 2017. Localized rapid warming of West Antarctic subsurface waters by remote winds. *Nature Climate Change* **7**, 595–603.

Spicer, J.I. & Gaston, K.J. 1999. *Physiological Diversity and its Ecological Implications*. Oxford: Blackwell Science.

Stammerjohn, S.E., Martinson, D.G., Smith, R.C., Yuan, X. & Rind, D. 2008. Trends in Antarctic annual sea-ice retreat and advance and their relation to El Niño-Southern Oscillation and Southern Annular Mode variability. *Journal of Geophysics Research Oceans* **113**, C03S90.

Stammerjohn, S.E., Massom, R., Rind, D. & Martinson, D. 2012. Regions of rapid sea-ice change: an inter-hemispheric seasonal comparison. *Geophysics Research Letters* **39**, L06501.

Stanwell-Smith, D.P. & Peck, L.S. 1998. Temperature and embryonic development in relation to spawning and field occurrence of larvae of 3 Antarctic echinoderms. *Biological Bulletin Woods Hole* **194**, 44–52.

Stanwell-Smith, D.P., Peck, L.S., Clarke, A., Murray, A. & Todd, C. 1999. Distribution, abundance and seasonality of pelagic marine invertebrate larvae in the maritime Antarctic. *Philosophical Transactions of the Royal Society of London B* **354**, 471–484.

Starmans, A., Gutt, J. & Arntz, W.E. 1999. Mega-epibenthic communities in Arctic and Antarctic shelf areas. *Marine Biology* **135**, 269–280.

Stearns, S.C. 1989. Trade-offs in life-history evolution. *Functional Ecology* **3**, 259–268.

Stearns, S.C. 1992. *The Evolution of Life Histories*. Oxford: Oxford University Press.

Stevens, M.I., Greenslade, P., Hogg, I.D. & Sunnucks, P. 2006. Southern hemisphere springtails: could any have survived glaciation of Antarctica? *Molecular Biology and Evolution* **23**, 874–882.

Stevens, M.M., Jackson, S., Bester, S.A., Terblanche, J.S. & Chown, S.L. 2010. Oxygen limitation and thermal tolerance in two terrestrial arthropod species. *Journal of Experimental Biology* **213**, 2209–2218.

Stewart, A. & Thompson, A. 2013. Connecting Antarctic cross-slope exchange with Southern Ocean overturning. *Journal of Physical Oceanography* **43**, 1453–1471.

Stillman, J.H. & Paganini, A.W. 2015. Biochemical adaptation to ocean acidification. *Journal of Experimental Biology* **218**, 1946–1955.

Storch, D., Fernandez, M., Navarrete, S.A. & Pörtner, H.O. 2011. Thermal tolerance of larval stages of the Chilean kelp crab Taliepus dentatus. *Marine Ecology Progress Series* **429**, 157–167.

Storch, D., Heilmayer, O., Hardewig, I. & Pörtner, H.O. 2003. *In vitro* protein synthesis capacities in a cold stenothermal and a temperate eurythermal pectinid. *Journal of Comparative Physiology* **173B**, 611–620.

Storey, K.B. & Storey, J.M. 2013. Molecular biology of freeze tolerance in animals. *Comparative Physiology* **3**, 1283–1308.

Strandberg, U., Käkelä, A., Lydersen, C., Kovacs, K., Grahl-Nielsen, O., Hyvärinen, H. & Käkelä, R. 2008. Stratification, composition, and function of marine mammal blubber: the ecology of fatty acids in marine mammals. *Physiological and Biochemical Zoology* **81**, 473–485.

Strathmann, R.R. 1985. Feeding and nonfeeding larval development and life-history evolution in marine invertebrates. *Annual Review of Ecology, Evolution and Systematics* **16**, 339–361.

Strathmann, R.R., Kendall, L.R. & Marsh, A.G. 2006. Embryonic and larval development of a cold adapted Antarctic ascidian. *Polar Biology* **29**, 495–501.

Strobel, A., Bennecke, S., Leo, E., Mintenbeck, K., Pörtner, H.O. & Mark, F.C. 2012. Metabolic shifts in the Antarctic fish Notothenia rossii in response to rising temperature and PCO_2. *Frontiers in Zoology* **9**, 28 only.

Strobel, A., Graeve, M., Pörtner, H.O. & Mark, F.C. 2013. Mitochondrial acclimation capacities to ocean warming and acidification are limited in the Antarctic Nototheniid fish, Notothenia rossii and Lepidonotothen squamifrons. *PLoS ONE* **8**, e68865.

Suckling, C.C., Clark, M.S., Beveridge, C., Brunner, L., Hughes, A., Cook, E., Davies, A.J. & Peck, L.S. 2014a. Experimental influence of pH on the early life-stages of sea urchins II: increasing parental exposure gives rise to different responses. *Invertebrate Reproduction and Development* **58**, 161–175.

Suckling, C.C., Clark, M.S., Peck, L.S. & Cook, E. 2014b. Experimental influence of pH on early life-stages of sea urchins I: different rates of introduction give different responses. *Invertebrate Reproduction and Development* **58**, 148–159.

Suckling, C.C., Clark, M.S., Richard, J., Morley, S.A., Thorne, M.A.S., Harper, E.M. & Peck, L.S. 2015. Adult acclimation to combined temperature and pH stressors significantly enhances reproductive outcomes compared to short-term exposures. *Journal of Animal Ecology* **84**, 773–784.

Suda, C.N.K., Vani, G.S., de Oliveira, M.F., Rodrigues, E. Jr., Rodrigues, E. & Lavrado, H.P. 2015. The biology and ecology of the Antarctic limpet Nacella concinna. *Polar Biology* **38**, 1949–1969.

Sun, T., Lin, F.-H., Campbell, R.L., Allingham, J.S. & Davies, P.L. 2014. An antifreeze protein folds with an interior network of more than 400 semi-clathrate waters. *Science* **343**, 795–798.

Sun, Y., Jin, K., Mao, X.O., Zhu, Y. & Greenberg, D.A. 2001. Neuroglobin is upregulated by and protects neurons from hypoxic-ischemic injury. *Proceedings of the National Academy of Science USA* **98**, 15306–15311.

Sunday, J.M., Bates, A.E. & Dulvy, N.K. 2011. Global analysis of thermal tolerance and latitude in ectotherms. *Proceedings of the Royal Society of London Series B* **278**, 1823–1830.

Sunday, J.M., Bates, A.E. & Dulvy, N.K. 2012. Thermal tolerance and the global redistribution of animals. *Nature Climate Change* **2**, 686–690.

Sunday, J.M., Calosi, P., Dupont, S., Munday, P.L., Stillman, J.H. & Reusch, T.B.H. 2014. Evolution in an acidifying ocean. *Trends in Ecology and Evolution* **29**, 117–125.

Sutherland, W.J., Clout, M., Depledge, M., Dicks, L.V., Dinsdale, J., Entwistle, A.C., Fleishman, E., Gibbons, D.W., Keim, B., Lickorish, F.A., Monk, K.A., Ockendon, N., Peck, L.S., Pretty, J., Rockstrom, J., Spalding, M.D., Tonneijck, F.H. & Wintle, B.C. 2015. A horizon scan of global conservation issues for 2015. *Trends in Ecology and Evolution* **30**, 17–24.

Sørensen, J.G. & Loeschcke, V. 2007. Studying stress responses in the post-genomic era: its ecological and evolutionary role. *Journal of Experimental Biology* **32**, 447–456.

Talmage, S.C. & Gobler, C.J. 2010. Effects of past, present, and future ocean carbon dioxide concentrations on the growth and survival of larval shellfish. *Proceedings of the National Academy of Science USA* **107**, 17246–17251.

Tandler, A. & Beamish, F.W.H. 1979. Mechanical and biochemical components of apparent specific dynamic action in largemouth bass, Micropterus salmoides Lacepede. *Journal of Fish Biology* **14**, 342–350.

Targett, T.E. 1990. Feeding, digestion & growth in Antarctic fishes: ecological factors affecting rates and efficiencies. In *Second International Symposium on the Biology of Antarctic Fishes*, G. di Prisco et al. (eds). Naples: IIGB Press, 37–39.

Tarling, G.A., Peck, V.L., Ward, P., Ensor, N.S., Achterberg, E.P., Tynan, E., Poulton, A.J., Mitchell, E. & Zubkov, M.V. 2016. Effects of acute ocean acidification on spatially-diverse polar pelagic foodwebs: Insights from on-deck microcosms. *Deep Sea Research Part II: Topical Studies in Oceanography* **127**, 75–92.

Taylor, S.E., Egginton, S., Taylor, E.W., Franklin, C.E. & Johnston, I.E. 1999. Estimation of intracellular pH in muscle of fishes from different thermal environments. *Journal of Thermal Biology* **24**, 199–208.

Teixidó, N., Garrabou, J., Gutt, J. & Arntz, W.E. 2004. Recovery in Antarctic benthos after iceberg disturbance: trends in benthic composition, abundance and growth forms. *Marine Ecology Progress Series* **278**, 1–16.

Telesca, L., Michalek, K., Sanders, T., Peck, L.S., Thyrring, J. & Harper, E.M. 2018. Blue mussel shell shape plasticity and natural environments: a quantitative approach. *Scientific Reports* **8**, 2865.

Terblanche, J.S., Hoffmann, A.A., Mitchell, K.A., Rako, L., le Roux, P.C. & Chown, S.L. 2011. Ecologically relevant measures of tolerance to potentially lethal temperatures. *Journal of Experimental Biology* **214**, 3713–3725.

Thatje, S., Hillenbrand, C.-D. & Larter, R. 2005. On the origin of Antarctic marine benthic community structure. *Trends in Ecology and Evolution* **20**, 534–540.

Thatje, S., Hillenbrand, C.-D., Mackensen, A. & Larter, R. 2008. Life hung by a thread: endurance of Antarctic fauna in glacial periods. *Ecology* **89**, 682–692.

Thiel, M. & Gutow, L. 2005. The ecology of rafting in the marine environment II: the rafting organisms and community. *Oceanography and Marine Biology an Annual Review* **43**, 279–418.

Thomas, D.N. & Diekmann, G.S. 2010. *Sea-Ice*, 2nd edition. Oxford: Wiley Blackwell.

Thomson, B.A.W. & Riddle, M.J. 2005. Bioturbation behaviour of the spatangoid urchin Abatus ingens in Antarctic marine sediments. *Marine Ecology Progress Series* **290**, 135–143.

Thorson, G. 1936. The larval development, growth and metabolism of Arctic marine bottom invertebrates compared with those of other seas. *Meddelelser om Grønland* **100**, 1–155.

Thorson, G. 1946. Reproduction and larval development of Danish marine bottom invertebrates, with special reference to the planktonic larvae in the sound (Øresund). *Meddelelser fra Kommissionen for Danmarks Fiskeri- og Havundersøgelser. Serie Plankton* **4**, 1–523.

Thorson, G. 1950. Reproductive and larval ecology of marine bottom invertebrates. *Biological Reviews* **25**, 1–45.

Todgham, A.E., Crombie, T.A. & Hofmann, G.E. 2017. The effect of temperature adaptation on the ubiquitin-proteasome pathway in notothenioid fishes. *Journal of Experimental Biology* **220**, 369–378.

Todgham, A.E., Hoaglund, E.A. & Hofmann, G.E. 2007. Is cold the new hot? Elevated ubiquitin-conjugated protein levels in tissues of Antarctic fish as evidence for cold-denaturation of proteins *in vivo*. *Journal of Comparative Physiology B* **177**, 857–866.

Tomanek, L. 2010. Variation in the heat shock response and its implication for predicting the effect of global climate change on species' biogeographical distribution ranges and metabolic costs. *Journal of Experimental Biology* **213**, 971–979.

Torre, L., Servetto, N., Leonel, E.M., Momo, F., Tatian, M., Abele, D. & Sahade, R. 2012. Respiratory responses of three Antarctic ascidians and a sea pen to increased sediment concentrations. *Polar Biology* **35**, 1743–1748.

Tremblay, N. & Abele, D. 2016. Response of three krill species to hypoxia and warming: an experimental approach to oxygen minimum zones expansion in coastal ecosystems. *Marine Ecology* **37**, 179–199.

Turner, J., Barrand, N.E., Bracegirdle, T. J., Convey, P., Hodgson, D.A., Jarvis, M., Jenkins, A., Marshall, G., Meredith, M.P., Roscoe, H., Shanklin, J., French, J., Goosse, H., Guglielmin, M., Gutt, J., Jacobs, S., Kennicutt II, M.C., Masson-Delmotte, V., Mayewski, P., Navarro, F., Robinson, S., Scambos, T., Sparrow, M., Summerhayes, C., Speer, K. & Klepikov, A. 2013. Antarctic climate change and the environment: an update. *Polar Record* **50**, 237–259.

Turner, J., Bindschadler, R., Convey, P., di Prisco, G., Fahrbach, E., Gutt, J., Hodgson, D., Mayewski, P. & Summerhayes, C. 2009. *Antarctic Climate Change and the Environment*. Cambridge, UK: Scientific Committee on Antarctic Research.

Turner, J., Lu, H., White, I., King, J.C., Phillips, T., Hosking, J.S., Bracegirdle, T.J., Marshall, G.J., Mulvaney, R. & Deb, P. 2016. Absence of 21st century warming on Antarctic Peninsula consistent with natural variability. *Nature* **535**, 411–415.

Tyler, P.A., Reeves, S., Peck, L.S., Clarke, A. & Powell, D. 2003. Seasonal variation in the gametogenic ecology of the Antarctic scallop Adamussium colbecki. *Polar Biology* **26**, 727–733.

Tyler, P.A., Young, C.M. & Clarke, A. 2000. Temperature and pressure tolerances of embryos and larvae of the Antarctic sea urchin Sterechinus neumayeri (Echinodermata: Echinoidea): potential for deep-sea invasion from high latitudes. *Marine Ecology Progress Series* **192**, 173–180.

Underwood, A.J. & Keough, M.J. 2001. Supply-side ecology - the nature and consequences of variations in recruitment of intertidal organisms. In *Marine Community Ecology*, M. Hay et al. (eds). Sunderland: Sinauer Associates, 183–200.

UNEP 2012. *Global Environment Outlook 5*. Malta: United Nations Environment Program, Progress Press Ltd.

Urban, H.J. & Mercuri, G. 1998. Population dynamics of the bivalve Laternula elliptica from Potter Cove, King George Island, South Shetland Islands. *Antarctic Science* **10**, 153–160.

Urban, H.J. & Silva, P. 1998. Upper temperature tolerance of two Antarctic mollusks (Laternula elliptica and Nacella concinna) from Potter Cove, King George Island, Antarctic Peninsula, Reports on Polar Research. *Alfred Wegener Institut for Polar and Marine Research, Bremerhaven* **299**, 230–236.

Vahl, O. & Sundet, J.H. 1985. Is sperm really so cheap? In *Marine Biology of Polar Regions and Effects of Stress on Marine Organisms*, J.S. Gray & M.E. Christiansen (eds). New York: John Wiley and Sons, 281–285.

Van Den Thillart, G. & Smit, H. 1984. Carbohydrate metabolism of goldfish Carassius auratus effects of long-term hypoxia acclimation on enzyme patterns of red muscle, white muscle and liver. *Journal of Comparative Physiology* **154**, 477–486.

Van der Have, T.M. & de Jong, G. 1996. Adult size in ectotherms: temperature effects on growth and differentiation. *Journal of Theoretical Biology* **183**, 329–340.

Van Dijk, P.L.M., Tesch, C., Hardewig, I. & Pörtner, H.O. 1999. Physiological disturbances at critically high temperatures: a comparison between stenothermal Antarctic and eurythermal temperate eelpouts (Zoarcidae). *Journal of Experimental Biology* **202**, 3611–3621.

Vance, R.R. 1973. On reproductive strategies in marine benthic invertebrates. *American Naturalist* **107**, 339–352.

Vanella, F.A., Boy, C.C., Lattuca, M.E. & Calvo, J. 2010. Temperature influence on post-prandial metabolic rate of sub-Antarctic teleost fish. *Comparative Biochemistry and Physiology* **156A**, 247–254.

Venables, H.J., Clarke, A. & Meredith, M.P. 2013. Winter-time controls on summer stratification and productivity at the western Antarctic Peninsula. *Limnology and Oceanography* **58**, 1035–1047.

Verberk, W.C., Overgaard, J., Ern, R., Bayley, M., Wang, T., Boardman, L. & Terblanche, J.S. 2016. Does oxygen limit thermal tolerance in arthropods? A critical review of current evidence. *Comparative Biochemistry and Physiology* **192A**, 64–78.

Verde, C., Giordano, D., di Prisco, G. & Andersen, Ø. 2012a. The hemoglobins of polar fish: evolutionary and physiological significance of multiplicity in Arctic fish. *Biodiversity* **13**, 228–233.

Verde, C., Giordano, D., di Prisco, G., Russo, R., Riccio, A. & Coppola, D. 2011. Evolutionary adaptations in antarctic fish: the oxygen transport system. *Oecologia Australis* **15**, 40–50.

Verde, C., Parisi, E. & di Prisco, G. 2006. The evolution of thermal adaptation in polar fish. *Gene* **385**, 137–145.

Verde, C., di Prisco, G., Giordano, D., Russo, R., Anderson, D. & Cowan, D. 2012b. Antarctic psychrophiles: models for understanding the molecular basis of survival at low temperature and responses to climate change. *Biodiversity* **13**, 249–256.

Verde, C., Vergara, A., Mazzarella, L. & di Prisco, G. 2008. The hemoglobins of fishes living at polar latitudes - current knowledge on structural adaptations in a changing environment. *Current Protein and Peptide Science* **9**, 578–590.

Vermeij, G.J. 1978. *Biogeography and Adaptation: Patterns of Marine Life*. Cambridge, MA: Harvard University Press.

Vermeij, G.J. 1987. *Evolution and Escalation: An Ecological History of Life*. Princeton, NJ: Princeton University Press.

Vermeij, G.J. 1993. *A Natural History of Shells*. Princeton, NJ: Princeton University Press, .

Vernet, M., Martinson, D., Iannuzzi, R., Stammerjohn, S., Kozlowski, W., Sines, K., Smith, R. & Garibotti, I. 2008. Primary production within the sea-ice zone west of the Antarctic Peninsula: I-sea ice, summer mixed layer, and irradiance. *Deep-Sea Research Part II* **55**, 2068–2085.

Vernon, H.M. 1900. The death temperature of certain marine organisms. *Journal of Physiology* **25**, 131–136.

Wakeling, J.M. & Johnston, I.A. 1998. Muscle power output limits fast-start performance in fish. *Journal of Experimental Biology* **201**, 1505–1526.

Walker, R.S., Gurven, M., Burger, O. & Hamilton, M.J. 2007. The trade-off between number and size of offspring in humans and other primates. *Proceedings of the Royal Society of London* **275**, 827–833.

Waller, C.L. 2013. Zonation in a cryptic Antarctic intertidal macrofaunal community. *Antarctic Science* **25**, 62–68.

Waller, C.L., Barnes, D.K.A. & Convey, P. 2006a. Ecological contrasts across an Antarctic land-sea interface. *Austral Ecology* **31**, 656–666.

Waller, C.L., Worland, M.R., Convey, P. & Barnes, D.K.A. 2006b. Ecophysiological strategies of Antarctic intertidal invertebrates faced with freezing stress. *Polar Biology* **29**, 1077–1083.

Waller, R.G., Tyler, P.A. & Smith, C.R. 2008. Fecundity and embryo development of three Antarctic deep-water scleractinians: Flabellum thouarsii, F. curvatum and F. impensum. *Deep-Sea Research Part II-Topical Studies in Oceanography* **55**, 2527–2534.

Wallin, M. & Strömberg, E. 1995. Cold-stable and cold-adapted microtubules. *International Reviews in Cytology* **157**, 1–31.

Wang, T., Lefevre, S., Iversen, N. K., Findorf, I., Buchanan, R. & McKenzie, D.J. 2014. Anaemia only causes a small reduction in the upper critical temperature of sea bass: is oxygen delivery the limiting factor for temperature tolerance in fishes? *Journal of Experimental Biology* **217**, 4275–4278.

Watson, S.-A., Peck, L.S. & Morley, S.A. 2013. Low global sensitivity of metabolic rate to temperature in calcified marine invertebrates. *Oecologia* **175**, 47–54.

Watson, S.-A., Peck, L.S., Morley, S.A. & Munday, P.L. 2017. Latitudinal trends in shell production cost from the tropics to the poles. *Science Reports* **3**, e1701362.

Watson, S.-A., Peck, L.S., Tyler, P.A., Southgate, P.C., Tan, K.S., Day, R.W. & Morley, S.A. 2012. Marine invertebrate skeleton size varies with latitude, temperature, and carbonate saturation: implications for global change and ocean acidification. *Global Change Biology* **18**, 3026–3038.

Watson, S.-A., Southgate, P., Tyler, P.A. & Peck, L.S. 2009. Early larval development of the Sydney rock oyster Saccostrea glomerata under near-future predictions of CO_2-driven ocean acidification. *Journal of Shellfish Research* **28**, 431–437.

Weinstein, R.B. & Somero, G.N. 1998. Effects of temperature on mitochondrial function in the Antarctic fish Trematomus bernacchii. *Journal of Comparative Physiology* **168B**, 190–196.

Welladsen, H.M., Southgate, P.C. & Heimann, K. 2010. The effects of exposure to near-future levels of ocean acidification on shell characteristics of Pinctada fucata (Bivalvia: Pteriidae). *Molluscan Research* **30**, 125–130.

Wells, R.M.G. 1978. Respiratory adaptation and energy-metabolism in Antarctic nototheniid fishes. *New Zealand Journal of Zoology* **5**, 813–815.

Wells, R.M.G. 1987. Respiration of Antarctic fish from McMurdo Sound. *Comparative Biochemistry and Physiology* **88**, 417–424.

Weslawski, M. & Opalinski, K.W. 1997. Winter and summer metabolic rates of Arctic amphipods. Preliminary results. In *Polish Polar Studies, 24th Polar Symposium*, P. Glowacki (ed.). Warszawa: Institute of Geophysics of the Polish Academy of Sciences, 307–317.

Whitaker, T.M. 1982. Primary production of phytoplankton off Signy Island, South Orkneys, the Antarctic. *Proceedings of the Royal Society of London* **214B**, 169–189.

White, C.R., Alton, L.A. & Frappell, P.B. 2012. Evolutionary adaptation and thermal plasticity are insufficient to completely overcome the acute thermodynamic effects of temperature. *Proceedings of the Royal Society of London* **279**, 1740–1747.

White, J.W., Morgan, S.G. & Fisher, J.L. 2014. Planktonic larval mortality rates are lower than widely expected. *Ecology* **95**, 3344–3353.

White, M.G. 1970. Aspects of the breeding biology of Glyptonotus antarcticus (eights) (Crustacea, Isopoda) at Signy Island, South Orkney Islands. In *Antarctic Ecology*, vol. 1, M.W. Holdgate (ed.). London: Academic Press, 279–285.

White, M.G. 1984. Marine Benthos. In *Antarctic Ecology*, vol. 2, R.M. Laws (ed.). London: Academic Press, 421–461.

Whiteley, N.M. 2011. Physiological and ecological responses of crustaceans to ocean acidification. *Marine Ecology Progress Series* **30**, 257–271.

Whiteley, N.M., Taylor, E.W. & El Haj, A.J. 1996. A comparison of the metabolic cost of protein synthesis in stenothermal and eurythermalisopod crustaceans. *American Journal of Physiology* **271**, R1295–R1303.

Whiteley, N.M., Taylor, E.W. & El Haj, A.J. 1997. Seasonal and latitudinal adaptation to temperature in crustaceans. *Journal of Thermal Biology* **22**, 419–427.

Wiedenmann, J., Cresswell, K.A. & Mangel, M. 2009. Connecting recruitment of Antarctic krill and sea-ice. *Limnology and Oceanography* **54**, 799–811.

Williams, G.C. 1957. Pleiotropy, natural selection and the evolution of senescence. *Evolution* **11**, 398–411.

Wilson, R.S., Franklin, C.E., Davison, W. & Kraft, P. 2001. Stenotherms at subzero temperatures: thermal dependence of swimming performance in Antarctic fish. *Journal of Comparative Physiology* **171B**, 263–269.

Wilson, R.S., Kuchel, L.J., Franklin, C.E. & Davison, W. 2002. Turning up the heat on subzero fish: thermal dependence of sustained swimming in an Antarctic notothenioid. *Journal of Thermal Biology* **27**, 381–386.

Winberg, G.G. 1956. *Rate of Metabolism and Food Requirements of Fishes*. Minsk: Belorussian State University (Translated from Russian by the Fisheries Research Board of Canada Translation Series No. 194, 1960).

Wittenberg, B.A. & Wittenberg, J.B. 2003. Myoglobin function assessed. *Journal of Experimental Biology* **206**, 2011–2020.

Wohlschlag, D.E. 1963. An Antarctic fish with unusually low metabolism. *Ecology* **44**, 557–564.

Wohlschlag, D.E. 1964. Respiratory metabolism and ecological characteristics of some fishes in McMurdo Sound, Antarctica. *Antarctic Research Series of the American Geophysical Union* **1**, 33–62.

Wood, H.L., Spicer, J.I., Lowe, D.M. & Widdicombe, S. 2010. Interaction of ocean acidification and temperature; the high cost of survival in the brittlestar Ophiura ophiura. *Marine Biology* **157**, 2001–2002.

Woods, A.H. & Moran, A.L. 2008. Temperature-oxygen interactions in Antarctic nudibranch egg masses. *Journal of Experimental Biology* **211**, 798–804.

Wooton, R. 2012. *Ecology of Teleost Fishes*. Berlin: Springer Science and Business Media.

Wägele, H. 1988. Riesenwuchs contra spektakuläre Farbenpracht – Nudibranchia der Antarktis. *Natur und Museum* **118**, 46–53.

Wägele, J.W. 1987. On the reproductive biology of Ceratoserolis trilobitoides (Crustacea: Isopoda): latitudinal variation of fecundity and embryonic development. *Polar Biology* **7**, 11–24.

Wägele, J.W. 1990. Growth in captivity and aspects of reproductive biology of the Antarctic fish parasite Aega antarctica (Crustacea, Isopoda). *Polar Biology* **10**, 521–527.

Węslawski, J.M. & Legezynska, J. 2002. Life cycles of some Arctic amphipods. *Polish Polar Research* **23**, 253–264.

Xiao, Q., Xia, J.-H., Zhang, X.-J., Li, Z., Wang, Y., Zhou, L. & Gui, J.-F. 2014. Type-IV antifreeze proteins are essential for epiboly and convergence in gastrulation of zebrafish embryos. *International Journal of Biological Sciences* **10**, 715–732.

Yang, S.-Y., Wojnar, J.M., Harris, P.W.R., DeVries, A.L., Evans, C.W. & Brimble, M.A. 2013. Chemical synthesis of a masked analogue of the fish antifreeze potentiating protein (AFPP). *Organic and Molecular Chemistry* **11**, 4935–4942.

Young, J.S. 2004. *Effects of temperature on elements of the motor control of behaviour in eurythermal and stenothermal crustaceans*. PhD thesis, University of Cambridge.

Young, J.S., Peck, L.S. & Matheson, T. 2006a. The effects of temperature on walking in temperate and Antarctic crustaceans. *Polar Biology* **29**, 978–987.

Young, J.S., Peck, L.S. & Matheson, T. 2006b. The effects of temperature on peripheral neuronal function in eurythermal and stenothermal crustaceans. *Journal of Experimental Biology* **209**, 1976–1987.

Yu, P.C., Sewell, M.A., Matson, P.G., Rivest, E.B., Kapsenberg, L. & Hofmann, G.E. 2013. Growth attenuation with developmental schedule progression in embryos and early larvae of Sterechinus neumayeri raised under elevated CO_2. *PLoS ONE* **8**, e52448.

Zacher, K., Wulff, A., Molis, M., Hanelt, D. & Wiencke, C. 2007. Ultraviolet radiation and consumer effects on a field-grown intertidal macroalgal assemblage in Antarctica. *Global Change Biology* **13**, 1201–1215.

Zachos, J., Pagani, M., Sloan, L., Thomas, E. & Billups, K. 2001. Trends, rhythms, and aberrations in global climate 65 Ma to present. *Science* **292**, 686–693.

Zachos, J.C., Dickens, G.R. & Zeebe, R.E. 2008. An early Cenozoic perspective on greenhouse warming and carbon-cycle dynamics. *Nature* **451**, 279–283.

Zhang, H., Bosch-Marce, M. & Shimoda, L.A. 2008. Mitochondrial autophagy is an HIF-1-dependent adaptive metabolic response to hypoxia. *Journal of Biological Chemistry* **283**, 10892–10903.

Zhang, H., Shin, P.K.S. & Cheung, S.G. 2016. Physiological responses and scope for growth in a marine scavenging gastropod, Nassarius festivus (Powys, 1835), are affected by salinity and temperature but not by ocean acidification. *ICES Journal of Marine Science* **73**, 814–824.

Ziveri, P., Passaro, M., Incarbona, A., Milazzo, M., Rodolfo-Metalpa, R. & Hall-Spencer, J.M. 2014. Decline in coccolithophore diversity and impact on coccolith morphogenesis along a natural CO_2 gradient. *Biological Bulletin* **226**, 282–290.

Oceanography and Marine Biology: An Annual Review, 2018, **56**, 237-310
© S. J. Hawkins, A. J. Evans, A. C. Dale, L. B. Firth, and I. P. Smith, Editors
Taylor & Francis

THE CARBON DIOXIDE VENTS OF ISCHIA, ITALY, A NATURAL SYSTEM TO ASSESS IMPACTS OF OCEAN ACIDIFICATION ON MARINE ECOSYSTEMS: AN OVERVIEW OF RESEARCH AND COMPARISONS WITH OTHER VENT SYSTEMS

SHAWNA ANDREA FOO[1]*, MARIA BYRNE[2], ELENA
RICEVUTO[3] & MARIA CRISTINA GAMBI[3]

[1]*Department of Global Ecology, Carnegie Institution for Science, Stanford, CA, United States*
[2]*School of Medical Sciences and School of Life and Environmental Sciences,
The University of Sydney, Sydney, New South Wales, Australia*
[3]*Stazione Zoologica Anton Dohrn, Department of Integrative
Marine Ecology, Villa Dohrn-Benthic Ecology Center, Ischia, Italy*
Corresponding author: Shawna Andrea Foo
e-mail: sfoo@carnegiescience.edu

Abstract

As the ocean continues to take up carbon dioxide (CO_2), it is difficult to predict the future of marine ecosystems. Natural CO_2 vent sites, mainly of volcanic origin, that provide a pH gradient are useful as a proxy to investigate ecological effects of ocean acidification. The effects of decreased pH can be assessed at increasing levels of organisation, from the responses of individuals of a species up through populations and communities to whole ecosystems. As a natural laboratory, CO_2 vent sites incorporate a range of environmental factors, such as gradients of nutrients, currents and species interactions that cannot be replicated in the laboratory or mesocosms, with the caveat that some vent systems have confounding factors such as hydrogen sulphide and metals. The first CO_2 vent sites to be investigated in an ocean acidification context were the vents at the Castello Aragonese on the island of Ischia, Italy. The gas released is primarily CO_2 with no evidence of toxic substances. They have been the focus of a wealth of studies, which are reviewed here and in context with research at other vent systems. Investigations of the species that occur along the pH gradients at Ischia show that, as the pH decreases, there is a reduction in calcifying species, reflecting the trends seen at other vent systems and in laboratory studies. The species assemblages at the Castello vents living at near future (2100) ocean acidification conditions (mean pH 7.8), show the resilience of many species to elevated CO_2, including many calcifying species (e.g. sea urchins, serpulids, bryozoans, foraminifera and corals). These taxa show different physiological and ecological mechanisms for acclimatisation and adaptation to low pH. As the oceans continue to acidify to pH levels <7.8, species assemblages are likely to become dominated by fleshy algae and smaller-bodied, generalist invertebrates. These observations suggest that ocean acidification will result in a simplification of marine food webs and trophic complexity.

Introduction

Ocean acidification

Over the past two centuries, increased anthropogenic emissions of CO_2 have resulted in a net increase in atmospheric CO_2 (IPCC 2014, Gattuso et al. 2015). The oceans constitute a major sink, absorbing approximately 40% of these emissions, causing a phenomenon known as 'ocean acidification' (OA) (Zeebe et al. 2008, IPCC 2014). Since the Industrial Revolution, the mean pH of surface ocean waters has decreased by 0.1 pH units, corresponding to a 26% increase in hydrogen ion concentration (Rhein et al. 2013). With the ocean changing at a pace much faster than past greenhouse events that led to mass extinctions (Zeebe et al. 2016), the fate of marine biodiversity and ecosystems is unclear.

Two factors co-vary with the CO_2-driven decrease in ocean pH: (1) hypercapnia, the increase in organism partial pressure of CO_2 (pCO_2), which can hinder metabolism and lead to impaired growth and reproduction (Pörtner 2008, Melzner et al. 2009, Byrne et al. 2013); and (2) reduced calcium carbonate ($CaCO_3$) mineral saturation. The reduction in carbonate minerals is a problem for marine calcifiers that build skeletons and shells (Guinotte & Fabry 2008, Kerr 2010, Howes et al. 2015). The impacts of changing ocean conditions on marine biota have been investigated in many laboratory studies (e.g. reviews by Byrne 2011, Harvey et al. 2013, Kroeker et al. 2013a, Przeslawski et al. 2015). Short-term studies cannot, however, incorporate evolutionary adaptation to environmental change, a feature that some fast-generational species have been shown to be capable of (Bush et al. 2016). It is also logistically difficult to consider multiple stressors simultaneously to more accurately reflect field conditions as well as to include species interactions. The latter have been shown to be affected by OA in laboratory experiments (e.g. Wahl et al. 2015, Jellison et al. 2016) and at vent systems (Nagelkerken et al. 2017).

CO_2 vents are being used as natural test systems and as proxies for future OA to assess potential outcomes for marine species and ecosystems. These systems provide insights into the biodiversity that survives and thrives in a lower pH ocean. As natural proxies, CO_2 vent systems provide a window into a future ocean and so are of great value. Vent systems are also useful to identify which organisms and functional groups are likely to be susceptible to decreased pH (Hall-Spencer et al. 2008), providing important data to compare with laboratory OA experiments. Vent systems show the impacts of OA-like conditions on a range of communities (e.g. rocky reefs, seagrass beds, coral reefs) and the taxa that are likely to decline as the pH decreases (Hall-Spencer et al. 2008, Gambi et al. 2010, Fabricius et al. 2011, Kroeker et al. 2011).

Volcanic CO_2 vents, a natural system for studying the effects of ocean acidification

The natural pH gradients at CO_2 vent sites have been used as model systems to disentangle the effects of OA at the ecosystem level (Kroeker et al. 2011, Connell et al. 2013, Fabricius et al. 2014). These low-pH sites provide insights into how ecosystems, communities and species will respond to a decrease in pH, as well as the acclimatisation and adaptation potential of the species that inhabit these regions (Hall-Spencer et al. 2008, Calosi et al. 2013b). The vents incorporate a range of environmental factors such as nutrients, currents and species interactions that cannot be replicated in the laboratory (Barry et al. 2010, Garrard et al. 2012). With the caveat that these are open systems, and thus biota with complex life histories (i.e. widely dispersing propagules or larvae) and mobile fauna may not necessarily spend their entire life cycle in the vents, these systems are especially important to identify the plant and animal species that have occupied these areas for years and decades. They are also a convenient and tractable field and experimental resource to assess the multifaceted effects of decreased pH, for instance in recent transplant experiments (e.g. Turner et al. 2016, Kumar et al. 2017a, Porzio et al. 2017).

Several CO_2 vent sites around the world in temperate (e.g. Italy, Greece, Spain, New Zealand, Portugal) and tropical (e.g. Papua New Guinea, Mexico, Japan) regions have been used for OA studies. At most of these sites, the gas released appears to be largely CO_2. The toxic gases hydrogen sulphide and methane have been detected at some sites in Papua New Guinea (Fabricius et al. 2011, Hassenrück et al. 2016), Vulcano, Italy (Sedwick & Stuben 1996, Boatta et al. 2013), Panarea, Italy (Gugliandolo et al. 2006), White Island, New Zealand (Tarasov 2006, Brinkman & Smith 2015) and Japan (Agostini et al. 2015). In addition, trace elements (e.g. As, Hg, Mn, Zn) are present in some vent systems (Tarasov 2006, Boatta et al. 2013). These toxic substances should be taken into consideration when assessing outcomes for species. As trace element solubility may be related to changes in seawater chemistry and decreases in pH, caution is required when using CO_2 vent sites as analogues for OA research (Vizzini et al. 2013).

In seawater, dissolved CO_2 forms carbonic acid and causes a decrease in carbonate ion concentration and an increase in bicarbonate ion concentration. This results in a release of hydrogen ions to maintain equilibrium, thus lowering pH, and thus the vent sites create a natural OA-like phenomenon by simulating this effect naturally. The vent systems reduce pH along a gradient, with some sites exhibiting pH levels predicted for the future. With continued oceanic uptake of CO_2, near future (by 2100) changes in ocean pH under best and worse case scenarios (RCP 2.6, 8.5) are expected to be -0.07 ± 0.001 to -0.33 ± 0.003 pH units, respectively (Caldeira & Wickett 2005, IPCC 2014, Gattuso et al. 2015, Howes et al. 2015). Far future scenarios (by 2300) predict the surface pH to reduce by 0.8–1.4 pH units (Caldeira & Wickett 2005).

Ischia: a model vent system

Of the vent systems that have been explored as proxies of a future ocean, the volcanic vent areas of the Castello Aragonese in the island of Ischia, Italy in the Mediterranean are very well characterised with respect to analysis of gas emissions. They have been investigated by volcanologists for decades (Rittmann 1930, Tedesco 1996). Extensive analysis of gas emissions at this vent system confirms that the majority of the gas composition is CO_2, with undetectable concentrations of trace elements and no hydrogen sulphide (Tedesco 1996, Hall-Spencer et al. 2008, Kroeker et al. 2011). Characterisation of the gas and its isotope composition across CO_2 vent sites distributed on the north-eastern side of Ischia, including one of the vents of the Castello Aragonese, show that the gas is composed of 90%–95% CO_2, 3%–6% N_2, 0.6%–0.8% O_2 and 0.2%–0.8% CH_4 at ambient temperature (Tedesco 1996). The gas chemistry analyses performed at the Castello vents are scattered over a relatively long time span (Tedesco 1996, Hall-Spencer et al. 2008), including the most recent analysis of the northern and southern vent sites in August 2017 (F. Italiano, G. Pecoraino & M.C. Gambi, unpublished data). All have shown the gas is primarily CO_2 (around 95%) with no hydrogen sulphide.

Given the pioneering and extensive research at the Castello CO_2 vents in Ischia, Italy, and the comparatively pure CO_2 nature of benthic gassing, our review focusses on studies that have been conducted in this vent system. The review begins with a description of the CO_2 vents of Ischia, a summary of the volcanic history in the area which generates the CO_2 bubbles, as well as an overview of studies that have characterised the gas and pH gradients. pH levels are reported as pH total (pH_T) or National Bureau of Standards pH (pH_{NBS}) depending on what was reported in the study. This is followed by a history of Ischia providing a hypothesised timeline for vent activity and, in more recent times, observations from scientists provide further bounds on vent activity and how it has changed over time. Research conducted at the Ischian vents is then reviewed, focussing on biodiversity, settlement and recruitment. This is considered in context with the plethora of studies that have used the vents as proxies to assess the potential impacts of climate driven OA from the species to the ecosystem level. Much of the Ischian CO_2 vents research has investigated the effects of CO_2 on *Posidonia oceanica* seagrass habitat and its associated epiphytic biodiversity, and this is covered extensively. For the taxonomic group levels, we present the effects of low pH on individual species. Species discussed in the review and listed in S1 are named according to the WoRMS World Register of Marine Species (WoRMS Editorial Board 2017). The use of the vents as a natural experimental system is discussed, detailing the transplant experiments conducted in Ischia. We also make comparisons with other vent systems and laboratory-based acidification research. Research undertaken using the Castello vent system is summarised in Tables 1 and 2, with pH values listed for each experimental site. The locations of the corresponding sites are shown in Figure 1 with mean pH values in Figure 2.

We summarise the key insights generated from the work at the Ischia CO_2 vents and in context with findings generated at other vent systems. Finally, we consider the way forward in using CO_2 vents as natural laboratories to study the impacts of acidification on benthic biota from species to ecosystem level as well as their limitations.

The CO_2 vents of Ischia: the geological setting and the zone of the Castello Aragonese

The submerged vent system at Ischia is located north-west of the Bay of Naples, Italy, characterised by submersed venting in both shallow (1–10 m) and deep waters (48 m) linked to local volcanic and tectonic features (Tedesco 1996, de Alteriis & Toscano 2003, Gambi 2014). Ischia, together with the 'Campi Flegrei' (Phlaegrean Fields), is part of the volcanic district lying west to the Bay of Naples in the Tyrrhenian Sea. The formation of the area dates to the Plio-Quaternary complex derived from the opening of the Tyrrhenian Sea (Peccerillo 2005). On the island of Ischia, secondary volcanism is still quite active and intense. The island has a well-documented eruptive history. Its volcanic-tectonic horst (i.e. the raised region of the Earth's crust bounded by faults), is one of the first described in the geological literature (Rittmann 1930, de Alteriis & Toscano 2003, de Alteriis et al. 2006, 2010). Various submerged volcanic, monogenic edifices are still evident around the island and the nearby continental shelf (de Alteriis et al. 2006). The last documented eruption was in 1301–1302 AD (de Alteriis & Toscano 2003), while the last major earthquake was a destructive event in 1883 AD, with a more recent minor event (4.0 Richter magnitude) in August 2017.

A series of recent geological surveys have revealed the complex geomorphology of the submerged portion of Ischia, confirming articulated benthic features, as well as the historical and archaeological records (see de Alteriis et al. 2006, 2010 for a bathymetric map and geological overview). The main block of the island was uplifted following a series of eruptions approximately 155,000 years ago. The Castello Aragonese represents the remains of a volcanic edifice formed from 150,000 to 75,000 years ago, defined as a synaptic volcano (Rittmann & Gottini 1981). It is adjacent to a fault corresponding to one of the many fractures present on the south-eastern sector of Ischia (Rittmann 1930). The faults favour the passage of CO_2 bubbles, derived from complex geological and geochemical magmatic processes (Vezzoli 1988), thus giving passage to the CO_2 bubbles that arise from the sea floor.

Several new vent systems along the same fault line have been discovered and are being used for research (Gambi 2014). The CO_2 vents of the Castello Aragonese have been investigated in over 80 studies and are the most intensely studied vents in the world.

Characterisation of the CO_2 vents of the Castello Aragonese

The vents occur at the rocky shores on the north and south sides of the Castello Aragonese located on the north-eastern coast of Ischia (40°43.84′N 13°57.08′E) (Figure 1). At the north site, gas is emitted at 0.7×10^6 L per day (area = 2000 m^2) with less than five vents per square metre. On the south site, gas is emitted at 1.4×10^6 L per day (area = 3000 m^2) with more than five vents per square metre (Pecoraino et al. 2005, Hall-Spencer et al. 2008).

Table 1   Effects of the CO_2 vents and resultant pH gradient on biodiversity at the Castello Aragonese

Study species	Vent site and pH levels measured	Results	References
Changes in overall species composition, settlement and recruitment			
Crustose coralline algae (CCA)	Cited values for S1–3 and N1–3 from Kroeker et al. 2011	After 14 months of placement of settlement tiles *in situ*, settlement of CCA in low pH (S2, N2) was decreased in abundance; however, the skeleton had a similar Mg^{++} content in comparison with CCA present in ambient conditions (S1, N1). At extreme low pH (S3, N3), CCA was significantly smaller with changes in Mg^{++} content.	Kamenos et al. 2016
Amphiglena mediterranea, Platynereis dumerilii and *Syllis prolifera* (polychaetes)	Cited values for S1–3 and N1–3 from Kroeker et al. 2011	Observations of colonisation in invertebrate collectors showed that the three species were found across all pH levels, with higher abundance in extreme low pH (S3, N3).	Ricevuto et al. 2014
All species	$pH_T \pm SD$ S1: 8.1 ± 0.1 S2: 7.8 ± 0.3 S3: 6.6 ± 0.5 N1: 8.0 ± 0.1 N2: 7.8 ± 0.2 N3: 7.2 ± 0.4	32 months after clearance of all species on natural rocky plots, species recovery was highly diverse in ambient pH levels (S1, N1). On the other hand, acidification decreased the variability of the communities where assemblages in extreme low pH (S3, N3) were dominated by algae.	Kroeker et al. 2013b[*]
All species	$pH_T \pm SD$ S1: 8.06 ± 0.09 S2: 7.75 ± 0.31 S3: 6.59 ± 0.51 N1: 7.95 ± 0.06 N2: 7.77 ± 0.19 N3: 7.21 ± 0.34	Recruitment on settlement tiles showed that in the earlier months (<3.5 months), calcareous species were recruited and grew at similar rates in both ambient (S1, N1) and low pH (S2, N2). As time progressed (>6.5 months), they were overgrown by fleshy seaweeds in the low-pH conditions.	Kroeker et al. 2013c[*]

Continued

Table 1 (Continued)   Effects of the CO_2 vents and resultant pH gradient on biodiversity at the Castello Aragonese

Study species	Vent site and pH levels measured	Results	References
Algal assemblages	$pH_T \pm SD$ S1: 8.06 ± 0.09 S2: 7.75 ± 0.30 S3: 6.59 ± 0.51 N1: 7.95 ± 0.06 N2: 7.77 ± 0.19 N3: 7.20 ± 0.36	After 4 months, the number of algal species that had settled on tiles decreased by 4% in low pH (S2, N2) and by 18% in the extreme low pH (S3, N3) zones as compared to control pH (S1, N1).	Porzio et al. 2013[*]
All species	Cited values for S1–3 and N1–3 from Cigliano et al. 2010 and Kroeker et al. 2011	In low pH (S2, N2), the settlement and colonisation of scouring pads by molluscs was greatly reduced in both number of species and abundance. On the other hand, polychaetes, amphipods, tanaids and isopods were found with a similar abundance across all pH levels.	Ricevuto et al. 2012
All species	$pH_T \pm SD$ S1: 8.1 ± 0.1 S2: 7.8 ± 0.3 S3: 6.6 ± 0.5 N1: 8.0 ± 0.1 N2: 7.8 ± 0.2 N3: 7.2 ± 0.4	Across the three pH zones, characterisation of species assemblages in rocky, vegetated reefs showed that the number of individuals did not differ, although there were fewer taxa and a lower biomass in the extreme pH zones (S3, N3).	Kroeker et al. 2011[*]
Calcifying and non-calcifying invertebrates	$pH_T \pm SD$ S1: 8.12 ± 0.01 S2: 7.62 ± 0.14 S3: 7.33 ± 0.13 N1: 8.09 ± 0.03 N2: 7.72 ± 0.12 N3: 7.39 ± 0.10	Assessment of settlement on artificial collectors (scouring pads) after one-month exposure showed that calcifiers were restricted to collectors in the control pH stations (S1, N1, including foraminifera, gastropods and most Bivalvia species) with a clear decrease in settlement diversity as pH decreased. The polychaete *Syllis prolifera* showed a higher abundance in extreme low pH (S3, N3).	Cigliano et al. 2010[*]

Continued

Table 1 (Continued) Effects of the CO_2 vents and resultant pH gradient on biodiversity at the Castello Aragonese

Study species	Vent site and pH levels measured	Results	References
Calcifying and non-calcifying taxa	$pH_T \pm SE$ S1: 8.14 ± 0.01 S2: 7.83 ± 0.06 S3: 6.57 ± 0.06 N1: 8.14 ± 0.02 N2: 7.87 ± 0.05 N3: 7.09 ± 0.11 P1: 8.17 ± 0.00 P2: 8.17 ± 0.01 P3: 8.00 ± 0.03 P4: 7.60 ± 0.05	The diversity of the rocky shore shifted to communities which lacked calcifying species as the pH was decreased, where calcifying species became reduced or absent as pH decreased beyond pH 7.8 (S2, N2).	Hall-Spencer et al. 2008
Effects on *Posidonia oceanica* and associated biodiversity			
Epiphyte community on *P. oceanica*	$pH_{NBS} \pm SE$ LA: 8.10 ± 0.04 S2: 7.80 ± 0.02 N2: 7.78 ± 0.01	Epiphyte communities in low pH (S2, N2) differed in composition and abundance compared to ambient pH (LA) levels. Coralline algae were in a lower abundance but only on the south sides of the vents, indicating other influential factors.	Nogueira et al. 2017
P. oceanica	Cited values for S1–3 from Hall-Spencer et al. 2008	There was a change in ∼50% of gene expression between low- (S2, S3) and ambient-pH (S1) conditions related to down-regulation in metal detoxification genes in low pH.	Lauritano et al. 2015
13 *P. oceanica* associated invertebrates	Cited values for S1 and S2 from Garrard et al. 2014	The chemotactic response in 54% of the species to volatile olfactory chemicals from *P. oceanica* leaves changed in low pH (S2).	Zupo et al. 2015
All species	Cited values for S1–3 and N1–3 from Cigliano et al. 2010	Settlement pattern of epibionts on *P. oceanica* mimics showed that in extreme low-pH (S3, N3) conditions; there were no calcifying species. Mimics were dominated by hydroids, tunicates and filamentous algae.	Donnarumma et al. 2014

Continued

Table 1 (Continued) Effects of the CO_2 vents and resultant pH gradient on biodiversity at the Castello Aragonese

Study species	Vent site and pH levels measured	Results	References
All species	$pH_{NBS} \pm SD$ S1: 8.12 ± 0.04 S2: 7.78 ± 0.39 N1: 8.13 ± 0.05 N2: 7.82 ± 0.31	Assessment of the invertebrate assemblage present on *P. oceanica* showed that many taxa (e.g. gastropods, amphipods, bivalves, isopods and polychaetes) were tolerant to low pH (S2, N2), with almost double the number of invertebrates found at low pH in comparison to the control (S1, N1), likely due to buffering effect of the *Posidonia* canopy and an increase in food supply.	Garrard et al. 2014
Epiphyte community on *P. oceanica*	$pH_T \pm SD$ P1: 8.17 ± 0.01 P2: 8.17 ± 0.03 P2.5: 8.11 ± 0.03 P3: 8.00 ± 0.16 P4: 7.66 ± 0.32	In the field, coralline algae were the most dominant contributor to calcium carbonate mass on *P. oceanica* at ambient pH (P1–3) levels but was absent at pH 7.7 (P4). Bryozoans were the only calcifier present on seagrass blades at this pH level.	Martin et al. 2008
Site characterisation			
Site characterisation	$pH_{SWS} \pm IQR$ *Surface pH* S1: 8.08, +0.08, −0.02 S2: 7.68, +0.22, +0.01 S3: 7.48, +0.69, −0.05 *Benthic pH* S1: 7.84, +0.04, −0.01 S2: 7.80, +0.21, −0.04 S3: no data	pH measurements were found to be lower and less stable than those from Hall-Spencer et al. 2008.	Kerrison et al. 2011
Assessment of pH variability	$pH_T \pm SD$ S2: 7.85 ± 0.27	pH variability was abiotically driven and due to CO_2 venting and subsequent mixing.	Hofmann et al. 2011[*]

Note: The studies presented are grouped by those that investigated biodiversity and species composition present at the vent sites, effects of the pH gradient on *Posidonia oceanica* and its associated biodiversity as well as studies with a focus on characterisation of pH, ordered from the study most recently published within each section. For each study, the specific species or group of focus is indicated, the pH values measured at the vent sites are shown, the results are summarised and the relevant citation is included. pH levels are reported as pH total (pH_T), National Bureau of Standards pH (pH_{NBS}) or pH on the seawater scale (pH_{SWS}), and measures of variability are reported depending on what was provided in the study. The location of the vent sites indicated correspond to those labelled in Figure 1. References with an asterisk (*) refer to studies that were used to determine average pH and standard deviation for Figure 2.

Table 2 Use of the CO_2 vents of the Castello Aragonese as a natural laboratory

Study species	Vent site and pH levels measured	Results	References
Group specific studies			
Polychaete assemblages	$pH_T \pm SD$ S1: 8.1 ± 0.1 S2: 7.8 ± 0.3 S3: 6.6 ± 0.5 N1: 8.0 ± 0.1 N2: 7.8 ± 0.2 N3: 7.2 ± 0.4	The number of species and abundance decreased in the extreme low pH (S3, N3) conditions, with decreases in both calcifying and non-calcifying species.	Gambi et al. 2016a*
Ostreopsis ovata	Cited values for S1–3 and N1–3 from Cigliano et al. 2010	Abundance of the harmful dinoflagellate did not vary with reduced pH, where it was tolerant to a wide range of pH values.	Di Cioccio et al. 2014
Sponges	Averaged values for S1–3 and N1–3 from Hall-Spencer et al. 2008, Cigliano et al. 2010 and Rodolfo-Metalpa et al. 2010	Only one species of sponge (*Crambe crambe*) was present at extreme low pH (S3, N3), with seven at low pH (S2, N2) and 11 at ambient pH (S1, N1).	Goodwin et al. 2013
Macroalgal communities	Cited values for S1–3 from Hall-Spencer et al. 2008	At low pH (S2), there was a 5% decrease in species richness in comparison with ambient pH (S1). At extreme low pH (S3), there was a decrease of 72% in species richness.	Porzio et al. 2011
Foraminifera	Cited values for S1–3 and N1–3 from Hall-Spencer et al. 2008	Reduction from 24 to 4 species when pH was reduced from ambient (S1, N1) to extreme low pH (S3, N3).	Dias et al. 2010
Species' response			
Sargassum vulgare (macroalgae)	Cited values for S3 from Hall-Spencer et al. 2008	Comparative transcriptome analysis showed that the species could adapt to extreme low pH (S3) through an increase in energy production.	Kumar et al. 2017a
Polychaetes	Cited values for S1–3 and N1–3 from Kroeker et al. 2011	In low and extreme low pH (S2–3, N2–3), there was a shift towards species that brood the eggs in protected maternal environments.	Lucey et al. 2015

Continued

245

Table 2 (Continued)   Use of the CO_2 vents of the Castello Aragonese as a natural laboratory

Study species	Vent site and pH levels measured	Results	References
Platynereis dumerilii, Polyophthalmus pictus and *Syllis prolifera* (polychaetes)	$pH_{NBS} \pm SD$ R: $8.05 \pm 0.00/8.11 \pm 0.00$ LA: $8.17 \pm 0.00/8.09 \pm 0.00$ S2: 7.69 ± 0.30 S3: $6.83 \pm 0.42/7.14 \pm 0.40$ N3: $7.12 \pm 0.30/7.37 \pm 0.37$	There were significant effects of low pH (S2–3, N3) on the C and N isotopic signatures in the polychaetes' organic matter food sources with consequent changes in the C:N ratio and quality of food but no significant effects on trophic interactions. In low pH, all macrophytes including *Posidonia*, showed a higher N content.	Ricevuto et al. 2015b
Patella caerulea (limpet)	$pH_T \pm SD$ S1: 8.03 ± 0.05 S3: $6.46 \pm 0.35 / 6.51 \pm 0.38$	Comparison of shells from ambient (S1) and extreme low pH (S3) sites showed that limpets from extreme low pH had a two-fold increase in aragonite area, with an increase in apical thickness of the shell.	Langer et al. 2014*
Microdeutopus sporadhi (amphipod)	Cited values for N1–3, S1–3 from Cigliano et al. 2010	*Microdeutopus sporadhi* were only found in artificial collectors and only at low (S2, N2) and extreme low (S3, N3) pH sites on the south side.	Scipione 2013
Anemonia viridis (sea anemone)	Cited values for N1, N3 from Meron et al. 2012b	Low pH (N3) affected the composition and diversity of associated microbial communities, but there were no negative physiological changes or presence of microbial stress indicators.	Meron et al. 2012a
Balanophyllia europaea and *Cladocora caespitosa* (corals)	Cited values for B1–3 from Rodolfo-Metalpa et al. 2011	Low pH (B2–3) did not impact the composition of associated microbial communities in both coral species.	Meron et al. 2012b

Continued

Table 2 (Continued)   Use of the CO_2 vents of the Castello Aragonese as a natural laboratory

Study species	Vent site and pH levels measured	Results	References
Transplant experiments			
Sargassum vulgare (macroalgae)	Cited values for S1 and S3 from Porzio et al. 2011	Transplanted extreme low pH (S3) *Sargassum vulgare* to control conditions (S1) induced a stress response, where low pH adapted *S. vulgare* had less efficient photosynthesis apparatus.	Porzio et al. 2017
Sabella spallanzanii (polychaete)	$pH_{NBS} \pm SD$ L1: 8.23 ± 0.05 L2: 8.25 ± 0.06 L3: 8.25 ± 0.06 S2: 7.44 ± 0.35 S3: 7.14 ± 0.40 N3: 7.37 ± 0.37	The antioxidant system showed a significant decrease in enzymatic activities, with impairment of the ability to neutralise hydroxyl radicals after 30 days in low pH (S2–3, N3).	Ricevuto et al. 2016
Pinna nobilis (pen shell)	$pH_{NBS} \pm SE$ L1: 8.17 ± 0.01 L2: 8.14 ± 0.02 L3: 8.17 ± 0.01 S2: 7.57 ± 0.15 S3: 7.31 ± 0.17 N2: 7.78 ± 0.09 N3: 7.74 ± 0.15	After 45 days, transplanted juveniles showed significant decreases in growth, oxygen consumption and mineralisation in the extreme low pH (S3, N3) sites, with an increase in mortality.	Basso et al. 2015
Calpensia nobilis (bryozoan)	$pH_T \pm SD$ S1: 8.09 ± 0.08 S2: 7.76 ± 0.43	Colonies at low pH (S2) grew at half the rate of those transplanted at the control site (S1).	Lombardi et al. 2015*
Sabella spallanzanii (polychaete)	$pH_{NBS} \pm SE$ L1: 8.12 ± 0.01 L2: 8.13 ± 0.01 L3: 8.15 ± 0.01 S2: 7.49 ± 0.06 S3: 7.16 ± 0.08 N3: 7.21 ± 0.06	After 5 days in low pH (S2–3, N3), *Sabella spallanzanii* showed increases in metabolism, with a decrease in carbonic anhydrase concentrations, essential in acid base and respiratory functioning.	Turner et al. 2015
Mytilus galloprovincialis (mussel)	$pH_T \pm SD$ S1: 8.07 ± 0.04 S3: 7.25 ± 0.44	After 68 days exposure to extreme low pH (S3), the shells showed a significant change in carbon, oxygen and elemental skeleton composition; however, transplantation caused observable changes in shell ultrastructure and texture.	Hahn et al. 2012*

Continued

Table 2 (Continued) Use of the CO_2 vents of the Castello Aragonese as a natural laboratory

Study species	Vent site and pH levels measured	Results	References
Myriapora truncata (bryozoan)	$pH_T \pm SD$ S1: $8.09 \pm 0.08/8.09 \pm 0.01/8.10 \pm 0.07$ S2: $7.63 \pm 0.42/7.75 \pm 0.405/7.83 \pm 0.41$ S3: $7.28 \pm 0.48/7.29 \pm 0.407/7.32 \pm 0.47$	*Myriapora truncata* could form new and complete zooids in only the control site (S1). In the first 34 days, there were significant changes in the protein expression of colonies in the intermediate (S2) and low pH (S3) conditions; however, this declined after 57 and 87 days of exposure.	Lombardi et al. 2011a*
Schizoporella errata (bryozoan)	$pH_T \pm SD$ Unique control near S2: 8.1 ± 0.06 S2: 7.83 ± 0.41	At low pH (S2), growth of the zooidal basal and lateral walls was retarded as well as the presence of fewer avicularia.	Lombardi et al. 2011b*
Myriapora truncata (bryozoan)	Cited values for B1–2 from Rodolfo-Metalpa et al. 2010	After 45 days of exposure in low pH (B1–2), live colonies were less corroded than dead colonies, where organic tissues protected the skeleton.	Lombardi et al. 2011c
Mytilus galloprovincialis, *Patella caerulea* (molluscs) and *Balanophyllia europaea*, *Cladocora caespitosa* (corals)	pH_T mean/min $\sim$P2.5: 8.06 / 7.95 B1: 7.38 / 6.83 B2: 7.53 / 7.10 B3: 7.77 / 7.22	Animals relied on the organic layers outside the shell and skeleton to prevent dissolution in low pH (B1–3).	Rodolfo-Metalpa et al. 2011
Myriapora truncata (bryozoan)	$pH_T \pm SD$ B1: 7.43 ± 0.31 B2: 7.66 ± 0.22 Unique control sites not labelled on a map C1: 8.06 ± 0.07 C2: 8.07 ± 0.10	Skeletons of live colonies in the lowest pH site (B1) were less corroded than dead colonies, likely due to organic tissue protecting the skeleton.	Rodolfo-Metalpa et al. 2010

Reciprocal transplant experiments

Sargassum vulgare (macroalgae)	Cited values for S3 from Porzio et al. 2011	Measures of photosynthetic rates and oxidative stress indicated that the vent population is adapted to live at extreme low pH (S3).	Kumar et al. 2017b

Continued

Table 2 (Continued) Use of the CO_2 vents of the Castello Aragonese as a natural laboratory

Study species	Vent site and pH levels measured	Results	References
Simplaria sp. (polychaete)	$pH_{NBS} \pm SD$ N1: 8.03 ± 0.08 ~P3: 8.03 ± 0.05 S2: 7.61 ± 0.26	Populations showed decreases in survival, reproductive maturation and output when transplanted to low pH (S2) for 66 days, regardless of whether they had originally resided in a control (N1, P3) or low-pH environment.	Lucey et al. 2016
Isopods: 'Sensitive' to low pH: *Cymodoce truncata*, *Dynamene bicolor* (=*Dynamene torelliae*) and 'Tolerant' to low pH: *D. bifida*	Cited values for R from Ricevuto et al. 2015b and for S2–3 from Cigliano et al. 2010	After 5 days in conditions, both sensitive species could maintain similar metabolism levels in low pH (S2–3). For *C. truncata*, however, this came at a cost with decreases in the levels of carbonic anhydrase produced.	Turner et al. 2016
Platynereis dumerilii, *Polyophthalmus pictus* and *Syllis prolifera* (polychaetes)	$pH_{NBS} \pm SD$ L1: 8.2 ± 0.1 L2: 8.2 ± 0.1 L3: 8.2 ± 0.1 S2: 7.4 ± 0.4 S3: 7.1 ± 0.4 N3: 7.4 ± 0.4	Low pH (S2–3, N3) adapted *Platynereis dumerilii* and *Syllis prolifera* had enhanced basal antioxidant protection. *Polyophthalmus pictus* did not show the same compensation.	Ricevuto et al. 2015a
Polychaetes: 'Tolerant' to low pH: *Platynereis dumerilii*, *Amphiglena mediterranea*, *Syllis prolifera* and *Polyophthalmus pictus* and 'Sensitive' to low pH: *Lysidice ninetta*, *Lysidice collaris* and *Sabella spallanzanii*	$pH_{NBS} \pm SE$ L1: 8.15 ± 0.01 L2: 8.16 ± 0.01 L3: 8.15 ± 0.01 S2: 7.19 ± 0.05 S3: 7.07 ± 0.07 N3: 7.24 ± 0.04	Species showed different mechanisms of adjustment to decreased pH. *Amphiglena mediterranea* from the CO_2 vents were acclimatised but not adapted to low pH (S2–3, N3), but *Platynereis dumerilii* from the CO_2 vents were adapted to low pH and were physiologically and genetically different from control (L1–3) populations.	Calosi et al. 2013b

Note: The studies presented are grouped by those which investigated characteristics of species naturally present in the vents, transplant experiments and reciprocal transplant experiments ordered from the study most recently published within each section. For each study, the specific species or group of focus is indicated, the pH values measured at the vent sites are shown, the results are summarised, and the relevant citation is included. pH levels are reported as pH total (pH_T) or National Bureau of Standards pH (pH_{NBS}), and measures of variability are reported depending on what was provided in the study. The location of the vent sites indicated correspond to those labelled in Figure 1. References with an asterisk (*) refer to studies that were used to determine average pH and standard deviation for Figure 2.

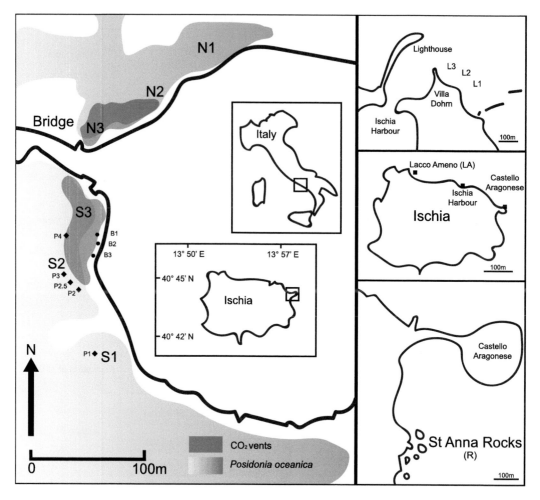

Figure 1 Map of the north and south vents on the rocky shores of the Castello Aragonese located on the north-eastern coast of Ischia, Italy and control sites around the Ischia island. Labels represent approximate locations of common sites used for experiments across varying pH levels. The left panel inset shows the location of Ischia island in Italy. The large left panel shows the vent sites around the Castello Aragonese where North 1–3 (N1–3) and South 1–3 (S1–3) sites represent areas of different pH levels and varying CO_2 bubbling intensity. On the south side, there are further experimental sites: Benthos 1–3 (B1–3) and *Posidonia* 1–4 (P1–4). The right panels show commonly used control sites with no vent activity and ambient pH: Lighthouse 1–3 (L1–3), Lacco Ameno (LA) and St. Anna Rocks (R). The middle and bottom right panels show these control locations in relation to the Castello Aragonese. See Tables 1 and 2 for pH levels recorded for each site and details of individual experiments.

Various studies (see Table 1) have characterised pH at the vent systems at Ischia, where the seawater pH reduces from normal seawater (sites N1, S1) across a gradient down to pH_T levels as low as 6.8 (site S3) (Figures 1, 2). The salinity of the vent sites is 38 psu, characteristic of the central region of the Tyrrhenian Sea (Lorenti et al. 2005). The vent activity does not heat the water, so the gas bubbles are at the same temperature as the surrounding waters (Hall-Spencer et al. 2008, Kerrison et al. 2011). The reduction in seawater pH is driven by an increased concentration of dissolved inorganic carbon (DIC) with moderately stable total alkalinities (TA), temperatures and salinities across ambient and reduced pH vent sites (Kroeker et al. 2011). Thus, as the pH decreases, there is an increase in the DIC-to-TA ratio, resulting in reductions in aragonite and calcite saturation states.

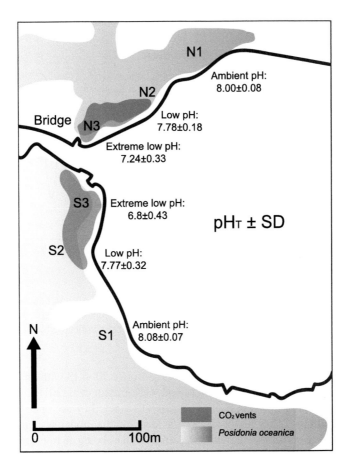

Figure 2 Average pH on the total scale (pH_T) $\pm$ SD recorded for the most commonly used experimental sites at Castello Aragonese. The values represent the means provided by those studies which recorded pH_T and provided the standard deviation, representative of the pH flux in each site. The studies included are indicated by an asterisk in the "Reference" column of Tables 1 and 2 (n for South sites 1–3 = 10, n for North sites 1–3 = 6). On both the north and south sides, as the sites converge towards the bridge, the pH values decrease from ambient to extremely low, corresponding with the intensity of the CO_2 bubbling.

Kerrison et al. (2011) monitored the pH and temperature at the south vent sites of the Castello and measured the surface and benthic pH (Table 1). They found that the pH was lower and much less stable than previously recorded (sites S1–S3; Hall-Spencer et al. 2008). In addition, pH at the surface and seabed differed with the lower bottom pH likely due to benthic sources of CO_2 such as respiration and calcification, as well as mixing of water at the surface. The high variability of gas release and thereby pH variability shows that vent activity changes over time (M.C. Gambi, pers. obs.). Thus, the pH at the vents of the Castello is dynamic and variable, a common feature of other vent systems (Dando et al. 1995, Fabricius et al. 2011, Boatta et al. 2013). This is an important consideration when comparing with outcomes of laboratory studies, which are usually conducted at stable pH and do not incorporate pH fluctuations.

The average pH levels at individual sites depends on location and time because pH can vary by 0.1 units within an hour (Figure 3). Hofmann et al. (2011) assessed pH at one location on the south side of the Castello (site S2, Figures 1, 2) using an autonomous sensor. In comparison with other deployments (e.g. tropical, polar), they showed that the greatest variability in pH over time was

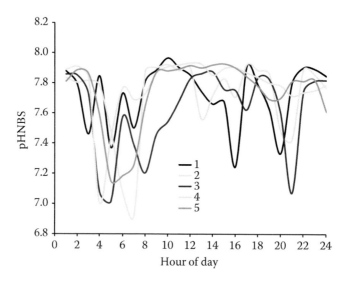

Figure 3 National Bureau of Standards pH (pH_{NBS}) logged every hour for 24 hours (September 2016) with a SeaFET™ at the CO_2 vents of Ischia. Each line represents a different consecutive day (a total of five days) where the variation in pH can vary by 0.1 units within an hour. Data collected by Dr Maria Cristina Gambi and Dr Shawna Foo.

documented at the Ischia S2 site, where pH_{NBS} ranged from 6.70–8.13 with an average of 7.84. This was due to variable CO_2 venting and mixing of water (Hofmann et al. 2011).

These vent systems are not a replicate of future OA since they are open systems with a large temporal variability in pH. They do, however, provide a unique model system to assess impacts of decreased pH on organisms and species interactions, as well as being an important natural laboratory for manipulative experiments (Garrard et al. 2012, Connell et al. 2013). The CO_2 vents of Ischia have been used extensively to investigate the effects of decreased pH on marine communities to help understand the changes in biodiversity associated with acidification. More recently, the vent sites have been used for transplant experiments to assess species level responses (e.g. Figure 4, see Table 2).

Archaeological features of the Castello area

Archaeological evidence from the Castello Aragonese has been useful in providing a possible timeline of vent activity. Ischia has a long colonisation history due to its strategic location in the Mediterranean, serving as a focal point for maritime trade and cultural exchange (Monti 1980, Buchner & Ridgway 1993). The island was the first western colony of Greek colonisation (Pithecusa, VIII century BC, Buchner & Ridgway 1993) and was after occupied by Romans from the fourth century BC where the island became known as Aenaria. The harbour of the Roman citadel was located in the stretch of water between the Castello Aragonese and the cliffs of St. Anna in Cartaromana Bay (Figure 5, Benini & Gialanella 2017a,b). As the presence of the Aenaria harbour indicates that the coastline was not originally submerged, the vents of the Castello cannot be older than the last known evidence of the harbour (approximately 1800 years ago).

The destruction and submergence of the Aenaria harbour (and presumably the citadel) was an abrupt event, probably caused by the Montagnone-Maschiatta eruption around the first century (Monti 1980), which was accompanied by strong earthquakes and volcano-settling tectonics in north-western Ischia. Recent studies suggest that the area underwent submersion of ~6 m in two phases. The first was between second century BC and the Angevin colonisation (thirteenth through

Figure 4 Photographs of *in situ* transplant experiments at low-pH vent field sites at the Castello Aragonese. CO_2 bubbles can be seen at the vent locations with examples of enclosures used to house polychaetes (bottom left and right panel) and sea urchin larvae (top left panel) over various lengths of time. Mesh of different sizes is used to retain the animals. Small buoys keep the ropes and enclosures suspended in the water column, or enclosures are weighed down on the benthos (Photos: Dr Maria Cristina Gambi).

the beginning of the fourteenth century AD), due to the eruptive event mentioned previously, coupled with other slower tectonic phenomena (based on dating of submerged building materials, Benini & Gialanella 2017a,b). A further submersion of ~2.5 m occurred during the Angevin period after the last eruption (AD 1301–1302), recorded on the island adjacent to Cartaromana Bay.

From archaeological information, it appears that most of the coastline of Cartaromana Bay where the vents currently are was not submerged until AD 140. Submersion of the area and consequent vent activity on the sea floor around the Castello could thus have been formed between 1800 to 700 years ago (if they formed after the eruption in 1302). Therefore, the evolutionary history of the ecosystem and the potential organism acclimatisation and adaptation trajectories could have occurred over hundreds of years in some areas, especially on the south side, as discussed below.

An anecdotal history of vent activity at the Castello and early benthic research

The island of Ischia was chosen by Anton Dohrn, the founder of the Stazione Zoologica of Naples, for his summer residence (Heuss 1991). In 1906, he built a villa on the top of San Pietro Hill at the entrance of the harbour, which was used as a 'buen retiro' for the family. The lower part (Casina a Mare), was used as a field station for scientific collections (Groeben 1985, Heuss 1991). In 1923, the Dohrn family annexed the villa to the Stazione Zoologica based in Naples (Bacci 1969), and the building was transformed into a marine ecology laboratory (Reparto di Ecologia), now known as the Villa Dohrn-Benthic Ecology Center.

Figure 5 Photograph of the Castello Aragonese. The photograph shows the Castello Aragonese islet from the south side with the bridge connecting the islet to the main island of Ischia. St. Anna Rocks can be seen in the front left of the photograph. The harbour of the ancient Roman citadel 'Aenaria' was located in the stretch of water between the Castello and the cliffs near St. Anna rocks. Aenaria was destroyed and submerged in two different phases (Benini & Gialanella 2017a,b), suggesting that the surrounding coastline and current vent areas were not submerged until AD 140 (Photo: Dr Maria Cristina Gambi).

Ecological observations around Ischia were scarce before the beginning of research at Villa Dohrn in the early 1970s. Early studies of the rocky shores of the Castello area represent a useful baseline of biodiversity in the region, including the areas around the Castello prior to specific studies at the vents related to OA (Hall-Spencer et al. 2008). This provides insight into the ecology and biodiversity associated with the vents and an understanding of the rate of change that the species may have experienced (Table 3).

Ecological studies around the Castello and observations of vent activity prior to OA research

Early biodiversity surveys around the coast of the Castello were focussed initially on sponges (Sarà 1959, Pulitzer-Finali 1970, Pulitzer-Finali & Pronzato 1976, see also Goodwin et al. 2013; Table 3 for list of recorded sponges) and sciophilious macroalgae (Boudouresque & Cinelli 1971). These studies did not document the exact location of where the samples were taken but are useful as a record of the occurrence of species nearby the vent zone (Table 3).

Invertebrates associated with the macroalgae or *Posidonia oceanica* seagrass beds, including polychaetes (Lanera & Gambi 1993, Gambi et al. 1997a, 1997b, 2000, Rouse & Gambi 1997), molluscs (Russo et al. 1983) and amphipods (Scipione 1999), were collected in the south side near areas that are now sites of intense venting and are recorded in the species checklist (Table 3).

Giraud et al. (1979) investigated *P. oceanica* meadows on the south side of the Castello where they inhabited the shallowest depths (0.5–1 m) with leaves often exposed at low tide (M.C. Gambi, pers. obs.). The meadows form a reef structure with a compacted 'matte' (thick layer of organic matter and sediment deposits) and density of up to 900 shoots per square metre (Giraud et al. 1979, Buia et al. 2003). Researchers did not note strong vent activity in the south side of the Castello; however, some modest bubbling was observed by several scientists at that time (F. Cinelli, J. Ott, G.F. Russo, M.C. Gambi, pers. comm.). Russo et al. (1983) sampled molluscs associated with the *P. oceanica* meadows on the south side, bordering the area of current CO_2 bubbling, and noted that the bubbling was relatively modest in intensity and distribution at the time (1981–1982; G.F. Russo, pers. comm.).

Today, in the shallowest area where *P. oceanica* occurs (0.5–0.8 m depth), the CO_2 bubbling is very intense ($pH_T \sim 6.8$), and the matte is completely dead with no living plants (Figure 6). The *P. oceanica* adjacent to this area ($pH_T \sim 7.8$), located at ~ 1–1.5 m depth, however, is dense with more than 1000 shoots per square metre (Buia et al. 2003, Hall-Spencer et al. 2008, Donnarumma et al. 2014, Garrard et al. 2014). Thus, although venting in the Castello system could have occurred for hundreds of years, at the site level, venting has changed locally, appearing and disappearing over time as also evident in recent observations.

The sheaths of the *P. oceanica* shoots around Ischia house polychaetes and isopod borers, regardless of pH (Gambi et al. 1997b, Gambi & Cafiero 2001). Monthly collections on the south side of the Castello (1997–1998) showed that isopod borers were present mainly at 1 m depth, where modest bubbling was present at that time, while polychaetes occurred mainly at 3 m depth

Figure 6 A view of the shallowest part of the vent system on the south side of the Castello. Living plants were still present in this area in the early 1980s; however, in 2006, the zone was characterised by a thick dead 'matte' of the seagrass *Posidonia oceanica* (Gambi M.C. pers. obs.). The progressive increase in the vent activity and associated acidification from the 1980s appears to have had effects on resident biota (Photo: Enric Ballesteros, June 2016).

(Gambi et al. 1997b, Gambi & Cafiero 2001). Further observations in 2006 at the same site noted an increase in venting activity (close to S3, mean pH_T 6.8). A disappearance in both the isopod borers and polychaetes, especially at 1 m depth, was also noted, although it is unknown whether this can be attributed to a change in pH (M.C. Gambi & M. Lorenti, pers. obs.).

Further evidence that vent activity on the south side increased in the mid-1990s was provided when gastropods with corroded shells were observed in September 1997 (e.g. *Gibbula* spp., *Hexaplex trunculus*) (M. Taviani & M.C. Gambi, pers. obs.). It was suspected that the bubbling was causing acidification (Buia et al. 2003), and pH_{NBS} values of 7.76–7.98 were recorded in seawater sampled in September 1997 (M.C. Gambi, unpublished data). Thus, on the south side of the Castello, and particularly at the shallow *P. oceanica* meadow (1–2 m depth), modest venting was noted since the 1970s and 1980s. The intensity of the bubbling increased during the 1990s, and by 2006, the pH_T was 6.57 (Hall-Spencer et al. 2008). Thus, the biota living at the southern vent areas have experienced venting, albeit with some variation, for a minimum of 40 years, if not for hundreds of years since the submersion of the area reconstructed from geological and archaeological records (Benini & Gialanella 2017a,b).

Venting is more recent on the north side of the Castello. The absence of bubbling on the north side up until the mid-1980s was confirmed by scientists working in this area (Ott 1980, Mazzella et al. 1981, F. Cinelli., C.-F. Boudoursque, R. Pronzato, G.F. Russo, pers. comm., M.C. Gambi, pers. obs.). The first reports of relatively high venting at this site was in 2006 (Hall-Spencer et al. 2008). Thus, the biota on the north side have experienced venting for at least the past 10 years and a possible maximum of 30 years.

Since 2016, CO_2 vent activity has visibly increased in both sides of the Castello. A site in the north that was used as a control (i.e. no CO_2 vents) for the experiments on the rocky reefs described in Kroeker et al. (2011, 2013b,c) is now characterised by modest bubbling (M.C. Gambi & K.J. Kroeker, pers. obs.).

The progressive increase in vent activity and associated acidification since the 1980s appears to have had effects on resident biota. The shallowest *P. oceanica* meadow (reef bed), which flourished in the 1980s, was a dead zone in 2006. The dead matte on the south side is now colonised during the summer by a dense cover of macroalgae (mainly *Dictyota* spp.) (Figure 6) which is grazed by juveniles of the herbivorous fish *Sarpa salpa* (M.C. Gambi, pers. obs.). Another possible consequence of the increase in CO_2 is the disappearance of *P. oceanica* borers (e.g. polychaetes, isopods). Furthermore, the brown macroalga *Sargassum vulgare* which was limited to a few large specimens in the 1980s on the south side (M.C. Buia & M.C. Gambi, pers. obs.), is now frequently cited as one of the most abundant algal species in the most acidified portion of the rocky reef on the south side (Porzio et al. 2011). Recently, the alga shows signs of decline due to increased vent activity (M.C. Gambi, pers. obs.).

Effects of the CO_2 vents on biodiversity and influence on species composition

Across the Castello vent systems at Ischia, numerous studies have compared the species that occur at vent sites with those that occur at nearby non-vent sites (Table 1). Hall-Spencer et al. (2008) performed the first detailed characterisation of the biodiversity at the vents, noting a paucity of calcifying species as pH decreased. In contrast, nearby control areas were richer in biodiversity (e.g. 69 macrofaunal species in ambient pH_T 8.14 *versus* 11 under extreme low pH of 6.57). As the pH decreased, the diversity of species on the rocky shore shifted to non-calcifying species, with reductions and disappearance of the calcifiers such as barnacles (e.g. *Chthamalus stellatus*) and coralline algae (Hall-Spencer et al. 2008). Calcifying species, which are susceptible to skeletal dissolution in OA conditions, are reduced in abundance and diversity compared to controls at other vent sites around the world (Fabricius et al. 2011, 2015, Goffredo et al. 2014, Vizzini et al. 2017).

Several macroalgae, seagrass, foraminifera, coral, polychaete, crustacean, mollusc and bryozoan species that are tolerant of moderate acidification (mean pH~7.8) have been identified at the Ischia system, although their abundance is reduced to 30% with respect to controls (Hall-Spencer et al. 2008, Table 1). At sites with pH levels that commensurate with far future OA projections (pH < 7.8), crustose coralline algae (CCA), calcifying foraminifera and sea urchins are absent (Hall-Spencer & Rodolfo-Metalpa 2012). Along the pH gradient from 8.08–6.8, the number of taxa and biomass decreases, although the number of individuals do not differ due to a higher number of small-bodied, tolerant species at the low-pH sites (Kroeker et al. 2011). It appears that the biomass and density at the low-pH sites may be maintained by an increase in the abundance of small, acidification-tolerant taxa. Dwarfing has been seen in several gastropod species (*Tritia corniculum*, previously known as *Nassarius corniculus*, and *Tritia neritea*, previously known as *Cyclope neritea*) at the CO_2 seeps in Vulcano (Garilli et al. 2015, Harvey et al. 2016), and also in long-term laboratory studies (Dworjanyn and Byrne, 2018).

As the ocean continues to decrease in pH, studies at the Ischia and other CO_2 vent sites indicate there will be an increase in the dominance of generalists and small-bodied species resulting in a simplification of food webs and trophic complexity (Vizzini et al. 2017). This, for instance, is evident for the polychaete assemblages on rocky substrate at the vents (Gambi et al. 2016a). Similarly, in Papua New Guinea, reductions in coral diversity as well as changes in species interactions were observed with reductions in pH (Fabricius et al. 2011). There was also a 3-fold decrease in zooplankton biomass at the Papua New Guinea seeps with an average pH_T of 7.8, in comparison to ambient nearby sites with a pH_T of 8.0 (Smith et al. 2016). In CO_2 vents at Columbretes Islands in Spain, coralligenous outcrops and rhodolith beds, usually dominant at pH 8.1, are replaced by kelp forests (*Laminaria rodriguezii*) at pH 7.9 (Linares et al. 2015).

The lower biodiversity observed at the low-pH sites can be linked with indirect effects of OA due to loss of some habitat formers (e.g. corals) (Gaylord et al. 2015, Sunday et al. 2017). In Papua New Guinea, the loss of structurally complex corals (e.g. *Acropora* species) at low-pH sites resulted in a decrease in habitat complexity and the loss of reef-associated macro invertebrates such as crustaceans and crinoids (Fabricius et al. 2014). Replacement of branching corals by boulder-shaped corals at CO_2 seeps contribute to the reduction in zooplankton biomass observed in Papua New Guinea (Smith et al. 2016). Similar simplification of coral reef structure was found at low-pH vent systems of 6.7–7.3 in Puerto Morelos, Mexico in the Caribbean (Crook et al. 2012).

Interestingly, the shifts in habitat caused by OA can favour increases in abundance of some species. At White Island, New Zealand, reefs changed from a mixture of kelp, urchin barrens and turf algae to one dominated by turf algae at the low-pH site (pH 7.86) (Nagelkerken et al. 2016). At seeps in Vulcano, there was a shift from seagrass-dominated habitat to one dominated by macroalgae and sand at the low-pH site (pH 7.76) (Nagelkerken et al. 2016). Fishes preferred the turf and sand habitats. The increase in fish abundance found at both the White Island and Vulcano seep sites was explained by their significant positive association with turf and sand rather than preference for low pH (Nagelkerken et al. 2016).

In Ischia, the CO_2 vents have been important sites for discovering new species, including an acoel worm (*Philactinoposthia ischiae*; Nilsson et al. 2011), two fabriciid polychaetes (*Brifacia aragonensis, Parafabricia mazzellae*; Giangrande et al. 2014) and a *Posidonia oceanica* boring isopod (*Limnoria mazzellae*; Cookson & Lorenti 2001). Some taxa, which are rare in other areas of the Mediterranean, have also been observed nearby the Castello vents, including the ectoproct *Loxosomella pes* (previously known as *Loxosoma pes*) (Nielsen 2008), the polychaete *Platynereis massiliensis* (Lucey et al. 2015, Valvassori et al. 2015, Wäge et al. 2017) and the amphipod *Microdeutopus sporadhi* (Scipione 2013). The latter species is tolerant of low-pH conditions, but it is not known if it is a vent specialist (Scipione 2013). Thus, the vents can support a somewhat novel biodiversity.

As seen in other vent systems (Hall-Spencer & Allen 2015), the CO_2 vent habitat at Ischia favours the settlement and success of introduced species (Gambi et al. 2016b). Among the 22 introduced species that have been documented around the coast of Ischia, eight of these have been recorded at the vents. These include a toxic dinoflagellate *Ostreopsis ovata* (Di Cioccio et al. 2014), two macroalgal species (*Caulerpa cylindracea* and *Asparagopsis taxiformis*), polychaetes (*Lysidice collaris, Branchiomma boholense,* and *Novafabricia infratorquata*) and crustaceans (*Mesanthura* cf. *romulea* and *Percnon gibbesi*) (Table 3, Gambi et al. 2016b). These species, however, are as abundant in the vents as they are in other surrounding ambient zones, except for *Novafabricia infratorquata*, which has been recorded only in the vent zone. A biodiversity checklist for the Ischian vent system provides a useful baseline to evaluate further changes in the species and communities inhabiting the system as venting intensity changes (Table 3).

Studies of settlement and recruitment at the CO_2 vents

The difference in the abundance of many species, particularly calcifiers at the low-pH sites at Ischia, could be due to the effect of low pH on larval settlement or post-settlement success (Cigliano et al. 2010). It is important to understand why calcareous species are not successful at low pH with respect to bottlenecks across life-history stages. There are three main approaches to analyse settlement success: (1) artificial settlement collectors placed in the water column or attached to the benthos (Cigliano et al. 2010, Ricevuto et al. 2012, 2014); (2) settlement tiles attached to rocky walls (Kroeker et al. 2013c, Porzio et al. 2013, Kamenos et al. 2016) and (3) clearance of substrata and inspection of recolonisation (Kroeker et al. 2013b).

Cigliano et al. (2010) placed artificial collectors at three north (pH_T means N1: 8.09, N2: 7.72, N3: 7.39) and three south (S1: 8.12, S2: 7.62, S3: 7.33) vent sites, quantifying settlement of organisms after one month. Calcifying taxa such as foraminiferans, polychaetes and molluscs exhibited significantly less recruitment at reduced pH vent sites in comparison with ambient control pH sites. The polychaete *Syllis prolifera* was most abundant at the lowest pH station (min–max pH_T 7.08–7.79). Calcifiers were restricted to the control pH stations (pH_T 8.09–8.15), including foraminiferan, gastropod and most bivalve species (Cigliano et al. 2010). At the intermediate sites (pH_T 7.41–7.99), the polychaetes *Amphiglena mediterranea* and *Platynereis dumerilii*, the isopod *Chondrochelia savignyi* and the amphipod *Caprella acanthifera* were most abundant. In contrast, more heavily calcified species were absent.

Ricevuto et al. (2012) analysed the settlement and colonisation of artificial collectors along the pH gradient at the vents, using previously characterised sites N1–3, S1–3 (Figure 2). Of the 10,927 individuals collected, the main taxa that had settled and survived were polychaetes, amphipods, isopods, bivalves and gastropods. Comparisons of recruitment between control and low-pH stations showed that gastropod species decreased in number and abundance with decreasing pH, while some species of polychaetes, amphipods, tanaids and isopods exhibited similar abundance across all pH levels. Juveniles of *Mytilus galloprovincialis*, however, were still abundant at extreme low pH.

Ricevuto et al. (2014) attached artificial collectors to the substratum along the vent sites and monitored colonisation of polychaetes after 30 days in different months (March, May–June, October). Three polychaete species (*Amphiglena mediterranea, Platynereis dumerilii* and *Syllis prolifera*) were present at all pH sites across all seasons and were most abundant at the lowest pH (mean pH_T 6.9) conditions. *Syllis prolifera* and *Platynereis dumerilii* are often associated with disturbed areas (Bellan et al. 1988, Musco et al. 2009), while all three species thrive in algal-dominated habitats (Giangrande 1988). Both factors are likely to have contributed to their abundance at the low-pH site.

Kroeker et al. (2013c) placed recruitment tiles across the pH gradients at the Castello vents and monitored them for 14 months. Within 3.5 months, calcareous species, including barnacles, limpets and CCA, recruited and grew at similar rates in both ambient (mean pH_T of S1: 8.06 and N1: 7.95) and low (mean pH_T of S2: 7.75 and N2: 7.77) conditions. On the other hand, at the extremely low-pH sites, poor settlement of calcareous species indicated physiological intolerance at an early life-history stage. The use of settlement tiles in parallel across all sites controlled for substrate complexity and colonisation history, allowing tracking of how biodiversity changed over time. After 3.5 months, calcareous species were overgrown by fleshy seaweeds in the low-pH conditions. Thus, differences in growth rates and resultant changes in competitive dynamics between fleshy seaweeds and calcareous species amplified the direct effects of acidification (Kroeker et al. 2013c). The results of this study contrast with laboratory studies indicating that recruitment of calcareous species is limited at the same low-pH levels (Kuffner et al. 2007, Russell et al. 2009). Interestingly, some of the species recruiting to the settlement tiles calcify as adults but not as larvae or spores (e.g. CCA and serpulid polychaetes). Thus, these species may be able to recruit to the vent sites and develop, and at least are initially tolerant of low pH. Similarly, the non-calcifying larvae of sea stars and early recruits can tolerate low pH (Nguyen et al. 2012, Nguyen & Byrne 2014). That said, while coral larvae are robust to low pH, the early calcifying juveniles succumb to this stressor, and acidification also impairs the settlement signal for the larvae (Anlauf et al. 2011, Doropoulos et al. 2012). At the vent sites, pH fluctuations over large ranges may create a respite opportunity for larvae to settle while the pH is closer to ambient (Kroeker et al. 2013c).

A further study examined the size and magnesium (Mg^{++}) content of CCA (*Lithophyllum*, *Titanoderma* and *Phymatolithon* species) on the same tiles used in Kroeker et al. (2013c) for 14 months in control and low-pH vent sites (mean pH_T 8.1, 7.8, 6.9) (Kamenos et al. 2016). The largest CCA from the low pH 7.8 site was comparable in size to that of CCA in the control site (Kamenos et al. 2016). Furthermore, the calcite of the CCA grown at pH 8.1 and 7.8 had a similar Mg^{++} content, which is surprising as their high-magnesium calcite skeleton is the most soluble form of biogenic $CaCO_3$ in acidification conditions (Kamenos et al. 2013). At the pH 6.9 site, the CCA were significantly smaller and exhibited a change in mineralogy with a shift from a high-magnesium calcite skeleton to hydrated calcium sulphate (gypsum) (Kamenos et al. 2016). As gypsum contains no Mg^{++}, this skeletal mineral is considered more stable in low pH. This change in skeletal mineralogy is similar to laboratory studies of the mussel *Mytilus edulis*, which did not produce aragonite when exposed to low pH (1000 ppm, Fitzer et al. 2014).

At CO_2 vent seeps in Papua New Guinea, CCA was absent on settlement tiles deployed for 13 months at pH levels ~7.8 (Fabricius et al. 2015). The different findings at this site for CCA compared with that at the Ischian vents may be due to differences in grazing pressure and light level, both of which were higher on settlement tiles in Papua New Guinea (Fabricius et al. 2015). Differences in seawater carbonate saturation levels could also contribute to the different responses. The Papua New Guinea vent sites have higher calcite and aragonite saturation levels than the vent sites in Ischia (seawater chemistry tables from Fabricius et al. 2011, Kroeker et al. 2013c). Determination of the magnesium-calcite levels in seawater, however, would be most relevant to CCA (Nash & Adey 2017) and was not done for either study.

Porzio et al. (2013) placed settlement tiles in three different pH zones at the north and south vent sites (control 8.06, medium pH_T 7.75 and low pH_T 6.59 for north sites and control 7.95, medium pH_T 7.77 and low pH_T 7.2 for the south sites) and documented the settlement of algal species. After four months, the number of species recorded on tiles was 4% lower in the medium pH and 18% lower in the low-pH zone compared to controls. Calcareous algae, such as *Titanoderma mediterraneum*, *Hydrolithon* sp. and *Corallina* sp., dominated at normal and medium pH sites, indicating that calcifying algae can persist in pH levels predicted for the end of the century. The low-pH site assemblages were dominated by noncalcifying macroalgae and turf species, primarily *Feldmannia* sp. (Porzio et al. 2013).

Kroeker et al. (2013b) investigated recruitment following clearance of substrata, in ambient (pH_T 8), low (pH_T 7.8) and extremely low pH (pH_T 7.2) sites at Ischia. After 32 months at control pH levels, recruitment and success of species was highly variable, resulting in diverse species assemblages. At the low-pH sites, community diversity was low, and algae dominated (Kroeker et al. 2013b). While there were no differences in the size and abundance of the sea urchins *Paracentrotus lividus* and *Arbacia lixula* at the low-pH zone compared to the control zones, lower feeding rates were observed leading to a reduction in the number of feeding halos (Kroeker et al. 2013b). This reduced grazing is suggested to have facilitated recovery rates and dominance of fleshy algal species at the low-pH site. Thus, altered species interactions appear to drive the differences seen in species assemblages (Kroeker et al. 2013b).

Some general patterns can be summarised from settlement experiments at the Castello. The results from deployments of artificial collectors and settlement tiles showed that most calcifying species are absent in extreme low and low-pH areas. Biodiversity is greatly reduced at pH levels <7.8. Settlement and recruitment dynamics of many species vary over time, in parallel with the variability in the intensity of CO_2 venting activity. This result is also likely to be influenced by variable predation and competitive dynamics as well as the open nature of the vent systems and surrounding mobile species and propagules.

The assemblages observed on the tiles and thereby the vent benthos are, therefore, a result of complex interactions between species which change as the pH is reduced (Kroeker et al. 2013c). In CO_2 vents at Methana, Greece, although sea urchins occurred in lower density at pH_{NBS} 7.7, the same levels of macroalgal biomass was maintained as compared to ambient pH levels due to herbivorous fish replacing the roles of sea urchins, thus maintaining functional redundancy (Baggini et al. 2015). Different species of macroalgae were more abundant at different times in the year at the Methana CO_2 seeps. For example, *Cystoseira corniculata* was more abundant in autumn, whereas *Sargassum vulgare* was more abundant in spring. Thus, seasonal patterns can also alter the resulting assemblages and community response to local decreases in pH (Baggini et al. 2014).

As discussed in Gaylord et al. (2015), ocean acidification can offer useful insights into ecological processes in a low-pH environment when interpreted through the lens of some of the basic ecological theories and hypotheses. The settlement and recruitment studies performed at the Castello vents highlight some of the ecological mechanisms underlying assemblage structure associated with species interactions which change as a direct or indirect effect of reduced pH (Kroeker et al. 2013a,b). The pattern of community succession on the tiles was driven by altered competitive interactions between CCA and fleshy algae, rather than direct negative effects of OA on the CCA. These experiments along the pH gradients are important in understanding the direct and indirect effects of decreased pH in shaping the structure and functioning of benthic communities. The clearing experiments showed that decreased pH-driven reduction in sea urchin grazing modulated habitat patchiness and heterogeneity and at the seascape level from low- to high-pH conditions. These insights form the basis to generate hypotheses and focussed questions on the effects of decreased pH on marine habitats and to design further manipulative experiments in the field and under laboratory or mesocosm conditions.

Effects of CO_2 vents on *Posidonia oceanica* habitat

The shallow, sandy bottom around the Castello supports the shallowest and most dense meadows of the seagrass *Posidonia oceanica* around the island of Ischia (Giraud et al. 1979, Buia et al. 2003). There is a clear gradient of shoot density in this seagrass meadow along the pH gradient (Hall-Spencer et al. 2008, Donnarumma et al. 2014, Garrard et al. 2014), with the highest density in the extreme low-pH area (approx. 1000 shoots per m^2) compared with the control sites (<500 shoots per m^2). Despite the greater density and coverage of seagrass in extreme low-pH vent areas, daily leaf elongation rates were lower than that of seagrass growing at ambient pH (Buia et al. 2009). Similar increases in seagrass abundance occur at CO_2 seeps in Papua New Guinea (Russell et al. 2013). The stimulation in production and biomass could be tightly linked with inorganic carbon physiology of the plant. Species capable of using both HCO_3^- and CO_2 have been shown to increase in abundance in low pH in the laboratory (Cornwall et al. 2017); however, this has not been determined for *P. oceanica*.

Posidonia oceanica leaves at the vent sites have a higher N content (lower C:N ratio) compared with those growing at control sites, and so have a higher nutritional value for herbivores (Ricevuto et al. 2015b, Scartazza et al. 2017). This may explain why *P. oceanica* is much more heavily grazed inside the vents by the herbivorous fish *Sarpa salpa* than outside the vents (Donnarumma et al. 2014). Additionally, increased CO_2 reduces the ability of plants to produce protective chemicals (phenolics), a change that may also contribute to their greater vulnerability to herbivory (Arnold et al. 2012).

The expression of 35 antioxidant and stress-related genes of *P. oceanica* from the vent sites in Ischia, and another Italian vent site in Panarea (Aeolian Archipelago, Sicily), exhibited differences (Lauritano et al. 2015). For both vent sites, approximately 50% of the genes analysed showed significant changes in expression in comparison to control seagrass from local control sites, with a significant down-regulation in metal detoxification genes. This indicates that *P. oceanica* living at the vents had potentially impaired metal detoxification processes. There was an up-regulation of free radical detoxification and heat shock protein genes for *P. oceanica* but only at the Panarea vent sites (Lauritano et al. 2015). The differences in the stress responses of the same species at different vent sites show that inherent differences between vent sites are important to consider (Vizzini et al. 2013).

Effects on seagrass associated biodiversity

Posidonia oceanica meadows host a diverse range of species (Mazzella et al. 1992), and many studies have investigated differences in the assemblages living in association with this seagrass at vent and non-vent sites. Martin et al. (2008) investigated algal epiphytes and found that coralline algae was the most dominant contributor to calcium carbonate mass on *P. oceanica* growing at pH_T 8.1 but was absent from the seagrass growing at pH_T 7.7, as also noted by Hall-Spencer et al. (2008). A few species of bryozoan (e.g. *Electra posidoniae*) were the only calcifiers present on seagrass blades at low-pH levels (Martin et al. 2008). This was likely due to their skeletal mineralogy, which has a low Mg^{++} content, making them less susceptible to chemical dissolution in OA-like conditions (Smith 2014). Overall, there was a 90% reduction in epiphytic calcium carbonate production in seagrass at the pH 7.7 sites compared with those at control sites (Martin et al. 2008). Therefore, while seagrass can thrive at low-pH levels, there are significant effects of growing in low-pH conditions on epiphyte communities, with a significant reduction in calcifying species.

Analysis of the settlement pattern of biota on *P. oceanica* mimics showed a reduction or disappearance of calcifying forms. Calcareous invertebrates (e.g. barnacles and bryozoans) and CCA were abundant at control pH levels but were not present at the lowest pH (S3: 7.33, N3: 7.39) vent sites (Donnarumma et al. 2014). The biota that settled onto the leaf mimics at the low-pH site was dominated by non-calcareous taxa (e.g. hydroids and tunicates) and filamentous algae (Donnarumma et al. 2014). It appears that a mean pH of 7.8, a pH level expected by 2100 (IPCC 2014), may represent the limit for the successful recruitment of many seagrass epibionts that calcify (Donnarumma et al. 2014).

Seagrass epiphyte biodiversity differs on the north and south sides of the Castello. A study comparing epiphyte communities of *P. oceanica* found that the abundance of CCA decreased in low pH (S2: 7.8, N2:7.78) but only on the south side (Nogueira et al. 2017). This difference may be influenced by different hydrodynamics and mixing of the water by wind at the north and south sides of the Castello, thereby decreasing the resident time of the low-pH water (Donnarumma et al. 2014). This would be expected to influence the differences between the seagrass epiphyte communities found on either side. The south site experiences a greater fluctuation in pH and reaches a lower minimum pH level than the north (Figure 2), potentially breaching the pH limit of CCA. This may explain why the abundance of CCA is decreased in low pH only on the south side.

Behavioural experiments with polychaetes, gastropods, isopods, crustaceans and amphipods indicate that low pH (pH~7.8) affects chemoreception of the volatile chemicals produced by

P. oceanica, a factor that may be important in influencing the structure of seagrass epifaunal communities (Zupo et al. 2015). This was examined by extracting volatile organic compounds (VOCs) from *P. oceanica* and incorporating them into agarose blocks to create a VOC gradient. Most of the species (7 out of 13, 54%) exhibited a different chemotactic response in low-pH water compared to control water (Zupo et al. 2015). For the polychaete (eg. *Platynereis dumerilii* and *Psamathe fusca*, previously known as *Kefersteinia cirrhata*), low pH reduced their ability to detect the VOCs. On the other hand, the amphipod *Gammarella fucicola* showed a higher attraction to *Posidonia oceanica* at low pH (Zupo et al. 2015).

Garrard et al. (2014) compared the mobile invertebrate assemblages present within *P. oceanica* meadows at two pH zones on the north and south sides of the Castello (pH$_{NBS}$ 8.1 and 7.8). Interestingly, many invertebrate taxa (e.g. gastropods, amphipods, bivalves, isopods and polychaetes) were tolerant to a low pH of 7.8, with almost double the number of these taxa found at low pH in comparison to the controls. This included many species that calcify. This might have been due to positive indirect effects of lower pH, as the higher density of the seagrass shoots at the low-pH site provides shelter from predation (Pages et al. 2012), as well as an increase in habitable space.

Importantly, the diffusion boundary layer (DBL) can produce conditions at the surface of an organism different to the rest of the seawater (Vogel 1994, Hurd & Pilditch 2011). For photosynthetic organisms such as *P. oceanica*, pH is usually higher due in the DBL during the day due to photosynthesis, which can provide a buffer for associated organisms from overlying lower pH seawater (Cornwall et al. 2014). Decreased pH and resultant increases in dissolved inorganic carbon (DIC) also stimulate plant production and biomass due to elevated bicarbonate ions and CO_2 (Cornwall et al. 2017), thus increasing food supply for herbivores and substrate for epibionts. In Vulcano CO_2 seeps of pH 7.71, the DBL around the seaweed *Padina pavonica* was not enough to act as a refugia for calcifying foraminifera (Pettit et al. 2015) or other invertebrate calcifiers (Vizzini et al. 2017). Thus, at far future pH levels, the DBL may not be sufficient to buffer calcifiers.

Effects of CO_2 vent sites on the abundance of calcifying and non-calcifying species

Many studies conducted at the Ischian vents have focussed on identifying the vulnerabilities of calcifying groups in contrast with non-calcifying groups when exposed to OA conditions (Table 2).

Flora

In a survey of 101 macroalgal species, most were present at low-pH$_T$ levels of 7.83 with only a 5% decrease in species richness at the low-pH site compared with controls (Porzio et al. 2011). The decrease was due to a disproportionate decrease in calcifying algae. Of the 20 species of calcareous algae found at ambient pH, 15 were present at the pH 7.83 sites with none at the extreme low-pH sites (pH 6.57). Interestingly, two species of CCA with high magnesium-calcite skeletons, *Hydrolithon cruciatum* and *Peyssonnelia squamaria*, increased in abundance at the pH 7.83 sites compared to controls. This was suggested to be due to a decrease in competition from other species less tolerant of low-pH conditions, or due to reduced predation pressure (Porzio et al. 2011). At pH 6.57, species richness decreased by 72%. At this pH, a few dominant macroalgal species, *Sargassum vulgare*, *Cladostephus spongiosus* and *Chondracanthus acicularis*, were present (Porzio et al. 2011). Similar results were seen in the CO_2 seeps at Papua New Guinea, where fleshy macroalgae and seagrasses were dominant at pH$_T$ 7.73–8.00 (Koch et al. 2013). Macroalgal diversity decreased with increasing CO_2 venting activity at seeps in Methana, Greece (Baggini et al. 2014). Laboratory studies showed enhanced productivity of turf-forming algae at pH 7.6, which can lead to shifts in species dominance from corals to seaweeds (Diaz-Pulido et al. 2010, Connell et al. 2013).

Fauna

The distribution, diversity and composition of foraminifera changes significantly at the Ischian vent sites as the pH decreases from control levels to extreme low pH (Dias et al. 2010). Reductions in the diversity of foraminiferans from 24 species to 4 was evident at the extreme low-pH_T site (8.14 vs 6.57). The assemblage at the vent sites changed from calcareous (e.g. *Triloculina, Pyrgo, Miliolinella, Quinqueloculina* and *Peneroplis*) to agglutinated species (e.g. *Trochammina, Miliammina, Textularia* and *Ammoglobigerina*). The tests of the latter species are composed of small pieces of sediment cemented together (Dias et al. 2010). Similar reductions in the abundance of foraminifera occurs at the CO_2 vent sites in Papua New Guinea, where calcifying species were absent at pH levels below 7.9 (Uthicke et al. 2013).

These results contrast with those from the deep-water (74–207 m) CO_2 vents in the northern Gulf of California, Mexico where calcifying, benthic foraminifera were found across all pH levels, down to the lowest pH levels of 7.55 (Pettit et al. 2013). As temperature is a major controller of foraminifera distribution (Murray 2006), and the shallow waters of the Ischian and Papua New Guinean vents can experience rapid increases in temperature in summer, it has been suggested that the stable cool water temperature environment in the Gulf supports a higher species diversity of foraminifera, despite the low-pH levels (Pettit et al. 2013). Thus, the simultaneous stress of low pH and high temperature may impair the ability of foraminifera to inhabit vent areas in warmer habitats. This would need to be investigated in multistressor pH-temperature experiments. A recent laboratory study found that *Ammonia* sp. were able to maintain normal calcification over a range of pH values down to 6.8 (Toyofuku et al. 2017). This was achieved through regulation of biochemical mechanisms and active outward proton pumping. These findings suggest that physiological adaptation to low pH is possible for some foraminiferan species. Foraminifera with diatom symbionts can be more resilient to low-pH conditions (pH 7.75) in laboratory experiments (Doo et al. 2014), although this was not evident at pH levels < 7.9 at the Papua New Guinea vents (Uthicke et al. 2013).

The number and abundance of polychaete species associated with the rocky vegetated reefs decreased in the extreme low-pH conditions (pH_T 6.6), including both calcifying and non-calcifying species with the disappearance of sessile polychaetes (Gambi et al. 2016a). The loss of sessile species could be because they are unable to respond as quickly as motile species, and that their ability to make their calcareous tube at low pH is impaired (Gambi et al. 2016a).

Sponges also decrease in diversity at the low-pH sites, with 11 species present at ambient controls (pH_T 8.13) and only one species (*Crambe crambe*) at the lowest pH sites (pH_T 6.68; Goodwin et al. 2013). An absence of predators and a high reproductive output may contribute to the ability of *C. crambe* to live at extremely low-pH levels.

The clearest effect of the low-pH vent sites on flora and fauna is the reduction and disappearance of calcifying species (e.g. molluscs, polychaete spirorbids, foraminifera) (Hall-Spencer et al. 2008, Martin et al. 2008, Cigliano et al. 2010, Porzio et al. 2011, Goodwin et al. 2013). Several species of algae appear to benefit from the CO_2 (e.g. *Sargassum vulgare, Cladostephus spongiosus, Chondracanthus acicularis*), potentially due to the disappearance of their main invertebrate consumers (e.g. grazing gastropods, sea urchins) at low pH (Porzio et al. 2011). However, the fact that some species with high magnesium-calcite skeletons (e.g. CCA species), a trait which would normally be considered extremely vulnerable to low pH, can persist in these extreme environments demonstrates that individuals in some species have substantial plasticity in their response to low pH.

Studies of species present in the vents

Investigation of species naturally present in the CO_2 vents provides insights into the potential acclimatisation and adaptive abilities of species to live with long-term exposure to low pH (Table 2).

Ochrophyta

A comparative transcriptomic study of *Sargassum vulgare* growing at vent and control sites found that transcripts involved with metabolic and cellular processes were differentially expressed in low-pH environments, with increased expression of genes involved in energy metabolism, photosynthesis and ion homeostasis (Kumar et al. 2017b). There was no change, however, in expression of genes involved in carbon acquisition processes, or antioxidant enzymes and heat shock proteins. These results suggest that *S. vulgare* has been able to survive low-pH vent conditions through an increase in energy production to maintain ion homeostasis and other cellular processes (Kumar et al. 2017b). *Sargassum* species appear to thrive in increased CO_2 conditions, as also observed in laboratory studies (Poore et al. 2016).

Cnidaria

The sea anemone *Anemonia viridis* displayed similar abundances of its symbiotic algal species (of the genus *Symbiodinium*) at low pH (7.0) and control pH sites, but the microbial communities differed. Changes in the microbial communities at low pH did not appear to affect the physiology and health of the sea anemones (Meron et al. 2012a). However, only four specimens were present at the low-pH site and limited to the north side. For two corals, *Balanophyllia europaea* and *Cladocora caespitosa*, low pH did not affect the associated microbial community (Meron et al. 2012b). In Papua New Guinea, the microbial communities in the corals *Acropora millepora* and *Porites cylindrica* significantly differed between control and low pH (Upa-Upasina, pH 7.81) vent sites (Morrow et al. 2015). For *Acropora millepora*, a 50% reduction in the symbiotic marine bacteria *Endozoicomonas* was seen in vent populations, a taxon that has been shown to be common in healthy corals (Mouchka et al. 2010).

Mollusca

Mollusc species that inhabit the low-pH vent areas at Ischia appear to have specific acclimatisation and/ or adaptive strategies to facilitate their survival. Langer et al. (2014) compared the shells of the limpet *Patella caerulea* from vent and control sites using Raman microscopy. Limpets living at low pH (mean pH 6.46) had a 2-fold increase in aragonite area of their shells and an increase in shell apical thickness. They appear to counteract low-pH conditions through enhanced shell production by depositing additional aragonitic layers on the inner side of the shell, although this would be expected to incur an energetic cost. This increase in aragonite in low-pH limpet shells is not due to preferential usage of aragonite over calcite but governed by shell formation mechanisms where calcite is usually only formed during shell elongation in limpets at ambient pH (Langer et al. 2014). Aragonite is more soluble than calcite in low-pH seawater, so aragonitic shells are considered more vulnerable to ocean acidification than calcitic ones (Carter 1980). At the Ischian vents, the limpets thickened their shells through additional deposition of aragonitic layers, indicating this mineral form is not necessarily disadvantageous under OA conditions. For *Mytilus edulis*, which normally produces both calcite and aragonite in the shell, juveniles did not produce aragonite when grown under high $p\mathrm{CO_2}$ due to the susceptibility of this biomineral to low pH (Fitzer et al. 2014), a finding that contrasts with the results found at the Ischian vents.

Rodolfo-Metalpa et al. (2010) transplanted several mollusc species including limpets and bivalves into the vent systems to determine the effect of low pH on survival (see section on 'Transplant experiments'). Interestingly, only one of these species is usually found within the reduced-pH vent sites, the limpet *Patella caerulea*. From this study, they concluded that the external organic periostracum cover over the shell of *Mytilus galloprovincialis* plays a major role in protecting the shell from dissolution (Rodolfo-Metalpa et al. 2010). *Patella caerulea*, however, lacks a periostracum and thus uses other means to survive pH as low as 6.8.

Polychaeta

Reproductive and life-history traits across several polychaete species from the Ischian vents indicate that this environment may select for species with parental care, as shown for the nereidid *Platynereis massiliensis* which is abundant at low-pH sites (Lucey et al. 2015). In contrast, its common free-spawning sibling species, *P. dumerilii*, is much rarer in the vent areas (Lucey et al. 2015, Wäge et al. 2017). Additionally, Gambi et al. (2016a) found that brooding polychaete species dominated the rocky reefs of the most acidified vent sites. Thus, brood protection of offspring may be an important trait facilitating persistence at the vent sites (Lucey et al. 2015, Gambi et al. 2016a).

A shift towards species that protect their offspring could be expected as the pH of the ocean decreases. Specific adaptations such as brooding allow species to endure low pH, as seen for other species which have evolved this life-history mode likely in response to living in harsh environments (e.g. *Cryptasterina* sea stars; Byrne et al. 2003, Puritz et al. 2012).

Use of the vents as a natural experimental system

The vent systems at the Castello Aragonese in Ischia have been used as a natural environment to study the effects of OA and to ground truth laboratory experiments and mesocosm studies which show that calcifiers are negatively impacted by decreased pH (e.g. abalone, Byrne et al. 2011; echinoids, Brothers et al. 2016, Carey et al. 2016; bivalves, Parker et al. 2013, Waldbusser et al. 2015). Transplant experiments have been used to document responses of calcifying species moved to the vent sites. Reciprocal transplant experiments allow the examination of specific acclimatisation strategies used to survive in low pH, as well as discerning effects of the transplant and effects of low pH. Studies that have conducted these experiments are summarised in Table 2.

Transplant experiments

Transplant experiments involve the relocation of species from control sites to vent sites (Table 2, Figure 4). Although these animals were not always sourced from the low-pH areas, these experiments provide insights that are not possible with those performed in the laboratory by incorporation of multiple factors (e.g. currents, water flow) that are difficult to replicate in the laboratory.

One-way transplants from control conditions to the vents

One-way transplant experiments involve sourcing of animals from sites with ambient seawater and placing them into low-pH vent and control-pH sites (not necessarily the original site the species were sourced from). However, most studies have this limitation, as only one study (Rodolfo-Metalpa et al. 2011) included procedural controls translocating within the source site to assess the potential for a disturbance response. This control treatment is required to assess the impact of transplant, an effect that has been shown to be observable and important (Hahn et al. 2012).

Several transplant experiments at the vent sites of Ischia have investigated the impacts of low pH (7.43, 7.66) on the bryozoan *Myriapora truncata* with a focus on growth and calcification (Rodolfo-Metalpa et al. 2010, Lombardi et al. 2011a, c). After 45 days, colonies transplanted to pH 7.66 were able to maintain the same calcification rates as observed in controls, but not at pH_T 7.43 where calcification was significantly decreased (Rodolfo-Metalpa et al. 2010). After 128 days, surviving colonies showed a significant loss of skeleton and lower levels of Mg^{++} in the skeleton (Lombardi et al. 2011c). During the second exposure to low pH, the bryozoans experienced a three-month, late summer heat wave (25–28°C compared to normal temperatures of 19–24°C). Calcification was impaired at both low pH and the control pH. It appeared that, for this species, warming breached stress tolerance thresholds (Rodolfo-Metalpa et al. 2010).

For another bryozoan, *Schizoporella errata*, transplanted from control (pH$_T$ 8.09) to a mean pH$_T$ of 7.76, growth of the zooidal basal and lateral walls was retarded, and there were fewer avicularia (defensive polymorphs) (Lombardi et al. 2011b). The smaller number of avicularia found in the bryozoans at the low-pH site suggests a change in resource allocation from defence to sustain growth (Lombardi et al. 2011b). Thus, this species of bryozoan could reallocate resources in response to OA, favouring growth over defence.

Transplants of the bryozoan *Calpensia nobilis* to vent sites (mean pH$_T$ 7.83) grew at half the rate of those transplanted to the control site. In those placed at the lower-pH site, the zooids had longer and wider primary openings. These colonies thus reallocated metabolic energy to strengthen existing zooids to enhance survival in low-pH conditions, a response that may provide sufficient plasticity to live in a changing ocean (Lombardi et al. 2015).

For bivalves, molluscs and corals, surface organic layers are essential to reduce or prevent skeletal dissolution in low-pH conditions. Comparison between shell dissolution rates of the bivalve *Mytilus galloprovincialis*, which has a periostracum covering its shell, to the gastropod *Patella caerulea*, which lacks a periostracum, as well as the solitary coral *Balanophyllia europaea*, which has a skeleton completely covered in tissue, to the colonial hermatypic coral *Cladocora caespitosa*, which has exposed skeletal parts (Rodolfo-Metalpa et al. 2011), show that these organic layers protect the skeleton at low-pH$_T$ sites of 7.4 and 7.7. However, when low pH was coupled with high temperatures due to a heat wave (+~8 °C), there was high mortality and huge decreases in calcification rates, demonstrating that ocean warming may be the more important stressor (Rodolfo-Metalpa et al. 2011).

Juvenile pen shells (*Pinna nobilis*) transplanted to control and vent sites for 45 days showed significant decreases in physiological responses (including growth, oxygen consumption and mineralisation) in pH$_{NBS}$ sites <7.6. The mortality rate increased from 10%–30% in ambient pH to 60%–70% at pH$_{NBS}$ 7.6 (Basso et al. 2015).

Studies of *Paracentrotus lividus* and *Arbacia lixula* translocated to the CO$_2$ vent sites in Vulcano for four days showed that they were able to regulate their acid–base homeostasis in pH$_{NBS}$~7.7 (Calosi et al. 2013a), parallel with observations in the laboratory for *Paracentrotus lividus* (Catarino et al. 2012). Whether similar results with the same species of sea urchin occur at Ischian vent systems remains to be seen. Female *A. lixula* inhabiting the Ischian vent systems show an adjusted egg phenotype which is resilient to low pH suggesting that specific adaptive strategies are needed to reside at the vents (Foo et al. 2018). At CO$_2$ seeps in Greece (Methana), *Arbacia lixula* and *Paracentrotus lividus* were differentially affected by acidified waters in comparison to ambient sites (Bray et al. 2014). This was evident in the trace element composition of the skeleton, which differed from that of sea urchins living at ambient pH. The vast increase in Mn^{++} (541% in *P. lividus*, 243% in *Arbacia lixula*) in the skeleton (Bray et al. 2014) indicated bioaccumulation of this element from surrounding vent water. There were also alterations in Zn^{++} and Sr^{++} levels in the skeleton (Bray et al. 2014).

Assessment of biochemical and metabolic functioning of the fanworm *Sabella spallanzanii* transplanted to low-pH vent sites for five days showed that in the short term, energy metabolism was increased, and carbonic anhydrase concentration was decreased (Turner et al. 2015). A further transplant experiment with *S. spallanzanii* found that after 30 days, the antioxidant system of the polychaete showed a significant decrease in enzymatic activities, with impairment in the ability to neutralize hydroxyl radicals (Ricevuto et al. 2016). Together, these two studies indicate the vulnerability of this polychaete to acidification, and why the species is only found at ambient-pH sites (Hall-Spencer et al. 2008, Calosi et al. 2013b).

For the macroalga *Sargassum vulgare*, transplantation from the extreme low-pH$_T$ site (6.7) to control sites (8.1) for three weeks induced a stress response (Porzio et al. 2017). Photochemical activity and expression of rubisco, an enzyme important for carbon fixation, decreased by 30%. Non-photochemical dissipation mechanisms increased by 50%, and photosynthetic pigment content increased by 40%, potentially compensating for the decrease in photochemical efficiency. It appears that the photosynthesis apparatus of *S. vulgare* living at the vent sites is adapted to the CO_2 levels found in the vent systems (Porzio et al. 2017).

Although the vent sites around the Castello Aragonese provide a natural system as a proxy to study the impacts of OA, care needs to be taken in interpreting the results of experiments. Live *Mytilus galloprovincialis* transplanted from control sites (pH_T 8.07) to vent sites (pH_T 7.25) exhibited the effect of transplantation (Hahn et al. 2012). After 68 days, the shells showed a strongly elevated change in carbon and oxygen isotope ratios, which was attributed to transplantation shock. These ratios can change depending on seawater DIC chemistry (Hahn et al. 2012). Comparisons between pre- and post-transplantation growth revealed a period of microstructural disarrangement in between periods of ordered growth, and this was attributed to the transplant process (Hahn et al. 2012).

With the exception of the study by Rodolfo-Metalpa et al. (2011), most of the transplant studies did not include a true control where animals were transplanted to the original site they were sourced from. Conducting reciprocal transplant experiments with control to control transplants are needed to address this problem to discern between the shock effects of handling and translocation and those caused by low pH. Furthermore, longer term studies (months to years) are needed to reduce the effect of the shock artefact as many studies report effects after a few days or weeks at the vents. Longer studies are needed to better reflect 'living in the future'.

Reciprocal transplants

Reciprocal transplant experiments involve the translocation of species from control conditions to low-pH vent conditions, and vice versa, as well as transplantation within each pH level (Table 2). Thus, these experiments consist of four (or more) treatments, that is, transplanting control populations to control conditions, control populations to vent conditions, vent populations to vent conditions and vent populations to control conditions. This fully crossed design is needed to disentangle the effects of habitat *versus* transplantation shock on response variables by the inclusion of multiple control treatments.

Transplants of the macroalga *Sargassum vulgare* from pH 8.2 to pH 6.7 sites for two weeks resulted in an increase in photosynthetic rates, metabolism and oxidative damage (Kumar et al. 2017a). For the reciprocal transplant where *S. vulgare* was moved from low pH to pH 8.2, the plants were able to maintain similar photosynthetic rates and oxidative activity with no oxidative damage compared to the procedural control transplants. It thus appears that populations at the vent sites are locally acclimatised (Kumar et al. 2017a). The results from this study contrast to those of Porzio et al. (2017), where a decrease in photosynthetic efficiency was observed when *S. vulgare* was transplanted from 6.7 to ambient pH. This reciprocal transplant study was longer (three weeks *versus* two weeks), a difference that is likely to have influenced the different outcome. Similarly, the macroalgae *Cystoseira compressa* and *Padina pavonica* transplanted to CO_2 vent sites in Vulcano exhibited an increase in photosynthesis at low pH_{NBS} 7.86 after four days. However, photosynthesis was more dependent on light and nutrient availability than on pH and carbonate conditions (Celis-Pla et al. 2015). Moreover, four days is likely to be too short for the plants to achieve an acclimatisation response.

Juvenile isopods (*Cymodoce truncata* and *Dynamene bicolor*, previously known as *Dynamene torelliae*) collected from ambient pH sites and transplanted for five days to both control and vent (mean pH 7.29) conditions maintained similar metabolism levels (Turner et al. 2016). For *Cymodoce truncata*, however, this came at a cost with a significant reduction in the amount of carbonic anhydrase produced, an enzyme essential in acid–base functioning. *Dynamene bifida* juveniles transplanted from control to low pH upregulated ATP production by 20%, indicating that they can adjust their metabolism in low-pH environments (Turner et al. 2016). This may also have indicated a stress response.

Calosi et al. (2013b) performed several short-term reciprocal transplant experiments with polychaetes living in and around the vents and measured metabolic rates after five days. Translocation of *Amphiglena mediterranea* from the CO_2 vents to control environments resulted in metabolic rates returning to a normal level, comparable to those normally inhabiting ambient conditions. This indicates *A. mediterranea* are acclimatised but not adapted to low pH. In contrast, *Platynereis dumerilii* transplanted from the CO_2 vents to control conditions exhibited an increase in metabolic rate of 44%. As transfer to ambient conditions induced a stress response, this could indicate that they were adapted to low pH. Elevated metabolism is likely to compensate for the low-pH environment to which they are normally subjected but could also reflect the short-term nature of the observations.

Ricevuto et al. (2015a) performed a reciprocal transplant experiment with three polychaete species, *Platynereis dumerilii*, *Polyophthalmus pictus* and *Syllis prolifera*. After 30 days, *Platynereis dumerilii* and *Syllis prolifera* placed in the vent sites (average pH_{NBS} 7.3) showed an enhanced ability to neutralise oxyradicals. On the other hand, *Polyophthalmus pictus* did not show the same response and was more sensitive to low pH (Ricevuto et al. 2015a).

A set of fitness traits was measured in a calcified, tubicolous spirorbid polychaete (*Simplaria* sp.) reciprocally transplanted into ambient and low-pH (pH_{NBS} 7.61) sites for 66 days (Lucey et al. 2016). Both groups of this species showed decreases in survival, reproductive output, maturation and population growth, regardless of whether they had originally resided in a control or low-pH environment. This suggests that local adaptation to the low-pH vent environment had not occurred.

Across six polychaete species, reciprocal transplant experiments indicate species-specific responses. Not all species show a capability to survive the low-pH conditions of the vents. For those already living at the vent sites, it appears that several species are acclimatised (e.g., *Amphiglena mediterranea*, *Simplaria* sp., *Syllis prolifera*) as indicated by their return to 'normal' functioning after transplantation to control conditions. Others show they have genetically or functionally adapted (e.g. *Platynereis massiliensis* vs *P. dumerilii*) as transplantation of the vent residents to control pH induced a stress response.

It will be important to repeat these studies to involve a longer residence time in the novel habitats (low and high pH). Short-term transfer results (e.g. days–weeks) can only provide a snapshot of the species' response and be potentially confounded by a transfer stress response. Longer term studies are needed to fully tease out the differences between shock responses and the capacity for adaptation and acclimatisation.

Insights from vent studies

A wealth of data from research at the Castello vents of Ischia and other vent systems show that many calcifying species are present at the sites with a mean pH∼7.8, including sea urchins, bryozoans, foraminifera, serpulid polychaetes, molluscs and corals (Hall-Spencer et al. 2008, Hall-Spencer & Rodolfo-Metalpa 2012, Meron et al. 2012b, Porzio et al. 2013, Donnarumma et al. 2014, Garrard et al. 2014). Below this pH at the Castello vents, there is a large reduction in calcifying species. Thus, as the ocean continues to decrease in pH, insights from the vents indicate that non-calcifying plants and animals or weak calcifiers will have the advantage (Cigliano et al. 2010, Kroeker et al. 2011, Ricevuto et al. 2012, Scipione 2013). Beyond pH levels <7.8, rocky reef habitats will likely be dominated by soft forms of algae and turf species, with a low community diversity.

Although some species are susceptible to low pH, growing at half the rate of conspecifics living in ambient conditions (e.g. *Calpensia nobilis*; Lombardi et al. 2015), they are able to persist despite negative effects on physiology. This is seen in the reallocation of resources to growth over other functions (e.g. bryozoan *Schizoporella errata* and several polychaetes species; Lombardi et al. 2011b, Ricevuto et al. 2015a). Thus, we can expect to see an increase in the abundance of smaller-bodied or slower-growing individuals in populations in a future ocean.

With respect to allocation of resources and energetic considerations, many vent species exhibit great plasticity in their responses to near future levels of low pH. Some 'winners' are able to maintain normal levels of calcification, growth and skeletal mineralogy in the face of low pH (Kamenos et al. 2016), acclimatising and genetically adapting to these conditions (Calosi et al. 2013b, Ricevuto et al. 2015a, Kumar et al. 2017a) and possibly becoming vent specialists (Nilsson et al. 2011, Scipione 2013, Giangrande et al. 2014).

Considerations for future experiments and caveats

Although CO_2 vents can provide key insights into the impacts of decreased pH on marine ecosystems, and are unique model systems to address questions on the potential impacts of OA on biological and ecological responses, as analogues of OA, there are caveats to their use:

1. *Global warming-ocean acidification mismatch*: Arguably one of the most important caveats to the use of vent systems as analogues for future OA is that they do not simultaneously incorporate the impacts of global warming (e.g. Hughes et al. 2017), as noted for species that occur at the vents of Ischia (Rodolfo-Metalpa et al. 2011). Along the vent pH gradients, there is a mismatch in the temperature-pH conditions with respect to those projected for the future. Under best and worse case scenarios (RCP 2.6, 8.5), an increase of 0.71 ± 0.45 up to 2.73 ± 0.72 °C by 2100 is predicted for the ocean, with more extreme changes for some shallow-water and coastal ecosystems (Gattuso et al. 2015), but the vents present future OA conditions at present-day temperatures. The few studies where a heat wave occurred during the experimental exposure of transplanted organisms showed strong negative synergistic effects of warming and OA (Rodolfo-Metalpa et al. 2010, 2011).

2. *Vents are open systems*: In comparison with the surrounding ocean and benthic ecosystems, CO_2 vent systems occur at very small spatial scales, much smaller than the foraging range of visiting fishes and the catchment source of the planktonic propagules characteristic of most marine invertebrates. Thus, the open nature of vent systems is a concern. Most associated mobile fauna do not spend their entire life cycle in the vents, limiting the potential for evolutionary adaptation to occur. The exception is the resident species that have the direct development/brooding life history (e.g. *Platynereis massiliensis, Simplaria sp.*, Lucey et al. 2015, 2016).

3. *Seawater chemistry*: This review focussed on the Ischian vent system of the Castello Aragonese, where gas emissions are almost 100% CO_2 with no toxic hydrogen sulphide gas (Tedesco 1996, Hall-Spencer et al. 2008, F. Italiano, G. Pecoraino & M.C. Gambi unpublished data). In other systems (e.g. White Island, PNG), the presence of H_2S is noted (Fabricius et al. 2011, Brinkman & Smith 2015) indicating that the seawater chemistry changes might not just be due to CO_2. For the Methana system in Greece, the massive bioaccumulation of a trace element (Mn^{++}) highly toxic to sea urchins (see Pinsino et al. 2010, 2011) in their skeleton (Bray et al. 2014) indicates that the vent water contains metals and trace elements. Thus, the negative effects reported for the biota at Methana (Baggini et al. 2014) may be due to changes in water chemistry in addition to local acidification. For Vulcano, sediments near the vents exceed sediment quality guidelines in their levels of trace elements of concern (e.g. As, Hg, Cu), creating a complex chemistry because low pH can increase their bioavailability (Vizzini et al. 2013). Resident seagrass and epiphytes are enriched for trace elements (e.g. Cd, Hg and Zn) (Vizzini et al. 2013). Seep volcanic systems often have metal and trace element input (Dekov & Savelli 2004), and so *a priori* investigation of trace element distribution and bioaccumulation is fundamental in selection of sites for use as analogues of OA.

4. *Variability in venting intensity*: The temporal variability in the carbonate chemistry of vent sites is much more pronounced than that for natural oceanic systems (Hofmann et al. 2011), and this may affect their utility. However, fluctuating pH levels are characteristic of some natural systems such as tide pools, coral reefs and areas of natural upwelling, and thus laboratory experiments with stable pH levels may not accurately reflect the natural effects of OA on the species or community (Kroeker et al. 2013c). Thus, vent systems provide an opportunity to look at the effects of flux on individual, population or community level responses in a way laboratory experiments that use constant pH cannot (Wahl et al. 2015). Fluctuating pH levels can influence community responses to acidification; however, it is not

understood whether these are driven by the minimum and maximum pH/pCO$_2$ reached, or if the rate of change is the greater determinant of biodiversity changes. What is essential is to be able to relate responses to flux through detailed documentation of carbonate chemistry with a fine time resolution series to link responses to the degree of acidification and how it changes.

The way forward and concluding remarks

The volcanic vent areas around the Castello Aragonese at Ischia are the first and among the best characterised and studied CO$_2$ vent sites in the world. They have provided important insights into the effects of long-term exposure of low pH on species and communities and are useful to complement laboratory and mesocosm OA studies. The gas at the Ischian vents has been well characterised over a long time period, confirming the dominance of CO$_2$ and lack of H$_2$S (Tedesco 1996, Hall-Spencer et al. 2008, F. Italiano, G. Pecoraino & M.C. Gambi, unpublished data). This feature makes this vent system a useful window to assess the diversity and ecosystem functionality of a future ocean.

Thus far, studies in Ischia have focussed on specific, well-characterised sites. Understanding whether species biodiversity is different in areas of overlap (i.e. the region of transition between different pH zones) as well as a complete characterisation of the pH gradients across the south and north vent sites will provide essential information on the pH limits of the species that occupy these areas. Transplantation experiments need appropriate procedural controls to fully assess translocation artefacts. Longer duration experiments will also be useful to discern between shock responses and acclimatisation. In parallel with field experiments, these experiments in conjunction with full characterisation of environmental conditions over time and in combination with 'omics' approaches will allow identification of genetic responses, and how acclimatisation, adaptation or epigenetics contribute to the ability to live in a reduced-pH ocean (Foo & Byrne 2016).

Consideration of other factors such as light and nutrients is also important to determine if these can modulate, attenuate or even reduce effects of OA on marine species. Teasing out the impacts and interactions of these multiple factors will require controlled laboratory experiments to be conducted in parallel with these field experiments.

In general, there are similar trends in species and ecosystem responses across different volcanic vent systems and reflect those observed in Ischia. However, differences in seawater carbonate chemistry, trace elements and hydrodynamic factors are likely to influence the discrepancies that have been observed in the responses of biodiversity. Investigations of the species that occur at the low-pH sites at the Castello Aragonese in Ischia show that, as the pH decreases, there is a reduction in calcifying species. This reflects the trends seen at other vent systems and in laboratory studies. At the other end of the low-pH bottleneck our marine ecosystems are passing through, there will be a reduction in species diversity, with structural and functional simplification of communities. Research at vent systems indicate that a resilient array of species will prevail resulting in different ecosystem functionality in a future ocean.

Acknowledgements

We wish to thank the archaeologists, Alessandra Benini and Costanza Gialanella, for information on the dating of the remains of the ancient Roman harbour of Aenaria. We also thank Maria Beatrice Scipione for checking the taxonomy of the amphipods of an earlier version of Table 3. Thanks to all colleagues who offered their personal observations on the Castello areas previous to the high-venting activity and OA studies. We also thank the referees for their thorough and useful comments which have greatly improved the review. Research was supported by an Australian Endeavour Research Fellowship (SAF) and the Stazione Zoologica Anton Dohrn of Napoli.

Table 3 A taxonomic checklist of the marine species recorded in the CO_2 vent sites of the Castello Aragonese, Ischia, as well as nearby ambient zones (approximately within 500 m range)

Genus	Species	Taxonomic authority	Site: Control (C) ambient pH, Intermediate (I) low pH, Acidified (A) extreme low pH	Literature citation
Phylum Cyanophyta				
Calothrix	sp.		C	Mazzella et al. 1981
Lyngnya	sp.		C	Mazzella et al. 1981
Phylum Dynophyta				
Ostreopsis	*ovata*	Fukuyo, 1981	C, I, A	Di Cioccio et al. 2014, Gambi et al. 2016b
Phylum Ochrophyta				
Class				
Bacillariophyceae				
Amphora	*marina*	W. Smith, 1857	C	Mazzella et al. 1981
Cocconeis	*britannica*	Naeg. Ex Kütz. 1849	C	Mazzella et al. 1981
Cocconeis	*scutellum*	Ehrenberg, 1838	C	Mazzella et al. 1981
Coscinodiscus	sp.		C	Mazzella et al. 1981, Gambi et al. 2000
Cyclotella	sp.		C	Gambi et al. 2000
Diploneis	sp.		C	Gambi et al. 2000
Entomoneis	sp.		C	Gambi et al. 2000
Fragilaria	sp.		C	Mazzella et al. 1981
Navicula	*ramosissima*	C. Agardh Cleve, 1895	C	Mazzella et al. 1981
Nitzschia	*vidovichii*	(Grunow) Grunow in Van Heurck, 1881	C	Mazzella et al. 1981
Podocystis	*adriatica*	(Kützing) Ralfs in Pritchard, 1861	C	Gambi et al. 2000
Striatella	sp.		C	Mazzella et al. 1981, Gambi et al. 2000
Class Phaeophyceae				
Cladosiphon	*cylindricus*	(Sauvageau) Kylin, 1940	C	Gambi et al. 2000
Cladostephus	*spongiosum f. verticillatum*	(Lightfoot) Prud'Homme van Reine, 1972	C, I, A	Hall-Spencer et al. 2008, Porzio et al. 2011
Colpomenia	*sinuosa*	(Mertens ex Roth) Derbès & Solier in Castagne, 1851	C, I, A	Hall-Spencer et al. 2008
Compsonema	sp.		C	Boudouresque & Cinelli 1971
Cutleria	*adspersa*	Mertens ex Roth De Notaris, 1842	C	Boudouresque & Cinelli 1971
Cutleria	sp.		C, I, A	Hall-Spencer et al. 2008
Cystoseira	*amentacea v. amentacea*	(C. Agardh) Bory de Saint-Vincent, 1832	C, I	Hall-Spencer et al. 2008

Continued

Table 3 (Continued) A taxonomic checklist of the marine species recorded in the CO_2 vent sites of the Castello Aragonese, Ischia, as well as nearby ambient zones (approximately within 500 m range)

Genus	Species	Taxonomic authority	Site: Control (C) ambient pH, Intermediate (I) low pH, Acidified (A) extreme low pH	Literature citation
Cystoseira	*amentacea* var. *stricta*	Montagne, 1846	C	Porzio et al. 2011
Cystoseira	*compressa f. compressa*	(Esper) Gerloff et Nizamuddin, 1975	C, I	Hall-Spencer et al. 2008
Cystoseira	*crinita*	Duby, 1830	C	Gambi et al. 2000, Gambi et al. 2003
Dictyopteris	*polypodioides*	(A.P. De Candolle) J.V. Lamouroux, 1809	C, I	Boudouresque & Cinelli 1971, Hall-Spencer et al. 2008, Porzio et al. 2011
Dictyota	*dichotoma v. dichotoma*	(Hudson) J.V. Lamouroux, 1809	C, I, A	Boudouresque & Cinelli 1971, Hall-Spencer et al. 2008, Porzio et al. 2011
Dictyota	*dichotoma var. intricata*	(C. Agardh) Greville, 1830	A	Porzio et al. 2011
Dictyota	*fasciola*	(Roth) J.V. Lamouroux, 1809	C, A	Hall-Spencer et al. 2008
Dictyota	*spiralis*	Montagne, 1846	C, I, A	Hall-Spencer et al. 2008, Porzio et al. 2011
Ectocarpus	*siliculosus*	Dillwyn Lyngbye, 1819	C	Boudouresque & Cinelli 1971
Ectocarpus	sp.		C, A	Nogueira et al. 2017
Feldmannia	*irregularis*	(Kützing) Hamel, 1939	I	Porzio et al. 2011
Halopteris	*filicina*	(Grateloup) Kützing, 1843	C, I, A	Hall-Spencer et al. 2008, Porzio et al. 2011
Halopteris	*scoparia*	(Linnaeus) Sauvageau, 1904	C, I, A	Hall-Spencer et al. 2008, Porzio et al. 2011, Ricevuto et al. 2015a,b
Kuckuckia	*cf spinosa*	(Kützing) Kornmann in Kuckuck, 1958	C	Hall-Spencer et al. 2008
Lobophora	*variegata*	(J.V. Lamouroux) Womersley ex E.C. Oliveira, 1977	C, I	Porzio et al. 2011
Myrionema	*orbiculare*	J. Agardh, 1848	C	Mazzella et al. 1981, Nogueira et al. 2017

Continued

Table 3 (Continued)   A taxonomic checklist of the marine species recorded in the CO_2 vent sites of the Castello Aragonese, Ischia, as well as nearby ambient zones (approximately within 500 m range)

Genus	Species	Taxonomic authority	Site: Control (C) ambient pH, Intermediate (I) low pH, Acidified (A) extreme low pH	Literature citation
Padina	*pavonica*	(Linnaeus) Thivy,1960	C, I	Hall-Spencer et al. 2008, Porzio et al. 2011
Parviphycus	sp.		C, A	Nogueira et al. 2017
Pseudolithoderma	*adriaticum*	(Hauck) Verlaque, 1988	C	Boudouresque & Cinelli 1971
Ralfsia	*verrucosa*	(Areschoug) Areschoug, 1845	C, I, A	Hall-Spencer et al. 2008
Sargassum	*vulgare*	C. Agardh, 1820	A	Hall-Spencer et al. 2008, Porzio et al. 2011, Ricevuto et al. 2015b
Scytosiphon	*lomentaria*	(Lyngbye) Link, 1833	C	Gambi et al. 2000
Sphacelaria	*cirrosa*	(Roth) C. Agardh, 1824	C, I, A	Boudouresque & Cinelli 1971, Gambi et al. 2000, Hall-Spencer et al. 2008, Porzio et al. 2011
Sphacelaria	*fusca*	(Hudson) S.F. Gray, 1821	C, I, A	Hall-Spencer et al. 2008
Sphacelaria	*rigidula*	Kützing, 1843	C	Porzio et al. 2011
Sphacelaria	sp.		C	Porzio et al. 2011
Sphacelaria	*tribuloides*	Meneghini, 1840	C, I	Porzio et al. 2011
Phylum Rhodophyta				
Acrosorium	*ciliolatum*	(Harvey) Kylin, 1924	C	Hall-Spencer et al. 2008
Aglaothamnion	*bipinnatum*	(P. Crouan & H. Crouan) Feldmann & G. Feldmann, 1948	I, A	Porzio et al. 2011
Aglaothamnion	*hookeri*	(Dillwyn) Maggs & Hommersand, 1993	C	Boudouresque & Cinelli 1971
Aglaothamnion	*tenuissimum*	(Bonnemaison) Feldmann-Mazoyer, 1941	C	Hall-Spencer et al. 2008
Amphiroa	*rigida*	J.V. Lamouroux, 1816	C, I	Hall-Spencer et al. 2008, Porzio et al. 2011
Amphiroa	*rubra*	(Philippi) Woelkerling, 1983	I	Porzio et al. 2011

Continued

Table 3 (Continued) A taxonomic checklist of the marine species recorded in the CO_2 vent sites of the Castello Aragonese, Ischia, as well as nearby ambient zones (approximately within 500 m range)

Genus	Species	Taxonomic authority	Site: Control (C) ambient pH, Intermediate (I) low pH, Acidified (A) extreme low pH	Literature citation
Anotrichium	*tenue*	(C. Agardh) Nägeli, 1862	C, I	Porzio et al. 2011
Antithamnion	*cruciatum*	(C. Agardh) Nägeli, 1847	C, I, A	Boudouresque & Cinelli 1971, Porzio et al. 2011
Antithamnion	*heterocladum*	Funk, 1955	C	Boudouresque & Cinelli 1971
Antithamnion	sp.		I	Porzio et al. 2011
Apoglossum	*ruscifolium*	(Turner) J. Agardh, 1898	C, I	Hall-Spencer et al. 2008
Asparagopsis	sp.		C, I	Porzio et al. 2011
Asparagopsis	*taxiformis*	(Delile) Trevisan, 1845	C, I, A	Hall-Spencer et al. 2008, Gambi et al. 2016b
Audouinella	*caespitosa*	(J. Agardh) P.S. Dixon in Parke & P.S. Dixon, 1976	C	Boudouresque & Cinelli 1971
Audouinella	*daviesii*	(Dillwyn) Woelkerling, 1971	C	Boudouresque & Cinelli 1971
Audouinella	*leptonema*	(Rosenvinge) Garbary, 1979	C	Boudouresque & Cinelli 1971
Callithamniella	*tingitana*	Schousboe ex Bornet Feldmann-Mazoyer, 1938	C	Boudouresque & Cinelli 1971
Callithamnion	*corynbosum*	(Smith) Lyngbye, 1819	C, A	Nogueira et al. 2017
Centroceras	*clavulatum*	(C. Agardh) Montagne, 1846	C	Porzio et al. 2011
Ceramium	*circinatum*	(Kützing) J. Agardh, 1851	C	Porzio et al. 2011
Ceramium	*codii*	(H. Richards) Mazoyer, 1938	C, I	Porzio et al. 2011
Ceramium	*diaphanum*	(Lightfoot) Roth, 1806	C, A	Boudouresque & Cinelli 1971, Porzio et al. 2011
Ceramium	*flaccidum*	(Harvey ex Kützing) Ardissone, 1871	C, I	Porzio et al. 2011
Ceramium	sp.		C, I	Hall-Spencer et al. 2008, Nogueira et al. 2017
Ceramium	*virgatum v. virgatum*	Roth, 1797	C, I	Boudouresque & Cinelli 1971, Hall-Spencer et al. 2008

Continued

Table 3 (Continued)   A taxonomic checklist of the marine species recorded in the CO_2 vent sites of the Castello Aragonese, Ischia, as well as nearby ambient zones (approximately within 500 m range)

Genus	Species	Taxonomic authority	Site: Control (C) ambient pH, Intermediate (I) low pH, Acidified (A) extreme low pH	Literature citation
Champia	*parvula*	(C. Agardh) Harvey, 1853	C, I	Porzio et al. 2011
Chondracanthus	*acicularis*	(Roth) Fredericq, 1993	A	Porzio et al. 2011
Chondracanthus	*teedei*	(Mertens ex Roth) Kützing, 1843	C, I, A	Hall-Spencer et al. 2008
Choreonema	*thuretii*	Bornet F. Schmitz, 1889	I	Porzio et al. 2011
Contarinia	*peyssonneliaeformis*	Zanardini, 1843	C	Boudouresque & Cinelli 1971
Contarinia	*squamariae*	(Meneghini) Denizot, 1968	I	Porzio et al. 2011
Corallina	*elongata*	J. Ellis & Solander, 1786	C, I	Boudouresque & Cinelli 1971, Hall-Spencer et al. 2008, Porzio et al. 2011
Corallina	*officinalis*	Linnaeus, 1758	C	Hall-Spencer et al. 2008, Porzio et al. 2011
Corallinaceae	spp.		C, I	Porzio et al. 2011
Crouania	*attenuata*	(C. Agardh) J. Agardh, 1842	C	Porzio et al. 2011
Crouania	*ischiana*	(Funk) C.F. Boudouresque & M.M. Perret-Boudouresque, 1987	C	Boudouresque & Cinelli 1971
Dasya	*baillouviana*	(S.G. Gmelin) Montagne in Barker-Webb & Berthelot 1841	C	Porzio et al. 2011
Dasya	*corymbifera*	J. Agardh, 1841	C	Boudouresque & Cinelli 1971, Nogueira et al. 2017
Dasya	*hutchinsiae*	Harvey, 1833	I	Porzio et al. 2011
Dasya	*rigidula*	(Kützing) Ardissone, 1878	C	Porzio et al. 2011
Erythrocladia	*polystromatica*	P.J.L. Dangeard, 1932	C	Cinelli et al. 1981
Erythrotrichia	*carnea*	Dillwyn J. Agardh, 1883	C	Boudouresque & Cinelli 1971
Erythrotrichia	*investiens*	Zanardini Bornet, 1892	C	Boudouresque & Cinelli 1971

Continued

Table 3 (Continued)   A taxonomic checklist of the marine species recorded in the CO_2 vent sites of the Castello Aragonese, Ischia, as well as nearby ambient zones (approximately within 500 m range)

Genus	Species	Taxonomic authority	Site: Control (C) ambient pH, Intermediate (I) low pH, Acidified (A) extreme low pH	Literature citation
Gelidium	bipectinatum	G. Furnari, 1999	C, I	Porzio et al. 2011
Gelidium	minusculum	(Weber-van Bosse) R.E. Norris, 1992	C	Porzio et al. 2011
Griffithsia	phyllamphora	J. Agardh, 1842	C, I	Porzio et al. 2011
Gulsonia	nodulosa	(Ercegovic) Feldmann & G. Feldmann, 1967	C, I	Hall-Spencer et al. 2008
Herposiphonia	secunda	(C. Agardh) Ambronn, 1880	C	Boudouresque & Cinelli 1971, Porzio et al. 2011
Herposiphonia	secunda f. tenella	(C. Agardh) M.J. Wynne, 1985	C, I	Porzio et al. 2011
Herposiphonia	sp.		I	Porzio et al. 2011
Heterosiphonia	crispella	C. Agardh M.J. Wynne, 1985	C, I	Porzio et al. 2011
Hildenbrandia	rubra	(Sommerfelt) Meneghini, 1841	C, I, A	Hall-Spencer et al. 2008, Martin et al. 2008, Porzio et al. 2011
Hydrolithon	boreale	(Foslie) Chamberlain, 1994	C, I	Martin et al. 2008, Porzio et al. 2011
Hydrolithon	cruciatum	(Bressan) Y.M. Chamberlain, 1994	C, I	Martin et al. 2008, Porzio et al. 2011
Hydrolithon	farinosum	(J.V. Lamouroux) Penrose & Y.M. Chamberlain, 1993	C, I	Boudouresque & Cinelli 1971, Mazzella et al. 1981, Porzio et al. 2011
Jania	longifurca	Zanardini, 1844	C	Hall-Spencer et al. 2008
Jania	rubens	(Linnaeus) J.V. Lamouroux, 1816	C, I	Boudouresque & Cinelli 1971, Hall-Spencer et al. 2008, Porzio et al. 2011, Ricevuto et al. 2015a,b, Nogueira et al. 2017
Laurencia	obtusa	(Hudson) J.V. Lamouroux 1813	C, I, A	Hall-Spencer et al. 2008, Porzio et al. 2011
Lithophyllum	incrustans	Philippi, 1837	C, I	Hall-Spencer et al. 2008, Porzio et al. 2011

Continued

Table 3 (Continued) A taxonomic checklist of the marine species recorded in the CO_2 vent sites of the Castello Aragonese, Ischia, as well as nearby ambient zones (approximately within 500 m range)

Genus	Species	Taxonomic authority	Site: Control (C) ambient pH, Intermediate (I) low pH, Acidified (A) extreme low pH	Literature citation
Lithophyllum	*pustulatum*	J.V. Lamouroux Foslie, 1904	C, I	Boudouresque & Cinelli 1971
Lithophyllum	*pustulatum var. confine*	(P.L. Crouan & H.M. Crouan Y.M.) Chamberlain, 1991	C	Boudouresque & Cinelli 1971
Lithophyllum	sp.		C, I	Porzio et al. 2011, Kamenos et al. 2016
Lithothamnium	sp.		C	Boudouresque & Cinelli 1971
Lomentaria	*verticillata*	Funk, 1944	C	Boudouresque & Cinelli 1971
Lophosiphonia	*cristata*	Falkenberg, 1901	C, I	Porzio et al. 2011, Nogueira et al. 2017
Lophosiphonia	sp.		C	Gambi et al. 2000
Melobesia	*membranacea*	Esper J.V. Lamouroux, 1812	C	Boudouresque & Cinelli 1971
Meredithia	*microphylla*	(J. Agardh) J. Agardh, 1892	C	Porzio et al. 2011
Mesophyllum	*lichenoides*	(J. Ellis) Me. Lemoine, 1928	C	Hall-Spencer et al. 2008
Mesophyllum	sp.		C, I	Porzio et al. 2011
Monosporus	*pedicellatus*	(Smith) Solier in Castagne, 1845	C	Porzio et al. 2011
Myriogramme	*minuta*	Kylin, 1924	C	Boudouresque & Cinelli 1971
Neogoniolithon	*brassica-florida*	(Harvey) Setchell & L.R. Mason, 1943	C, I	Porzio et al. 2011
Nithophyllum	*punctatum*	(Stackhouse) Greville, 1830	C, I	Porzio et al. 2011
Osmundea	*truncata*	(Kützing) K.W. Nam & Maggs, 1994	C, I, A	Porzio et al. 2011
Palisada	*perforata*	(Bory de Saint-Vincent) K.W. Nam, 2007	C	Porzio et al. 2011
Parviphycus (*Gelidiella*)	*pannosus* (*pannosa*)	(Feldmann) G. Furnari, 2010	C, I	Porzio et al. 2011
Peyssonnelia	*armorica*	(P. et H. Crouan) Weber van Bosse, 1916	C, A	Boudouresque & Cinelli 1971, Porzio et al. 2011
Peyssonnelia	*bornetii*	Boudouresque & Denizot, 1973	I	Porzio et al. 2011
Peyssonnelia	*dubyi*	P. et H. Crouan, 1844	C	Boudouresque & Cinelli 1971, Porzio et al. 2011

Continued

Table 3 (Continued)   A taxonomic checklist of the marine species recorded in the CO_2 vent sites of the Castello Aragonese, Ischia, as well as nearby ambient zones (approximately within 500 m range)

Genus	Species	Taxonomic authority	Site: Control (C) ambient pH, Intermediate (I) low pH, Acidified (A) extreme low pH	Literature citation
Peyssonnelia	*orientalis*	(Weber-van Bosse) Cormaci & G. Furnari, 1987	C	Cinelli et al. 1981
Peyssonnelia	*polymorpha*	(Zanardini) F. Schmitz in Falkenberg, 1879	C, I	Porzio et al. 2011
Peyssonnelia	*rosa-marina*	Boudouresque & Denizot, 1973	C	Boudouresque & Cinelli 1971
Peyssonnelia	cfr. *rubra*	(Greville) J. Agardh, 1851	I	Porzio et al. 2011
Peyssonnelia	*squamaria*	(S.G. Gmelin) Decaisne, 1841	C, I	Boudouresque & Cinelli 1971, Hall-Spencer et al. 2008, Porzio et al. 2011, 2013
Phyllophora	*crispa*	(Hudson) P.S. Dixon, 1964	C	Porzio et al. 2011
Phyllophora	*sicula*	(Kützing) Guiry & L.M. Irvine, 1976	I	Porzio et al. 2011
Phymatolithon	*cfr lenormandii*	(Areschoug) Adey, 1966	C, I	Porzio et al. 2011
Phymatolithon	*lenormandii*	(Areschoug) Adey, 1966	I	Porzio et al. 2011
Pneophyllum	*confervicola*	Kützing Y.M. Chamberlain, 1983	C, I	Martin et al. 2008
Pneophyllum	*fragile*	Kützing, 1843	C,I	Martin et al. 2008, Porzio et al. 2011
Pneophyllum	*zonale*	P.L. Crouan & H.M. Crouan Y.M. Chamberlain, 1983	C,I	Martin et al. 2008
Polysiphonia	*banyulensis*	Coppejans, 1975	C	Cinelli et al. 1981
Polysiphonia	*denudata*	(Dillwyn) Greville ex Harvey, 1833	C	Porzio et al. 2011
Polysiphonia	*fibrata*	(Dillwyn) Harvey, 1833	I	Porzio et al. 2011
Polysiphonia	cf. *stricta*	(Mertens ex Dillwyn) Greville, 1824	C, I	Hall-Spencer et al. 2008
Polysiphonia (*Lophosiphonia*)	*scopulorum*	Harvey, 1855	C, I, A	Cinelli et al. 1981, Porzio et al. 2011
Pterocladiella	*capillacea*	(S.G. Gmelin) Santelices & Hommersand, 1997	C, A	Boudouresque & Cinelli 1971, Porzio et al. 2011

Continued

Table 3 (Continued)   A taxonomic checklist of the marine species recorded in the CO_2 vent sites of the Castello Aragonese, Ischia, as well as nearby ambient zones (approximately within 500 m range)

Genus	Species	Taxonomic authority	Site: Control (C) ambient pH, Intermediate (I) low pH, Acidified (A) extreme low pH	Literature citation
Pterothamnion	*plumula*	(J. Ellis) Thuret, 1863	C, I	Hall-Spencer et al. 2008
Ptilothamnion	*sphaericum*	(P. Crouan & H. Crouan ex J. Agardh) Maggs & Hommersand, 1993	I	Porzio et al. 2011
Rhodymenia	*ardissonei*	Kuntze Feldmann, 1937	C	Boudouresque & Cinelli 1971
Schottera	*nicaeensis*	J.V. Lamouroux ex Duby Guiry & Hollenberg, 1975	C	Boudouresque & Cinelli 1971
Spermothamnion	*repens*	(Dillwyn) Magnus, 1873	C, A	Gambi et al. 2000, Porzio et al. 2011
Spermothamnion	*strictum*	(C. Agardh) Ardissone, 1883	C, I, A	Porzio et al. 2011
Stylonema	*alsidii*	K.M. Drew, 1956	C, A	Nogueira et al. 2017
Stylonema	*cornu-cervi*	Reinsch, 1875	C	Boudouresque & Cinelli 1971
Titanoderma	spp.		I, A	Donnarumma et al. 2014
Titanoderma mediterraneum		(Foslie) Woelkerling, 1988	C, I	Porzio et al. 2013
Phylum Chlorophyta				
Acetabularia	*acetabulum*	(Linnaeus) P.C. Silva, 1952	C	Hall-Spencer et al. 2008
Acrochaetium	*hauckii*	Schiffner, 1916	C	Boudouresque & Cinelli 1971
Acrochaetium	sp.1		C	Boudouresque & Cinelli 1971
Acrochaetium	sp.2		C	Boudouresque & Cinelli 1971
Bryopsis	*duplex*	De Notaris, 1844	C	Boudouresque & Cinelli 1971
Bryopsis	*plumosa*	(Hudson) C. Agardh, 1823	C, I	Hall-Spencer et al. 2008, Porzio et al. 2011
Caulerpa	*cylindracea (racemosa)*	Sonder, 1845	C, I, A	Buia et al. 2003, Hall-Spencer et al. 2008, Gambi et al. 2016b
Caulerpa	*prolifera*	(Forsskål) J.V. Lamouroux, 1809	C, I	Gambi et al. 2000, Hall-Spencer et al. 2008

Continued

Table 3 (Continued)   A taxonomic checklist of the marine species recorded in the CO_2 vent sites of the Castello Aragonese, Ischia, as well as nearby ambient zones (approximately within 500 m range)

Genus	Species	Taxonomic authority	Site: Control (C) ambient pH, Intermediate (I) low pH, Acidified (A) extreme low pH	Literature citation
Chaetomorpha	*aerea*	Dillwyn Kützing, 1849	C, A	Nogueira et al. 2017
Chaetomorpha	*linum*	(O.F. Müller) Kützing, 1845	C, I, A	Porzio et al. 2011
Chaetomorpha	sp.		C, I, A	Hall-Spencer et al. 2008
Cladophora	*dalmatica*	Kützing, 1843	C, I	Porzio et al. 2011, Nogueira et al. 2017
Cladophora	*flexuosa*	(O.F. Müller) Kützing, 1843	C	Porzio et al. 2011
Cladophora	*laetevirens*	(Dillwyn) Kützing, 1843	C, I, A	Porzio et al. 2011, Nogueira et al. 2017
Cladophora	*lehmanniana*	Lindenberg Kützing, 1843	C, A	Nogueira et al. 2017
Cladophora	*pellucida*	(Hudson) Kützing, 1843	I, A	Porzio et al. 2011
Cladophora	*prolifera*	(Roth) Kützing, 1843	C, I, A	Hall-Spencer et al. 2008, Porzio et al. 2011, Ricevuto et al. 2015a,b
Cladophora	*rupestris*	(Linnaeus) Kützing, 1843	C, I, A	Hall-Spencer et al. 2008, Porzio et al. 2011
Cladophora	*sericea*	(Hudson) Kützing, 1843	C	Porzio et al. 2011
Cladophora	*socialis*	Kützing, 1849	C	Porzio et al. 2011
Codium	*bursa*	(Olivi) C. Agardh, 1817	C, I, A	Hall-Spencer et al. 2008
Codium	*effusum*	(Rafinesque) Delle Chiaje, 1829	C, I	Hall-Spencer et al. 2008
Codium	*vermilara*	(Olivi) Delle Chiaje, 1829	C, I	Hall-Spencer et al. 2008
Derbesia	sp.		C	Boudouresque & Cinelli 1971
Derbesia	*tenuissima*	(Moris & De Notaris) P.L. Crouan & H.M. Crouan, 1867	C	Boudouresque & Cinelli 1971
Flabellia	*petiolata*	(Turra) Nizamuddin, 1987	C, I	Hall-Spencer et al. 2008, Porzio et al. 2011, Ricevuto et al. 2015a,b

Continued

Table 3 (Continued)   A taxonomic checklist of the marine species recorded in the CO_2 vent sites of the Castello Aragonese, Ischia, as well as nearby ambient zones (approximately within 500 m range)

Genus	Species	Taxonomic authority	Site: Control (C) ambient pH, Intermediate (I) low pH, Acidified (A) extreme low pH	Literature citation
Halimeda	*tuna*	(J. Ellis & Solander) J.V. Lamouroux, 1816	C, I	Hall-Spencer et al. 2008, Porzio et al. 2011
Parvocaulis	*parvulus*	(Solms-Laubach) S. Berger et al., 2003	C, I	Porzio et al. 2011
Pedobesia	*simplex*	(Meneghini ex Kützing) M.J. Wynne & F. Leliaert, 2001	C	Hall-Spencer et al. 2008
Pseudochlorodesmis	*furcellata*	(Zanardini) Børgesen, 1925	C, I	Boudouresque & Cinelli 1971, Porzio et al. 2011
Rhizoclonium	*tortuosum*	(Dillwyn) Kützing, 1845	C, I, A	Hall-Spencer et al. 2008
Ulva	*laetevirens (rigida)*	Areschough, 1854	C	Gambi et al. 2000
Valonia	*utricularis*	(Roth) C. Agardh, 1823	C, I, A	Boudouresque & Cinelli 1971, Hall-Spencer et al. 2008, Porzio et al. 2011
Phylum Magnoliophyta				
Cymodocea	*nodosa*	(Ucria) Ascherson, 1870	C	Kraemer & Mazzella 1996, Lanera & Gambi 1993
Posidonia	*oceanica*	(Linnaeus) Delile, 1813	C, I, A	Giraud et al. 1979, Ott 1980, Kraemer et al. 1997, Buia et al. 2003, Hall-Spencer et al. 2008, Garrard et al. 2014
Zostera (Zosterella)	*noltei*	Hornemann, 1832	C	Kraemer & Mazzella 1996
Phylum Granuloreticulosa Class Foraminifera				
Ammoglobigerina	*globigeriniformis*	(Parker & Jones, 1865)	A	Dias et al. 2010
Ammonia	*inflata*	(Seguenza, 1862)	C	Dias et al. 2010
Ammonia	*tepida*	(Cushman, 1926)	I	Dias et al. 2010
Brizalina	sp.		C	Dias et al. 2010
Cibicides	*advenum*	(d'Orbigny, 1839)	C	Dias et al. 2010
Cibicides	*refulgens*	Montfort, 1808	C	Dias et al. 2010
Cycloforina	*tenuicollis*	(Wiesner, 1923)	C	Dias et al. 2010

Continued

Table 3 (Continued)   A taxonomic checklist of the marine species recorded in the CO_2 vent sites of the Castello Aragonese, Ischia, as well as nearby ambient zones (approximately within 500 m range)

Genus	Species	Taxonomic authority	Site: Control (C) ambient pH, Intermediate (I) low pH, Acidified (A) extreme low pH	Literature citation
Elphidium	*aculeatum*	(d'Orbigny, 1846)	C, I, A	Cigliano et al. 2010, Dias et al. 2010
Elphidium	*complanatum*	(d'Orbigny, 1839)	C, I	Cigliano et al. 2010
Elphidium	*depressulum*	(Cushman, 1933)	C, I, A	Cigliano et al. 2010
Elphidium	sp. cf. *E. advenum*	(Cushman, 1922)	C	Dias et al. 2010
Lepidiodeuteramina	sp.		C	Dias et al. 2010
Massilina	*gualtieriana*	(d'Orbigny, 1839)	C	Dias et al. 2010
Massilina	*secans*	(d'Orbigny, 1826)	C	Cigliano et al. 2010
Miliammina	*fusca*	(Brady, 1870)	I, A	Dias et al. 2010
Miliolinella	*elongata*	Kruit, 1955	C	Dias et al. 2010
Miliolinella	*labiosa*	(d'Orbigny, 1839)	C	Dias et al. 2010
Milionella	*subrotundata*	(Montagu, 1803)	C	Cigliano et al. 2010
Parrina	*bradyi*	(Millett, 1898)	C	Dias et al. 2010
Peneroplis	*pertusus*	(Forsskal, 1775)	C	Dias et al. 2010
Peneroplis	*planatus*	(Fitchel & Moll, 1798)	C	Dias et al. 2010
Planorbulina	*mediterranensis*	d'Orbigny, 1826	C	Dias et al. 2010
Pyrgo	sp.		C	Dias et al. 2010
Quinqueloculina	*aff. parvula*	Schlumberger, 1895	C	Cigliano et al. 2010
Quinqueloculina	*berthelotiana*	d'Orbigny, 1839	C	Cigliano et al. 2010, Dias et al. 2010
Quinqueloculina	*jugosa*	Cushman, 1944	C	Dias et al. 2010
Quinqueloculina	*parvula*	Schlumberger, 1894	C	Cigliano et al. 2010
Quinqueloculina	*seminula*	(Linné, 1758)	C	Dias et al. 2010
Reophax	sp.		I	Dias et al. 2010
Rosalina	*vilardeboana*	d'Orbigny, 1839	C	Dias et al. 2010
Textularia	sp. cf. *T. bocki*	Höglund, 1947	A	Dias et al. 2010
Triloculina	*plicata*	Terquem, 1876	C	Cigliano et al. 2010
Triloculina	*schreiberiana*	d'Orbigny, 1839	C	Cigliano et al. 2010
Triloculina	sp.		C	Dias et al. 2010
Triloculina	*tricarinata*	d'Orbigny, 1826	C	Dias et al. 2010
Trochammina	*inflata*	(Montagu, 1803)	I, A	Dias et al. 2010
Vertebralina	*striata*	d'Orbigny, 1826	C	Cigliano et al. 2010
Phylum Nematoda	spp.		C, I, A	Cigliano et al. 2010
Phylum Porifera				
Agelas	*oroides*	(Schmidt, 1864)	C	Hall-Spencer et al. 2008
Aplysilla	*rosea*	(Barrois, 1876)	C	Pulitzer-Finali & Pronzato 1976
Cacospongia	sp.		C, I, A	Hall-Spencer et al. 2008

Continued

Table 3 (Continued)   A taxonomic checklist of the marine species recorded in the CO_2 vent sites of the Castello Aragonese, Ischia, as well as nearby ambient zones (approximately within 500 m range)

Genus	Species	Taxonomic authority	Site: Control (C) ambient pH, Intermediate (I) low pH, Acidified (A) extreme low pH	Literature citation
Chondrilla	*nucula*	Schmidt, 1862	C, I	Hall-Spencer et al. 2008, Goodwin et al. 2013
Chondrosia	*reniformis*	Nardo, 1847	C, I	Sarà 1959, Hall-Spencer et al. 2008, Goodwin et al. 2013
Crambe	*crambe*	(Schmidt, 1862)	C, I, A	Goodwin et al. 2013
Dysidea	sp.		C, I, A	Hall-Spencer et al. 2008
Eurypon	*cf cinctum*	Sarà, 1960	C, I	Goodwin et al. 2013
Geodia	*cydonium*	(Jameson, 1811)	C	Sarà 1959
Haliclona (Reniera)	*mediterranea*	Griessinger, 1971	C, I, A	Hall-Spencer et al. 2008, Goodwin et al. 2013, Gambi M.C. personal collection (2017)
Hemimycale	*columella*	(Bowerbank, 1874)	C,A	Goodwin et al. 2013
Ircinia	*variabilis*	(Schmidt, 1862)	C, I	Hall-Spencer et al. 2008, Pulitzer-Finali & Pronzato 1976
Microciona	*cf toxitenuis*	(Topsent, 1925)	A	Goodwin et al. 2013
Petrosia	*ficiformis*	(Poiret, 1789)	C, I	Goodwin et al. 2013
Phorbas	*fictitius*	(Bowerbank, 1866)	A	Goodwin et al. 2013
Phorbas	*tenacior*	(Topsent, 1925)	C, I	Goodwin et al. 2013
Sarcotragus	sp.		I	Goodwin et al. 2013
Sarcotragus	*spinosulus*	Schmidt, 1862	C	Nielsen 2008
Scalarispongia	*scalaris*	(Schmidt, 1862)	C, I	Goodwin et al. 2013
Spirastrella	*cunctatrix*	Schmidt, 1868	C, I, A	Hall-Spencer et al. 2008, Goodwin et al. 2013
Sycon	*raphanus*	Schmidt, 1862	C	Sarà 1959
Phylum Entoprocta				
Loxosomella	*pes*	(Schmidt, 1878)	C	Nielsen 2008
Phylum Cnidaria				
Class Antozoa				
Actinia	*equina*	(Linnaeus, 1758)	C, I, A	Hall-Spencer et al. 2008
Anemonia	*viridis*	(Forsskål, 1775)	C, A	Hall-Spencer et al. 2008, Meron et al. 2012a
Balanophyllia	*europaea*	(Risso, 1826)	C	Hall-Spencer et al. 2008
Caryophyllia	*smithii*	Stokes & Broderip, 1828	C	Hall-Spencer et al. 2008

Continued

Table 3 (Continued)   A taxonomic checklist of the marine species recorded in the CO_2 vent sites of the Castello Aragonese, Ischia, as well as nearby ambient zones (approximately within 500 m range)

Genus	Species	Taxonomic authority	Site: Control (C) ambient pH, Intermediate (I) low pH, Acidified (A) extreme low pH	Literature citation
Cladocora	*caespitosa*	(Linnaeus, 1767)	C	Hall-Spencer et al. 2008
Parazoanthus	*axinellae*	(Schmidt,1862)	C, I	Hall-Spencer et al. 2008
Class Hydrozoa				
Aglaophenia	*pluma*	(Linnaeus, 1758)	C, I	Hall-Spencer et al. 2008
Eudendrium	sp.		C, I	Hall-Spencer et al. 2008
Plumularia	*obliqua (Plumularia) (posidoniae)*	(Picard, 1952)	C, A	Nogueira et al. 2017
Sertularia	*distans*	(Lamouroux, 1816)	C, A	Nogueira et al. 2017
Phylum Plathelminthes				
Anoplodium	*parasita*	Schneider, 1858		Kroll & Jangoux 1989
Philactinoposthia	*ischiae*	Nilsson et al., 2011	I, A	Nilsson et al. 2011
Phylum Sipunculida				
Aspidosiphon	*muelleri*	Diesing, 1851	C, I, A	Kroeker et al. 2011
Golfingia (Golfingia)	*vulgaris*	(Blainville, 1827)	C, I, A	Kroeker et al. 2011
Phascolion (Phascolion)	*strombus*	(Montagu, 1804)	C, I, A	Kroeker et al. 2011
Phascolosoma (Phascolosoma)	*stephensoni*	(Stephen, 1942)	A	Gambi M.C. personal collection (2016)
Phylum Annelida				
Class Polychaeta				
Amblosyllis	*madeirensis*	Langerhans, 1879	C	Gambi et al. 2016a
Amphicorina	*armandi*	(Claparede, 1864)	I	Gambi et al. 2016a
Amphicorina	*persinosa*	(Ben-Eliahu, 1975)	C	Gambi et al. 2016a
Amphiglena	*mediterranea*	(Leydig, 1851)	C, I, A	Gambi et al. 1997a, Rouse & Gambi 1997, Cigliano et al. 2010, Kroeker et al. 2011, Garrard et al. 2014, Gambi et al. 2016a
Autolytinae	sp.		C	Gambi et al. 2016a
Autolytinae	spp.		C	Kroeker et al. 2011
Axiothella	sp.		C	Lanera & Gambi 1993
Branchiomma	*boholense (as Branchiomma) (bairdi)*	(Grube, 1878)	C, I, A	Arias et al. 2013, Gambi et al. 2016b, Del Pasqua et al. 2018

Continued

Table 3 (Continued)   A taxonomic checklist of the marine species recorded in the CO_2 vent sites of the Castello Aragonese, Ischia, as well as nearby ambient zones (approximately within 500 m range)

Genus	Species	Taxonomic authority	Site: Control (C) ambient pH, Intermediate (I) low pH, Acidified (A) extreme low pH	Literature citation
Branchiomma	*bombyx*	(Dalyell, 1853)	C	Kroeker et al. 2011
Branchiomma	*lucullanum*	(Delle Chiaje, 1828)	C, I	Kroeker et al. 2011
Branchiosyllis	*exilis*	(Gravier, 1900)	C, I	Gambi et al. 2016a
Brania	*armini*	(Langerhans, 1881)	I	Gambi et al. 2016a
Brania	*pusilla*	(Dujardin, 1851)	C, I, A	Gambi et al. 2016a
Brifacia	*aragonensis*	Giangrande et al. 2014	C, I, A	Giangrande et al. 2014
Capitellidae	sp.		I, A	Kroeker et al. 2011
Capitomastus	*minimus*	Langerhans, 1880	C	Lanera & Gambi 1993
Ceratonereis	*costae*	(Grube, 1840)	C, I	Cigliano et al. 2010, Kroeker et al. 2011
Ceratonereis	*hircinicola*	(Eisig, 1870)	C	Kroeker et al. 2011, Gambi et al. 2016a
Chone	*duneri*	Malmgren, 1867	C	Lanera & Gambi 1993
Chrysopetalum	*debile*	(Grube, 1855)	C	Gambi et al. 1997a
Cirratulidae	spp.		C, I	Kroeker et al. 2011
Ctenodrilidae	gen. sp.		C, I, A	Kroeker et al. 2011
Diopatra	*neapolitana*	Delle Chiaje, 1841	C	Lanera & Gambi 1993
Dodecaceria	*concharum*	Örsted, 1843	C, I	Kroeker et al. 2011, Gambi et al. 2016a
Dorvillea (Schistomeringos)	*rudolphii*	(Delle Chiaje, 1828)	C, A	Kroeker et al. 2011, Gambi et al. 2016a
Erinaceusyllis	*belizensis*	(Russel, 1989)	C	Gambi et al. 2016a
Euclymene	*oerstedi*	(Claparède, 1863)	C	Lanera & Gambi 1993
Eunice	*vittata*	(Delle Chiaje, 1828)	C	Lanera & Gambi 1993
Eusyllis	*lamelligera*	Marion & Bobretzsky, 1875	C, I	Gambi et al. 2016a
Exogone (Exogone)	*dispar*	(Webster, 1879)	C, I, A	Gambi et al. 1997a, Lanera & Gambi 1993, Kroeker et al. 2011, Gambi et al. 2016a
Exogone (Exogone)	*naidina*	Oertsed, 1845	C, I, A	Cigliano et al. 2010, Gambi et al. 2016a
Exogone (Exogone)	*rostrata*	Naville, 1933	C	Gambi et al. 2016a
Exogone (Paraexogone)	*meridionalis*	Cognetti, 1955	C, I, A	Cigliano et al. 2010
Fabricia	*stellaris stellaris*	(O.F. Mueller, 1774)	C, I, A	Lanera & Gambi 1993, Kroeker et al. 2011, Giangrande et al. 2014

Continued

Table 3 (Continued)   A taxonomic checklist of the marine species recorded in the CO_2 vent sites of the Castello Aragonese, Ischia, as well as nearby ambient zones (approximately within 500 m range)

Genus	Species	Taxonomic authority	Site: Control (C) ambient pH, Intermediate (I) low pH, Acidified (A) extreme low pH	Literature citation
Glycera	*lapidum*	Quatrefages, 1865	C	Lanera & Gambi 1993, Gambi et al. 1997a
Gyptis	*propinqua*	Marion & Bobretzky, 1875	C, I	Cigliano et al. 2010
Haplosyllis	*granulosa*	(Lattig, Martin & San Martin, 2007)	C, I, A	Gambi et al. 2016a
Harmothoe	sp.		A	Kroeker et al. 2011, Gambi et al. 2016a
Hesionidae	sp.		C, I	Kroeker et al. 2011
Hydroides	*pseudouncinatus*	Zibrowius, 1968	C, I	Kroeker et al. 2011, Gambi et al. 2016a
Laetmonice	*hystrix*	(Savigny, 1820)	C	Lanera & Gambi 1993
Lepidonotus	*clava*	(Montagu, 1808)	C, I, A	Kroeker et al. 2011
Levinsenia	*gracilis*	(Tauber, 1879)	C, I, A	Lanera & Gambi 1993, Gambi et al. 2016a
Lumbrineris	*latreilli*	Audouin & Milne-Edwards, 1834	C	Lanera & Gambi 1993
Lumbrineris	sp.		C, I	Kroeker et al. 2011, Gambi et al. 2016a
Lysidice	*collaris*	Grube, 1870	C, I, A	Gambi & Cafiero 2001, Kroeker et al. 2011, Gambi et al. 2016a
Lysidice	*ninetta*	Audouin & Milne-Edwards, 1833	C, I, A	Gambi & Cafiero 2001, Kroeker et al. 2011, Gambi et al. 2016a
Lysidice	*unicornis*	(Grube, 1840)	C, I	Lanera & Gambi 1993, Gambi et al. 1997b, Gambi and Cafiero 2001, Kroeker et al. 2011, Gambi et al. 2016a
Myrianida	cf *edwarsi*	Saint-Joseph, 1887	C, I, A	Gambi et al. 2016a
Neodexiospira	*pseudocorrugata*	(Bush, 1904)	C	Lanera & Gambi 1993, Cigliano et al. 2010
Nephtys	*cirrhosa*	Ehlers, 1868	C	Lanera & Gambi 1993
Nephtys	*hombergi*	Savigny, 1818	C	Lanera & Gambi 1993
Nereis	*funchalensis*	(Langerhans, 1880)	I	Gambi et al. 2016a
Nereis	*zonata*	Malmgren, 1867	C, I	Gambi et al. 2016a

Continued

Table 3 (Continued)   A taxonomic checklist of the marine species recorded in the CO_2 vent sites of the Castello Aragonese, Ischia, as well as nearby ambient zones (approximately within 500 m range)

Genus	Species	Taxonomic authority	Site: Control (C) ambient pH, Intermediate (I) low pH, Acidified (A) extreme low pH	Literature citation
Novafabricia	*infratorquata*	(Fitzhugh, 1983)	I	Giangrande et al. 2014
Novafabricia	*posidoniae*	Giangrande & Licciano, 2006	C, I, A	Giangrande et al. 2014
Odontosyllis	*fulgurans*	(Audouin & Milne Edwards, 1834)	C	Gambi et al. 2016a
Orbinia	*cuvieri*	(Audouin & Milne Edwards, 1833)	C	Lanera & Gambi 1993
Palola	*siciliensis*	(Grube, 1840)	C, I, A	Gambi & Cafiero 2001, Kroeker et al. 2011, Gambi et al. 2016a
Paradoneis	*armata*	Glemarec, 1966	C	Lanera & Gambi 1993
Paradoneis	*ilvana*	Castelli, 1985	C	Lanera & Gambi 1993
Paraehlersia	*ferrugina*	(Langerhans, 1881)	C, I	Gambi et al. 2016a
Parafabricia	*mazzellae*	Giangrande et al. 2014	C, I, A	Giangrande et al. 2014
Parapionosyllis	*brevicirra*	(Day, 1954)	C	Gambi et al. 2016a
Parasabella	*saxicola*	(Grube, 1861)	C	Gambi et al. 2016a
Perinereis	*cultrifera*	(Grube, 1840)	C, I	Kroeker et al. 2011, Gambi et al. 2016a
Pholoe	sp.		I	Kroeker et al. 2011, Gambi et al. 2016a
Phyllodocidae	sp.		C	Kroeker et al. 2011
Platynereis	*dumerilii*	(Audouin & Milne-Edwards, 1834)	C, I, A	Gambi et al. 2000, Lanera & Gambi 1993, Gambi et al. 1997a, Cigliano et al. 2010, Kroeker et al. 2011
Platynereis	*massiliensis*	(Moquin-Tandon, 1869)	A, I	Calosi et al. 2013b, Lucey et al. 2015, Valvassori et al. 2015
Polydora	sp.		C, I, A	Kroeker et al. 2011
Polyophthalmus	*pictus*	(Dujardin, 1839)	C, I, A	Gambi et al. 1997a, Lanera & Gambi 1993, Cigliano et al. 2010, Kroeker et al. 2011, Gambi et al. 2016a
Pontogenia	*chrysocoma*	(Baird, 1865)	C, I	Garrard et al. 2014
Prionospio (*Prionospio*)	*malmgreni*	Claparède, 1870	C	Gambi et al. 1997a, Lanera & Gambi 1993

Continued

Table 3 (Continued) A taxonomic checklist of the marine species recorded in the CO_2 vent sites of the Castello Aragonese, Ischia, as well as nearby ambient zones (approximately within 500 m range)

Genus	Species	Taxonomic authority	Site: Control (C) ambient pH, Intermediate (I) low pH, Acidified (A) extreme low pH	Literature citation
Proceraea	cf *paraurantiaca*	Nygren, 2004	C	Gambi et al. 2016a
Prosphaerosyllis	*xarifae*	(Hartmann-Schroeder, 1960)	C, I	Gambi et al. 2016a
Protodorvillea	*kefersteini*	(McIntosh, 1869)	C	Lanera & Gambi 1993, Gambi et al. 1997a
Psamathe (Kefersteinia) (cirrhata)	*fusca*	Johnston, 1836	C	Lanera & Gambi 1993
Rubifabriciola	*tonerella*	(Banse, 1959)	I, A	Giangrande et al. 2014
Sabella	*spallanzanii*	(Gmelin, 1791)	C, I, A	Hall-Spencer et al. 2008, Calosi et al. 2013b
Salvatoria	*alvaradoi*	(San Martin, 1984)	C	Lanera & Gambi 1993
Salvatoria	*clavata*	(Claparède, 1863)	C, I, A	Lanera & Gambi 1993, Gambi et al. 2016a
Salvatoria	*limbata*	(Claparede, 1868)	C, I	Gambi et al. 2016a
Scolelepis	sp.		C	Lanera & Gambi 1993
Scoletoma	*impatiens*	(Claparède, 1868)	I	Kroeker et al. 2011, Gambi et al. 2016a
Serpula	*vermicularis*	Linnaeus, 1767	C, I	Kroeker et al. 2011, Gambi et al. 2016a
Simplaria	*pseudomilitaris*	(Thiriot-Quievreux, 1965)	C	Lanera & Gambi 1993
Simplaria	sp.		C, I, A	Lucey et al. 2016
Sphaerodoropsis	cf. *sphaerulifer*	(Moore, 1911)	C	Lanera & Gambi 1993
Sphaerosyllis (Sphaerosyllis)	*hystrix*	Claparède, 1863	C	Lanera & Gambi 1993
Sphaerosyllis (Sphaerosyllis)	*thomasi*	San Martin, 1984	C	Lanera & Gambi 1993
Sphaerosyllis	*austriaca*	Banse, 1959	C, I, A	Gambi et al. 2016a
Sphaerosyllis	*hystrix*	Claparede, 1863	C, I, A	Gambi et al. 2016a
Sphaerosyllis	*pirifera*	Claparede, 1868	C, I, A	Gambi et al. 2016a
Spio	*decoratus*	Bobretzky, 1870	C, I, A	Cigliano et al. 2010
Spionidae	gen. sp.		C, I	Kroeker et al. 2011
Spirobranchus	*polytrema*	(Philippi, 1844)	C	Kroeker et al. 2011, Gambi et al. 2016a
Spirobranchus	*triqueter*	(Linnaeus, 1758)	C	Hall-Spencer et al. 2008, Kroeker et al. 2011, Gambi et al. 2016a

Continued

Table 3 (Continued) A taxonomic checklist of the marine species recorded in the CO$_2$ vent sites of the Castello Aragonese, Ischia, as well as nearby ambient zones (approximately within 500 m range)

Genus	Species	Taxonomic authority	Site: Control (C) ambient pH, Intermediate (I) low pH, Acidified (A) extreme low pH	Literature citation
Spirorbis	*marioni*	Caullery & Mesnil, 1897	C	Cigliano et al. 2010
Subadyte	*pellucida*	(Ehlers, 1864)	C	Kroeker et al. 2011, Gambi et al. 2016a
Syllides	*fulvus*	(Marion & Bobretzsky, 1875)	A	Gambi et al. 2016a
Syllis	*armillaris*	(O.F. Mueller, 1771)	C	Gambi et al. 2016a
Syllis	*beneliahue*	(Campoy & Aquezar, 1982)	C	Gambi et al. 2016a
Syllis	*compacta*	Gravier, 1900	C	Gambi et al. 2016a
Syllis	*corallicola*	Verrill, 1900	C	Gambi et al. 2016a
Syllis	*ferrani*	Alos & San Martin, 1987	C	Gambi et al. 2016a
Syllis	*garciai*	(Campoy, 1982)	C	Lanera & Gambi 1993, Gambi et al. 2016a
Syllis	*gerlachi*	(Hartmann-Schroeder, 1960)	C, I, A	Gambi et al. 2016a
Syllis	*gracilis*	Grube, 1840	C, I	Gambi et al. 2016a
Syllis	*krohnii*	Ehlers, 1864	C, I	Gambi et al. 2016a
Syllis	*prolifera*	Krohn, 1852	C, I, A	Gambi et al. 1997a, Lanera & Gambi 1993, Cigliano et al. 2010, Kroeker et al. 2011, Gambi et al. 2016a
Syllis	*rosea*	Langerhans, 1879	I	Gambi et al. 2016a
Syllis	*variegata*	Grube, 1860	C	Lanera & Gambi 1993, Gambi et al. 2016a
Synmerosyllis	*lamelligera*	Saint-Joseph, 1887	C, I	Gambi et al. 2016a
Terebellidae	spp.		C	Kroeker et al. 2011
Trichobranchus	*glacialis*	Malmgren, 1866	C	Cigliano et al. 2010
Trypanosyllis	*coeliaca*	Claparede, 1868	C	Gambi et al. 2016a
Vermiliopsis	*striaticeps*	(Grube, 1862)	C	Kroeker et al. 2011, Gambi et al. 2016a
Phylum Mollusca **Class** **Polyplacophora**				
Acanthochitona	*crinita*	(Pennant, 1777)	C, I	Hall-Spencer et al. 2008
Lepidochitona	*cinerea*	(Linnaeus, 1767)	C	Hall-Spencer et al. 2008

Continued

Table 3 (Continued)   A taxonomic checklist of the marine species recorded in the CO_2 vent sites of the Castello Aragonese, Ischia, as well as nearby ambient zones (approximately within 500 m range)

Genus	Species	Taxonomic authority	Site: Control (C) ambient pH, Intermediate (I) low pH, Acidified (A) extreme low pH	Literature citation
Class Cephalopoda				
Octopus	*vulgaris*	Cuvier, 1797	C, I	Hall-Spencer et al. 2008
Class Gasteropoda				
Alvania	*cimex*	(Linnaeus, 1758)	C	Kroeker et al. 2011
Alvania	*geryonia*	(Nardo, 1847)	I	Kroeker et al. 2011
Alvania	*lineata*	Risso, 1826	C, A	Garrard et al. 2014
Alvania	*mamillata*	Risso, 1826	I	Kroeker et al. 2011
Alvania	*subcrenulata*	(Bucquoy, Dautzenberg & Dollfus, 1884)	C	Cigliano et al. 2010
Alvania	spp.		C, I, A	Kroeker et al. 2011
Aplysia	*depilans*	Gmelin, 1791	C, I, A	Hall-Spencer et al. 2008
Bittium	*latreilli*	(Payraudeau, 1826)	C, I, A	Kroeker et al. 2011, Garrard et al. 2014
Bittium	*reticulatum*	(Da Costa, 1778)	C, I	Hall-Spencer et al. 2008
Bittium	sp.		C, I	Kroeker et al. 2011
Bolma	*rugosa*	(Linnaeus, 1767)	C	Hall-Spencer et al. 2008
Buccinulum	*corneum*	(Linnaeus, 1758)	C, I	Hall-Spencer et al. 2008, Kroeker et al. 2011
Bulla	*striata*	Bruguiere, 1789	I	Kroeker et al. 2011
Calliostoma (*Calliostoma*)	*laugieri laugieri*	Payraudeau, 1826	C	Kroeker et al. 2011, Chiarore & Patti 2013
Calliostoma	sp.		C, I	Kroeker et al. 2011
Cerithium	sp.		C	Kroeker et al. 2011
Cerithium	*vulgatum*	Bruguière, 1792	C, I	Hall-Spencer et al. 2008, Kroeker et al. 2011
Clanculus (*Clanculopsis*)	*jussieui*	Payraudeau, 1826	A	Cigliano et al. 2010
Columbella	*rustica*	(Linnaeus, 1758)	C, I, A	Kroeker *et al.* 2011, Garrard *et al.* 2014
Dikoleps	*nitens*	(Philippi, 1884)	C, A	Cigliano *et al.* 2010
Diodora	sp.		C, I	Kroeker *et al.* 2011
Elysia	*timida*	(Risso, 1818)	C, I, A	Hall-Spencer *et al.* 2008
Etonina	*fulgida*	(J. Adams, 1797)	A	Chiarore & Patti 2013
Fasciolaria	*lignaria*	(Linnaeus, 1758)	C, I	Hall-Spencer *et al.* 2008

Continued

Table 3 (Continued)   A taxonomic checklist of the marine species recorded in the CO_2 vent sites of the Castello Aragonese, Ischia, as well as nearby ambient zones (approximately within 500 m range)

Genus	Species	Taxonomic authority	Site: Control (C) ambient pH, Intermediate (I) low pH, Acidified (A) extreme low pH	Literature citation
Flabellina	sp.		C, I, A	Hall-Spencer *et al.* 2008
Fossarus	*ambigus*	(Linnaeus, 1758)	C	Cigliano *et al.* 2010
Fusinus Fusinus	*pulchellus*	(Philippi, 1844)	C	Kroeker *et al.* 2011
Gibberula	*miliaria*	(Linnaeus, 1858)	C, A	Garrard *et al.* 2014
Gibberula	*philippii*	(Monterosato, 1878)	C, A	Garrard *et al.* 2014
Gibbula (Colliculus)	*adansonii adansonii*	(Payraudeau, 1826)	C, I	Kroeker *et al.* 2011
Gibbula (Forskalena)	*fanulum*	(Gmelin,1791)	C	Cigliano *et al.* 2010
Gibbula (Gibbula)	*varia*	(Linnaeus, 1758)	C	Cigliano *et al.* 2010
Gibbula	sp.		C, I	Kroeker *et al.* 2011
Haliotis	sp.		C, I	Hall-Spencer *et al.* 2008
Hexaplex	*trunculus*	(Linnaeus, 1758)	C, I, A	Hall-Spencer *et al.* 2008, Chiarore & Patti 2013
Hydrobia	sp.		C	Cigliano *et al.* 2010
Jujubibus	*striatus striatus*	(Linnaeus, 1758)	C, I	Hall-Spencer *et al.* 2008, Kroeker *et al.* 2011, Garrard *et al.* 2014
Jujubinus	*gravinae*	(Dautzenberg, 1881)	C, I	Cigliano *et al.* 2010
Littorina (Melaraphe)	*neritoides*	(Linnaeus, 1758)	C, I, A	Hall-Spencer *et al.* 2008
Mangelia	*costulata*	Risso, 1826	C, A	Garrard et al. 2014
Melanella	*polita*	(Linnaeus, 1758)	C	Cigliano *et al.* 2010
Mitrella	*scripta*	(Linnaeus, 1758)	C, I	Kroeker *et al.* 2011
Mitrella	*scripta*	(Linnaeus, 1858)	C, A	Garrard et al. 2014
Muricopsis	*cristata*	(Brocchi, 1814)	C	Kroeker *et al.* 2011
Ocinebrina	*edwardsii*	(Payraudeau, 1826)	C, I	Hall-Spencer *et al.* 2998, Kroeker *et al.* 2011
Patella	*coerulea*	Linnaeus, 1758	C, I, A	Hall-Spencer *et al.* 2008, Rodolfo-Metalpa *et al.* 2011, Chiarore & Patti 2013
Patella	*rustica*	Linnaeus, 1758	C, I, A	Hall-Spencer *et al.* 2008
Patella	sp.		C, I	Kroeker *et al.* 2011
Patella	*ulyssiponensis*	Gmelin, 1791	C, I, A	Hall-Spencer *et al.* 2008
Phorcus	*richardi*	(Payreaudeau, 1826)	I	Cigliano *et al.* 2010
Phorcus (Osilinus)	*turbinatus (turbinatus)*	(Von Born, 1778)	C, I	Hall-Spencer *et al.* 2008, Cigliano *et al.* 2010

Continued

Table 3 (Continued)   A taxonomic checklist of the marine species recorded in the CO_2 vent sites of the Castello Aragonese, Ischia, as well as nearby ambient zones (approximately within 500 m range)

Genus	Species	Taxonomic authority	Site: Control (C) ambient pH, Intermediate (I) low pH, Acidified (A) extreme low pH	Literature citation
Pusia	*tricolor*	(Gmelin, 1791)	C, A	Garrard et al. 2014
Pusillina	*margiminia*	(Nordsieck, 1972)	I	Cigliano et al. 2010
Pusillina	sp.		C, A	Cigliano et al. 2010
Rissoa	*guerinii*	Recluz, 1843	C, A	Garrard et al. 2014
Rissoa	*italiensis*	Verduin, 1985	C, A	Garrard et al. 2014
Rissoa	*lia*	(Monterosato, 1884)	A	Chiarore & Patti 2013
Rissoa	*membranacea var. labiosa*	(Montagu, 1803)	A	Chiarore & Patti 2013
Rissoa	sp.		I	Kroeker et al. 2011
Rissoa	*variabilis*	(Megerle von Mühlfeld, 1824)	C, A	Garrard et al. 2014
Rissoa	*variabilis*	(Von Muehtfeldt, 1824)	C, I, A	Cigliano et al. 2010, Kroeker et al. 2011
Rissoa	*ventricosa*	Desmarest, 1814	A	Cigliano et al. 2010
Rissoa	*violacea*	Desmarest, 1814	C, A	Cigliano et al. 2010, Chiarore & Patti 2013
Scissurella	*costata*	d'Orbigny, 1824	A	Chiarore & Patti 2013
Serpulorbis	*arenaria*	(Linnaeus, 1767)	C	Hall-Spencer *et al.* 2008
Setia	*pulcherrima*	(Jeffreys, 1848)	A	Chiarore & Patti 2013
Stramonita	*haemastoma*	(Linnaeus, 1766)	C, I	Hall-Spencer *et al.* 2008
Tricolia	*pullus pullus*	(Linnaeus, 1758)	C, I	Hall-Spencer *et al.* 2008, Cigliano *et al.* 2010
Tritia (Nassarius)	*incrassata*	(Stroem, 1768)	C, I	Kroeker *et al.* 2011
Tritia (Nassarius) sp.	sp.		C	Kroeker *et al.* 2011
Tritia corniculum (Nassarius) (corniculum)	*corniculum*	(Olivi, 1792)	A	Garrard *et al.* 2014
Vermetus (Vermetus)	*triquetrus*	Ant. Bivona, 1832	C	Hall-Spencer *et al.* 2008
Class Bivalvia				
Abra	*alba*	(W. Wood, 1802)	C, A	Garrard et al. 2014
Anomia	*ephippium*	Linnaeus, 1758	C, I, A	Hall-Spencer *et al.* 2008, Kroeker *et al.* 2011, Chiarore & Patti 2013
Arca	*noae*	Linnaeus, 1758	C, I	Hall-Spencer *et al.* 2008
Cardiidae	sp.		I	Kroeker *et al.* 2011
Chama	*gryphoides*	Linnaeus, 1758	C	Kroeker *et al.* 2011
Gastrochaena	*dubia*	(Pennant, 1777)	C	Kroeker *et al.* 2011

Continued

Table 3 (Continued)   A taxonomic checklist of the marine species recorded in the CO_2 vent sites of the Castello Aragonese, Ischia, as well as nearby ambient zones (approximately within 500 m range)

Genus	Species	Taxonomic authority	Site: Control (C) ambient pH, Intermediate (I) low pH, Acidified (A) extreme low pH	Literature citation
Glans	*elegans*	(Réquien, 1848)	C, I	Kroeker *et al.* 2011
Hiatella	*rugosa*	(Linnaeus, 1767)	C	Kroeker *et al.* 2011
Lima	*lima*	(Linnaeus, 1758)	C	Hall-Spencer *et al.* 2008
Mimachlamys	*varia*	(Linnaeus, 1758)	A	Chiarore & Patti 2013
Musculus	*subpictus*	(Cantraine, 1835)	C, A	Garrard *et al.* 2014
Mytilus	*galloprovincialis*	Lamarck, 1819	C, I, A	Hall-Spencer *et al.* 2008, Ricevuto *et al* 2012, Chiarore & Patti 2013
Ostrea	sp.		C	Hall-Spencer *et al.* 2008
Sub-phylum Crustacea				
Order Mysidacea				
Anchialina	*agilis*	(G.O. Sars, 1877)	C	Wittmann 2001
Anchialina	*oculata*	Hoenigman, 1960	C	Wittmann 2001
Erythrops	*elegans*	(G.O. Sars, 1863)	C	Wittmann 2001
Gastrosaccus	*mediterraneus*	Bacescu, 1970	C	Wittmann 2001
Hemimysis	*lamornae*	Bacescu, 1937	C	Wittmann 2001
Leptomysis	*buergii*	Bacescu, 1966	C	Wittmann & Stagl 1996
Leptomysis	*lingvura*	(Gourret, 1888)	C	Ariani et al. 1993
Leptomysis	*posidoniae*	Wittmann, 1986	C	Ariani et al. 1993, Wittmann & Stagl 1996
Pyroleptomysis	*rubra*	Wittmann, 1985	C	Wittmann 2001
Siriella	*clausii*	G.O. Sars, 1877	C	Wittmann 2001
Siriella	*gracilipes*	(Nouvel, 1942)	C	Wittmann 2001
Order Ostracoda	spp.		C, I, A	Cigliano et al. 2010
Order Copepoda	spp.		C, I, A	Cigliano et al. 2010
Order Decapoda				
Acanthonyx	*lunulatus*	(Risso, 1816)	C, I	Kroeker et al. 2011
Alpheus	*dentipes*	Guérin-Méneville, 1832	C, I, A	Kroeker et al. 2011
Alpheus	*glaber*	(Olivi, 1792)	C, I	Kroeker et al. 2011
Anapagurus	*laevis*	(Bell, 1845)	C, I	Lanera 1987, Kroeker et al. 2011
Athanas	*nitescens*	(Leach, 1814)	C, I	Kroeker et al. 2011
Calcinus	*tubularis*	(Linnaeus, 1767)	C, I	Kroeker et al. 2011, Garrard et al. 2014
Cestopagurus	*timidus*	(Roux, 1830)	C, I	Kroeker et al. 2011, Garrard et al. 2014

Continued

Table 3 (Continued)   A taxonomic checklist of the marine species recorded in the CO_2 vent sites of the Castello Aragonese, Ischia, as well as nearby ambient zones (approximately within 500 m range)

Genus	Species	Taxonomic authority	Site: Control (C) ambient pH, Intermediate (I) low pH, Acidified (A) extreme low pH	Literature citation
Clibanarius	*erythropus*	(Latreille, 1818)	C, I	Kroeker et al. 2011
Eriphia	*verrucosa*	(Forsskål, 1775)	C, I	Kroeker et al. 2011
Eualus	*cranchii*	(Leach, 1817)	C, A	Garrard et al. 2014
Galathea	*bolivari*	Zariquiey Álvarez, 1950	C, A	Garrard et al. 2014
Hippolyte	*inermis*	Leach, 1815	I	Kroeker et al. 2011
Hippolyte	*leptocerus*	(Heller, 1863)	C, I, A	Kroeker et al. 2011, Garrard et al. 2014
Hippolyte	*varians*	Leach, 1814	C, I, A	Kroeker et al. 2011
Maja (Maya)	*crispata (verrucosa)*	(Risso, 1827)	C	Tertschnig 1989
Munida	*intermedia*	A. Milne-Edwards & Bouvier, 1899	C, I	Kroeker et al. 2011
Pachygrapsus	*marmoratus*	(Fabricius, 1787)	C, I, A	Lanera 1987, Hall-Spencer et al. 2008, Kroeker et al. 2011
Palaemon	*serratus*	(Pennant, 1777)	C, I, A	Hall-Spencer et al. 2008
Percon	cf *gibbesi*	(H. Milne Edwards, 1853)	I, A	Gambi et al. 2016b
Pilumnus	*hirtellus*	(Linnaeus, 1761)	C, I, A	Kroeker et al. 2011
Processa	*acutirostris*	Nouvel & Holthuis, 1957	I	Kroeker et al. 2011
Processa	*canaliculata*	Leach, 1815	C, A	Garrard et al. 2014
Order Cirripedia				
Balanus	*perforatus*	Bruguiére, 1789	C, I	Hall-Spencer et al. 2008
Balanus	spp.		C, I, A	Donnarumma et al. 2014
Chthamalus	*stellatus*	(Poli, 1791)	C, I, A	Hall-Spencer et al. 2008
Euraphia	*depressa*	(Poli, 1791)	C, I	Hall-Spencer et al. 2008
Order Amphipoda				
Ampelisca (Ampelisca)	*rubella*	A. Costa, 1864	C A	Scipione 1999 Kroeker et al. 2011
Ampelisca	*serraticauda*	Chevreux, 1888	C, A	Garrard et al. 2014
Ampithoe	*helleri*	G. Karaman, 1975	C	Scipione 1999, Garrard et al. 2014
Ampithoe	*ramondi*	Audouin, 1826	C, I, A	Scipione 1999, Cigliano et al. 2010, Garrard et al. 2014
Ampithoe	sp.		C, I, A	Kroeker et al. 2011

Continued

Table 3 (Continued) A taxonomic checklist of the marine species recorded in the CO_2 vent sites of the Castello Aragonese, Ischia, as well as nearby ambient zones (approximately within 500 m range)

Genus	Species	Taxonomic authority	Site: Control (C) ambient pH, Intermediate (I) low pH, Acidified (A) extreme low pH	Literature citation
Aora	*spinicornis*	Afonso, 1976	C	Scipione 1999
Aora	spp.		C, A	Kroeker et al. 2011, Garrard et al. 2014
Apherusa	*chiereghinii*	Giordani Soika, 1950	C, A	Scipione 1999, Cigliano et al. 2010
Apocorophioum	*acutum*	(Chevreux, 1908)	C	Scipione 1999
Apolochus	cf *picadurus*	(J.L. Barnard, 1962)	C, A	Garrard et al. 2014
Apolochus	sp.		C, I, A	Kroeker et al. 2011
Atylus	*guttatus*	(A. Costa, 1851)	C	Scipione 1999
Caprella	*acanthifera*	Leach, 1814	C, I, A	Scipione 1999, Cigliano et al. 2010, Garrard et al. 2014
Caprella	sp.		C	Cigliano et al. 2010
Colomastix	sp.		C, A	Kroeker et al. 2011
Cymadusa	*crassicornis*	(A. Costa, 1857)	C	Scipione 1999
Dexamine	*spiniventris*	(A. Costa, 1853)	C, I, A	Scipione 1999, Cigliano et al. 2010
Dexamine	*spinosa*	(Montagu, 1813)	C, A	Lanera 1987, Scipione 1999, Garrard et al. 2014
Elasmopus	*rapax*	A. Costa, 1853	C, I	Cigliano et al. 2010
Elasmopus	sp.		C, I, A	Kroeker et al. 2011
Ericthonius	*difformis*	H. Milne Edwards, 1830	A	Garrard et al. 2014
Ericthonius	*punctatus*	(Bate, 1857)	C	Lanera 1987, Scipione 1999, Cigliano et al. 2010, Garrard et al. 2014
Eusiroides	sp.		C	Kroeker et al. 2011
Gammarella	*fucicola*	(Leach, 1814)	A	Cigliano et al. 2010
Gammaropsis	*palmata*	(Stebbing & Robertson, 1891)	C, I, A	Scipione 1999, Cigliano et al. 2010
Gitana	sp.		C	Kroeker et al. 2011
Hyalidae	sp.		C, I, A	Kroeker et al. 2011
Hyale	*camptonyx*	(Heller, 1866)	C, I, A	Cigliano et al. 2010
Iphimedia	*minuta*	G.O. Sars, 1882	C	Cigliano et al. 2010
Ischyrocerus	*inexpectatus*	Ruffo, 1959	C, I, A	Cigliano et al. 2010
Jassa	*marmorata*	Holmes, 1903	C, I	Cigliano et al. 2010
Lembos	*websteri*	Bate, 1857	C, I, A	Scipione 2013, Garrard et al. 2014
Leptocheirus	*guttatus*	(Grube, 1864)	I, A	Scipione 2013
Leptocheirus	*pectinatus*	(Norman, 1869)	I, A	Scipione 2013

Continued

Table 3 (Continued)   A taxonomic checklist of the marine species recorded in the CO_2 vent sites of the Castello Aragonese, Ischia, as well as nearby ambient zones (approximately within 500 m range)

Genus	Species	Taxonomic authority	Site: Control (C) ambient pH, Intermediate (I) low pH, Acidified (A) extreme low pH	Literature citation
Leptocheirus	*pilosus*	Zaddach, 1844	C, A	Cigliano et al. 2010, Scipione 2013
Leucothoe	sp.		C, I, A	Kroeker et al. 2011
Liljeborgia	*dellavallei*	Stebbing, 1906	C, I, A	Kroeker et al. 2011, Garrard et al. 2014
Lysianassa	*costae*	Milne-Edwards, 1830	A	Cigliano et al. 2010
Lysianassa	*pilicornis*	(Heller, 1866)	C, A	Garrard et al. 2014
Maera	spp.		C, I	Kroeker et al. 2011, Garrard et al. 2014
Metaphoxus	*simplex*	(Spence Bate, 1857)	C, I, A	Kroeker et al. 2011, Garrard et al. 2014
Microdeutopus	*chelifer*	(Bate, 1862)	C, I, A	Scipione 2013
Microdeutopus	*obtusatus*	Myers, 1973	C, I, A	Scipione 2013
Microdeutopus	*sporadhi*	Myers, 1969	I, A	Scipione 2013
Monocorophium	sp.		C, I, A	Kroeker et al. 2011
Orchomene	*humilis*	(A. Costa, 1853)	C	Scipione 1999
Peltocoxa	*marioni*	Catta, 1875	C	Scipione 1999
Pereionotus	*testudo*	(Montagu, 1808)	C	Scipione 1999
Photis	*longicaudata*	(Bate & Westwood, 1862)	C	Scipione 1999
Phtisica	*marina*	Slabber, 1769	C	Scipione 1999
Podocerus	sp.		C, I	Kroeker et al. 2011
Protohyale	*schmidtii*	(Heller, 1866)	A	Cigliano et al. 2010, Garrard et al. 2014
Pseudoprotella	*phasma*	(Montagu, 1804)	C	Scipione 1999
Quadrimaera	*inaequipes*	(A. Costa, 1857)	C	Cigliano et al. 2010, Garrard et al. 2014
Stenothoe	*cavimana*	Chevreux, 1908	I, A	Cigliano et al. 2010
Stenothoe	*monoculoides*	(Montagu, 1813)	A	Cigliano et al. 2010
Tethylembos	*viguieri*	(Chevreux, 1911)	C, A	Garrard et al. 2014
Tritaeta	sp.		C, A	Kroeker et al. 2011
Order Tanaidacea				
Chondrochelia (*Leptochelia*)	*savignyi*	(Kroyer, 1842)	C, I, A	Cigliano et al. 2010, Lanera 1987, Garrard et al. 2014
Parapseudes	*dulongii*	(Audouin, 1826)	A	Cigliano et al. 2010
Parapseudes	*latifrons*	(Grube,1864)	I, A	Cigliano et al. 2010
Parapseudidae	spp.		C, I	Kroeker et al. 2011
Pseudoleptochelia	*anomala*	(Sars, 1882)	C, A	Garrard et al. 2014
Pseudoparatanais incertae sedis			C, I	Kroeker et al. 2011

Continued

Table 3 (Continued)   A taxonomic checklist of the marine species recorded in the CO_2 vent sites of the Castello Aragonese, Ischia, as well as nearby ambient zones (approximately within 500 m range)

Genus	Species	Taxonomic authority	Site: Control (C) ambient pH, Intermediate (I) low pH, Acidified (A) extreme low pH	Literature citation
Order Isopoda				
Apanthura	corsica	Amar, 1953	C, A	Garrard et al. 2014
Carpias	spp.		C, I	Kroeker et al. 2011
Cymodoce	hanseni	Dumay, 1972	C, A	Lanera 1987, Cigliano et al. 2010, Garrard et al. 2014
Cymodoce	truncata	Leach, 1814	C	Cigliano et al. 2010, Turner et al. 2016
Dynamene	bicolor (Dynamene) (torelliae)	(Rathke, 1837)	C	Turner et al. 2016
Dynamene	bifida	Torelli, 1930	I, A	Cigliano et al. 2010, Turner et al. 2016
Dynamene	tubicauda	Holdich, 1968	C, A	Garrard et al. 2014
Gnathia	sp.		C, I, A	Kroeker et al. 2011
Jaeropsis	spp.		C, I, A	Kroeker et al. 2011
Joeropsis	brevicornis	Koehler, 1885	C, A	Garrard et al. 2014
Mesanthura	cf. romulea	Poore & Lew Ton, 1986	C, I	Kroeker et al. 2011, Gambi et al. 2016b
Paranthura	nigropunctata	(Lucas, 1849)	C	Wagele 1982, Cigliano et al. 2010
Paranthuridae	sp.		C, I	Kroeker et al. 2011
Zenobiana	prismatica	(Risso, 1826)	C	Cigliano et al. 2010
Order Cumacea	spp.		C, A	Cigliano et al. 2010
Phylum Chaetognata				
Sagitta	sp.		C, I, A	Cigliano et al. 2010
Phylum Bryozoa				
Callopora	lineata	(Linnaeus, 1767)	C, I	Martin et al. 2008
Diaperoecia	spp.		C, A	Nogueira et al. 2017
Elettra	posidoniae	Gautier, 1954	C,I, A	Martin et al. 2008, Donnarumma et al. 2014, Nogueira et al. 2017
Fenestrulina	joannae	(Calvet, 1902)	C, A	Nogueira et al. 2017
Microporella	ciliata	(Pallas, 1766)	C, I	Martin et al. 2008
Patinella	spp.		C, A	Nogueira et al. 2017
Tubulipora	sp.		C, I	Martin et al. 2008
Phylum Echinodermata				
Class Holothuroidea				
Holothuria	forskali	Delle Chiaje, 1823	C, I	Hall-Spencer et al. 2008
Holothuria	tubulosa	Gmelin, 1788	C, I	Kroll & Jangoux 1989, Hall-Spencer et al. 2008

Continued

Table 3 (Continued)   A taxonomic checklist of the marine species recorded in the CO_2 vent sites of the Castello Aragonese, Ischia, as well as nearby ambient zones (approximately within 500 m range)

Genus	Species	Taxonomic authority	Site: Control (C) ambient pH, Intermediate (I) low pH, Acidified (A) extreme low pH	Literature citation
Class Echinoidea				
Arbacia	*lixula*	(Linnaeus, 1758)	C, I	Hall-Spencer et al. 2008, Kroeker et al. 2013a
Paracentrotus	*lividus*	(Lamarck, 1816)	C, I	Tertschnig 1989, Hall-Spencer et al. 2008, Kroeker et al. 2013a
Class Asteroidea				
Coscinasterias	*tenuispina*	(Lamarck, 1816)	C, I	Hall-Spencer et al. 2008
Echinaster	*sepositus*	(Retzius, 1783)	C, I	Hall-Spencer et al. 2008
Class Ophiuroidea				
Amphipholis	*squamata*	(Delle Chiaje, 1828)	A, C	Gambi M.C. personal collection (2012)
Phylum Pisces **Class Osteichthyes**				
Chromis	*chromis*	(Linnaeus, 1758)	C, I	Guidetti & Bussotti 1998, Hall-Spencer et al. 2008
Coris	*julis*	(Linnaeus, 1758)	C	Guidetti & Bussotti 1998
Dentex	*dentex*	(Linnaeus, 1758)	C	Guidetti & Bussotti 1998
Diplodus	*annularis*	(Linnaeus, 1758)	C, I	Guidetti & Bussotti 1998, Hall-Spencer et al. 2008
Diplodus	*sargus sargus*	(Linnaeus, 1758)	C	Guidetti & Bussotti 1998
Diplodus	*vulgaris*	(Geofroy Saint-Hilaire, 1817)	C	Guidetti & Bussotti 1998
Labrus	*mixtus*	Linnaeus, 1758	C, I	Hall-Spencer et al. 2008
Labrus	*viridis*	Linnaeus, 1758	C, I	Hall-Spencer et al. 2008
Lithognathus	*mormyrus*	(Linnaeus, 1758)	C	Guidetti & Bussotti 1998
Mullus	*surmuletus*	(Linnaeus, 1758)	C	Guidetti & Bussotti 1998
Muraena	*helena*	Linnaeus, 1758	C, I, A	Hall-Spencer et al. 2008

Continued

Table 3 (Continued)   A taxonomic checklist of the marine species recorded in the CO_2 vent sites of the Castello Aragonese, Ischia, as well as nearby ambient zones (approximately within 500 m range)

Genus	Species	Taxonomic authority	Site: Control (C) ambient pH, Intermediate (I) low pH, Acidified (A) extreme low pH	Literature citation
Oblada	*melanura*	(Linnaeus, 1758)	C	Guidetti & Bussotti 1998
Sarpa	*salpa*	(Linnaeus, 1758)	C, I, A	Guidetti & Bussotti 1998, Hall-Spencer et al. 2008, Donnarumma et al. 2014
Scorpaena	*porcus*	Linnaeus, 1758	C, I, A	Hall-Spencer et al. 2008
Sparus	*aurata*	(Linnaeus, 1758)	C	Guidetti & Bussotti 1998
Sphyraena	*sphyraena*	(Linnaeus, 1758)	C	Guidetti & Bussotti 1998
Spondyliosoma	*cantharus*	(Linnaeus, 1758)	C	Guidetti & Bussotti 1998
Symphodus	*cinereus*	(Bonnaterre, 1788)	C	Guidetti & Bussotti 1998
Symphodus	*melanocercus*	(Risso, 1810)	C	Guidetti & Bussotti 1998
Symphodus	*ocellatus*	(Forsskal, 1775)	C	Guidetti & Bussotti 1998
Symphodus	*roissali*	(Risso, 1810)	C, I	Hall-Spencer et al. 2008
Symphodus	*rostratus*	(Bloch, 1791)	C	Guidetti & Bussotti 1998
Symphodus	*tinca*	Linnaeus, 1758	C	Guidetti & Bussotti 1998
Tripterygion	*tripteronotus*	(Risso, 1810)	C, I	Hall-Spencer et al. 2008

Note: Species have been organised by Phylum and then either by Class or Order. Taxonomic authority with the name and year of the original description has been given where available but are not indicated for those species only identified to genus level, indicated by "sp." (singular) or "spp." (plural). Species are named according to the WoRMS World Register of Marine Species (WoRMS Editorial Board 2017) where the previously reported name is given in brackets under the Genus and Species names. For each species, the pH zone where the animal is found at is specified, where C is control pH (North 1 (N1), South 1 (S1) or other control areas within 500 m of the S1 and N1 zones), I is intermediate pH (N2, S2) and A is acidified extreme low pH (N3, S3) equivalent to sites labelled in Figure 1 and pH levels indicated in Figure 2. The relevant literature citation regarding that species is also indicated. This biodiversity checklist for the Ischian vent system provides a useful baseline to evaluate further changes in the species and communities.

References

Agostini, S., Wadaa, S., Kona, K., Omorib, A., Kohtsuka, H., Fujimura, H., Tsuchiya, Y., Sato, T., Shinagawaa, H., Yamada, Y. & Inaba, K. 2015. Geochemistry of two shallow CO_2 seeps in Shikine Island (Japan) and their potential for ocean acidification research. *Regional Studies in Marine Science* **2**, 45–53.

Anlauf, H., D'Croz, L. & O'Dea, A. 2011. A corrosive concoction: the combined effects of ocean warming and acidification on the early growth of a stony coral are multiplicative. *Journal of Experimental Marine Biology and Ecology* **397**, 13–20.

Ariani, A.P., Wittmann, K.J. & Franco, E. 1993. A comparative study of static bodies in mysid crustaceans: Evolutionary implications of crystallographics characteristics. *The Biological Bulletin* **185**, 393–404.

Arias, A., Giangrande, A., Gambi, M.C. & Anadon, N. 2013. Biology and new records of the invasive species *Branchiomma bairdi* (Annelida: Sabellidae) in the Mediterranean Sea. *Mediterranean Marine Science* **14**, 162–171.

Arnold, T.M., Mealey, C., Leahey, H., Miller, A.W., Hall-Spencer, J., Milazzo, M. & Maers, K. 2012. Ocean acidification and the loss of phenolic substances in marine plants. *PLoS ONE* **7**, e35107.

Bacci, G. 1969. A future for ecological research at the Zoological Station of Naples. *Pubblicazioni della Stazione Zoologica di Napoli* **37**, 7–15.

Baggini, C., Issaris, Y., Salomidi, M. & Hall-Spencer, J.M. 2015. Herbivore diversity improves benthic community resilience to ocean acidification. *Journal of Experimental Marine Biology and Ecology* **469**, 98–104.

Baggini, C., Salomidi, M., Voutsinas, E., Bray, L., Krasakopoulou, E. & Hall-Spencer, J.M. 2014. Seasonality affects macroalgal community response to increases in pCO_2. *PLoS ONE* **9**, e106520.

Barry, J.P., Hall-Spencer, J.M. & Tyrrell, T. 2010. *In situ* perturbation experiments: natural venting sites, spatial/temporal gradients in ocean pH, manipulative *in situ* pCO_2 perturbations. In *Guide to Best Practices for Ocean Acidification Research and Data Reporting*, U. Riebesell (ed.). Luxembourg: Publications Office of the European Union, 123–136.

Basso, L., Hendriks, I., Rodriguez-Navarro, A., Gambi, M.C. & Duarte, C.M. 2015. High pCO_2 conditions affect growth, survival and metabolism of juvenile pen shells (*Pinna nobilis*) at a natural CO_2 vent system (Italy). *Estuaries and Coasts* **38**, 1986–1999.

Bellan, G., Desrosiers, G. & Willsie, A. 1988. Use of an annelid pollution index for monitoring a moderately polluted littoral zone. *Marine Pollution Bulletin* **19**, 662–665.

Benini, A. & Gialanella, C. 2017a. Ischia in età romana: notizie preliminari sugli scavi subacquei di Cartaromana. *Proceedings Symposium Archeologia Subacquea 2.0. Una proposta per il futuro del patrimonio culturale sommerso in Italia*. In press. (In Italian).

Benini, A. & Gialanella, C. 2017b. Ischia tra terra e mare. Notizie preliminari sugli scavi di Cartaromana. *Atti del Convegno "Nautae longe a patria sua vivunt", Il Mediterraneo e la Storia II, (Ischia 2015). Acta Istituti Romani Finlandiae*. **45**, 13–27. (In Italian).

Boatta, F., D'Alessandro, W., Gagliano, A.L., Liotta, M., Milazzo, M., Rodolfo-Metalpa, R., Hall-Spencer, J.M. & Parello, F. 2013. Geochemical survey of Levante Bay, Vulcano Island (Italy), a natural laboratory for the study of ocean acidification. *Marine Pollution Bulletin* **73**, 485–494.

Boudouresque, C.-F. & Cinelli F. 1971. Le peuplement des biotopes sciaphiles superficiels de mode battue de l'île d'Ischia (Golfe de Naples, Italie). *Pubblicazioni Stazione Zoologica di Napoli* **39**, 1–43.

Bray, L., Pancucci-Papadopoulou, M.A. & Hall-Spencer, J.M. 2014. Sea urchin response to rising pCO_2 shows ocean acidification may fundamentally alter the chemistry of marine skeletons. *Mediterranean Marine Science* **15**, 510–519.

Brinkman, A.T.J. & Smith, A.M. 2015. Effect of climate change on crustose coralline algae at a temperate vent site, White Island, New Zealand. *Marine and Freshwater Research* **66**, 360–370.

Brothers, C.J., Harianto, J., McClintock, J.B. & Byrne, M. 2016. Sea urchins in a high-CO_2 world: the influence of acclimation on the immune response to ocean warming and acidification. *Proceedings of the Royal Society B* **283**, 20161501.

Buchner, G. & Ridgway, D. 1993. *Pithekoussai and the First Western Greeks*. Pithkoussai I Monumenti Antichi dei Lincei, Serie monografica, Cambridge: Cambridge University Press, vol. **4**, Roma, 851 pp.

Buia, M.C., Ferrara, A., Rodolfo-Metalpa, R., Lorenti, M. & Hall-Spencer, J.M. 2009. Effects of seawater acidification on *Posidonia oceanica* system. *Proceedings Mediterranean Seagrass Workshop 2009*, September 6-10 2009. Hvar, Croatia: **22** (Abstract).

Buia, M.C., Gambi, M.C., Lorenti, M., Dappiano, M. & Zupo, V. 2003. Aggiornamento sulla distribuzione e sullo stato ambientale dei sistemi a fanerogame marine (*Posidonia oceanica* e *Cymodocea nodosa)* delle isole Flegree. *Accademia Scienze Lettere ed Arti Napoli, Memorie Società Scienze Fisiche e Matematiche* **5**, 163–186.

Buia, M.C., Mazzella, L., Russo, G.F. & Scipione, M.B. 1995. Observations on the distribution of *Cymodocea nodosa* (Ucria) Aschers. Prairies around the island of Ischia (Gulf of Naples). *Rapports Commission internationale exploration scientifique de la Mer Méditerranée*, **29**, 205–208.

Bush, A., Mokany, K., Catullo, R., Hoffmann, A., Kellermann, V., Sgro, C., McEvey, S. & Ferrier, S. 2016. Incorporating evolutionary adaptation in species distribution modelling reduces projected vulnerability to climate change. *Ecology Letters* **19**, 1468–1478.

Byrne, M. 2011. Impact of ocean warming and ocean acidification on marine invertebrate life history stages: vulnerabilities and potential for persistence in a changing ocean. *Oceanography and Marine Biology Annual Review* **49**, 1–42.

Byrne, M., Hart, M.W., Cerra, A. & Cisternas, P. 2003. Reproduction and larval morphology of broadcasting and viviparous species in the *Cryptasterina* species complex. *The Biological Bulletin* **205**, 285–294.

Byrne, M., Ho, M.A., Wong, E., Soars, N., Selvakumaraswamy, P., Sheppard Brennand, H., Dworjanyn, S.A. & Davis, A.R. 2011. Unshelled abalone and corrupted urchins, development of marine calcifiers in a changing ocean. *Proceedings of the Royal Society B* **278**, 2376e2383.

Byrne, M., Lamare, M., Winter, D., Dworjanyn, S.A. & Uthicke, S. 2013. The stunting effect of a high CO_2 ocean on calcification and development in sea urchin larvae, a synthesis from the tropics to the poles. *Philosophical Transactions of the Royal Society B* **368**, 20120439.

Caldeira, K. & Wickett, M.E. 2005. Ocean model predictions of chemistry changes from carbon dioxide emissions to the atmosphere and ocean. *Journal of Geophysical Research* **110**, C09S04.

Calosi, P., Rastrick, S.P., Graziano, M., Thomas, S.C., Baggini, C., Carter, H.A., Hall-Spencer, J.M., Milazzo, M. & Spicer, J.I. 2013a. Distribution of sea urchins living near shallow water CO_2 vents is dependent upon species acid-base and ion-regulatory abilities. *Marine Pollution Bulletin* **73**, 470–484.

Calosi, P., Rastrick, S.P.S., Lombardi, C., de Guzman, H.J., Davidson, L., Jahnke, M., Giangrande, A., Hardege, J.D., Schulze, A., Spicer, J.I. & Gambi, M.C. 2013b. Adaptation and acclimatization to ocean acidification in marine ectotherms: an *in situ* transplant experiment with polychaetes at a shallow CO_2 vent system. *Philosophical Transactions of the Royal Society B* **368**, 20120444.

Carey, M., Harianto, J. & Byrne, M. 2016. Sea urchins in a high-CO_2 world: partitioned effects of body size, ocean warming and acidification on metabolic rate. *Journal of Experimental Biology* **219**, 1178–1186.

Carter, J.G. 1980. Environmental and biological controls of bivalve shell mineralogy and structure. In *Skeletal Growth of Aquatic Organisms*, D.C. Rhoads & R.A. Lutz (eds). New York: Plenum. 69–115.

Catarino, A.I., Bauwens, M. & Dubois, P. 2012. Acid-base balance and metabolic response of the sea urchin *Paracentrotus lividus* to different seawater pH and temperatures. *Environmental Science and Pollution Research* **19**, 2344–2353.

Celis-Pla, P.S.M., Hall-Spencer, J.M., Horta, P.A., Milazzo, M., Korbee, N., Cornwall, C.E. & Figueroa, F.L. 2015. Macroalgal responses to ocean acidification depend on nutrient and light levels. *Frontiers in Marine Science* **2**, 26.

Chiarore, A. & Patti, F.P. 2013. Molluschi associati all'alga bruna *Sargassum vulgare* (C. Agardh, 1820) (Fucales, Sargassaceae) rinvenuti lungo le coste dell'isola di Ischia (Napoli): Checklist preliminare. *Lavori S.I.M.* 31, 10–11 (In Italian).

Cigliano, M., Gambi, M.C., Rodolfo-Metalpa, R., Patti, F.P. & Hall-Spencer, J.M. 2010. Effects of ocean acidification on invertebrate settlement at natural volcanic CO_2 vents. *Marine Biology* **157**, 2489–2502.

Cinelli, F., Boudouresque, C.-F., Mazzella, L. & Richard, M. 1981. Algae marine rare o nuove per la flora italica. *Quaderni del Laboratorio di Tecnologia della Pesca, Ancona*, **3**, 467–480. (In Italian)

Connell, S.D., Kroeker, K.J., Fabricius, K.E., Kline, D.I. & Russell, B.D. 2013. The other ocean acidification problem: CO_2 as a resource among competitors for ecosystem dominance. *Philosophical Transactions of the Royal Society B* **368**, 20120442.

Cookson, L.J. & Lorenti, M. 2001. A new species of limnoriid seagrass borer (Isopoda) from the Mediterranean. *Crustaceana* **74**, 339–346.

Cornwall, C.E., Boyd, P.W., McGraw, C.M., Hepburn, C.D., Pilditch, C.A., Morris, J.N. Smith, A.M. & Hurd, C.L. 2014. Diffusion boundary layers ameliorate the negative effects of ocean acidification on the temperate coralline macroalga *Arthrocardia corymbosa*. *PLoS ONE* **9**, e97235.

Cornwall, C.E., Revill, A.T., Hall-Spencer, J.M., Milazzo, M., Raven J.A. & Hurd, C.L. 2017. Inorganic carbon physiology underpins macroalgal responses to elevated CO_2. *Scientific Reports* 7, 46297.

Crook, E.D., Potts, D., Rebolledo-Vieyra, M., Hernandez, L. & Paytan, A. 2012. Calcifying coral abundance near low-pH springs: implications for future ocean acidification. *Coral Reefs* 31, 239–245.

Dando, P., Hughes, J., Leahy, Y., Niven, S., Taylor, L. & Smith, C. 1995. Gas venting rates from submarine hydrothermal areas around the island of Milos, Hellenic Volcanic Arc. *Continental Shelf Research* 15, 913–929.

de Alteriis, G., Insinga, D., Morabito, S., Morra, V., Chiocci, F.L., Terrasi, F., Lubritto, C., Di Benedetto, C. & Pazzanese, M. 2010. Age of submarine debris avalanches and tephrostratigraphy offshore Ischia Island, Tyrrhenian Sea, Italy. *Marine Geology* 278, 1–18.

de Alteriis, G., Tonielli, R., Passaro, S. & De Lauro, M. 2006. *Isole Flegree (Ischia e Procida). Serie: Batimetria dei fondali marini della Campania*. Liguori Editore, Napoli, 73 pp. (in Italian)

de Alteriis, G. & Toscano, F. 2003. Introduzione alla geologia dei mari circostanti le isole Flegree di Ischia, Procida e Vivara. *Accademia Scienze Lettere ed Arti Napoli, Memorie Società Scienze Fisiche e Matematiche* 5, 3–25. (in Italian)

Dekov, M. & Savelli, C. 2004. Hydrothermal activity in the SE Tyrrhenian Sea: an overview of 30 years of research. *Marine Geology* 204, 161–185.

Del Pasqua, M., Schulze, A., Tovar-Hernández, M.A., Keppel, E., Lezzi, M., Gambi, M.C. & Giangrande, A. 2018. Racing for the Mediterranean: Clarifying the taxonomic status of *Branchiomma bairdi* and *Branchiomma boholense* (Annelida: Sabellidae) using molecular and morphological evidence. *PLoS ONE* 13, e0197104.

Di Cioccio, D., Buia, M.C. & Zingone, A. 2014. Ocean acidification will not deliver us from *Ostreopsis*. Proceedings ISSHA Conference. In *Harmful Algae 2012. Proceedings of the 15th International Conference on Harmful Algae. International Society for the Study of Harmful Algae*, H.G. Kim, B. Reguiera, G.M. Hallegraeff, C.K. Lee & J.K. Choi (eds), Changwon, Korea, 85–88.

Dias, B.B., Hart, M.B., Smart, C.W. & Hall-Spencer, J.M. 2010. Modern seawater acidification: the response of foraminifera to high-CO_2 conditions in the Mediterranean Sea. *Journal of the Geological Society* 167, 843–846.

Diaz-Pulido, G., Gouezo, M., Tilbrook, B., Dove, S. & Anthony, K.R.N. 2010. High CO_2 enhances the competitive strength of seaweeds over corals. *Ecology Letters* 14, 156–162.

Donnarumma, L., Lombardi, C., Cocito, S. & Gambi, M.C. 2014. Settlement pattern of *Posidonia oceanica* epibionts along a gradient of ocean acidification: an approach with mimics. *Mediterranean Marine Science* 15, 498–509.

Doo, S.S., Kazuhiko, F., Byrne, M. & Uthicke, S. 2014. Fate of calcifying tropical symbiont-bearing large benthic foraminifera: living sands in a changing ocean. *The Biological Bulletin* 226, 169–186.

Doropoulos, C., Ward, S., Diaz-Pulido, G., Hoegh-Guldberg, O. & Mumby, P.J. 2012. Ocean acidification reduces coral recruitment by disrupting intimate larval-algal settlement interactions. *Ecology Letters* 15, 338–346.

Dworjanyn, S.A., Byrne, M. 2018. Impacts of ocean acidification on sea urchin growth across the juvenile to mature adult life-stage transition is mitigated by warming. *Proceedings of the Royal Society B: Biological Sciences* 285, 20172684.

Fabricius, K.E., De'ath, G., Noonan, S. & Uthicke, S. 2014. Ecological effects of ocean acidification and habitat complexity on reef-associated macroinvertebrate communities. *Proceedings of the Royal Society B* 281, 20132479.

Fabricius, K.E., Kluibenschedl, A., Harrington, L., Noonan, S. & De'ath, G. 2015. *In situ* changes of tropical crustose coralline algae along carbon dioxide gradients. *Scientific Reports* 5, 9537.

Fabricius, K.E., Langdon, C., Uthicke, S., Humphrey, C., Noonan, S., De'ath, G., Okazaki, R., Muehllehner, N., Glas, M.S. & Lough, J.M. 2011. Losers and winners in coral reefs acclimatized to elevated carbon dioxide concentrations. *Nature Climate Change* 1, 165–169.

Fitzer, S.C., Cusack, M., Phoenix, V.R. & Kamenos, N.A. 2014. Ocean acidification reduces the crystallographic control in juvenile mussel shells. *Journal of Structural Biology* 188, 39–45.

Foo, S.A. & Byrne, M. 2016. Acclimatization and adaptive capacity of marine species in a changing ocean. *Advances in Marine Biology* 74, 69–116.

Foo, S.A., Byrne, M. & Gambi, M.C. 2018. Residing at low pH matters, resilience of the egg jelly coat of sea urchins living at a CO_2 vent site. *Marine Biology* 165, 97.

Gambi, M.C. 2014. Emissioni sommerse di CO_2 lungo le coste dell'isola d'Ischia. Rilievi su altre aree come possibili laboratori naturali per lo studio dell'acidificazione e cambiamento climatico a mare. *Notiziario S.I.B.M.* **66**, 67–79 (in Italian).

Gambi, M.C. & Cafiero, G. 2001. Functional diversity in the *Posidonia oceanica* ecosystem: an example with polychaete borers of the scales. In *Mediterranean Ecosystems: Structure and Processes*, F. M. Faranda et al. (eds). Italy: Springer-Verlag, 399–405.

Gambi, M.C., De Lauro, M., Iannuzzi, F. (eds). 2003. Ambiente marino costiero e territorio delle isole Flegree (Ischia, Procida e Vivara). Risultati di uno studio multidisciplinare. *Accademia Scienze Lettere ed Arti Napoli, Memorie Società Scienze Fisiche e Matematiche*, Liguori Ed. Napoli, Italy, **5**, 1–425 (In Italian)

Gambi, M.C., Hall-Spencer, J.M., Cigliano, M., Cocito, S., Lombardi, C., Lorenti, M., Patti, F.P., Porzio, L., Rodolfo-Metalpa, R., Scipione, M.B. & Buia, M.C. 2010. Using volcanic marine CO_2 vents to study the effects of ocean acidification on benthic biota: highlights from Castello Aragonese d'Ischia (Tyrrhenian Sea). *Biologia Marina Mediterranea* **17**, 86–88.

Gambi, M.C., Lorenti, M., Bussotti, S. & Guidetti, P. 1997b. Borers in *Posidonia oceanica* scales: taxonomical composition and occurrence. *Biologia Marina Mediterranea* **4**, 384–387.

Gambi M.C, Ramella l., Sella G., Protto P., Aldieri E. 1997a. Variation in genome size in benthic polychaetes: systematic and ecological relationships. *Journal of the Biological Association U.K.*, **77**, 1045–1057.

Gambi, M.C., Lorenti, M., Patti, F.P. & Zupo, V. 2016b. An annotated list of alien marine species of the Ischia Island. *Notiziario S.I.B.M.* **70**, 64–68.

Gambi, M.C., Musco, L., Giangrande, A., Badalamenti, F., Micheli, F. & Kroeker, K.J. 2016a. Distribution and functional traits of polychaetes in a CO_2 vent system: winners and losers among closely related species. *Marine Ecology Progress Series* **550**, 121–134.

Gambi, M.C., Zupo, V., Buia, M.C. & Mazzella, L. 2000. Feeding ecology of the polychaete *Platynereis dumerilii* (Audouin & Milne Edwards) (Nereididae) in the seagrass *Posidonia oceanica* system: role of the epiphytic flora. *Ophelia* **53**, 189–202.

Garrard, S.L., Gambi, M.C., Scipione, M.B., Patti, F.P., Lorenti, M., Zupo, V., Paterson, D.M. & Buia, M.C. 2014. Indirect effects may buffer negative responses of seagrass invertebrate communities to ocean acidification. *Journal of Experimental Marine Biology and Ecology* **461**, 31–38.

Garrard, S.L., Hunter, R.C., Frommel, A.Y., Lane, A.C., Phillips, J.C., Cooper, R., Dineshram, R., Cardini, U., McCoy, S.J., Arnberg, M., Rodrigues Alves, B.G., Annane, S., de Orte, M.R., Kumar, A., Aguirre-Martınez, G.V., Maneja, R. H., Basallote, M.D., Ape, F., Torstensson, A. & Bjoerk, M.M. 2012. Biological impacts of ocean acidification: a postgraduate perspective on research priorities. *Marine Biology* **160**, 1789–1805.

Garilli, V., Rodolfo-Metalpa, R., Scuderi, D., Brusca, L., Parrinello, D., Rastrick, S.P.S., Foggo, A., Twitchett, R.J., Hall-Spencer, J.M. & Milazzo, M. 2015. Physiological advantages of dwarfing in surviving extinctions in high-CO_2 oceans. *Nature Climate Change* **5**, 678–682.

Gattuso, J.P., Magnan, A., Billé, R., Cheung, W.W.L., Howes, E.L., Joos, F., Allemand, D., Bopp, L., Cooley, S.R., Eakin, C.M., Hoegh-Guldberg, O., Kelly, R.P., Pörtner, H.O., Rogers, A.D., Baxter, J.M., Laffoley, D., Osborn, D., Rankovic, A., Rochette, J., Sumaila, U.R., Treyer, S. & Turley, C. 2015. Contrasting futures for ocean and society from different anthropogenic CO_2 emissions scenarios. *Oceanography* **349**, 4722–4721.

Gaylord, B., Kroeker, K.J., Sunday, J.M., Anderson, K.M., Barry, J.P., Brown, N.E., Connell, S.D., Dupont, S., Fabricius, K.E., Hall-Spencer, J.M., Klinger, T., Milazzo, M., Munday, P.L., Russell, B.D., Sanford, E., Schreiber, S.J., Thiyagarajan, V., Vaughan, H., Widdicombe, S. & Harley, C.D.G. 2015. Ocean acidification through the lens of ecological theory. *Ecology* **96**, 3–15.

Giangrande, A. 1988. Polychaete zonation and its relation to algal distribution down a vertical cliff in the western Mediterranean (Italy): a structural analysis. *Journal of Experimental Marine Biology and Ecology* **120**, 263–276.

Giangrande, A., Gambi, M.C., Micheli, F. & Kroeker, K.J. 2014. Fabriciidae (Annelida, Sabellida) from a naturally acidified coastal system (Italy) with description of two new species. *Journal Marine Biological Association of the United Kingdom* **94**, 1417–1427.

Giraud, G., Boudouresque, C.F., Cinelli, F., Fresi, E. & Mazzella, L. 1979. Observations sur l'herbier de *Posidonia oceanica* (L.) Delile autor de l'ile d'Ischia (Italie). *Giornale Botanico Italiano* **113**, 261–274. (in French)

Goffredo, S., Prada, F., Caroselli, E., Capaccioni, B., Zaccanti, F., Fantazzini, P., Fermani, S., Reggi, M., Levy, O., Fabricius, K.E., Dubinsky, Z. & Falini, G. 2014. Biomineralization control related to population density under ocean acidification. *Nature Climate Change* **4**, 293–297.

Goodwin, C., Rodolfo-Metalpa, R., Picton, B. & Hall-Spencer, J.M. 2013. Effects of ocean acidification on sponge communities. *Marine Ecology* **35**, 41–49.

Groeben, C. 1985. Anton Dohrn – the statesman of Darwinism. *The Biological Bulletin* **168**, 4–25.

Gugliandolo, C., Italiano, F. & Maugeri, T.L. 2006. The submarine hydrothermal system of Panarea (Southern Italy): biogeochemical processes at the thermal fluids-sea bottom interface. *Annals of Geophysics*, **49**, 783–792.

Guidetti, P. & Bussotti, S. 1998. Juveniles of littoral fish species in shallow seagrass beds: preliminary quali-quantitative data. *Biologia Marina Mediterranea* **5**, 347–350.

Guinotte, J.M. & Fabry, V.J. 2008. Ocean acidification and its potential effects on marine invertebrates. *Annual New York Academy of Sciences* **1134**, 320–343.

Hahn, S., Rodolfo-Metalpa, R., Griesshaber, E., Schmahl, W.W., Buhl, D., Hall-Spencer, J.M., Baggini, C., Fehr, K.T. & Immenhauser, A. 2012. Marine bivalve shell geochemistry and ultrastructure from modern low pH environments: environmental effect *versus* experimental bias. *Biogeosciences* **9**, 1897–1914.

Hall-Spencer, J.M. & Allen, R. 2015. The impact of CO_2 emissions on "nuisance" marine species. *Research and Reports in Biodiversity Studies* **4**, 33–46.

Hall-Spencer, J.M. & Rodolfo-Metalpa, R. 2012. Effects of ocean acidification on Mediterranean coastal habitats: lessons from carbon dioxide vents off Ischia. In *Life in the Mediterranean Sea: A Look at Habitat Changes. Nova Science Publishers*, N. Stambler (ed.). Israel: Bar Ilan University, 671–684.

Hall-Spencer, J.M., Rodolfo-Metalpa, R., Martin, S., Ransome, E., Fine, M., Turner, S.M., Rowley, S.J., Tedesco, D. & Buia, M.C. 2008. Volcanic carbon dioxide vents show ecosystem effects of ocean acidification. *Nature* **454**, 96–99.

Harvey, B.P., Gwynn-Jones, D. & Moore, P.J. 2013. Meta-analysis reveals complex marine biological responses to the interactive effects of ocean acidification and warming. *Ecology and Evolution* **3**, 1016–1030.

Harvey, B.P., McKeown, N.J., Rastrick, S.P.S., Bertolini, C., Foggo, A., Graham, H., Hall-Spencer, J.M., Milazzo, M., Shaw, P.W., Small, D.P. & Moore, P.J. 2016. Individual and population-level responses to ocean acidification. *Scientific Reports* **6**, 20194.

Hassenrück, C., Fink, A., Lichtschlag, A., Tegetmeyer, H.E., de Beer, D. & Ramette, A. 2016. Quantification of the effects of ocean acidification on sediment microbial communities in the environment: the importance of ecosystem approaches. *FEMS Microbiology Ecology* **92**, fiw027.

Heuss, T. 1991. *Anton Dohrn: A Life for Science*. Germany: Springer.

Hofmann, G.E., Smith, J.E., Johnson, K.S., Send, U., Levin, L.A., Micheli, F., Paytan, A., Price, N.N., Peterson, B., Takeshita, Y., Matson, P.G., Derse Crook, E., Kroeker, K.J., Gambi, M.C., Rivest, E.B., Frieder, C.A., Yu, P.C. & Martz, T.R. 2011. High-frequency dynamics of ocean pH: a multi-ecosystem comparison. *PLoS ONE* **6**, e28983.

Howes, E.L., Joos, F., Eakin, C.M. & Gattuso, J-P. 2015. An updated synthesis of the observed and projected impacts of climate change on the chemical, physical and biological processes in the oceans. *Frontiers in Marine Science* **2**, 36.

Hughes, T.P., Kerry, J.T., Álvarez-Noriega, M., Álvarez-Romero, J.G., Anderson, K.D., Baird, A.H., Babcock, R.C., Beger, M., Bellwood, D.R., Berkelmans, R., Bridge, T.C., Butler, I.R., Byrne, M., Cantin, N.E., Comeau, S., Connolly, S.R., Cumming, G.S., Dalton, S.J., Diaz-Pulido, G., Eakin, C.M., Figueira, W.F., Gilmour, J.P., Harrison, H.B., Heron, S.F., Hoey, A.S., Hobbs, J.-P.A., Hoogenboom, M.O., Kennedy, E.V., Kuo, C.-Y., Lough, J.M., Lowe, R.J., Liu, G., McCulloch, M.T., Malcolm, H.A., McWilliam, M.J., Pandolfi, J.M., Pears, R.J., Pratchett, M.S., Schoepf, V., Simpson, T., Skirving, W.J., Sommer, B., Torda, G., Wachenfeld, D.R., Willis, B.L. & Wilson, S.K. 2017. Global warming and recurrent mass bleaching of corals. *Nature* **543**, 373–377.

Hurd, C.L. & Pilditch, C.A. 2011. Flow-induced morphological variations affect diffusion boundary-layer thickness of *Macrocystis pyrifera* (Heterokontophyta, Laminariales). *Journal of Phycology* **47**, 341–351.

IPCC, 2014. *Climate Change 2014: Synthesis Report. Contribution of Working Groups I, II and III to the Fifth Assessment Report of the Intergovernmental Panel on Climate Change*, R.K. Pachauri & L.A. Meyer (eds). Switzerland: IPCC.

Jellison, B.M., Ninokawa, A.T., Hill, T.M., Sanford, E. & Gaylord, B. 2016. Ocean acidification alters the response of intertidal snails to a key sea star predator. *Proceedings of the Royal Society B* **283**, 20160890.

Kamenos, N.A., Burdett, H.L., Aloisio, E., Findlay, H.S., Martin, S., Longbone, C., Dunn, J., Widdicombe, S. & Calosi, P. 2013. Coralline algal structure is more sensitive to rate, rather than the magnitude, of ocean acidification. *Global Change Biology* **19**, 3621–3628.

Kamenos, N.A., Perna, G., Gambi, M.C., Micheli, F. & Kroeker, K.J. 2016. Coralline algae in a naturally acidified ecosystem persist by maintaining control of skeletal mineralogy and size. *Proceedings of the Royal Society B* **283**, 20161159.

Kerr, R.A. 2010. Ocean acidification unprecedented, unsettling. *Science* **328**, 1500–1501.

Kerrison, P., Hall-Spencer, J.M., Suggett, D.J., Hepburn, L.J. & Steinke, M. 2011. Assessment of pH variability at a coastal CO_2 vent for ocean acidification studies. *Estuarine Coastal and Shelf Science* **94**, 129–137.

Koch, M., Bowes, G., Ross, C. & Zhang, X.-H. 2013. Climate change and ocean acidification effects on seagrasses and marine macroalgae. *Global Change Biology* **19**, 103–132.

Kraemer, G.P. & Mazzella, L. 1996. Nitrogen assimilation and growth dynamics of the Mediterranean seagrasses *Posidonia oceanica, Cymodocea nodosa* and *Zostera noltii*. In *Seagrass Biology: Proceedings of an International Workshop*, J. Kuo, R.C. Phillips, D. Walker & H. Kirkman (eds). Rottnest Island, Western Australia, 181–190.

Kraemer, G.P., Mazzella, L. & Alberte, R.S. 1997. Nitrogen assimilation and partitioning in the Mediterranean seagrass *Posidonia oceanica*. *Marine Ecology* **18**, 175–188.

Kroeker, K.J., Kordas, R.L., Crim, R., Hendriks, I.E., Ramjo, L., Singh, G.S., Duarte, C.M. & Gattuso, J.-P. 2013a. Impacts of ocean acidification on marine organisms: quantifying sensitivities and interaction with warming. *Global Change Biology* **19**, 1884–1896.

Kroeker, K.J., Micheli, F. & Gambi, M.C. 2013b. Community dynamics and ecosystem simplification in a high-CO_2 ocean. *Proceedings of the National Academy of Sciences USA* **110**, 12721–12726.

Kroeker, K.J., Micheli, F. & Gambi, M.C. 2013c. Ocean acidification causes ecosystem shifts via altered competitive interactions. *Nature Climate Change* **3**, 156–159.

Kroeker, K.J., Micheli, F., Gambi, M.C. & Martz, T.R. 2011. Divergent ecosystem responses within a benthic marine community to ocean acidification. *Proceedings of the National Academy of Sciences USA* **108**, 14515–14520.

Kroll, A. & Jangoux, M. 1989. Les Gregarines (Sporozoa) et les Umagillides (Turbellaria) parasites du coelome et du systeme hemal de *Holothuria tubulosa*, Gmelin (Echinodermata, Holothuroidea). *Vie Marine, HS* **10**, 193–204 (In French).

Kuffner, I.B., Andersson, A.J., Jokiel, P.L., Rodgers, K.S. & Mackenzie, F.T. 2007. Decreased abundance of crustose coralline algae due to ocean acidification. *Nature Geoscience* **1**, 114–117.

Kumar, A., Abdelgawad, H., Castellano, I., Lorenti, M., Delledonne, M., Beemster, G.T.S., Asard, H., Buia, M.C. & Palumbo, A. 2017a. Physiological and biochemical analyses shed light on the response of *Sargassum vulgare* to ocean acidification at different time scales. *Frontiers in Plant Science* **8**, 570.

Kumar, A., Castellano, I., Patti, F.P., Delledonne, M., Abdelgawad, H., Beemster, G.T.S., Asard, H., Palumbo, A. & Buia, M.C. 2017b. Molecular response of *Sargassum vulgare* to acidification at volcanic CO_2 vents - insights from de novo transcriptomic analysis. *Molecular Ecology*, **26**(8), 2276–2290.

Lanera, P. 1987. Studio sulla fauna associata a prati della fanerogama marina *Cymodocea nodosa* (Ucria) Aschers. dell'isola d'Ischia. *Master thesis* Dep.t of Zoology University Federico II Naples (Italy) 119 pp. (In Italian).

Lanera, P. & Gambi, M.C. 1993. Polychaete distribution in some *Cymodocea nodosa* meadows around the island of Ischia (Gulf of Naples, Italy). *Oebalia* **19**, 89–103.

Langer, G., Nehrke, G., Baggini, C., Rodolfo-Metalpa, R., Hall-Spencer, J.M. & Bijma, J. 2014. Limpets counteract ocean acidification induced shell corrosion by thickening of aragonitic shell layers. *Biogeosciences* **11**, 7363–7368.

Lauritano, C., Ruocco, M., Dattolo, E., Buia, M.C., Silva, J., Santos, R., Olive, I., Costa, M.M. & Procaccini, G. 2015. Response of key stress-related genes of the seagrass *Posidonia oceanica* in the vicinity of submarine volcanic vents. *Biogeosciences* **12**, 4185–4194.

Linares, C., Vidal, M., Canals, M., Kersting, D.K., Amblas, D., Aspillaga, E., Cebrian, E., Delgado-Huertas, A., Diaz, D., Garrabou, J., Hereu, B., Navarro, L., Teixido, N. & Ballesteros, E. 2015. Persistent natural acidification drives major distribution shifts in marine benthic ecosystems. *Proceedings of the Royal Society B* **282**, 20150587.

Lombardi, C., Cocito, S., Gambi, M.C., Cisterna, B., Flach, F., Taylor, P.D., Keltie, K., Freer, A. & Cusack, M. 2011a. Effects of ocean acidification on growth, organic tissue and protein profile of the Mediterranean bryozoan *Myriapora truncata*. *Aquatic Biology* **13**, 251–262.

Lombardi, C., Cocito, S., Gambi, M.C. & Taylor, P.D. 2015. Morphological plasticity in a calcifying modular organism: evidence from an *in situ* transplant experiment in a natural CO_2 vent system. *Royal Society Open Science* **2**, 140413.

Lombardi, C., Gambi, M.C., Vasapollo, C., Taylor, A.C. & Cocito, S. 2011b. Skeletal alterations and polymorphism in a Mediterranean bryozoan at natural CO_2 vents. *Zoomorphology* **130**, 135–145.

Lombardi, C., Rodolfo-Metalpa, R., Cocito, S., Gambi, M.C. & Taylor, P.D. 2011c. Structural and geochemical alterations in the Mg calcite bryozoan *Myriapora truncata* under elevated seawater pCO_2 simulating ocean acidification. *Marine Ecology* **32**, 211–221.

Lorenti, M., Buia, M.C., Di Martino, V. & Modigh, M. 2005. Occurrence of mucous aggregates and their impact on *Posidonia oceanica* beds. *Science of the Total Environment* **353**, 369–379.

Lucey, N.M., Lombardi, C., DeMarchi, L., Schulze, A., Gambi, M.C. & Calosi, P. 2015. To brood or not to brood. Are marine organisms that protect their offspring more resilient to ocean acidification? *Scientific Report* **5**, 12009.

Lucey, N.L., Lombardi, C., Florio, M., DeMarchi, L., Nannini, M., Rundle, S., Gambi, M.C. & Calosi, P. 2016. No evidence of local adaptation to low pH in a calcifying polychaete population from a shallow CO_2 vent system. *Evolutionary Applications* **9**, 1054–1071.

Martin, S., Rodolfo-Metalpa, R., Ransome, E., Rowley, S., Buia, M.C., Gattuso, J.P. & Hall-Spencer, J.M. 2008. Effects of naturally acidified seawater on seagrass calcareous epibionts. *Biology Letters* **4**, 689–692.

Mazzella, L., Buia, M.C., Gambi, M.C., Lorenti, M., Russo, G.F. & Scipione, M.B. 1992. Plant-animal trophic relationships in the *Posidonia oceanica* ecosystem of the Mediterranean Sea: a review. In *Plant-Animal Interactions in the Marine Benthos*, D.M. John et al. (eds). Oxford: Clarendon Press, 165–188.

Mazzella, L., Cinelli, F., Ott, J.A. & Klepal, W. 1981. Studi sperimentali *in situ* sull'epifitismo della *Posidonia oceanica* Delile. *Quaderni del Laboratorio di Tecnologia della Pesca, Ancona* **3**, 481–492. (in Italian)

Melzner, F., Gutowska, M.A., Langenbuch, M., Dupont, S., Lucassen, M., Thorndyke, M.C., Bleich, M. & Portner, H.O. 2009. Physiological basis for high CO_2 tolerance in marine ectothermic animals: pre-adaptation through lifestyle and ontogeny? *Biogeosciences* **6**, 2313–2331.

Meron, D., Buia, M.C., Fine, M. & Banin, E. 2012a. Changes in microbial communities associated with the sea anemone *Anemonia viridis* in a natural pH gradient. *Microbial Ecology* **65**, 269–276.

Meron, D., Rodolfo-Metalpa, R., Cunning, R., Baker, A.C., Fine, M. & Banin, E. 2012b. Changes in coral microbial communities in response to a natural pH gradient. *ISME Journal* **6**, 1775–1785.

Monti, P. 1980. *Ischia. Archeologia e storia*. Tipografia F.lli Porzio Napoli, 1–832. (in Italian)

Morrow, K.M., Bourne, D.G., Humphrey, C., Botte, E.S., Laffy, P., Zaneveld, J., Uthicke, S., Fabricius, K.E. & Webster, N.S. 2015. Natural volcanic CO_2 seeps reveal future trajectories for host–microbial associations in corals and sponges. *International Society for Microbial Ecology* **9**, 894–908.

Mouchka, M.E., Hewson, I. & Harvell, C.D. 2010. Coral-associated bacterial assemblages: current knowledge and the potential for climate-driven impacts. *Integrative and Comparative Biology* **50**, 662–674.

Murray, J.W. 2006. *Ecology and Applications of Benthic Foraminifera*. Cambridge, UK: Cambridge University Press.

Musco, L., Terlizzi, A., Licciano, M. & Giangrande, A. 2009. Taxonomic structure and the effectiveness of surrogates in environmental monitoring: a lesson from polychaetes. *Marine Ecology Progress Series* **383**, 199–210.

Nagelkerken, I., Goldenberg, S.U., Ferreira, C.M., Russell, B.D. & Connell, S.D. 2017. Species interactions drive fish biodiversity loss in a high-CO_2 world. *Current Biology* **27**, 1–8.

Nagelkerken, I., Russell, B.D., Gillanders, B.M. & Connell, S.D. 2016. Ocean acidification alters fish populations indirectly through habitat modification. *Nature Climate Change* **6**, 89–93.

Nash, M.C. & Adey, W. 2017. Multiple phases of Mg-calcite in crustose coralline algae suggest caution for temperature proxy and ocean acidification assessment: lessons from the ultrastructure and biomineralization in *Phymatolithon* (Rhodophyta, Corallinales). *Journal of Phycology* **53**, 970–984, doi:10.111/jpy.12559.

Nguyen, H. & Byrne, M. 2014. Early benthic juvenile *Parvulastra exigua* (Asteroidea) are tolerant to extreme acidification and warming in its intertidal habitat. *Journal of Experimental Marine Biology and Ecology* **453**, 36–42.

Nguyen, H., Doo, S., Soars, N. & Byrne, M. 2012. Noncalcifying larvae in a changing ocean: warming not acidification/hypercapnia, is the dominant stressor on development of the sea star *Meridiastra calcar*. *Global Change Biology* **18**, 2466–2476.

Nielsen, C. 2008. A review of the solitary entoprocts reported from sponges from Napoli (Italy), with designation of a neotype of *Loxosoma pes* Schimdt, 1878. *Journal of Natural History* **42**, 1573–1579.

Nilsson, K.S., Wallberg, A. & Jondelius, U. 2011. New species of Acoela from the Mediterranean, the Red Sea, and the South Pacific. *Zootaxa* **2867**, 1–31.

Nogueira, P., Gambi, M.C., Vizzini, S., Califano, G., Tavares, A.M., Santos, R. & Martínez-Crego, B. 2017. Altered epiphyte community and sea urchin diet in *Posidonia oceanica* meadows in the vicinity of volcanic CO_2 vents. *Marine Environmental Research* **127**, 102–111.

Ott, J.A. 1980. Growth and production in *Posidonia oceanica* (L.) Delile. *Marine Ecology* **1**, 47–64.

Pages, J.F., Farina, S., Gera, A., Arthur, R., Romero, J. & Alcoverro, T. 2012. Indirect interactions in seagrasses: fish herbivores increase predation risk to sea urchins by modifying plant traits. *Functional Ecology* **26**, 1015–1023.

Parker, L.M., Ross, R.M., O'Connor, W.A., Portner, H.O., Scanes, E. & Wright, J.M. 2013. Predicting the response of molluscs to the impact of ocean acidification. *Biology* **2**, 651–692.

Peccerillo, A. 2005. *Plio-Quaternary Volcanism in Italy: Petrology, Geochemistry, Geodynamics*. Italy: Springer.

Pecoraino, G., Brusca, L., D'Alessandro, W., Giammanco, S., Inguaggiato, S. & Longo, M. 2005. Total CO_2 output from Ischia Island volcano (Italy). *Geochemical Journal* **39**, 451–458.

Pettit, L.R., Hart, M.B., Medina-Sánchez, A.N., Smart, C.W., Rodolfo-Metalpa, R., Hall-Spencer, J.M. & Prol-Ledesma, R.M. 2013. Benthic foraminifera show some resilience to ocean acidification in the northern Gulf of California, Mexico. *Marine Pollution Bulletin* **73**, 452–462.

Pettit, L.R., Smart, C.W., Hart, M.B., Milazzo, M. & Hall-Spencer, J.M. 2015. Seaweed fails to prevent ocean acidification impact on foraminifera along a shallow-water CO_2 gradient. *Ecology and Evolution* **5**, 1784–1793.

Pinsino, A., Matranga, V., Trinchella, F. & Roccheri, M.C. 2010. Sea urchin embryos as an *in vitro* model for the assessment of managanese toxicity: developmental and stress response effects. *Ecotoxicology* **19**, 555–562.

Pinsino, A., Roccheri, M.C., Costa, C. & Matranga, V. 2011. Managanese interferes with calcium, perturbs ERK signaling and produces embryos with no skeleton. *Toxicology Science* **123**, 217–230.

Poore, A.G.B., Graham, S.E., Byrne, M. & Dworjanyn, S.D. 2016. Effects of ocean warming and lowered pH on algal growth and palatability to a grazing gastropod. *Marine Biology* **163**, 99.

Pörtner, H.O. 2008. Ecosystem effects of ocean acidification in times of ocean warming: a physiologist's view. *Marine Ecology Progress Series* **373**, 203–217.

Porzio, L., Buia, M.C. & Hall-Spencer, J.M. 2011. Effects of ocean acidification on macroalgal communities. *Journal Experimental Marine Biology and Ecology* **400**, 278–287.

Porzio, L., Buia, M. C., Lorenti, M., De Maio, A. & Arena, C. 2017. Physiological responses of a population of *Sargassum vulgare* (Phaeophyceae) to high pCO_2/low pH: implications for its long-term distribution. *Science of the Total Environment* **576**, 917–925.

Porzio, L., Garrard, S.L. & Buia, M.C. 2013. The effect of ocean acidification on early algal colonization stages at natural CO_2 vents. *Marine Biology* **160**, 2247–2259.

Przeslawski, R., Byrne, M. & Mellin, C. 2015. A review and meta-analysis of the effects of multiple abiotic stressors on marine embryos and larvae. *Global Change Biology* **21**, 2122–2140.

Pulitzer-Finali, G. 1970. Report on a collection of sponges from the Bay of Naples. I. Sclerospongiae, Lithistida, Tetractinellida, Epipolasida. *Pubblicazioni della Stazione Zoologica di Napoli* **38**, 328–354.

Pulitzer-Finali, G. & Pronzato, R. 1976. Report on a collection of sponges from the Bay of Naples. II. Keratosa. *Pubblicazioni della Stazione Zoologica di Napoli* **40**, 83–104.

Puritz, J.B., Keever, C.S., Addison, A.A., Byrne, M., Hart, M.W., Grosberg, R.K. & Toonen, R.J. 2012. Extraordinarily rapid life-history divergence between *Cryptasterina* sea star species. *Proceedings of the Royal Society B* **279**, 3914–3922.

Rhein, M., Rintoul, S.R., Aoki, S., Campos, E., Chambers, D., Feely, R.A., Gulev, S., Johnson, G.C., Josey, S.A., Kostianoy, A., Mauritzen, C., Roemmich, D., Talley, L.D. & Wang, F. 2013. Observations: Ocean. In *Climate Change 2013: The Physical Science Basis*. Contribution of Working Group I to the Fifth Assessment Report of the Intergovernmental Panel on Climate Change. T.F. Stocker et al. (eds). Cambridge, UK and New York: Cambridge University Press, 255–316.

Ricevuto, E., Benedetti, M., Regoli, F., Spicer, J.I. & Gambi, M.C. 2015a. Antioxidant capacity of polychaetes occurring at a natural CO_2 vent system: results of an *in situ* reciprocal transplant experiment. *Marine Environmental Research* **112**, 44–51.

Ricevuto, E., Kroeker, K.J., Ferrigno, F., Micheli, F. & Gambi, M.C. 2014. Spatio-temporal variability of polychaete colonization at volcanic CO_2 vents (Italy) indicates high tolerance to ocean acidification. *Marine Biology* **161**, 2909–2919.

Ricevuto, E., Lanzoni, I., Fattorini, D., Regoli, F. & Gambi, M.C. 2016. Arsenic speciation and susceptibility to oxidative stress in the fanworm *Sabella spallanzanii* (Gmelin) (Annelida, Sabellidae) under naturally acidified conditions: an *in situ* transplant experiment in a Mediterranean CO_2 vent system. *Science of the Total Environment* **544**, 765–773.

Ricevuto, E., Lorenti, M., Patti, F.P., Scipione, M.B. & Gambi, M.C. 2012. Temporal trends of benthic invertebrate settlement along a gradient of ocean acidification at natural CO_2 vents (Tyrrhenian Sea). *Biologia Marina Mediterranea* **19**, 49–52.

Ricevuto, E., Vizzini, S. & Gambi, M.C. 2015b. Ocean acidification effects on stable isotope signatures and trophic interactions of polychaete consumers and organic matter sources at a CO_2 shallow vent system. *Journal of Experimental Marine Biology and Ecology* **468**, 105–117.

Rittmann, A. 1930. Geologie der Insel Ischia. *Zeitung Vulkanologie* **6**, 1–265. (in German)

Rittmann, A. & Gottini, V. 1981. L'isola d'Ischia - Geologia. *Bollettino del Servizio Geologico d'Italia* **101**, 131–274. (in Italian)

Rodolfo-Metalpa, R., Houlbrèque, F., Tambutté, E., Boisson, F., Baggini, C., Patti, F.P., Jeffree, R., Fine, M., Foggo, A., Gattuso, J.-P. & Hall-Spencer, J.M. 2011. R. Coral and mollusc resistance to ocean acidification adversely affected by warming. *Nature Climate Change* **1**, 308–312.

Rodolfo-Metalpa, R., Lombardi, C., Cocito, S., Hall-Spencer, J.M. & Gambi, M.C. 2010. Effects of ocean acidification and high temperatures on the bryozoan *Myriapora truncata* at natural CO_2 vents. *Marine Ecology* **31**, 447–456.

Rouse, G.W. & Gambi, M.C. 1997. Cladistic relationships within *Amphiglena* Claparède (Polychaeta: Sabellidae) with a new species and a redescription of *A. mediterranea* (Leydig). *Journal of Natural History* **31**, 999–1018.

Russell, B.D., Connell, S.D., Uthicke, S., Muehllehner, N., Fabricius, K.E. & Hall-Spencer, J.M. 2013. Future seagrass beds: can increased productivity lead to increased carbon storage? *Marine Pollution Bulletin* **73**, 463–469.

Russell, B.D., Thompson, J-A.I., Falkenberg, L.J. & Connell, S.D. 2009. Synergistic effects of climate change and local stressors: CO_2 and nutrient-driven change in subtidal rocky habitats. *Global Change Biology* **15**, 2153–2162.

Russo, G.F., Fresi, E., Vinci, D. & Chessa, L.A. 1983. Malacofauna di strato foliare delle praterie di *Posidonia oceanica* (L.) Delile intorno all'isola d'Ischia (Golfo di Napoli): analisi strutturale del popolamento estivo in rapporto alla profondità ed alla esposizione. *Nova Thalassia* **6**, 655–661. (In Italian).

Sarà, M. 1959. Poriferi del litorale dell'isola d'Ischia e loro ripartizione per ambienti. *Pubblicazioni della Stazione Zoologica di Napoli* **31**, 421–472. (in Italian)

Scartazza, A., Moscatello, S., Gavrichkova, O., Buia, M.C., Lauteri, M., Battistelli, A., Lorenti, M., Garrard, S.L., Calfapietra, C. & Brugnoli, E. 2017. Carbon and nitrogen allocation strategy in *Posidonia oceanica* is altered by seawater acidification. *Science of the Total Environment* **607**, 954–964.

Scipione, M.B. 1999. Amphipod biodiversity in the foliar stratum of shallow-water *Posidonia oceanica* beds in the Mediterranean Sea. In *Crustacean and the Biodiversity Crisis*, F.R. Schram & J.C. van Vaupel Kelin (eds). Brill, The Netherlands: Leiden, 649–662.

Scipione, M.B. 2013. On the presence of the Mediterranean endemic *Microdeutopus sporadhi* Myers, 1969 (Crustacea: Amphipoda: Aoridae) in the Gulf of Naples (Italy) with a review on its distribution and ecology. *Mediterranean Marine Science* **4**, 56–63.

Sedwick, P. & Stuben, D. 1996. Chemistry of shallow submarine warm springs in an arc-volcanic setting: Vulcano Island, Aeolian Archipelago, Italy. *Marine Chemistry* **53**, 147–161.

Smith, A.M. 2014. Growth and calcification of marine bryozoans in a changing ocean. *The Biological Bulletin* **226**, 203–210.

Smith, J.N., De'ath, G., Richter, C., Cornils, A., Hall-Spencer, J.M. & Fabricius, K.E. 2016. Ocean acidification reduces demersal zooplankton that reside in tropical coral reefs. *Nature Climate Change* **6**, 1124–1129.

Sunday, J.M., Fabricius, K.E., Kroeker, K.J., Anderson, K.M., Brown, N.E., Barry, J.P., Connell, S.D., Dupont, S., Gaylord, B., Hall-Spencer, J.M., Klinger, T., Milazzo, M., Munday, P.L., Russell, B.D., Sanford, E., Thiyagarajan, V., Vaughan, M.L.H., Widdicombe, S. & Harley, C.D.G. 2017. Ocean acidification can mediate biodiversity shifts by changing biogenic habitat. *Nature Climate Change* **7**, 81–85.

Tarasov, V.G. 2006. Effects of shallow-water hydrothermal venting on biological communities of coastal marine ecosystems of the Western Pacific. *Advances in Marine Biology* **50**, 267–421.

Tedesco, D. 1996. Chemical and isotopic investigation of fumarolic gases from Ischia Island (Southern Italy): evidence of magmatic and crustal contribution. *Journal of Vulcanology and Geothermal Research* **74**, 233–242.

Tertschnig, W.P. 1989. Predation of the sea urchin *Paracentrotus lividus* by the spider crab *Maja crispata*. *Vie Marine, HS* **19**, 95–103.

Toyofuku, T., Matsuo, M.Y., Nooijer, L.J., Nagai, Y., Kawada, S., Fujita, K., Reichart, G.-J., Nomaki, H., Tsuchiya, M., Sakaguchi, H. & Kitazato, H. 2017. Proton pumping accompanies calcification in foraminifera. *Nature Communications* **8**, 14145.

Turner, L.M., Ricevuto, E., Massa-Gallucci, A., Gambi, M.C. & Calosi, P. 2015. Energy metabolism and cellular homeostasis trade-offs provide the basis for a new type of sensitivity to ocean acidification in a marine polychaete at a high-CO_2 vent: adenylate and phosphagen energy pools *versus* carbonic anhydrase. *Journal of Experimental Biology* **218**, 2148–2151.

Turner, L.M., Ricevuto, E., Massa-Gallucci, A., Lorenti, M., Gambi, M.C. & Calosi, P. 2016. Metabolic responses to high pCO_2 conditions at a CO_2 vent site in the juveniles of a marine isopod assemblage. *Marine Biology* **163**, 211.

Uthicke, S., Momigliano, P. & Fabricius, K.E. 2013. High risk of extinction of benthic foraminifera in this century due to ocean acidification. *Scientific Reports* **3**, 1769.

Valvassori, G., Massa-Gallucci, A. & Gambi, M.C. 2015. Reappraisal of *Platynereis massiliensis* (Moquin-Tandon) (Annelida, Nereididae), a neglected sibling species of *Platynereis dumerilii* (Audouin & Milne Edwards). *Biologia Marina Mediterranea* **22**, 113–116.

Vezzoli, L. 1988. *The island of Ischia*. Quaderni de "La Ricerca Scientifica" CNR Progetto Finalizzato Geodinamica, Monografie finali **10**, 1–134.

Vizzini, S., Di Leonardo, R., Costa, V., Tramati, C.D., Luzzu, F. & Mazzola, A. 2013. Trace element bias in the use of CO_2 vents as analogues for low pH environments: implications for contamination levels in acidified oceans. *Estuarine, Coastal and Shelf Science* **134**, 19–30.

Vizzini, S., Martínez-Crego, B., Andolina, C., Massa-Gallucci, A., Connell, S.D. & Gambi, M.C. 2017. Ocean acidification as a driver of community simplification via the collapse of higher-order and rise of lower-order consumers. *Nature Scientific Report* **7**, 4018.

Vogel, S. 1994. *Life in Moving Fluids: The Physical Biology of Flow*. Princeton, NJ: Princeton University Press, 467 pp.

Wäge, J., Valvassori, G., Hardege, J.D., Shulze, A. & Gambi, M.C. 2017. The sibling polychaetes *Platynereis dumerilii* and *Platynereis massiliensis* in the Mediterranean Sea: are phylogeographic patterns related to exposure to ocean acidification? *Marine Biology* **164**, 199.

Wagele, J.W. 1982. Neubeschreibung und Verleich der mediterranen *Paranthura* Arten (Crustacea, Isopoda, Anthuridea). *Marine Ecology* **3**, 109–132.

Wahl, M., Saderne, V. & Sawall, Y. 2015. How good are we at assessing the impact of ocean acidification in coastal systems? Limitations, omissions and strengths of commonly used experimental approaches with special emphasis on the neglected role of fluctuations. *Marine and Freshwater Research* **67**, 25–36.

Waldbusser, G.G., Hales, B., Langdon, C.J., Haley, B.A., Schrader, P., Brunner, E.L., Gray, M.W., Miller, C.A., Gimenez, I. & Hutchinson, G. 2015. Ocean acidification has multiple modes of action on bivalve larvae. *PLoS ONE* **10**, e0128376.

Wittmann, K.J. 2001. Centennial changes in the near-shore Mysid Fauna of the Gulf of Naples (Mediterranean Sea) with description of *Heteromysis riedli* sp. n. (Crustacea, Mysidacea). *Marine Ecology* **22**, 85–110.

Wittmann, K.J. & Stagl, V. 1996. Die Mysidaceen-Sammlung an NturhistorischenMuseum in Wien: eine kritische Sichtung im Spiegel der Sammlungsgeschichte (Crustacea, Malacostraca). *Annales des Naturhistorischen Museums in Wien. Serie B furt Botanik und Zoologie* 157–191. (In German).

WoRMS Editorial Board, 2017. World Register of Marine Species. Online. http://www.marinespecies.org at VLIZ (accessed 27 October 2017).

Zeebe, R.E., Ridgwell, A. & Zachos, J.C. 2016. Anthropogenic carbon release rate unprecedented during the past 66 million years. *Nature Geoscience* **9**, 325–329.

Zeebe, R.E., Zachos, J.C., Caldeira, K. & Tyrrell, T. 2008. Oceans: carbon emissions and acidification. *Science* **321**, 51–52.

Zupo, V., Maibam, C., Buia, M.C., Gambi, M.C., Patti, F.P., Scipione, M.B., Lorenti, M. & Fink, P. 2015. Chemoreception of the seagrass *Posidonia oceanica* by benthic invertebrates is altered by seawater acidification. *Journal of Chemical Ecology* **41**, 766–779.

Oceanography and Marine Biology: An Annual Review, 2018, **56**, 311-370
© S. J. Hawkins, A. J. Evans, A. C. Dale, L. B. Firth, and I. P. Smith, Editors
Taylor & Francis

IMPACTS AND ENVIRONMENTAL RISKS OF OIL SPILLS ON MARINE INVERTEBRATES, ALGAE AND SEAGRASS: A GLOBAL REVIEW FROM AN AUSTRALIAN PERSPECTIVE

JOHN K. KEESING[1,2]*, ADAM GARTNER[3], MARK WESTERA[3], GRAHAM J. EDGAR[4,5], JOANNE MYERS[1], NICK J. HARDMAN-MOUNTFORD[1,2] & MARK BAILEY[3]

[1]*CSIRO Oceans and Atmosphere, Indian Ocean Marine Research Centre, M097, 35 Stirling Highway, Crawley, 6009, Australia*
[2]*University of Western Australia Oceans Institute, Indian Ocean Marine Research Centre, M097, 35 Stirling Highway, Crawley, 6009, Australia*
[3]*BMT Pty Ltd, PO Box 462, Wembley, 6913, Australia*
[4]*Aquenal Pty Ltd, 244 Summerleas Rd, Kingston, 7050, Australia*
[5]*Institute for Marine and Antarctic Studies, University of Tasmania, Private Bag 49, Hobart, 7001, Australia*
Corresponding author: John K. Keesing
e-mail: john.keesing@csiro.au

Abstract

Marine invertebrates and macrophytes are sensitive to the toxic effects of oil. Depending on the intensity, duration and circumstances of the exposure, they can suffer high levels of initial mortality together with prolonged sublethal effects that can act at individual, population and community levels. Under some circumstances, recovery from these impacts can take years to decades. However, effects are variable because some taxa are less sensitive than others, and many factors can mitigate the degree of exposure, meaning that impacts are moderate in many cases, and recovery occurs within a few years. Exposure is affected by a myriad of factors including: type and amount of oil, extent of weathering, persistence of exposure, application of dispersants or other clean-up measures, habitat type, temperature and depth, species present and their stage of development or maturity, and processes of recolonisation, particularly recruitment. Almost every oil spill is unique in terms of its impact because of differing levels of exposure and the type of habitats, communities and species assemblages in the receiving environment. Between 1970 and February 2017, there were 51 significant oil spills in Australia. Five occurred offshore with negligible likely or expected impacts. Of the others, only 24 of the spills were studied in detail, while 19 had only cursory or no assessment despite the potential for oil spills to impact the marine environment. The majority were limited to temperate waters, although 10 of the 14 spills since 2000 were in tropical coastal or offshore areas, seven were in north Queensland in areas close to the Great Barrier Reef. All four spills that have occurred from offshore petroleum industry infrastructure have occurred since 2009. In Australia, as elsewhere, a prespill need exists to assess the risk of a spill, establish environmental baselines, determine the likely exposure of the receiving environment, and test the toxicity of the oil against key animal and plant species in the area of potential impact. Subsequent to any spill, the baseline provides a reference for targeted impact monitoring.

Introduction

Background to this review

The aim of this paper is to review the impacts of oil and oil spills on marine invertebrates, seagrasses and macroalgae, particularly in an Australian context. Reviewing oil spill literature is a daunting task, and there is always the risk of omission. For example, Harwell & Gentile (2006) reviewed almost 300 papers, just on the effects of the 1989 Exxon Valdez oil spill in Alaska. Other reviews on the effects of oil on marine invertebrates have been undertaken (Johnson 1977, Loya & Rinkevich 1980, Suchanek 1993). For marine algae and seagrasses, a detailed review was provided by O'Brien & Dixon (1976), while the toxicity of oil to aquatic plants was recently reviewed by Lewis & Pryor (2013). Apart from a report by Runcie et al. (2005) on seagrasses, no detailed review exists on the impacts of oil on marine flora and fauna from an Australian perspective.

Our assessment fully evaluates the small number of Australian studies that we are aware of and captures both the primary and grey literature. We then draw on a subset of the international literature to capture the status of existing knowledge from field and laboratory studies and the experience of overseas oil spills to identify the largest gaps in our ability to infer likely impacts of any future spills in Australia.

The scope for our review includes marine macrophytes and all life stages of marine invertebrates with a benthic adult phase. We also include zooplankton, but we do not consider impacts on phytoplankton or microbial communities. Other studies review and experimentally evaluate how microbial communities respond to oil (MacNaughton et al. 1999, Nyman & Green 2015); the review by Lewis & Pryor (2013) covers microalgae to some degree. The complex nature of oil solubility and hydrocarbon fractionation in seawater, and thus the bioavailability of toxicants, make it difficult to compare results between studies using different oils, ratios in oil–water mixes and methods of creating water-accommodated (soluble) fractions (WAFs or WSFs—note we use both terms where different authors used them) (Redman 2015, Redman & Parkerton 2015); consequently, we have not attempted to make such comparisons in reviewing literature in this paper. However, we emphasise that the circumstances and level of exposure to oil are key elements needed when assessing risk, the likelihood of biological impact and impact detection.

The scale and nature of oil spills has changed in recent decades. Previously, major spills were primarily the result of shipping-related incidents, but recent spills from deep-water offshore fixed installations have resulted in markedly different circumstances of exposure of the environment to releases of oil. Over three months from 21 August 2009, a damaged production well (Montara) released gas, condensate and ~4750 tonnes of crude oil into the Timor Sea, creating arguably Australia's most significant oil spill. Not long after the Montara incident, the Deepwater Horizon oil spill (20 April 2010) occurred in the Gulf of Mexico, with resultant discharge of an estimated 7 million tonnes of crude oil, the largest accidental volume ever released. While both incidents highlighted a general lack of preparedness for such events, the environmental monitoring programmes triggered by these oil spills, especially Deepwater Horizon, resulted in extensive research on the impacts of spills. That both events occurred in warm-water environments is also pertinent to advancing our understanding of the implications of oil spills in an Australian context, since the greatest expansion in oil and gas exploration and extraction has been on the country's tropical north-west shelf. Both these spills provide an opportunity to expand on previous research prompted by major incidents that occurred in temperate/cold-water environments (Exxon Valdez and Torrey Canyon by way of examples).

The review is structured as follows. The remainder of this introductory section provides some context of how oil behaves and breaks down over time in seawater and its fate in coastal environments. The second section provides a summary of the marine oil spills of significance that have occurred around Australia since 1970 and then follows a detailed review, structured by geography and/or habitat type, of the impacts of these spills and the studies (if any) undertaken on them. The third section provides a detailed review of the direct and indirect impacts, both acute and chronic, for each of the major taxonomic groups of invertebrates, seagrasses and macroalgae, and the factors that affect the severity of impact from exposure to oil. The fourth section evaluates the evidence for recovery of species and assemblages from the impact of oil spills, both in the short term and long term, and the factors that affect recovery. This section also sets out some key attributes for good impact assessment based on our review. Finally, we provide a concluding section which summarises the key points from the review and highlights important knowledge gaps that we feel should be addressed.

Behaviour and fate of crude oil in coastal environments

Our review is principally focussed on the interaction between crude oils and marine organisms within intertidal and subtidal marine coastal areas. As such, an important consideration is how the behaviour and fate of crude oils vary in space and time, as this in turn influences the likelihood and type of interaction spilt oils may have on these organisms.

Following an oil spill, evaporation will typically remove about one-third of the volume of a medium crude oil slick within the first 24 hours, but a significant residue will always remain (National Research Council [NRC] 2003). Many crude oils will emulsify readily, a process that can greatly reduce subsequent weathering rates (NRC 2003). Crude oils also have the potential to adsorb onto intertidal sediments (Lee et al. 2015), with the risk of subsequent erosion of oiled sediments from the shoreline and deposition in nearshore habitats. Dissolution from slicks and adsorbed oil can persist for weeks to years (NRC 2003). This is particularly relevant to understanding impacts on benthic invertebrates that principally occupy the seafloor and intertidal zones.

While nearshore oil spills obviously pose a high risk to benthic animals and plants, in an offshore spill, a significant portion of toxic components are likely to be lost before the oil reaches coastal waters or the seafloor (Haapkylä et al. 2007). Most crude oils spread very thinly on open waters to average thicknesses of ~0.1 mm (Lee et al. 2013). The application of dispersants further enhances the transport of oil as small droplets are entrained into the water column. Dispersant application can dilute oil concentrations to less than 100 ppm under turbulent (1 m) wave conditions (Lee et al. 2013). Within 24 hours of release, it is expected that oil will disperse and mix into the water column and be diluted to concentrations well below 10 ppm, with dilution continuing as time proceeds. As biodegradation and dilution takes place over the following weeks, dispersed oil concentrations could be expected to decline to less than 1 ppm (Lee et al. 2013). The response of organisms exposed to oil maybe acute or chronic. For the purposes of this review, the Australian and New Zealand Environment and Conservation Council and the Agriculture and Resource Management Council of Australia and New Zealand (ANZECC & ARMCANZ 2000) definitions are used, whereby acute is defined as a rapid adverse effect (e.g. death) caused by a substance in a living organism. Acute can be used to define either the exposure or the response to an exposure (effect). Chronic is defined as lingering or continuing for a long time, often for periods from several weeks to years. Chronic can also be used to define either the exposure of a species or its response to an exposure (effect). Chronic exposure typically includes a biological response of relatively slow progress and long continuance, often affecting a life stage (ANZECC & ARMCANZ 2000).

Oil spills in Australia and the impact of oil on marine invertebrates, algae and seagrass

Scientific research conducted on Australian oil spills

According to the Australian Maritime Safety Authority (AMSA 2017), there were at least 44 large or notable oil spills in Australia between 1970 and 2012; we have documented a further seven between 2013 and 2017 (Table 1) with the potential to impact marine habitats and the floral and faunal assemblages occurring in them. Most of these were the result of shipping incidents (42) but also include refinery (1) and bulk storage spills (3), deliberate illegal dumping onshore (1), leaks from platform or seabed oil and gas infrastructure (3) and an oil platform wellhead blowout (1).

Most crude oil and fuel spills in Australian waters have had little or no assessment of possible or realised impacts of oil. Of the 51 spills that have occurred from 1970 to 2017, six were in offshore waters, with no significant impacts (or no expected impact), 24 involved studies of potential effects, and 19 had no or only cursory assessment of impact and two are currently being assessed (Table 2). Much of the work undertaken to date to assess the impacts of oil spills in Australia has not been reported in the peer reviewed literature, thus limiting its utility and accessibility. For example, of the 24 spills cited in Table 2 with published outcomes, documentation on impacts or the assessment of potential impacts in peer reviewed journals occurred in just 10 cases. One of the possible reasons there are limited published studies of oil spill impacts in Australia is that few have been able to be conducted with the foresight, or in some cases fortuity, to have rigorous experimental designs in place; notable exceptions are Edgar & Barrett (2000) and MacFarlane & Burchett (2003). However, several published studies have incorporated appropriate control-impact only comparisons by examining oiled and non-oiled locations, and then repeating measurements for a period of time post impact (e.g. Smith & Simpson 1995, 1998, Schlacher et al. 2011).

Offshore spills and coral reefs

Of the 51 spills listed in Table 1, 10 spills occurred far offshore from the mainland, with only two of these having the potential to cause significant environmental impacts to benthic or intertidal habitats (the 2002

Table 1 Records of significant or notable historical Australian oil spills since 1970

Date	Vessel/source	Location, state	Oil amount (tonnes)
3 March 1970	OCEANIC GRANDEUR	Torres Strait, Qld	1100
1 January 1975	LAKE ILLAWARRA	Hobart, Tas.	81[b]
4 February 1975	ESSO DEN HAAG	Port Stanvac, SA	<41[b]
26 May 1974	SYGNA	Newcastle, NSW	700
14 July 1975	PRINCESS ANNE MARIE	Offshore, WA	14,800
24 July 1976	FU LONG II	Geraldton, WA	Unknown[b]
18 December 1976	BETHIOUA	Tamar River, Tas.	356[b]
8 January 1977	AUSTRALIS	Sydney Cove, NSW	Unknown[b]
7 March 1977	YUN HAI	Newcastle, NSW	100[b]
31 March 1977	STOLT SHEAF	Ballast Point, NSW	20[b]
15 March 1978	Unknown	Teewah to Ballina	Unknown[b]
10 September 1979	WORLD ENCOURAGEMENT	Botany Bay, NSW	95
29 January 1981	Unknown	Botany Bay, NSW	50–100[c]
29 October 1981	ANRO ASIA	Bribie Island, Qld	100
22 January 1982	ESSO GIPPSLAND	Port Stanvac, SA	Unknown
August 1982	Oil barge	Parramatta River, NSW	Unknown[e]
5 February 1985	ARTHUR PHILLIP	Botany Bay, NSW	7–50[d]
August 1987	Storage tank leak	Withnell Bay, WA	50[e]
3 December 1987	NELLA DAN	Macquarie Island	125
6 February 1988	SIR ALEXANDER GLEN	Port Walcott, WA	450
20 May 1988	KOREAN STAR	Cape Cuvier, WA	600
28 July 1988	AL QURAIN	Portland, Vic.	184
21 May 1990	ARTHUR PHILLIP	Cape Otway, Vic.	Unknown

Continued

Table 1 (Continued)   Records of significant or notable historical Australian oil spills since 1970

Date	Vessel/source	Location, state	Oil amount (tonnes)
14 February 1991	SANKO HARVEST	Esperance, WA	700
21 July 1991	KIRKI	Offshore, WA	17,280
December 1991	Fuel storage depot	Cape Flattery, Qld	8[e]
30 August 1992	ERA	Port Bonython, SA	300
1993	Unknown	Point Sampson, WA	1[e]
December 1994	Illegal dumping	Cairns, Qld	6[e]
10 July 1995	IRON BARON	Hebe Reef, Tas	325
1996	Unknown	Point Sampson, WA	1[e]
2 March 1998	Fuel storage depot	Christmas Island	60
28 June 1999	Mobil refinery	Port Stanvac, SA	230
26 July 1999	MV TORUNGEN	Varanus Island, WA	25
3 August 1999	LAURA D'AMATO	Sydney, NSW	250
16 December 1999	METAXATA	Ballina, NSW	1400[f]
18 December 1999	SYLVAN ARROW	Wilson's Promontory, Vic.	<2
2 September 2001	PAX PHOENIX	Holbourne Island, Qld	<1
25 December 2002	PACIFIC QUEST	Border Island, Qld	Unknown
24 January 2006	GLOBAL PEACE	Gladstone, Qld	25
11 March 2009	PACIFIC ADVENTURER	Cape Moreton, Qld	270
21 August 2009	MONTARA Wellhead	NW Australian coast	4750
3 April 2010	SHEN NENG[a]	Great Keppel Island, Qld	4
9 January 2012	MV TYCOON	Christmas Island	102
6 June 2013	FV JOSEPH M[f]	Lady Elliot Island, Qld	<1
29 July 2013	Refuelling accident[f]	Port of Brisbane, Qld	Unknown
7 September 2013	Esso COBIA platform[f,g]	Bass Strait, Victoria, 50 km offshore	<1
3 October 2013	Seafaris tourist catamaran[f]	Cape Tribulation, Qld	Unknown (vessel caught fire and sank with 12–15 t of diesel on board)
17 July 2015	Alleged ship responsible being investigated[f,h]	Cape Upstart, Qld	10–15
April 2016	WOODSIDE COSSACK[d] Well[i,j]	Pilbara, 150 km offshore Karratha, WA	10.5
1 February 2017	ESSO WEST TUNA platform[k,l]	Bass Strait, Victoria, 45 km offshore	Unknown

Source: Unless specified, source for records up to 2012 is Australian Maritime Safety Authority (AMSA). 2017. Major historical incidents. Australian Maritime Safety Authority. Available at https://www.amsa.gov.au/environment/protecting-our-environment/major-historical-incidents/ (accessed 6 August 2017).

[a] Manuell (1979).

[b] Dexter (1984).

[c] Anink et al. (1985).

[d] Duke & Burns (1999).

[e] Nelson (2000).

[f] AMSA records of pollution events since 2012 (involving oil spills) rated as category 3 (moderate) or above provided by the Australian Maritime Safety Authority.

[g] Australian Broadcasting Commission (ABC) (2013).

[h] Australian Broadcasting Commission (ABC) (2017a).

[i] Australian Broadcasting Commission (ABC) (2017b).

[j] *The West Australian* (2017).

[k] NOPSEMA (2017).

[l] *The Guardian* (2017).

Table 2 Summary of habitats and flora/fauna impacted or potentially impacted by oil spills in Australia since 1970

Year	Vessel/source	Intertidal habitats potentially impacted	Subtidal habitats potentially impacted	Studies on impacts	Description of spill and/or response
Nearshore subantarctic					
1987	Nella Dan	Rocky shore	Reef	Molluscs, echinoderms, crustaceans: Pople et al. (1990); Simpson et al. (1995); Smith & Simpson (1995, 1998) algae: Pople et al. (1990); Simpson et al. (1995); Smith & Simpson (1995, 1998)	AMSA (2017); Lipscombe (2000)
Nearshore temperate					
1974	Sygna	Beach	Unknown	None	AMSA (2017)
1975	Esso den Haag	Unknown	Unknown	None	Manuell (1979)
1976	Bethioua	Unknown	Unknown	None	Manuell (1979)
1976	Fu Long II	Unknown	Unknown	None	Manuell (1979)
1978	Unknown	Unknown	Unknown	None	Manuell (1979)
1981	Anro Asia	Beach	Unknown	None	AMSA (2017)
1981	Not known	Mangroves	Mangroves	Mangroves: Allaway et al. (1985); Allaway (1987) amphipods: Dexter (1984)	Dexter (1984); Allaway et al. (1985); Allaway (1987)
1982	Esso Gippsland	Beach	Unknown, possibly seagrass	None	AMSA (2017)
1988	Korean Star	Rocky shore, beach	Unknown, possibly reef	None	AMSA (2017); Lipscombe (2000)
1990	Arthur Phillip	Beach, mangrove	Probably reef, seagrass	Penguins: (AMSA 2017)	AMSA (2017)
1991	Sanko Harvest	Beach	Probably reef, seagrass	Seals: Gales (1991)	AMSA (2017)
1992	Era	Mangroves	Unknown	Mangroves: Wardrop et al. (1996) fish: Connolly & Jones (1996) sea birds: AMSA (2017)	AMSA (2017); Lipscombe (2000)
1995	Iron Baron	Beach, reef, estuary	Reef	Invertebrates, fish, algae: Edgar & Barrett (2000) penguins: Goldsworthy et al. (2000a,b); Giese et al. (2000)	AMSA (2017); Edgar & Barrett (2000); Lipscombe (2000)
1999	Port Stanvac Refinery	Beach	Unknown, possibly seagrass	None	AMSA (2000a); AMSA (2017); Lipscombe (2000)

Continued

Table 2 (Continued) Summary of habitats and flora/fauna impacted or potentially impacted by oil spills in Australia since 1970

Year	Vessel/source	Intertidal habitats potentially impacted	Subtidal habitats potentially impacted	Studies on impacts	Description of spill and/or response
2006	GLOBAL PEACE	Mangrove, seagrass, beach		Mangrove: Andersen et al. (2008) seagrass: Taylor et al. (2006); Taylor & Rasheed (2011) crabs: Andersen et al. (2008); Melville et al. (2009)	AMSA (2017)
2009	PACIFIC ADVENTURER	Beach, rocky shore		Invertebrates: Schlacher et al. (2011); Stevens et al. (2012); Finlayson et al. (2015) fish: Rissik & Esdaile (2011) wetlands: Rissik et al. (2011); Bi et al. (2011)	AMSA (2017); Schlacher et al. (2011)
2010	SHEN NENG I	Reef, beach	Reef	Physical and antifouling impacts: GBRMPA (2011)	AMSA (2017); GBRMPA (2011)
Nearshore tropical					
1970	OCEANIC GRANDEUR	Beach, reef, and mangroves	Unknown, possibly reef	None but extensive deaths of pearl oysters reported: Loya & Rinkevich (1980)	AMSA (2017); Manuell (1979)
1988	SIR ALEXANDER GLEN	Beach, mangrove and rocky shore	Unknown	Mangroves: Duke & Burns (1999)	AMSA (2017)
1998	Christmas Island	Beach, probably rocky shore	Unknown, probably reef	None	Dept Mines & Energy WA (1999); Lipscombe (2000)
1999	MV TORUNGEN	Beach, reef	Possibly reef	Oysters: AMSA (2017)	AMSA (2017)
2012	MV TYCOON	Beach, rocky shore, reef	Unknown, probably reef	None	AMSA (2017)
2013	FV JOSEPH M	Coral reef, beach	Coral reef	None, no impact expected	AMSA records[a]
2013	Seafaris catamaran	Coral reef, beach	Coral reef	Unknown as of November 2017	AMSA records[a]
2015	Alleged ship responsible being investigated	Coral reef, beach	Coral reef	Sea birds, turtle	AMSA records[a], ABC (2017a)

Continued

317

Table 2 (Continued) Summary of habitats and flora/fauna impacted or potentially impacted by oil spills in Australia since 1970

Continued

Year	Vessel/source	Intertidal habitats potentially impacted	Subtidal habitats potentially impacted	Studies on impacts	Description of spill and/or response
Harbour/estuary temperate					
1975	Lake Illawarra	Unknown	Unknown	None	Manuell (1979)
1977	Australis	Unknown	Unknown	None	Manuell (1979)
1977	Stolt Sheaf	Unknown	Unknown	None	Manuell (1979)
1977	Yun Hai	Unknown	Unknown	None	Manuell (1979)
1979	World Encouragement	Mangroves		Mangroves: Allaway (1982, 1987); Duke and Burns (1999)	AMSA (2017)
1982	Parramatta River barge	Mangrove		Mangroves: reference to crabs: Anink et al. (1985); Duke & Burns (1999)	Duke & Burns (1999)
1985	Arthur Phillip	Mangrove	Unknown	Mangroves, crustaceans and molluscs: Anink et al. (1985)	Anink et al. (1985)
1988	Al Qurain	Harbour rock walls	None	Penguins: AMSA (2017)	AMSA (2017)
1999	Laura D'Amato	Beach, rocky shore	Unknown, probably rocky reef	Molluscs and other invertebrates: MacFarlane & Burchett (2003) gastropods: Reid & MacFarlane (2003) amphipods: Jones (2003)	AMSA (2000b); Lipscombe (2000); Underwood (2002)
Harbour/estuary tropical					
1987	Withnell Bay storage tank	Mangrove		Mangroves: Duke & Burns (1999)	Duke & Burns (1999)
1991	Cape Flattery storage tank	Mangrove		Mangroves: Duke & Burns (1999)	Duke & Burns (1999)
1993	Point Sampson boat harbour	Mangrove		Mangroves: Duke & Burns (1999)	Duke & Burns (1999)
1994	Cairns illegal dumping	Mangrove		Mangroves: Duke & Burns (1999)	Duke & Burns (1999)
1996	Point Sampson boat harbour	Mangrove		Mangroves: Duke & Burns (1999)	Duke & Burns (1999)
2013	Port of Brisbane	Mangrove		Cormorants, pelicans oiled: AMSA records[a]	AMSA records[a]

Table 2 (Continued)   Summary of habitats and flora/fauna impacted or potentially impacted by oil spills in Australia since 1970

Year	Vessel/source	Intertidal habitats potentially impacted	Subtidal habitats potentially impacted	Studies on impacts	Description of spill and/or response
Offshore					
1975	Princess Anne Marie	None, spill offshore	Unknown, probably none	None	AMSA (2017); Manuell (1979)
1991	Kirki	None, spill offshore	None, spill offshore	None	AMSA (2017); Walker (1991)
1999	Metaxata	None	None	None	AMSA (2000c); Nelson (2000)
1999	Sylvan Arrow	None, spill offshore	Unknown, probably none	None	AMSA (2017)
2001	Pax Phoenix	None, spill offshore	Unknown, probably none	None	AMSA (2017)
2002	Pacific Quest	None, spill offshore	Unknown	None	AMSA (2017)
2009	Montara Wellhead	Reef	Reef	Fish, birds, reptiles, mammals: PTTEP Australasia (2013) corals, seagrass: Heyward et al. (2012); Heyward et al. (2013)	Australian Government (2011); AMSA (2017); Hunter (2010)
2013	Esso Cobia offshore platform	Beach	Unknown, probably sandy seabed	None (minor spill no expected impact)	AMSA records[a]
2016	Woodside Cossack4 well	Coral reef	Unknown, probably subtidal reef or sediments	None	ABC (2017b); The West Australian (2017)
2017	Esso West Tuna platform	Beach	Unknown, probably sandy seabed	Unknown as of December 2017	NOPSEMA (2017); The Guardian (2017)

[a] Australian Maritime Safety Authority (AMSA) records of pollution events since 2012 (involving oil spills) rated as category 3 (moderate) or above provided by the Australian Maritime Safety Authority.

PACIFIC QUEST event at the southern end of the Great Barrier Reef in eastern Australia and the 2009 MONTARA platform wellhead blowout off the north-west coast of Australia). Of these, only the MONTARA spill was subject to any detailed environmental assessment (PTTEP Australasia 2009, 2013). There have been three other spills or leaks from offshore oil and gas infrastructure: one minor spill from the ESSO COBIA platform in Bass Strait in 2013 and a recent (2017) incident from another Bass Strait platform (ESSO WEST TUNA) for which few details are available as yet; plus, the 2016 WOODSIDE COSSACK4 well leak of 10,500 litres in the Pilbara, which was not assessed for any impacts (see Tables 1 and 2).

The MONTARA H1 wellhead blowout released gas, condensate and ~4750 tonnes of crude oil into the Timor Sea 260 km offshore from 21 August to 3 November 2009. Assessments of potential impacts of this release focussed on seabirds, reptiles, commercial fish species and coral reefs (PTTEP Australasia 2013). No apparent recent mortality or disturbance impacts were detected at the nearest coral reefs (Vulcan Shoal) 27 km from the spill site. However, assessment may have begun too late to detect impacts on corals and other invertebrate fauna associated with coral reef habitats. Surveys of coral reef habitat in proximity to the MONTARA wellhead began in April 2010, some eight months after the start of the spill (Heyward 2010). Most studies examined in this review found that the most significant mortalities were immediate, if they occurred at all.

After the MONTARA spill, decreased seagrass cover was observed at Vulcan Shoal, although whether this was a natural disturbance or an effect of crude oil exposure could not be determined. In 2010, six months after the uncontrolled release, Vulcan Shoal was found to support an extensive and lush seagrass meadow (*Thalassodendron ciliatum* (32.9% cover)) (Heyward et al. 2013). However, significant declines in seagrass cover occurred in 2011 (5.2% cover), and <1% cover of seagrass was recorded on this shoal by 2013 (Heyward et al. 2013). The seagrass remnants seen in the 2011 survey (fibrous rhizomes embedded in sand and rubble areas across the central and northern plateau areas) were also further reduced by 2013 (Heyward et al. 2013). Sediment samples confirmed that Vulcan Shoal had hydrocarbons present, and this exposure, while low, was higher than other shoals examined in this study (Heyward et al. 2013). As the 2010 survey did not detect a decline in seagrass cover immediately after the spill, Heyward et al. (2013) suggested that a delayed effect from the uncontrolled release (i.e. 6–16 months post spill) seemed unlikely. Hard corals on Vulcan Shoal were normal in appearance and had not decreased in abundance during the same period, suggesting that the cause of the seagrass loss was either selective to seagrass or perhaps physical in nature (Heyward et al. 2013). There was also no evidence of accentuated algal epiphyte growth on remaining seagrasses or other nearby biota (corals) that could potentially be associated with oil spills to explain a lagged response. It was concluded that a storm or other source of strong seabed shear forces might have been responsible for the reduced seagrass cover in the 2011–2013 period (Heyward et al. 2013).

In addition to their presence in sediments at Vulcan shoals, hydrocarbons were also detected at lower levels in sediments sampled at Ashmore, Cartier and Seringapatam Reefs, the closest of which is about 120 km from the MONTARA well site. Approximately 50% of the sediment samples contained hydrocarbons five months after the spill, and ~35% of the sediment samples contained hydrocarbons 15 months after the spill (Heyward et al. 2012). Gas chromatography mass spectrometry analyses and reconstructed ion chromatogram analyses showed that the hydrocarbons had patterns typical of degraded oil, indicating that significant weathering processes had taken place. It was not possible to match the oil found in the sediments to the MONTARA reservoir (Heyward et al. 2012). The levels recorded were orders of magnitude lower than what would constitute a risk to the environment according the ANZECC & ARMCANZ (2000) guidelines; their presence in the sediments suggests that hydrocarbons from spills can be incorporated into sediments, as can naturally released hydrocarbons. There are few sites of proven active seepage of hydrocarbons in Australia with the exception of the Timor Sea (Logan et al. 2008, Wasmund et al. 2009). However, the hydrocarbon seeps identified in the Timor Sea (Burns et al. 2010) do not appear to have a chronic or measurable effect on the quality of the marine environment (Sim et al. 2012), and consequently, it could be concluded that they do not pose a risk to animal and plant communities in the area.

The absence of any prior surveys or baseline data, and the delayed start of the post-spill survey, precluded the authors drawing firm conclusions about whether the MONTARA spill resulted in any impacts at Vulcan Shoal or other surveyed locations (Heyward et al. 2010, Heyward 2011). The salient lessons from the MONTARA spill relevant to benthic macroinvertebrate and macrophyte assessment more generally include the need for baseline prespill monitoring in areas vulnerable to impact and the necessity for a more rapid response to assess impacts. In addition, Heyward et al. (2010) found considerable ecological differences among shoals and reefs surveyed, indicating potentially different histories of disturbance. These would likely have confounded efforts to compare impacted and non-impacted reefs if there had been a significant effect of oil at one or more sites. This finding reinforces the need for any prespill baseline monitoring programmes to include a historical series of measurements at reference sites expected to be unimpacted in the event of a spill.

As evidenced by the PACIFIC QUEST and MONTARA spills, shipping along the Great Barrier Reef and oil and gas extraction off north-western Australia provides a potential for oil spills to impact offshore coral reefs. Corals and their larvae are sensitive to the toxic effects of oil (see later in this review). On the Great Barrier Reef, as elsewhere in the Indo-Pacific region, most coral species spawn annually over a period of just a few days following the late spring/early summer full moon (Kojis & Quinn 1982, Harrison et al. 1984, Babcock et al. 1986), and the buoyant nature of coral gametes and larvae (Harrison & Wallace 1990) places them at an elevated risk of contact with oil following spills that coincide with spawning events (and up to four weeks after). The toxicity of oil has also been demonstrated to be increased by photo-oxidation (exposure to UV light) (Negri et al. 2015), thus exacerbating the risk to coral gametes and larvae. On the Australian west coast, a larger primary mass spawning period has been reported in autumn, centred around the March full moon (Simpson et al. 1991). There is also a smaller but ecologically significant multispecific spawning period, involving fewer species and colonies, reported during late spring or early summer, often following the November full moon (Rosser & Gilmour 2008, Gilmour et al. 2009, Rosser 2013) in north-western Australia.

Subantarctic spill

Australia has experienced one Subantarctic oil spill on Macquarie Island, from the NELLA DAN in 1987, which resulted in significant impacts on marine invertebrates but seemingly minimal impacts on algae. These impacts have been well documented by Pople et al. (1990), Simpson et al. (1995) and Smith & Simpson (1995, 1998) who examined intertidal and subtidal habitats one year and seven years post impact. Pople et al. (1990) described high mortalities of marine invertebrates immediately post impact. These studies compared impacted and non-impacted (control) exposed rocky shores at different intertidal heights, as well as more sheltered kelp holdfast habitat. After one year, significant differences existed between sites, with habitats at all impacted sites having lower abundances of marine invertebrates. Gastropods (limpets and trochids) and echinoderms (holothurians and seastars) were heavily impacted on the exposed shores (Pople et al. 1990), while isopods were most impacted amongst more sheltered kelp holdfast habitats (Smith & Simpson 1995). Pople et al. (1990) noted a small amount of mortality of the green algae *Spongomorpha pacifica* (=*Acrosiphonia pacifica*) near the site of the heaviest oiling. Despite finding a significant difference in cover of the red alga *Palmaria georgica* between oil and control sites, they attributed the high cover of this alga at control sites to habitat differences rather than to oiling. After seven years, invertebrate abundances at impacted sites on exposed shores were comparable to control sites; however, significant differences remained within the more sheltered kelp holdfast habitat. Some impacted sites showed continuing presence of diesel and invertebrate communities dominated by opportunistic polychaete and oligochaete worms, whereas kelp holdfast invertebrate faunal assemblages at control sites were dominated by crustaceans such as isopods and other peracarids (Smith & Simpson 1998).

Spills in mangroves and estuaries

A moderate proportion (12/51) of oil spills in Australia have impacted mangrove habitats (Table 2); however, associated studies rarely considered impacts on fauna. After a small spill of oil from the ARTHUR PHILLIP in 1985, Anink et al. (1985) recorded up to 743 μg g^{-1} of hydrocarbons in sediment and significant impacts on mangroves in Botany Bay. They noted that invertebrate mortality (crabs) was not evident until around four weeks post spill, and that the same observation had been made in the Parramatta River spill in 1982, with dead crabs first observed three weeks after the spill. Andersen et al. (2008) found evidence that burrowing crabs in the high intertidal area in mangroves suffered high mortality (reduced incidence of crab holes) one month following the GLOBAL PEACE spill in Gladstone harbour, but that after six months, numbers had recovered (Melville et al. 2009). Species richness and abundance of marine invertebrates in the lower intertidal zone did not differ between impacted and control sites one-month post spill. Seagrass meadows were found to be not impacted by the GLOBAL PEACE spill in a study where good prespill spatial and seasonal baseline data existed (Taylor et al. 2006, Taylor & Rasheed 2011).

Spills on temperate and subtropical shores and reefs

Following the 1995 IRON BARON spill in Tasmania, Edgar & Barrett (2000) undertook the first rigorous assessment of the impacts of an oil spill on a subtidal temperate reef in Australia. In contrast to the obvious mortality to invertebrates in the high intertidal zone (such as amphipods) caused from this spill, Edgar & Barrett (2000) found no evidence of oil impacts on species richness and abundance of macroalgae, reef invertebrates and fish across gradients of exposure to heavy oil which had been mixed to a sufficient depth to contact the reef. However, other Australian studies showed varying responses to oil spills when examining impacts at different intertidal levels. When assessing the impact of the PACIFIC ADVENTURER oil spill one week and three months post spill, Schlacher et al. (2011) found significantly less diversity and abundance of invertebrates on impacted beaches low on the shore, with no detectable impacts on the upper shore. Three months post spill, there was no evidence of oil remaining on the beach, but the lower density and diversity of invertebrates on impacted beaches remained (Schlacher et al. 2011). For the same spill, assessment of impacts on rocky shores (Stevens et al. 2012) showed a very significant impact on diversity and abundance at both mid and high intertidal areas, with the most severe effects in the high intertidal. Significant effects remained five months post spill; by four years post spill, midshore communities of invertebrates had returned to preimpact levels, but higher intertidal communities continued to be affected (Finlayson et al. 2015).

MacFarlane & Burchett (2003) studied the impact of the 1999 LAURA D'AMATO oil spill on rocky intertidal reefs in Sydney Harbour. Pulmonate limpets (*Siphonaria*) and trochid snails (*Austrocochlea*) suffered significant mortality at the most heavily oiled site. Less impacted sites also showed impacts of oiling on densities of some species, but these changes were of a similar magnitude to seasonal variability observed at some sites surveyed for eight years prior to the spill. During 2000, MacFarlane & Burchett (2003) documented some recovery at all sites 12 months post spill. They did not note any significant effects on macroalgae from the spill. A follow-up laboratory study (Reid & MacFarlane 2003) confirmed the toxicity of oil similar to that released from the LAURA D'AMATO on *Austrocochlea*. The same oil spill caused near total mortality of the amphipod *Exoediceros* sp. on sandy beaches affected by the spill (Jones 2003). Rates of recovery tended to be most rapid at the less impacted site, which recovered after four months compared to more heavily impacted sites, some of which showed no recovery after nine months. Rapid recovery of communities was found by Dexter (1984), who documented a decline in polychaetes and amphipods on a beach impacted by a 1981 spill in Botany Bay but with full recovery to preimpact levels after three to five months.

Oil spill simulation field studies in Australia

Few studies simulating the effects of an oil spill in natural habitats have been undertaken in Australia. Clarke & Ward (1994) found applications of Bass Strait crude oil, crude oil plus dispersant and diesel (which simulated an unplanned spill in salt marshes in Jervis Bay, NSW), caused high rates of mortality among high intertidal gastropods (*Littorina*, *Bembicium*, *Salinator* and *Ophicardelus*) and crabs. Similar effects were found amongst the different contaminants used. Clarke & Ward (1994) found that treated plots had been recolonised from adjacent areas after 12 months. Residual lower densities in some treated plots after this time possibly resulted from greater predation of gastropods in the treated areas, where salt marsh plant cover had been reduced by the simulated spill treatments.

McGuinness (1990) carried out similar experiments in both mangrove and salt marsh plots in Botany Bay, NSW, using weathered (one part oil agitated with two parts seawater) Dubai light crude oil. High mortality of the gastropods *Assiminea*, *Cassidula* (=*Melosidula*) and *Salinator* was measured, with increased mortality in salt marsh compared to mangroves. However, plots recovered within weeks, due in large part to rapid recolonisation from adjacent habitats (McGuinness 1990).

Both these studies found that recovery was rapid due to recolonisation from adjacent plots. In the event of an actual oil spill affecting a large area of salt marsh or mangrove, recovery is unlikely to be as rapid or as effective as observed by McGuinness (1990) or Clarke & Ward (1994). In Gladstone, Queensland, Burns et al. (2000) treated plots of mangroves with Gippsland crude and Bunker C oils to compare the weathering effects of each and the effect of dispersants but did not examine the influence on mangrove invertebrates (Burns et al. 1999, 2000). Similarly, in the temperate Gulf St Vincent in South Australia, Wardrop et al. (1987) measured the rates of mangrove defoliation and recovery after application of Arabian light crude and Tirrawarra crude oils with and without dispersant. They found increased levels of initial defoliation on mangroves in treatments with dispersants but no long-term negative impacts, although they did not examine any effects on biota associated with mangroves.

Thompson et al. (2007) conducted a study in Antarctica using synthetic lubricants to assess the effect of recolonisation in defaunated plots in marine sediments. They found that abundances of recently colonised animals in plots were the same five weeks post oiling, but community composition was quite different, with numbers of certain crustaceans (amphipods, tanaidaceans and cumaceans) reduced in treated plots compared to controls (Thompson et al. 2007).

Measured responses of Australian invertebrates, seagrasses and macroalgae to oil toxicity in laboratory studies

Marine invertebrate toxicity responses

Experimental assays of oil and dispersant toxicities in Australia are limited to a few studies. Gulec et al. (1997) measured the 96-hour 50% lethal concentrations (96-hour LC50 s) for the amphipod *Allorchestes compressa* of 311,000 ppm (WAF of Bass Strait crude oil diluted in seawater), 16.2 ppm (oil dispersed 9:1 with Corexit 9500 and diluted in seawater) and 14.8 ppm (as previously for Corexit 9527) indicating that the oil alone (without dispersant) was much less toxic. It should be noted that these are not the concentrations of hydrocarbons but the concentrations of WAF. In the case of Gulec et al. (1997), the WAF was obtained by stirring one part crude oil and nine parts water for 24 hours and then, after settling for one hour, siphoning off the WAF from below the oil. In the previous example, 311,000 ppm is 311 mL WAF and 689 mL seawater in each litre. They also demonstrated that a WAF using burnt oil was less toxic than unburnt oil (Gulec & Holdway 1999). *Octopus pallidus* hatchlings exposed to WAFs of Bass Strait crude oil and dispersed oil had a 48-hour LC50 of 0.39 ppm and 1.83 ppm, respectively, suggesting that they are much more sensitive to oil than amphipods (Long & Holdway 2002). The trochid snail *Austrocochlea porcata* had a 96-hour LC50 of 12 ppm (Reid & MacFarlane 2003). Although not conducted on Australian invertebrate species, Neff et al. (2000)

compared the toxicities of three types of light to medium density north-west Australian crude oils and diesel fuel on penaeid prawns, mysid shrimps and sea urchin larvae.

Despite the importance of coral reefs in Australia, there have been few studies of the response of corals and coral larvae to oil in Australia. Branches of *Acropora muricata* (=*Acropora formosa*) exposed to marine fuel oil showed complete living tissue disintegration after 48 hours (Harrison et al. 1990). Harrison et al. (1990) further reported that the *A. muricata* colonies expelled zooxanthellae and showed enhanced production of mucus within the first hour of exposure to marine fuel oil, mortality after 12 hours exposure, and an increase in the concentration of pigmented bacteria on the mucus after 48 hours.

Most broadcast spawning corals have buoyant eggs and larvae (Harrison & Wallace 1990), making them potentially highly vulnerable to oil spills (Harrison 1999). The WAF of heavy crude oil from the Wandoo reservoir off north-western Australia did not inhibit fertilisation of *Acropora millepora* gametes. However, fertilisation was disrupted by the dispersant Corexit 9527 and the dispersed oil fraction (Negri & Heyward 2000). Negri & Heyward (2000) further showed that larval metamorphosis was inhibited by exposure to crude oil, the dispersant Corexit 9527 and the dispersed oil fraction. Moreover, although the crude oil and dispersant inhibited larval metamorphosis individually, this toxicity was magnified when the larvae were exposed to combinations of both (i.e. dispersed oil fraction). A similar result was reported in three different species (*Acropora tenuis, Platygyra sinensis, Coelastrea aspera* [=*Goniastrea aspera*]); significantly increased larval mortality rates in the dispersant (Ardrox 6120) and dispersed bunker fuel oil 467TM treatments were recorded compared to seawater controls (Lane & Harrison 2000). Increased exposure time to oil resulted in increased toxic effects, and the dispersed oil was the most toxic of the contaminants tested due to an increase in bioavailability of hydrocarbons as a result of the dispersal (Lane & Harrison 2000). Of the three species, significant mortality was only measured in *C. aspera* larvae when exposed to treatments of different concentrations of the oil WAF alone. However, the experiment was hampered by low rates of survivorship in the controls.

Natural gas condensate can be considered as a light crude oil, and the WAF of Browse Basin condensate (north-western Australia) was shown to inhibit metamorphosis of coral (*Acropora tenuis*) and sponge (*Rhopaloeides odorabile*) larvae (Negri et al. 2015). Coral larvae were 100 times more sensitive (inhibition at 100 μg L^{-1} polycyclic aromatic hydrocarbons [PAHs]) than sponge larvae (10,000 μg L^{-1}). When the WAF was exposed to ultraviolet (UV) light, the sensitivity of coral larvae, but not sponge larvae, increased by 40% (Negri et al. 2015). Oil exposed to ultraviolet radiation in surface waters may photo-oxidise (Pelletier et al. 1997), which may substantially increase the toxicity and bioavailability of hydrocarbon components such as PAHs (Ehrhardt & Burns 1993, Dutta & Harayama 2000, Neff 2002).

Unfortunately, studies of experimental laboratory and field-based oil toxicity are not directly comparable given the use of different field/laboratory conditions, weathering techniques (with/ without predistilling, and WAF varying between two and nine parts seawater), LC50 calculations (ppm total hydrocarbons versus % WAF), and time periods (24–96 hours). The use of regression analyses to predict LC50 s from experimental data also makes comparison difficult, and applications to field situations even more challenging. As described previously, Long & Holdway (2002) calculated a 48-hour LC50 for newly hatched octopus of 0.39 ppm total hydrocarbons. However, they also reported a 48-hour no observed effect concentration of 0.36 ppm and a lowest observed effect concentration of 0.71 ppm. Given that exposure times and circumstances in the field can vary greatly, the approach taken by Long & Holdway (2002) of also reporting 24-hour LC50 s is useful when trying to extrapolate likely ecological field effects from laboratory toxicity experiments. These points highlight the need for ecologically meaningful approaches and improved consideration of how laboratory toxicity trials are related to the field situation.

Three Australian studies have found sublethal behavioural impacts on seastars and a gastropod. Ryder et al. (2004) found that the herbivorous seastar *Parvulastra exigua* (=*Patiriella exigua*) from Port Phillip Bay avoided oiled sediment in the laboratory and in doing so was able to avoid its narcotising effects. The ability of the predatory Port Phillip Bay seastar *Coscinasterias muricata* to locate prey was significantly reduced when exposed to WAFs of Bass Strait crude oil with and without added dispersant; however, seastars exposed to a burnt oil WAF maintained the same ability to locate prey as control animals (Georgiades et al. 2003). Gulec et al. (1997) studied suppression of burying behaviour of the marine sand snail *Conuber conicum* (=*Polinices conicus*) after 30 minutes of exposure to dilutions of Bass Strait crude oil WAF, crude oil plus Corexit 9500 and crude oil plus Corexit 9527. Burying was suppressed in 50% of snails (EC50) at 190,000 ppm, 65.4 ppm and 56.3 ppm, respectively (Gulec et al. 1997).

Seagrass toxicity responses

Australia has some of the most extensive seagrass meadows in the world, which are of great importance to coastal stability and as a food source and nursery for coastal invertebrates, fishes, mammals and reptiles (Larkum et al. 1989, Butler & Jernakoff 1999). Despite the importance of these habitats, there have been few studies on the toxic response to oil of seagrasses in Australia. The first such study was carried out by Hatcher & Larkum (1982), who found exposure of *Posidonia australis* to 6800 ppm of Bass Straight crude oil and 120 ppm of the dispersant Corexit 8667 resulted in decreased photosynthesis and increased respiration, but that leaf production was not affected. The dispersed oil mixture induced a greater physiological stress than the oil on its own. However, Ralph & Burchett (1998) found photosynthesis and pigments in *Halophila ovalis* showed only a minor negative response when exposed to 1% weight/volume (w/v) Bass Strait crude oil and an oil: Corexit 9527 mix (10:1).

More recently, Macinnis-Ng & Ralph (2003) showed Champion crude oil (0.25% w/v) had an effect on photosynthesis and pigments of *Zostera capricorni* in the laboratory; however, *in situ* field trials in enclosures demonstrated reduced impacts. It was concluded that laboratory trials may overemphasise toxic effects. In another study on *Z. capricorni*, Wilson & Ralph (2012) showed no impact at concentrations less than 0.4% of the water-soluble fraction (WSF) of Tapis crude oil. Thus, while such studies highlight the potential for toxicity following exposure to oil and oil and dispersant mixes, consensus on impacts to the overall health of seagrass is lacking due to a high degree of variability in oiling scenarios and potentially response among seagrass species. This remains an issue for both the Australian and international context (Fonseca et al. 2017).

Macroalgae toxicity responses

Measurements of toxicity of Australian macroalgae to oil and dispersants are limited to the measurements made by Burridge & Shir (1995). In these experiments, they examined the effects on germination of the brown alga *Phyllospora comosa* and found that the 48-hour EC50 (maximal effective concentration that induces a response halfway between the baseline and maximum germination inhibition after 48 hours) was 130 ppm for Bass Strait crude oil and 6800 ppm for diesel fuel.

Application of two dispersants (Corexit 7664, 8667A) resulted in substantially decreased toxicity of the crude oil to a 48-hour EC50 of 2500 and 4000 ppm, respectively, while little change in toxicity was evident with use of two other dispersants (Corexit 9500 and 9527). All four dispersants increased toxicity of diesel (48-hour EC50 of 340–420 ppm) (Burridge & Shir 1995).

Impacts of oil on marine invertebrates, seagrass and macroalgae

Acute and toxic impacts of oil on marine invertebrates

Impacts on zooplankton and invertebrate larval stages

Larval life stages of marine invertebrates are likely to represent the most vulnerable period for exposure to toxicants. However, in long-lived iteroparous species with short larval periods, the impact of an oil spill on a population or a species may be minimal. On the other hand, semelparous species with long larval periods might suffer a major impact.

Larval assays of toxicity to contaminants in seawater have been used extensively to determine both the toxicity of different oils and their fractions, and what concentrations and time exposures constitute lethal or sublethal or minimum observable effects. Sea urchin and bivalve mollusc larvae have been most commonly used, as they are easy to culture or are commercially available. Assays have used either contaminated seawater collected following a spill or the preparation of known concentrations in the laboratory. The former method is likely to give results that can be directly related to an oil spill event. For example, Beiras & Saco-Alvarez (2006) used seawater sampled from the shore at various times following the PRESTIGE oil spill in Spain in 2002 to test for toxicity against sea urchin, *Paracentrotus lividus*, larvae. They found that even after a 4-fold dilution this WAF was toxic to the larvae immediately after the oil spill, and toxicity to larvae from the undiluted contaminated seawater persisted in seawater collected two months after the spill.

The planktonic larvae of marine invertebrates are highly sensitive to the toxic effects of hydrocarbons in a WAF. Chia (1971) reported on an oil spill in northern Washington State in the United States, which killed numerous adult marine invertebrates, and noted that the diesel spill had occurred during the spawning season for many species. He then tested the larvae of 14 species of echinoderms (seastars and a sea urchin), gastropods, a bivalve and a chiton mollusc, annelid worms, a barnacle and an ascidian in a 0.5% oil–water mixture. He found that while all larvae in control conditions survived, all larvae in the oil–water mixture died within three hours to three days, other than those of one seastar *Crossaster*, which all had died after eight days (Chia 1973). Unlike the other invertebrates, *Crossaster* has large yolky eggs with lecithotrophic (non-feeding) development, which may be a reproductive strategy that is resilient to oiling. Around the same time, Wells (1972) also demonstrated the toxic effect of oil on lobster (Crustacea) larvae (Wells 1972, Wells & Sprague 1976, Stejskal 2000). Byrne & Calder (1977) and Nicol et al. (1977) further demonstrated that oil disrupted embryonic development, causing mortality in a bivalve mollusc and a sea urchin, while PAHs have been demonstrated to inhibit settlement of sponges or cause mortality of sponge recruits (Cebrian & Uriz 2007). For zooplankton, Elmgren et al. (1983) found ostracods and harpacticoid copepods were significantly impacted by the 1977 TSESIS spill in Sweden. Almeda et al. (2013) examined the effects of crude oil on mortality on a wide range of copepod species in the Gulf of Mexico and determined an LC50 of 31.4 ppm WAF. In other studies, Almeda et al. (2014a,b) demonstrated that dispersed oil is more highly toxic than crude oil alone.

Following the Gulf of Mexico DEEPWATER HORIZON spill in 2010, settlement of the commercially important crab *Callinectes sapidus* along the Mississippi coast was assessed (Fulford et al. 2014). Natural settlement rates of this species vary considerably each year, and they did not detect any change attributable to the oil spill. They also noted that concentrations of oil known to be toxic to crab larvae were not experienced in locations important for crab settlement, at least in their study area.

Studies consistently indicate that the impact of oil on larvae is largely dependent on exposure concentration (Almeda et al. 2013, Fulford et al. 2014). Thus, the risk of exposure to toxic concentrations is a critical element when evaluating likelihood of impact. This emphasises the importance of studies that determine at what concentrations hydrocarbons cause mortality or have significant sublethal impacts on receptor species and the likelihood of exposure to those concentrations. Because of the varying responses observed by different species to different types of oils, and the degree of weathering, the context becomes very important. This underlines the importance of studies that not only provide data in a context relevant to Australia, but does so in at least a regional or preferably local context. For this reason, we advocate tests that determine the toxicity of Australian oils against taxa most likely to be exposed to them. In addition, we suggest that tests seek to simulate the concentration levels likely to occur at the time of exposure. This is especially emphasised for prespill assessments for large unconfined spills, where an advancing plume may create an exposure risk to animals in intertidal and subtidal habitats.

Impacts on adult invertebrates

Early reports on the impacts of oil spills generally relate to intertidal exposures following shipping or refinery accidents. These indicated that a very wide range of marine invertebrate taxa are affected by oil, with very high potential mortalities (Mitchell et al. 1970, Spooner 1970, Chia 1971, Woodin et al. 1972, Chan 1977). Given the different taxa that characterise species assemblages across the wide range of benthic habitat types and latitudes, we provide examples below of the available information on how different groups of marine invertebrates are affected by oil exposure. Although the following analysis suggests a range of responses between taxa, it is likely the severity of a spill and the levels of exposure (not the type of animal) are probably more important in determining the level of mortality that occurs.

Sponges Sponges form an important and often dominant component of the fauna of many Australian benthic marine habitats (Fromont et al. 2012), and this is especially true of the Australian North West Shelf where extensive oil and gas exploration and production is occurring. There are few reports of sponge mortality from oil spills in the international literature, and those that do exist make it unclear whether sponges are highly vulnerable to oil toxicity or not. After the 1986 oil spill in Galeta, Panama, sponges growing on oil-covered mangrove roots died (Burns et al. 1993). Similarly, following the 2002 PRESTIGE oil spill, *Hymeniacidon perlevis* and *Tethya* sp. (defining species in the lower intertidal area of their study site in France) were killed and had not reappeared by 2011, as shown by annual monitoring to 2011 (Castège et al. 2014). However, it is possible that the high-pressure hot-water cleaning of the site contributed to the mortality and/or the absence of recovery in this instance (Castège et al. 2014). Rocky shore monitoring in Milford Haven, Wales, six months before and after the 1996 SEA EMPRESS spill did not detect any effects on abundance of *Hymeniacidon perlevis* in areas affected by the spill but not cleaned, suggesting this sponge at least is not especially vulnerable to the toxic effects of oil (Moore 1997). Batista et al. (2013) determined that *Hymeniacidon heliophila* was a good indicator of PAH pollution in Brazil (Batista et al. 2013). However, apart from the study of Harvey et al. (1999), who found an absence of genotoxic effects on *Halichondria panicea* after the 1996 SEA EMPRESS spill in Wales (Harvey et al. 1999), we are not aware of any laboratory studies on the toxic effects or harmful concentrations of oil to adult sponges. This is arguably a high-priority need in the Australian context.

Bryozoans Bryozoans can comprise a significant proportion of benthic biomass, and this is particularly true on Australia's North West Shelf (Keesing et al. 2011). However, this group is rarely reported when considering the impacts of oil spills, although Burns et al. (1993) found that along with hydroids, bryozoans were the least impacted and the fastest taxa to recover on the roots of mangrove trees following the 1986 Galeta refinery oil spill in Panama. In the absence of any other studies, an understanding of the response of Australian bryozoa to oil is needed.

Cnidarians As a group, anthozoan scleractinian corals are sensitive to oil pollution and can suffer high mortality on both intertidal and subtidal reefs affected by oil spills (Jackson et al. 1989, Guzmán & Holst 1993, Guzmán et al. 1994), as well as chronic sublethal effects (see reviews by Johnson 1977, Loya & Rinkevich 1980, Fucik et al. 1984, Suchanek 1993). Direct coating of corals by oil have been reported to result in stress and mortality of coral colonies (Loya & Rinkevich 1979, 1980, Yender & Michel 2010). Rinkevich & Loya (1977) reported that a greater number of *Stylophora pistillata* colonies died on hydrocarbon-affected reefs after one year (∼42% mortality) compared to control reefs (∼10% mortality) in Eilat, Israel. Coral mortality was also recorded following a major oil spill in Panama (70% Venezuelan, 30% Mexican Isthmus crude oil), and the total coral cover in heavily oiled reefs declined by between ∼55% and ∼75% at 3–6 m and <3 m depths, respectively (Jackson et al. 1989). A significant decline in coral cover was recorded at ≤3 m depth, and most corals at >3 m depth that remained alive after the spill showed signs of stress that included tissue swelling, conspicuous production of mucus and recently dead areas devoid of live coral tissue (Jackson et al. 1989).

Laboratory studies have also demonstrated stress and mortality of coral colonies exposed to oil. Johannes et al. (1972) exposed 22 species of coral at Eniwetok Atoll to Santa Maria crude oil for 1.5 hours and showed a complete breakdown of tissue on the areas to which crude oil adhered, whereas no visible effects were evident in the areas to which the oil did not adhere to the coral colonies. Reimer (1975) exposed colonies of *Pocillopora* cf. *damicornis* to Bunker-C and marine diesel oils and reported that all colonies lost almost all living tissue after 16 days, while the control colonies sustained >95% living tissue. Colonies of other genera (*Psammocora* and *Porites*) were more resilient, surviving for longer, but died after 114 days. However, this was done under extreme conditions with corals in some cases being fully submersed in oil. Similarly, branches of *Acropora muricata* (=*Acropora formosa*) exposed to marine fuel oil showed complete living tissue disintegration after 48 hours (Harrison et al. 1990). Harrison et al. (1990) further reported that the *A. muricata* colonies expelled zooxanthellae and showed enhanced production of mucus after 12–24 hours. While no significant impact on the survivorship from the WAF of Egyptian crude oil was reported in *Stylophora pistillata* and *Pocillopora damicornis* following exposure, significant mortality was reported following exposure to dispersants (Emulgal C-100, Dispolen 36S, Inipol IP-90, Petrotech PTI-25, Slickgone NS, Bioreico R-93) and the dispersed oil fraction (Shafir et al. 2007).

Laboratory and field studies have demonstrated that branching corals appear to have a higher susceptibility to hydrocarbon exposure than massive corals or corals with large polyps. After exposure to Santa Maria crude oil, branching *Acropora* and branching *Pocillopora* showed the highest susceptibility to coating and retention, and oil was still visible on the corals after four weeks (Johannes et al. 1972). Corals with large fleshy polyps and abundant mucus (*Fungia* and *Lobophyllia* [=*Symphyllia*]) retained almost no oil and showed no damage, and *Astreopora*, *Favia*, *Favites*, *Montipora*, *Plesiastrea*, *Porites*, *Psammocora* and *Turbinaria* showed intermediate effects (Johannes et al. 1972). The increased susceptibility of branching coral species was further reinforced by Guzmán et al. (1991), whereby following a major oil spill in Panama (70% Venezuelan, 30% Mexican Isthmus crude oil), the branching coral *Acropora palmata* suffered far greater mortality compared to the massive species that were common on the local reefs.

Based on the literature, those genera capable of producing large amounts of mucus may be less susceptible to damage from exposure to oil. Production of mucus is generally a stress response (Beeden et al. 2008, Erftemeijer et al. 2012) and may aid removal of oil and toxicants and provide these corals with a level of resilience. However, excess production of mucus resulting from hydrocarbon exposure may also lead to enhanced bacterial growth and degradation of the coral tissue (Loya & Rinkevich 1980, Peters et al. 1981) as well as being energetically costly (see Davies & Hawkins 1998 for review).

The reported responses of other cnidarians to oil vary considerably in the literature. Jackson et al. (1989) recorded that the hydrozoan *Millepora* and zoanthids *Palythoa* and *Zoanthus sociatus* were significantly affected (along with scleractinian corals) after the 1986 Galeta refinery spill. Cohen et al. (1977) found that toxicity of crude oil on octocorals (*Heteroxenia fuscescens*) was only evident at very high levels of exposure (12 ppt), but sublethal effects occurred at lower concentrations. There is some evidence that hydroids may be resilient to oil spill impacts, but this is equivocal. Suchanek (1993) reviewed laboratory studies that indicated the hydroid *Tubularia* and the scyphozoan *Aurelia* were both sensitive to oil; conversely, Burns et al. (1993) found that hydroids (and bryozoans) growing on mangrove roots were minimally impacted following the 1986 Galeta refinery oil spill.

Anthozoan actinians (anemones) were severely impacted by the 1986 Galeta oiling and much slower to recover than other taxa, with reduced densities persisting five years after the event (Burns et al. 1993). However, Castège et al. (2014) found two anemones (*Actinia equina* and *Anemonia viridis*) were among a group of invertebrate species that were minimally impacted or recovered quickly (within one year) following the 2002 PRESTIGE oil spill that affected the French coastline. Similarly, the anemone *Anthopleura elegantissima* was one of the few species that survived the 1957 TAMPICO MARU spill in Mexico that killed most marine invertebrates (Mitchell et al. 1970). The widely varying responses of cnidarians, and their importance among benthic marine communities in

Australia (and especially the tropics; e.g. Keesing et al. 2011), indicates that a specific examination of their response to potential levels of exposure in Australia is required.

Crustaceans   Motile crustaceans as a group are among the most vulnerable marine invertebrates to oil spills and suffer high mortalities, behavioural disorders, and reduced recruitment (Krebs & Burns 1977). Crabs are highly conspicuous components of intertidal assemblages and are among the first casualties to be reported after a spill (e.g. Spooner 1970, Woodin et al. 1972, Chan 1977). Following the 1969 spill in West Falmouth, Massachusetts, USA, fiddler crabs (*Minuca pugnax* [=*Uca pugnax*]) suffered high mortality. Stomatopods were heavily impacted in intertidal seagrass beds after the 1986 refinery spill at Galeta, Panama (Jackson et al. 1989). Following the 1977 Tsesis oil spill in the Baltic Sea, numbers of amphipods (*Pontoporeia*) were reduced by 95% as oil began to be deposited onto the benthos (Elmgren et al. 1983). Amphipods, isopods and crabs were also heavily impacted by the 1978 Amoco Cadiz spill in France (Chassé 1978, O'Sullivan 1978, Conan 1982), with populations of numerous amphipod species remaining at low levels or absent 10 years later (Dauvin 1987, Dauvin & Gentil 1990).

Although high initial mortalities can occur following spills, some crustaceans apparently recover quickly. After the 2002 Prestige oil spill in France, several shrimp and crab species (*Athanas nitescens*, *Carcinus maenas*, *Eriphia verrucosa* [=*Eriphia spinifrons*], *Galathea squamifera*) were among a group of invertebrate species on a rocky intertidal shore that were minimally impacted or recovered quickly (within one year; Castège et al. 2014). Conversely, after the West Falmouth spill, the fiddler crab population had still not recovered seven years post spill because the habitat was still contaminated (Krebs & Burns 1977).

Compared with motile crustaceans, adult barnacles as a group have been regarded as very resistant to the effects of oil (Suchanek 1993). Following the Amoco Cadiz spill, Chassé (1978) found barnacles *Chthamalus* and *Balanus* did not suffer mortality, although this may have been due to their position at a lower intertidal height on the shore, as extensive gastropod mortalities were found higher on the shore. In Brazil, Lopes et al. (1997) studied the impacts of an oil pipeline spill and found that among the crustaceans, crabs and isopods suffered heavy mortality, but barnacles (*Chthamalus* and *Tetraclita*) were not significantly affected (Lopes et al. 1997). However, the circumstances of a spill (and not the type of animal) are probably more important in determining the levels of mortality experienced. For example, following the 1971 diesel spill in Washington State, USA, substantial mortality of *Balanus glandula* and *Semibalanus cariosus* (=*Balanus cariosus*) were recorded (Woodin et al. 1972), while after the 2002 Prestige spill in Spain, mortality of the barnacle *Chthamalus montagui* depended on extent of oiling (Penela-Arenaz et al. 2009).

Some Australian studies have reported significant oil spill impacts on amphipods (Edgar & Barrett 2000, Jones 2003) and crabs (Anink et al. 1985, Clarke & Ward 1994, Andersen et al. 2008) from accidental or planned oil spills in Sydney Harbour, Jervis Bay and Tasmania. Given their vulnerability to oil and the commercial importance of crustacean invertebrates, more attention should be given to assessing the toxicity and sublethal responses of crustaceans to oil in areas of anticipated risk across different Australian regions.

Tunicates   Tunicates were among the heavily impacted taxa within the invertebrate communities on the roots of mangroves subject to oiling following the 1986 refinery spill in Galeta, Panama. Like anemones, tunicate populations had not recovered after five years (Burns et al. 1993). Castège et al. (2014) reported a similar time (two to five years) for the tunicate *Botryllus schlosseri* to reappear at their study site in France after the 2002 Prestige oil spill. Few reports exist on the impacts of oil on tunicates. Nevertheless, given their importance as filter feeders and the abundance of some species among intertidal and benthic assemblages in eastern Australia (e.g. *Pyura stolonifera*, Dakin 1960), and among subtidal habitats on the Australian North West Shelf (Keesing et al. 2011), more work on their vulnerability to oil spills is warranted.

Worms We group a diverse range of worm-like phyla here, and not surprisingly, they have a diverse range of sensitivities to oil. In his review, Johnson (1977) considered several studies and concluded that adult polychaetes were in general highly resistant to oil toxicity. The polychaete *Capitella capitata* opportunistically proliferates in anthropogenically disturbed sediments, including those impacted by oil, even where very high mortality of other invertebrates occurs. An extreme example of this is described by Sanders (1978) after the Florida oil spill in West Falmouth, USA (Sanders 1978).

However, some polychaetes have been reported to suffer significant mortalities following oil spills, including after the Amoco Cadiz spill in France (O'Sullivan 1978) and the 1974 Bouchard 65 spill in Massachusetts, USA, where a large number of *Alitta virens* (=*Nereis virens*) were killed among numerous other marine invertebrates (Hampson & Moul 1978). Elmgren et al. (1983) reported polychaetes (*Bylgides sarsi* [=*Harmothoe sarsi*]) were reduced by 95% in sediments contaminated by the 1977 Tsesis oil spill in the Baltic Sea and that turbellarians and kinorhynchs were also significantly affected, but nematodes were not greatly affected. Beyrem et al. (2010) examined the response of lagoon sediment nematode assemblages from Tunisia to lubricating oil contamination in laboratory experiments and found this caused reductions in both abundance and species diversity, although with differing responses amongst individual species. The reason that some species (e.g. *Daptonema trabeculosum*) were very sensitive to the oil, while others (e.g. *Spirinia gerlachi*) were resilient could not be established (Beyrem et al. 2010). Nematodes in deep-water (∼1200 m) samples were found to respond positively to the 2010 Deepwater Horizon spill in the Gulf of Mexico (Montagna et al. 2013). It was hypothesised that the nematodes may have responded to enhancement of the bacterial flora through oil-induced organic sediment enrichment and reduction of competitive species in taxa negatively impacted by the spill, such as copepods. The ratio of nematodes to copepods has previously been proposed as an indication of pollution impact (Raffaelli & Mason 1981), although the effectiveness of its use has been subject to extensive debate (e.g. see Amjad & Gray 1983, Raffaelli 1987). Dexter (1984) reported reductions in polychaetes after an oil spill in Australia, but probable impacts of oil spills on worms in the Australian situation are not generally known.

Echinoderms Echinoderms are among the most vulnerable of marine invertebrates to oil spills, and many early studies documenting oil spills indicated extensive mortalities of echinoderms after a spill (e.g. Mitchell et al. 1970, Chia 1971, Woodin et al. 1972, Chan 1977, Jackson et al. 1989). Castège et al. (2014) found three species of echinoderms (the ophiuroid *Amphipholis squamata* and the echinoids *Echinus esculentus* and *Psammechinus miliaris*) were among a group of invertebrates that disappeared from a French rocky shore after the 2002 Prestige oil spill and took two to four years to recover. On the other hand, the seastar *Asterina gibbosa* and a holothurian (*Holothuria* sp.) were either minimally or not impacted by the spill. Conan (1982) reported severe mortality of 1 million heart urchins following the 1978 Amoco Cadiz spill in France. Ballou et al. (1989) simulated the effects of crude oil and dispersed oil on a coral reef in Panama and found that all *Echinometra lucunter* and *Lytechinus variegatus* were killed in the experimental treatment areas. Jackson et al. (1989) also reported high mortality (∼80%) of *Echinometra lucunter* in Panama. Echinoderms also experience a range of significant sublethal impacts from oil exposure on their movement, reproduction and feeding (Johnson 1977).

Australia's marine ecosystems harbour a high level of echinoderm diversity, including in areas where petroleum exploration and extraction, as well as a high level of commercial shipping, also occur. For example, one live-bearing intertidal species of seastar in the Great Australian Bight (*Parvulastra parvivipara*) has a species distribution of <200 km (Edgar 2012), a very small range for a marine species. A seastar with similar brooding habit, *Asterina phylactica*, suffered significant mortality following the 1996 Sea Empress spill in Wales (Moore 2006). On the Australian North West Shelf, ecologically important heart urchins can be superabundant (Keesing & Irvine 2012). The density and diversity of crinoids on the Great Barrier Reef is extraordinarily high (Bradbury et al. 1987), and yet we could find no studies on oil toxicity to crinoids for any country. Two studies have been conducted on sublethal impacts of oil on seastars in south-eastern Australia (Georgiades et al. 2003, Ryder et al. 2004) but no tropical studies or studies on other classes of echinoderms—this is an important priority for future work.

Molluscs   Gastropods, particularly herbivores, are consistently reported as experiencing very high mortality due to oil spills, either due to the oil itself or inappropriate use of dispersants (e.g. Smith 1968, Mitchell et al. 1970, Chia 1971, Woodin et al. 1972, Southward & Southward 1978, Le Hir & Hily 2002). Mortality rates are dependent on degree of exposure, which in turn is often associated with shore height in intertidal populations. Following the 1978 Amoco Cadiz spill in France, Chassé (1978) and O'Sullivan (1978) documented high mortality of gastropods *Littorina*, *Gibbula* and *Monodonta* and to a lesser extent limpets of the genus *Patella*. In that study, mussels escaped mortality due to their position lower in the intertidal zone where barnacles were also unaffected. However, in other studies where heavy intertidal oiling has occurred, mussels have also been shown to suffer high mortality (e.g. Mitchell et al. 1970). In the Amoco Cadiz spill, Conan (1982) refers to the massive mortality of 14.5 million bivalves of other families (Cardiidae, Solenidae, Macridae and Veneridae). Contrary to the situation with the Amoco Cadiz spill, very heavy mortality of *Patella* followed the 1967 Torrey Canyon spill in Cornwall in the United Kingdom; however, most of this was attributed the toxicity of the first-generation dispersants used (Smith 1968, Southward & Southward 1978, Hawkins & Southward 1992, Hawkins et al. 2017b).

Following the 1971 diesel spill in Washington State, USA, Woodin et al. (1972) recorded substantial mortality of numerous molluscs including chitons (*Mopalia* sp. and *Katharina tunicata*), bivalves (*Clinocardium nuttallii* and *Macoma* spp.) and gastropods (*Acmaea* spp.), while oysters (*Magallana gigas* [=*Crassostrea gigas*]), mussels (*Mytilus edulis*) and the gastropods (*Littorina scutulata* and *Littorina sitkana*) experienced little or no mortality. Predatory whelks (*Thais* spp.) were found with moribund appearance but recovered when returned into clean seawater. In another study on the same spill, Chia (1971) found extensive mortality of marine invertebrates including limpets and chitons but noted that two species of littorinid periwinkles seemed unaffected. Conversely, following a large 1986 refinery spill in Panama, Garrity & Levings (1990) found that both neritids and littorinids were severely impacted and almost absent from the affected sites more than two years post spill, although the severity of effects on molluscs (in terms of immediate mortality) varied spatially with the amount of oil deposited (Garrity & Levings 1990). Subtidal impacts on gastropods (e.g. mortality of the abalone *Haliotis rufescens* and other subtidal gastropods) have also been reported after the Tampico Maru tanker was shipwrecked and spilled oil for eight to nine months in Baja California, Mexico in 1957 (North et al. 1965, Mitchell et al. 1970).

Some molluscs have been reported to be resilient to the effects of oil. For example, the gastropod *Cerithium* has been reported to continue to feed on oiled intertidal flats (Spooner 1970, Chan 1977). After the 1977 Tsesis oil spill in the Baltic Sea, Elmgren et al. (1983) showed that despite oil contamination of sediment killing 95% of amphipods and polychaetes, the clam *Limecola balthica* (=*Macoma balthica*) experienced minimal mortality despite becoming highly contaminated (to a level of 2 mg g^{-1} total hydrocarbons).

As toxic PAHs can readily bind to sediment, phytoplankton and other particulate organic matter, they can be readily ingested by filter feeding invertebrates such as mussels and oysters. Bivalve molluscs in particular are effective at bioaccumulation of these toxicants (La Peyre et al. 2014), and so can potentially suffer from a range of sublethal impacts. Given the range of often contradictory and inconsistent responses of molluscs to oil, and their diversity and importance in intertidal and subtidal assemblages around Australia, there is a need for studies on the response of local species to the types of oil they may be exposed to in different parts of Australia.

Acute and toxic impacts of oil on marine macrophytes

Seagrasses and macroalgae are considered vulnerable to oil spills because they are located in nearshore subtidal and intertidal habitats, which are the areas where uncontained oil tends to accumulate following its release. Over 75% of oil spills in the United States are estimated to occur within coastal waters, specifically estuaries, enclosed bays and wetlands (Kennish 1992). Despite this risk, and the demonstrated toxicity of oils to marine plants (Hatcher & Larkum 1982, Thorhaug et al. 1986, Ralph & Burchett 1998, Wilson & Ralph 2012, Lewis & Pryor 2013), cases of large-scale losses of seagrasses and macroalgae following an oil spill are restricted to a few isolated incidents (Nelson-Smith 1973, Floc'h & Diouris 1980, Jackson et al. 1989). This, in part, appears to be a reflection of environment, with geomorphologic and nearshore hydrodynamic conditions likely to influence the behaviour of oils such that the effects in a sheltered bay with low energy and little water exchange will be different to the effects on an exposed rocky coast with high energy and strong current (Taylor & Rasheed 2011). Understanding such considerations is important, as responses to oil spills need to be tailored to suit conditions at the local scale.

As most of the constituents of petroleum oil have low solubility in water, buoyant surface plumes are generally transported directly to intertidal habitats, often bypassing subtidal habitats and reducing the levels of exposure. Accordingly, marine flora that occur in the shallow intertidal zone are more susceptible to large-scale oil spills than flora that occur subtidally (Lewis & Pryor 2013). Following the 1986 Galeta oil spill in Panama (>8 million litres of crude oil spilled into a complex region of mangroves, seagrasses and coral reefs just east of the Caribbean entrance to the Panama Canal), Jackson et al. (1989) reported loss of entire beds of the seagrass *Thalassia testudinum* (a genus also common in tropical areas of Australia) on heavily oiled reef flats, while subtidal meadows in the same area survived. Substantial macroalgal losses following oiling of intertidal habitats were also reported by Floc'h & Diouris (1980) and Nelson-Smith (1973), while more recently there have been anecdotal accounts of large-scale losses of macroalgae in intertidal seaweed farms in Timor following the MONTARA oil spill (Mason 2011). It is important to note, however, that some of the decline reported by Floc'h & Diouris (1980) was attributed elsewhere to indirect effects associated with grazing of herbivores, rather than from direct hydrocarbon toxicity associated with crude oil impacts on macroalgae (Laubier 1980).

Although in many reported cases subtidal seagrass or macroalgal assemblages have not been severely affected following oil spills (Kenworthy et al. 1993, Dean et al. 1998, Edgar et al. 2003, Taylor et al. 2006, Taylor & Rasheed 2011), subtidal assemblages can be subjected to, and impacted by, direct contact with oil under certain environmental conditions. For example, subtidal beds of the seagrass *Thalassia testudinum* were decimated following a crude oil spill in Puerto Rico in 1973, where strong weather conditions caused the entrainment of oil into the seagrass meadow (Nadeau & Bergquist 1977). In this instance, the ship's master intentionally released 37,579 barrels (5.98 million litres) of Venezuelan (Tijuana) crude oil (following a vessel grounding), of which an estimated 24,000 barrels (3.82 million litres) was stranded within nearby coastal ecosystems (Hoff & Michel 2014). While the crude oil was considered to be of low toxicity, the impacts to the seagrass were so severe that even the rhizome layer was affected (Nadeau & Bergquist 1977).

Despite the potential for more severe consequences, sublethal responses (e.g. localised necrosis) have more commonly been reported following exposure to seagrasses and macroalgae in subtidal depths (Jackson et al. 1989, Taylor & Rasheed 2011). This suggests that environmental factors can act to reduce the potential for toxicity of oil in subtidal environments. For example, Taylor & Rasheed (2011) reported that following a spill of ~25 tonnes (~28,000 litres) of bunker oil in the subtropical Port of Gladstone, Queensland, in 2006, no significant differences were detected in seagrass shoot biomass or seagrass extent (i.e. area affected) between exposed and 'clean' seagrass meadows. This lack of impact was suggested to result from several extenuating factors: the oil spill occurred on a high neap tide; consequently, the intertidal meadows were probably not exposed until two to three days post spill, by which time the oil would have spread and thinned, with evaporation, photooxidation and dissolution processes likely reducing the quantity of remaining oil in the area (Taylor & Rasheed 2011) and reducing its acute toxicity.

Effects of direct contact of oil on seagrasses and macroalgae

Smothering, fouling and asphyxiation are some of the physical effects that have been documented from oil contamination in marine plants (Blumer 1971, Cintron et al. 1981). In macroalgae, oil can act as a physical barrier for the diffusion of CO_2 across cell walls (O'Brien & Dixon 1976), although the mucilaginous slime covering many large brown algae is considered a protective device against coating by oil. For example, following the 1969 PLATFORM A blowout off Santa Barbara, California, oil retained in the canopy of giant kelp *Macrocystis* beds adhered tightly to blades reaching the surface, but removal of the oil layer revealed healthy tissue beneath (Mitchell et al. 1970). Observable damage to the macroalgae was negligible, likely because of secretion of mucus (Mitchell et al. 1970). A lack of damage to *Macrocystis* was also reported following exposure to Arabian crude oil spilled from the tanker METULA, wrecked in the Magellan Straits (Wardley-Smith 1974). In addition to gas exchange problems, macroalgae, and likely some small annual seagrass species, can become seriously overweighted and subject to breakage by waves (Nelson-Smith 1973) or other hydrodynamic processes following exposure to oil. A strong oil adsorptive capacity has been noted for many macroalgal species (including *Ascophyllum nodosum*, *Fucus* spp., *Pelvetia canaliculata*, *Hesperophycus californicus* [=*Hesperophycus harveyanus*], *Mastocarpus stellatus* [=*Gigartina stellata*] and *Gelidium crinale*; O'Brien & Dixon 1976), and the seagrass *Phyllospadix torreyi* (Foster et al. 1971), and this is likely to be exacerbated where the oil strands on a rising tiding contacting partially desiccated plants.

Toxicity effects on seagrasses

When seagrass leaves are exposed to petroleum oil, sublethal quantities of the WAF can be incorporated into the tissue, causing a reduction in tolerance to other stress factors (Zieman et al. 1984). The toxic components of oil are thought to be the PAHs, which are lipophilic and tend to accumulate in the thylakoid membranes of chloroplasts (Ren et al. 1994). Consequently, thylakoid membrane oxidation impacting on photosynthesis is a symptom of oil toxicity (Ren et al. 1994, Marwood et al. 1999). The type of oil spilled has different effects on different aquatic plants, and the use of dispersants can also contribute to overall impact (Thorhaug 1988).

Direct contact with the above-ground biomass of seagrass following oil exposure can result in morphological changes (Jacobs 1988). Seagrass blades can become bleached, blackened, yellowed or detached from the plant following direct oil contamination (den Hartog & Jacobs 1980, Jackson et al. 1989, Dean et al. 1998), while other effects from direct contact include a decrease in the density of vegetative and flowering shoots (den Hartog & Jacobs 1980, Jackson et al. 1989, Dean et al. 1998). There are, however, many reported cases where hydrocarbon exposure has not led to any form of physical change in seagrasses (Kenworthy et al. 1993, Dean et al. 1998, Taylor et al. 2006, Taylor & Rasheed 2011). For example, approximately one year after exposure during the Gulf War spill, Kenworthy et al. (1993) found no measurable impact on species composition, distribution, abundance, net production or growth of the annual seagrass species *Halodule uninervis*, *Halophila ovalis* or *Halophila stipulacea*.

Studies of exposure of oil and dispersants to seagrasses have revealed a range of responses with some species more tolerant than others. *Thalassia testudinum* has been used in a number of studies from the Caribbean (Baca & Getter 1984, Thorhaug et al. 1986, Ballou et al. 1987, Thorhaug & Marcus 1987a,b) and is generally less sensitive to dispersed oil exposure (LD50—lethal dose causing 50% mortality of 1.25 mL L^{-1}) than other seagrasses such as *Halodule wrightii* and *Syringodium filiforme* (LD50 of 0.75 mL L^{-1}). In other studies, seagrasses have been shown to be less sensitive to unmixed oil exposure than to dispersed oil. In support of the field observations of Kenworthy et al.

(1993) discussed previously, Durako et al. (1993) found that a 1% weight/volume mix of crude oil in seawater had no detectable effect on *Halodule uninervis*, *Halophila stipulacea* and *Halophila ovalis*. In Australia, the same concentrations of oil and an oil/dispersant mix also had negligible impacts on *H. ovalis* (Ralph & Burchett 1998). From these studies, it appears that a WSF of between 0.25%–1% (w/v) solution of crude oil in seawater is sufficient to induce physiological responses; however, valid generalisations concerning the magnitude of phytotoxicity for crude and refined oils are difficult to make due to the uneven nature and wide range of reported toxic effect concentrations (Hatcher & Larkum 1982, Thorhaug et al. 1986, Thorhaug 1988, Ralph & Burchett 1998, Lewis & Pryor 2013). Further, some species are more sensitive to oil exposure than others (Thorhaug et al. 1986), although the reasons for this remain unclear. It is, therefore, difficult to compare results from studies testing different oils on different species; any one of these variables, or a complex interaction between them, could be responsible for the variable results (Runcie et al. 2005).

Phytotoxic effect of petroleum oil on seagrasses can apparently lead to a range of sublethal responses including reduced growth rates (Howard & Edgar 1994), bleaching, decrease in the density of shoots, reduced flowering success (den Hartog & Jacobs 1980, Dean et al. 1998) and blackened leaves that may detach from the plant (den Hartog & Jacobs 1980). Direct exposure, however, does not always induce toxic effects (Kenworthy et al. 1993, Dean et al. 1998), even under laboratory conditions (Wilson & Ralph 2012). The disparity among research findings may be due to different experimental methods, including the range of indicators used, varying exposure and temperature regimes, and the specific petrochemicals evaluated (Lewis & Pryor 2013). Further, morphological variation between seagrass species is considerable, and species resilience to petrochemical impacts is likely to reflect morphological traits.

Published studies of exposure to oils exist for nine of the world's 72 seagrass species (Lewis & Pryor 2013), with congeners of all nine species present in Australian waters. Thorhaug et al. (1986) showed clear differences in the response of different species of tropical seagrass to oil, but other research to date has largely been conducted on single species (Hatcher & Larkum 1982, Ralph & Burchett 1998, Macinnis-Ng & Ralph 2003). While it is difficult to generalise for these reasons, the seagrass *Thalassia testudinum* has been consistently more tolerant to crude oils and dispersed oils than other species that occur in tropical waters, such as *Halodule wrightii* and *Syringodium filiforme* (Thorhaug et al. 1986). *Halophila* spp. have also shown a degree of tolerance to petrochemical exposure (up to 1% (w/v) solution of Bass Strait crude oil) under laboratory conditions (Ralph & Burchett 1998) and during field testing (Kenworthy et al. 1993).

Stress conditions associated with oil exposure to seagrasses can reduce the rate of photosynthesis, block photosynthetic electron transport or disturb the pigment-protein apparatus (Maxwell & Johnson 2000). When light energy is absorbed by a plant, some is used for photochemical reactions, but a proportion is emitted as heat or fluorescence. If the functional state of the photosynthetic apparatus changes, the amount of fluorescence emitted also changes. This information can be used to quantify a stressor (Maxwell & Johnson 2000). In this way, chlorophyll-a fluorescence can provide important physiological data on the effect of hydrocarbon exposure on photosynthetic activity (Ralph & Burchett 1998, Macinnis-Ng & Ralph 2003, Wilson & Ralph 2012).

Photosynthetic stress of seagrasses is typically monitored using chlorophyll-a fluorescence (specifically the effective quantum yield of Photosystem II ($\Delta F/F_m'$) using pulse amplitude modulation techniques), but also via analyses of chlorophyll-a pigment concentrations. The value of $\Delta F/F_m'$ provides information regarding photosynthetic activity and thus physiological health of the seagrass. In laboratory experiments, Wilson & Ralph (2012) reported minimal, if any, long-term change to $\Delta F/F_m'$ in the seagrass *Zostera capricorni* when exposed to crude oil but found short-term increases similar to stimulatory effects observed in other oil exposure studies (Karydis & Fogg 1980, Chan & Chiu 1985). Oil taken up by the leaf blades was suggested to lead to a short-term stimulatory effect (hours) before returning to typical rates of photosynthesis (Wilson & Ralph 2012). In other studies, toxic impacts (declines in photosynthesis) of oils were detected using $\Delta F/F_m'$ in seagrass

(Ralph & Burchett 1998, Macinnis-Ng & Ralph 2003), freshwater macrophytes (Marwood et al. 2001), phytoplankton (Marwood et al. 1999) and corals (Jones & Heyward 2003). For seagrasses, Marwood et al. (2001) concluded that oil can cause the inactivation of Photosystem II reaction centres due to oxidation or degradation of D1 proteins.

Toxicity effects on macroalgae

It has been known for at least 60 years that crude and refined oils are phytotoxic to algae (Currier & Peoples 1954, Van Overbeek & Blondeau 1954; cited in Lewis & Pryor 2013), with water-soluble hydrocarbon molecules more toxic to macroalgae than larger molecules (Van Overbeek & Blondeau 1954, Kauss et al. 1973, cited in O'Brien & Dixon, 1976). There are several reported cases of toxicity associated with oil inducing morphological change in macroalgae (Lewis & Pryor 2013). The sequential effects of an oil spill on intertidal algae have been recorded in a heavily oiled cove near to the wreck of the AMOCO CADIZ in Brittany, France, in 1978 (Floc'h & Diouris 1980). This release involved ~1.6 million barrels (~250,000 tonnes) of light crude oil produced from Ras Tanura, Saudi Arabia. Fresh oil initially acted like a contact herbicide, causing intensive bleaching of the most sensitive thalli of green algae (Floc'h & Diouris 1980). This was followed by localised necrosis in red algae and widespread necrosis in brown algae (Floc'h & Diouris 1980). More recently, Stekoll & Deysher (2000) reported physical injury to the macroalga *Fucus distichus* (=*Fucus gardneri*) following the 1989 EXXON VALDEZ oil spill. This resulted in lower biomass, lower percent cover, impaired reproductive capability and alterations to population structure (as evidenced by differences in densities and proportions amongst size classes). However, in this case, injuries may also have been caused by the oil spill clean-up (Stekoll & Deysher 2000).

Oil concentrations with observed toxic effects on algae vary greatly among species and studies, ranging from 0.002 to 10,000 ppm (Lewis & Pryor 2013). Indeed, algae show a variable response to the effects of oil under the same conditions in comparative studies. For example, of 10 species of Norwegian algae (four Phaeophyta, three Rhodophyta and three Chlorophyta) monitored for two years under mesocosm conditions with continuous exposure to 0.129 ppm of diesel (129 μg L^{-1} hydrocarbon concentration or 30 μg L^{-1} WAF), only two species, *Fucus evanescens* (a kelp) and *Phymatolithon lenormandii* (encrusting coralline rhodophyte), declined in abundance (Bokn et al. 1993). On the other hand, in north-eastern India, Premila & Rao (1997) found that 0.02 ppm of crude oil was sufficient to inhibit growth of 10 species (one Phaeophyta, four Rhodophyta and five Chlorophyta) and that 72-hour LC50 values among these varied from 0.28 to 1.9 ppm, with green algae the most sensitive. As with seagrasses, the sensitivity of macroalgae to spills appears to decline from high- to low-intertidal levels on shores, with filamentous red algae the most susceptible (O'Brien & Dixon 1976). Subtidal macroalgae do not always die or exhibit reduced growth rates following oil spills (Peckol et al. 1990, Lewis & Pryor 2013), other than in the most severe situations (Jackson et al. 1989).

Reported toxic responses to oil have included a variety of physiological changes to enzyme systems, photosynthesis, respiration and nucleic acid synthesis (Lewis & Pryor 2013). Photosynthetic uptake has been the most commonly used physiological index in studies on the toxic effects of oil on algae (Lewis & Pryor 2013). Disruption of cellular membranes by certain types of hydrocarbons can have an adverse effect on photosynthesis (O'Brien & Dixon 1976). The double lipophilic layer is susceptible to swelling by hydrocarbons penetrating the outer protein layers. Darkening of cells and distortion of membranes has been reported by Kauss et al. (1973, cited in O'Brien & Dixon 1976), indicating possible interference with membrane structures (O'Brien & Dixon 1976). It has also been suggested that hydrocarbons dissolve in the lipid phase of the grana of chloroplasts (O'Brien & Dixon 1976). An increase in distance between individual chlorophyll molecules caused by membrane distortion and other disturbances to submicroscopic structures appears to impair photosynthetic ability (O'Brien & Dixon 1976).

Despite the well-established pool of literature on macroalgae exposure to petroleum oils, very few investigations have reported effects on species that are common in Australian waters (Lewis & Pryor 2013). Rather, most studies on macroalgae have occurred in response to northern hemisphere oil spills (Lewis & Pryor 2013), with a focus on the phaeophyte *Fucus*, a common temperate intertidal species. Noticeably absent from the literature on toxicity are the attached, meadow forming species within the *Sargassum* genus, which have widespread distribution across Australia's tropical north (especially in the Kimberley; Huisman et al. 2009) and also temperate south, and is thought to have in excess of 200 species (Edgar 2012). While no evidence indicates that *Sargassum* spp. and *Fucus* spp. respond similarly to hydrocarbon exposure, they do occur in the same taxonomic order (Fucales) and share similar morphologies, including flat, branched fronds that are either straplike or tapering, as well as characteristic vesicles. *Ulva* spp. and *Padina* spp., which belong to different taxonomic orders (Ulvales and Dictyotales, respectively) but also commonly occur in Australian waters, have both been reported with inhibited growth following oil exposure (Lewis & Pryor 2013).

Chronic and indirect effects of oil exposure

Impacts on reproduction and growth

Invertebrates   Reduction in the success or extent of reproductive activity following exposure to oil has been shown for several different invertebrate species, indicating that sublethal effects of oil can threaten reproductive success of a population impacted by oil.

Berdugo et al. (1977) showed a reduction in fecundity, brood size and rate of egg production in planktonic copepods following exposure to oil. Similarly, sublethal concentrations of crude oil were found to decrease brood numbers in amphipods when females were exposed during the incubation period (Linden 1976). Elmgren et al. (1983) also found that several months following the Tsesis oil spill in Sweden, female amphipods (*Monoporeia affinis* [=*Pontoporeia affinis*]) showed a significant increase in proportion of abnormal eggs.

Blumer et al. (1970) found that while an oil spill did not cause mortality, mussels failed to reproduce following the spill (with mussels from an area not affected by the oil spill reproducing normally). Other studies (Renzoni 1973, 1975; Nicol et al. 1977) showed that exposure to No. 2 fuel oil WAF affected sperm motility and reduced fertilisation in sand dollars and bivalves. Vashchenko (1980) reared sea urchin larvae from gametes obtained from adult *Mesocentrotus nudus* (=*Strongylocentrotus nudus*) maintained for 45 days in seawater containing 30 mg L^{-1} of diesel. While larvae from control urchins developed normally, those produced from gametes of urchins maintained in the diesel-contaminated water had a higher proportion of abnormal and non-viable larvae after three days, including those larvae reared from control eggs or sperm and treatment eggs or sperm (i.e. if one set of gametes came from a control urchin, the larvae still did not develop normally) (Vashchenko 1980).

Karinen et al. (1985) exposed Dungeness crabs (*Metacarcinus magister* [=*Cancer magister*]) to various concentrations of crude-oil-contaminated sediment and found that moulting was affected, mating was often unsuccessful, and that egg-carrying females produced significantly lower numbers of larvae than control crabs. These larvae also had shorter survival times than larvae from control crabs (Karinen et al. 1985).

Decreased reproductive success of both brooding and broadcast spawning corals has been shown following exposure to oil. The hydrocarbon effects on adult colonies range from the premature expulsion of larvae (Cohen et al. 1977, Rinkevich & Loya 1977, Villanueva et al. 2011), significant reduction in the number of colonies with gonads in their polyps (Rinkevich & Loya 1977), and significantly smaller gonads (Guzmán & Holst 1993). Despite these effects, Negri & Heyward (2000) suggested that the early life stages of corals may in fact be more sensitive to hydrocarbons and showed that larval metamorphosis was inhibited by exposure to crude oil and dispersed oil. Epstein et al. (2000) exposed *Stylophora pistillata* larvae to Egyptian crude oil and reported no mortality from the WAF of the oil, no significant mortality from the dispersants (Emulgal C-100,

Biosolve, Inipol IP-90, Petrotech PTI-25, Bioreico R-93), but the dispersed oil fraction resulted in significant mortality. Furthermore, the authors reported a significant reduction in settlement rates from the WAF and dispersants, and no successful settlement was evident in the dispersed oil treatment. Anomalies in planula morphology and behaviour were also recorded in the dispersant and dispersed oil treatments (Epstein et al. 2000). Larval settlement and survival in *Porites astreoides* and *Orbicella faveolata* (=*Montastraea faveolata*) decreased with increasing concentrations of the WAF of DEEPWATER HORIZON crude oil, the dispersant Corexit 9500 and the dispersed oil (Goodbody-Gringley et al. 2013). Malampaya natural gas condensate has also been shown to result in larval mortality in *Seriatopora hystrix* and *S. guttata* (=*S. guttatus*) but not in *Stylophora pistillata*, *Pocillopora damicornis* and *P. verrucosa* (Villanueva et al. 2008).

There is also evidence that exposure to oil affects coral growth. Birkeland et al. (1976) exposed colonies of *Porites furcata* to Bunker C oil and reported a significant difference in growth between the control and treatment colonies. Guzmán et al. (1991) found a reduction in growth rates in three species (*Porites astreoides*, *Pseudodiploria strigosa* [=*Diploria strigosa*], *Orbicella annularis* [=*Montastraea annularis*]) following the major 1986 oil spill near Galeta in Panama, with growth rates of *Porites astreoides* being lower during the three years after the spill than before the spill (Guzmán et al. 1994). Following the Panama oil spill, *Siderastrea siderea* showed no reduction in growth rate (Guzmán et al. 1991), but the growth rate of *S. siderea* differed significantly before and after the spill (Guzmán et al. 1994). Decreased juvenile growth rates have also been reported in *Seriatopora hystrix*, *S. guttata* (=*S. guttatus*) and *Stylophora pistillata* following exposure to the WAF of Malampaya natural gas condensate (Villanueva et al. 2008).

Seagrasses   While considerable variation in reproductive strategy occurs at the species level, perennial seagrasses typically rely on lateral shoot growth for local persistence and expansion, with use of seeds a secondary strategy (Larkum & den Hartog 1989). However, production of new plants from dispersed seeds can provide an important auxiliary mechanism for re-establishing meadows (Orth et al. 1994). While it could be presumed that responses to oil exposure among seagrasses with different life cycles and reproductive strategies will vary, there have been no systematic studies to specifically test if such differences occur. From the few reported studies available, species that rely heavily on recruitment from seeds appear to suffer greater impacts from residual oil in sediments than species with vegetative reproduction (Dean et al. 1998). Dean et al. (1998) reported that seeds collected from an oiled site had a higher germination rate than seeds from its paired reference site, but the seedlings produced from the oiled site also had higher rates of mitotic abnormalities.

Oiling also appears to affect flowering success (Houghton et al. 1993, Dean et al. 1998). In an assessment of seagrass health following the EXXON VALDEZ crude oil spill in Alaska, Dean et al. (1998) reported that the density of flowering shoots was more than twice as high at reference sites compared to oiled sites. The paucity of flowering shoots at oiled sites translated directly to a lack of inflorescences produced. These results were comparable to those of Houghton et al. (1993), who independently conducted surveys of eelgrass following the EXXON VALDEZ oil spill. It is difficult to predict the implications of a reduction in flowering success associated in oil spills for long-term viability of seagrass meadows, as in both instances, declines in flowering success were short-term (natural flowering success had recovered within two years post spill) and did not lead to reductions in shoot density in subsequent years (Houghton et al. 1993, Dean et al. 1998).

There has been no examination of the effect of oils on the various life stages of seagrasses, and tolerances of young versus mature plants are unknown.

Macroalgae   Macroalgal reproduction varies considerably among phyla and species but, in general, involves mass production of gametes, zygotes and spores forming propagules that can be readily dispersed to nearby habitats. Gametes can be either male or female and are released by the development of reproductive bodies on the plant (Holmquist 1997, Peterson et al. 2002).

Once released, spores or eggs attach to a surface or substrate and then are fertilised and mature into zygotes, juvenile and eventually adult plants. Sensitivities to toxicants at different macroalgal life stages is common, and gametes, larva and zygote stages have all proven more responsive to oil exposure than adult growth stages (Thursby et al. 1985, Lewis & Pryor 2013). Of the response parameters reported by Lewis & Pryor (2013), germination, reproduction and growth rate have been more sensitive than other indicators of macroalgal health, with inhibition the dominant effect.

Impacts on movement, attachment and feeding of invertebrates

Animals not exposed to oil at concentrations high enough to kill them may still suffer mortality not directly related to the oil impact. For example, a reduced ability to move away from oiled areas or escape predators may reduce survival. Percy & Mullin (1977) found impaired movement in amphipods *Onisimus affinis* and hydromedusa *Halitholus cirratus* when exposed to low concentrations of crude oil. Johnson (1977) provides numerous examples of the narcotising effect of hydrocarbons in oil causing reduced mobility and respiration in decapod crustaceans, leading to reduced survival and increased vulnerability to predation (including during the process of moulting). O'Sullivan (1978) found that after the 1978 Amoco Cadiz oil spill in France, limpets remained attached to the reef, but their grip was weakened. This may have affected their ability to survive strong breaking waves or resist predators. Similarly, in an experiment in the Arctic, Mageau et al. (1987) demonstrated impairment of movement and attachment in the urchin *Strongylocentrotus droebachiensis* from loss of movement of tube feet and spines following exposure to dispersed crude oil.

Numerous studies have shown that sublethal exposures of marine invertebrates to oil result in reduced feeding rates or ability to feed effectively. Feeding rates of the predatory seastar *Asterias rubens* on mussels were depressed when exposed to crude oil at 200 ppm (Crapp 1971). A similar reduction in feeding rate and a reduced growth rate was found in *Evasterias troschelii* feeding on mussels in Alaska after the seastars had been exposed to a very dilute (0.12 ppm) crude oil seawater mixture (O'Clair & Rice 1985). Corals have also been shown to respond to oil pollution by abnormal feeding reactions such as mouth opening or reduced pulsation rates (Reimer 1975, Cohen et al. 1977, Rinkevich & Loya 1979).

Diseases and physiological responses

A range of tumour and blood-type diseases of crustaceans and molluscs have been reported in animals exposed to oil spills (see Hodgins et al. 1977 for review of oil-induced disease responses in fish and invertebrates). Deoxyribonucleic acid (DNA) damage and potential mutations from oil exposure were found in mussels following the Prestige oil spill in Spain (Perez-Cadahia et al. 2004, Laffon et al. 2006) and the oil spill off Spain from the ship Aegean Sea (Sole et al. 1996). However, after the 1996 Sea Empress spill in Wales, DNA damage and potential mutations were found to be more prevalent in fishes than invertebrates (mussels and sponges) (Harvey et al. 1999).

Changes in the cellular physiological condition of coral have been shown following exposure to oil. Downs et al. (2006) measured the effects of a spill of intermediate grade oil on *Porites lobata*. Samples were collected three months after initial exposure. Changes were recorded at the cellular level in terms of lesions found adducted to coral macromolecules. The cellular physiology was consistent with the pathological profile that results from the interaction of corals with PAHs (e.g. an injury resulting from exposure to a xenobiotic). Similar results were recorded by Rougee et al. (2006) when exposing *Pocillopora damicornis* in the laboratory to fuel oil. Moreover, following a major oil spill in Panama, Burns & Knap (1989) found that corals stressed by hydrocarbons had altered protein-to-lipid ratios, suggesting a disruption in the lipid synthesis system. Coral mucus is rich in wax esters, triglycerides and other lipids and is an energy-rich link in the coral reef food chain (Benson & Muscatine 1974). As such, in addition to the adverse change in coral physiology, the disruption in the lipid synthesis system has the potential to cascade to other components in the ecosystem that are dependent on the corals for food and substrate (Burns & Knap 1989).

A concentration dependent change in the photosynthetic rate of the zooxanthellae was reported in *Stylophora pistillata* following exposure to the water-soluble fraction (WSF) of Iranian crude oil (Rinkevich & Loya 1983). Cook & Knap (1983) exposed colonies of *Pseudodiploria strigosa* (=*Diploria strigosa*) to varying concentrations (18–20 ppm) of Arabian light crude oil and the dispersant Corexit 9527, and while the hydrocarbon exposure alone did not affect carbon fixation, the dispersed oil resulted in significant impacts to photosynthesis. Knap et al. (1985) noted these effects were temporary however, and there was no evidence of a long-term effect on the coral skeleton. Similarly, Dodge et al. (1985) found no detrimental effects on calcification rates of *P. strigosa* when exposed to the same experimental conditions as that applied by Cook & Knap (1983) (i.e. Arabian light crude/Corexit 9527). There is also some evidence that at low levels of exposure to oil a slight enhancement of photosynthesis and calcification rates can occur (Rinkevich & Loya 1983, Dodge et al. 1985) even where it is supressed at higher hydrocarbon concentrations (Rinkevich & Loya 1983). This stimulatory effect in response to low levels of pollutants is known as hormesis (Dodge et al. 1985). Jones & Heyward (2003) reported a reduction in the photochemical efficiency of zooxanthellae in *Plesiastrea versipora* exposed to produced formation water (3%–50% w/v). Consequently, it appears that sublethal exposure to hydrocarbons can have an adverse effect on the photosynthetic efficiency of zooxanthellae, but that the effect can be inconsistent and reversible.

Corals secrete mucus as a protective mechanism, and research has shown that its production varies with exposure to oil. Peters et al. (1981) exposed *Manicina areolata* colonies to No. 2 fuel oil for three months and noted an increase in mucous sensory cell activity after two, four and six weeks. Coral mucus, therefore, not only serves to protect coral tissue from oil but also as a mechanism of clearing oil from contaminated corals. Production of mucus may also serve as a major pathway of energy loss, given reports that 40% of the primary production in *Acropora* is lost as mucus (Loya & Rinkevich 1980). In stressed corals, this energy loss may constitute a large energy drain, which could further compromise the health of the colonies. Moreover, coral mucus serves as a food source to other reef organisms (Cole et al. 2008, Rotjan & Lewis 2008), and in the event of oil exposure, the mucus may therefore serve as a pathway for contamination (Shigenaka 2001).

Stimulation of epiphytic algal growth

The effects of oil on epiphytic micro- and macroalgal communities are complex. Petroleum oils are both toxic and sources of labile carbon for energy and growth (Altenburger et al. 2004). In theory, any stimulatory effects on growth of epiphytic algae on seagrasses can lead to indirect seagrass loss (Cambridge et al. 2007). If enhanced epiphytic growth covers the greater portion of a seagrass leaf surface, insufficient light becomes available to enable photosynthesis (Masini & Manning 1997) and, if persistent, can lead to declines in seagrass shoot density and ultimately seagrass loss (Cambridge et al. 2007). den Hartog (1986) suggested that the impact of reduced light as a result of enhanced epiphytic or phytoplankton loading could potentially be greater than the impact of the petrochemical toxicity on seagrasses, while Jackson et al. (1989) reported high levels of epiphytic algal growth on subtidal seagrass (*Thalassia testudinum*) for several months following the 1986 oil spill at Galeta in Panama. The level of epiphytic growth in the Galeta example, however, was insufficient to induce changes in seagrass health (Jackson et al. 1989). Despite the potential for this indirect cause–effect pathway, there is limited evidence to confirm the seriousness of this effect. Further, assuming there is a legitimate link, there is no evidence to confirm that the level of stimulation would result in the same magnitude of epiphytic growth (and loss of seagrass) induced by other nutrient sources (Cambridge et al. 2007).

Changes to community assemblages and trophic structure

The sudden loss of a particular group of animals in any disturbed habitat often leads to a change in community structure caused by founder effects, and this has certainly occurred after some oil spills. Numerous studies following oil spills report a positive response by, and occasionally a proliferation of, marine algae (Mitchell et al. 1970, Chan 1977, Chassé 1978, Southward & Southward 1978, Newey & Seed 1995, Le Hir & Hily 2002, Marshall & Edgar 2003, Barillé-Boyer et al. 2004), an outcome attributed either to a loss of herbivores or to an increase in nutrients (or both). The 1967 TORREY CANYON spill resulted in extensive mortality to herbivorous limpets and other invertebrates on the Cornwall coast in the United Kingdom when 40,000 tonnes of Kuwait crude oil washed ashore. The bare space was rapidly colonised by ephemeral chlorophytes (*Blidingia* and *Ulva* [=*Enteromorpha*]) and then dense stands of the phaeophyte *Fucus,* which resulted in the death of most remaining barnacles. Limpet recruitment then began to occur beneath the plant canopy and displaced the *Fucus* after about five years, but the species composition, biomass and abundance was greatly different and fluctuated dramatically for several years, and the shoreline did not begin to resemble prespill characteristics until the 1980s (Hawkins & Southward 1992, Hawkins et al. 2017a). Following the 1993 BRAER oil spill in Scotland, Newey & Seed (1995) found a proliferation of *Ulva* and *Porphyra* in the impacted area where grazers (such as limpets) were greatly reduced. Castège et al. (2014) found an overshoot recovery of grazers ~18 months after the 2002 PRESTIGE oil spill in France, and this was attributed to rapid algal growth in the oil-impacted area. Also within France, a dramatic increase in algae (*Ulva* and *Grateloupia*) in intertidal rock pools after the 1999 ERIKA oil spill was recorded, which resulted in 100% mortality of the grazing sea urchins *Paracentrotus lividus* and *Psammechinus miliaris* (Barillé-Boyer et al. 2004). In some of these cited cases, the algae were able to respond not just to a reduction in grazers, but also to an abundance of bare spaces caused by the death of encrusting cover invertebrate species such as barnacles (Newey & Seed 1995) and sponges (Castège et al. 2014).

Oil spills have the potential to have community-scale effects. Despite this, most studies on the topic tend to track impacts on singular species over relatively short temporal scales. Peterson et al. (2003), however, suggest that expectations of short-term singular species effects from oiling ignore the complexity of ecological interactions and should be replaced by a better understanding of the sequence of delayed indirect effects over much longer timescales (Peterson 2001). Changes in community composition in macroalgal communities following the 1989 EXXON VALDEZ incident provides an interesting case in point. Following this oil spill event, there was considerable loss of macroalgal cover (*Fucus distichus* [=*Fucus gardneri*]) within the intertidal zone, which triggered a cascade of indirect effects (Peterson et al. 2003). Freeing of space on the rocks and the losses of important grazing (limpets and periwinkles) and predatory (whelks) gastropods combined to promote initial blooms of ephemeral green algae in 1989 and 1990 and an opportunistic barnacle (*Chthamalus dalli*) in 1991 (Peterson et al. 2003). Absence of structural algal canopy also led to declines in associated invertebrates and inhibited recovery of *Fucus distichus* itself, whose recruits survive better under the protective cover of the adult plants. Those *F. distichus* plants that subsequently settled on *Chthamalus dalli* became dislodged during storms because of the structural instability of the attachment of this opportunistic barnacle (Peterson et al. 2003). After apparent recovery of *Fucus distichus*, previously oiled shores exhibited another mortality event in 1994, which was suggested to be a cyclic instability associated with simultaneous senility of a single-aged stand (Driskell et al. 2001).

The 1986 Galeta oil spill in Panama provides another interesting case study of selective effects at the community scale. Seagrasses and associated algal epiphytes provide important feeding, nursery and refuge habitats for marine fauna (Bostrom & Bonsdorff 2000, Smit et al. 2005). The Galeta spill resulted in large declines in seagrass (*Thalassia testudinum*) in intertidal areas, although subtidal seagrass meadows remained, albeit largely covered in epiphytic algae (Jackson et al. 1989). Within the dead but intact intertidal seagrass root-rhizome mats, Jackson et al. (1989) reported losses in some macroinvertebrate fauna (including amphipods, tanaidaceans and ophiuroids), while relative abundances of other macroinvertebrate taxa (bivalves, gastropods and polychaetes) were consistent with surrounding unoiled seagrasses.

The selective nature of oil toxicity caused by the resilience of some groups can, in turn, also alter community structure. Following a large 2006 oil spill off Estonia, Kotta et al. (2008) compared the abundance of guilds of herbivores, suspension feeders and deposit-feeding invertebrates immediately after the spill and 18 months later and found that herbivores (especially amphipods and isopods) were decimated, while deposit feeders and suspension feeders were not impacted. Three years after the 1978 AMOCO CADIZ spill in France, Conan (1982) found that opportunistic polychaetes had come to dominate sand/mud habitats, but clam populations had not recovered and had unstable recruitment. As well as selective mortality affecting communities after an oil spill, opportunistic recruitment by less sensitive species can affect community dynamics. For example, after the 1989 EXXON VALDEZ spill in Alaska, Jewett et al. (1999) found the abundance and biomass of subtidal epifauna and infauna at oiled sites among seagrass beds was higher than at control sites, partly due to the response of mussels and polychaetes to organic enrichment at the oiled sites. These effects were found to persist for at least six years.

Peterson (2001) and Peterson et al. (2003) documented the long-term effects of the 1989 EXXON VALDEZ spill and concluded that profound chronic effects remained more than 10 years later, particularly in trophic interactions and effects on populations of birds and sea otters, some of which related back to their invertebrate prey or foraging areas. Suspension-feeding clams and mussels can only slowly metabolise hydrocarbons, and when continuously exposed to sedimented oil, they concentrate the hydrocarbons, leading to chronically elevated tissue contamination (Peterson et al. 2003). In the case of the 1989 EXXON VALDEZ spill, persistent bioaccumulation of hydrocarbons within clams (*Leukoma staminea* [=*Protothaca staminea*]) and mussels (*Mytilus trossulus*) meant that foraging sea otters that consumed the bivalves suffered chronic exposure to hydrocarbons for many years (Peterson et al. 2003). Carls et al. (2001) estimated it could take 30 years for mussel beds to be free of hydrocarbon contamination because of oil trapped in the sediments beneath the beds. However, Payne et al. (2008) argued (on the basis of sampling for 11 years longer than Carls et al. 2001) that levels of contamination in mussels had reduced to very low levels by 2006, and that bioaccumulation by mussels only presented a problem when substrate with adsorbed oil was disturbed. A continuing problem is that foraging birds and sea otters often disturb sediment in the EXXON VALDEZ spill area, exposing the sedimented oil (Peterson et al. 2003, Payne et al. 2008).

While petroleum hydrocarbon uptake and accumulation can occur in seagrasses and macroalgae, uptake does not appear to be a major entry point into food chains (Lewis & Pryor 2013), although there remains some research interest in the topic. For example, after the NASSIA tanker accident (14 March 1994), there were elevated concentrations of oil (total petroleum hydrocarbons [TPHs] up to 196.76 μg g^{-1}) reported in five different species of marine macroalgae occurring in the Bosporus Straits, Turkey (Binark et al. 2000). There is also evidence that the bioaccumulation rate is likely to be species specific (Getter et al. 1985, Binark et al. 2000), with uptake also shown to be affected by salinity and temperature (Wolfe et al. 1998b). The influence of dispersants on uptake of some fractions of petroleum hydrocarbons (e.g. [^{14}C]naphthalene) has ranged from a limited effect to as much as a 10-fold increase for some marine algae (Wolfe et al. 1998a) and seagrasses (Lewis & Pryor 2013).

The toxic effects of tissue-accumulated hydrocarbon concentrations, such as those referred to previously, are almost unknown (Lewis & Pryor 2013). In laboratory experiments undertaken by Navas et al. (2006), *Daphnia magna* were fed with algae previously exposed to fuel, but no toxic effects were detected despite the presence of chemicals (elevated ethoxyresorufin-O-deethylase (EROD) activities of up to 16 pMol mg^{-1} min^{-1} in RTG-2 cells) that could cause sublethal effects to organisms (Navas et al. 2006). The scarcity of threshold effect data, in general, reduces the value of most oil-related bioaccumulation studies to the risk assessment process and prevents the calculation of tissue screening concentrations and toxic units that have proven useful for non-oil contaminants and aquatic biota (Shephard 1998).

In addition to bioaccumulation, novel pathways of injury to marine fauna from offshore oil spills have also recently been reported following the DEEPWATER HORIZON spill (Powers et al. 2013). The pelagic brown alga *Sargassum* forms a floating platform, or mat, which supports biodiversity and secondary production. According to Powers et al. (2013), the vast pool of oil resulting from the DEEPWATER HORIZON oil spill came into contact with a large portion of the Gulf of Mexico's floating *Sargassum* mats, leading to three pathways for oil spill-related injury: (1) *Sargassum* accumulated oil on the surface, exposing animals to high concentrations of contaminants; (2) application of dispersant sank *Sargassum*, thus removing the floating mat habitat and potentially transporting oil and dispersant vertically; and (3) low oxygen surrounded the floating mat habitat, potentially stressing animals that reside in the alga. Such pathways represent direct, sublethal and indirect effects of oil and dispersant release that minimise the ecosystem services provided by floating *Sargassum* (Powers et al. 2013).

Loss of genetic diversity

Loss of genetic diversity in the razor clam *Ensis siliqua* four years after the 2002 PRESTIGE oil spill in Spain was attributed to this event (Fernandez-Tajes et al. 2012). These authors compared prespill data of genetic diversity with that from the population that recovered following the oil spill. The latter had been produced from a very much smaller population than existed before the oil spill. This result contrasted that of Pineira et al. (2008), who, after the same spill, did not find any evidence for reduced genetic diversity in the littorinid snail *Littorina saxatilis*. The difference in results is probably explained by the short time (only 18 months) after the spill that Pineira et al. (2008) made their study and/or the widely differing life histories of the two molluscs. *Littorina saxatilis* is ovoviviparous, brooding its young internally before hatching, and the species has very limited dispersive capacity, while *Ensis siliqua* is a free-spawning species with external fertilisation. The latter life-history strategy should permit greater gene flow within and between populations, although this did not occur in this case, at least in the initial years following the spill.

Factors influencing impacts

Impacts of residual oil

One important mechanism that prolongs the effects of an oil spill occurs when oil is sedimented into the substrate. Oil is readily incorporated into muddy or sandy substrates where it can both restrict the recovery of infauna and burrowing fauna (such as crabs) and cause secondary continuous exposure via erosion of the sedimented layers (e.g. Hayes et al. 1993). Burns et al. (1993) found that five years after the 1986 refinery oil spill in Galeta, Panama, sedimented oil continued to leach from mangrove sediments and to bioaccumulate in bivalve molluscs. Similarly, more than seven years after a long history of acute and chronic oil spill pollution in Curacao, coarse rubble habitats cemented together with tar had a lower abundance and diversity of gastropod molluscs than unpolluted areas of similar habitat (Nagelkerken & Debrot 1995).

Sedimented oil also has an effect on re-colonisation and bioturbation. Wells & Sprague (1976) found that sedimented oil disrupted the burrowing behaviour of post-larval American lobster *Homarus americanus*. Dow (1978) showed how five successive-year classes of the burrowing bivalve *Mya arenaria* were killed following the 1971 spill from a storage facility in Long Cove, Maine, USA, as new recruits burrowed into sediment contaminated to 250 ppm oil at 15–25 cm below the surface. Gilfillan & Vandermeulen (1978) found the same species was still subject to significant lethal and sublethal effects six years after the 1970 ARROW oil spill contaminated lagoon sediments (87–3800 ppm) in Nova Scotia, Canada, with reduced growth and metabolic rates and fewer mature adults than at an unimpacted site. In Argentina, Ferrando et al. (2015) used cores extracted from muddy Argentinean sediments to show how oil contamination results in reduced bioturbation following the mortality of infaunal species. This effect will exacerbate oil spill impacts by reducing irrigation and oxygenation of subsurface layers, resulting in anoxic effects as an indirect effect of oiling.

Hydrocarbons can adsorb on aquatic plant-rooted soils and sediments where they can persist for up to 30 years (Lewis & Pryor 2013). Research has demonstrated that substrate oiling has adversely impacted other aquatic primary producers such as freshwater phytoplankton, wetland plants (Gilfillan et al. 1989, Naidoo et al. 2010) and mangroves (Klekowski et al. 1994). Phytotoxicities of substrate-bound hydrocarbons to seagrasses and macroalgae are important to understand, since the protection of below-ground root-rhizome systems are likely to be important for plant regeneration after oil spills. It is also possible that smaller ephemeral seagrasses, which tend to have shorter rhizomes that persist for weeks to months, may be differentially affected compared to larger seagrasses, which tend to have longer and more persistent rhizomes that exist for months to years. Furthermore, evaluation of above-substrate toxic effects may underestimate below-substrate toxic effects (Lin et al. 2002).

Composition of oil and weathering

Refined oils, diesel and heavy bunker fuel oils are apparently more toxic than crude oil. Anderson et al. (1974) compared the toxicity of a heavy fuel oil (Bunker C), a light fuel oil (No. 2, similar to diesel) and two crude oils to three species of shrimps and mysids—the two fuel oils were more toxic than the crude oils, and the heavy fuel oil was more toxic than the lighter distillate (Anderson et al. 1974). Within Australia, crude oils from different oil fields show a range in density (Neff et al. 2000), so the type and source of oil in an unplanned spill is a very important factor in determining the extent of impact and level of exposure to hydrocarbons. There has been very little work done specifically on the toxicity of natural gas condensates (except see Negri et al. 2015) that are particularly relevant to Australia's North West Shelf, but these are known to show toxicity to coral larvae and to affect coral reproduction.

Apart from their differing toxicities, different oil types behave differently in a spill according to their density and other properties. Edgar et al. (2003) attributed the minimal impacts of the 2000 JESSICA spill in the Galapagos, in part to the thinning effect that the diesel fuel had on the heavy bunker fuel when the two mixed following the spill. In that spill, other circumstances also mitigated the impacts of the spill (e.g. waves, evaporation and currents moving the oil offshore).

Weathering of oil is the process of evaporation of some of the volatile fractions from floating spilled oil and its dilution, modification and breakup by wave mixing, UV radiation, chemical reactions and biological degradation. Oil that has time to weather to a significant degree before it reaches and influences intertidal or benthic habitat will have less acute toxic impacts than freshly spilled oil. Apart from the oil's composition (or type), the main factors that will influence weathering are wave mixing, which will be affected by wind speed and temperature which effects of rates of evaporation and biodegradation (Neff et al. 2000, Venosa & Zhu 2003, Lee et al. 2013).

Although not conducted on Australian invertebrate species, Neff et al. (2000) compared the toxicities of light (Wonnich, Campbell) and medium density (Agincourt) north-west Australian crude oils, and Australian diesel oil, on penaeid prawns, mysid shrimps and sea urchin larvae. They used the oils and several dilutions of predistilled WAFs to simulate weathering and assess and compare toxicity. In general, weathering significantly reduced the toxicity of all oils, with variable toxicity between test animals and when different oils were applied (although weathering had a minimal effect on changing the toxicity of the oil to the crustaceans). The heavier Agincourt crude had minimal toxicity on all but the prawns, while for Wonnich, Campbell and diesel oils the percentage of the WAF that resulted in LC50 after 96 hours (determined by regression) ranged from 30%–48% WAF for prawns and mysids. For sea urchin larvae, results were expressed as percentage of the WAF that resulted in abnormal development on 50% of the larvae after 60 hours (60-hour EC50). These varied between 11%–68% of the WAF for Wonnich and Campbell oils, depending on the type of sea urchin larvae. For diesel, the 60-hour EC50 varied from 27% to 100% (non-toxic) depending on the type of sea urchin larvae used.

Use of dispersants and shoreline clean-up

The use of chemical dispersants in oil spill clean-up offers the benefit of reducing the threat of oil reaching shorelines and intertidal communities. However, debate exists over the merit of using dispersants to help break up oil spills and mitigate the impacts of oil toxicity, as opposed to allowing weathering and natural break-up of the oil slicks or other methods such as burning. Although this is not the only consideration in the use of dispersants, in general, most studies have concluded that dispersed oils are more toxic to marine invertebrates than the oil on its own. Most of the ecological damage caused by the 1967 TORREY CANYON spill in England was due to the use of dispersants and other cleaning measures (Smith 1968, Southward & Southward 1978, Hawkins & Southward 1992), with sites where dispersant was not used recovering after three years compared to as much as 15 years at sites where dispersant was used (Hawkins et al. 2017a,b). It should be stated, however, that modern dispersants are less toxic (Hawkins et al. 2017a), undergo more rigorous toxicity testing and are subject to an approvals process not in place in 1967 when first-generation dispersants were used. Almost all studies we examined found that in a WAF, dispersed oil is more toxic than the oil alone. For example, Fisher & Foss (1993) determined that a dispersed oil–water fraction using two commercial oil dispersants (Corexit 7664 and Corexit 9527) was 10 times more toxic to embryos of grass shrimp (*Palaemon pugio* [=*Palaemonetes pugio*]) than the oil–water fraction on its own (Fisher & Foss 1993). More recently, Almeda et al. (2013) found that dispersed oil was more than three times more toxic to mesozooplankton than crude oil alone. In further experiments, Almeda et al. (2014b) compared growth and survival of nauplii larvae of the barnacle *Amphibalanus improvisus* and tornaria larvae of the enteropneust (acorn worm) *Schizocardium* sp. when exposed to crude oil and dispersed crude oil (using Corexit 9500A). They found that the dispersed oil had a greater toxicity (Almeda et al. 2014b), and they reached the same conclusion when studying the same impacts on the copepods *Acartia tonsa*, *Temora turbinata* and *Parvocalanus crassirostris* (Almeda et al. 2014a). They concluded that the application of dispersants was likely to have a greater effect on post-spill recruitment of marine invertebrates than crude oil alone. The widely consistent demonstration that dispersed oils are more toxic to marine invertebrates than oil–water mixtures alone has led to calls for oil spill clean-ups to employ burning off the volatile fraction instead of using dispersants (Georgiades et al. 2003).

The situation with seagrass and macroalgae and dispersants is far less clear, and conflicting findings have complicated the decision-making process of whether or not to disperse oil (Wilson & Ralph 2012, Lewis & Pryor 2013). Research outcomes include dispersed oil posing a greater threat than non-dispersed oil, dispersed oil posing less of a threat than non-dispersed oil, and that neither oil nor dispersed oil impact seagrass or macroalgae (Wilson & Ralph 2012, Lewis & Pryor 2013). Further complicating the matter, responses to dispersed oils can be highly variable between species (Thorhaug et al. 1986).

Seagrasses have been shown to absorb more aliphatic and aromatic oil fractions when the oil is dispersed, therefore increasing its toxicity (den Hartog 1986). Dispersants are thought to affect the waxy cuticle of the seagrass blade and, in doing so, increase the penetrability of the dispersed oil to the photosynthetic organs, particularly the thylakoid membrane (Wolfe et al. 1998a). Most commonly, it appears that non-dispersed oil leads to less photosynthetic stress compared with the addition of a chemical dispersant. Wilson & Ralph (2012) and Hatcher & Larkum (1982) both found that dispersed oil had a greater impact on seagrasses than untreated oil. However, Ralph & Burchett (1998) found no comparative difference in photosynthetic activity of *Halophila ovalis* when exposed to crude oil and dispersed crude oil (both treatments induced similar declines). Macinnis-Ng & Ralph (2003) also found little difference between the impacts of oil and dispersed oil on *Zostera capricorni* in field experiments (but showed a greater impact from oil compared to dispersed oil in laboratory experiments).

It is also important to consider the type of dispersant used. Wilson & Ralph (2012) found that the toxicity of dispersed crude oil to *Z. capricorni* was specifically related to the brand of dispersant and the petrochemical loading in the water column. While both Corexit 9527 and Ardrox 6120 dispersed oil treatments produced negative impacts to *Z. capricorni,* the Ardrox 6120 dispersed treatment had a more sustained negative impact on the seagrass. For *Halophila ovalis*, $\Delta F/F_m'$ impacts were detected within the first four hours but were followed by full recovery. These laboratory experiments suggest that dispersed crude oil is more toxic to *Zostera capricorni* and *Halophila ovalis* than non-dispersed crude oil, and that Ardrox 6120 dispersed crude oil is slightly more toxic than Corexit 9527 dispersed crude oil.

Although the increased toxicity of dispersed oils has been demonstrated in numerous studies, the practise may be warranted in situations offshore, where reduced shoreline oiling and reduced risk of sedimented oil would result from using dispersants despite the higher toxicity to marine life in deeper offshore waters. This consideration is particularly relevant in the Australian context, for oil spills that occur far offshore, such as the MONTARA wellhead blowout.

Mechanical cleaning of oil-impacted shorelines has also been shown to be potentially more destructive than the effects of the oil spill on its own (Broman et al. 1983). Rolan & Gallagher (1991) reported that even more than eight years after the 1978 Esso BERNICIA oil spill in the Shetland Islands, Scotland, mechanically cleaned shorelines (i.e. by removal of oiled substrate) had not recovered, whereas those areas that were inaccessible to mechanical cleaning had recovered within one year despite the persistence of some residual oil (Rolan & Gallagher 1991). After the 2002 PRESTIGE oil spill in France, Castège et al. (2014) attributed high-pressure hot-water cleaning as one reason why communities did not recover normally, including the permanent loss of some sponges (e.g. *Hymeniacidon perlevis* and *Tethya* sp.). Le Hir & Hily (2002) also found that recovery of the rocky shore after the 1999 ERIKA spill in France was retarded by high-pressure cleaning as it removed macroalgae that had survived the spill, thereby altering the succession dynamics and recovery of grazers to the area. In a review of the recovery periods described for 34 oil spills, Sell et al. (1995) concluded that clean-up and treatment usually resulted in longer recovery periods. Hawkins & Southward (1992) stated that one of the most important lessons from the 1967 TORREY CANYON spill was that dispersants should not be used against spills impacting exposed rocky shorelines (see also Hawkins et al. 2017a).

Importance of water temperature and exposure time

Because temperature influences the dissolved fraction of oil in water, higher water temperatures dramatically affect the toxicity of both monocyclic aromatic hydrocarbons (MAHs) and PAHs. This needs to be considered when conducting experiments and applying laboratory results to the real world, as impacts may vary in temperate and tropical regions and between seasons. Jiang et al. (2012) studied the effect of temperature and exposure time of several zooplankton species to a WAF using crude oil from China. Regardless of temperature, increasing exposure time from 24 hours to 72 hours generally doubled the toxicity (e.g. halved the LC50 concentration). Increasing the temperature from 8.5°C to 16.5°C, and then to 31.2°C, also doubled the toxicity at each temperature step. For example, for the copepod *Labidocera euchaeta*, 24-hour LC50 concentrations changed from ~22 mg L^{-1} to 13 mg L^{-1} to 4 mg L^{-1} at 8.5°C, 16.5°C and 31.2°C, respectively. Consistent with this, Fisher & Foss (1993) compared toxicity of fuel oil to embryos of grass shrimp *Palaemon pugio* (=*Palaemonetes pugio*) at different temperatures and found that the effects of toxicity onset earlier at higher temperatures.

Depth, wave exposure and habitat influences

Intertidal and shallow subtidal habitats   Our assessment of the literature confirms some general patterns. Typically, communities on exposed rocky shores are less impacted by oil spills than on sheltered shores, and intertidal communities on the higher part of the shore are usually more impacted than those on the lower shore and subtidal communities. Salt marsh, mangrove and other intertidal sedimentary habitats are probably no more vulnerable to the initial impacts from oil spills than sheltered intertidal rocky shores, but they recover much more slowly due to the residual effects of sedimented oil, resupply of disturbed sediments and loss of perennial habitat-forming macrophytes.

Thus, subtidal communities on exposed rocky coasts subject to oil spills (as sometimes occur from shipping accidents away from ports), generally escape significant impacts. Examples of this experience are described (Edgar & Barrett 2000, Lougheed et al. 2002, Edgar et al. 2003) for shipping accidents in Australia and the Galapagos. Nevertheless, exceptions exist, such as when quantities of oil released are very great (e.g. the 2002 PRESTIGE oil spill in Spain) and/or when the oil discharges continue for many months (e.g. the 1957 TAMPICO MARU in Mexico; North et al. 1965, Mitchell et al. 1970).

For intertidal rocky shores, Sell et al. (1995) reviewed case studies of 21 spills and showed that exposed rocky shores recovered more quickly than sheltered or moderately exposed shores. This pattern is consistent with the general patterns described previously and is reflected in a number of studies. Lopes et al. (1997) studied the impact of a burst oil pipeline spill in Brazil, where such accidents are common (171 spills between 1974 and 1994). Oiled areas of rocky intertidal shore resulted in immediate mortality of crabs, littorinid snails and isopods, while areas monitored adjacent to the impacted area showed no significant change in cover of barnacles (*Chthamalus* and *Tetraclita*) and mussels (*Brachidontes*). Although doubt exists over the extent of actual oiling of the area surveyed (the impacted area was chosen because of extensive before impact data, with the authors noting that adjacent areas were more heavily oiled), the study indicates a lower level of sensitivity of barnacles and mussels to oil spills on rocky shores where wave action can rapidly disperse, dilute and naturally remove oil. Kotta et al. (2008) also found sheltered and deeper sites were more impacted by oiling than shallow, exposed sites following a large 2006 oil spill in the Baltic Sea off Estonia. On Macquarie Island, invertebrate abundances at impacted sites on exposed shores had returned to prespill impact levels and were comparable to control sites after seven years, but this was not the case among the more sheltered kelp holdfast habitat, which had not recovered after that time (Smith & Simpson 1995).

Several studies of oil spill impacts on intertidal rocky shores have found differential mortality of invertebrates at different positions of tidal height on the shore. This difference is often expressed as different types of animals being more or less sensitive. However, our view from this review is that, while taxa clearly differ to some degree in sensitivity to oiling, observed differences in the field following oil spills are generally due more to zonation of intertidal animals rather than differential sensitivities. Regardless of taxa, invertebrates in the lower intertidal zone of rocky shores generally suffer lower mortality than those in the upper intertidal zone. For example, mussels and barnacles escaped mortality due to their position lower in the intertidal zone following the 1978 AMOCO CADIZ spill in France, while gastropods higher on the shore were killed (Chassé 1978). However, where heavy oiling of mussels has occurred (e.g. the 1957 TAMPICO MARU spill in Mexico), they too suffered high mortality (Mitchell et al. 1970). Similarly, after the 2002 PRESTIGE spill in Spain, the extent of mortality of barnacles (often regarded as being resistant to the effects of oil) was found to be dependent on the extent of oiling at different locations (Penela-Arenaz et al. 2009). In this case, the barnacle involved was *Chthamalus montagui*, which occupies the higher intertidal area (Penela-Arenaz et al. 2009). Thus, their observations support our assessment that tidal height is more important than species sensitivities, especially in spills involving heavily oiling.

Marine invertebrates occupying intertidal sedimentary habitats such as mangroves, salt marshes, mud flats, sand flats and beaches are especially vulnerable to oil spills. For example, massive mortality of 14.5 million bivalves of the families Cardiidae, Solenidae, Macridae and Veneridae occurred after the AMOCO CADIZ spill in France (Conan 1982), while Spanish beaches affected by the 2002 PRESTIGE spill initially lost up to 67% of species richness (de la Huz et al. 2005). Similarly, elsewhere in this review we discuss direct mortality of intertidal amphipods, isopods and burrowing crabs by oil spills and the long recovery times in intertidal habitats. Perhaps the most extreme examples are impacts following the 1969 FLORIDA barge oil spill in West Falmouth, Massachusetts, USA (Krebs & Burns 1977), and the 1986 refinery spill at Galeta, Panama (Burns et al. 1993). Incorporation of oil into intertidal habitats prolongs the exposure of animals to residual oil, both through burrow activities and as the oil is re-exposed due to erosion or other disturbances. Nevertheless, some evidence suggests that invertebrates on intertidal beach habitats respond differently to those on intertidal rocky shores with respect to zone of greatest impact. Following the 2009 PACIFIC ADVENTURER oil spill in Australia, Schlacher et al. (2011) found greater impacts lower on the shore on beaches, rather than on the upper shore. For the same spill, impacts were greater on the high intertidal part of the rocky shore than in the lower intertidal (Stevens et al. 2012).

Lastly, where subtidal sedimentary and seagrass habitats are affected by oil spills, high mortalities can also occur. For example, subtidal heart urchins and amphipods were decimated by the spill from the AMOCO CADIZ that sank 5 km offshore from the Brittany coast of France (Conan 1982). The significance of the effect of oil on subtidal sedimentary habitats is likely to vary according to the size of the spill, depth and degree of mixing, and interactive effects involving exposure. Penela-Arenaz et al. (2009) reported that the heart urchin *Echinocardium cordatum* in Spain was not affected by the PRESTIGE oil spill. The sheer scale of the IXTOC I wellhead blowout in the Gulf of Mexico (475,000 tonnes, of which 120,000 tonnes sank to the bottom) over 290 days during 1979 and 1980 is thought to have resulted in significant subtidal impacts along the Gulf's sandy shores, where all crabs present along hundreds of kilometres of coastline were inferred to have been killed (Jernelöv & Linden 1981). Post-spill monitoring of subtidal seagrass beds (>1 m deep) following the 1986 refinery spill at Galeta, Panama (Jackson et al. 1989) indicated that, relative to control areas, amphipods, tanaidaceans, crabs and ophiuroids were severely impacted by the spill, with very slow recovery, while bivalves, gastropods and polychaetes were either less impacted or recovered over 18 months.

Deepwater habitats   Few studies inform us about the impact and/or recovery of oil spills on deep benthic habitats and their fauna. Guidetti et al. (2000) compared fauna in impacted and non-impacted subtidal areas in 75–80 m of water eight years after the 1999 HAVEN oil spill in the north-west Mediterranean Sea off Italy. Tar aggregates remained in the impacted areas, but comparison of numbers of macroinvertebrates (including polychaetes, sipunculids, bivalves and crustaceans—tanaidaceans, isopods and amphipods) between areas with and without tar aggregates showed no significant differences. Following the 2010 DEEPWATER HORIZON blowout in the Gulf of Mexico, Felder et al. (2014) found that shrimp, crab and lobster species associated with rhodolith and macroalgal habitat in 55–80 m water depth decreased dramatically in both diversity and abundance. They concluded this was an indirect effect of the oil killing the algae and rhodoliths, rather than actual mortality from the oil. Montagna et al. (2013) found that in very deep water (~1200 m), copepods were affected negatively by the same spill, although nematodes responded positively (probably due to nutrient enrichment after the spill and subsequent increased bacterial production). After the 2002 PRESTIGE oil spill off Spain, Sanchez et al. (2006) surveyed depths of 70–500 m where tar aggregates 200–300 kg km^{-2} existed over an area ~25 km in diameter. This area had good historical data from fisheries surveys, which was used to identify reductions in the abundance of the Norwegian lobster *Plesionika heterocarpus*, which were attributed to direct mortality from the oil spill.

Capacity for recovery

Evidence for short-term recovery (months to a few years)

Sell et al. (1995) reviewed studies on 34 oil spills to form a generalised view that times taken for community assemblages to recover after an oil spill are about three years for rocky shores and five years for salt marshes. However, numerous exceptions were identified. Very few of the case studies were carried through to full 'recovery' (in this case defined as 'where a natural biota has been established and is within the range of dominance, diversity, abundance and zonation expected for that habitat'). Sell et al. (1995) found that, in most cases, recovery on heavily oiled, exposed rocky shores was well advanced by two years and, in the one case available, fully recovered within three years, except where the shoreline had been subject to mechanical and/or chemical cleaning treatment. In such cases, recovery was not complete after four years in the studies examined. For shores that are heavily oiled, moderately exposed and sheltered, few case studies are available where post-spill clean-up had not been carried out. Mostly these sites were in a recovering stage two to three years post spill, except in three cases where they were still in the recovery phase 10 years post spill. Sell et al. (1995) noted that in at least one of these cases, the clean-up procedure was particularly intense. For moderate to lightly oiled rocky shores, recovery was much more rapid, that is, within one to five years (mostly two), except where shoreline clean-up had been undertaken. For heavily oiled salt marsh habitat, Sell et al. (1995) found periods of three to six years for full recovery (with some exceptions); treated shorelines took longer to recover.

The review by Sell et al. (1995) provides an excellent 'big picture' assessment, although after 20 years is overdue for an update. However, results are simplified because their approach does not discriminate important differences in spill circumstances, particularly gradients of impact and differential impacts on different taxa within the same spill. Moore (2006) also reviewed the impacts of several oil spills and concluded that where oil had been removed from the environment and the clean-up process had not resulted in physical damage to biota that recovery was generally rapid, in the order of a few years, but that recovery was much longer when long-term oil contamination remained present.

Eighteen months after a large oil spill off Estonia in 2006, Kotta et al. (2008) found that herbivores (especially amphipods and isopods) had not recovered, while deposit feeders and suspension feeders (which had not been impacted) remained stable. Elmgren et al. (1983) found that three years after the TSESIS oil spill, *Pontoporeia* amphipod numbers were still depressed, while the polychaete *Bylgides sarsi* (=*Harmothoe sarsi*) had returned to preimpact densities, and *Limecola balthica* (=*Macoma balthica*), which had not been affected by the spill, became significantly more abundant. Elmgren et al. (1983) estimated it would take 5–10 years (or even longer) for the relative densities of species in the affected area to return to normal. Garrity & Levings (1990) found significant impacts on mollusc populations on intertidal mudflats in Galeta, Panama after a 1986 refinery spill. They monitored impact and control sites, and after three years, recruitment in the spill areas remained reduced relative to unimpacted sites; some species of littorinids had not re-established after three years. Garrity & Levings (1990) estimated it would take 5–10 years (or perhaps longer) for the relative densities of species in the affected area to return to normal. After the 1999 ERIKA spill in France resulted in 100% mortality of the sea urchins *Paracentrotus lividus* and *Psammechinus miliaris*, Barillé-Boyer et al. (2004) found it took two years until the first urchins reappeared and three years for densities to return to preimpact levels ($\sim$60 m^{-2}).

In general, studies of simulated oil spills have demonstrated rapid recovery from recolonisation from adjacent plots (McGuinness 1990, Egres et al. 2012), but this is likely to be unrealistic in an extensive real-world spill. One large-scale (900 m^2) simulated oil spill in Panama caused total mortality of the sea urchins present, and those that reappeared seven months later (as recovery started to take place) were smaller and most likely recruits rather than migrants (Ballou et al. 1989). Egres et al. (2012) experimentally oiled sandy intertidal sand flat plots with diesel in Brazil to simulate a small spill of fuel oil. They documented high mortalities of animals (e.g. gastropods, oligochaetes and ostracods) in the plots (just 0.35 m^2, each oiled with 2.5 L) and then rapid recovery of the treated plots within two days, resulting from immigration by animals from outside the treated plots. While studies such as this reinforce the toxic nature of hydrocarbons in the field, they do not greatly inform determination of recovery rates except to say that where oiling is patchy at scales of metres, and not heavy nor persistent enough to be incorporated into sedimentary processes, recolonisation of soft-substrate habitats can occur rapidly.

For seagrass or macroalgal habitat subject to only minor disturbance from hydrocarbon exposure, the duration of recovery can be rapid—within months to less than a year (Dean et al. 1998). Following the 1989 Exxon Valdez oil spill, eelgrass (*Zostera marina*) beds in heavily oiled bays were exposed to moderate concentrations of hydrocarbons. A year after the spill, Dean et al. (1998) quantified injuries (shoot density and flowering shoots) to eelgrass to be slight and not significantly detectable relative to undisturbed sites (noting the power of the experimental design was low and may have masked statistical outcomes). Five years after the Exxon Valdez oil spill, follow-up surveys indicated no significant differences between oiled and reference sites (measured via biomass, seed density, seed germination or the incidence of normal mitosis in seedlings), and no signs of the elimination of eelgrass beds (Dean et al. 1998). Dean et al. (1998) concluded that the eelgrass populations recovered from possible damage by 1991, which coincided with sharp declines in hydrocarbon concentrations.

Evidence for long-term recovery (>5 years)

Despite some studies finding good recovery after periods of three to six years, others have demonstrated incomplete recovery or lagging sublethal effects that persisted for decades. Long-term monitoring to 1990 following the 1967 Torrey Canyon spill indicated that recovery of communities, to the extent that they reflect prespill species assemblages and their natural levels of spatial and temporal variability, takes close to 15 years (Hawkins & Southward 1992). Incomplete recovery (to the extent it should be regarded as permanent impact) has occurred in some of the worst examples of oil spills, although this point is hotly debated as in the case of the Exxon Valdez spill (Peterson et al. 2003, Harwell & Gentile 2006, Payne et al. 2008, Bodkin et al. 2014).

Studies reviewed here suggest that in intertidal sedimentary habitats such as salt marshes and mangroves, effects of oiling can last decades. Krebs & Burns (1977) found significant effects of the Florida barge oil spill, which affected salt marsh habitat in West Falmouth, Massachusetts, USA, remained after seven years with fiddler crabs (*Minuca pugnax* [=*Uca pugnax*]) at reduced densities compared to prespill levels and residual chronic effects on crab health and behaviour. Twenty years after the same spill, Teal et al. (1992) found that sedimented oil still remained at 10–15 cm below the surface in the heaviest oiled areas in sufficient levels to affect crab utilisation of habitat. They concluded that if these sediments were disturbed such that the oil was again exposed at the surface, it would lead to toxic concentrations of oil reoccurring. Another survey 30 years after the event (Reddy et al. 2002) found similar results, with oil still present in cores between 6–28 cm from the surface. Carls et al. (2001) measured the rate of decline in hydrocarbons in mussel beds and the underlying sediment affected by the 1989 Exxon Valdez oil spill and concluded that it would take another 30 years to recover. Harwell & Gentile (2006) reviewed studies of the impacts of the Exxon Valdez spill and concluded that with the exception of killer whales, other species (including the invertebrates) that had suffered a decline in abundance following the spill had recovered in abundance within six years. However, Peterson (2001) and Peterson et al. (2003) concluded that persistent low-level exposure to residual sedimented oil from the Exxon Valdez continued to cause impacts to several species, with sublethal effects likely to continue for many years as continued exposure from oil in contaminated sediments resulted in bioaccumulation of hydrocarbons by macroinvertebrates and then through the food chain to birds and mammals.

Beyond the issue of sublethal effects, numerous examples in the literature indicate partial recovery within a few years, as suggested by Sell et al. (1995), but with long-term effects persisting for many more years. The 1957 Tampico Maru wreck in Baja California, Mexico partially blocked a cove and spilled oil for eight to nine months. Despite this incident, the intertidal gastropod *Littorina keenae* (=*Littorina planaxis*) survived, but subtidal gastropods including abalone (*Haliotis* spp.) were killed, with reduced numbers seven years later (North et al. 1965). Conan (1982) noted that after the 1978 Amoco Cadiz spill of 223,000 tonnes of crude oil in France, delayed effects on mortality, growth and recruitment were still observed up to three years after the spill. This author estimated it would take three to six generations (5–10 years in the case of bivalves) before populations recovered fully. In a longer-term investigation, Dauvin & Gentil (1990) found that populations of most amphipod species had recovered after 10 years to densities similar to before the AMOCO CADIZ spill; however, some formerly abundant species were not found in any post-spill surveys and had probably been extirpated. After the 2002 PRESTIGE oil spill in France, Castège et al. (2014) showed that despite the recovery of many taxa, nine years after the spill the sponges *Hymeniacidon perlevis* and *Tethya* sp., which had dominated some parts of the lower intertidal area before the spill, had not returned.

There are limited studies on the recovery of coral communities following oils spills. Guzmán et al. (1994) reported that there was no evidence of coral recovery five years after the 1986 major oil spill in Panama. An understanding of the capacity and timeline to recovery can be drawn from other impacts to coral reefs such as coral bleaching from elevated sea surface temperatures and physical damage from cyclones. Studies have shown significant recovery of corals from these types of impacts in less than 15 years in some cases, longer in others (Baker et al. 2008). Recovery may be delayed if there is residual oil in the sediment, especially where PAHs continue to be released.

Where exposure to oil has resulted in severe impacts to a seagrass or macroalgal habitat (e.g. Galeta oil spill in Panama), there has been little follow-up reporting on recovery. This has considerably limited our understanding of the possible capacity, or required durations, of seagrass and macroalgal assemblages to return to predisturbance conditions. Based on other investigations of where seagrass meadows have been severely disturbed (i.e. considerable reductions in shoot density and loss of leaf biomass), recovery may not commence for many years after the stressor has ceased (McMahon et al. 2011).

In terms of community-scale impacts (i.e. those involving both the plant assemblages and their associated fauna), the persistence of toxic subsurface oil and chronic exposures in sediments (even at sublethal levels) can result in delayed population reductions and cascades of indirect effects that can significantly hamper ecosystem recovery for many years (Peterson et al. 2003). Peterson et al. (2003) reported chronic exposures to *Fucus distichus* (=*Fucus gardneri*) assemblages after the EXXON VALDEZ spill, which led to a complex pattern of shifts in community composition represented by opportunistic species, many of which did not perform the same ecosystem functions as the original community.

Assessing impact and recovery from oil spills

A number of excellent best practice guidelines for oil spill response and monitoring have been published recently in Australia (Hook et al. 2016) and internationally (IPIECA-IOGP 2015, 2016); however, a number of key lessons learned are apparent from the literature reviewed here and are worth emphasising when assessing the impacts of oil spills on marine animals and plants.

Use best practice study design

Best practice is a before-after-control-impact (BACI) with well-established time series of baseline data (Green 2005) or 'beyond' BACI, which extends the method to multiple control sites and temporal replication of preimpact sampling (Underwood 1991, 1994). Good examples in the Australian context are Edgar & Barrett (2000) and MacFarlane & Burchett (2003), although often the availability of preimpact baseline is fortuitous, having been established for other purposes. The lack of baseline data at affected and unaffected sites prior to an oil spill is regarded as the greatest impediment to understanding the effects of oil spills (Bodkin et al. 2014). Some of the most informative studies were where multiple years of prespill baseline data existed in impacted and unimpacted areas. Ideally, this should include multiple impacted and non-impacted sampling locations (e.g. Southward & Southward, 1978). Most often this was fortuitous, such as the overlap of high human population density, regional research laboratories and shipping lanes that exists in Europe. Nevertheless, availability of baseline data does not necessarily involve serendipity. Quantitative baseline surveys encompassing thousands of species using the Edgar & Barrett (2000) method have now been undertaken at over 2000 sites around the Australian continent through the Reef Life Survey programme, including Vulcan Shoal and >100 other locations on the North-West Shelf (Edgar & Stuart-Smith 2014, Edgar et al. 2016). As a consequence, future spills affecting shallow reef systems in Australia have a high probability of prespill abundance data for hundreds of species in the near vicinity, plus comparable reference sites enabling a BACI statistical design, while also providing a monitoring framework for continental-scale state-of-the-environment reporting (Stuart-Smith et al. 2017).

Other planned approaches involve establishment of baselines in areas of intensive or frequent industrial use, as well as control sites, as occurred with the long-term seagrass monitoring programme in Gladstone. This was established several years before the 2006 GLOBAL PEACE spill in Gladstone and subsequently provided an excellent prespill baseline (Taylor et al. 2006, Taylor & Rasheed 2011). In the absence of a baseline, a 'bullseye' sampling design that establishes a gradient between impact and control is recommended (Green 2005). Smith & Simpson (1995, 1998) and Schlacher et al. (2011) have demonstrated it is possible to make very good control-impact-only comparisons by repeated measure studies of oiled and non-oiled locations.

Sampling design with sufficient power to detect change needs to be used

This would seem to be self-evident; however, Peterson et al. (2001) showed that some of the assessments undertaken following the EXXON VALDEZ spill in Alaska had insufficient power to detect an impact where it had occurred, and that this had contributed to the controversy over the severity of spill impacts (Peterson et al. 2001).

Impacts should be assessed outside the zone of maximum damage

This is likely to be the high-impact zone, but it may be small in area relative to the areas of more moderate or lesser impact. Assessing these areas, in addition to the main impact sites and control sites, is likely to be more informative about gradients of impact and recovery times, relative sensitivities of different species and sublethal impacts.

Indirect and sublethal impacts are important

Numerous studies have pointed to the ongoing long-term sublethal effects of oil spills (Johnson 1977, Peterson 2001), and interactions between exposure to oiling, environmental conditions, habitat, biota and behaviour.

Monitoring should continue until recovery is complete or change has stabilised

Very few studies have followed the course of recovery from an oil spill through to full recovery. Most ceased while the communities were in a recovery mode or well on the way to recovery (Sell et al. 1995). Researchers occasionally returned after 20 or even annually over 50 years (e.g. Hawkins et al. 2017a), but generally these were in the worst cases where recovery may never be complete and damage perhaps permanent (e.g. Teal et al. 1992, Reddy et al. 2002 in the case of the 1969 West Falmouth, USA spill) or where interactions of recovery with other impacts such as imposex from tributyl-tin anti-fouling paints and climate-driven fluctuations and recent warming had become the focus of the research (Southward et al. 2004, Hawkins et al. 2017a,b). Intervals between monitoring surveys can increase with time.

Archive samples for future analysis

Changes in analytical methods, improved detection limits and technological improvements in instrumentation over the course of a long-running monitoring programme (such as occurred with the 1989 EXXON VALDEZ spill; Payne et al. 2008) can confuse interpretation and contribute to debate about impact and recovery.

Consider a wide range of taxa

Different species have varying sensitivities to impact, and with the limited state of knowledge, it is currently difficult to pick winners in terms of the best sentinel species. Unpredictable impacts involving species interactions are likely in the most severe spills and may otherwise go unnoticed.

Conclusions

Marine invertebrates, algae and seagrasses are sensitive to the toxic effects of oil. Depending on the intensity, duration and circumstances of the exposure, invertebrates can suffer high initial mortality together with prolonged sublethal effects, which can act at both the individual and population level. Under some circumstances, recovery from these impacts can take years to decades. Although crude oil exposure to seagrasses and macroalgae can be toxic, toxicity varies among species, and, in general, marine algae and seagrasses are less sensitive to oil than marine invertebrates, with generally faster recovery rates. Some commonly occurring species of seagrasses in Australia may be less sensitive than those that occur elsewhere based on the studies reviewed.

While a range of generalities can be stated about the response of marine invertebrates, algae and seagrasses to oil spills, almost every oil spill is unique in terms of its impact because of differing levels of exposure. The variety of factors that contribute to exposure include: type of oil, amount of oil, extent of weathering, whether the exposure is transient or persistent, whether dispersants or other clean-up measures were used, the type of habitats and depths affected, the species present and their stage of development or maturity, the species assemblages present, and how the process of recolonisation proceeds in terms of recruitment and other dynamics. The importance of each of the factors and how they affect the degree of impact have been explored in this review.

The type and source of oil in an unplanned spill is a very important factor in determining the extent of the impact and the level of exposure to toxic hydrocarbons. Refined oils, diesel and heavy bunker fuel oils are more toxic than crude oils, and the type of oils spilled usually depends on whether the accident involves a cargo ship, an oil tanker, a refinery spill or a wellhead blowout. In Australia, crude oils from different oil fields possess a wide range of densities.

The degree of exposure from an oil spill will depend, in part, on the degree of weathering and dilution (e.g. from wave action) that takes place from the time oil makes contact with the shoreline or benthos. In general, the part of weathering that involves evaporation will drive off some toxicants, but it also serves to concentrate others and affects the consistency of the oil. In terms of toxicity to marine invertebrates, the extent of dilution determines exposure concentrations and is more important than a short-term period of weathering. Almost without exception among the studies we reviewed, the use of mechanical cleaning of oil from the substrate resulted in higher mortality and longer recovery times for marine invertebrates than the oil on its own. Although this was also true for the use of dispersants in the nearshore environment, their role in achieving dilution and preventing excessive oil stranding leading to chronic exposure to sedimented oil probably weighs in favour of their use where sedimentary habitats are threatened. For exposed rocky shores, the evidence weighs in favour of not using dispersants.

Likewise, for seagrass and macroalgae, the evidence suggests chemically dispersed oils are more toxic than crude oils, although results differ between studies and are not as clear-cut as for marine invertebrates. Research outcomes range from dispersed oil posing a greater threat than non-dispersed oil, to dispersed oil posing less of a threat than non-dispersed oil, to neither oil nor dispersed oil impacting seagrass or macroalgae. Further complicating the matter, responses to dispersed oils can be highly variable between species. As dispersed oils are significantly more toxic than oil alone, the use of dispersants offshore to reduce sea surface oil, improve dilution, aid weathering and decrease concentrations before shorelines are impacted should be considered on a case-by-case basis and is likely to be warranted in some situations, especially in Australia where a large number of offshore installations exist.

The type of habitat impacted by the oil, including substratum type, depth and wave exposure, is a very significant factor in modulating impacts on marine invertebrates and plants in those habitats. Typically, communities on exposed rocky shores are less impacted by oil spills than on sheltered shores, and intertidal communities on the higher part of the shore are usually more impacted than those on the lower shore and subtidal communities. Salt marsh, mangrove and other intertidal sedimentary habitats are probably no more vulnerable to the initial impacts from oil spills than sheltered intertidal rocky shores, but they recover much more slowly due to the residual effects of sedimented oil which is re-supplied as sediments are subsequently disturbed.

Thus, oil spills that result in significant exposure to intertidal sedimentary habitats cause high levels of mortality to marine invertebrates, with prolonged chronic effects when oil is incorporated into sedimentary layers, which causes disruption to burrowing and bioturbation. Oil is easily sedimented into these types of habitats, with resultant risk of re-exposure of oil through erosion or by animals foraging and burrowing. Several studies reviewed here showed that oil incorporated in sediments can release lethally toxic levels of hydrocarbons at least two decades after the initial exposure. Loss of salt marsh and mangrove habitat resulting from oil spills in sedimentary habitats also impacts marine invertebrates—first directly from exposure and loss of habitat, and later from increased predation (e.g. from birds) as the plant cover remains thinned for some time. In sedimentary habitats, the taxa predominantly impacted (at least in the studies reviewed) were small crustaceans such as amphipods, crabs, bivalves and gastropods. However, some taxa such as cerithiid gastropods (creepers) and some nematodes and polychaetes appear more resilient to oiling. For subtidal sedimentary habitats and seagrass beds, amphipods, crabs, bivalves, gastropods and sea urchins are affected, but unless the spill is particularly heavy (as has occurred in some spills in Europe and Panama), subtidal habitats are less frequently impacted than intertidal habitats. In Australia, spills have impacted mangrove and beach habitats, resulting primarily in mortality to crustaceans and gastropod molluscs.

For intertidal reef or rocky shore habitats, the initial contact from heavy oiling causes high mortality, especially of small crustaceans such as amphipods and isopods, gastropod molluscs and echinoderms. Uncertainty exists over the sensitivity of sponges with the very few studies that have been undertaken giving conflicting results. In tropical areas, scleractinian and hydrozoan corals are affected along with anemones and zoanthids. Macroalgae are also susceptible to direct contact, while seagrasses have greater resilience due to rhizomes below the sediment surface. Longer-term effects of exposure to oil appear to be less on exposed rocky intertidal habitats than on those that are sheltered, apparently because exposed intertidal habitats are vigorously washed by turbulent waves, and oil is less likely to be trapped within the substratum. Overall, with some caveats, intertidal rocky shores appear more resilient to the long-term effects of oil spills but not to initial exposure mortalities. Subtidal reef habitats also suffer mortalities, but these are less than on intertidal reefs and limited to very severe spills of oil from ships or refineries close to shore. There have been no major impacts to subtidal reefs detected from oil spills in Australia to date.

The extensive literature on the effects of oil on coral shows that the potential damage to coral reefs is of particular concern, and there is substantial evidence that exposure to oil spills can be toxic to corals. Some corals are less resilient than others, and changes may occur to the gross percent cover of corals or the composition of different coral species. Branching coral (e.g. *Acropora* and *Pocillopora*) are generally more susceptible to impacts from oil than massive corals or corals with large polyps (e.g. *Favia* and *Fungia*). Studies have shown a range of impacts including changes in growth rate, feeding behaviour, cellular physiological condition, photosynthetic ability of zooxanthellae, fecundity, settlement, bioaccumulation and mortality. However, some studies also demonstrate recovery of corals from these effects over time. Toxicological and physiological studies are important to understand direct effects on corals in terms of growth and reproductive capacity. Corals are most vulnerable to oil spills during coral spawning, and subsequent larval and juvenile life stages, when they drift on or near the surface. Hydrocarbons exposed to ultraviolet radiation in surface waters may photo-oxidise, which may increase the toxicity and bioavailability of hydrocarbon components such as PAHs and could impact the larval phase of corals.

Several studies claim their results show that barnacles and mussels are less sensitive to oiling than other taxa. We found little evidence for this and suggest these findings more likely reflect differences in exposure, and that impacts of oiling on rocky intertidal shores are consistently more severe in the high intertidal zone compared with the lower intertidal. Intertidal herbivorous gastropods, small crustaceans and sea urchins appear particularly vulnerable to oil spills, with high mortalities recorded in many studies. It is possible that this reflects the abundance of this type of animal in intertidal habitats relative to other taxa (including predatory gastropods, for example), which may have been initially uncommon enough to allow statistically powerful comparisons to be made between oiled and control sites. Such species often exhibit low densities and high variance-to-mean ratios, making comparisons between sites difficult. Thus, it is possible that a broader range of invertebrates than those commonly studied to determine impact response might also be just as vulnerable to oil spills. Decapod crustaceans (crabs and shrimps) are another group that appear particularly vulnerable based on the studies reviewed. This was particularly true of burrowing species in sedimentary habitats, which can in turn drive oxygenation of shallow sediments and thus play a keystone role as habitat engineers.

Most of the available literature on the effects of oil on seagrasses and macroalgae show results of acute responses and/or acute exposure, which can be highly variable depending on the experimental set-up, oil type, and species assessed. Commonly reported acute toxic effects of oiling on seagrasses and macroalgae include reduced rates of photosynthesis and respiration, which can lead to morphological changes in the plant structure (loss of leaves/thalli, reduced biomass, etc.), and ultimately death. However, acute toxicity tests may not necessarily offer the best understanding of a hydrocarbon effect, and there is a need for more thorough assessment of long-term implications for oil spills in nearshore and intertidal marine habitats.

The indirect effects of oil spill impacts on marine invertebrates include changes to dominance patterns in community assemblages. In the studies we reviewed, two types in particular were evident. First, on soft sediment habitats, some species of nematodes and polychaetes dominated recovery processes and achieved very high abundances relative to prespill levels. This was due to either (or both) their relative insensitivity to the oil and/or their quickness to recolonise, including in response to the organic enrichment that occurs when oil is sedimented. Release from competition when formerly dominant taxa are removed by oiling may also contribute. Second, on rocky intertidal shores and subtidal reefs, heavy mortality of grazing amphipods, gastropods and urchins was followed by a proliferation of opportunistic algae taking advantage of the lack of grazers, the space cleared by the death of sessile invertebrates, and organic enrichment.

Although observations from field studies suggest that exposure to oil can affect reproductive success among seagrasses and macroalgae, there remains a large gap in the literature around this topic. Further, it is difficult to make generalisations as there is a considerable variation among different seagrasses, from large perennial species to small ephemeral species, and responses to different oil types are also likely to vary accordingly. To better understand the effects of hydrocarbons on the diversity of reproductive strategies and life-cycle stages, a more systematic approach to assessing these effects is required beyond the limited information presently available from field observations.

In terms of recovery, it appears that where oil exposure has led to minimal disturbance to a seagrass or macroalgal assemblage, the duration of recovery can be rapid—within months. However, where exposure to oil has resulted in severe impacts to seagrasses or macroalgae, there has been little follow-up reporting on recovery. This has limited our understanding of possible duration times required for seagrass and macroalgal assemblages to return to abundances/biomass similar to predisturbance conditions.

The key factor in all these considerations of oil impacts on marine animals and plants is the level of exposure. Where oiling is slight because of low concentration exposure (as might occur far from the spill site or in a small spill), the impacts of oil spills on marine invertebrates, seagrasses and macroalgae appear low or at least short-lived. That said, sublethal impacts (e.g. impaired motility in crabs, lower adhesion strength in limpets) have been found at very low concentrations, and these may be sufficient to cause the animals to be unable to feed or avoid predators, thus affecting their likelihood of survival. Similarly, low concentrations of hydrocarbons can result in a range of other sublethal effects such as reproductive impairment that causes effects at the population level (reproductive and/or recruitment failure, disease, DNA damage and loss of genetic diversity). Bioaccumulation of hydrocarbons in crustaceans and in bivalve molluscs impacts reproductive success and results in transfer of hydrocarbons higher up the food chain. In general, oil appears more highly toxic to larval invertebrates than adults, so this impact needs to be considered.

In the Australian context at least, the level of exposure to oil and subsequent impact will differ greatly between offshore wellhead accidents (e.g. MONTARA 2009), shipping accidents close to shore (e.g. PACIFIC ADVENTURER 2009), and refinery/oil storage depot spills (e.g. Port Stanvac refinery 1999), and prespill planning and baseline assessment needs to be considered differently. Even though the volumes of oil likely to be involved are much greater from wellhead blowouts, the risk of direct impacts, at least to intertidal and shallow subtidal reefs and sedimentary habitats, seems low in comparison to other types of spills. For this reason, prespill precautionary assessments should not just seek to establish baselines against which to assess impact but should determine the risk to exposure of a range of oil–water fractions and hydrocarbon concentrations of marine organisms, and to test the response of a range of receptor species to those concentrations. Both lethal and sublethal responses need to be assessed, and perhaps most importantly given the differing toxicities of different types of crude and refined oils, the assessments need to be done using the oil with highest risk in terms of local geography.

In this review, we examined the records of assessment of 51 significant oil spills in Australia between 1970 and 2017. Of these, six occurred offshore with no likely or expected impact on benthic invertebrates or macrophytes. Despite the potential for oil spills to impact marine invertebrates, only 24 cases had potential direct effects of oil studied, and 19 cases had only cursory or no assessment of impact. Of those 24 spills where impact assessments are available in published or unpublished reports, only eight considered impacts on invertebrates, with many others focussing on birds or the primary plant habitat affected but with little or no consideration of the invertebrate communities they support. With the exception of the 2009 MONTARA wellhead spill and the 1999 TORUNGEN spill, detailed assessment of spill impacts on invertebrates in Australia have been limited to temperate waters. Despite the majority of spills in Australia since 1970 being in temperate Australia, this is not the current trend. Ten of the 14 spills this century have been in the tropics, seven along the north Queensland coast in proximity to the Great Barrier Reef, and all four spills associated with offshore oil and gas infrastructure have been since 2009.

We also found very few assessments of the toxicity and sublethal effects of oil on Australian marine invertebrates and macroalgae, but that seagrasses were better studied. Studies undertaken have nearly all been confined to south-eastern Australia. While they are useful in the local context, a high priority remains to test the responses of Australian marine invertebrates across the range of habitats and geography and types of oil they might be exposed to. Given the nutrient-deficient status of Australian seas relative to the North American and European locations where most studies have been undertaken, the concentrations of oil needed for lethal impacts on marine plants and invertebrates may well be lower in Australia. In particular, we identified a number of taxa of habitat-forming, sessile, filter-feeding invertebrates (sponges, bryozoa, tunicates) that need assessment of their response to oil. In addition, more information is needed about Australian species in different parts of Australia for taxonomic groups that are known from overseas studies to be vulnerable to exposure to oil—these include molluscs, crustaceans, echinoderms, and most macroalgae and tropical seagrasses. Lastly, in this review we outlined some of the lessons learned in assessment of oil spill impacts from the studies examined and provided some recommendations to be considered in responding to oil spills.

Acknowledgements

Much of the material examined in this review was assembled as part of a series of unpublished white papers prepared for the Australian Petroleum Production and Exploration Association (APPEA). Christina Street and Meryn Scott from CSIRO provided assistance with compiling the literature used in this review, and Christine Hanson from BMT assisted with editing. We thank Paul Irving from the Australian Maritime Safety Authority (AMSA) for providing access to AMSA records of recent pollution events between 2013 and 2015 and Jon Moore for providing helpful comments on the manuscript.

References

Allaway, W.G. 1982. Mangrove die-back in Botany Bay. *Wetlands (Australia)* **2**, 2–7.

Allaway, W.G. 1985. Exploitation and destruction of mangroves in Australia. Mangrove ecosystems of Asia and the Pacific: status exploitation and management. In *Proceedings of the Research for Development Seminar*, 18–25th May 1985, Australian Institute of Marine Science, Townsville, Australia.

Allaway, W.G. 1987. Exploitation and destruction of mangroves in Australia. In *Mangrove Ecosystems of Asia and the Pacific, Status, Exploitation and Management*, C.D. Field & A.J. Dartnall (eds). Townsville, Australia: Australian Institute of Marine Sciences, 183–192.

Allaway, W.G., Cole, M. & Jackson, J.E. 1985. *Oil spills and mangrove die-back in Botany Bay*. Final Report to the Coastal Council of New South Wales, Australia.

Almeda, R., Baca, S., Hyattm C. & Buskey, E.J. 2014a. Ingestion and sublethal effects of physically and chemically dispersed crude oil on marine planktonic copepods. *Ecotoxicology* **23**, 988–1003.

Almeda, R., Bona, S., Foster, C.R. & Buskey, E.J. 2014b. Dispersant Corexit 9500a and chemically dispersed crude oil decreases the growth rates of meroplanktonic barnacle nauplii (*Amphibalanus improvisus*) and tornaria larvae (*Schizocardium* sp.). *Marine Environmental Research* **99**, 212–217.

Almeda, R., Wambaugh, Z., Wang, Z., Hyatt, C., Liu, Z. & Buskey, E.J. 2013. Interactions between zooplankton and crude oil: toxic effects and bioaccumulation of polycyclic aromatic hydrocarbons. *PLoS One* **8**, e67212.

Altenburger, R., Walter, H. & Grote, M. 2004. What contributes to the combined effect of a complex mixture? *Environmental Science & Technology* **38**, 6353–6362.

Amjad, S. & Gray, J.S. 1983. Use of the nematode-copepod ratio as an index of organic pollution. *Marine Pollution Bulletin* **14**, 178–181.

Andersen, L.E., Melville, F. & Jolley, D. 2008. An assessment of an oil spill in Gladstone, Australia—impacts on intertidal areas at one month post-spill. *Marine Pollution Bulletin* **57**, 607–615.

Anderson, J.W., Neff, J.M., Cox, B.A., Tatem, H.E. & Hightower, G.M. 1974. Characteristics of dispersions and water-soluble extracts of crude and refined oils and their toxicity to estuarine crustaceans and fish. *Marine Biology* **27**, 75–88.

Anink, P.J., Roberts, D.E., Hunt, D.R. & Jacobs, N.F. 1985. Oil spill in Botany Bay: short-term effects and long-term implications. *Wetlands (Australia)* **5**, 32–41.

Australian and New Zealand Environment and Conservation Council (ANZECC) & Agriculture and Resource Management Council of Australian and New Zealand (ARMCANZ). 2000. *Australian and New Zealand Guidelines for Fresh and Marine Water Quality. Volume 1: The Guidelines.* Canberra, Australian Capital Territory: Australian and New Zealand Environment and Conservation Council and Agriculture and Resource Management Council of Australia and New Zealand, October 2000.

Australian Broadcasting Commission (ABC). 2013. http://www.abc.net.au/news/2013-09-19/oil-spill-triggers-esso-infrastructure-fears/4967316

Australian Broadcasting Commission (ABC). 2017a. http://www.abc.net.au/news/2017-06-22/panama-company-faces-massive-fine-over-great-barrier-reef-spill/8633984

Australian Broadcasting Commission (ABC). 2017b. http://www.abc.net.au/news/2017-05-19/woodside-oil-spill-no-lasting-damage-company-says/8543084

Australian Government. 2011. Final Government Response to the Report of the Montara Commission of Inquiry. Australian Government, Canberra.

Australian Maritime Safety Authority (AMSA). 2000a. *The response to the Port Stanvac oil spill.* Report of the Incident Analysis Team, prepared for the Australian Maritime Safety Authority, Canberra, April 2000.

Australian Maritime Safety Authority (AMSA). 2000b. *The response to the Laura D'Amato oil spill*. Report of the Incident Analysis Team, Prepared for the Australian Maritime Safety Authority, Canberra, April 2000.

Australian Maritime Safety Authority (AMSA). 2000c. *Annual report: national plan to combat pollution of the sea by oil and other noxious and hazardous substances 1999–2000*. Prepared by the Australian Maritime Safety Authority, Canberra.

Australian Maritime Safety Authority (AMSA). 2017. Major historical incidents. Australian Maritime Safety Authority. Online. https://www.amsa.gov.au/environment/protecting-our-environment/major-historical-incidents/ (accessed 6 August 2017).

Babcock, R.C., Bull, G.D., Harrison, P.L., Heyward, A.J., Oliver, J.K., Wallace, C.C. & Willis, B.L. 1986. Synchronous spawnings of 105 scleractinian coral species on the Great Barrier Reef. *Marine Biology* **90**, 379–394.

Baca, B.J. & Getter, C.D. 1984. The toxicity of oil and chemically dispersed oil to the seagrass *Thalassia testudinum*. In *Oil Spill Chemical Dispersants, STP 840*, T.E. Allen (ed.). Philadelphia, PA: American Society for Testing and Materials, 314–323.

Baker, A.C., Glynn, P.W. & Riegl, B. 2008. Climate change and coral reef bleaching: an ecological assessment of long-term impacts, recovery trends and future outlook. *Estuarine, Coastal and Shelf Science* **80**, 435–471.

Ballou, T.G., Dodge, R.E., Hess, S.C., Knap, A.H. & Sleeter, T.D. 1987. *Effects of Dispersed and Undispersed Crude Oil on Mangroves, Seagrasses and Corals*. Contribution 1187. Washington, DC: American Petroleum Institute, 1–229.

Ballou, T.G., Hess, S.C., Dodge, R.E., Knap, A.H. & Sleeter, T.D. 1989. Effects of untreated and chemically dispersed oil on tropical marine communities: A long-term field experiment. In *International Oil Spill Conference*, February 1989, Vol. **1989**(1), 447–454.

Barillé-Boyer, A.L., Gruet, Y., Barillé, L. & Harin, N. 2004. Temporal changes in community structure of tide pools following the 'Erika' oil spill. *Aquatic Living Resources* **17**, 323–328.

Batista, D., Tellini, K., Nudi, A.H., Massone, T.P., Scofield, A.L. & Wagener, A.L. 2013. Marine sponges as bioindicators of oil and combustion derived PAH in coastal waters. *Marine Environmental Research* **92**, 234–243.

Beeden, R., Willis, B.L., Raymundo, L.J., Page, C.A. & Weil, E. 2008. Underwater cards for assessing coral health on Indo-Pacific reefs. Coral Reef Targeted Research and Capacity Building for Management. Melbourne: Currie Communications.

Beiras, R. & Saco-Alvarez, L. 2006. Toxicity of seawater and sand affected by the Prestige fuel-oil spill using bivalve and sea urchin embryogenesis bioassays. *Water Air and Soil Pollution* **177**, 457–466.

Benson, A.A. & Muscatine, L. 1974. Wax in coral mucus—energy transfer from corals to fishes. *Limnology and Oceanography* **19**, 810–814.

Berdugo, V., Harris, R.P. & O'Hara, S.C. 1977. The effect of petroleum hydrocarbons on reproduction of an estuarine planktonic copepod in laboratory cultures. *Marine Pollution Bulletin* **8**, 138–143.

Beyrem, H., Louati, H., Essid, N., Aissa, P. & Mahmoudi, E. 2010. Effects of two lubricant oils on marine nematode assemblages in a laboratory microcosm experiment. *Marine Environmental Research* **69**, 248–253.

Bi, H., Rissik D., Macova, M., Hearn, L., Mueller, J.F. & Escher, B. 2011. Recovery of a freshwater wetland from chemical contamination after an oil spill. *Journal of Environmental Monitoring* **13**, 713–720.

Binark, N., Güven, K.C., Gezgin, T. & Ünlü, S. 2000. Oil pollution of marine algae. *Bulletin of Environmental Contamination and Toxicology* **64**, 866–872.

Birkeland, C., Reimer, A.A. & Young, J.R. 1976. *Survey of marine communities in Panama and experiments with oil*. US Environmental Protection Agency, Report No. EPA-60013-76-028, Rhode Island.

Blumer, M. 1971. Scientific aspects of the oil spill problem. *Environmental Affairs* **1**, 54–73.

Blumer, M., Souza, G. & Sass, J. 1970. Hydrocarbon pollution of edible shellfish by an oil spill. *Marine Biology* **5**, 195–202.

Bodkin, J.L., Esler, D., Rice, S.D., Matkin, C.O. & Ballachey, B.E. 2014. The effects of spilled oil on coastal ecosystems: lessons from the *Exxon Valdez* spill. In *Coastal Conservation*, J.L. Lockwood (ed.). Cambridge, UK: Cambridge University Press.

Bokn, T.L., Moy, F.E. & Murray, S.N. 1993. Long term effects of the water-accommodated fraction (WAF) of diesel oil on rocky shore populations maintained in experimental mesocosms. *Botanica Marina* **36**, 313–319.

Bostrom, C. & Bonsdorff, E. 2000. Zoobenthic community establishment and habitat complexity—the importance of seagrass shoot-density, morphology and physical disturbance for faunal recruitment. *Marine Ecology Progress Series* **205**, 123–138.

Bradbury, R.H., Reichelt, R.E., Meyer, D.L. & Birtles, R.A. 1987. Patterns in the distribution of the crinoid community at Davies Reef on the Central Great Barrier Reef. *Coral Reefs* **5**, 189–196.

Broman, D., Ganning, B. & Lindblad, C. 1983. Effects of high pressure, hot water shore cleaning after oil spills on shore ecosystems in the Northern Baltic proper. *Marine Environmental Research* **10**(3), 173–187.

Burns, K.A., Brinkman, D.L., Brunskill, G.J., Logan, G.A., Volk, H., Wasmund, K. & Zagorskis, I. 2010. Fluxes and fate of petroleum hydrocarbons in the Timor Sea ecosystem with special reference to active natural hydrocarbon seepage. *Marine Chemistry* **118**, 140–155.

Burns, K.A., Codi, S. & Duke, N.C. 2000. Gladstone, Australia field studies: weathering and degradation of hydrocarbons in oiled mangrove and salt marsh sediments with and without the application of an experimental bioremediation protocol. *Marine Pollution Bulletin* **41**, 392–402.

Burns, K.A., Codi, S., Pratt, C. & Duke, N.C. 1999. Weathering of hydrocarbons in mangrove sediments: testing the effects of using dispersants to treat oil spills. *Organic Geochemistry* **30**, 1273–1286.

Burns, K.A., Garrity, S.D. & Levings, S.C. 1993. How many years until mangrove ecosystems recover from catastrophic oil-spills. *Marine Pollution Bulletin* **26**, 239–248.

Burns, K.A. & Knap, A.H. 1989. The Bahai Las Minas oil spill—hydrocarbon uptake by reef building corals. *Marine Pollution Bulletin* **20**, 391–398.

Burridge, T.R. & Shir, M. 1995. The comparative effects of oil, dispersants, and oil/dispersant conjugates on germination of marine macroalga *Phyllospora comosa.* (Fucales: Phaeophyta). *Marine Pollution Bulletin* **31**, 446–452.

Butler, A. & Jernakoff, P. 1999. *Seagrass in Australia. Strategic Review and Development of an R&D Plan.* Collingwood, Australia: CSIRO Publishing.

Byrne, C.J. & Calder, J.A. 1977. Effect of the water-soluble fractions of crude, refined and waste oils on the embryonic and larval stages of the quahog clam *Mercenaria* sp. *Marine Biology* **40**, 235–251.

Cambridge, M., How, J.R., Lavery, P.S. & Vanderklift, M.A. 2007. Retrospective analysis of epiphyte assemblages in relation to seagrass loss in a eutrophic coastal embayment. *Marine Ecology Progress Series* **346**, 97–107.

Carls, M.G., Babcock, M.M., Harris, P.M., Irvine, G.V., Cusick, J.A. & Rice, S.D. 2001. Persistence of oiling in mussel beds after the Exxon Valdez oil spill. *Marine Environmental Research* **51**, 167–190.

Castège, I., Milon, E. & Pautrizel, F. 2014. Response of benthic macrofauna to an oil pollution: lessons from the 'Prestige' oil spill on the rocky shore of Guéthary (south of the Bay of Biscay, France). *Deep Sea Research Part II: Topical Studies in Oceanography* **106**, 192–97.

Cebrian, E. & Uriz, M.J. 2007. Contrasting effects of heavy metals and hydrocarbons on larval settlement and juvenile survival in sponges. *Aquatic Toxicology* **81**, 137–143.

Chan, E.I. 1977. Oil pollution and tropical littoral communities: biological effects of the 1975 Florida Keys oil spill. In *International Oil Spill Conference Proceedings*, American Petroleum Institute, Washington, DC, March 1977, **1997**(1), 539–542.

Chan, K. & Chiu, S.Y. 1985. The effects of diesel oil and oil dispersants on growth, photosynthesis, and respiration of *Chlorella salina. Archives of Environmental Contamination and Toxicology* **14**, 325–331.

Chassé, C. 1978. The ecological impact on and near shores by the Amoco Cadiz oil spill. *Marine Pollution Bulletin* **9**, 298–301.

Chia F.S. 1971. Diesel oil spill at Anacortes. *Marine Pollution Bulletin* **2**, 105–106.

Chia, F.S. 1973. Killing of marine larvae by diesel oil. *Marine Pollution Bulletin* **4**, 29–30.

Cintron, G., Lugo, A.E., Marinez, R., Cintron, B.B. & Encarnacion, L. 1981. *Impact of oil in the tropical marine environment.* Prepared by Division of Marine Research, Department of Natural Resources, Puerto Rico.

Clarke, P.J. & Ward, T.J. 1994. The response of southern-hemisphere salt-marsh plants and gastropods to experimental contamination by petroleum-hydrocarbons. *Journal of Experimental Marine Biology and Ecology* **175**, 43–57.

Cohen, Y., Nissenbaum, A. & Eisler, R. 1977. Effects of Iranian crude oil on the Red Sea octocoral *Heteroxenia fuscescens. Environmental Pollution* **12**, 173–186.

Cole, A.J., Pratchett, M.S. & Jones, G.P. 2008. Diversity and functional importance of coral-feeding fishes on tropical coral reefs. *Fish and Fisheries* **9**, 286–307.

Cook, C.B. & Knap, A.H. 1983. Effects of crude oil and chemical dispersant on photosynthesis in the brain coral *Diploria strigosa. Marine Biology* **78**, 21–27.

Conan, G. 1982. The long-term effects of the Amoco Cadiz oil-spill. *Philosophical Transactions of the Royal Society of London Series B-Biological Sciences* **297**, 323–333.

Connolly, R.M. & Jones, G.K. 1996. Determining effects of an oil spill on fish communities in a mangrove-seagrass ecosystem in southern Australia. *Australasian Journal of Ecotoxicology* **2**, 3–15.

Crapp, G.B. 1971. Chronic oil pollution. In *Symposium on the Ecological Effects of Oil Pollution*. Institute of Petroleum, London, England.

Currier, H. & Peoples, S. 1954. Phytotoxicity of hydrocarbons. *Hilgardia* **23**(6), 155–173.

Dakin, W.J. 1960. *Australian Seashores: A Guide for Beach-Lover, the Naturalist, the Shore Fisherman, and the Students*. Sydney, Australia: Angus & Robertson. Second edition.

Dauvin, J.C. 1987. Evolution à long terme (1978–1986) des populations d'amphipodes des sables fins de la pierre noire (baie de morlaix, manche occidentale) aprs`s la catastrophe de l'Amoco Cadiz. *Marine Environmental Research* **21**, 247–273.

Dauvin, J.C. & Gentil, F. 1990. Conditions of the peracarid populations of subtidal communities in northern Brittany ten years after the Amoco Cadiz oil spill. *Marine Pollution Bulletin* **21**, 123–130.

Davies, M.S. & Hawkins, S.J. 1998. Mucus from marine molluscs. *Advances in Marine Biology* **34**, 1–71.

Dean, T.A., Stekoll, M.S., Jewett, S.C., Smith, R.O. & Hose, J.E. 1998. Eelgrass (*Zostera marina L.*) in Prince William Sound, Alaska: effects of the Exxon Valdez oil spill. *Marine Pollution Bulletin* **36**, 201–210.

de la Huz, R., Lastra, M., Junoy, J., Castellanos, C. & Viéitez, J.M. 2005. Biological impacts of oil pollution and cleaning in the intertidal zone of exposed sandy beaches: preliminary study of the 'Prestige' oil spill. *Estuarine Coastal Shelf Science* **65**, 19–29.

den Hartog, C. 1986. Effects of oil pollution on seagrass beds. In *Proceedings of the First International Conference on the Impact of Oil Spill in the Persian Gulf*, 20–27 May 1984, University of Tehran, Iran, 255–268.

den Hartog, C. & Jacobs, R.P.W.M. 1980. Effects of the Amoco Cadiz oil spill on an eelgrass community at Roscoff (France) with special reference to the mobile benthic fauna. *Helgolaender Meeresuntersuchungen* **33**, 182–191.

Department of Mines and Energy. 1999. *Explosives and Dangerous Goods Act 1961: Summary of Accident Reports 1998*. Perth, Australia: Government of Western Australia.

Dexter, D.M. 1984. Temporal and spatial variability in the community structure of the fauna of four sandy beaches in south-eastern New South Wales. *Marine and Freshwater Research* **35**, 663–672.

Dodge, R.E., Knap, A.H., Wyers, S.C., Frith, H.R., Sleeter, T.D. & Smith, S.R. 1985. The effect of dispersed oil on the calcification rate of the reef-building coral Diploria strigosa. *Proceedings of the Fifth International Coral Reef Symposium*, Tahiti 6, 453–457.

Dow, R.L. 1978. Size-selective mortalities of clams in an oil-spill site. *Marine Pollution Bulletin* **9**, 45–48.

Downs, C.A., Richmond, R.H., Mendiola, W.J., Rougee, L. & Ostrander, G.K. 2006. Cellular physiological effects of the MV Kyowa Violet fuel-oil spill on the hard coral, *Porites lobata*. *Environmental Toxicology and Chemistry* **25**, 3171–3180.

Driskell, W.B., Ruesink, J.L., Lees, D.C., Houghton, J.P. & Lindstrom, S.C. 2001. Long-term signal of disturbance: *Fucus gardneri* after the Exxon Valdez oil spill. *Ecological Applications* **11**, 815–827.

Duke, N.C. & Burns, K.A. 1999. *Fate and effects of oil and dispersed oil on mangrove ecosystems in Australia*. Final Report to the Australian Petroleum Production and Exploration Association. Australian Institute of Marine Science, Townsville, Australia.

Durako, M.J., Kenworthy, W.J., Fatemy, S.M.R., Valavi, H. & Thayer, G.W. 1993. Assessment of the toxicity of Kuwait crude-oil on the photosynthesis and respiration of seagrasses of the Northern Gulf. *Marine Pollution Bulletin* **27**, 223–227.

Dutta, T.K. & Harayama, S. 2000. Fate of crude oil by the combination of photooxidation and biodegradation. *Environmental Science and Technology* **34**, 1500–1505.

Edgar, G.J. 2012. *Australian Marine Life: The Plants and Animals of Temperate Waters*. Sydney, Australia: New Holland.

Edgar, G.J. & Barrett, N.S. 2000. Impact of the Iron Baron oil spill on subtidal reef assemblages in Tasmania. *Marine Pollution Bulletin* **40**, 36–49.

Edgar, G.J., Bates, A.E., Bird, T.J., Jones, A.H., Kininmonth, S., Stuart-Smith, R.D. & Webb, T.J. 2016. New approaches to marine conservation through the scaling up of ecological data. *Annual Review of Marine Science* **8**, 435–461.

Edgar, G.J., Kerrison, L., Shepherd, S.A. & Toral-Granda, M.V. 2003. Impacts of the *Jessica* oil spill on intertidal and shallow subtidal plants and animals. *Marine Pollution Bulletin* **47**, 276–283.

Edgar, G.J. & Stuart-Smith, R.D. 2014. Systematic global assessment of reef fish communities by the Reef Life Survey program. *Scientific Data* **1**, 140007:1–8.

Egres, A.G., Martins, C.C., de Oliveira, V.M. & de Cunha Lana, P. 2012. Effects of an experimental in situ diesel oil spill on the benthic community of unvegetated tidal flats in a subtropical estuary (Paranagua Bay, Brazil). *Marine Pollution Bulletin* **64**, 2681–2691.

Ehrhardt, M.G. & Burns, K.A. 1993. Hydrocarbons and related photo-oxidation products in Saudi Arabian Gulf coastal waters and hydrocarbons in underlying sediments and bioindicator bivalves. *Marine Pollution Bulletin* **27**, 187–197.

Elmgren R., Hansson S., Larsson U., Sundelin B. & Boehm P.D. 1983. The "Tsesis" oil-spill—acute and long-term impact on the benthos. *Marine Biology* **73**, 51–65.

Epstein, N., Bak, R.P.M. & Rinkevich, J. 2000. Toxicity of third generation dispersants and dispersed Egyptian crude oil on Red Sea coral larvae. *Marine Pollution Bulletin* **40**, 497–503.

Erftemeijer, P.L.A., Riegl, B., Hoeksema, B.W. & Todd, P.A. 2012. Environmental impacts of dredging and other sediment disturbances on corals: a review. *Marine Pollution Bulletin* **64**, 1737–1765.

Felder, D.L., Thoma, B.P., Schmidt, W.E., Sauvage, T., Self-Krayesky, S.L., Chistoserdov, A., Bracken-Grissom, H.D. & Fredericq, S. 2014. Seaweeds and decapod crustaceans on Gulf Deep Banks after the Macondo oil spill. *Bioscience* **64**, 808–819.

Fernandez-Tajes, J., Arias-Perez, A., Fernandez-Moreno, M. & Mendez, J. 2012. Sharp decrease of genetic variation in two Spanish localities of razor clam *Ensis siliqua*: natural fluctuation or Prestige oil spill effects? *Ecotoxicology* **21**, 225–233.

Ferrando, A., Gonzalez, E., Franco, M., Commendatore, M., Nievas, M., Militon, C., Stora, G., Gilbert, F., Esteves, J.L. & Cuny, P. 2015. Oil spill effects on macrofaunal communities and bioturbation of pristine marine sediments (Caleta Valdes, Patagonia, Argentina): experimental evidence of low resistance capacities of benthic systems without history of pollution. *Environmental Science Pollution Research* **22**, 15294–15306.

Finlayson, K., Stevens, T., Arthur, J.M. & Rissik, D. 2015. Recovery of a subtropical rocky shore is not yet complete, four years after a moderate sized oil spill. *Marine Pollution Bulletin* **93**, 27–36.

Fisher, W.S. & Foss, S.S. 1993. A simple test for toxicity of number 2 fuel oil and oil dispersants to embryos of grass shrimp, *Palaemonetes pugio*. *Marine Pollution Bulletin* **26**, 385–391.

Floc'h, J.Y. & Diouris, M. 1980. Initial effects of the Amoco Cadiz oil on intertidal algae. *Ambio* **9**, 284–286.

Fonseca, M., Piniak, G.A. & Cosentino-Manning, N. 2017. Susceptibility of seagrass to oil spills: a case study with eelgrass, *Zostera marina* in San Francisco Bay, USA. *Marine Pollution Bulletin* **115**, 29–38.

Foster, M.S., Neushul, M. & Zingmark, R. 1971. The Santa Barbara oil spill. Part 2. Initial effects on intertidal and kelp bed organisms. *Environmental Pollution* **2**, 115–134.

Fromont, J., Althaus, F., McEnnulty, F.R., Williams, A., Salotti, M., Gomez, O. & Gowlett-Holmes, K. 2012. Living on the edge: the sponge fauna of Australia's southwestern and northwestern deep continental margin. *Hydrobiologia* **687**, 127–142.

Fucik, K.W., Bright, T.J. & Goodman, K.S. 1984. Measurements of damage, recovery, and rehabilitation of coral reefs exposed to oil. In *Restoration of Habitats Impacted by Oil Spills*, J. Cairns & A.L. Buikema (eds). London: Butterworth, 182 pp.

Fulford, R.S., Griffitt, R.J., Brown-Peterson, N.J., Perry, H. & Sanchez-Rubio, G. 2014. Impacts of the Deepwater Horizon Oil Spill on blue crab, *Callinectes sapidus*, larval settlement in Mississippi. In *Impacts of Oil Spill Disasters on Marine Fisheries in North America*, B.A. Alford et al. (eds). Boca Raton, FL: CRC Press, 247–261.

Gales, N.J. 1991. *New Zealand fur seals and oil: an overview of assessment, treatment, toxic effects and survivorship*. The 1991 Sanko Harvest Oil Spill. Unpublished Report. Western Australian Department of Conservation and Land Management, Perth, Australia.

Garrity, S.D. & Levings, S.C. 1990. Effects of an oil spill on the gastropods of a tropical intertidal reef flat. *Marine Environmental Research* **30**, 119–153.

Georgiades, E.T., Holdway, D.A., Brennan, S.E., Butty, J.S. & Temara, A. 2003. The impact of oil-derived products on the behaviour and biochemistry of the eleven-armed asteroid *Coscinasterias muricata* (Echinodermata). *Marine Environmental Research* **55**, 257–276.

Getter, C.D., Ballou, T.G. & Koons, C.B. 1985. Effects of dispersed oil on mangroves: synthesis of a seven-year study. *Marine Pollution Bulletin* **16**, 318–324.

Giese, M., Goldsworthy, S.D., Gales, R., Brothers, N. & Hamill, J. 2000. Effects of the Iron Baron oil spill on little penguins (*Eudyptula minor*). III. Breeding success of rehabilitated oiled birds. *Wildlife Research* **27**, 583–591.

Gilfillan, E.S., Page, D.S., Bass, A.E., Foster, J.C., Fickert, P.M., Ellis, W.G., Rusk, S. & Brown, C. 1989. Use of Na/K ratios in leaf tissues to determine effects of petroleum on salt exclusion in marine halophytes. *Marine Pollution Bulletin* **20**, 272–276.

Gilfillan, E.S. & Vandermeulen, J.H. 1978. Alterations in growth and physiology of soft-shell clams, *Mya arenaria*, chronically oiled with Bunker C from Chedabucto Bay, Nova Scotia, 1970–1976. *Journal of the Fisheries Research Board of Canada* **35**, 630–636.

Gilmour, J.P., Smith, L.D. & Brinkman, R.M. 2009. Biannual spawning, rapid larval development and evidence of self-seeding for scleractinian corals at an isolated system of reefs. *Marine Biology* **156**, 1297–1309.

Goldsworthy, S.D., Gales, R.P., Giese, M. & Brothers, N. 2000a. Effects of the Iron Baron oil spill on little penguins (*Eudyptula minor*). I. Estimates of mortality. *Wildlife Research* **27**, 559–571.

Goldsworthy, S.D., Giese, M., Gales, R.P., Brothers, N. & Hamill, J. 2000b. Effects of the Iron Baron oil spill on little penguins (*Eudyptula minor*). II. Post-release survival of rehabilitated oiled birds. *Wildlife Research* **27**, 573–582.

Goodbody-Gringley, G., Wetzel, D.L., Gillon, D., Pulster, E., Miller, A. & Ritchie, K.B. 2013. Toxicity of Deepwater Horizon source oil and the chemical dispersant, Corexit (R) 9500, to coral larvae. *PloS One* **8**, e45574.

Great Barrier Reef Marine Park Authority (GBRMPA). 2011. *Grounding of the Shen Neng 1 on Douglas Shoal, April 2010.* Impact Assessment Report. Great Barrier Reef Marine Park Authority, Townsville, Australia.

Green, R.H. 2005. Marine coastal monitoring: designing an effective offshore oil and gas environmental effects monitoring program. In *Proceedings of the Offshore Oil and Gas Environmental Effects Monitoring Workshop: Approaches and Technologies*, S.L. Armsworthy et al. (eds). Bedford Institute of Oceanography, Dartmouth, Nova Scotia, May 26–30, 2003. Columbus, OH: Battelle Press, 373–397.

Guidetti, P., Modena, M., La Mesa, G. & Vacchi, M. 2000. Composition, abundance and stratification of macrobenthos in the marine area impacted by tar aggregates derived from the Haven oil spill (Ligurian Sea, Italy). *Marine Pollution Bulletin* **40**, 1161–1166.

Gulec, I. & Holdway, D.A. 1999. The toxicity of laboratory burned oil to the amphipod *Allorchestes compressa* and the snail *Polinices conicus*. *Spill Science & Technology Bulletin* **5**, 135–139.

Gulec, I., Leonard, B. & Holdway, D.A. 1997. Oil and dispersed oil toxicity to amphipods and snails. *Spill Science & Technology Bulletin* **4**, 1–6.

Guzmán, H.M., Burns, K.A. & Jackson, J.B.C. 1994. Injury, regeneration and growth of Caribbean reef corals after a major oil spill in Panama. *Marine Ecology Progress Series* **105**, 231–241.

Guzmán, H.M. & Holst, I. 1993. Effects of chronic oil sediment pollution on the reproduction of the Caribbean Reef coral *Siderastrea siderea*. *Marine Pollution Bulletin* **26**, 276–282.

Guzmán, H.M., Jackson, J.B.C. & Weil, E. 1991. Short term ecological consequences of a major oil spill on Panamanian subtidal reef corals. *Coral Reefs* **10**, 1–12.

Haapkylä, J., Ramade, F. & Salvat, B. 2007. Oil pollution on coral reefs: a review of the state of knowledge and management needs. *Vie Et Milieu-Life and Environment* **57**, 95–111.

Hampson, G.R. & Moul, E.T. 1978. No. 2 fuel oil spill in Bourne, Massachusetts—immediate assessment of effects on marine invertebrates and a 3-year study of growth and recovery of a salt marsh. *Journal of the Fisheries Research Board of Canada* **35**, 731–744.

Harrison, P.L. 1999. Oil pollutants inhibit fertilization and larval settlement in the scleractinian reef coral *Acropora tenuis* from the Great Barrier Reef, Australia. Sources, fates and consequences of pollutants in the Great Barrier Reef and Torres Strait, Great Barrier Reef Marine Park Authority, 11–12.

Harrison, P.L., Babcock, R.C., Oliver, J.K., Bull, G.D., Wallace, C.C. & Willis, B.L. 1984. Mass spawning in the tropical reef corals. *Science* **223**, 1186–1189.

Harrison, P.L., Collins, J.C., Alexander, C.G. & Harrison, B.L. 1990. The effects of fuel oil and dispersant on the tissues of a staghorn coral *Acropora formosa*: a pilot study. In *Proceedings of the Second National Workshop on the Role of Scientific Support Coordinator*, Hastings, Victoria, 51–61.

Harrison, P.L. & Wallace, C.C. 1990. Reproduction, dispersal and recruitment of scleractinian corals. In *Ecosystems of the World 25: Coral Reefs*, Z. Dubinsky (ed.). Amsterdam: Elsevier, 133–207.

Harvey, J.S., Lyons, B.P., Page, T.S., Stewart, C. & Parry, J.M. 1999. An assessment of the genotoxic impact of the Sea Empress oil spill by the measurement of DNA adduct levels in selected invertebrate and vertebrate species. *Mutation Research/Genetic Toxicology Environment Mutagenesis* **441**, 103–114.

Harwell, M.A. & Gentile, J.H. 2006. Ecological significance of residual exposures and effects from the Exxon Valdez oil spill. *Integrated Environmental Assessment Management* **2**, 204–246.

Hatcher, A.I. & Larkum, A.W.D. 1982. The effects of short term exposure to Bass Strait crude oil and Corexit 8667 on benthic community metabolism in *Posidonia australis* Hook. f. dominated microcosms. *Aquatic Botany* **12**, 219–227.

Hawkins, S.J., Evans, A.J., Mieszkowska, N., Adams, L.C., Bray, S., Burrows, M.T., Firth, L.B., Genner, M.J., Leung, K.M.Y., Moore, P.J., Pack, K., Schuster, H., Sims, D.W., Whittington, M. & Southward, E.C. 2017a. Distinguishing globally-driven changes from regional-and local-scale impacts: the case for long-term and broad-scale studies of recovery from pollution. *Marine Pollution Bulletin*, http://dx.doi.org/10.1016/j.marpolbul.2017.01.068

Hawkins, S.J., Evans, A.J., Moore, J., Whittington, M., Pack, K., Firth, L.B., Adams, L.C., Moore, P.J., Masterson-Algar, P., Mieszkowska, N. & Southward, E.C. 2017b. From the Torrey Canyon to today: 50 year retrospective from the oil spill and interaction with climate driven fluctuations on Cornish rocky shores. In *International Oil Spill Conference Proceedings* **2017**, 74–103.

Hawkins, S.J. & Southward, A.J. 1992. The Torrey Canyon oil spill: recovery of rocky shore communities. Chapter 13 In *Restoring the Nation's Marine Environment. NOAA Symposium on Habitat Restoration*, Washington, DC, 1990, G.W. Thayer (ed.). Maryland Sea Grant College, 473–542.

Hayes, M.O., Michel, J., Montello, T.M., Aurand, D.V., Almansi, A.M., Almoamen, A.H., Sauer, T.C. & Thayer, G.W. 1993. Distribution and weathering of shoreline oil one year after the Gulf War oil spill. *Marine Pollution Bulletin* **27**, 135–142.

Heyward, A. 2010. *Environmental Study S6.4 Corals*. Montara Surveys: Final Report on Benthic Surveys at Ashmore, Cartier and Seringapatam Reefs. Prepared for PTTEP Australasia (Ashmore Cartier) Pty Ltd by Australian Institute of Marine Science, Perth, Western Australia.

Heyward, A. 2011. *Monitoring Study S6b Corals reefs, Montara: 2011 shallow reef surveys at Ashmore, Cartier and Seringapatam reefs*. Prepared for PTTEP Australasia (Ashmore Cartier) Pty Ltd by Australian Institute of Marine Science, Perth, Western Australia.

Heyward, A., Jones, R., Travers, M., Burns, K., Suosaari, G., Colquhoun, J., Case, M., Radford, B., Meekan, M., Markey, K., Schenk, T., O'Leary, R.A., Brooks, K., Tinkler, P., Cooper, T. & Emslie, M. 2012. *Montara: 2011 Shallow reef surveys at Ashmore, Cartier and Seringapatam reefs final report*. Prepared for PTTEP Australasia (Ashmore Cartier) Pty Ltd by Australian Institute of Marine Science, Perth, April 2012.

Heyward, A., Moore, C., Radford, B. & Colquhoun, J. 2010. *Environmental Study S5*. Monitoring Program for the Montara Well Release Timor Sea: Final Report on the Nature of Barracouta and Vulcan Shoals. Prepared for PTTEP Australasia (Ashmore Cartier) Pty Ltd by Australian Institute of Marine Science, Perth, Western Australia.

Heyward, A., Speed, C.W., Meekan, M., Cappo, M., Case, M., Colquhoun, J., Fisher, R., Meeuwig, J. & Radford, B. 2013. Montara: Vulcan, Barracouta East and Goeree Shoals Survey 2013. Prepared for PTTEP Australasia (Ashmore Cartier) Pty Ltd by Australian Institute of Marine Science, August 2013.

Hodgins, H.O.M., Bruce, B., Hawkes & Joyce, W. 1977. Marine fish and invertebrate diseases, host disease resistance, and pathological effects of petroleum. Disease resistance and pathology. In *Effects of Petroleum on Arctic and Subarctic Marine Animals*, D.C. Mallins (ed.). New York: Academic Press, 95–128.

Hoff, R. & Michel, J. 2014. Mangrove case studies. In *Oil Spills in Mangroves: Planning and Response Considerations*. R. Hoff & J. Michel (eds). Seattle, Washington: National Oceanic and Atmospheric Administration, 1–19.

Holmquist, J.G. 1997. Disturbance and gap formation in a marine benthic mosaic: influence of shifting macroalgal patches on seagrass structure and mobile invertebrates. *Marine Ecology Progress Series* **158**, 121–130.

Hook, S., Batley, G., Holloway, M., Ross, A. & Irving, P. (eds). 2016. *Oil Spill Monitoring Handbook*. Melbourne: CSIRO Publishing.

Houghton, J.P., Lees, D.C. & Driskell, W.B. 1993. Evaluation of the condition of Prince William Sound shorelines following the Exxon Valdez oil spill and subsequent shoreline treatment. Prepared for National Oceanic and Atmospheric Administration, Ocean Assessment Division, Washington, DC.

Howard, R.K. & Edgar, G.J. 1994. Seagrass meadows. In *Marine Biology*, L.S. Hammond & R.N. Synnot (eds). Melbourne, Australia: Longman Cheshire.

Hunter, T. 2010. montara oil spill and the national marine oil spill contingency plan: disaster response or just a disaster. *Australian and New Zealand Maritime Law Journal* **24**, 46–58.

Huisman, J.M., Leliaert, F., Verbruggen, H. & Townsend, R.A. 2009. Marine benthic plants of Western Australia's shelf-edge atolls. *Records of the Western Australian Museum Supplement* **77**, 50–87.

IPIECA-IOGP. 2015. *Impacts of oil spills on marine ecology: good practice guidelines for incident management and emergency response personnel*. IOGP Report 525. International Association of Oil & Gas Producers, London, United Kingdom.

IPIECA-IOGP. 2016. *Impacts of oil spills on shorelines: good practice guidelines for incident management and emergency response personnel*. IOGP Report 534. International Association of Oil & Gas Producers, London, United Kingdom.

Jackson, J.B.C., Cubit, J.D., Keller, B.D., Batista, V., Burns, K., Caffey, H.M., Caldwell, R.L., Garrity, S.D., Getter, C.D., Gonzalez, C., Guzmán, H.M., Kaufmann, K.W., Knap, A.H., Levings, S.C., Marshall, M.J., Steger, R., Thompson, R.C. & Weil, E. 1989. Ecological effects of a major oil spill on Panamanian coastal marine communities. *Science* **243**, 37–44.

Jacobs, R.P.W.M. 1988. Oil and the seagrass ecosystem of the Red Sea. *Oil and Chemical Pollution* **5**, 21–45.

Jernelöv, A. & Linden, O. 1981. Ixtoc I: a case study of the world's largest oil spill. *Ambio* **10**, 299–306.

Jewett, S.C., Dean, T.A., Smith, R.O. & Blanchard, A. 1999. Exxon Valdez oil spill: impacts and recovery in the soft-bottom benthic community in and adjacent to eelgrass beds. *Marine Ecology Progress Series* **185**, 59–83.

Jiang, Z., Huang, Y., Chen, Q., Zeng, J. & Xu, X. 2012. Acute toxicity of crude oil water accommodated fraction on marine copepods: the relative importance of acclimatization temperature and body size. *Marine Environmental Research* **81**, 12–17.

Johannes, R.E., Maragos, J. & Coles, S.L. 1972. Oil damages corals exposed to air. *Marine Pollution Bulletin* **3**, 29–30.

Johnson, F. 1977. Sublethal biological effects of petroleum hydrocarbon exposures: bacteria, algae, and invertebrates. In *Effects of Petroleum on Arctic and Subarctic Marine Environments and Organisms. Vol. 2. Biological Effects*, D.C. Malins (ed.). New York: Academic Press, 271–318.

Jones, A.R. 2003. Ecological recovery of amphipods on sandy beaches following oil pollution: an interim assessment. *Journal of Coastal Research*, Special Issue **35** *Proceedings of the Brazilian Symposium on Sandy Beaches: Morphodynamics, ecology, uses, hazards and management (spring 2003)*, 66–73.

Jones, R.J. & Heyward, A.J. 2003. The effects of Produced Formation Water (PFW) on coral and isolated symbiotic dinoflagellates of coral. *Marine and Freshwater Research* **54**, 153–162.

Karinen, J.F., Rice, S.D. & Babcock, M.M. 1985. *Reproductive success in Dungeness crab (Cancer magister) during long-term exposures to oil contaminated sediments*. Final Report Outer Continental Shelf Environmental Assessment Program. NOAA, National Marine Fisheries Service, Auke Bay, Alaska.

Karydis, M. & Fogg, G.E. 1980. Physiological effects of hydrocarbons on the marine diatom *Cyclotella cryptica*. *Microbial Ecology* **6**, 281–290.

Kauss, P., Hutchinson, T.C., Soto, C., Hellebust, J. & Griffiths, M. 1973. The toxicity of crude oil and its components to freshwater algae. In *International Oil Spill Conference* **1973**(1), 703–714. American Petroleum Institute.

Keesing, J.K. & Irvine, T.R. 2012. Aspects of the biology of an abundant spatangoid urchin, *Breynia desorii* in the Kimberley region of north-western Australia. In *Echinoderms in a Changing World: Proceedings of the 13th International Echinoderm Conference*, January 5–9 2009, C. Johnson (ed.). University of Tasmania, Hobart Tasmania, Australia: CRC Press, 165–174.

Keesing, J.K., Irvine, T.R., Alderslade, P., Clapin, G., Fromont, J., Hosie, A.M., Huisman, J.M., Phillips, J.C., Naughton, K.M., Marsh, L.M., Slack-Smith, S.M., Thomson, D.P. & Watson, J.E. 2011. Marine benthic flora and fauna of Gourdon Bay and the Dampier Peninsula in the Kimberley region of North-Western Australia. *Journal of the Royal Society Western Australia* **94**, 285–301.

Kennish, M.J. 1992. *Ecology of Estuaries*. Florida: CRC Press.

Kenworthy, W.J., Durako, M.J., Fatemy, S.M.R., Valavi, H. & Thayer, G.W. 1993. Ecology of seagrasses in northeastern Saudi Arabia one year after the Gulf War oil spill. *Marine Pollution Bulletin* **27**, 213–222.

Klekowski, E.J., Corredor, J., Morell, J.M. & Del Castillo, C.A. 1994. Petroleum pollution and mutation in mangroves. *Marine Pollution Bulletin* **28**, 166–169.

Knap, A.H., Wyers, S.C., Dodge, R.E., Sleeter, T.D., Frith, H.R., Smith, S.R. & Cook, C.B. 1985. The effects of chemically and physically dispersed oil on the brain coral Diploria strigosa (Dana) – a summary review. In *International Oil Spill Conference* **1985**(1), 547–551. American Petroleum Institute.

Kojis, B.L. & Quinn, N.J. 1982. Reproductive ecology of two faviid corals (Coelenterata: Scleractinia). *Marine Ecology Progress Series* **8**, 251–255.

Kotta, J., Aps, R. & Herkul, K. 2008. Predicting ecological resilience of marine benthic communities facing a high risk of oil spills. In *Environmental Problems in Coastal Regions VII, Book 99*, C.A. Brebbia (ed.). Southampton, UK: Wit Press **99**, 101–110.

Krebs, C.T. & Burns, K. 1977. Long-term effects of an oil spill on populations of the salt-marsh crab *Uca pugnax*. *Science* **197**, 484–487.

Laffon, B., Rabade, T., Pasaro, E. & Mendez, J. 2006. Monitoring of the impact of *Prestige* oil spill on *Mytilus galloprovincialis* from Galician Coast. *Environment International* **32**, 342–348.

Lane, A. & Harrison, P.I. 2000. Effects of oil contaminants on survivorship of larvae of the scleractinian reef corals *Acropora tenuis, Goniastrea aspera* and *Platygyra sinensis* from the Great Barrier Reef. In *Proceedings of 9th International Coral Reef Symposium*, 23–27 October 2000, Bali, Indonesia.

La Peyre, J., Casas, S. & Miles, S. 2014. Oyster responses to the Deepwater Horizon oil spill across coastal Louisiana: examining oyster health and hydrocarbon bioaccumulation. In *Impacts of Oil Spill Disasters on Marine Habitats and Fisheries in North America*, J.B. Alford et al. (eds). Boca Raton, FL: CRC/ Taylor & Francis Group, 269–294.

Larkum, A.W.D. & den Hartog, C. 1989. Evolution and biogeography of seagrasses. In *Biology of Seagrasses: A Treatise on the Biology of Seagrasses with Special Reference to the Australian Region*, A.W.D. Larkum, A.J. McComb & S.A. Shepherd (eds). Amsterdam: Elsevier.

Larkum, A.W.D., McComb, A.J. & Shepherd, S.A. 1989. *Biology of Seagrasses: A Treatise on the Biology of Seagrasses with Special Reference to the Australian Region*. Amsterdam: Elsevier Science Publishers.

Laubier, L. 1980. The amoco cadiz oil spill: an ecological impact study. *Ambio* **9**, 268–276.

Lee, K., Boufadel, M., Chen, B., Foght, J., Hodson, P., Swanson, S. & Venosa, A. 2015. *The Behaviour and Environmental Impacts of Crude Oil Released Into Aqueous Environments*. Ottawa, Canada: The Royal Society of Canada.

Lee, K., Nedwed, T., Prince, R. & Palandro, D. 2013. Lab tests on the biodegradation of chemically dispersed oil should consider the rapid dilution that occurs at sea. *Marine Pollution Bulletin* **73**, 314–318.

Le Hir, M. & Hily, C. 2002. First observations in a high rocky-shore community after the Erika oil spill (December 1999, Brittany, France). *Marine Pollution Bulletin* **44**, 1243–1252.

Lewis, M. & Pryor, R. 2013. Toxicities of oils, dispersants and dispersed oils to algae and aquatic plants: review and database value to resource sustainability. *Environmental Pollution* **180**, 345–367.

Lin, Q., Mendelssohn, I.A., Suidan, M.T., Lee, K. & Venosa, A.D. 2002. The dose response relationship between No. 2 fuel oil and the growth of the salt marsh grass, *Spartina alterniflora*. *Marine Pollution Bulletin* **44**, 897–902.

Linden, O. 1976. Effects of oil on the reproduction of the amphipod *Gammarus oceanicus*. *Ambio* **5**, 36–37.

Lipscombe, R. 2000. Australia's tyranny of distance in oil spill response. *Spill Science & Technology Bulletin* **6**, 13–25.

Logan, G.A., Jones, A.T., Ryan, G.J., Wettle, M., Thankappan, M., Groesjean, V., Rollet, N. & Kennard, J.M. 2008. *Review of Australian offshore natural hydrocarbon seepage studies*. Geoscience Australia, Report No. 2008/17, Canberra, ACT, Australia.

Long, S.M. & Holdway, D.A. 2002. Acute toxicity of crude and dispersed oil to *Octopus pallidus* (Hoyle, 1885) hatchlings. *Water Research* **36**, 2769–2776.

Lopes, C.F., Milanelli, J.C.C., Prosperi, V.A., Zanardi, E. & Truzzi, A.C. 1997. Coastal monitoring program of Sao Sebastiao Channel: assessing the effects of 'Tebar V' oil spill on rocky shore populations. *Marine Pollution Bulletin* **34**, 923–927.

Lougheed, L.W., Edgar, G.J. & Snell, H.L. 2002. *Biological impacts of the Jessica oil spill on the Galápagos environment*. Final Report. Charles Darwin Foundation, Puerto Ayora, Galápagos, Ecuador.

Loya, Y. & Rinkevich, B. 1979. Abortion effect in corals induced by oil pollution. *Marine Ecology Progress Series* **1**, 77–80.

Loya, Y. & Rinkevich, B. 1980. Effects of oil pollution on coral-reef communities. *Marine Ecology Progress Series* **3**, 167–180.

MacFarlane, G.R. & Burchett, M.D. 2003. Assessing effects of petroleum oil on intertidal invertebrate communities in Sydney Harbour: preparedness pays off. *Australasian Journal of Ecotoxicology* **9**, 29–38.

Macinnis-Ng, C.M.O. & Ralph, P.J. 2003. In situ impact of petrochemicals on the photosynthesis of the seagrass *Zostera capricorni*. *Marine Pollution Bulletin* **46**, 1395–1407.

MacNaughton, S.J., Stephen, J.R., Venosa, A.D., Davis, G.A., Chang, Y.J. & White, D.C. 1999. Microbial population changes during bioremediation of an experimental oil spill. *Applied and Environmental Microbiology* **65**, 3566–3574.

Mageau, C., Engelhardt, F.R., Gilfillan, E.S. & Boehm, P.D. 1987. Effects of short-term exposure to dispersed oil in arctic invertebrates. *Arctic* **40**, 162–171.

Manuell, R.W. 1979. Oil spill prevention and control in Australia. In *International Oil Spill Conference. American Petroleum Institute*, Los Angles **1**, 293–297.

Marshall, P.A. & Edgar, G.J. 2003. The effect of the Jessica grounding on subtidal invertebrate and plant communities at the Galápagos wreck site. *Marine Pollution Bulletin* **47**, 284–295.

Marwood, C.A., Smith, R.E.H., Solomon, K.R., Charlton, M.N. & Greenberg, B.M. 1999. Intact and photomodified polycyclic aromatic hydrocarbons inhibit photosynthesis in natural assemblages of Lake Erie phytoplankton exposed to solar radiation. *Ecotoxicology and Environmental Safety* **44**, 322–327.

Marwood, C.A., Solomon, K.R. & Greenberg, B.M. 2001. Chlorophyll fluorescence as a bioindicator of effects on growth in aquatic macrophytes from mixtures of polycyclic aromatic hydrocarbons. *Environmental Toxicology and Chemistry* **20**, 890–898.

Masini, R.J. & Manning, C.R. 1997. The photosynthetic responses to irradiance and temperature of four meadow-forming seagrasses. *Aquatic Botany* **58**, 21–36.

Mason, C. 2011. *Submission by the West Timor Care Foundation (WTCF), Kupang (West Timor), Republic of Indonesia, to the Draft Government Response to the Report of the montara Commission of Inquiry.* Prepared for WTCF, Kupang, West Timor.

Maxwell, K. & Johnson, G.N. 2000. Chlorophyll fluorescence—a practical guide. *Journal of Experimental Botany* **51**, 659–668.

McGuinness, K.A. 1990. Effects of oil spills on macro-invertebrates of saltmarshes and mangrove forests in Botany Bay, New South Wales, Australia. *Journal of Experimental Marine Biology and Ecology* **142**, 121–135.

McMahon, K., Lavery, P. & Mulligan, M. 2011. Recovery from the impact of light reduction on the seagrass *Amphibolis griffithii*, insights for dredging management. *Marine Pollution Bulletin* **62**, 270–283.

Melville, F., Andersen, L.E. & Jolley, D.F. 2009. The Gladstone (Australia) oil spill—impacts on intertidal areas: baseline and six months post-spill. *Marine Pollution Bulletin* **58**, 263–271.

Mitchell, C.T., Anderson, E.K., Jones, L.G. & North, W.J. 1970. What oil does to ecology. *Journal Water Pollution Control Federation* **42**, 812–818.

Montagna, P.A., Baguley, J.G., Cooksey, C., Hartwell, I., Hyde, L.J., Hyland, J.L., Kalke, R.D., Kracker, L.M., Reuscher, M. & Rhodes, A.C. 2013. Deep-sea benthic footprint of the Deepwater Horizon blowout. *PLoS One* **8**, e70540.

Moore, J. 1997. *Rocky shore transect monitoring in Milford Haven, October 1996. Impacts of the Sea Empress oil spill.* A Report to the Countryside Council for Wales from OPRU, Neyland, Pembrokeshire. Report No. OPRU/12/97. 30 pp. plus appendices.

Moore, J. 2006. Long term ecological impacts of marine oil spills. *Proceedings of the Interspill Conference. March 2006*, London, 21–23.

Nadeau, R.J. & Berquist, E.T. 1977. Effects of the March 18, 1973, oil spill near Cabo Rojo, Puerto Rico on tropical marine communities. In *Proceedings, International Oil Spill Conference*, Vol **1977**(1), 535–538. American Petroleum Institute, Washington, DC.

Nagelkerken, I.A. & Debrot, A.O. 1995. Mollusc communities of tropical rubble shores of Curaçao: long-term (7+ years) impacts of oil pollution. *Marine Pollution Bulletin* **30**, 592–598.

Naidoo, G.N., Naidoo, Y. & Achar, P. 2010. Response of the mangroves *Avicennia marina* and *Bruguiera gymnorrhiza* to oil contamination. *Flora* **205**, 357–362.

National Offshore Petroleum Safety and Environmental Management Authority (NOPSEMA). 2017. https://www.nopsema.gov.au/news-and-media/hydrocarbon-sheen-in-the-bass-strait-west-tuna-platform/

National Research Council. 2003. Oil in the Sea III: Inputs, Fates, and Effects. National Research Council (US). *Prepared by Committee on Oil in the Sea: Inputs, Fates, and Effects*, Washington, DC: National Academies Press.

Navas, J.M., Babín, M., Casado, S., Fernández, C. & Tarazona, J.V. 2006. The Prestige oil spill: a laboratory study about the toxicity of the water-soluble fraction of the fuel oil. *Marine Environmental Research* **62**, S352–S355.

Neff, J. 2002. *Bioaccumulation in Marine Organisms—Effect of Contaminants from Oil Well Produced Water.* Amsterdam: Elsevier.

Neff, J.M., Ostazeski, S., Gardiner, W. & Stejskal, I. 2000. Effects of weathering on the toxicity of three offshore Australian crude oils and a diesel fuel to marine animals. *Environmental Toxicology and Chemistry* **19**, 1809–1821.

Negri, A.P., Brinkman, D.L., Flores, F., Botté, E.S., Jones, R.J. & Webster, N.S. 2015. Acute ecotoxicology of natural oil and gas condensate to coral reef larvae. *Scientific Reports* **6**, 21153.

Negri, A.P. & Heyward, A.J. 2000. Inhibition of fertilization and larval metamorphosis of the coral *Acropora millepora* (Ehrenberg, 1834) by petroleum products. *Marine Pollution Bulletin* **41**, 420–427.

Nelson, P. 2000. Australia's national plan to combat pollution of the sea by oil and other noxious and hazardous substances–overview and current issues. *Spill Science & Technology Bulletin* **6**, 3–11.

Nelson-Smith, A. 1973. *Oil Pollution and Marine Ecology.* New York: Plenum Press.

Newey, S. & Seed, R. 1995. The effects of the Braer oil-spill on rocky intertidal communities in South Shetland, Scotland. *Marine Pollution Bulletin* **30**, 274–280.

Nicol, J.A.C., Donahue, W.H., Wang, R.T. & Winters, K. 1977. Chemical composition and effects of water extracts of petroleum on eggs of the sand dollar *Mellita quinquiesperforata*. *Marine Biology* **40**, 309–316.

North, J.W., Neushul, M.J. & Clendenning, K.A. 1965. Successive biological changes observed in a marine cove exposed to a large spillage of oil. In *Pollutions Marines par les Microorganismes et les Produits Pétroliers (Symposium de Monaco, Avril 1964)*. Paris: Commission Internationale Pour l'Exploration Scientifique de la Mer Méditerranée, 335–354.

Nyman, J.A. & Green, C.G. 2015. A brief review of the effects of oil and dispersed oil on coastal wetlands including suggestions for future research. In *Impacts of Oil Spill Disasters on Marine Habitats and Fisheries in North America*, J.B. Alford et al. (eds). Boca Raton, FL: CRC/Taylor & Francis Group.

O'Brien, P. & Dixon, P. 1976. The effects of oils and oil components on algae: a review. *British Phycological Journal* **11**, 115–141.

O'Clair, C.E. & Rice, S.D. 1985. Depression of feeding and growth-rates of the seastar *Evasterias troschelii* during long-term exposure to the water-soluble faction of crude-oil. *Marine Biology* **84**, 331–340.

Orth, R.J., Luckenbach, M. & Moore, K.A. 1994. Seed dispersal in a marine macrophyte: implications for colonization and restoration. *Ecology* **75**, 1927–1939.

O'Sullivan, A.J. 1978. The Amoco Cadiz oil-spill. *Marine Pollution Bulletin* **9**, 123–128.

Payne, J.R., Driskell, W.B., Short, J.W. & Larsen, M.L. 2008. Long term monitoring for oil in the Exxon Valdez spill region. *Marine Pollution Bulletin* **56**, 2067–2081.

Peckol, P., Levings, S.C. & Garrity, S.D. 1990. Kelp response following the World Prodigy oil spill. *Marine Pollution Bulletin* **21**, 473–476.

Pelletier, M.C., Burgess, R.M., Ho, K.T., Kuhn, A., McKinney, R.A. & Ryba, S.A. 1997. Phototoxicity of individual polycyclic aromatic hydrocarbons and petroleum to marine invertebrate larvae and juveniles. *Environmental Toxicology and Chemistry* **16**, 2190–2199.

Penela-Arenaz, M., Bellas, J. & Vazquez, E. 2009. Effects of the Prestige oil spill on the biota of NW Spain: 5 years of learning. *Advanced Marine Biology* **56**, 365–396.

Percy, J.A. & Mullin, T.C. 1977. Effects of crude oil on the locomotory activity of Arctic marine invertebrates. *Marine Pollution Bulletin* **8**, 35–40.

Perez-Cadahia, B., Laffon, B., Pasaro, E. & Mendez, J. 2004. Evaluation of PAH bioaccumulation and DNA damage in mussels (*Mytilus galloprovincialis*) exposed to spilled Prestige crude oil. *Comparative Biochemistry and Physiology Part C Toxicology & Pharmacology* **138**, 453–460.

Peters, E.C., Meyers, P.A., Yevich, P.P. & Blake, N.J. 1981. Bioaccumulation and histopathological effects of oil on a stony coral. *Marine Pollution Bulletin* **12**, 333–339.

Peterson, B.J., Rose, C.D., Rutten, L.M. & Fourqurean, J.W. 2002. Disturbance and recovery following catastrophic grazing: studies of a successional chronosequence in a seagrass bed. *Oikos* **97**, 361–370.

Peterson, C.H. 2001. The "Exxon Valdez" oil spill in Alaska: acute, indirect and chronic effects on the ecosystem. *Advances in Marine Biology* **39**, 1–103.

Peterson, C.H., McDonald, L.L., Green, R.H. & Erickson, W.P. 2001. Sampling design begets conclusions: the statistical basis for detection of injury to and recovery of shoreline communities after the 'Exxon Valdez' oil spill. *Marine Ecology Progress Series* **210**, 255–283.

Peterson, C.H., Rice, S.D., Short, J.W., Esler, D., Bodkin, J.L., Ballachey, B.E. & Irons, D.B. 2003. Long-term ecosystem response to the Exxon Valdez oil spill. *Science* **302**, 2082–2086.

Pineira, J., Quesada, H., Rolan-Alvarez, E. & Caballero, A. 2008. Genetic impact of the Prestige oil spill in wild populations of a poor dispersal marine snail from intertidal rocky shores. *Marine Pollution Bulletin* **56**, 270–281.

Pople, A., Simpson, R.D. & Cairns, S.C. 1990. An incident of Southern Ocean oil pollution: effects of a spillage of diesel fuel on the rocky shore of Macquarie-Island (Sub-Antarctic). *Australian Journal of Marine and Freshwater Research* **41**, 603–620.

Powers, S.P., Hernandez, F.J., Condon, R.H., Drymon, J.M. & Free, C.M. 2013. Novel pathways for injury from offshore oil spills: direct, sublethal and indirect effects of the Deepwater Horizon oil spill on pelagic *Sargassum* communities. *PLoS ONE* **8**, e74802.

Premila, V.E. & Rao, M.U. 1997. Effect of crude oil on the growth and reproduction of some benthic marine algae of Visakhapatnam coastline. *Indian Journal of Marine Science* **26**, 195–200.

PTTEP Australasia. 2009. *Monitoring plan for the Montara well release Timor Sea—as agreed between PTTEP Australasia (Ashmore Cartier) Pty. Ltd. and the Department of the Environment, Water, Heritage and the Arts.* Report No. PTTEPAA 04, Victoria, Australia, October 2009.

PTTEP Australasia. 2013. *Montara environmental monitoring program: report of research.* A New Body of World Class Research on the Timor Sea. West Perth, Western Australia.

Raffaelli, D. 1987. The behaviour of the nematode/copepod ratio in organic pollution studies. *Marine Environmental Research* **23**, 135–152.

Raffaelli, D.G. & Mason, C.F. 1981. Pollution monitoring with meiofauna, using the ratio of nematodes to copepods. *Marine Pollution Bulletin* **12**, 158–163.

Ralph, P.J. & Burchett, M.D. 1998. Impact of petrochemicals on the photosynthesis of *Halophila ovalis* using chlorophyll fluorescence. *Marine Pollution Bulletin* **36**, 429–436.

Reddy, C.M., Eglinton, T.I., Hounshell, A., White, H.K., Xu, L., Gaines, R.B. & Frysinger, G.S. 2002. The West Falmouth oil spill after thirty years: the persistence of petroleum hydrocarbons in marsh sediments. *Environmental Science & Technology* **36**, 4754–4760.

Redman, A.D. 2015. Role of entrained droplet oil on the bioavailability of petroleum substances in aqueous exposures. *Marine Pollution Bulletin* **97**, 342–348.

Redman, A.D. & Parkerton, T.F. 2015. Guidance for improving comparability and relevance of oil toxicity tests. *Marine Pollution Bulletin* **98**, 156–170.

Reid, D.J. & MacFarlane, G.R. 2003. Potential biomarkers of crude oil exposure in the gastropod mollusc, *Austrocochlea porcata*: laboratory and manipulative field studies. *Environmental Pollution* **126**, 147–155.

Reimer, A.A. 1975. Effects of crude oil on corals. *Marine Pollution Bulletin* **6**, 39–43.

Ren, L., Huang, X.D., McConkey, B.J., Dixon, D.G. & Greenberg, B.M. 1994. Photoinduced toxicity of three polycyclic aromatic hydrocarbons (fluoranthene, pyrene and naphthalene) to the duckweed *Lemna gibba*. *Ecotoxicology and Environmental Safety* **28**, 160–170.

Renzoni, A. 1973. Influence of crude oil, derivatives and dispersants on larvae. *Marine Pollution Bulletin* **4**, 9–13.

Renzoni, A. 1975. Toxicity of three oils to bivalve gametes and larvae. *Marine Pollution Bulletin* **6**, 125–128.

Rinkevich, B. & Loya, Y. 1977. Harmful effects of chronic oil pollution on a Red Sea scleractinian coral population. In *Proceedings of the Third International Coral Reef Symposium*, Miami, FL **2**, 585–591.

Rinkevich, B. & Loya, Y. 1979. Laboratory experiments on the effects of crude oil on the Red Sea coral *Stylophora pistillata*. *Marine Pollution Bulletin* **10**, 328–330.

Rinkevich, B. & Loya, Y. 1983. Response of zooxanthellae photosynthesis to low concentrations of petroleum hydrocarbons. *Bulletin of the Institute of Oceanography and Fisheries* **9**, 109–115, Academy of Scientific Research and Technology, Cairo, Egypt.

Rissik, D. & Esdaile, J. 2011. Recovery of fish communities in two freshwater wetlands on Moreton Island impacted by the 'Pacific Adventurer' oil spill. *Royal Society of Queensland* **117**, 477–483.

Rissik, D., Hough, P. & Gruythuysen, J. 2011. The 'Pacific Adventurer' oil spill and two freshwater wetlands on Moreton Island, South East Queensland: effects and rehabilitation. *Royal Society of Queensland* **117**, 467–475.

Rolan, R.G. & Gallagher, R. 1991. Recovery of intertidal biotic communities at Sullom Voe following the Esso Bernicia oil spill of 1978. In *International Oil Spill Conference* Vol. **1991**(1), 461–465. American Petroleum Institute, Dallas.

Rosser, N.L. 2013. Biannual coral spawning decreases at higher latitudes on Western Australian reefs. *Coral Reefs* **32**, 455–460.

Rosser, N.L. & Gilmour, J.P. 2008. New insights into patterns of coral spawning on Western Australian reefs. *Coral Reefs* **27**, 345–349.

Rotjan, R.D. & Lewis, S.M. 2008. Impact of coral predators on tropical reefs. *Marine Ecology Progress Series* **367**, 73–91.

Rougee, L., Downs, C.A., Richmond, R.H. & Ostrander, G.K. 2006. Alteration of normal cellular profiles in the scleractinian coral (*Pocillopora damicornis*) following laboratory exposure to fuel oil. *Environmental Toxicology and Chemistry* **25**, 3181–3187.

Runcie, J., Macinnis-Ng, C. & Ralph, P. 2005. *The toxic effects of petrochemicals on seagrasses: literature review.* Report Prepared by Institute for Water and Environmental Resource Management and Department of Environmental Sciences University of Technology, Sydney, for Australian Maritime Safety Authority, Canberra.

Ryder, K., Temara, A. & Holdway, D.A. 2004. Avoidance of crude-oil contaminated sediment by the Australian seastar, *Patiriella exigua* (Echinodermata: Asteroidea). *Marine Pollution Bulletin* **49**, 900–909.

Sanchez, F., Velasco, F., Cartes, J.E., Olaso, I., Preciado, I., Fanelli, E., Serrano, A. & Gutierrez-Zabala, J.L. 2006. Monitoring the Prestige oil spill impacts on some key species of the Northern Iberian Shelf. *Marine Pollution Bulletin* **53**, 332–349.

Sanders, H.L. 1978. Florida oil-spill impact on Buzzards Bay benthic fauna—West-Falmouth. *Journal of the Fisheries Research Board of Canada* **35**, 717–730.

Schlacher, T.A., Holzheimer, A., Stevens, T. & Rissik, D. 2011. Impacts of the 'Pacific Adventurer' oil spill on the macrobenthos of subtropical sandy beaches. *Estuaries and Coasts* **34**, 937–949.

Sell, D., Conway, L., Clark, T., Picken, G.B., Baker, J.M., Dunnet, G.M., McIntyre, A.D. & Clark, R.B. 1995. Scientific criteria to optimize oil spill cleanup. In *International Oil Spill Conference Proceedings,* Long Beach, February–March 1995, 595–610.

Shafir, S., Van Rijn, J. & Rinkevich, B. 2007. Short and long term toxicity of crude oil and oil dispersants to two representative coral species. *Environmental Science and Technology* **41**, 5571–5574.

Shephard, B.K. 1998. Quantification of ecological risks to aquatic biota from bioaccumulated chemicals. In *National Sediment Bioaccumulation Conference Proceedings.* Bethesda, MD: EPA 31–52.

Shigenaka, G. 2001. *Toxicity of Oil to Reef-Building Corals: A Spill Response Perspective.* NOAA Technical Memorandum NOS OR&R 8. Seattle, Washington: National Oceanic and Atmospheric Administration.

Sim, C., Masini, R.J., Daly, T., Tacey, W., Kemps, H.A. & McAlpine, K.W. 2012. *Petroleum hydrocarbon content of shoreline sediment and intertidal biota at selected sites in the Kimberley bioregion.* Western Australia, Marine Technical Report Series. Office of the Environmental Protection Authority, Report No. MTR4, Perth, Western Australia, May 2012.

Simpson, C.J., Pearce, A.F. & Walker, D.I. 1991. Mass spawning of corals on Western Australian reefs and comparisons with the Great Barrier Reef. *Journal of the Royal Society of Western Australia* **74**, 85–91.

Simpson, R.D., Smith, S.D.A. & Pople, A.R. 1995. The effects of a spillage of diesel fuel on a rocky shore in the Sub-Antarctic region (Macquarie Island). *Marine Pollution Bulletin* **31**, 367–371.

Smith, A.J., Brearley, A., Hyndes, G.A., Lavery, P.S. & Walker, D.I. 2005. Carbon and nitrogen isotope analysis of an *Amphibolis griffithii* seagrass bed. *Estuarine, Coastal and Shelf Science* **65**, 545–556.

Smith, J.E. 1968. *Torrey Canyon pollution and marine life.* A Report by the Plymouth Laboratory of the Marine Biological Association of the United Kingdom. Cambridge University Press, London, 196 pp.

Smith, S.D.A. & Simpson, R.D. 1995. Effects of the Nella Dan oil spill on the fauna of *Durvillaea antarctica* holdfasts. *Marine Ecology Progress Series* **121**, 73–89.

Smith, S.D.A. & Simpson, R.D. 1998. Recovery of benthic communities at Macquarie Island (Sub-Antarctic) following a small oil spill. *Marine Biology* **131**, 567–581.

Sole, M., Porte, C., Biosca, X., Mitchelmore, C.L., Chipman, J.K., Livingstone, D.R. & Albaiges, J. 1996. Effects of the "Aegean Sea" oil spill on biotransformation enzymes, oxidative stress and DNA-adducts in digestive gland of the mussel (*Mytilus edulis* L). *Comparative Biochemistry and Physiology Part C Toxicology & Pharmacology* **113**, 257–265.

Southward, A.J. & Southward, E.C. 1978. Recolonization of rocky shores in Cornwall after use of toxic dispersant to clean up the torrey canyon spill. *Journal of the Fisheries Research Board of Canada* **35**, 682–706.

Southward, A.J., Langmead, O., Hardman-Mountford, N.J., Aiken, J., Boalch, G.T., Dando, P.R., Genner, M.J. et al. 2004. Long-term oceanographic and ecological research in the western English Channel. *Advances in Marine Biology* **47**, 1–105. http://dx.doi.org/10.1016/S0065-2881(04)47001-1.

Spooner, M. 1970. Oil spill in Tarut Bay Saudi Arabia. *Marine Pollution Bulletin* **1**, 166–167.

Stejskal, I.V. 2000. Obtaining approvals for oil and gas projects in shallow water marine areas in Western Australia using an environmental risk assessment framework. *Spill Science & Technology Bulletin* **6**, 69–76.

Stekoll, M.S. & Deysher, L. 2000. Response of the dominant alga *Fucus gardneri* (Silva) (Phaeophyceae) to the Exxon Valdez oil spill and clean-up. *Marine Pollution Bulletin* **40**, 1028–1041.

Stevens, T., Boden, A., Arthur, J.M., Schlacher, T.A., Rissik, D. & Atkinson, S. 2012. Initial effects of a moderate-sized oil spill on benthic assemblage structure of a subtropical rocky shore. *Estuarine and Coastal Shelf Science* **109**, 107–115.

Stuart-Smith, R.D., Edgar, G.J., Barrett, N.S., Bates, A.E., Baker, S.C., Bax, N.J., Becerro, M.A., Berkhout, J., Blanchard, J.L., Brock, D.J., Clark, G.F., Cooper, A.T., Davis, T.R., Day, P.B., Duffy, J.E., Holmes, T.H., Howe, S.A., Jordan, A., Kininmonth, S., Knott, N.A., Lefcheck, J.S., Ling, S.D., Parr, A., Strain, E., Sweatman, H. & Thomson, R. 2017. Assessing national biodiversity trends for rocky and coral reefs through the integration of citizen science and scientific monitoring programs. *BioScience* **67**, 134–146.

Suchanek, T.H. 1993. Oil impacts on marine invertebrate populations and communities. *American Zoologist* **33**, 510–523.

Taylor, H.A. & Rasheed, M.A. 2011. Impacts of a fuel oil spill on seagrass meadows in a subtropical port, Gladstone, Australia—the value of long-term marine habitat monitoring in high risk areas. *Marine Pollution Bulletin* **63**, 431–437.

Taylor, H.A., Rasheed, M.A. & Thomas, R. 2006. Port Curtis post oil spill seagrass assessment, Gladstone—February 2006. Department of Primary Industries and Fisheries (DPI&F) Information Series QI06046 (DPI&F, Cairns), 19 pp.

Teal, J.M., Farrington, J.W., Burns, K.A., Stegeman, J.J., Tripp, B.W., Woodin, B. & Phinney, C. 1992. The West Falmouth oil-spill after 20 years—fate of fuel-oil compounds and effects on animals. *Marine Pollution Bulletin* **24**, 607–614.

The Guardian. 2017. https://www.theguardian.com/environment/2017/feb/03/oil-spill-near-exxonmobil-drilling-platform-in-bass-strait-to-be-investigated

The West Australian. 2017. https://thewest.com.au/business/oil-gas/woodside-oil-well-in-pilbara-leaked-into-ocean-for-two-months-ng-b88480665z

Thompson, B.A.W., Goldsworthy, P.M., Riddle, M.J., Snape, I. & Stark, J.S. 2007. Contamination effects by a 'conventional' and a 'biodegradable' lubricant oil on infaunal recruitment to Antarctic sediments: a field experiment. *Journal of Experimental Marine Biology and Ecology* **340**, 213–226.

Thorhaug, A. 1988. Dispersed oil effects on mangroves, seagrasses and corals in the wider Caribbean. In *Proceedings of the Sixth International Coral Reef Symposium*. Vol. **2** 1988. Townsville, Australia, 337–339.

Thorhaug, A. & Marcus, J. 1987a. Effects of seven dispersants on growth of three subtropical/tropical Atlantic seagrasses. In *Proceedings, Conference on Oil Pollution Organized Under the Auspices of the International Association on Water Pollution Research and Control. Netherlands Organization for Applied Scientific Research*. Amsterdam, Netherlands: Springer, 201–205.

Thorhaug, A. & Marcus, J. 1987b. Oil spill clean-up: the effect of three dispersants on three subtropical/tropical sea grasses. *Marine Pollution Bulletin* **18**, 124–126.

Thorhaug, A., Marcus, J. & Booker, F. 1986. Oil and dispersed oil on subtropical and tropical seagrasses in laboratory studies. *Marine Pollution Bulletin* **17**, 357–361.

Thursby, G.B., Steel, R.L. & Kane, M.E. 1985. Effects of organic chemicals on growth and reproduction in the marine red algae, *Champia parvula*. *Environmental Toxicology Chemistry Journal* **4**, 797–805.

Underwood, A.J. 1991. Beyond BACI: experimental designs for detecting human environmental impacts on temporal variations in natural populations. *Marine and Freshwater Research* **42**, 569–587.

Underwood, A.J. 1994. On beyond BACI: sampling designs that might reliably detect environmental disturbances. *Ecological Applications* **4**, 3–15.

Underwood, A.J. 2002. Establishing the true environmental impact of a spill. In *Spillcon 2002: 9th International Oil Spill Conference*, Sydney, Australia, September 2002.

Van Overbeek, J. & Blondeau, R. 1954. Mode of action of phytotoxic oils. *Weeds* **3**(1), 55–65.

Vashchenko, M.A. 1980. Effects of oil pollution on the development of sex cells in sea urchins. *Helgoländer Meeresuntersuchungen* **33**, 297–300.

Venosa, A.D. & Zhu, X. 2003. Biodegradation of crude oil contaminating marine shorelines and freshwater wetlands. *Spill Science & Technology Bulletin* **8**(2), 163–178.

Villanueva, R.D., Montano, M.N.E. & Yap, H.T. 2008. Effects of natural gas condensate—water accommodated fraction on coral larvae. *Marine Pollution Bulletin* **56**, 1422–1428.

Villanueva, R.D., Yap, H.T. & Montano, M.N.E. 2011. Reproductive effects of the water-accommodated fraction of a natural gas condensate in the Indo-Pacific reef-building coral *Pocillopora damicornis*. *Ecotoxicology and Environmental Safety* **74**, 2268–2274.

Walker, D.I. 1991. Major oil spill off the coast of Western Australia. *Marine Pollution Bulletin* **22**, 424–478.

Wardley-Smith, J. 1974. Magellan Straits spill. *Marine Pollution Bulletin* **5**, 163–164.

Wardrop, J.A., Butler, A.J. & Johnson, J.E. 1987. A field study of the toxicity of two oils and a dispersant to the mangrove *Avicennia marina*. *Marine Biology* **96**(1), 151–156.

Wardrop, J.A., Wagstaff, B., Pfennig, P., Leeder, J. & Connolly, R.M. 1996. *The distribution, persistence and effects of petroleum hydrocarbons in mangroves impacted by the 'Era' oil spill (September 1992)*. Final Phase One Report. Published by the Office of the Environment Protection Authority, South Australian Department of Environment and Natural Resources, 120 pp.

Wasmund, K., Kurtboke, D.I., Burns, K.A. & Bourne, D.G. 2009. Microbial diversity in sediments associated with a shallow methane seep in the tropical Timor Sea of Australia reveals a novel aerobic methanotroph diversity. *Fems Microbiology Ecology* **68**, 142–151.

Wells, P.G. 1972. Influence of Venezuelan crude oil on lobster larvae. *Marine Pollution Bulletin* **3**, 105–106.

Wells, P.G. & Sprague, J.B. 1976. Effects of crude oil on American lobster (*Homarus Americanus*) larvae in the laboratory. *Journal of the Fisheries Board of Canada* **33**, 1604–1614.

Wilson, K.G. & Ralph, P.J. 2012. Laboratory testing protocol for the impact of dispersed petrochemicals on seagrass. *Marine Pollution Bulletin* **64**, 2421–2427.

Wolfe, M.F., Schlosser, J.A., Schwartz, G.J.B., Singaram, S., Mielbrecht, E.E., Tjeerdema, R.S. & Sowby, M.L. 1998a. Influence of dispersants on the bioavailability and trophic transfer of petroleum hydrocarbons to primary levels of a marine food chain. *Aquatic Toxicology* **42**, 211–227.

Wolfe, M.F., Schwartz, G.J.B., Singaram, S., Mielbrecht, E.E., Tjeerdema, R.S. & Sowby, M.L. 1998b. Effects of salinity and temperature on the bioavailability of dispersed petroleum hydrocarbons to the golden-brown algae, *Isochrysis galbana*. *Archives of Environmental Contamination and Toxicology* **35**, 268–273.

Woodin, S.A., Nyblade, C.F. & Chia, F.S. 1972. Effect of diesel oil spill on invertebrates. *Marine Pollution Bulletin* **3**, 139–143.

Yender, R.A. & Michel, J. 2010. Oil toxicity to corals. In *Oil Spills in Coral Reefs: Planning & Response Considerations*. R.A. Yender & J. Michel (eds). Seattle, Washington: National Oceanic and Atmospheric Administration, 25–35.

Zieman, J.C., Orth, R., Phillips, R.C., Thayer, G.W. & Thorhaug, A. 1984. The effects of oil on seagrass ecosystems. In *Restoration of Habitats Impacted by Oil Spills*, J.J. Cairns & A.L. Buikema (eds). Boston, MA: Butterworth Publishers, 36–64.

Oceanography and Marine Biology: An Annual Review, 2018, **56**, 371-448
© S. J. Hawkins, A. J. Evans, A. C. Dale, L. B. Firth, and I. P. Smith, Editors
Taylor & Francis

SYMBIOTIC POLYCHAETES REVISITED: AN UPDATE OF THE KNOWN SPECIES AND RELATIONSHIPS (1998–2017)

DANIEL MARTIN[1]* & TEMIR A. BRITAYEV[2]*

[1]Centre d'Estudis Avançats de Blanes (CEAB – CSIC), 17300 Blanes, Girona, Catalunya, Spain
[2]A.N. Severtzov Institute of Ecology and Evolution (RAS), Moscow, Russia
*Corresponding authors: Daniel Martin
e-mail: dani@ceab.csic.es
Temir A. Britayev
e-mail:britayev@yandex.ru

Abstract

Here we consider the growing knowledge on symbiotic polychaetes since this particular group of worms, and their relationships with their hosts, were reviewed by Martin & Britayev (1998). The current number of symbiotic polychaetes (excluding myzostomids) reported has almost doubled since 1998 (618 versus 373 species) and are now known to be involved in 1626 relationships (966 in 1998), representing 245 and 660 newly reported species and relationships, respectively. Overall, 490 (292 in 1998) species involved in 1229 (713 in 1998) relationships are commensals, and 128 (81 in 1998) involved in 397 (253 in 1998) relationships are parasitic. New commensal and parasitic species and/ or relationships have been respectively reported for eight (Chaetopteridae, Siboglinidae, Fabriciidae, Aphroditidae, Orbiniidae, Pholoididae, Scalibregmatidae, Sigalionidae) and five (Fabriciidae, Typhloscolecidae, Phyllodocidae, Polynoidae, Hesionidae, Serpulidae) polychaete families. Three additional taxa (cephalopod molluscs, gorgonocephalid ophiuroids and ascidian tunicates) are now known to harbour commensal polychaetes, and a further three taxa (decapod crustaceans, chaetognaths and brachiopods) are now known to host parasitic polychaetes. Here we discuss, family by family, the main characteristics and nature of symbiotic polychaetes and their relationships. We conclude that some of the biases identified in 1998 are still uncorrected. Despite the noticeable increase of taxonomic studies describing new species and reporting new relationships, there is still a lack of ecological and biological studies, either descriptive or experimental (e.g. based on behavioural observations of living organisms), addressing the actual nature of the associations. We have also identified that most studies are restricted to a specialised academic world. The next logical step would be to transfer this knowledge to non-specialised audiences. In other words, to contribute to the preservation of our seas, it is our duty to raise awareness of the potential ecological and economic impacts of these symbiotic associations and to allow other eyes to enjoy the intrinsic beauty of symbiotic worms.

Introduction

The term symbiosis was defined as a long-term association between two or more different species that are intimately interacting at different levels (de Bary 1879), which usually involves behavioural, physiological and/or morphological adaptations (Goff 1982). Parasitisms $(+/-)$, commensalisms $(=/+)$ and mutualisms $(+/+)$ are traditionally distinguished according to the effects on the fitness of their partners. However, 'symbiosis' has a broader sense because the limits between these three types of interactions are often unclear. In fact, modern perspectives consider these associations as forming a continuum in which the end of one type and the beginning of another is often difficult to distinguish (Parmentier & Michet 2013). Moreover, the real nature of many of these associations is still unknown.

Marine invertebrate symbioses are, with a few exceptions, more poorly known than terrestrial ones despite the paradigmatic function of symbiotic consortia in habitats such as coral reefs in tropical areas of the world's oceans (Hoegh-Guldberg 1999). Different approaches may contribute to progress in the knowledge of their role, from specialised studies on the functional characteristics of a given partnership, to comprehensive reviews on the extent of the association. Among marine invertebrates, the annelid polychaetes are ubiquitous organisms including numerous representatives living in symbiosis with virtually all known major taxa in the marine realm.

The existing literature on symbiotic polychaetes was abundant but dispersed, and it was not until 1998 that we undertook an exhaustive analysis (Martin & Britayev 1998). In this review, we considered more than 400 references, which allowed us to discuss the characteristics of the symbiotic polychaetes (excluding myzostomids) known to date. The Myzostomida is a highly specialised order of annelids with a highly derived body plan that obscures their phylogenetic affinities to other metazoans (Eeckahaut & Lanterbecq 2005) and merits a specific approach (see some examples from the last 10 years at Bleidorn et al. 2009; Lanterbecq et al. 2009, 2010; Bo et al. 2014; Summers & Rouse 2014; Terrana & Eeckahaut 2017). It was considered as outside the scope of the review by Martin & Britayev (1998), and to be consistent, we are here following the same criteria. The numbers resulting from the 1998 review were outstanding, with a total of 373 species (292 commensals, 81 parasites) and 966 relationships (713 commensals, 253 parasites). However, we did not consider our contribution as conclusive but as an updated baseline to support an expected increase of the interest of this topic. We stated, "The aim of the review is, therefore, to attract the attention of scientists to–and to encourage further studies on–the ecology of this particular and diverse group of symbionts." From 1998 on, numerous studies addressing the taxonomy, but also the biology and the ecology, of symbiotic polychaetes have been published, which suggests that our objective has been (or is being) achieved.

Having this objective in mind, we realised that the amount of newly produced information has grown enough to deserve an update. Therefore, we have here compiled a full list of the symbiotic polychaetes and their hosts and relationships that have been newly reported from 1998 to date. Moreover, we realised that a few pre-1998 records were overlooked in the first review, and these have also been considered in the present one. In 1998, the weak delineations between the different types of symbiotic relationships led us to use the term commensal as opposite to parasitic for statistical purposes. Accordingly, the term 'commensals' encompassed both real commensals and mutualistic symbionts, which are usually difficult to distinguish in the absence of functional analyses of the associations. In the present update, we are using the same assumption to compare the taxonomic distribution of the symbiotic polychaetes found in 1998 with the equivalent information that exists now. To prevent taxonomic and nomenclatural mistakes, both the ancient database and all new records added since 1998 have been checked through the World Register of Marine Species (WoRMS, Horton et al. 2017), with a few exceptions in which we used either alternative online resources, such as the Marine Species Identification Portal (http://species-identification.org/about.php) and the Encyclopaedia of Life (http://www.eol.org), or the original papers in which the species were described or mentioned (Tables 1 and 2).

The compilation of all new records and the corresponding papers dealing with symbiotic polychaetes also allows us firstly to update the overall statistics concerning these organisms; secondly to summarise the main new taxonomic information both for the polychaetes and their hosts; and thirdly to discuss, family by family, the main characteristics and the nature of the established relationships, whenever available. The final conclusive section contains an overview of the main general trends on symbiotic polychaetes and their relationships, as extracted from the updated information, and stresses some facts that we perceive as the way forward for future studies on this topic.

Table 1 List of newly reported commensal polychaetes and their respective hosts from 1998 to 2017.

Symbiont		Host	Source references
Amphynomidae			
Benthoscolex cubanus Hartman, 1942	E	Unknown sea urchins	(Hartman 1942)
Antonbruunidae			
Antonbruunia gerdesi Quiroga & Sellanes, 2009	B	*Calyptogena gallardoi* Sellanes & Krylova, 2005	(Quiroga & Sellanes 2009)
Antonbruunia sociabilis Mackie, Oliver & Nygren, 2015	B	*Thyasira scotiae* Oliver & Drewery, 2014	(Mackie et al. 2015)
Aphroditidae			
Pontogenia chrysocoma (Baird, 1865)	Ec	*Bonellia viridis* Rolando, 1822 (1821 volume)	(Anker et al. 2005)
	D	*Upogebia deltaura* (Leach, 1815)	(Schembri & Jaccarini 1978)
Capitellidae			
Capitella floridana Hartman, 1959	Ce	*Lolliguncula brevis* (Blainville, 1823)	(Boletzky & von Dohle 1967, Zeidberg et al. 2011)
Capitella hermaphrodita Boletzky & Dohle, 1967	Ce	*Loligo vulgaris* Lamarck, 1798	(Boletzky & von Dohle 1967, Zeidberg et al. 2011)
Notomastus latericeus Sars, 1851	Ec	*Echiurus echiurus echiurus* (Pallas, 1766)	(Anker et al. 2005)
	Ec	*Lissomyema exilii* (Muller F., 1883)	(Anker et al. 2005)
Chaetopteridae			
Spiochaetopterus sp.	Cn	*Montipora* cf. *aequituberculata* Bernard, 1897	(Bergsma 2009)
	Cn	*Montipora* cf. *hispida* (Dana, 1846)	(Bergsma 2009)
Chrysopetalidae			
Bhawania goodei Webster, 1884	Ec	*Lissomyema exilii*	(Anker et al. 2005)
Iheyomytilidicola tridentatus Miura & Hashimoto, 1996	B	*Bathymodiolus aduloides* Hashimoto & Okutani, 1994	(Miura & Hashimoto 1996, Dreyer et al. 2004)
Iheyomytilidicola lauensis Aguado & Rouse, 2011	B	*Bathymodiolus* sp.	(Aguado & Rouse 2011)
Laubierus alvini Aguado & Rouse, 2011	B	*Bathymodiolus* sp.	(Aguado & Rouse 2011)
Laubierus mucronatus Blake, 1993	B	*Bathymodiolus heckerae* Turner, Gustafson, Lutz & Vrijenhoek, 1998	(Ward et al. 2004)
Miura spinosa Blake, 1993	B	Unidentified	(Blake 1993, Dreyer et al. 2004)
Natsushima bifurcata Miura & Laubier, 1990	B	*Acharax johnsoni* (Dall, 1891)	(Miura & Hashimoto 1996)
Natsushima graciliceps Miura & Hashimoto, 1996	B	*Solemya* sp.	(Miura & Hashimoto 1996, Dreyer et al. 2004)
Natsushima sashai Aguado & Rouse, 2011	B	*Acharax* sp.	(Aguado & Rouse 2011)
Shinkai fontefridae Aguado & Rouse, 2011	B	*Phreagena kilmeri* (F. R. Bernard, 1974)	(Aguado & Rouse 2011)
	B	*Archivesica gigas* (Dall, 1896)	(Aguado & Rouse 2011)

Continued

Table 1 (Continued)   List of newly reported commensal polychaetes and their respective hosts from 1998 to 2017.

Symbiont		Host	Source references
Shinkai longipedata Miura & Ohta, 1991	B	Unidentified vesicomyid	(Aguado & Rouse 2011)
Shinkai semilonga Miura & Hashimoto, 1996	B	*Calyptogena solidissima* Okutani, Hashimoto & Fujikura, 1992	(Miura & Hashimoto 1996, Dreyer et al. 2004)
Shinkai robusta Quiroga & Sellanes, 2009	B	*Calyptogena gallardoi* Sellanes & Krylova, 2005	(Quiroga & Sellanes 2009)
Thyasiridicola branchiatus Miura & Hashimoto, 1996	B	*Conchocele bisecta* (Conrad, 1849)	(Miura & Hashimoto 1996, Dreyer et al. 2004)
Vesicomyicola trifulcatus Dreyer, Miura & van Dover, 2004	B	*Vesicomya* sp.	(Dreyer et al. 2004)
Dorvilleidae			
Iphitime cuenoti Fauvel, 1914	D	*Bathynectes maravigna* (Prestandrea, 1839)	(Lattig et al. 2017)
	D	*Pagurus* sp.	(Fage & Legèndre 1925, Hartman 1954, Williams & McDermott 2004)
Eunicidae			
Aciculomarphysa comes Hartmann-Schröder & Zibrowius, 1998	Cn	Unidentified	(Hartmann-Schröder & Zibrowius 1998, Molodtsova & Budaeva 2007)
Eunice antipathum (Pourtalès, 1867)	Cn	*Distichopathes filix* (Pourtalès, 1867)	(Hartmann-Schröder & Zibrowius 1998, Molodtsova & Budaeva 2007)
	Cn	*Elatopathes abietina* (Pourtalès, 1874)	(Hartmann-Schröder & Zibrowius 1998, Molodtsova & Budaeva 2007)
Eunice aphroditois (Pallas, 1788)	G	*Onustus longleyi* (Bartsch, 1931)	(Whorff 1991)
Eunice cf. *dubitata* Fauchald, 1974	Cn	*Madrepora oculata* Linnaeus, 1758	(Buhl-Mortensen & Mortensen 2004, Oppelt et al. 2017)
Eunice kristiani Hartmann-Schröder & Zibrowius, 1998	Cn	cf. *Cupressopathes cylindrica* (Brook, 1889)	(Hartmann-Schröder & Zibrowius 1998, Molodtsova & Budaeva 2007)
Eunice marianae Hartmann-Schröder & Zibrowius, 1998	Cn	cf. *Cupressopathes cylindrica*	(Hartmann-Schröder & Zibrowius 1998, Molodtsova & Budaeva 2007)
Eunice norvegica Linnaeus, 1767	Cn	*Lophelia pertusa*	(Zibrowius 1980, Roberts 2005, Mueller et al. 2013, Oppelt et al. 2017)
	Cn	*Madrepora oculata*	(Zibrowius 1980, Roberts 2005)
	Cn	*Solenosmilia variabilis*	(Zibrowius 1980, Roberts 2005)
Eunice sp.	Cn	*Madrepora oculata*	(Jensen & Frederiksen 1992)
Eunice sp.	Cn	*Madrepora arbuscula* (Moseley, 1881)	(Buhl-Mortensen & Mortensen 2005)
Eunice sp.	Cn	*Madrepora minutiseptum* Cairns & Zibrowius, 1997	(Cairns & Zibrowius 1997)
Eunice sp.	Cn	*Madrepora* cf. *porcellana* (Moseley, 1881)	(Cairns & Zibrowius 1997)

Continued

Table 1 (Continued)   List of newly reported commensal polychaetes and their respective hosts from 1998 to 2017.

Symbiont		Host	Source references
Eunice sp.	Cn	*Lobophyllia hemprichii* (Ehrenberg, 1834)	(Chisholm & Kelley 2001)
Eunice sp.	G	*Onustus longleyi*	(Whorff 1991)
Lysidice unicornis (Grube, 1840)	Ec	*Lissomyema exilii*	(Anker et al. 2005)
Fabriciidae			
Brandtika asiatica Jones, 1974a	G	*Mekongia jullieni* Deshayes, 1876	(Jones 1974a, Fitzhugh & Rouse 1999)
	B	*Hyriopsis delaportei* Haas, 1914	(Jones 1974a, Fitzhugh & Rouse 1999)
Monroika africana (Monro, 1936)	G	*Hydrobia plena* Bequaert & Clench, 1936	(Jones 1974a, Fitzhugh & Rouse 1999)
Histriobdellidae			
Stratiodrilus arrielai Amaral & Morgado, 1997	D	*Aegla perobae* Hebling & Rodrigues, 1977	(Amaral & Morgado 1997)
Stratiodrilus circensis Steiner & Amaral, 1999	D	*Aegla* cf. *schmitii* Hobbs III, 1979	(Steiner & Amaral 1999)
	D	*Aegla leptodactyla* Buckup & Rossi, 1977	(Daudt & Amato 2007)
Stratiodrilus robustus Steiner & Amaral, 1999	D	*Trichodactylus* sp.	(Steiner & Amaral 1999)
Stratiodrilus tasmanica Haswell, 1900	D	*Astacopsis franklinii* var. *tasmanicus* Smith, 1912	(Haswell 1900)
Stratiodrilus vilae Amato, 2001	D	*Parastacus brasiliensis* (von Martens, 1869)	(Amato 2001)
	D	*Parastacus defossus* Faxon, 1898	(Amato 2001)
Hesionidae			
Hesione picta Müller in Grube, 1858	O	*Ophionereis reticulata* (Say, 1825)	(De Assis et al. 2012)
Leocrates chinensis Kinberg, 1866	Cn	Corals	(Pettibone 1970)
Leocrates claparedii (Costa in Claparède, 1868)	Cn	Corals	(Pettibone 1970)
Nereimyra punctata (Müller, 1788)	D	*Pagurus cuanensis*	(Samuelsen 1970, Williams & McDermott 2004)
Oxydromus angustifrons (Grube, 1878)	E	*Salmacis sphaeroides* (Linnaeus, 1758)	(Chim et al. 2013)
	E	*Temnopleurus toreumaticus* (Leske, 1778)	(Chim et al. 2013)
	E	Unknown sand dollars	(Mohammad 1971)
Oxydromus bunbuku (Uchida, 2004)	E	*Brissus agassizii* Döderlein, 1885	(Uchida 2004)
	E	*Brissus latecarinatus* (Leske, 1778)	(Uchida 2004)
Oxydromus flexuosus (Delle Chiaje, 1827)	Ec	*Maxmuelleria lankesteri* (Herdman, 1898)	(Anker et al. 2005)
Oxydromus herrmanni (Giard, 1882)	H	*Balanoglossus robinii* Giard, 1882	(Giard 1882)

Continued

Table 1 (Continued)   List of newly reported commensal polychaetes and their respective hosts from 1998 to 2017.

Symbiont		Host	Source references
	H	*Balanoglossus salmoneus* Giard, 1882	(Giard 1882)
Oxydromus pallidus Claparède, 1864	Ec	*Lissomyema exilii*	(Anker et al. 2005)
Oxydromus sp.	E	*Linopneustes longispinus* (A. Agassiz, 1878)	(Emson et al. 1993, Miller & Wolf 2008)
Oxydromus sp. 1	A	*Luidia maculata* Müller & Troschel, 1842	(Britayev & Antokhina 2012)
	A	*Archaster angulatus* Müller & Troschel, 1842	(Britayev & Antokhina 2012)
Oxydromus sp. 2	A	*Luidia maculata*	(Britayev & Antokhina 2012)
	A	*Archaster angulatus*	(Britayev & Antokhina 2012)
	A	*Archaster typicus* Müller & Troschel, 1842	(Britayev & Antokhina 2012)
Oxydromus sp. 3	E	*Clypeaster* cf. *reticulatus* (Linnaeus, 1758)	(Britayev & Antokhina 2012)
Oxydromus sp. 4	E	*Salmacis bicolor* L. Agassiz in L. Agassiz & Desor, 1846	(Britayev & Antokhina 2012, as *Ophiodromus* sp. 4, Britayev et al. 2013)
	E	*Rhynobrissus hemiasteroides* A. Agassiz, 1879	(Britayev & Antokhina 2012)
Parahesione sp.	D	*Upogebia* sp.	(Britayev & Antokhina 2012)
Struwela noodti Hartmann-Schröder, 1959	E	*Mellita longifissa* Michelin, 1858	(Hartmann-Schröder 1959)
Struwela sp.	E	*Encope grandis* L. Agassiz, 1841	(Campos et al. 2009)
Struwela sp.	E	*Encope micropora* L. Agassiz, 1841	(Campos et al. 2009)
Struwela sp.	E	*Mellita granti* Mortensen, 1948	(Campos et al. 2009)
Struwela sp.	E	*Mellita longifissa*	(Campos et al. 2009)
Lumbrineridae			
Lumbrineris flabellicola Fage, 1936	Cn	*Caryophyllia (Acanthociathus) spinicarens* (Moseley, 1881)	(Miura & Shirayama 1992, Cairns & Zibrowius 1997)
	Cn	*Caryophyllia (Acanthocyathus) grayi* (Milne Edwards & Haime, 1848)	(Cairns & Zibrowius 1997)
	Cn	*Caryophyllia (Caryophyllia) quadragenaria* Alcock, 1902	(Miura & Shirayama 1992)
	Cn	*Caryophyllia* sp.[a]	(Martinell & Domènech 2009)
	Cn	*Ceratotrochus* sp.[a]	(Martinell & Domènech 2009)
	Cn	*Conotrochus brunneus* (Moseley, 1881)	(Cairns & Zibrowius 1997)
	Cn	Dendrophyllid coral	(Cairns & Zibrowius 1997)
	Cn	*Flabellium (Flabellum)* sp.	(Zibrowius et al. 1975)
	Cn	*Flabellum (Flabellum) lamellosum* Alcock, 1902	(Cairns & Zibrowius 1997)
	Cn	*Flabellum (Flabellum) patens* Moseley, 1881	(Cairns & Zibrowius 1997)
	Cn	*Flabellum* sp.[a]	(Martinell & Domènech 2009)

Continued

Table 1 (Continued) List of newly reported commensal polychaetes and their respective hosts from 1998 to 2017.

Symbiont		Host	Source references
	Cn	*Rhizotrochus typus* Milne Edwards & Haime, 1848	(Cairns & Zibrowius 1997)
	Cn	*Trochocyathus* sp.[a]	(Martinell & Domènech 2009)
	Cn	*Truncatoflabellum candeanum* (Milne Edwards & Haime, 1848)	(Zibrowius et al. 1975)
Nereididae			
Cheilonereis peristomialis Benham, 1916	D	*Diacanthurus rubricatus* (Henderson, 1888)	(Hand 1975, Williams & McDermott 2004)
	D	*Paguristes subpilosus* Henderson, 1888	(Hand 1975, Williams & McDermott 2004)
	D	*Sympagurus dimorphus* (Studer, 1883)	(Hand 1975, Williams & McDermott 2004)
Neanthes fucata (Savigny in Lamarck, 1818)	D	*Pagurus bernhardus* (Linnaeus, 1758)	(Clark 1956, Spooner et al. 1957, Gilpin-Brown 1969, Goerke 1971, Jensen & Bender 1973, Williams & McDermott 2004)
	D	*Pagurus cuanensis*	(Spooner et al. 1957, Williams 2002, Williams & McDermott 2004)
Orbiniidae			
Naineris setosa (Verrill, 1900)	Ec	*Lissomyema exilii*	(Anker et al. 2005)
Pholoididae			
Pholoe minuta (Fabricius, 1780)	Ec	*Echiurus echiurus echiurus*	(Anker et al. 2005)
Pilargidae			
Sigatargis commensalis Misra, 1999	H	*Saccoglosus* sp.	(Misra 1999)
Phyllodocidae			
Eumida ophiuricola Britayev, Doignon & Eeckahaut, 1999	O	*Ophiothrix (Acanthophiothrix) purpurea* von Martens, 1867	(Britayev et al. 1999)
Polynoidae			
Acanthicolepis asperrima (M. Sars, 1861)	Cn	*Madrepora oculata*	(Fiege & Barnich 2009)
Acanthicolepis zibrowii Barnich & Fiege, 2010	Cn	*Madrepora oculata*	(Barnich & Fiege 2010, Núñez et al. 2011)
Adyte assimilis (McIntosh, 1874)	E	*Echinus* sp.	(Okuda 1950)
Anotochaetonoe michelbhaudi Britayev & Martin, 2005	P	*Phyllochaetopterus* sp.	(Britayev & Martin 2005)
Antipathipolyeunoa sp.	Cn	Unidentified anthipatharian	(Serpetti et al. 2017)
	P	*Spiochaetopterus* sp.	(Britayev & Martin 2005)
Arctonoe pulchra (Johnson, 1897)	E	Unidentified	(Clark 1956)
Arctonoe vittata (Grube, 1855)	A	*Evasterias retifera* Djakonov, 1938	(Tokaji et al. 2014)
	D	*Labidochirus splendescens* (Owen, 1839)	(Hoberg et al. 1982, Williams & McDermott 2004)
	D	*Pagurus capillatus* (Benedict, 1892)	(Hoberg et al. 1982, Williams & McDermott 2004)

Continued

Table 1 (Continued)   List of newly reported commensal polychaetes and their respective hosts from 1998 to 2017.

Symbiont		Host	Source references
	D	*Pagurus trigonocheirus* (Stimpson, 1858)	(Hoberg et al. 1982, Williams & McDermott 2004)
	G	*Haliotis discus hanai* Ino, 1953	(Park et al. 2016)
	G	*Fusitriton oregonensis* (Redfield, 1846)	(Pettibone 1953)
	G	*Scelidotoma gigas* (Martens, 1881)	(Tokaji et al. 2014, Park et al. 2016)
	N	*Cadlina japónica* Baba, 1937	(Park et al. 2016)
	N	*Triopha catalinae* (Cooper, 1863)	(Park et al. 2016)
Asterophylia culcitae Britayev & Fauchald, 2005	Ho	*Choriaster granulatus* Lütken, 1869	(Britayev & Antokhina 2012)
	Ho	*Culcita novaeguineae* Müller & Troschel, 1842	(Britayev & Fauchald 2005, Britayev & Antokhina 2012)
	A	*Echinaster luzonicus* (Gray, 1840)	(Britayev & Antokhina 2012)
	Ho	*Euretaster insignis* (Sladen, 1882)	(Britayev & Antokhina 2012)
	Ho	*Fromia monilis* (Perrier, 1869)	(Britayev & Antokhina 2012)
	Ho	*Linckia laevigata* (Linnaeus, 1758)	(Britayev & Antokhina 2012)
	Ho	*Protoreaster nodosus* (Linnaeus, 1758)	(Britayev & Antokhina 2012)
	Cr	unstalked crinoids	(Britayev & Antokhina 2012)
Australaugenira iberica Ravara & Cunha, 2016	Cn	*Acanella* sp.	(Ravara & Cunha 2016)
Australaugenira michaelseni Pettibone, 1969	Cn	*Dendronephthya* sp.	(Britayev & Antokhina 2012)
Australaugeneria rutilans (Grube, 1878)	Cn	*Dendronephthya* sp.	(Britayev & Antokhina 2012)
Branchipolynoe pettiboneae Miura & Hashimoto, 1991	B	*Bathymodiolus brevior* Cosel, Métivier & Hashimoto, 1994	(Blake et al. 2006)
	B	*Bathymodiolus platifrons* Hashimoto & Okutani, 1994	(Blake et al. 2006)
	B	*Bathymodiolus japonicus* Hashimoto & Okutani, 1994	(Hashimoto & Okutani 1994)
	B	*Bathymodiolus elongatus* Cosel, Métivier & Hashimoto, 1994	(Blake et al. 2006)
Brichionoe sp.	E	Unidentified	(Serpetti et al. 2017)
	O	Unidentified	(Serpetti et al. 2017)
	Go	Unidentified	(Serpetti et al. 2017)
Eunoe bathydomus (Ditlevsen, 1917)	Ho	*Deima validum validum* Théel, 1879	(Shields et al. 2013)
Eunoe depressa Moore, 1905	D	*Elassochirus cavimanus* (Miers, 1879)	(Hoberg et al. 1982, Williams & McDermott 2004)
	D	*Labidochirus splendescens*	(Hoberg et al. 1982, Williams & McDermott 2004)
	D	*Pagurus aleuticus* (Benedict, 1892)	(Hoberg et al. 1982, Williams & McDermott 2004)
	D	*Pagurus capillatus*	(Hoberg et al. 1982, Williams & McDermott 2004)
	D	*Pagurus confragosus* (Benedict, 1892)	(Hoberg et al. 1982, Williams & McDermott 2004)

Continued

Table 1 (Continued) List of newly reported commensal polychaetes and their respective hosts from 1998 to 2017.

Symbiont		Host	Source references
	D	*Pagurus ochotoensis*	(Hoberg et al. 1982, Williams & McDermott 2004)
	D	*Pagurus rathbuni* (Benedict, 1892)	(Hoberg et al. 1982, Williams & McDermott 2004)
	D	*Pagurus setosus* (Benedict, 1892)	(Hoberg et al. 1982, Williams & McDermott 2004)
	D	*Pagurus trigonocheirus*	(Hoberg et al. 1982, Williams & McDermott 2004)
Eunoe nodosa (M. Sars, 1861)	D	*Elassochirus cavimanus*	(Hoberg et al. 1982, Williams & McDermott 2004)
	D	*Pagurus aleuticus*	(Hoberg et al. 1982, Williams & McDermott 2004)
	D	*Pagurus confragosus*	(Hoberg et al. 1982, Williams & McDermott 2004)
	D	*Pagurus ochotensis* Brandt, 1851	(Hoberg et al. 1982, Williams & McDermott 2004)
	D	*Pagurus trigonocheirus*	(Hoberg et al. 1982, Williams & McDermott 2004)
Eunoe purpurea Treadwell, 1936	Cn	*Bathypathes* cf. *alternata* Brook, 1889	(Barnich et al. 2013)
	Cn	*Madrepora oculata*	(Barnich et al. 2013)
Eunoe senta (Moore, 1902)	D	*Labidochirus splendescens*	(Hoberg et al. 1982, Williams & McDermott 2004)
	D	*Pagurus capillatus*	(Hoberg et al. 1982, Williams & McDermott 2004)
	D	*Pagurus trigonocheirus*	(Hoberg et al. 1982, Williams & McDermott 2004)
Eunoe spinulosa Verrill, 1879	Cn	*Acanella arbuscula* (Johnson, 1862)	(Buhl-Mortensen & Mortensen 2004)
Eunoe sp.	Ho	*Pannychia taylorae* O'Loughlin in O'Loughlin et al., 2013	(Serpetti et al. 2017)
	Ho	Unidentified	(Serpetti et al. 2017)
	Ec	Unidentified	(Serpetti et al. 2017)
Gattyana ciliata Moore, 1902	D	*Labidochirus splendescens*	(Hoberg et al. 1982, Williams & McDermott 2004)
	D	*Pagurus capillatus*	(Hoberg et al. 1982, Williams & McDermott 2004)
	D	*Pagurus rathbuni*	(Hoberg et al. 1982, Williams & McDermott 2004)
	D	*Pagurus trigonocheirus*	(Hoberg et al. 1982, Williams & McDermott 2004)
Gattyana cirrhosa (Pallas, 1766)	Ec	*Echiurus echiurus echiurus*	(Anker et al. 2005)
	P	*Neoamphitrite figulus* (Dalyell, 1853)	(Pettibone 1953)
	P	*Thelepus* sp.	(Pettibone 1953)
Gaudichaudius cimex (Quatrefages, 1866)	D	*Clibanarius padavensis* de Man, 1888	(Parulekar 1969, Williams & McDermott 2004)
	D	*Diogenes alias* McLaughlin & Holthuis, 2001	(Achari 1974, Pettibone 1986b, Williams & McDermott 2004)

Continued

Table 1 (Continued) List of newly reported commensal polychaetes and their respective hosts from 1998 to 2017.

Symbiont		Host	Source references
Gorgoniapolynoe caeciliae (Fauvel, 1913)	Cn	*Acanthogorgia armata* Verrill, 1878	(Barnich et al. 2013)
	Cn	*Corallium secundum* Dana, 1846	(Stock 1986)
Gorgoniapolynoe sp.	Cn	*Narella vermifera* Cairns & Bayer, 2007 [2008]	(Cairns & Bayer 2008)
Gorgoniapolynoe sp.	Cn	*Narella macrocalyx* Cairns & Bayer, 2007 [2008]	(Cairns & Bayer 2008)
Gorgoniapolynoe sp.	Cn	*Narella alata* Cairns & Bayer, 2007 [2008]	(Cairns & Bayer 2008)
Gorgoniapolynoe sp.	Cn	*Candidella helminophora* (Nutting, 1908)	(Nutting 1908, Cairns 2009)
Gorekia crassicirris (Willey, 1902)	E	*Abatus (Pseudabatus) nimrodi* (Koehler, 1911)	(Schiaparelli et al. 2011)
	E	*Brachysternaster chescheri* Larrain, 1985	(Schiaparelli et al. 2011)
Halosydna johnsoni (Darboux, 1899)	P	*Amphitrite* sp.	(Johnson 1897)
	P	*Telepus* sp.	(Johnson 1897)
	P	*Terebella* sp.	(Johnson 1897)
Harmothoe dannyi Barnich, Beuck & Freiwald, 2013	Cn	*Stylaster erubescens* Pourtalès, 1868	(Barnich et al. 2013)
Harmothoe extenuata (Grube, 1840)	D	*Pagurus bernhardus*	(Brightwell 1951, Williams & McDermott 2004)
	G	*Xenophora crispa* (König, 1825)	(Intès & Le Loeuff 1975)
Harmothoe gilchristi Day, 1960	Cn	*Cladocora caespitosa* (Linnaeus, 1767)	(Barnich & Fiege 2000)
	Cn	*Isidella* sp.	(Barnich & Fiege 2000)
	Cn	*Dendrophyllia ramea* (Linnaeus, 1758)	(Barnich & Fiege 2000)
	Cn	*Solenosmilia variabilis* Duncan, 1873	(Miranda & Brasil 2014)
	Cn	*Lophelia pertusa*	(Miranda & Brasil 2014)
	Cn	*Enallopsammia rostrata* (Pourtalès, 1878)	(Miranda & Brasil 2014)
	Cn	*Madrepora oculata*	(Miranda & Brasil 2014)
Harmothoe imbricata (Linnaeus, 1767)	Ec	*Lissomyema exilii*	(Anker et al. 2005)
Harmothoe ruthae Miranda & Brasil, 2014	Cn	*Enallopsammia rostrata*	(Miranda & Brasil 2014)
	Cn	*Solenosmilia variabilis*	(Miranda & Brasil 2014)
	Cn	*Lophelia pertusa*	(Miranda & Brasil 2014)
	Cn	*Madrepora oculata*	(Miranda & Brasil 2014)
	Cn	*Errinia* sp.	(Miranda & Brasil 2014)
Harmothoe vesiculosa Ditlevsen, 1917	Cn	*Lophelia pertusa* (Linnaeus, 1758)	(Fiege & Barnich 2009)
	Cn	*Madrepora oculata*	(Fiege & Barnich 2009)
Harmothoe oculinarum (Storm, 1879)	P	*Eunice norvegica* (Linnaeus, 1767)	(Mortensen 2001, Buhl-Mortensen & Mortensen 2004, Fiege & Barnich 2009)
Hesperonoe adventor (Skogsberg in Fisher & MacGinitie, 1928)	Ec	*Echiurus echiurus alascanus* Fisher, 1946	(Anker et al. 2005)

Continued

Table 1 (Continued)   List of newly reported commensal polychaetes and their respective hosts from 1998 to 2017.

Symbiont		Host	Source references
Hesperonoe coreensis Hong, Lee & Sato, 2017	D	*Upogebia major* (De Haan, 1841)	(Sato et al. 2001, Hong et al. 2017)
Hesperonoe hwanghaiensis Uschacov & Wu, 1959	D	*Upogebia major*	(Sato et al. 2001, Hong et al. 2017)
Hesperonoe japonensis Hong, Lee & Sato, 2017	D	*Upogebia major*	(Hong et al. 2017)
	D	*Austinogebia narutensis* (Sakai, 1986)	(Hong et al. 2017)
Hesperonoe laevis Hartman, 1961	Ec	*Listriolobus pelodes* Fisher, 1946	(Anker et al. 2005)
Heteralentia ptycholepis (Grube, 1878)	Cr	*Phanogenia gracilis* (Hartlaub, 1893)	(Britayev & Antokhina 2012)
Hololepidella boninensis Nishi & Tachikawa, 1998	O	*Ophiocoma* sp.	(Nishi & Tachikawa 1998)
Hololepidella laingensis Britayev, Doignon & Eeckahaut, 1999	Cr	*Amphimetra tessellata* (Müller, 1841)	(Mekhova & Britayev 2015)
	Cr	*Comatula solaris* Lamarck, 1816	(Mekhova & Britayev 2015)
	Cr	*Capillaster gracilicirrus* AH Clark, 1912	(Mekhova & Britayev 2015)
	Cr	*Zygometra comata* AH Clark, 1911	(Mekhova & Britayev 2015)
	A	*Acanthaster planci* (Linnaeus, 1758)	(Britayev & Antokhina 2012)
	A	*Culcita novaeguineae*	(Britayev & Antokhina 2012)
	A	*Linckia laevigata*	(Britayev & Antokhina 2012)
	Cr	*Capillaster multiradiatus* (Linnaeus, 1758)	(Britayev et al. 1999)
	Cr	*Dichometra flagellata*	(Britayev et al. 1999)
	Cr	*Lamprometra palmata* (Müller, 1841)	(Britayev et al. 1999)
	Cr	*Himerometra robustipinna* (Carpenter, 1881)	(Britayev et al. 1999, 2016)
	Cr	*Anneissia bennetti* (Müller, 1841)	(Britayev et al. 1999)
	Cr	*Comaster* sp.	(Britayev et al. 1999)
	Cr	*Comaster multifidus* (Müller, 1841)	(Britayev et al. 1999)
Hololepidella millari Britayev, Doignon & Eeckahaut, 1999	A	*Acanthaster planci*	(Britayev & Antokhina 2012)
	A	*Choriaster granulatus*	(Britayev & Antokhina 2012)
	A	*Culcita novaeguineae*	(Britayev & Antokhina 2012)
	A	*Halityle regularis* Fisher, 1913	(Britayev & Antokhina 2012)
	A	*Protoreaster nodosus*	(Britayev & Antokhina 2012)
	A	*Euretaster insignis*	(Britayev & Antokhina 2012)
	Cr	*Clarkcomanthus alternans* (Carpenter, 1881)	(Britayev et al. 1999)
	A	*Linckia laevigata*	(Britayev et al. 1999)
	A	*Linckia guildingi*	(Britayev et al. 1999)
Hololepidella nigropunctata (Horst, 1915)	O	*Ophiothrix (Acanthophiothrix) purpurea*	(Britayev et al. 1999)
Hololepidella sp.	Cn	*Galaxea astreata* (Lamarck, 1816)	(Britayev & Antokhina 2012, Britayev et al. 2015)

Continued

Table 1 (Continued)   List of newly reported commensal polychaetes and their respective hosts from 1998 to 2017.

Symbiont		Host	Source references
Lepidasthenia brunnea Day, 1960	P	*Phyllochaetopterus* sp.	(Britayev & Martin 2005)
Lepidasthenia esbelta Amaral & Nonato, 1982	P	*Thelepus* sp.	(Amaral & Nonato 1982, Barnich et al. 2012b)
Lepidasthenia loboi Salazar-Vallejo, González & Salazar-Silva	P	*Thelepus antarcticus* Kinberg, 1867	(Salazar-Vallejo et al. 2015)
Lepidonotus sublevis Verrill, 1873	D	*Pagurus annulipes* (Stimpson, 1860)	(Dauer 1991, Williams & McDermott 2004)
	D	*Pagurus longicarpus* Say, 1817	(Dauer 1991, McDermott 2001, Williams & McDermott 2004)
Lepidonotus sp.	Ec	*Ochetostoma erythrogrammon* Leuckart & Ruppell, 1828	(Anker et al. 2005, Goto & Kato 2011)
Leucia violacea (Storm, 1879)	Cn	*Lophelia pertusa*	(Fiege & Barnich 2009)
	Cn	*Madrepora oculata*	(Fiege & Barnich 2009)
Medioantenna variopinta Di Camillo, Martin & Britayev, 2011	Cn	*Solanderia secunda* (Inaba, 1892)	(Nishi & Tachikawa 1999, Di Camillo et al. 2011)
Neohololepidella antipathicola Hartmann-Schröder & Zibrowius, 1999	Cn	*Elatopathes abietina* (Pourtalès, 1874)	(Hartmann-Schröder & Zibrowius 1998, Molodtsova & Budaeva 2007)
	Cn	*Distichopathes filix*	(Hartmann-Schröder & Zibrowius 1998, Molodtsova & Budaeva 2007)
Neolagisca jeffreysi (MaciIntosh, 1876)	Cn	*Madrepora oculata*	(Fiege & Barnich 2009 Núñez et al. 2011)
Neopolynoe acanellae (Verril, 1881)	Cn	*Thouarella (Thouarella) variabilis* Wright & Studer, 1889	D. Martin, personal observations
	Po	*Chondrocladia* sp.	(Bock et al. 2010)
Neopolynoe paradoxa (Anon, 1888)	Cn	*Lophelia pertusa*	(Jensen & Frederiksen 1992)
Ophthalmonoe pettiboneae Petersen & Britayev, 1997	P	*Chaetopterus* sp.	(Britayev & Martin 2005)
Paradyte crinoidicola (Potts, 1910)	Cr	*Amphimetra ensifer* (AH Clark, 1909)	(Britayev & Antokhina 2012)
	Cr	*Amphimetra tessellata* (Müller, 1841)	(Britayev & Antokhina 2012)
	Cr	*Cenometra bella* (Hartlaub, 1890)	(Britayev & Antokhina 2012)
	Cr	*Clarkcomanthus alternans*	(Britayev et al. 1999)
	Cr	*Comanthus gisleni* Hoggett, Birtles & Vail, 1986	(Britayev & Antokhina 2012)
	Cr	*Comaster multifidus*	(Britayev et al. 1999)
	Cr	*Comaster nobilis* (Carpenter, 1884)	(Britayev & Antokhina 2012)
	Cr	*Comatella nigra* (Carpenter, 1888)	(Britayev & Antokhina 2012)
	Cr	*Dichometra flagellata* (Müller, 1841)	(Britayev et al. 1999)
	Cr	*Liparometra regalis* (Carpenter, 1888)	(Britayev & Antokhina 2012)
	Cr	*Anneissia bennetti*	(Britayev et al. 1999)
	Cr	*Anneissia pinguis* (AH Clark, 1909)	(Britayev & Antokhina 2012)
	Cr	*Phanogenia gracilis*	(Britayev & Antokhina 2012)
	Cr	*Stephanometra tenuipinna* (Hartlaub, 1890)	(Britayev & Antokhina 2012)

Continued

Table 1 (Continued) List of newly reported commensal polychaetes and their respective hosts from 1998 to 2017.

Symbiont		Host	Source references
Paradyte levis (Marenzeller, 1902)	P	*Dendronephthya* sp.	(Britayev & Antokhina 2012)
Parahalosydnopsis tubicola (Day, 1973)	P	*Loimia verrucosa* Caullery, 1944	(Polgar et al. 2015)
Parahololepidella greefi (Augener, 1919)	Cn	*Tanacetipathes* cf. *spinescens* (Gray, 1857)	(Britayev et al. 2014)
Parapolyeunoa flynni (Benham, 1921)	Cn	*Errina aspera* (Linnaeus, 1767)	(Barnich et al. 2012a)
	Cn	*Inferiolabiata labiata* (Moseley, 1879)	(Barnich et al. 2012a)
	Cn	*Thouarella* sp.	(Barnich et al. 2012a)
Polyeunoa laevis McIntosh, 1885	Cn	*Primnosis* sp.	(Barnich et al. 2012a,b, Stiller 1996)
	Cn	*Dasystenella* sp.	(Barnich et al. 2012a,b, Stiller 1996)
Neopolynoe antarctica (Kinberg, 1858)	P	Unidentified terebellid	(Uschakov 1962, Hartmann-Schröder 1989, Barnich et al. 2012a)
	Cn	*Thuiaria* sp.	(Hartmann-Schröder 1989, Barnich et al. 2012a)
Unidentified	Cn	*Thouarella (Thouarella) bipinnata* Cairns, 2006	(Cairns 2006)
Unidentified	Cn	*Thouarella (Euthouarella) laxa* Versluys, 1906	(Versluys 1906, Watling et al. 2011)
Unidentified	Cn	*Thouarella cristata* Cairns 2011	(Versluys 1906, Cairns 2011, Watling et al. 2011)
Unidentified	Cn	*Minuisis pseudoplanum* Grant, 1976	(Alderslade 1998, Watling et al. 2011)
Unidentified	Cn	*Minuisis granti* Alderslade, 1998	(Alderslade 1998, Watling et al. 2011)
Unidentified	Cn	*Narella hypsocalyx* Cairns 2012	(Cairns 2012)
Unidentified	Cn	*Narella vulgaris* Cairns 2012	(Cairns 2012)
Unidentified	Cn	*Narella mosaica* Cairns 2012	(Cairns 2012)
Unidentified	Cn	*Narella dampieri* Cairns 2012	(Cairns 2012)
Pottsiscalisetosus praelongus (Marenzeller, 1902)	A	*Luidia maculata*	(Britayev & Antokhina 2012)
	A	*Archaster angulatus*	(Britayev & Antokhina 2012)
Showascalisetosus shimizui Imajima, 1997	Cn	*Stylaster* sp.	(Imajima 1997)
Unidentified	Ho	*Oneirophanta mutabilis mutabilis* Théel, 1879	(Schiaparelli et al. 2010)
Unidentified	Ho	*Laetmogone* sp.	(Schiaparelli et al. 2010)
Unidentified	D	*Paguristes eremita*	(Stachowitsch 1977, Williams & McDermott 2004)
Unidentified	D	*Pagurus cuanensis*	(Anker et al. 2005)
Unidentified	G	*Onustus longleyi*	(Whorff 1991)
Unidentified	P	*Eunice norvegica*	(Mortensen 2001)
Serpulidae			
Circeis spirillum (Linnaeus, 1758)	D	*Pagurus bernhardus*	(Samuelsen 1970, Jensen & Bender 1973, Williams & McDermott 2004)

Continued

Table 1 (Continued)   List of newly reported commensal polychaetes and their respective hosts from 1998 to 2017.

Symbiont		Host	Source references
	D	*Pagurus cuanensis*	(Samuelsen 1970, Williams & McDermott 2004)
	D	*Pagurus prideaux*	(Samuelsen 1970, Williams & McDermott 2004)
Josephella commensalis[a] Baluk & Radwański, 1997	Cn	*Tarbellastraea reussiana*[a] (Milne-Edwards & Haime, 1850)	(Bałuk & Radwański 1997)
Propomatoceros sulcicarinata[a] Ware, 1974	Cn	*Glomerula lombricus*[a] (Defrance, 1827)	(Garberoglio & Lazo 2011)
	Cn	*Mucroserpula mucroserpula*[a] Regenhardt, 1961	(Garberoglio & Lazo 2011)
Pseudovermilia madracicola Ten Hove, 1989	Cn	*Madracis senaria* Wells, 1973	(Hoeksema et al. 2017)
Spirobranchus corniculatus (Grube, 1862)	Cn	*Acropora digitifera* (Dana, 1846)	(Rowley 2008)
	Cn	*Acropora humilis* (Dana, 1846)	(Rowley 2008)
	Cn	*Acropora hyacinthus* (Dana, 1846)	(Rowley 2008)
	Cn	*Acropora loripes* (Brook, 1892)	(Rowley 2008)
	Cn	*Astreopora cucullata* Lamberts, 1980	(Rowley 2008)
	Cn	*Coeloseris mayeri* Vaughan, 1918	(Rowley 2008)
	Cn	*Cyphastrea chalcidicum* (Forskål, 1775)	(Rowley 2008)
	Cn	*Cyphastrea microphthalma* (Lamarck, 1816)	(Rowley 2008)
	Cn	*Cyphastrea serailia* (Forskål, 1775)	(Rowley 2008)
	Cn	*Dipsastraea amicorum* (Milne Edwards & Haime, 1849)	(Rowley 2008)
	Cn	*Dipsastraea favus* (Forskål, 1775)	(Ben-Tzvi et al. 2006)
	Cn	*Dipsastraea laxa* (Klunzinger, 1879)	(Ben-Tzvi et al. 2006)
	Cn	*Dipsastraea pallida* (Dana, 1846)	(Rowley 2008)
	Cn	*Dipsastraea speciosa* (Dana, 1846)	(Rowley 2008)
	Cn	*Favites abdita* (Ellis & Solander, 1786)	(Rowley 2008)
	Cn	*Favites pentagona* (Esper, 1795)	(Rowley 2008)
	Cn	*Galaxea astreata*	(Rowley 2008)
	Cn	*Merulina ampliata* (Ellis & Solander, 1786)	(Rowley 2008)
	Cn	*Millepora platyphylla* Hemprich & Ehrenberg, 1834	(Rowley 2008)
	Cn	*Millepora tenera* Boschma, 1949	(Rowley 2008)
	Cn	*Montipora aequituberculata*	(Rowley 2008)
	Cn	*Montipora foliosa* (Pallas, 1766)	(Rowley 2008)
	Cn	*Montipora foveolata* (Dana, 1846)	(Rowley 2008)
	Cn	*Montipora grisea* Bernard, 1897	(Rowley 2008)
	Cn	*Montipora informis* Bernard, 1897	(Rowley 2008)
	Cn	*Montipora monasteriata* (Forskål,1775)	(Rowley 2008)
	Cn	*Montipora spongodes* Bernard, 1897	(Rowley 2008)
	Cn	*Montipora spumosa* (Lamarck, 1816)	(Rowley 2008)

Continued

Table 1 (Continued)   List of newly reported commensal polychaetes and their respective hosts from 1998 to 2017.

Symbiont		Host	Source references
	Cn	*Montipora tuberculosa* (Lamarck, 1816)	(Rowley 2008)
	Cn	*Montipora undata* Bernard, 1897	(Rowley 2008)
	Cn	*Montipora venosa* (Ehrenberg, 1834)	(Rowley 2008)
	Cn	*Montipora verrucosa* (Lamarck, 1816)	(Rowley 2008)
	Cn	*Mycedium elephantotus* (Pallas, 1766)	(Rowley 2008)
	Cn	*Porites annae* Crossland, 1952	(Rowley 2008)
	Cn	*Porites lichen* Dana, 1846	(Rowley 2008)
	Cn	*Porites lobata* Dana, 1846	(Rowley 2008)
	Cn	*Porites nigrescens* Dana, 1848	(Rowley 2008)
	Cn	*Porites rus* (Forskål, 1775)	(Rowley 2008)
	Cn	*Seriatopora hystrix* Dana, 1846	(Rowley 2008)
	Cn	*Stylocoeniella armata* (Ehrenberg,1834)	(Rowley 2008)
	Cn	*Stylophora pistillata* Esper, 1797	(Rowley 2008)
	B	*Tridacna squamosa* Lamarck, 1819	(Schoot et al. 2016)
Spirobranchus giganteus (Pallas, 1766)	Cn	*Acropora palmata* (Lamarck, 1816)	(Hoeksema & ten Hove 2016, Hoeksema et al. 2017)
	Cn	*Agaricia agaricites* (Linnaeus, 1758)	(Hoeksema & ten Hove 2016, Hoeksema et al. 2017)
	Cn	*Agaricia humilis* Verrill, 1901	(Hoeksema & ten Hove 2016, Hoeksema et al. 2017)
	Cn	*Agaricia lamarcki* Milne Edwards and Haime, 1851	(Hoeksema & ten Hove 2016, Hoeksema et al. 2017)
	Cn	*Colpophyllia natans* (Houttuyn, 1772)	(Hoeksema & ten Hove 2016, Hoeksema et al. 2017)
	Cn	*Dendrogyra cylindrus* Ehrenberg, 1834	(Hoeksema & ten Hove 2016, Hoeksema et al. 2017)
	Cn	*Dichocoenia stokesii* Milne Edwards and Haime, 1848	(Hoeksema & ten Hove 2016, Hoeksema et al. 2017)
	Cn	*Eusmilia fastigiata* (Pallas, 1766)	(Hoeksema & ten Hove 2016, Hoeksema et al. 2017)
	Cn	*Helioseris cucullata* (Ellis and Solander, 1786)	(Hoeksema et al. 2017)
	Cn	*Madracis auretenra* Locke, Weil & Coates, 2007	(Hoeksema & ten Hove 2016, Hoeksema et al. 2017)
	Cn	*Madracis decactis* (Lyman, 1859)	(Hoeksema & ten Hove 2016, Hoeksema et al. 2017)
	Cn	*Madracis formosa* Wells, 1973	(Hoeksema & ten Hove 2016, Hoeksema et al. 2017)
	Cn	*Madracis pharensis* (Heller, 1868)	(Hoeksema & ten Hove 2016, Hoeksema et al. 2017)
	Cn	*Meandrina meandrites* (Linnaeus, 1758)	(Hoeksema & ten Hove 2016, Hoeksema et al. 2017)
	Cn	*Oculina valenciennesi* Milne Edwards and Haime, 1850	(Hoeksema & ten Hove 2016, Hoeksema et al. 2017)

Continued

Table 1 (Continued) List of newly reported commensal polychaetes and their respective hosts from 1998 to 2017.

Symbiont		Host	Source references
	Cn	*Orbicella faveolata* (Ellis and Solander, 1786)	(Hoeksema & ten Hove 2016, Hoeksema et al. 2017)
	Cn	*Orbicella franksi* (Gregory, 1895)	(Hoeksema & ten Hove 2016, Hoeksema et al. 2017)
	Cn	*Porites branneri* Rathbun, 1887	(Hoeksema & ten Hove 2016, Hoeksema et al. 2017)
	Cn	*Porites furcata* Lamarck, 1816	(Hoeksema & ten Hove 2016, Hoeksema et al. 2017)
	Cn	*Porites porites* (Pallas, 1766)	(Hoeksema & ten Hove 2016, Hoeksema et al. 2017)
	Cn	*Rhizopsammia goesi* (Lindström, 1877)	(Hoeksema & ten Hove 2016, Hoeksema et al. 2017)
	Cn	*Stephanocoenia intersepta* (Lamarck, 1816)	(Hoeksema & ten Hove 2016, Hoeksema et al. 2017)
	Cn	*Tubastraea coccinea* Lesson, 1829	(Hoeksema & ten Hove 2016, Hoeksema et al. 2017)
	Po	*Callyspongia (Cladochalina) vaginalis* (Lamarck, 1814)	(García-Hernández & Hoeksema 2017)
	Po	*Clathria curacaoensis* Arndt, 1927	(García-Hernández & Hoeksema 2017)
	Po	*Desmapsamma anchorata* (Carter, 1882)	(García-Hernández & Hoeksema 2017)
	Po	*Monanchora arbuscula* (Duchassaing & Michelotti, 1864)	(García-Hernández & Hoeksema 2017)
	Po	*Ircinia felix* (Duchassaing & Michelotti, 1864)	(García-Hernández & Hoeksema 2017)
	Po	*Neofibularia nolitangere* (Duchassaing & Michelotti, 1864)	(García-Hernández & Hoeksema 2017)
	Po	4 unidentified species	(García-Hernández & Hoeksema 2017)
	Po	*Callyspongia vaginalis*	(García-Hernández & Hoeksema 2017)
Spirorbis (Spirorbis) cuneatus Gee, 1964	D	*Paguristes eremita*	(Stachowitsch 1977, Williams & McDermott 2004)
	D	*Pagurus cuanensis*	(Stachowitsch 1977, Williams & McDermott 2004)
Spirorbis sp.	D	*Pagurus bernhardus*	(Augener 1926, Williams & McDermott 2004)
Spirorbis sp.	D	*Pagurus granosimanus*	(Walker 1988, Williams & McDermott 2004)
Spirorbis sp.	D	*Pagurus prideaux*	(Augener 1926, Williams & McDermott 2004)
Spirorbis sp.	D	*Pagurus samuelis*	(Walker & Carlton 1995, Williams & McDermott 2004)
Spirorbis sp.	D	*Petrochirus diogenes* (Linnaeus, 1758)	(Pearse 1932, Williams & McDermott 2004)
Spirorbis (Spirorbis) spirorbis (Linnaeus, 1758)	D	*Pagurus bernhardus*	(Eliason 1962, Samuelsen 1970, Williams & McDermott 2004)

Continued

Table 1 (Continued)   List of newly reported commensal polychaetes and their respective hosts from 1998 to 2017.

Symbiont		Host	Source references
Spirorbis (Spirorbis) tridentatus Levinsen, 1883	D	*Anapagurus chiroacanthus* (Lilljeborg, 1856)	(Samuelsen 1970, Williams & McDermott 2004)
	D	*Pagurus bernhardus*	(Samuelsen 1970, Williams & McDermott 2004)
	D	*Pagurus cuanensis*	(Samuelsen 1970, Williams & McDermott 2004)
Paradexiospira (Spirorbides) vitrea (Fabricius, 1780)	D	*Pagurus cuanensis*	(Samuelsen 1970, Williams & McDermott 2004)
Vermiliopsis sp.	Cn	*Helioseris cucullata*	(Hoeksema et al. 2017)
Vermiliopsis sp.	Cn	*Tubastraea coccinea* (Lesson, 1829)	(Hoeksema & ten Hove 2016)
	Cn	*Stylocoeniella armata* (Ehrenberg, 1834)	(Rowley 2008)
	Cn	*Stylophora pistillata* Esper, 1797	(Rowley 2008)
Scalibregmatidae			
Scalibregma inflatum Rathke, 1843	Ec	*Echiurus echiurus echiurus*	(Anker et al. 2005)
Siboglinidae			
Osedax rubiplumus Rouse, Goffredi & Vrijenhoek, 2004	P	*Osedax rubiplumus* Rouse, Goffredi & Vrijenhoek, 2004	(Rouse et al. 2004)
Osedax frankpressi Rouse, Goffredi & Vrijenhoek, 2004	P	*Osedax frankpressi* Rouse, Goffredi & Vrijenhoek, 2004	(Rouse et al. 2004)
Osedax japonicus Fujikura, Fujiwara & Kawato, 2006	P	*Osedax japonicus* Fujikura, Fujiwara & Kawato, 2006	(Fujikura et al. 2006)
Osedax mucofloris Glover et al., 2005	P	*Osedax mucofloris* Glover et al., 2005	(Glover et al. 2005)
Osedax roseus Rouse et al., 2008	P	*Osedax roseus* Rouse et al., 2008	(Rouse et al. 2008)
Osedax "spiral"	P	*Osedax* "spiral"	(Braby et al. 2007)
Osedax "rosy"	P	*Osedax* "rosy"	(Braby et al. 2007)
Osedax "yellow collar"	P	*Osedax* "yellow collar"	(Braby et al. 2007)
Osedax "orange collar"	P	*Osedax* "orange collar"	(Braby et al. 2007)
Osedax "nude-palp-A"	P	*Osedax* "nude-palp-A"	(Jones et al., 2008)
Osedax "nude-palp-B"	P	*Osedax* "nude-palp-B"	(Jones et al. 2008)
Osedax "nude-palp-C"	P	*Osedax* "nude-palp-C"	(Rouse et al. 2009)
Osedax "nude-palp-D"	P	*Osedax* "nude-palp-D"	(Vrijenhoek et al. 2009)
Osedax "nude-palp-E"	P	*Osedax* "nude-palp-E"	(Vrijenhoek et al. 2009)
Osedax "nude-palp-F"	P	*Osedax* "nude-palp-F"	(Vrijenhoek et al. 2009)
Osedax "white-collar"	P	*Osedax* "white-collar"	(Vrijenhoek et al. 2009)
Osedax "yellow patch"	P	*Osedax* "yellow patch"	(Vrijenhoek et al. 2009)
Osedax "green-palp"	P	*Osedax* "green-palp"	(Vrijenhoek et al. 2009)
Sigalionidae			
Pholoides sinepapillatus Miranda & Brasil, 2014	Cn	*Solenosmilia variabilis.*	(Miranda & Brasil 2014)
	Cn	*Errinia* sp.	(Miranda & Brasil 2014)
Spionidae			
Carazziella spongilla Sato-Okoshi, 1998	Po	*Spongilla alba* Carter, 1849	(Sato-Okoshi 1998)

Continued

Table 1 (Continued) List of newly reported commensal polychaetes and their respective hosts from 1998 to 2017.

Symbiont		Host	Source references
Polydora colonia Moore, 1907	Po	*Clathria (Clathria) prolifera* (Ellis & Solander, 1786)	(David & Williams 2012)
	Po	*Halichondria bowerbanki* Burton, 1930	(David & Williams 2012)
Polydora nanomon Orensky & Williams, 2009	D	*Calcinus tibicen*	(Orensky & Williams 2009)
	D	*Phimochirus holthuisi* (Provenzano, 1961)	(Orensky & Williams 2009)
	D	*Paguristes* sp.	(Orensky & Williams 2009)
	D	Unidentified pagurid	(Orensky & Williams 2009)
cf. *Polydora* sp.[a]	E	*Echinocorys ovatus*[a] (Leske, 1778)	(Wisshak & Neumann 2006)
Polydorella kamakamai Williams, 2004	Po	*Clathria (Thalysias) cervicornis* (Thiele, 1903)	(Williams 2004)
Polydorella dawydoffi Radashevsky, 1996	Po	*Chalinula nematifera* (de Laubenfels, 1954)	(Williams 2004)
	Po	*Xestospongia* sp.	(Britayev & Antokhina 2012)
Polydorella smurovi Tzetlin & Britayev, 1985	Po	*Negombata magnifica* (Keller, 1889)	(Naumann et al. 2016)
Polydorella sp.	Po	*Haliclona djeedara* Fromont & Abdo, 2014	(Fromont & Abdo 2014)
Polydorella sp.	Po	*Haliclona durdong* Fromont & Abdo, 2014	(Fromont & Abdo 2014)
Syllidae			
Genus A	Po	orange finger sponge	(San Martín et al. 2010)
Alcyonosyllis aidae Álvarez-Campos, San Martín & Aguado, 20013	Cn	*Dendronephthya* sp.	(Álvarez-Campos et al. 2013)
Alcyonosyllis bisetosa (Hartmann-Schröder, 1960)	Cn	Unidentified alcyonarian	(Glasby & Aguado 2009, Hartmann-Schröder 1960)
Alcyonosyllis glasbyi San Martín & Nishi, 2003	Cn	*Melithaea flabellifera* (Kükenthal, 1908)	(Kumagai & Aoki 2003, San Martín & Nishi 2003)
Alcyonosyllis gorgoniacolo (Sun &Yang, 2004)	Cn	Orange-red gorgonia	(Sun & Yang 2004, Glasby & Aguado 2009, Lattig & Martin 2009)
Alcyonosyllis hinterkircheri Glasby & Aguado, 2009	Cn	*Goniopora* cf. *stokesi* Milne Edwards & Haime, 1851	(Glasby & Aguado 2009)
Alcyonosyllis phili Glasby & Watson, 2001	Cn	*Dendronephthya* sp.	(Glasby & Watson 2001)
	Cn	*Melithaea* sp. 1	(Glasby & Watson 2001)
	Cn	*Melithaea* sp. 2	(Glasby & Watson 2001)
	Cn	Nephtheidae gen. sp.	(Glasby & Watson 2001)
	Cn	Unidentified gorgonian	(Glasby & Aguado 2009)
	Cn	Unidentified sea fan	(Glasby & Aguado 2009)
	Cn	Unidentified nephtheid	(Britayev & Antokhina 2012)
	Cn	*Dendronephthya* sp.	(Britayev & Antokhina 2012)
	Cn	Unidentified hard coral	(Aguado & Glasby 2015)
	Cn	Unidentified gorgonian	(Britayev & Antokhina 2012)

Continued

Table 1 (Continued)   List of newly reported commensal polychaetes and their respective hosts from 1998 to 2017.

Symbiont		Host	Source references
	Cn	*Carijoa* sp.	(Britayev & Antokhina 2012)
Bollandiella antipathicola (Glasby, 1994)	Cn	*Cirrhipathes* cf. *rumphii* van Pesch, 1910	Bo and Martin, unpublished results
	Cn	*Stichopathes gravieri* Molodtsova, 2006	Martin & Molodtsova, unpublished results
Branchiosyllis lamelligera Verrill, 1900	Po	*Aplysina archeri*	(San Martín et al. 2013)
	Po	unidentified, massive purple and yellow	(San Martín et al. 2013)
Branchiosyllis tamenderensis Paresque, Fukuda & Matos, 2016	Po	*Tedania ignis* (Duchassaing & Michelotti, 1864)	(Paresque et al. 2016)
Eusyllis blomstrandi Malmgren, 1867	D	*Elassochirus cavimanus*	(Hoberg et al. 1982, Williams & McDermott 2004)
	D	*Labidochirus splendescens*	(Hoberg et al. 1982, Williams & McDermott 2004)
	D	*Pagurus aleuticus*	(Hoberg et al. 1982, Williams & McDermott 2004)
	D	*Pagurus capillatus*	(Hoberg et al. 1982, Williams & McDermott 2004)
	D	*Pagurus confragosus*	(Hoberg et al. 1982, Williams & McDermott 2004)
	D	*Pagurus ochotensis*	(Hoberg et al. 1982, Williams & McDermott 2004)
	D	*Pagurus setosus*	(Hoberg et al. 1982, Williams & McDermott 2004)
	D	*Pagurus trigonocheirus*	(Hoberg et al. 1982, Williams & McDermott 2004)
Haplosyllides aberrans (Fauvel, 1939)	D	*Platycaris latirostris* Holthuis, 1952	(Martin et al. 2008, 2009)
	D	*Palaemonella rotumana* (Borradaile, 1898)	I. Marin, personal communication
Haplosyllides ophiocomae Martin, Aguado & Britayev, 2009	O	*Ophiocoma pusilla* (Brock, 1888)	(Hartmann-Schröder 1978, Martin et al. 2009)
Haplosyllis amphimedonicola Paresque & Nogueira, 2014	Po	*Amphimedon viridis* Duchassaing & Michelotti, 1864	(Paresque & Nogueira 2014)
Haplosyllis aplysinae Lattig & Martin, 2011a	Po	*Aplysina insularis* (Duchassaing & Michelotti, 1864)	(Lattig & Martin 2011a)
	Po	*Aplysina lacunosa* (Lamarck, 1814)	(Lattig & Martin 2011a)
	Po	*Aplysina cauliformis* (Carter, 1882)	(Lattig & Martin 2011a)
	Po	*Aplysina bathyphila* Maldonado & Young, 1998	(Lattig & Martin 2011a)
	Po	*Aplysina fistularis* (Pallas, 1766)	(Lattig & Martin 2011a)
	Po	*Aplysina* sp.	(Lattig & Martin 2011a)
	Po	*Aplysina archeri*	(Fiore & Jutte 2010, non-confirmed)
Haplosyllis basticola Sardá, Ávila & Paul, 2002	Po	*Ianthella basta* (Pallas, 1766)	(Sardá et al. 2002, Lattig & Martin 2009)
	Po	*Anomoianthella lamella* Pulitzer-Finali & Pronzato, 1999	(Magnino et al. 1999a, Lattig et al. 2010b)
	Po	Unidentified iantellid	(Lattig et al. 2010b)

Continued

Table 1 (Continued)   List of newly reported commensal polychaetes and their respective hosts from 1998 to 2017.

Symbiont		Host	Source references
Haplosyllis chaetofusorata Lattig & Martin, 2011a	Po	*Verongula reiswigi* Alcolado, 1984	(Lattig & Martin 2011a)
	Po	*Verongula rigida* (Esper, 1794)	(Lattig & Martin 2011a, non-confirmed)
Haplosyllis cephalata Verrill, 1900	Po	Unidentified	(Lattig & Martin 2009)
	Po	*Ircinia campana* (Lamarck, 1814)	(Fiore & Jutte 2010, non-confirmed, Lattig & Martin 2011a)
Haplosyllis chamaeleon Laubier, 1960	Cn	*Paramuricea grayi* (Johnson, 1861)	(Lattig & Martin 2009)
Haplosyllis crassicirrata Aguado, San Martín & Nishi, 2006	Po	Grey with orange flesh	(Lattig et al. 2010a)
Haplosyllis ingensicola Lattig, Martin & Aguado, 2010a	Po	*Acanthostrongylophora ingens* (Thiele, 1899)	(Lattig et al. 2010a)
Haplosyllis djiboutiensis Gravier, 1900	Po	Unidentified	(Lattig & Martin 2009)
	Po	Unidentified	(Lattig & Martin 2009, Lee & Rho 1994)
	Po	Pink from sandy bottoms	(Lattig & Martin 2011b)
	Po	Unidentified sp. 1	(Lattig & Martin 2011b)
	Po	Unidentified sp. 2	(Lattig & Martin 2011b)
	Po	Unidentified sp. 3	(Lattig & Martin 2011b)
	Po	Unidentified sp. 4	(Lattig & Martin 2011b)
Haplosyllis eldagainoae Lattig & Martin, 2011	Po	*Theonella swinhoei* ray, 1868	(Magnino & Gaino 1998, Magnino et al. 1999b, Lattig & Martin 2011b)
Haplosyllis giuseppemagninoi Lattig & Martin, 2011	Po	*Liosina paradoxa* Thiele, 1899	(Magnino & Gaino 1998, Lattig & Martin 2011b)
Haplosyllis gula Treadwell, 1924	Po	*Neofibularia nolitangere* (Duchassaing & Michelotti, 1864)	(Lattig & Martin 2009)
Haplosyllis loboi Paola, San Martín & Martin, 2006	Po	Unidentified	(Paola et al. 2006, Paresque et al. 2016)
Haplosyllis navasi Lattig & Martin 2011a	Po	*Ircinia strobilina* (Lamarck, 1816)	(Lattig & Martin 2011a)
Haplosyllis nicoleae Lattig, Martin & Aguado, 2010a	Po	*Clathria (Thalysias) reinwardti* Vosmaer, 1880	(Lattig et al. 2010a)
	Po	*Biemna trirhaphis* (Topsent, 1897)	(Lattig et al. 2010a)
	Po	*Melophlus sarasinorum* Thiele, 1899	(Lattig et al. 2010a)
Haplosyllis niphatidae Lattig & Martin, 2011	Po	*Niphates erecta* Duchassaing & Michelotti, 1864	(Lattig & Martin 2011a)
	Po	*Niphates digitalis* (Lamarck, 1814)	(Lattig & Martin 2011a)
	Po	Pink coral reeef sponge	(Lattig & Martin 2011a)
Haplosyllis streptocephala (Grube, 1857)	Po	Unidentified	(Lattig & Martin 2009)
Haplosyllis villogorgicola Martin, Núñez, Riera & Gil, 2002	Cn	*Villogorgia bebrycoides* (Koch, 1887)	(Martin et al. 2002)
Haplosyllis sp.	Po	*Haliclona* sp.	Britayev, personal observations
Haplosyllis sp.	Po	*Topsentia* sp.	(Fiore & Jutte 2010, non-confirmed)

Continued

Table 1 (Continued)   List of newly reported commensal polychaetes and their respective hosts from 1998 to 2017.

Symbiont		Host	Source references
Haplosyllis sp.	Po	*Cliona* sp.	(Fiore & Jutte 2010, non-confirmed)
Haplosyllis sp.	Po	*Erylus* sp.	(Fiore & Jutte 2010, non-confirmed)
Haplosyllis sp.	Po	*Geodia* sp.	(Fiore & Jutte 2010, non-confirmed)
Haplosyllis sp.	T	Unidentified Didemnid	(Fiore & Jutte 2010, non-confirmed)
Haplosyllis sp.	A	*Luidia maculata*	(Britayev & Antokhina 2012)
Haplosyllis sp.	A	*Archaster angulatus*	(Britayev & Antokhina 2012)
Haplosyllis sp.	Po	*Mycale (Zygomycale) parishii* (Bowerbank, 1875)	(Duarte & Nalesso 1996)
Haplosyllis sp.	Po	*Cliona varians* (Duchassaing & Michelotti, 1864)	(Stofel et al. 2008)
Haplosyllis sp.	Po	*Mycale (Carmia) microsigmatosa* Arndt, 1927	(Ribeiro et al. 2003)
Haplosyllis sp.	Cn	*Mussismilia hispida* (Verrill, 1901)	(Nogueira 2000)
Haplosyllis sp.	Br	*Schizoporella errata* (Waters, 1878)	(Morgado & Tanaka 2001)
Haplosyllis sp.	Po	*Callyspongia* sp.	(Britayev & Antokhina 2012)
Haplosyllis sp.	Po	Unidentified	(Britayev & Antokhina 2012)
Inermosyllis sp.	A	*Archaster angulatus*	(Britayev & Antokhina 2012)
	A	*Luidia maculata*	(Britayev & Antokhina 2012)
Pionosyllis magnifica Moore, 1906	D	*Paralithodes camtschaticus* (Tilesius, 1815)	(López et al. 2001)
Ramisyllis multicaudata Glasby, Schroeder & Aguado, 2012	Po	*Petrosia* sp.	(Glasby et al. 2012)
Syllis cf. *armillaris* (O.F. Müller, 1776)	D	*Paragiopagurus boletifer* (de Saint Laurent, 1972)	(López et al. 2001)
	D	*Paguristes* sp.	(López et al. 2001)
Syllis exiliformis Imajima, 2003	Cn	*Verrucella* sp.	(Imajima 2003)
Syllis ferrani Alós & San Martín, 1978	D	*Pagurus excavatus* (Herbst, 1791)	(López et al. 2001)
Syllis mayeri Musco & Giangrande, 2005	Po	*Ircinia strobilina*	(Musco & Giangrande 2005)
Syllis mercedesae Lucas, San Martin & Parapar, 2012	S	*Phascolion (Phascolion) strombus strombus* (Montagu, 1804)	(Troncoso et al. 2000, Martins et al. 2013)
Syllis sp.	S	*Aspidosiphon* sp.	D. Martin, personal observations
Syllis sp.	S	Unidentified	D. Martin, personal observations
Syllis sp.	S	*Phascolion (Phascolion) caupo* Hendrix, 1975	(Hendrix 1975)
Syllis sp.	S	*Phascolion (Lesenka) collare* Selenka & de Man, 1883	(Knudsen 1944; D. Martin, personal observations)
Syllis pontxioi San Martín & López, 2000	D	*Clibanarius* sp.	(López et al. 2001)
	D	*Paguristes* sp.	(López et al. 2001)
Syllis ramosa McIntosh, 1879	Po	Hexactinellid sponges	(McIntosh 1879, 1885)
	Po	*Crateromorpha* sp.	(Imajima 1966, Read 2001)
Syllis unzima Simon, San Martín & Robinson, 2014	Ho	*Holothuria (Metriatyla) scabra* Jaeger, 1833	(Simon et al. 2014)

Continued

Table 1 (Continued)   List of newly reported commensal polychaetes and their respective hosts from 1998 to 2017.

Symbiont		Host	Source references
Trypanobia cryptica Aguado, Murray & Hutcings 2015	Po	Red sponge	(Aguado et al. 2015)
Trypanobia depressa (Augener, 1913)	Po	Red sponge	(Aguado et al. 2015)
Typosyllis sp.	Cn	*Cyphastrea microphthalma* (Lamarck, 1816)	(Randall & Eldredge 1976)
Typosyllis sp.	Cn	*Dipsastraea favus* (Forskål, 1775)	(Randall & Eldredge 1976)
Typosyllis sp.	Cn	*Dipsastraea speciosa* (Dana, 1846)	(Randall & Eldredge 1976)
Typosyllis sp.	Cn	*Favia* sp.	(Randall & Eldredge 1976)
Typosyllis sp.	Cn	*Favites pentagona* (Esper, 1795)	(Randall & Eldredge 1976)
Typosyllis sp.	Cn	*Favites valenciennesi* (Milne Edwards & Haime, 1849)	(Randall & Eldredge 1976)
Typosyllis sp.	Cn	*Favites* sp.	(Randall & Eldredge 1976)
Typosyllis sp.	Cn	*Symphyllia recta* (Dana, 1846)	(Randall & Eldredge 1976)

Source: Currently accepted names for both polychaetes and their hosts have been checked through the WoRMS website. (Horton T. et al. 2017. World Register of Marine Species (WoRMS). WoRMS Editorial Board.)

[a] Fossil species.

Po – Porifera, B – Bivalvia, G – Gastropoda, N – Nudibranchia, Ce – Cephalopoda, P – Polychaeta, Ec – Echiuroidea, S – Sipunculoidea, H – Hemichordata, D – Decapoda, A – Asteroidea, Cr – Crinoidea, E – Echinoidea, Ho – Holothuroidea, O – Ophiuroidea, Go – Gorgonocephalid, Cn – Cnidaria, Br – Bryozoa, T – Tunicata.

Symbiotic polychaetes by numbers

The known symbiotic polychaetes (excluding myzostomids) currently increased to 618 species involved in 1626 relationships, which implies that 245 species and 660 relationships have been newly reported since 1998. Overall, 490 species are involved in 1229 commensal relationships (Table 1), and 128 are involved in 397 parasitic relationships (Table 2). Since the first known symbiotic polychaete was described by Koch (1846) until 2017, this represents an average of 3.6 and 9.5 new species and relationships being described/reported per year. However, the annual averages were of 2.4 new symbiotic species (1.9 commensal, 0.5 parasitic) and about 6.4 relationships (4.7 commensal, 1.7 parasitic) until 1998. From that year until 2017, the number of newly reported species and relationships per year has been 12.9 (10.4 commensals, 2.5 parasites) and 34.8 (27.2 commensals, 7.6 parasites), which overall represents around five times more new reports of both species and relationships per year.

Concerning the degree of specificity of the associations in which the symbiotic polychaetes are involved, monoxenous relationships (i.e. those involving a single host species) followed the same trend as the total number, being almost double in 2017 than in 1998, still dominant and representing more than half of the total (Table 3). Despite the new knowledge produced in recent years, it is still not possible to assess whether many of these associations are really monoxenous or if this character corresponds to a lack of real information. For instance, some relevant 'monoxenous' species appear now to be present in one or several additional hosts, including *Bollandiella antipathicola* (Glasby, 1994), *Haplosyllis chamaeleon* Laubier, 1960 and *Haplosyllides aberrans* (Fauvel, 1939).

On the other hand, in 1998, there were only nine species (3.2%) involved in 10 or more relationships with different host species (i.e. polyxenous), and this number increased to almost 29 (6.3%) in 2017. Considering the current precision of the available tools for determining the taxonomic status of living organisms, we believe virtually all these multihost symbionts to be true polyxenous species. This is in contrast to some previous cases in which sibling species were involved, such as the paradigmatic *Haplosyllis spongicola* (Grube, 1855) species-complex (discussed in the commensal and parasitic Syllidae sections).

Table 2 List of newly reported parasitic polychaetes and their respective hosts from 1998 to 2017.

Parasite		Host	Source references
Cirratulidae			
Dodecaceria concharum Örsted, 1843	D	*Calcinus gaimardi* (H. Milne Edwards, 1848)	(Williams & McDermott 2004)
	D	*Dardanus arrosor* (Herbst, 1796)	(Cuadras & Pereira 1977, Williams & McDermott 2004)
	D	*Paguristes eremita* (Linnaeus, 1767)	(Stachowitsch 1977, Williams & McDermott 2004)
	D	*Pagurus cuanensis* Bell, 1846	(Stachowitsch 1977, Williams & McDermott 2004)
Dorvilleidae			
Veneriserva pygoclava meridionalis Micaletto, Gambi & Cantone, 2002	P	*Laetmonice producta* Grube, 1876	(Micaletto et al. 2002)
Hesionidae			
Oxydromus okupa Martin, Meca et al., 2017	B	*Scrobicularia plana* (da Costa, 1778)	(Martin et al. 2012, 2015, 2017a)
	B	*Macomopsis pellucida* (Splenger, 1798)	(Martin et al., 2012, 2015, 2017a)
Ichthyotomidae			
Ichthyotomus sanguinarius Eisig, 1906	F	*Conger conger* (Linnaeus, 1758)	(Culurgioni et al. 2006)
Oenonidae			
Haematocleptes terebellidis Wiren, 1886	P	*Ampharete falcata* Eliason 1955	(Mackie & Garwood 1995)
Labrorostratus caribensis Hernández-Alcántara, Crus-Prez & Solís-Weiss, 2015	P	Nereididae sp.	(Hernández-Alcántara et al. 2015)
Labrorostratus zaragozensis Hernández-Alcántara & Solís-Weiss, 1998	P	*Terebellides californica* Williams, 1984	(Hernández-Alcántara & Solís-Weiss 1998)
Phyllodocidae			
Eulalia bilineata (Johnston, 1840)	D	*Pagurus bernhardus*	(Jensen & Bender 1973)
Eulalia viridis (Linnaeus, 1767)	D	*Pagurus cuanensis*	(Samuelsen 1970)
Galapagomystides aristata Blake, 1985	U	-	(Jenkins et al. 2002)
Polynoidae			
Branchipolynoe seepensis Pettibone, 1986a	B	*Bathymodiolus azoricus* Cosel & Comtet, 1999	(Jollivet et al. 1998, Britayev et al. 2003a,b, 2007, Plouviez et al. 2008)
	B	*Batymodiolus heckerae*	(Becker et al. 2013)
	B	*Bathymodiolus puteoserpentis* Cosel, Métivier & Hashimoto, 1994	(Chevaldonné et al. 1998; Ward et al. 2004; Britayev et al. 2007)
	B	*Bathymodiolus* sp.	(Jollivet et al. 1998)
Eunoe opalina McIntosh, 1885	Ho	*Bathyplotes bongraini* Vaney, 1914	(Schiaparelli et al., 2010)

Continued

Table 2 (Continued) List of newly reported parasitic polychaetes and their respective hosts from 1998 to 2017.

Parasite		Host	Source references
Gastrolepidia clavigera Schmarda, 1861	Ho	*Holothuria (Acanthotrapeza) coluber* Semper, 1868	(Britayev et al. 1999; Marudhupandi et al. 2014)
	Ho	*Holothuria (Microthele) nobilis* (Selenka, 1867)	(Marudhupandi et al. 2014)
	Ho	*Holothuria* sp.	(Marudhupandi et al. 2014)
	Ho	*Synapta maculata* (Chamisso & Eysenhardt, 1821)	(Britayev & Antokhina 2012)
Thormora johnstoni (Kinberg, 1856)	P	*Palola viridis* Gray in Stair, 1847	(Hauenschild et al. 1968)
Sabellidae			
Terebrasabella heterouncinata Fitzhugh & Rouse, 1999	G	*Haliotis midae* Linnaeus, 1758	(Fitzhugh & Rouse 1999; Simon et al. 2002, 2004, 2005)
	G	*Haliotis rufescens* Swainson, 1822	(Day et al. 2000, Culver & Kuris 2004, Moore et al. 2007)
	G	*Tegula funebralis* (A. Adams, 1855)	(Culver & Kuris 2004, Moore et al. 2007)
	G	*Lottia gigantea* Gray in G. B. Sowerby I, 1834	(Culver & Kuris 2004)
	G	*Lottia pelta* (Rathke, 1833)	(Culver & Kuris 2004)
	G	*Fissurella volcano* Reeve, 1849	(Culver & Kuris 2004)
	G	*Megastraea undosa* (W. Wood, 1828)	(Culver & Kuris 2004)
	G	*Tectarius striatus* (King, 1832)	(Culver & Kuris 2004)
	G	*Cerithideopsis californica* (Haldeman, 1840)	(Culver & Kuris 2004)
	G	*Maxwellia santarosana* (Dall, 1905)	(Culver & Kuris 2004)
	G	*Nucella emarginata* (Deshayes, 1839)	(Culver & Kuris 2004)
	G	*Kelletia kelletii* (Forbes, 1850)	(Culver & Kuris 2004)
	G	*Olivella biplicata* (Sowerby I, 1825)	(Culver & Kuris 2004)
	G	*Californicus californicus* (Reeve, 1844)	(Culver & Kuris 2004)
Serpulidae			
Unknown	Bc	*Peregrinella multicarinata*[a] (Lamarck, 1819)	(Kiel 2008)
Spionidae			
Boccardia berkeleyorum Blake & Woodwick, 1971	D	*Pagurus granosimanus* (Stimpson, 1859)	(Blake & Woodwick 1971)
	D	*Calcinus gaimardii*	(Williams 2001a, Williams & McDermott 2004)
	D	*Dardanus lagopodes* (Forskål, 1775)	(Williams 2001a, Williams & McDermott 2004)
	D	Unidentified pagurid	(Blake & Evans 1973, Williams 2001a, Williams & McDermott 2004)

Continued

Table 2 (Continued)   List of newly reported parasitic polychaetes and their respective hosts from 1998 to 2017.

Parasite		Host	Source references
Boccardia columbiana Berkeley, 1927	D	*Pagurus granosimanus*	(Woodwick 1963, Williams & McDermott 2004)
	D	*Pagurus samuelis* (Stimpson, 1857)	(Woodwick 1963, Williams & McDermott 2004)
	D	Unidentified pagurid	(Blake & Evans 1973, Williams & McDermott 2004)
Boccardia proboscidea Hartman, 1940	D	Unidentified pagurid	(Blake & Evans 1973, Williams & McDermott 2004)
Boccardia tricuspa (Hartman, 1939)	D	*Pagurus granosimanus*	(Woodwick 1963, Williams & McDermott 2004)
	D	*Pagurus hirsutiusculus* (Dana, 1851)	(Blake & Woodwick 1971, Williams & McDermott 2004)
	D	*Pagurus samuelis*	(Woodwick 1963, Williams & McDermott 2004)
	D	Unidentified pagurid	(Blake & Evans 1973, Williams & McDermott 2004)
Boccardiella hamata (Webster, 1879)	D	Unidentified pagurid	(Blake & Evans 1973, Williams & McDermott 2004)
	D	*Pagurus pollicaris* Say, 1817	(Dean & Blake 1966)
Dipolydora alborectalis (Radashevsky, 1993)	D	Unidentified pagurid	(Radashevsky 1993, Williams & McDermott 2004)
Dipolydora armata (Langerhans, 1880)	Cn	*Leptastrea purpurea* (Dana, 1846)	(Okuda 1937)
	Cn	*Millepora complanata* Lamarck, 1816	(Lewis 1998)
	D	*Calcinus gaimardii*	(Williams 2001a, Williams & McDermott 2004)
	D	*Calcinus latens* (Randall, 1840)	(Williams 2001a, Williams & McDermott 2004)
	D	*Calcinus minutus* Buitendijk, 1937	(Williams 2001a, Williams & McDermott 2004)
	D	*Calcinus pulcher* Forest, 1958	(Williams 2001a, Williams & McDermott 2004)
	D	*Calcinus tubularis* (Linnaeus, 1767)	(Bick 2001, Williams & McDermott 2004)
	D	*Ciliopagurus strigatus* (Herbst, 1804)	(Williams 2001a, Williams & McDermott 2004)
	D	*Clibanarius englaucus* Ball & Haig, 1972	(Williams 2001a, Williams & McDermott 2004)
	D	*Clibanarius erythropus* (Latreille, 1818)	(Bick 2001, Williams & McDermott 2004)
	D	*Dardanus lagopodes*	(Williams 2001a, Williams & McDermott 2004)
	D	*Paguristes runyanae* Haig & Ball, 1988	(Williams 2001a, Williams & McDermott 2004)
	D	Unidentified pagurid	(Sato-Okoshi 1999, Williams & McDermott 2004)
Dipolydora bidentata (Zachs, 1933)	D	*Pagurus granosimanus*	(Blake & Woodwick 1971, Williams & McDermott 2004)

Continued

Table 2 (Continued)   List of newly reported parasitic polychaetes and their respective hosts from 1998 to 2017.

Parasite		Host	Source references
	D	*Pagurus samuelis*	(Blake & Woodwick 1971, Williams & McDermott 2004)
	D	Unidentified pagurid	(Radashevsky 1993, Williams & McDermott 2004)
Dipolydora commensalis (Andrews, 1981)	D	*Elassochirus cavimanus*	(Hoberg et al. 1982, Williams & McDermott 2004)
	D	*Labidochirus splendescens*	(Hoberg et al. 1982, Williams & McDermott 2004)
	D	*Pagurus aleuticus*	(Hoberg et al. 1982, Williams & McDermott 2004)
	D	*Pagurus granosimanus*	(Hatfield 1965, Williams & McDermott 2004)
	D	*Pagurus ochotensis*	(Hoberg et al. 1982, Williams & McDermott 2004)
	D	*Pagurus setosus*	(Hoberg et al. 1982, Williams & McDermott 2004)
	D	*Pagurus trigonocheirus*	(Hoberg et al. 1982, Williams & McDermott 2004)
Dipolydora concharum (Verrill, 1880)	D	Unidentified pagurid	(Radashevsky 1993, Williams & McDermott 2004)
Dipolydora elegantissima (Blake & Woodwick, 1972)	D	*Pagurus granosimanus*	(Blake & Woodwick 1971, Williams & McDermott 2004)
Dipolydora socialis (Schmarda, 1861)	D	*Calcinus gaimardii*	(Williams 2001a, Williams & McDermott 2004)
	D	*Calcinus minutus*	(Williams 2001a, Williams & McDermott 2004)
	D	*Dardanus lagopodes*	(Williams 2001a, Williams & McDermott 2004)
	D	*Paguristes runyanae*	(Williams 2001a, Williams & McDermott 2004)
	D	*Pagurus brachiomastus* (Thallwitz, 1892)	(Radashevsky 1993, Williams & McDermott 2004)
Dipolydora tridenticulata (Woodwick, 1964)	D	*Calcinus gaimardii*	(Williams 2001a, Williams & McDermott 2004)
	D	*Calcinus latens*	(Williams 2001a, Williams & McDermott 2004)
	D	*Calcinus minutus*	(Williams 2001a, Williams & McDermott 2004)
	D	*Dardanus lagopodes*	(Williams 2001a, Williams & McDermott 2004)
Scolelepis laonicola (Tzetlin, 1985)	P	*Scolelepis laonicola* (Tzetlin, 1985)	(Tzetlin 1985, Vortsepneva et al. 2006, 2008)
Polydora cf. *alloporis* Light, 1970	Cn	*Distichopora robusta* Lindner, Cairns & Guzman, 2004	(Lindner et al. 2004)
Polydora bioccipitalis Blake & Woodwick, 1971	D	*Pagurus hirsutiusculus*	(Blake & Woodwick 1971, Williams & McDermott 2004)
Polydora ciliata (Johnston, 1838)	D	*Dardanus arrosor*	(Cuadras & Pereira 1977, Williams & McDermott 2004)

Continued

Table 2 (Continued)   List of newly reported parasitic polychaetes and their respective hosts from 1998 to 2017.

Parasite		Host	Source references
	D	*Diogenes pugilator* (Roux, 1829)	(Codreanu & Mack-Fira 1961, Williams & McDermott 2004)
	D	*Pagurus bernhardus*	(Jensen & Bender 1973, Reiss et al. 2003, Williams & McDermott 2004)
	D	*Pagurus samuelis*	(Williams & McDermott 2004, Woodwick 1963)
Polydora mabinii Williams, 2001	D	*Calcinus latens*	(Williams 2001a, Williams & McDermott 2004)
Polydora maculata Day, 1963	D	Unidentified pagurid	(Day 1963, Williams & McDermott 2004)
Polydora neocaeca Williams & Radashevsky, 1999	D	*Pagurus longicarpus*	(Williams & McDermott 2004, Williams & Radashevsky 1999)
Polydora pygidialis Blake & Woodwick, 1971	D	*Pagurus granosimanus*	(Blake & Woodwick 1971, Williams & McDermott 2004)
Polydora rickettsi Woodwick, 1961	G	*Crepipatella peruviana* (Lamarck, 1822)	(Bertrán et al. 2005)
Polydora robi Williams, 2000	D	*Calcinus gaimardii*	(Williams 2002, Williams & McDermott 2004)
	D	*Calcinus latens*	(Williams 2000, 2001a,b, 2002, Williams & McDermott 2004)
	D	*Calcinus minutus*	(Williams 2000, 2001a,b, 2002, Williams & McDermott 2004)
	D	*Calcinus pulcher*	(Williams 2000, 2001a,b, 2002, Williams & McDermott 2004)
	D	*Ciliopagurus strigatus*	(Williams 2001a, Williams & McDermott 2004)
	D	*Clibanarius cruentatus* H. Milne Edwards, 1848)	(Williams 2000, 2001a,b, 2002, Williams & McDermott 2004)
	D	*Clibanarius englaucus*	(Williams 2000, 2001a,b, 2002, Williams & McDermott 2004)
	D	*Dardanus lagopodes*	(Williams 2000, 2001a,b, 2002, Williams & McDermott 2004)
	D	*Dardanus woodmasoni* (Alcock, 1905)	(Williams 2000, 2001a,b, 2002, Williams & McDermott 2004)
	D	*Dardanus* sp.	(Williams 2000, Williams & McDermott 2004)
	D	*Diogenes* sp.	(Williams 2000, 2001a,b, 2002, Williams & McDermott 2004)
Polydora umangivora Williams, 2001	D	*Calcinus gaimardii*	(Williams 2001a, Williams & McDermott 2004)
	D	*Dardanus lagopodes*	(Williams 2001a, Williams & McDermott 2004)
	D	*Paguristes runyanae*	(Williams 2001a, Williams & McDermott 2004)
Polydora villosa Radashevsky & Hsieh, 2000	Cn	*Montipora angulata* (Lamarck, 1816)	(Liu & Hsieh 2000)
	Cn	*Montipora hispida* (Dana, 1846)	(Liu & Hsieh 2000)

Continued

Table 2 (Continued) List of newly reported parasitic polychaetes and their respective hosts from 1998 to 2017.

Parasite		Host	Source references
	Cn	*Montipora informis*	(Liu & Hsieh 2000)
	Cn	*Porites lichen* Dana, 1848	(Liu & Hsieh 2000)
	Cn	*Porites lobata*	(Liu & Hsieh 2000)
	Cn	*Porites lutea* Quoy & Gaimard, 1833	(Liu & Hsieh 2000)
	Cn	*Hydnophora exesa* (Pallas, 1766)	(Radashevsky & Hsieh 2000)
	Cn	*Cyphastrea chalcidicum*	(Radashevsky & Hsieh 2000)
Polydora sp.	D	*Calcinus gaimardii*	(Williams 2001a, Williams & McDermott 2004)
Polydora sp.	D	*Paguristes eremita*	(Stachowitsch 1980, Williams & McDermott 2004)
Polydora sp.	D	*Pagurus cuanensis*	(Samuelsen 1970, Stachowitsch 1980, Williams & McDermott 2004)
Polydora sp.	D	*Pagurus bernhardus*	(Samuelsen 1970, Williams & McDermott 2004)
Polydora sp.	D	*Pagurus granosimanus*	(Walker 1988, Williams & McDermott 2004)
Polydora sp.	D	*Pagurus prideaux* Leach, 1815	(Samuelsen 1970, Williams & McDermott 2004)
Polydora sp.	D	*Pagurus longicarpus*	(Frey 1987, Williams & McDermott 2004)
Polydora sp.	D	*Pagurus pollicaris* Say, 1817	(Frey 1987, Williams & McDermott 2004)
Polydora sp.	D	*Pagurus samuelis*	(Walker & Carlton 1995, Williams & McDermott 2004)
Polydora sp.	D	Unidentified pagurid	(Sato-Okoshi 1999, Williams & McDermott 2004)
Polydora sp.	D	Unidentified pagurid	(Smyth 1989, 1990, Williams & McDermott 2004)
cf. *Polydora*	Cn	*Astreopora myriophthalma* (Lamarck, 1816)	(Wielgus et al. 2002, 2006, Wielgus & Levy 2006)
cf. *Polydora*	Cn	*Cyphastrea* spp.	(Wielgus et al. 2002, 2006, Wielgus & Levy 2006)
cf. *Polydora*	Cn	*Echinopora forskaliana* (Milne Edwards & Haime, 1849)	(Wielgus et al. 2002, 2006, Wielgus & Levy 2006)
cf. *Polydora*	Cn	*Leptastrea* sp.	(Wielgus et al. 2002, 2006, Wielgus & Levy 2006)
cf. *Polydora*	Cn	*Millepora* sp.	(Wielgus et al. 2002, 2006, Wielgus & Levy 2006)
cf. *Polydora*	Cn	*Montipora* sp.	(Wielgus et al. 2002, 2006, Wielgus & Levy 2006)
cf. *Polydora*	Cn	*Pavona varians* Verrill, 1864	(Wielgus et al. 2002, 2006, Wielgus & Levy 2006)
cf. *Polydora*	Cn	*Porites lutea*	(Wielgus et al. 2002, 2006, Wielgus & Levy 2006)
Pseudopolydora reishi Woodwick, 1964	D	*Ciliopagurus strigatus*	(Williams 2001a, Williams & McDermott 2004)

Continued

Table 2 (Continued)   List of newly reported parasitic polychaetes and their respective hosts from 1998 to 2017.

Parasite		Host	Source references
Syllidae			
Branchiosyllis lamellifera Verrill, 1900	Po	*Aplysina archeri* (Higgin, 1875)	(San Martín et al. 2013)
	Po	massive purple-yellow	(San Martín et al. 2013)
Haplosyllis gula Treadwell, 1924	P	*Eunice* sp.	(Lattig & Martin 2009, Potts 1912, Treadwell 1909)
	P	*Glycera* sp.	(Lattig & Martin 2009)
Imajimaea draculai (San Martín & López, 2002)	Cn	*Funiculina quadrangularis* (Pallas, 1766)	(Nygren & Pleijel 2010)
Proceraea exoryxae Martin, Nygren & Creuz-Rivera, 2017	Tu	*Phallusia nigra* Savigny, 1816	(Martin et al. 2017b)
Procerastea halleziana Malaquin, 1893	Cn	*Ectopleura crocea* (Agassiz, 1862)	(Genzano & San Martín 2002)
Procerastea cornuta Agassiz, 1862	Cn	Unidentified hydroid	(Pettibone 1963)
	Cn	Unidentified Coral	(Gardiner 1976)
Epigamia alexandri (Malmgren, 1867)	Cn	*Abietinaria turgida* (Clark, 1877)	(Britayev & San Martín 2001)
	Cn	*Orthopyxis integra* (MacGillivray, 1842)	T.A. Britayev, personal observations
Typhoscolecidae			
Sagitella kowalewskii Wagner, 1872	Ch	Unidentified	(Feigenbaum 1979)
Tomopteris (Johnstonella) helgolandica (Greeff, 1879)	Ch	Unidentified	(Feigenbaum 1979)
Travisiopsis dubia Støp-Bowitz, 1948	Ch	*Flaccisagitta enflata* (Grassi, 1881)	(Øresland & Bray 2005)
Typhloscolex muelleri Busch, 1851	Ch	*Flaccisagitta enflata*	(Øresland & Bray 2005)
Typhloscolex sp.	Ch	*Flaccisagitta enflata*	(Feigenbaum 1979)
Typhloscolex sp.	Ch	*Eukrohnia hamata* (Möbius, 1875)	(Øresland & Pleijel 1991)
Typhloscolex sp.	Ch	*Pseudosagitta maxima* (Conant, 1896)	(Øresland & Bray 2005)

Source: Currently accepted names for both polychaetes and their hosts have been checked through the WoRMS website. (Horton T. et al. 2017. World Register of Marine Species (WoRMS). WoRMS Editorial Board.)

[a] Fossil species.

Po – Porifera, B – Bivalvia, G – Gastropoda, P – Polychaeta, D – Decapoda, Ho – Holothuroidea, Cn – Cnidaria, Bc – Brachiopoda, Ch – Chaetognatha, F – Fish, Tu – Tunicata, U – Unknown.

Taxonomic distribution

Symbiotic polychaetes

Commensals

Among polychaetes, 28 families were reported in 1998 as including commensals, while the number has increased to 33 in 2017 (Tables 1 and 4). Eight families have been newly reported as including symbiotic representatives: Aphroditidae, Chaetopteridae, Orbiniidae, Pholoididae, Scalibregmatidae

Table 3   Specificity among symbiotic polychaetes, expressed as a number of species (No) associated with different number of hosts (Hosts) and relative importance (as %) in 1998 and 2017

Hosts	1998		2017	
	No	(%)	No	(%)
1	162	57.45	301	56.9
2	55	19.5	104	19.66
3	18	6.38	23	4.35
4	16	5.67	27	5.1
5	10	3.55	20	3.78
6	4	1.42	9	1.7
7	–	–	3	0.57
8	3	1.06	11	2.08
9	5	1.77	2	0.38
10	1	0.35	4	0.76
11	2	0.71	7	1.32
12	1	0.35	2	0.38
13	–	–	2	0.38
14	–	–	2	0.38
15	–	–	2	0.38
16	–	–	1	0.19
17	–	–	1	0.19
19	–	–	1	0.19
20	3	1.06	–	–
21	–	–	1	0.19
22	1	0.35	–	–
24	–	–	1	0.19
30	1	0.35	–	–
31	–	–	1	1
38	–	–	2	0.38
49	–	–	1	0.19
53	–	–	1	0.19

and Sigalionidae with only one commensal species, and Fabriciidae and Siboglinidae, with two and 18 commensals, respectively. In these cases, almost all species are involved in a single relationship. Despite the total number of families in 2017 coinciding with the number of newly reported plus those known in 1998, there are two significant taxonomic remarks. The single species of Sabellidae reported in 1998 belonged to the Fabriciinae, a subfamily that is now considered a valid family (Huang et al. 2011), while the Spirorbidae is currently considered a subfamily within the Serpulidae (Rzhavsky et al. 2013), so that all records (both the 1998 and the new ones) now expand the commensal serpulid list. In turn, a polychaete (considered to possibly be a species of *Polydora*, although there is no way to prove this identification) has been reported as a commensal of an echinoderm in the fossil register (Wisshak & Neumann 2006).

The families that experienced a higher increase in number of known commensal species are the Polynoidae (56) and Syllidae (48), which altogether represent more than 50% of all newly reported commensals. These two families also present the highest number of newly reported relationships, 171 and 114, respectively (i.e. more than 55% of the total). Among the remaining families, only the Siboglinidae represented around 10% (18) of the newly reported commensal species. In turn, the Serpulidae reached around 20% (99) of the newly reported relationships, mostly due to *Spirobranchus*

Table 4 Comparison of the total number of species and relationships, and relative importance (%), within the families reported as including commensal polychaetes until 1998 and 2017

Family	1998				2017			
	No Sp	(%)	No Rel	(%)	No Sp	(%)	No Rel	(%)
Polynoidae	163	57.60	426	61.74	219	44.97	597	48.86
Syllidae	21	7.42	33	4.78	69	14.17	147	12.01
Hesionidae	13	4.59	36	5.22	27	5.54	65	5.31
Chaetopteridae	–	–	–	–	1	0.21	2	0.16
Chrysopetalidae	11	3.89	13	1.88	23	4.72	29	2.37
Siboglinidae	–	–	–	–	18	3.70	18	1.47
Eunicidae	6	2.12	8	1.16	17	3.49	25	2.04
Spionidae	11	3.89	38	5.51	14	2.87	35	2.86
Histriobdellidae	9	3.18	13	1.88	13	2.67	20	1.64
Serpulidae	10	3.53	38	5.51	22	4.52	137	11.20
Nereididae	5	1.77	12	1.74	6	1.23	19	1.55
Capitellidae	1	0.35	2	0.29	6	1.23	6	0.49
Spintheridae	4	1.41	8	1.16	4	0.82	8	0.65
Dorvilleidae	4	1.41	6	0.87	11	2.26	28	2.29
Phyllodocidae	2	0.71	4	0.58	4	0.82	6	0.49
Sphaerodoridae	4	1.41	5	0.72	4	0.82	5	0.41
Pilargidae	3	1.06	3	0.43	4	0.82	4	0.33
Amphinomidae	3	1.06	7	1.01	3	0.62	8	0.65
Flabelligeridae	3	1.06	6	0.87	3	0.62	6	0.49
Lumbrineridae	2	0.71	21	3.04	2	0.41	35	2.86
Ctenodrilidae	2	0.71	5	0.72	2	0.41	5	0.41
Fabriciidae	–	–	–	–	2	0.41	3	0.25
Antonbruunidae	1	0.35	1	0.14	3	0.62	3	0.25
Sigalionidae	–	–	–	–	1	0.21	2	0.16
Aphroditidae	–	–	–	–	1	0.21	2	0.16
Sabellidae	1	0.35	1	0.14	1	0.21	1	0.08
Eulepthidae	1	0.35	1	0.14	1	0.21	1	0.08
Fauvelopsidae	1	0.35	1	0.14	1	0.21	1	0.08
Oenonidae	1	0.35	1	0.14	1	0.21	1	0.08
Orbiniidae	–	–	–	–	1	0.21	1	0.08
Pholoididae	–	–	–	–	1	0.21	1	0.08
Sabellariidae	1	0.35	1	0.14	1	0.21	1	0.08
Scalibregmatidae	–	–	–	–	1	0.21	1	0.08
Spirorbidae	2	0.71	3	0.43	–	–	–	–
Iphitimidae	7	2.47	20	2.90	–	–	–	–

Bold: Families not reported in 1998. No Sp: Number of symbiotic polychaete species; No Rel: Number of relationships.

corniculatus (Grube, 1862) and *S. giganteus* (Pallas, 1766), which were previously considered synonymous, but also to the various reports of Spirorbinae, now a subfamily within the Serpulidae but formerly considered as a family (Rzhavsky et al. 2013). In the case of the Lumbrineridae, the 14 newly reported relationships correspond to the same species previously reported in 1998, *Lumbrineris flabellicola* Fage, 1936. The Amphinomidae incorporates one newly reported relationship but no new commensal species, while the Ctenodrilidae, Eulepethidae, Fauvelopsidae, Flabelligeridae, Oenonidae, Sabellariidae, Sphaerodoridae and Spintheridae have no new commensal species or new relationships. Despite this, O'Reilly (2016) reported new records for the already known association between the sphaerodorid *Commensodorum commensalis* (Lützen, 1961) and the trichobranchid *Terebellides stroemii* M. Sars, 1835, which appear to be more frequent than previously reported, at least in Scottish waters. In turn, the Spionidae currently contain three more commensal species than in 1998, one of them within the fossil records, but three less commensal relationships, mainly because *Dipolydora commensalis* (Andrews, 1891) originally considered as commensal is, in fact, a parasite (Williams 2002). Finally, a significant increase both in species and relationships was observed for the Chrysopetalidae.

Parasites

We are here considering both boring and non-boring parasitic polychaetes together (Tables 2 and 5). Accordingly, six polychaete families were not previously reported as including parasitic polychaetes: Hesionidae and Serpulidae (1 species and relationship, respectively), Polynoidae (4 species and 30 relationships), Phyllodocidae (3 species and relationships), and Typhloscolecidae (5 species and 7 relationships), but also the Fabriciidae (8 species and 30 relationships), which comprise the seven species of *Caobangia*. This genus was placed previously within the Caobangiidae (Jones 1974b), then within the Sabellidae (Fitzhugh 1989), and now is considered as Fabriciidae (Kupriyanova & Rouse 2008, Read 2015). Overall, 14 families were reported in 1998 as including parasites, while the number had increased to 16 in 2017.

The Spionidae showed the highest increase in newly reported parasitic species (29 species, 52%), followed by Fabriciidae (8, 14%) and Typhloscolecidae (5, 9%). This is even more marked

Table 5 Comparison of the total number of species and relationships, and relative importance (%), within the families reported as including parasitic polychaetes until 1998 and 2017

Family	1998				2016			
	No Sp	(%)	No Rel	(%)	No Sp	(%)	No Rel	(%)
Spionidae	31	37.35	131	50.97	60	46.15	237	59.40
Oenonidae	19	22.89	28	10.89	21	16.15	31	7.77
Syllidae	11	13.25	45	17.51	15	11.54	25	6.27
Fabriciidae	–	–	–	–	8	6.15	30	7.52
Typhloscolecidae	–	–	–	–	5	3.85	7	1.75
Cirratulidae	2	2.41	4	1.56	3	2.31	8	2.01
Dorvilleidae	2	2.41	2	0.78	3	2.31	3	0.75
Phyllodocidae	–	–	–	–	3	2.31	3	0.75
Polynoidae	–	–	–	–	4	3.08	29	7.27
Sabellidae	2	2.41	2	0.78	2	1.54	15	3.76
Alciopidae	1	1.20	1	0.39	1	0.77	1	0.25
Chysopetallidae	2	2.41	2	0.78	1	0.77	1	0.25
Hesionidae	–	–	–	–	1	0.77	3	0.75
Ichthyotomidae	1	1.20	3	1.17	1	0.77	3	0.75
Lumbrineridae	1	1.20	2	0.78	1	0.77	2	0.50
Serpulidae	–	–	–	–	1	0.77	1	0.25
Caobangiidae	7	8.43	31	12.06	–	–	–	–
Capitellidae	2	2.41	2	0.78	–	–	–	–
Iphitimidae	1	1.20	1	0.39	–	–	–	–
Sphaerodoridae	1	1.20	3	1.17	–	–	–	–

Bold: Families not reported in 1998. No Sp: Number of symbiotic polychaete species; No Rel: Number of relationships.

in terms of relationships (106, 53%), being then followed by Fabriciidae and Polynoidae (around 15%). The latter was mainly due to relationships being considered as commensals in the 1998 review, which turned out to be parasitic when carefully analysed (Britayev & Lyskin 2002, Britayev et al. 2007). The Sabellidae did not show increase in the number of newly reported parasitic species but represent around 10% of the newly reported parasitic relationships. In turn, several families contained parasitic species in 1998 that are currently considered as commensals and, thus, do not have parasitic representatives nowadays (Capitellidae, Sphaerodoridae), similarly with the iphitimid Dorvilleidae. Among parasitic Syllidae, two species are newly reported, but the number of relationships decreased by 23, as many of the newly described species of the *Haplosyllis spongicola*-complex appear to be closer to mutualists than to parasites (see Lattig & Martin 2011a and references therein). Finally, the numbers of parasitic Alciopidae and Lumbrineridae did not vary from 1998 to 2017.

Host taxa

Commensalisms

Among the groups hosting commensal polychaetes (Tables 1 and 6), only cephalopods (with three species and four relationships), gorgonocephalid ophiuroids and ascidians (with one species and one relationship each) have not been previously reported. Cnidarians are the taxon that increased

Table 6 Comparison of the total number of species and relationships, and relative importance (%), within the taxa reported as including species associated with commensal polychaetes until 1998 and 2017

Host taxa	1998				2017			
	No Sp	(%)	No Rel	(%)	No Sp	(%)	No Rel	(%)
Cnidaria	117	20.56	139	19.50	279	29.78	329	27.93
Decapoda	68	11.95	81	11.36	100	10.67	165	14.01
Polychaeta	85	14.94	121	16.97	108	11.53	154	13.07
Porifera	46	8.08	63	8.84	120	12.81	137	11.63
Asteroidea	52	9.14	72	10.10	64	6.83	98	8.32
Holothuroidea	42	7.38	47	6.59	33	3.52	39	3.31
Ophiuroidea	48	8.44	55	7.71	46	4.91	6	0.51
Bivalvia	28	4.92	30	4.21	40	4.27	55	4.67
Crinoidea	23	4.04	28	3.93	44	4.70	56	4.75
Echinoidea	19	3.34	28	3.93	38	4.06	50	4.24
Gastropoda	10	1.76	11	1.54	16	1.71	20	1.70
Sipunculida	9	1.58	10	1.40	13	1.39	15	1.27
Echiuroidea	1	0.18	1	0.14	8	0.85	16	1.36
Hemichordata	6	1.05	8	1.12	8	0.85	11	0.93
Cephalopoda	–	–	–	–	3	0.32	4	0.34
Cirripedia	5	0.88	5	0.70	5	0.53	5	0.42
Foraminifera	3	0.53	4	0.56	3	0.32	4	0.34
Nudibranchia	2	0.35	2	0.28	2	0.21	4	0.34
Vestimentifera	2	0.35	4	0.56	2	0.21	4	0.34
Bryozoa	1	0.18	1	0.14	2	0.21	2	0.17
Tunicata	–	–	–	–	1	0.11	1	0.08
Polyplacophora	1	0.18	2	0.28	1	0.11	2	0.17
Isopoda	1	0.18	1	0.14	1	0.11	1	0.08

Bold: Taxa not reported in 1998. No Sp: Number of symbiotic polychaete species; No Rel: Number of relationships.

the most in terms of number of newly reported species and relationships: 162 and 190, respectively, which represents around 43% and 41% of the total. Sponges also increased significantly, including around 20% of the newly reported host species (74) and around 16% of the relationships (84). Around 9% of the newly reported host species (32) and around 18% of the relationships (84) occurred in decapods. Among foraminifers, isopods, nudibranchs, polyplacophorans and vestimentiferans, there were no newly reported host species and relationships. Three less host species but five more relationships were reported in ophiuroids, while in holothurians, there were nine less species and eight less relationships (i.e. reconsidered now as parasitic).

Parasitisms

Among the groups harbouring parasitic polychaetes (Tables 2 and 7), decapods, chaetognaths, and brachiopods have become newly reported as hosts. Decapods, holothurians, gastropods and cnidarians present the highest number of both newly reported species (41, 23, 15, 15) and relationships (97, 24, 12, 14), which represent around 36%, 20%, 13% and 13%, and 58%, 14%, 7% and 8% of the total. The lower increase in relationships versus species in cnidarians and gastropods can be explained by the fact that some newly reported species are involved in a relatively low number of relationships, while some species currently considered as commensals (but as parasites in 1998) are involved in more numerous relationships. In turn, amongst cephalopods, ophiuroids and sponges, there was a decrease both in number of newly reported host species and relationships, the last in connection with the respective relationships with the species of the *Haplosyllis spongicola* species-complex. Finally, the number of parasites in phanerogams, ctenophores, nemerteans, cirripedes, bryozoans, echiurans, holothurians and tunicates did not vary from 1998 to 2017.

Table 7 Comparison of the total number of species and relationships, and relative importance (%), within the taxa reported as including species associated with parasitic polychaetes until 1998 and 2017

Host taxa	1998				2017			
	No Sp	(%)	No Rel	(%)	No Sp	(%)	No Rel	(%)
Gastropoda	53	28.49	72	28.02	68	24.03	84	21.16
Bivalvia	42	22.58	78	30.35	50	17.67	84	21.16
Decapoda	–	–	–	–	41	14.49	97	24.43
Polychaeta	28	15.05	32	12.45	34	12.01	37	9.32
Cnidaria	9	4.84	10	3.89	24	8.48	24	6.05
Porifera	30	16.13	40	15.56	18	6.36	18	4.53
Bryozoa	4	2.15	4	1.56	4	1.41	4	1.01
Chaetognatha	–	–	–	–	4	1.41	7	1.76
Pisces	3	1.61	3	1.17	3	1.06	3	0.76
Cirripedia	2	1.08	3	1.17	2	0.71	3	0.76
Ophiuroidea	6	3.23	6	2.33	2	0.71	2	0.50
Tunicata	2	1.08	2	0.78	3	1.06	3	0.76
Brachiopoda	–	–	–	–	1	0.35	1	0.25
Ctenophora	1	0.54	1	0.39	1	0.35	1	0.25
Echiuroidea	1	0.54	1	0.39	1	0.35	1	0.25
Holothuroidea	1	0.54	1	0.39	24	8.48	25	6.30
Nemertini	1	0.54	1	0.39	1	0.35	1	0.25
Phanerogamia	1	0.54	1	0.39	1	0.35	1	0.25
Unknown	–	–	–	–	1	0.35	1	0.25
Cephalopoda	2	1.08	2	0.78	–	–	–	–

Bold: Taxa not reported in 1998. No Sp: Number of symbiotic polychaete species; No Rel: Number of relationships.

Overview of the newly reported relationships since 1998

Commensal polychaetes

Amphinomidae

The amphinomid *Benthoscolex cubanus* Hartman, 1942 was originally described from the guts of unnamed sea urchins collected by the ATLANTIS off Cuba between 219–430 m depth. The report was overlooked by Martin & Britayev (1998), who registered only its association with *Heterobrissus hystrix* (Agassiz, 1880) from Bahamas, where it occurs between 460–480 m depth. Although the Caribbean location and depth range may suggest that both reports refer to the same association, there is no way to confirm this hypothesis.

Antonbruunidae

Antonbruunia gerdesi Quiroga & Sellanes, 2009 and *A. sociabilis* Mackie, Oliver & Nygren 2015 are the second and third known species of the family. From an ecological point of view, *A. gerdesi* (from off Chile) differs from *A. viridis* Hartmann & Boss, 1965 from off Mozambique in inhabiting a chemosynthetic vesicomyid bivalve (*Calyptogena gallardoi* Sellanes & Krylova, 2005) from a 750–900 m deep cold seep instead of a lucinid bivalve (*Opalocina fosteri* Hartman & Boss, 1965) from shallow-water (<100 m depth) hypoxic sediments of black-brown, oozy mud and detritus. In turn, *Antonbruunia sociabilis* inhabits a thyassirid deep-sea bivalve from a sulphidic seep in the north-east Atlantic Ocean west of Scotland. The three antonbruunids also differ in their patterns of infestation. *Antonbruunia viridis* often lives in male-female pairs inside their host. In contrast, the Chilean species is most often solitary, while the Scottish one may have up to nine specimens living together (which is the reason of its specific epithet 'sociabilis'). The family was suggested to have close affinities with the former Nautiliniellidae and was later incorporated into the Calamyzinae (Chrysopetalidae) but now seems to be more related to the Pilargidae, based on molecular analyses (Mackie et al. 2015).

Antonbruunids do not seem to cause damage to their host bivalves or negatively influence their metabolism. Possible food sources could be gill particle strings and mantle pseudofaeces, but considering the chemosymbiotic nature of their hosts, it is possible that the worms could also feed on the bacterial communities associated with the bivalves (Mackie et al. 2015).

Aphroditidae

Pontogenia chrysocoma (Baird, 1865) occurs deep inside tubes built by the echiuran *Bonellia viridis* Rolando, 1822 (1821 volume) and the decapod *Upogebia deltaura* (Leach, 1816) in the Mediterranean. Presumably, the worm feeds on the faeces and pseudofaeces of both hosts. The relationship with the echiuran has been recently reported by Anker et al. (2005), but rather than supply new data, they just repeated the Schembri & Jaccarini (1978) account, where the decapod is also considered as owner of the tube and, thus, as host for the polychaete.

Capitellidae

Symbiotic capitellids now contain two species of *Capitella* living in association with cephalopods, which were not reported in Martin & Britayev (1998). These species, as well as the two previously known ones, *Capitella ovincola* Hartman, 1947 and *C. minima* (Langerhans, 1880) were initially considered parasitic, as they were observed to feed on the embryos inside the egg masses. However, a recent study carried out on *C. ovincola* demonstrates that the association provides mutual benefits to the partners, particularly food and shelter for the capitellids and an increased hatch rate for the squid embryos (Zeidberg et al. 2011). Therefore, we have considered this species within the commensals *sensu lato* for statistical purposes. The other three symbiotic capitellids associated with cephalopods may likewise be closer to mutualists than to parasites.

Chaetopteridae

Chaetopterids are better known as hosts than as symbionts (Britayev & Martin 2016). In fact, there is only one unidentified species of *Spiochaetopterus* reported to be in commensal association with a coral of the genus *Montipora* from French Polynesia (Bergsma 2009). This association belongs to the group of tube-dwelling organisms that cause major morphological changes in the host coral colonies by inducing the formation of aberrant skeletal structures. Usually these colonies exhibit encrusting or plating growth forms, while only those harbouring the symbionts were branching, showing finger-like structures. Almost 95% of these fingers harbour vermetid mollusc, amphipod and the chaetopterid symbionts, but only 30% of them were occupied by the polychaetes. The longest finger inhabited by *Spiochaetopterus* sp. was 122 mm long and resembled the structures induced by *Polydora villosa* Radashevsky & Hsieh, 2000 on different coral hosts (including species of *Montipora*), although these are much shorter. The fingers are formed when the coral encrusts the tubes built by the polychaetes, which must continually extend their tubes to avoid overgrowth by the rapid growth of the host.

The symbiont-induced fingers dramatically alter the colony topography by adding considerable material and an extra 3-dimensional structure to the reef landscape. Coral area and surface (and, thus, living tissues) increase with respect to non-fingered ones, as does the interstitial volume. Both affect coral physiology, fitness and life cycle as well as the available habitat for other coral associates or habitual inhabitants. Changes in host corals' morphology may have a reduced skeletal strength but also may increase the photosynthetic potential of the host and nutrient availability. These certainly seem to improve the ecological role of the infested colonies, which may harbour more interstitial fauna and more rapidly overgrow colonies of other space-competitor coral species than non-infested ones (see Bergsma 2009 and references therein). Thus, overall the association here is considered as commensalistic, or even mutualistic, rather than parasitic.

Chrysopetalidae

Commensal chrysopetalids are represented by a new report of *Bhawania goodei* Webster, 1884 as associated with a gallery building host, the echiurid *Lissomyema exilii* (F. Muller in Lampert, 1883) (Anker et al. 2005). Previous reports mentioned this species as associated only with sipunculans. In both cases, however, the worm moves freely inside the host galleries and thus could be considered a non-obligate commensal. However, most new reports (15 over 16) correspond to species of the recently synonymised Nautiliniellidae (Aguado et al. 2013). Like the previous known species of this group, all newly reported species are obligate symbionts of chemosynthetic bivalves mainly inhabiting deep-sea seeps, vents and other similar environments. Particularly, *Shinkai longipedata* (Miura & Ohta, 1991), from unidentified vesicomyid clams, has been reported as often living in pairs (male and female) between the gill lamellae and foot of its host (Aguado & Rouse 2011). The species is the largest chrysopetalid documented to date and lacks the proboscidial papillae characteristic of all non-symbiotic members of the family (Watson & Faulwetter 2017). Feeding on host mucus, gametes and pseudofaeces was suggested for obligate symbiont calamyzins (Becker et al. 2013). However, the recent discovery of the presence of a very small pair of platelet/stylet jaws with a sharp pointed distal structure in *S. longipedata* could indicate predation on bivalve tissues (Watson & Faulwetter 2017), thus casting some doubts on the qualification of the 'nautiliniellid' calamyzins as commensals.

Dorvilleidae

There are two newly reported relationships involving dorvilleids. Both correspond to species of *Iphitime*, previously belonging to the Iphitimidae. This family was synonymised with the Dorvilleidae, and *Iphitime* was even suggested to be a possible synonym of *Ophryotrocha* (Heggøy et al. 2007, Wiklund et al. 2012). The first new report is the association of *Iphitime cuenoti* Fauvel, 1914 (Figure 1A) with the bathyal crab *Bathynectes maravigna* (Prestandrea, 1839), in which the very low infestation prevented assessment of its prevalence (Lattig et al. 2017). Despite no other

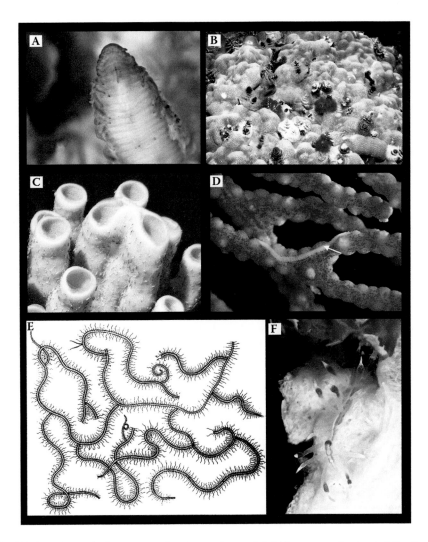

Figure 1 (A) Anterior end of a preserved male specimen of *Iphitime cuenoti* extracted for the branchial chamber of its host crab (Photo: D. Martin); (B) Numerous specimens of *Spirobranchus corniculatus* on its host coral (Photo: Paul Human); (C) Hundreds of small-sized specimens of *Polydorella* sp. inside their tubes on its host sponge (Photo: O.V. Savinkin); (D) A specimen of *Alcyonosyllis glasbyi* on its gorgonian host (preserved) (Photo: D. Martin); (E) *Syllis ramosa*, (redrawn from McIntosh (1879)); (F) Nudibranch-mimicking syllid on its alcyonacean host (Photo: O.V. Savinkin).

new symbiotic dorvilleids and hosts being reported in this paper, the authors provided the first phylogeographic approach to the distribution of *Iphitime cuenoti*. They hypothesised that the wide bathymetric distribution of the known hosts (i.e. from 100–600 m depth) could be responsible for the genetic homogeneity of the Mediterranean populations, together with the absence of a barrier effect of the oceanographic fronts in the north-west Mediterranean.

The second newly reported relationship refers to an association with pagurid decapods, which was overlooked by Martin & Britayev (1998). Most usually, pagurids harbour *Iphitime paguri* Fage & Legèndre, 1933, a species that has been recently redescribed as having *Pagurus prideaux* Leach, 1815 as the main host and *Pagurus bernhardus* (Linnaeus, 1758) as a secondary host (Høisæter & Samuelsen 2006). This species has distinct patterns of host utilisation depending on its life stage. Mature adults may be found intertwined in the apex of the shell inhabited by its host, together with some small larvae, while intermediate-sized worms occur in the gill chamber and in a sulcus on the carapax of the pagurid host (Høisæter & Samuelsen 2006).

Eunicidae

Newly reported symbiotic eunicids are mainly associated with cnidarians except for *Eunice aphroditois* (Pallas, 1788) and an unidentified species of *Eunice* associated with a gastropod, and *Lysidice unicornis* (Grube 1840), reported as inhabiting the galleries of an echiurid. The latter were considered as energy commensals, since they build their own burrows into the walls of the host galleries and likely take profit only from the water circulation produced by the hosts (Anker et al. 2005).

The previous reports of symbiotic eunicids associated with cnidarians from both tropical and cold-water reefs were more often presented as commensalisms or inquilinisms (i.e. a specific association in which the inquiline species lives on or inside the host to obtain shelter) (Martin & Britayev 1998). This is because of doubts about their possible negative influence on the host since they either excavate galleries or induce the formation of extra tissues that cover the worm tubes when attached to the coral branches. Nevertheless, the opportunity of rearing deep-sea corals in controlled aquaria allowed Roberts (2005) to modify this qualification towards a mutualism. Accordingly, the polychaete plays a basic functional role as reef-aggregating agent, combined with the ability of the coral to anastomose its branches, together with the polychaete tubes (which are calcified by the own corals) and debris. The calcified eunicid tubes contribute to connecting the coral colonies, modifying the balance between coral growth and erosion in favour of the former. Moreover, the allogeneic engineering behaviour of the eunicids accelerates reef growth, helping to prevent the reef from being covered by sediments in areas with active sediment transport. Finally, the polychaetes also contribute to the formation of new reef patches by connecting the fragments broken from the main colonies. More recently, however, the hypothesis of mucus segregated by symbiotic eunicids being involved in tube calcification has been discarded, as corallite thecae and calcified polychaete tubes did not differ in microstructure, growth rate and geochemical composition (Oppelt et al. 2017). Therefore, an interesting question remains still open: if mucus is not involved, which cue is triggering the corallites to produce the extra calcium carbonate required to cover the symbiont tubes?

The mutualistic character of the association between eunicids and cold reef-corals is even reinforced by the observations of the aggressive behaviour of the specimens of *Eunice norvegica* (Linnaeus, 1767) associated with *Lophelia pertusa* (Linnaeus, 1758), which attacked the spines and the test of the sea urchin, *Cidaris cidaris* (Linnaeus, 1758) climbing on a living host colony (Mortensen 2001). At the same time, the eunicids demonstrate kleptoparasitic behaviour, as they were observed to steal food captured by the polyps (Mortensen 2001).

Fabriciidae

The two fabriciid species of the genus *Brandtika* and *Monroika* considered here as commensals live in association with freshwater gastropods and bivalves. The species are always found living in sand grain tubes encrusting the shell surface of their hosts (Jones 1974a), a behaviour that clearly differs from that of the species of *Caobangia*, which are reported to actively bore into the shells of their mollusc hosts, mainly around the apex in the gastropods and all along the valves in the bivalves (Jones 1974b). They are characterised by lacking the typical annelid segmentation, showing a gut posteriorly bent to form an external trunk that runs around the pear-shaped body and ending by the anus, which opens dorsally at the anterior-most thoracic region. This peculiar body organisation closely resembles that of sipunculans or phoronids. Most developmental stages at least in *Caobangia billeti* Giard, 1893, however, are typical of those from annelids, and only one intermediate phase appears to be unusual. The peculiarity consists in a synchronous (instead of subsequent) development of segments, which gives rise to a comparatively faster settlement, metamorphosis and burrow formation. This may be perceived as a strategy with obvious adaptive meaning for a symbiotic species, particularly taking into account that the hosts are semi-terrestrial snails (Kolbasova & Tzetlin 2017).

Histriobdellidae

The five newly reported histriobdellids all belong to the genus *Stratiodrilus* and live in association with freshwater anomuran decapods (Table 1). No other relevant information was provided for these species (Haswell 1900, Amaral & Morgado 1997, Steiner & Amaral 1999, Amato 2001, Daudt & Amato 2007), preventing discussion of their associations compared to those involving previously known histriobdellids.

Hesionidae

The commensal species of this family are able to colonise a large variety of hosts, from hemichordates, polychaetes and sipunculans to different types of molluscs, crustaceans and echinoderms. The newly reported species follow a similar pattern, although echinoderms (mostly sea urchins) are the main hosts. Most of these new records correspond to species reported as potentially new to science. Among them, the Vietnamese and Caribbean *Oxydromus* associated with starfish or with sea urchins, one species of *Parahesione* from Vietnamese decapods, and one species of *Struwela* from sea urchins at the Sea of Cortez (Miller & Wolf 2008, Campos et al. 2009, Britayev & Antokhina 2012). No symbionts were reported in the recent revision of *Nereimyra* (Pleijel et al. 2012). Nevertheless, *N. punctata* (Müller 1788) was reported as living in association with a pagurid decapod, without any additional information of the characteristics of the association (Williams & McDermott 2004).

Lumbrineridae

The additional reports of symbiotic lumbrinerids all relate to *Lumbrineris flabellicola* Fage, 1936 living in association with additional cnidarian hosts, mainly scleractinians but also zoantharians and hydroids (Miura & Shirayama 1992, Cairns & Zibrowius 1997), which were overlooked by Martin & Britayev (1998). These increase the polyxenous character of the species, almost doubling the number of cnidarian host species (Table 1).

Special mention should be addressed to the most recent report of an equivalent association in the fossil record, where traces of the presence of the polychaete occurred within four fossil solitary corals of the families Caryophylliidae and Flabellidae (Martinell & Domènech 2009). In these corals, the symbiont seems not to be rejected by the hosts, as they did not produce overgrowths or deformation structures, supporting the hypothesis of commensalism based on modern records of the relationship. Accordingly, the host would be neutral, and the worm would benefit from a protected substrate to live on and a possibly easy-to-obtain source food. Contrary to recent reports, however, the fossil ones did not occur in the deep sea. They are mostly from shallow-water sediments, suggesting a possible shift on the ecological preferences of the partners (Martinell & Domènech 2009).

Nereididae

The report of *Cheilonereis peristomialis* Benham, 1916 as a commensal with pagurids was hidden within a paper on the associations of molluscs and decapods with sea anemones (Hand 1975). This was recently cited by Williams & McDermott (2004), who also reported two new hosts for *Neanthes fucata* (Savigny in Lamarck, 1818), a well-known hermit crab symbiont. The authors reported a similar behaviour for both nereidids, so that they may be considered as something between commensals and kleptoparasites. However, Hand (1975) also reported a curious behaviour of *Cheilonereis peristomialis* in relation to the host pagurid tapping on its associated sea anemones. The hermit crab, *Sympagurus dimorphus* (Studer, 1883), typically taps the anemone column with its chelae or dactyls to help encourage it to attach to its shell.

Accordingly, the worm emerged in a similar way to when it steals food from the crab, but instead it extended its anterior region onto the disk and among the tentacles of the anemone being tapped. The worm apparently touched the disk and tentacles of the anemone with its tentacles and palps, but there was no response from the anemone, nor was the worm attacked by nematocyst discharge. The worm repeatedly explored the anemone and then withdrew out of sight, only to emerge again. This behaviour of the worm started some time after the crab tapping began and ended when the anemone had brought its tentacles into full adherent contact with the shell. Whether this behaviour of the worm helped the pagurid in obtaining its new associated sea anemone was not clarified by Hand (1975), but, if so, it could be an additional factor allowing us to consider their association as mutualistic.

Orbiniidae

Naineris setosa (Verrill, 1900) was reported as facultative commensal of the echiurid *Lissomyema exilii* by Anker et al. (2005). However, these authors did not define the location of the polychaete inside the host galleries or the type of association.

Pholoididae

Pholoe minuta (Fabricius, 1780) was also reported from the galleries of its echiurid host *Echiurus echiurus echiurus* (Pallas, 1766) by Anker et al. (2005). It is suggested that the inhabiting polychaetes are agile enough to steal food from the extended proboscis. However, there are no direct observations of such behaviour in the case of the pholoidid.

Pilargidae

Sigatargis commensalis Misra, 1999 was found inside the burrows of an estuarine hemichordate. The symbiotic pilargid lives in close association with the host, firmly attached to its gastric region. No other details on the character of the association were indicated in the original description by Misra (1999) and the species has not ever been found again.

Phyllodocidae

The phyllodocids rarely occur as symbiotic species. For instance, no symbionts were reported in two revisions of *Eumida* (Eibye-Jacobsen 1991, Nygren & Pleijel 2011). However, *Eumida ophiuricola* Britayev, Doignon & Eeckahaut, 1999 is known to be in association with a brittle star. The species occurred occasionally on the surface of its host (shared with a symbiotic copepod) but was more frequent inside the bursa. The host species may also harbour a symbiotic polycladid inside the bursa, which was never shared by specimens of the two symbionts (Britayev et al. 1999).

Polynoidae

The polynoids are certainly the most diverse polychaete family in terms of number of symbionts and present some of the most spectacular examples of colouring and mimicry (Figure 2A–H). Among the 891 species and 173 genera recorded as valid (Read & Fauchald 2017), 219 species (i.e. 24.6%) belonging to 79 (i.e. 45.7%) genera are considered to live as commensals, 163 by Martin & Britayev (1998) and 56 in the present update, which are involved in a total of 597 associations. Among the newly reported commensal polynoids identified to species level (Table 1), 15 were described as new species, and 24 were either recorded as symbionts for the first time or omitted in the 1998 review, while the remaining ones were already known as symbionts, and we are here reporting new associations. Among those identified to the genus level, at least the species of *Hololepidella* associated with the coral *Galaxea* sp. is definitely an undescribed one (T.A. Britayev & D. Fiege, unpublished results).

The most diverse host taxa harbouring polynoids are cnidarians, decapods and polychaetes. Conversely, the least diverse are sponges, which may either be a non-appropriate habitat for polynoids

Figure 2 (A)–(F). Commensal Polynoidae: (A) Two specimens of *Australaugeneria* sp. on its host alcyonacean (Photo: Ned Deloach); (B) One specimen of *Acholoe squamosa* on a lateral of the arm of its host starfish (Photo: D. Martin); (C and D) *Asterophilia culcitae* on its host starfish *Linckia laevigata* (C) and *Culcita novaeguineae* (D) (Photos: O.V. Savinkin); (E) *Ophthalmonoe pettiboneae* inside a dissected tube of *Chaetopterus*, sharing it with a female/male couple of a symbiotic porcellanid crab (Photo: T.A. Britayev); (F) *Arctonoe vittata* below its host limpet (Photo: T.A. Britayev). (G) and (H) Parasitic Polynoidae *Gastrolepidia clavigera* on its host sea cucumbers (Photos: O.V. Savinkin). (G) On the dark blue *Stichopus chloronotus*; (H) On the dotted *Pearsonothuria graeffei*.

or a result from a comparatively low sampling effort addressed to these organisms. The number of relationships in which commensal polynoids are involved were also highly polyxenous in the case of cnidarian, decapod, crinoid and asteroid hosts but tended to be monoxenous for polychaete hosts. Among 13 of the species described as new, more than half occurred in association with a single host, so that their associations can be tentatively considered species-specific. On the other hand, relatively numerous commensal polynoids are associated with a wide range of hosts, including from five to 11 species (e.g. *Asterophilia culcitae* Britayev & Fauchald, 2005) (Figure 2C and D), *Hololepidella laingensis* Britayev, Doignon & Eeckahaut, 1999 or *H. millari* Britayev, Doignon & Eeckahaut,

1999. A few species live in association with organisms that are themselves also symbionts, such as an unidentified polynoid and *Harmothoe oculinarum* (Storm 1879) living with *Eunice norvegica* (Linnaeus 1767), which itself lives with *Lophelia pertusa* (Linnaeus 1758) (Mortensen 2001).

Most new observations confirmed the postulated prevalence of regular distributions among commensal polynoids. In other words, each host individual appears to be infested by one adult worm or by a male-female pair (Martin & Britayev 1998). Infestation by single worms occurs, for instance, in *Arctonoe vittata* (Grube, 1855) (Figure 2F), *Branchipolynoe pettiboneae* Miura & Hashimoto, 1991, *Gorekia crassicirris* (Willey, 1902) and *Ophthalmonoe pettiboneae* Petersen & Britayev, 1997 (Figure 2E), while host infestation by a pair of adults occurs in *Asterophilia culcitae* Britayev & Fauchald, 2005, *Eunoe bathydomus* (Ditlevsen 1917), *Medioantenna variopinta* Di Camillo, Martin & Britayev, 2011 and *Subadyte pellucida* (Ehlers, 1864) (Table 1). In some species (e.g. *Asterophilia culcitae*; Figure 2C and D), infestation by adult pairs was lower than that by a single worm. This was tentatively attributed to the formation of temporary couples when males migrated from host to host in search of females to fertilise (Britayev et al. 2007).

Regular distributions were suggested to result from host-to-host migration (or secondary dispersion) of post-settled symbionts, induced by intraspecific competition (Britayev 1991, Britayev & Smurov 1985). Long-distance migration has been experimentally demonstrated for the crinoid-associated *Paradyte crinoidicola* (Potts, 1910) by Mekhova & Britayev (2015). Both post-settled juvenile and adult scale worms rapidly recolonised depopulated crinoids, in which infestation rates fast approached those of control hosts. The long-distance migrations of *P. crinoidicola* likely resulted from a combination of both intraspecific aggressive interactions and mating partner searching behaviour. The first mechanism was indirectly inferred from the high frequency of body traumas. The second was indirectly supported by the observed sex-ratio deviation in favour of males in experimental series, indicating a higher migratory activity of males in comparison to females (Britayev & Mekhova 2014).

Secondary dispersion may be used to colonise either the same host species or an alternative one. In other words, this process may also lead to host switching during the symbiont's life cycle (Britayev 1991), an advantageous strategy for achieving optimal host exploitation by resource partitioning. A detailed experimental demonstration of this interesting but poorly documented strategy can be found in a population of *Arctonoe vittata* living in association with its well-known hosts, the starfish *Asterias amurensis* Lütken, 1871 (intermediate host) and the limpet *Niveotectura pallida* (Gould, 1861) (definitive host) (Tokaji et al. 2014). On the starfish, small worms suffer from predation either by other organisms, such as crabs, or by larger conspecifics. Strong intraspecific competition often drives the small worms away from a given starfish host, and they are forced to look for another. However, large specimens of *Arctonoe vittata* living in limpets in the experiments proved to be highly territorial and always attempt to monopolise their hosts. This suggested that intraspecific competition between limpet-associated scale worms in natural conditions might be intense. Accordingly, small scale worms improve their survival rate by settling on starfishes and, as they grow, later switching to limpet hosts where survival is higher, and reproduction takes place (Tokaji et al. 2014). This life-history pattern results from optimal host exploitation but cannot be perceived unless the behaviour of the scale worm is studied by considering the triple partnership. Many other scale worms (as well as other commensal polychaetes) have been reported as living in association with several hosts, often as separate records (Martin & Britayev 1998). However, as in the case of *A. vittata*, it is also quite likely that some (or even many) of them would show similar adaptive host-switching behaviour. In other cases, both juveniles and adults live always in association with the same hosts but occupy different positions. For instance, *Hesperonoe japonensis* Hong, Lee & Sato, 2017 (previously reported as *H. hwanghaiensis* Uschakov & Wu, 1959) lives inside the burrows of the ghost shrimps *Upogebia major* (De Haan, 1841) and *Austinogebia narutensis* (Sakai, 1986) (Hong et al. 2017). In the *Upogebia major* galleries, the juveniles are commonly attached to the ventral or lateral surface of the thorax or abdomen of the host. As soon as they become adults, they detach themselves from the host carapace and move freely on the inner surface of the host burrow (Sato et al. 2001). A similar behaviour was previously reported for the congeneric *Hesperonoe complanata* (Johnson, 1901) living in association with another ghost shrimp, *Neotrypaea californiensis* (Dana, 1854) (MacGinitie 1939, MacGinitie & MacGinitie 1968), and may also be presumed for the third Asian species of the genus, *Hesperonoe coreensis* Hong, Lee & Sato, 2017, living inside galleries of *Upogebia major*. *Hesperonoe japonensis* also has a uniform, bright red colouring (Sato et al. 2001), probably as a result of being rich in blood pigments like other symbionts living in relatively oxygen poor conditions (Martin & Britayev 1998).

The general scarcity of detailed information on host-symbiont relationships between symbiotic polychaetes particularly applies to scale worms. In many cases, the worms are referred to as commensals in taxonomic papers, where a few details on the association are included (see Table 1 for examples). In papers mainly dealing with the description of a scale worm as a new species, the possibility of a commensal habit is mentioned, but no evidence on the real existence of the postulated relationships is provided. For instance, Jourde et al. (2015) described *Malmgrenia louiseae* Jourde, Sampaio, Barnich, et al., 2015 as being collected together with several potential hosts (i.e. one echiuran and three synaptid holothurians). However, only in the case of *Leptosynapta* cf. *bergensis* (Östergren, 1905) was one worm found in contact with the potential host. Accordingly, despite the genus *Malmgrenia* including well-known symbiotic species, we have decided to mention *M. louiseae* in this section but not to incorporate the species in the list of commensals pending a well-supported demonstration of the symbiotic mode of life suggested by Jourde et al. (2015). In some cases, references to symbiotic polychaetes can be found in papers dealing with the taxonomy of the hosts. Among these, most probably involve deep-sea corals (Cairns 2006, 2009, 2011, 2012, Cairns & Bayer 2008), in which species-specificity and high prevalence of associations lead these authors to consider skeletal modifications induced by the polynoid symbionts as a taxonomically robust character for defining coral species. However, in many of these papers, the responsible symbionts have not been identified (Table 1).

On the other hand, during the past two decades, there has been a growing knowledge on particular species, leading to altering their classification as commensals. Among them, for instance, are *Gastrolepidia clavigera* Schmarda, 1861 and *Branchipolynoe seepensis* Pettibone, 1986, which are now considered closer to parasites (Britayev & Lyskin 2002, Britayev et al. 2003a,b, 2007), and, thus, the characteristics of their associations are discussed later in the 'Parasitic polychaetes' section.

The limits between commensalism and parasitism are often unclear in the case of scale worms, particularly when the host-symbiont interactions are based on limited observations. For example, the unusual association of the Antarctic scale worm *Gorekia crassicirris*, which inhabited the gut of the irregular sea urchin *Abatus (Pseudabatus) nimrodi* (Kœhler, 1908), was assessed as commensalism (inquilinism), although parasitism could not be excluded because there was no actual evaluation of the possible damage caused by the polychaete on its host (Schiaparelli et al. 2011). The association between *Hololepidella* sp. and the scleractinian coral *Galaxea astreata* (Lamarck, 1816) was also controversial (Britayev et al. 2015). The guts of these symbiotic scale worms contained mucus with unicellular algae (similar in size, shape and colour to the host coral zooxanthellae) and host cnidocysts, together with fragments of copepods resembling a parasitic species also infesting the same host colonies. The zooxanthellae and cnidocysts pointed to the symbionts feeding on host tissues, thus suggesting a parasitic interaction. Other scleractinian symbionts (e.g. crabs and shrimps) that also feed on host tissues show mutually beneficial relationships and are known as cleaning or defensive mutualists of corals (Stewart et al. 2006). This suggests that mutual relationships between *Hololepidella* sp. and *Galaxea* may also be possible, as supported by the remnants of copepods found in the guts of the polychaete. This dilemma, as well as those represented by many of the currently reported symbiotic interactions involving scale worms (and polychaetes in general) can only be assessed through detailed field observations and experimental studies.

The recent use of new methodologies, such as molecular analyses, has helped in the assessment of the phylogenetic and taxonomic relationships within the Polynoidae (e.g. Wiklund et al. 2005,

Norlinder et al. 2012, Shields et al. 2013). However, only one study of several deep-sea free-living and symbiotic species from the Southwest Indian Ridge (Serpetti et al. 2017) has attempted to assess the implications of symbiosis in the phylogeny of the family and the potential role of hosts in the speciation processes. The authors postulate that: (1) most basal species of the subfamily Polynoinae were obligate symbionts with specific morphological adaptations; (2) obligate and facultative commensalisms and free-living behaviour have evolved several times along the phylogeny of the group; (3) obligate octocoral commensal species appear to be monophyletic; and (4) to assess the possible role of behaviour in speciation processes, more symbiotic species associated with a wider range of hosts, as well as more numerous free-living representatives, have to be analysed. The authors base their phylogenetic inferences on a dataset of two mitochondrial (COI, 16S) and two nuclear (18S, 28S) ribosomal genetic markers. There are issues when mapping traits onto molecularly derived phylogenetic trees. Ideally, sequence data and all relevant morphological characters should be taken together with behavioural data to provide a complete view (Fitzhugh 2006). This would certainly be an interesting study to carry out in the near future.

Serpulidae

Most new reports of commensals within this family refer to two conspicuous species of *Spirobranchus*: *S. giganteus* and *S. corniculatus* (Figure 1B). Both are obligate associates of living hermatypic corals and were considered as synonyms. However, they appear now to be clearly separated by their morphology and by their known geographic distributions, the former living in Atlantic reefs, the latter in Indo-Pacific ones (ten Hove 1994, Willette et al. 2015).

    Spirobranchus giganteus apparently shows species-specific attraction for its different hosts together with a highly specialised behaviour (see Rowley 2008 and references therein). The larvae of the species seem to be attracted by exudates of its host corals, but, at the same time, they seem also to be attracted to conspecific clusters of adults on some colonies, hence other colonies can be devoid of worms. As with many other 'commensal' associations, when accurately examined, the relationship of this symbiotic serpulid with their hosts shows clear indications of mutualism, since the worms seem able to protect the coral hosts from predation. An additional benefit seems to be related to an increase of water circulation that benefits the health of the adjacent polyps and facilitates coral recovery in algal-dominated environments. This increase in water circulation close to the coral surface would also decrease susceptibility to bleaching, improve dispersal of waste products of the coral host, and increase nutrient availability from waste materials excreted by the associated partner.

    The two symbiotic species of *Spirobranchus* have also been recently reported in association with non-cnidarian organisms, although it is not clear whether they may be considered as true hosts. The species of *Spirobranchus* are known to be resistant to the toxins produced by some epibiotic sponges to overgrow, kill the living tissues and then occupy the coral surface (Hoeksema et al. 2016). This ability likely explains the association of *S. giganteus* with at least 10 different sponges in Curaçao (southern Caribbean). Worm tubes originate in the original coral host, which are overgrown by the sponges thus allowing the worms to continue growing, therefore being considered as secondary hosts (García-Hernández & Hoeksema 2017). Some Mediterranean species of *Spirobranchus* might apparently be involved in similar associations. However, the host sponges were not reported, and the true nature of their associations was not clarified (Koukouras et al. 1996). On the other hand, *S. corniculatus* has been reported as epibiont on the fluted giant clam *Tridacna squamosa* Lamarck, 1819 (Schoot et al. 2016). The tubes were found on the ventral side of the clam shell, close to the edge and often covered by the mantle, so that the worms could only extend out the branchial crown when the mantle was retracted. However, no information on the real nature of this association has been provided to date. In addition, there are some other reports of serpulids living in association with giant clams, but in these cases, there was no information on whether the species belonged or not to *Spirobranchus* (Vicentuan-Cabaitan et al. 2014, Neo et al. 2015).

Other serpulids, like some species of *Vermiliopsis*, also show specific associations with corals but are apparently difficult to detect, as they tend to be found under the host surface (Hoeksema & ten Hove 2016). Some other reports of cnidarian-associated serpulid-like species come from the fossil register. Among them, *Propomatoceros sulcicarinatus* Ware, 1975, whose tubes were located laying parallel near the tips of the coral branches and were embedded by the coral tissues when growing (Garberoglio & Lazo 2011). According to these authors, the association may be qualified as a symbiosis, probably close to mutualism. They based this postulation on existing knowledge of actual serpulid-coral relationships, as well as on the fact that *Propomatoceros* appears to be related to *Spirobranchus*, an obligate symbiont which may likely provide cues to understand the obligatory symbiosis in fossil species. Another commensal fossil was *Josephella commensalis* Bałuk & Radwański, 1997, whose presence stimulated the host to deposit acicular sclerites and granules over the worm tubes (Bałuk & Radwański 1997). The association was also qualified as commensalistic since the tubes were embedded in the coral tissues, gaining space and protection without harming the coral. However, the authors also reported enormous calcium secretions in some polyps caused by overpopulations of the commensal and leading to the death of part of the coral colony, so the qualification as commensal may be, at least, doubtful. Particularly, symbiotic relationships are not reported for the single living species of the genus *Josephella marenzelleri* Caullery & Mesnil, 1896 (ten Hove & Kupriyanova 2009).

The newly recorded symbiotic spirorbins belonging to the genus *Circeis* and *Spirorbis* (Table 1) also contribute to increase the list of commensal serpulids. All of them appear to be facultative commensals living in association with pagurid decapods (Williams & McDermott 2004). However, other authors considered most spirorbins as epibionts, as there are no observations on the possible interactions between the worms and their hosts. Before 1998, only *Circeis paguri* Knight-Jones & Knight-Jones, 1977, which creates tubes in the lumen of gastropod shells occupied by *Pagurus bernhardus* (or attaches to the telson of the host), was reported as an obligate commensal (Williams & McDermott 2004).

Scalibregmatidae

The report of *Scalibregma inflatum* Rathke, 1843 as a commensal of echiurids is based on resin casts of sediment cores, which always contained burrows of the polychaete tapping into the echiurid burrows. Anker et al. (2005) suggested that the polychaete may feed on faecal pellets of the host.

Siboglinidae

The report of the siboglinid species of *Osedax* as commensals refers to the fact that the female tubes contain the so-called 'harems'. In other words, these are numerous dwarf, paedomorphic males (often several tens) filled with developing sperm and yolk droplets that resemble siboglinid trochophore larvae, which occur clustered around the females' oviduct (Rouse et al. 2004). The presence of yolk may indicate that males do not need to feed. Also, they are not attached to the female body but to the tube. Thus, they cannot be considered as parasites. The relationship seems more to be a type of commensalism closer to a mutualism, where the males obtain protection from females, and females benefit by easy reproduction in the highly localised deep-sea whale falls that most commonly provides habitat for the species of *Osedax*.

The number of males increases with female size, which suggests that the latter may attract the former during larval settlement. It has also been suggested that there may be an environmental control of sex, with the larvae settling on whale bones and on females respectively becoming females and males. Apparently, molecular evidence shows that males recruit on females from a common larval pool, discarding selective recruitment on the own maternal individual (Rouse et al. 2008, Vrijenhoek et al. 2008, Whiteman 2008).

Sigalionidae

Pholoides sinepapillatus Miranda & Brasil, 2014 was reported as living in association with deep-sea corals. As no details on the type of relationship were provided in the original description, and no other mentions of the species are known to date, we have assumed that it might be something closer to the commensalism often reported for some species of the closely related family Polynoidae.

Spionidae

Newly reported commensal spionids are mostly associated with sponges and decapods. *Polydora colonia* Moore, 1907 is a widely distributed symbiont of sponges, recently reported as introduced in Mediterranean waters. Previous mentions of the species were always as living in association with sponges, but there were no clear statements on the nature of the association. David & Williams (2012) consistently observed the presence of sponge material in the gut of the polydoran. However, there was no firm evidence of selective feeding on the host sponge that could support defining the association as closer to parasitism. The observed sponge remains could come from a cleaning activity of the symbiont, which would have a completely different ecological meaning.

The other sponge-associated polydorans belong to the genus *Polydorella* (Figure 1C), which is characterised by combining sexual and asexual (paratomy) reproduction as a strategy to facilitate host colonisation. In the case of *P. kamakamai* Williams, 2004, most worms have very large sand grains in the gut, indicating that they may feed on large food particles removed from the sponge surface, an ability that could certainly benefit the host. More recently, clear indications of mutual benefits have been reported for the association of *Polydorella smurovi* Tzetlin & Britayev, 1985 with *Negombata magnifica* (Keller, 1889). These are the polychaete capturing and concentrating loose detrital matter on the sponge surface (among which there may be shed old sponge cells) to be consumed, thus contributing to clearing the sponge surface of debris (Naumann et al. 2016).

The polydoran associated with hermit crabs, *Polydora nanomon* Orensky & Williams, 2009, is an obligate commensal (the third known species). The species most commonly occurs in the shell apex where it makes a hole, enters the lumen of the uppermost whorl and connects to the columella with a tube of mucus and detritus. The species shows sexual dimorphism, with one large female sharing a host with several (up to four) smaller males. The females seem to be able to feed on the surface of the shell surrounding the openings of their holes. In turn, the males (whose feeding behaviour remains unknown) excavate burrows within the detrital matrix of the females' tube, where they appear to be able to actively crawl (Orensky & Williams 2009).

Additionally, in the fossil records, there is a report of U-shaped borings in sea urchins that closely resemble those of living species of *Polydora* in other calcareous substrates (Wisshak & Neumann 2006). The borings show traces of regeneration, indicating that the sea urchin was living when infested. The borings were located between the peristome and periproct of the host, an advantageous position for the symbiont to be able to collect the organic matter resuspended during the locomotion and feeding activity of the host, as well as to obtain protection as an additional benefit. In turn, the association could only be moderately harmful for the host. Wisshak & Neumann (2006) considered that this was the first evidence for a possible association of a boring polydoran with a sea urchin and with an echinoderm in general. The authors also suggested that the dominance of large, long-living sea urchins in the benthic community and the limited availability of other hard substrates could favour the choice of *Echinocorys* as a suitable host by the boring polychaetes.

Syllidae

The syllids, together with the polynoids, are the families that have experienced a major increase in number of newly reported commensal species and relationships. The Syllidae currently contains 1009 species (Read & Fauchald 2017), among which 69 (i.e. 6.8%) belonging to 16 genera are considered to live as commensals both by Martin & Britayev (1998) and in the present update (21 and 48 species, respectively). Contrary to Polynoidae, however, more than a half (i.e. around 32) of the new commensal reports since 1998 correspond to species described as new to science, basically as a result of the new species of *Alcyonosyllis*, but also of the extensive taxonomic studies on *Haplosyllis* (Table 1).

Species of *Alcyonosyllis* are most commonly associated with alcyonaceans, but one is found on a scleractinian coral. In all cases, they show marked colour patterns, the former mostly based on different combinations of reddish and whitish bands and spots, sometimes with traces of blue as in *A. phili* Glasby & Watson, 2001 (the type species); the latter a combination of greenish and whitish bands. This colouring is most often lost in preserved specimens, even if kept on the host gorgonian (Figure 1D). Most species show an intraspecific variation in colouring intensity, not always related to the intensity of that of the host. Although this could suggest a possible aposematic colouring, it is well accepted that the combination of host and symbiont colour patterns help the latter be camouflaged within the branches of the former. The best expression of camouflage, however, is attained by *A. hinterkircheri* Glasby & Aguado, 2009, whose colour pattern perfectly mimics that of its host scleractinian coral.

The species of *Haplosyllis* range from large-sized sponge predators like the type species, *H. spongicola*, which cannot be formally considered a symbiont and is more often found living solitary, to the small-sized symbionts (also often associated with sponges), which occur in high densities inside their hosts. The newly reported relationships are based on a series of meticulous taxonomic studies that have unravelled the so-called *H. spongicola* sibling species-complex. The sampling of numerous different hosts, together with careful morphological and morphometric observations, has tripled the number of known species, though this is likely an underestimate of the real diversity of this genus, which is still awaiting molecular analysis.

The new reports of symbiotic *Haplosyllis* mostly corresponds to sponge endosymbionts, whose associations tend now to be viewed more as mutualism than parasitism, despite evidence demonstrating that the worms feed on the sponges (Lattig & Martin 2011a). These authors suggested that the particular metabolism of the sponges might be favoured by the presence of the symbionts, which may help the host regenerate their cellular components. They also highlighted the fact that the worms can defend their host from the attack of predators and that the worm-sponge associations survived even if environmental conditions changed drastically, for instance, when a shallow-water reef sponge was experimentally transferred to dark, deep waters where the associated photosynthetic cyanobacteria cannot survive (Maldonado & Young 1998).

Some sponge endosymbiotic *Haplosyllis* seem also to show reproductive adaptations, and, for instance, it has been postulated that female stolons remain inside the host to favour its rapid colonisation, while males are able to leave the host to facilitate fertilisation (Lattig & Martin 2011a). A similar situation was also reported for another sponge endosymbiont, *Haplosyllides floridana* Augener, 1922 living in association with *Xestospongia muta* (Schmidt, 1870). From young juveniles to female epitokous forms, including large asexual specimens, all different reproductive phases of this species were found living as endosymbionts, with the exception of male stolons that seemed to be free swimmers (Martin et al. 2009). Another species of *Haplosyllides*, *H. aberrans* (Fauvel, 1939) from Vietnam, was originally described from a planktonic male stolon. However, its recent report as a symbiont defined the species as a parasite of shrimps that, in turn, are coral symbionts. Whether these findings represent a real parasitism or are the expression of the stress of an endosymbiont forced out of a hypothetical host sponge is still a matter of discussion (Martin et al. 2009). A similar question remains open for the third known species of the genus *H. ophiocomae* Martin, Aguado & Britayev, 2009, reported as associated with a brittle star.

One of the most extreme adaptations to sponge endosymbiotic life occurs within syllids. More specifically, we refer to the ramified species of the family, such as *Syllis ramosa* McIntosh, 1879 (Figure 1E) or *Ramisyllis multicaudata* Glasby, Shroeder & Aguado, 2012. In these species, a single individual with numerous branches occupies the host aquiferous canals, instead of the numerous individuals typical of *Haplosyllis* or *Haplosyllides*. Despite this amazing body shape, the functionality of the symbiosis appears to be similar to that of the two aforementioned genera. *Ramisyllis* has only one head (which is occupied in sucking sponge cells), and numerous pygidia protruding from the host sponge and moving actively on its surface (Glasby et al. 2012). These emergent fragments have shiny whitish dorsal cirri and are, thus, perfectly visible. Although it has not yet been demonstrated, this suggests a possible aposematic function that, to some extent, might replace the active defence strategy of the two aforementioned non-branching genera. *Ramisyllis* seems also to have separate sexes, as the stolons from an individual inside one host sponge are either all males or all females (Schroeder et al. 2017). Moreover, both male and female stolons have swimming chaetae, which suggests that they may abandon the host sponge for mating. This suggests a strategy that appears to be different to that of *Haplosyllis* and *Haplosyllides*, in which female stolons remained inside the host, while males were found to swim. In these species, all development phases from fertilisation to juveniles and adults may occur inside the host. In fact, for an endosymbiotic population, the consequences of such a behavioural segregation between male and female stolons may have a high fitness value, as it assures (and/or contributes to increase) the occupation of already colonised hosts, while only half of the reproductive bodies are on risk outside the host. In turn, for an endosymbiotic individual like *Ramisyllis*, host occupation is singular. As evolutionary selective pressures tend to assure colonisation of new hosts, having swimming stolons of both sexes, together with external fertilisation and, likely, planktonic larval development seem to be excellent options.

Among the species not associated with sponges, there may be highlighted the report of *Haplosyllis villogorgicola* Martin, Núñez, Riera & Gil, 2002, a mid-sized ectosymbiont that lives in anastomosed nests formed by the branches of its host gorgonian. This species therefore demonstrates a lifestyle somewhat intermediate between *Haplosyllis anthogorgicola* Utinomi, 1956 a small-sized endosymbiont living inside the tissues of its gorgonian host, and *H. chamaeleon*, a relatively large ectosymbiont that induces no detectable reaction in its host gorgonian. *Haplosyllis chamaeleon* was originally described as a Mediterranean endemic species strictly associated with *Paramuricea clavata* (Risso, 1826). However, it has also been discovered in Atlantic waters off the Iberian Peninsula, living in association with the closely related *Paramuricea grayi* (Johnson, 1861). Morphologically, both geographically separated populations of *Haplosyllis chamaeleon* are indistinguishable, but no attempts have been made to check their deoxyribonucleic acid (DNA).

Bollandiella is a monotypic genus represented by *B. antipathicola* (Glasby, 1994), living in association with an antipatharian in Japanese waters. However, studies in progress have revealed the presence of two other geographically separated populations. One of the new populations lives in association with a deep-sea antipatharian in the Cantabrian Sea, off the northern costs of the Iberian Peninsula (D. Martin & T. Molodtsova, unpublished data). The other lives in association with a whip-like antipatharian in Indonesian waters (M. Bo & D. Martin, unpublished data). Both findings may well represent additional new species of *Bollandiella* and thus open new possibilities in the study of this peculiar genus. In particular, the existence of specimens fixed and preserved in ethanol may facilitate elucidation of its phylogenetic position. The Indonesian population appeared to be sexually dimorphic as, in addition to the normal *Bollandiella*-like males living on the surface of the coral, there were numerous, highly simplified worms, all females, living in galleries inside the soft tissues of the host. These two morphotypes possess two simple chaetae per parapodial bundle that are, however, slightly different. There are only two possible ways to validate the hypothesised sexual dimorphism, through molecular or behavioural analysis, both being equally interesting for future studies.

An undescribed species of syllid associated with scleractinian corals was reported as belonging to *Typosyllis* by Randall & Eldredge (1976), a generic assignation that seems doubtful in light of current knowledge of cnidarian-associated syllids. The polychaete occurred in high numbers in different coral colonies, where they induced the formation of a system of tubercles and epitheca-like tubes joining the subphacelloid corallites, as well as, in some cases, deeply incised intercalicular grooves. However, the association apparently did not harm the host colonies, and other than the skeletal modifications, the colonies had a healthy aspect and grew normally (Randall & Eldredge 1976).

In addition to sponges and cnidarians, syllids may also be associated with many different groups, such as the facultative symbiosis of *Eusyllis blomstrandi* Malmgren, 1867 with different species of hermit crabs, the various species of *Syllis* associated with sipunculans, decapods and holothurians, and an undescribed species of *Inermosyllis* associated with a starfish (Table 1).

An interesting aspect of some of the new reports here recorded is that many correspond to morphospecies that have not yet been formally described. This, together with the fact that they seem able to utilise many taxa as potential hosts, certainly allows us to anticipate that the reported diversity of commensal syllids will continue to increase in the near future. A good example to illustrate this point is an undescribed species that has been photographed twice but never collected. The first specimen occurred in Vietnam (Figure 1F) on an alcyonacean of the genus *Dendronephthya*. The second, more recent, observation was from Brunei by B. Meyers from a non-identified alcyonacean (https://www.flickr.com/photos/brianmayes/14776372948/, https://www.flickr.com/photos/brianmayes/14962640162/). It is not listed in Table 1, as no specimens have been collected allowing its identification. However, we certainly believe it could be a new species, in accordance with Leslie Harris (Natural History Museum of Los Angeles), who checked the images captured from Brunei (B. Meyers, personal communication). In addition to the beauty of this syllid, it may represent the first documented case of an aposematic polychaete mimicking a well-protected organism. In fact, the whole body of the worm looks like the nudibranch *Samla bicolor* (Kelaart, 1858), with the dorsal cirri particularly simulating the cerata of the mollusc (some images are available at https://www.gbif.org/species/5796483). Contrary to what happens with cerata, however, we believe that cnidoblasts would not be accumulated in the tips of the syllid cirri. We thus suggest that, to some extent, this species may display a mimicry strategy similar to that of unarmed flies mimicking bees.

Parasitic polychaetes

Chrysopetalidae

No new species and relationships have been reported among the obligate parasitic calamyzin chrysopetalids. Thus, *Calamyzas* is still a monotypic genus including only *C. amphictenicola* Arwidsson, 1932 living in association ampharetid hosts. Recent observations, however, revealed that male and female couples shared the same hosts, where they were attached on the furrow at the base of the host gills (Watson & Faulwetter 2017) (Figure 3A). Observed *in vivo*, these specimens showed brown transversal stripes along the body, allowing them to be camouflaged among the striped gills of their ampharetid host (Watson & Faulwetter 2017) (Figure 3B). The careful morphological observations of these authors (based on microcomputed tomography) showed for the first time the presence of camerate neurochaetal shafts in *C. amphictenicola*, a feature that they suggest may be

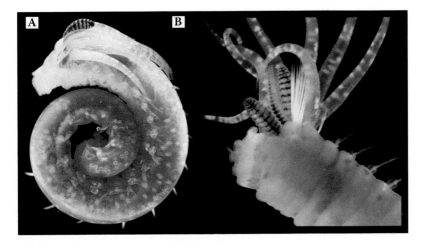

Figure 3 Living specimens of *Calamyzas amphictenicola* attached to their ampharetid hosts. (A) One female (up, dark striped, with oocytes inside) and one male (down, slightly pale) sharing the same host (Photo: Arne Nygren); (B) Three specimens sharing the same host (Photo: Frederick Pleijel).

related with the mobility of the parasite, as it has been found on different locations on the host body when attached for feeding.

Cirratulidae

The supposedly parasitic cirratulid belongs to *Dodecaceria*, likely *D. concharum* Örsted, 1843, although the species assignment is not clear in all cases. The type of relationships they have with pagurids seems to be accidental, as they are well known as borers of the shells that hermit crabs occupy. In turn, its classification as a parasite is also not clear, as there are no details on how its presence may negatively influence the hosts (Williams & McDermott 2004).

Dorvilleidae

The newly reported parasitic dorvilleid corresponds to a subspecies of *Veneriserva pygoclava* Rossi 1984, *V. p. meridionalis* Micaletto, Gambi & Cantone, 2002. Both inhabit the coelom of aphroditid polychaetes but differ in a few morphological characters, size and geographical distribution (Southern California and Antarctica, respectively). *Veneriserva p. meridionalis* was very numerous, so relevant statistics describing the peculiarities of the association with the host are reported in the original description. Accordingly, there was a significant relationship of the prevalence with the bathymetrical distribution of the host, while there were no size relationships between hosts and symbionts. The evident reduction of the jaw apparatus and intestine led Micaletto et al. (2002) to suggest that it may be a parasite, although its feeding mode was not clearly defined.

Hesionidae

The newly reported parasitic hesionid was initially cited as *Parasyllidea humesi* Pettibone, 1961 by Martin et al. (2012), then redescribed as *Oxydromus humesi* (Pettibone, 1961) by Martin et al. (2015), and recently described as a new species, *Oxydromus okupa* Martin, Meca & Gil in Martin et al. (2017a). The species lives endosymbiotically in the branchial chamber of two different tellinid bivalves. In the case of the host *Scrobicularia plana* (Da Costa, 1778), despite the absence of tissue damage (not in foot or in branchiae), the studied population showed a significant negative influence (i.e. a reduced tissue biomass). As this could be entirely attributable to the presence of the symbiont, Martin et al. (2012) considered the relationship as a parasitism.

Oxydromus okupa shows a territorial, aggressive behaviour and a regular distribution, with one symbiont per host (with a few exceptions in which male and female share a host, more rarely including a juvenile) as well as a characteristic host-entering behaviour in which the symbiont is able to use the inhalant siphon to gain the branchial chamber of the host (Figure 4A and B). An alternative entry may be in the vicinity of the foot, but attempts through other points along the valve outline are repulsed by the host (Martin et al. 2015). Males seem to be more motile (likely related with the success of fertilisation), while females tend to remain inside their hosts (Martin et al. 2017a), a behaviour previously reported for other polychaetes (scale worms) living in association with bivalves (deep-sea mytilids) (Britayev et al. 2007).

Ichthyotomidae

Ichthyotomus sanguinarius Eisig, 1906 is probably one of the most intriguing parasitic polychaetes ever described. The peculiar scissor-like jaws with spoon-like stylets and backwardly teeth (used to pierce the skin of the eel hosts and attach themselves firmly), the muscular pharynx (used to suck the blood of the host), and the haemophilic glands (used to store blood) points to its high degree of specialisation. There are further adaptations to the parasitic mode of life: the haemophilic glands produce a secretion preventing coagulation of the fish blood. Some glands in the ventral cirri produce a secretion which, together with the cup-shaped oral sucker, allow the worms to firmly stick to the host fins. In fact, although the parasites may leave one host and reattach themselves to another one using a very vigorous muscular action, some of them may even die in their efforts to free themselves

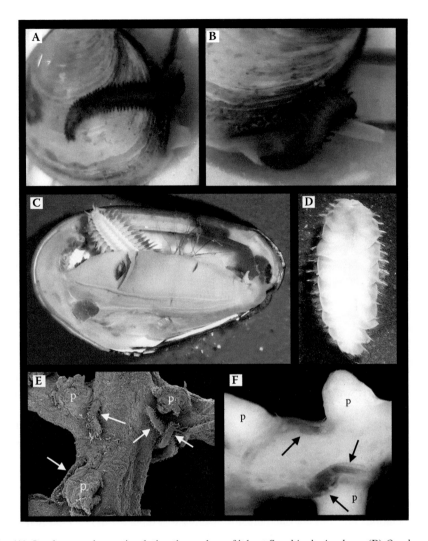

Figure 4 (A) *Oxydromus okupa* stimulating the syphon of it host *Scrobicularia plana*; (B) *Oxydromus okupa* entering the host through the exhalant siphon; (C) A specimen of *Bathymodiolus azoricus* opened to show the presence of *Branchipolynoe seepensis* in its branchial chamber; (D) The same *Branchipolynoe seepensis* isolated from the host, dorsal view; (E) Fragment of *Muricella ramosa* with numerous specimens of *Haplosyllis* sp. (arrows) protruding from the host coenenchyme near the polyps (p) (scanning electron microscope); (F) Fragment of *Carijoa* with three specimens of *Haplosyllis* sp. (arrows) protruding from the host coenenchyme near the polyps (p) (Photos: D. Martin).

from the host (Potts 1912). The species has a reduced gut and seems to use the sucked host blood as an additional source of oxygen. The species shows marked sexual dimorphism, and they seem to live in couples, the males being small and white and the females large and yellow.

Despite the peculiar characteristics of *Ichthyotomus sanguinarius*, its relative large size and the frequent capture of its host, the Mediterranean painted eel *Echelus myrus* (Linnaeus, 1758), the species had not been found again since the original description until Culurgioni et al. (2006) found at least 25 couples living attached to conger eels, *Conger conger* (Linnaeus, 1758) in Sardinian waters. Conger eels had already been reported as hosts (as *Conger vulgaris* Yarrell, 1832), together with another eel, *Dalophis imberbis* (Delaroche, 1809) (as *Sphagebranchus imberbis* Delaroche, 1809) (e.g. in Eisig 1906, Paris 1955). Some of the worms were found in the skin of the conger eels, a location that was also not previously reported. Unfortunately, Culurgioni et al. (2006) did not describe their specimens and only provided images of the posterior end of the worm, so the peculiar characteristics of the species could not be confirmed based on modern observations.

Oenonidae

With some rare exceptions, parasitic oenonids are very rare and often only known from their original descriptions, which consist of only one or a few specimens. The only relevant information in the original description (and single finding to date) of *Labrorostratus zaragozensis* (Hernández-Alcántara & Solís-Weiss, 1998) is that the species has a somewhat reduced jaw apparatus and belongs to the oenonid group in which the endoparasites are larger than their hosts. Conversely, the previously known association between *Haematocleptes terebellidis* (Wiren, 1886) and its host trichobranchid *Terebellides stroemii* seems to be an exception. Accordingly, it may have been overlooked by most marine ecologists and could be more widely distributed than previously thought (O'Reilly 2016).

Phyllodocidae

Galapagomystides aristata Blake, 1985 is a peculiar phyllodocid that lives in large numbers among tubeworms, clams and mussels at hydrothermal vents on the East Pacific Rise (see Jenkins et al. 2002 and references therein). The species has a peculiar morphology of the mid-gut, which is filled with blood and has large, numerous spherical inclusions in its tissues. Blood is very abundant, suggesting constant feeding or long-term blood storage, while the nature of the spherical inclusions led Jenkins et al. (2002) to speculate that they may have a role in the regulation and storage of iron resulting from degrading the exogenous haemoglobin from the blood food. Also, the worms have a ring of small spines at the end of the proboscis, which may help in attaching to the potential hosts. However, attempts to identify possible hosts have not been successful (Jenkins et al. 2002).

The two species of *Eulalia* reported as parasites of hermit crabs (Samuelsen 1970, Jensen & Bender 1973), probably correspond to accidental findings as there are no ways to demonstrate whether they were real parasites.

Polynoidae

Polynoids are the most numerous commensal/mutualistic symbiotic polychaetes both in terms of species and relationships, but not one parasitic species was reported in the 1998 review. However, the reanalysis of the existing bibliography revealed the existence of at least four parasitic polynoids. During the last two decades, the status of *Gastrolepidia clavigera* and *Branchipolynoe seepensis* has been reconsidered from commensal to parasitic, *Eunoe opalina* McIntosh, 1885 was newly described as symbiont, and *Thormora johnstoni* (Kinberg, 1856) was omitted in the 1998 review. In three of these four species, the qualification as parasite derives from careful behavioural observations. Therefore, we cannot discard that more associations routinely qualified as commensalistic could result to be parasitic when analysed in detail.

Gastrolepidia clavigera (Figure 2G and H) is a well-known scale worm, widely distributed in shallow Indo-West Pacific waters where it lives in association with sea cucumbers (Britayev & Zamyshliak 1996). Apparently, its presence did not affect the host negatively, while the colour mimicry and the use of host as a refuge prevailed in qualifying it as a commensal. However, detailed diet studies based on dissected gut contents demonstrated that the most common food particles were pieces of holothurian tissues, confirmed by the spicules belonging to the host species (Britayev & Lyskin 2002). Moreover, as holothurian spicules differ relative to body regions, it was even possible to determine, in some cases, which parts of host body were eaten by the symbiont. However, besides the host tissues, remnants of free-living crustaceans and copepods parasitising the holothurians were also found in the scale worm gut, which was suggested to provide indirect support of the control of one parasite on the density of the other.

The second 'commensal' species that has been reconsidered as likely being parasitic is *Branchipolynoe seepensis* (Figure 4C and D). The different reports of this 'species' from Atlantic seep and vent sites demonstrated the existence of a cryptic species-complex in parallel with the description of the peculiar larval dispersal strategies between their highly specific, scattered, deep-sea habitats (Chevaldonné et al. 1998, Jollivet et al. 1998, Plouviez et al. 2008). *Branchipolynoe seepensis* inhabited the mantle cavity of the hydrothermal mussels *Bathymodiolus azoricus* Von Cosel & Comtet 1999, *B. puteoserpentis* Von Cosel, Métivier & Hashimoto 1994 and *B. heckerae* Turner, Gustafson, Lutz & Vrijenhoek 1998 in Gustafson et al. (1998) and was conventionally considered as commensal, without an effective demonstration of the quality of the relationship (Jollivet et al. 2000). More recently, analyses of the location and gut contents of the symbionts indicated that they may either feed on filtered, agglutinated, suspended particles transported toward the host mouth or on suspended organic particles transported to the siphon opening (Britayev et al. 2003a,b, 2007). Thus, they may certainly be considered as kleptoparasites. Moreover, *Branchipolynoe seepensis* has a negative influence on their mytilid hosts, as supported by the finding of damaged gill filaments, foot deformation and shortening or elimination of labial palps in the infested individuals (Britayev et al. 2003b, 2007, Ward et al. 2004). The incidence of these traumas was about four to five times higher in infested than in non-infested molluscs, suggesting that they were caused by the symbionts. However, host tissues were not considered the habitual food source for the symbionts but as an accidental result of the 'normal' feeding activities of these powerfully jawed symbionts. On the other hand, by stealing host food, consuming oxygen and accidentally damaging the host tissues, the symbionts may have a negative influence on the host metabolism and, thus, on their growth. This was reflected by a modification of the shell proportions in infested and non-infested specimens of *Bathymodiolus azoricus*, likely due to the inhibition of growth in the former. In summary, the symbiont behaviour is now being considered as closer to parasitism than to commensalism (Britayev et al. 2003a,b, 2007).

The 'free-living' polynoid *Eunoe opalina* was recently redefined as parasite. It most often inhabits a depression excavated in the ventral surface of its host holothurian, in which the tegument has been partly removed (Schiaparelli et al. 2010). As for *Gastrolepidia clavigera*, remains of host tissues were found in the polynoid gut, which also contained other types of remains. Among the latter were foraminifers, diatoms, nematodes, tissues of other holothurians, chaetae of polychaetes and fragments of ophiuroids, bivalves and crustaceans, which was interpreted as an indication of alternation between parasitic and predatory behaviour. This alternation was suggested to be a successful strategy in Antarctic waters and may have a great impact in the simplified trophic webs characterising the region (Schiaparelli et al. 2010).

Omitted in the 1998 review, *Thormora johnstoni* lives in association with Palolo worms in Samoa, a relationship that has never been reported again since the original mention by Hauenschild et al. (1968). These authors reported *T. johnstoni* as being regularly found living solitarily inside host tubes. Infested hosts, in turn, habitually exhibited blood-shots in the body wall, especially in the posterior end. Shots were even present in hosts reared for several months in laboratory conditions. In turn, the polynoid guts contained a reddish liquid, probably a mixture of blood and coelomic fluid sucked from the Palolos, so that *T. johnstoni* was considered a temporary ectoparasite. In laboratory cultures, however, the polynoids were often observed in the immediate vicinity of the Palolo worms, being tolerated by their potential host in contrast to other intruders that disturbed the eunicids and were immediately repulsed. Whether this conforms to a possible benefit from the host derived from the presence of the symbiont or to an adaptation of the symbiont to exploit its host remains open to future research.

Sabellidae

The new reports of parasitic sabellids all correspond to *Terebrasabella heterouncinata* Fitzhugh & Rouse, 1999. Originally described from South Africa, the species was considered endemic and a pest of cultured abalones, affecting particularly the marketability of this valuable resource. The species could be considered among those defined as parasites from an anthropogenic point of view. However, careful studies carried out in South Africa (Simon et al. 2002, 2004, 2005) demonstrated that the presence of the sabellid had a strong negative influence on the growth rate of the abalone, thus leading to consideration of the species as a real parasite.

Moreover, *Terebrasabella heterouncinata* was revealed to have an invasive character. Introduced in California, probably through abalone cultures, it was able to colonise native habitats via effluent release from the aquaculture facilities. However, the species was successfully eradicated (Culver & Kuris 2000). Later, its invasive potential was assessed using Californian gastropods, revealing that it is the only known boring organism able to colonise the aperture of living gastropods. Their larvae seem to be chemically attracted by specific shell components, and once settled, they may induce a calcification response in the host that results in the formation of a worm tube (see Culver & Kuris 2004 and references therein). The susceptibility of the different potential hosts was, however, varied, and species such as *Neobernaya spadicea* (Swainson, 1823) were resistant to attack. Therefore, studies like that by Culver & Kuris (2004) are necessary to prevent the potential impacts of hypothetical invasion of pest species such as *Terebrasabella heterouncinata*.

Serpulidae

The mention of a parasitic serpulid comes from the fossil record. The host was a brachiopod that showed the presence of tubes attached to the inner surface of the valve. Shell materials covered these tubes indicating that symbionts settlement on the shell margin could occur when the host mantle was temporarily retracted, and then the symbionts continue to grow their tubes to avoid being covered by the shell materials or by mantle (Kiel 2008). Alternatively, we suggest that the symbiont growth could trigger a protective reaction in the host, this leading to exactly the same result. This second possibility seems to be more feasible than the originally suggested hypothesis and agrees with the equivalent host responses in other known associations between symbiotic polychaetes and different hosts with calcareous skeletons, such as corals (e.g. chaetopterids or spionids) or gorgonians (e.g. polynoids or syllids) already mentioned in this paper, so we cannot discard a non-parasitic relationship. The similarity of this association with those involving *Terebrasabella heterouncinata* is remarkable and may open evolutionary speculations, which are certainly interesting but far from the scope of this review.

Spionidae

Most Spionidae newly reported as parasites are associated with pagurids and cnidarians. In the former case, they live in shells occupied by the hermit crabs where they are the most prevalent borers despite the facultative character of the associations (Williams & McDermott 2004). Contrary to the considerable research carried out on the nature of the relationships of the commensal polydorans mentioned previously, the extent to which the relationships of the shell-boring polydorans are or are not of a parasitic nature remains doubtful, with a few exceptions. *Dipolydora commensalis*, for example, is a typical shell-boring spionid that was originally described as commensal. The worms were later reported as being either suspension- or deposit-feeders by removing particles from the branchial (respiratory) current of the hermit crabs or by collecting particles from the hermit crab leg setae (Dauer 1991). Thus, no damage to the host was attributed to the relationships. However, later studies demonstrated that the particles removed from leg chaetae were, in fact, hermit crab embryos, so the species must be considered as an egg predator, in a similar way to the closely related *Polydora robi* Williams, 2000. The influence on the reproductive success of host hermit crabs, however, has not yet been evaluated (Williams 2002).

The newly reported spionids associated with cnidarians are in most cases polydorans (although not formally identified, they probably belong to the genus *Polydora*) living in tropical corals. To some extent, most infestations are related to increasing organic pollution from waste discharges or fish-farming areas (Wielgus et al. 2006). Contrary to the symbiotic chaetopterids living in the same conditions, which we consider to be closer to commensals (see previous section), these coral-associated polydorans are here considered as parasites. In fact, they excavate numerous galleries that favour the weakening of the skeletal strength, thus probably increasing the possibility of storm-derived damages to reefs (Wielgus et al. 2002). Also, they secondarily induce malformations in the coral hosts, which increase the roughness and bumpiness. This may cause shadowing that reduces the amount of light reaching the zooxanthellae and may induce a photoacclimation response to this reduced irradiance (Wielgus & Levy 2006). Further, the presence of the spionids may induce local eutrophication increases (through the accumulation of faecal products) that may affect zooxanthellae and favour algal proliferation (Wielgus & Levy 2006).

Other boring spionids newly reported as cnidarian associates are *Dipolydora armata* (Langerhans, 1880) and *Polydora* cf. *alloporis* Light, 1970 (Lewis 1998, Lindner et al. 2004). The first species lives in complex branching tunnels within the coenosteum of its coral host. Although burrow openings are single tubular limbs lined with mucus, they are surrounded by swellings of host skeleton and soft tissues. As in other species with tubes attached or excavated in calcareous hosts, the continued growth of the host tissue and the worm tubes gives rise to joint structures, in this case developing into several millimetre-long spines, often irregular, whose surface lacks zooids. *Polydora* cf. *alloporis* lives in U-shaped galleries with a pair of closely placed circular holes that run along the centre of the branch core and occasionally come to the surface of the coenosteum of the calcareous hydrocoral host. The authors do not speculate on the relationships of these species with their hosts, but the fact they are borers led us to consider them as closer to parasites along with the other polydorans considered previously.

Probably, the most peculiar parasitic spionid is not, in fact, a real new report. Dwarf males of *Scolelepis laonicola* (Tzetlin, 1985) live as parasites of the 'normal' spionid-like females (Vortsepneva et al. 2008) and were originally described as a different genus and species, *Asetocalamyzas laonicola* Tzetlin, 1985 within the Calamyzidae, a monospecific taxon that is now considered as a subfamily within the Chrysopetallidae (Aguado et al. 2013). These peculiar parasitic males are highly simplified in comparison with the normal females but have extensive coelomic cavities, complete septae, well-developed segmental nephridia, circulatory and digestive systems, and a rudimentary nervous system. The most striking characteristic is, however, that host and parasite tissues are highly integrated around the area where the latter penetrates the former. In fact, their cuticles are continuous, and their blood vessels apparently interlace in the fusion zone. Accordingly, Vortsepneva et al. (2008) suggested that the parasite may alternatively feed through direct exchange with the female using the interlaced blood vessels (i.e. a placenta-like structure) or by sucking coelomic fluids, but the well-developed gut, filled with materials, in the parasitic males seems to be more consistent with the second method.

Syllidae

Contrary to earlier assessments, most associations involving symbiotic syllids were revealed to be closer to commensalisms/mutualisms than to parasitisms. This was despite the apparently detrimental interaction caused by the common presence of thousands of worms living on or inside the hosts, often feeding on them (as explained previously, in the section dedicated to commensal syllids). This may also be the case of the American specimens of *Branchiosyllis lamellifera* Verrill, 1900 living in association with sponges, as no data other than the presence of spicules adhered or inserted in the tegument were reported (San Martín et al. 2013).

A few newly reported associations involving syllids may certainly be parasitic, such as the small-sized autolytins associated with cnidarians. However, in many cases, there are no other relationships between host and parasite apart from the attachment of the worm tubes on the cnidarians. The presence of the symbiont seemed not to induce any reaction by the host, and it cannot be discarded that, despite the obligate nature of the association, the worm simply lives on, preys on and feeds upon the host, as Nygren & Pleijel (2010) suggested for *Imagimaea draculai* (San Martín & López 2002). To some extent, it remains difficult to distinguish between a small, highly specialised, species-specific predator and a true parasite. No behavioural observations (either *in situ* or experimental) for making this distinction have been reported to date. The most recent parasitic autolytin, however, lives in association with a tunicate. *Proceraea exoryxae* Martin, Nygren & Cruz-Rivera, 2017 excavates galleries in the tunica of the solitary ascidian *Phallusia nigra* Savigny, 1816 and is, thus, the first known example of a polychaete miner and the first syllid known to live in association with ascidians. The association has been found only once and in a single host, but numerous specimens of the parasite occurred in the galleries, including male and female stolons, which has been considered as an indirect support of a highly specialised behaviour (Martin et al. 2017b). Whether *Proceraea exoryxae* feeds only on the tunica or may also feed on the internal tissues of the host tunicate is an interesting topic for a future study that will help to precisely characterise this association.

Another syllid newly reported as a parasite, *Haplosyllis gula* Treadwell, 1924, occurs firmly attached to the branchiae or the dorsal cirri of its polychaete hosts. Its broad pharynx seems to be an adaptation to an ectoparasitic mode of life (Lattig & Martin 2009). However, the species was later reported as endosymbiont of sponges (Lattig & Martin 2011a). As for other species of *Haplosyllis* and *Haplosyllides*, enormous populations composed of hundreds or thousands of worms occur inside host sponges, while a few specimens are reported as attached to different appendages of other hosts (e.g. polychaetes, crustaceans). Whether these secondary relationships are steps in the life cycle of these species or just the result of the stress of being forced to be outside their host sponges is still a matter of discussion.

Two more small-sized species of *Haplosyllis* appear to be parasitic. Both are known to live in association with gorgonians. The first one with *Muricella ramosa* (Thompson & Henderson, 1905) from the Indian Ocean off Somalia (J. Gil & D. Martin, personal observation) (Figure 4E). The second one with *Carijoa* sp. from the South China Sea off Vietnam (T.A. Britayev & D. Martin, personal observation) (Figure 4F). Both are still undetermined but certainly belong to new species. They are also representatives of lifestyle progressively intermediate between those of the gorgonian associates *Haplosyllis anthogorgicola* and *H. villogorgicola*. Like the former, they live in galleries excavated in the coenenchyme of the host. The Somalian species induced the host to produce well-defined and consolidated gallery openings that indicates frequent use and some reaction form the host to the presence of the symbiont. The presence of the Vietnamese species apparently does not trigger any reaction from the host cnidarian, as the galleries have irregular openings that seem to be just surface breaks. In both cases, however, the worms tend to protrude from the coenenchyme near the host polyps, pointing to a possible kleptoparasitic behaviour as suggested for *H. anthogorgicola* (Martin et al. 2002).

Typhloscolecidae

Parasitic typhloscolecids are associated with chaetognaths (see Øresland & Bray 2005 and references therein). The typical behaviour consists of causing decapitation of the hosts, either by directly feeding on the host head or the surrounding tissue (i.e. in the neck or anterior body end areas) and seems to be widespread all around the world's oceans. The polychaetes may be found attached to different parts of the host body, most often on or around the head. However, completely decapitated hosts are never infested. Again, the polychaetes seem to feed by perforating the host tissues, then sucking coelomic fluids so that decapitations may be a secondary result of major tissue damage (Øresland & Bray 2005). Whether these worms are real parasites of just small, specialised predators is also an unsolved question.

Conclusions

The information summarised in this review partially responds to the question of whether there have been significant changes in our understanding of symbiotic polychaetes. Overall, 265 papers contain new reports of these organisms and/or their relationships from 1998 to 2017 (i.e. 18 years) versus the almost 425 in the 1998 review. There is certainly an increasing production of new knowledge, and this explains why five times more species and six times more relationships have been newly reported per year after 1998. Moreover, this trend continues progressing as, by the time we were trying to close the present review early in 2017, a few more papers dealing with symbiotic polychaetes were already published, just in time to be mentioned here (Lattig et al. 2017, Martin et al. 2017a,b, Serpetti et al. 2017).

By updating the information on symbiotic polychaetes with the most recent publications on that topic, we have been able to infer some general trends. One of the most consistent concerns the degree of specificity of the associations, as the number of symbionts living in association with a single host (i.e. monoxenous) still dominates. However, some things have not changed. For instance, the increasing amount of new knowledge does not allow us to fully assess whether some of the new (as well as the previously known) associations are really species-specific, or instead they qualified as such due to a lack of real information. In some cases, we suspect that the second possibility could be more likely, and this seems to be confirmed by certain newly published records (e.g. within the species of *Haplosyllis*). Moving to the other extreme, the number of polychaetes involved in multiple relationships has almost doubled, most being validated (i.e. not attributable to species-complexes). Therefore, there seems to be a trend in which symbiotic polychaetes are often involved in relationships that are much more complex than expected. We are almost certain that this trend will be confirmed as new knowledge is generated.

Another growing trend is the increasing number of mutualisms, reassigning species formerly considered as commensals or parasites. In fact, we found at least 22 species of polychaetes from different families (one from the fossil register) being (or being suggested to be) involved in mutualistic associations.

On the other hand, there is growing evidence that designation of an association as commensal or parasitic based on an anthropogenic perspective may be misleading or erroneous, at least without the support of careful behavioural observations and other experimental data. This concept has been recently developed from a paleoecological point of view (Zapalski 2011), but it certainly also applies to present interactions between living organisms, which are often defined on the basis of preserved material. For instance, some small, highly numerous, host-feeding worms (presumed parasites) have demonstrably provided clear benefits to their hosts in the form of tissue regeneration or defence (e.g. the case of some tropical symbiotic species of *Haplosyllis*), thus being considered as mutualists. In turn, some large, solitary worms, whose presence appeared not to harm the host were, in fact, causing damage (sometimes indirectly) and negatively affected the host metabolism, thus becoming examples of parasitism (e.g. some species of *Oxydromus* and *Branchipolynoe*).

Nevertheless, our 1998 review successfully raised the profile of these animals such that symbiotic polychaetes are now more and more attractive to researchers, who are paying much more attention to the peculiarities of the relationships in which they are involved (rather than simply reporting new species or relationships). Overall, there has been a noticeable increase of the studies dedicated to symbiotic polychaetes. However, descriptive and experimental studies (e.g. based on behavioural observations of living organisms) are still needed to address the true nature of the relationships. On the other hand, further work on how these relationships co-evolve from a phylogenetic perspective is required, as the single example addressing this question is limited to a few species of deep-sea polynoids collected during the same cruise (Serpetti et al. 2017). The phylogeographic perspective is also promising, but to date, it is also limited to a single study on deep-sea iphitimid dorvilleids (Lattig et al. 2017). In addition to asking taxonomically inclined scientists to broaden the scope of their studies, we need to convince scientists working on generalist ecological studies based in collecting and identifying organisms about the importance of paying attention to the possible presence of symbiotic organisms (polychaetes, of course, but also other taxa), which may occur in every sample they are working on, leading to much greater diversity than is superficially apparent.

We understand they may not be directly interested in studying them and the associations in which they are involved. However, such information may prompt dedicated studies by specialists: there are no excuses in the era of easy access to global communications and information. For instance, today this can be easily achieved by simply distributing images of their findings via social media (e.g. https://twitter.com/WPolyDb/status/908852928244355073).

Marine symbioses in general require environmental stability over large timescales to get established (Brasier & Lindsay 1998), and those involving polychaetes are certainly not an exception. In turn, this allows the established consortia to colonise and persist in unfavourable environments (Uriz et al. 2012). Although often cryptic, symbionts are important contributors to the biodiversity of the benthic ecosystems (e.g. Lattig & Martin 2011a and references therein, Wirtz et al. 2009). Therefore, they are undoubtedly highly relevant from ecological and conservation perspectives in our changing ocean conditions. This certainly justifies gathering new knowledge on the role they may play in the marine ecosystems, as well as on their potential responses to currently projected environmental changes – a perspective that has been seldom considered in the studies dealing with symbiotic polychaetes.

Finally, we would like to conclude by stressing one aspect that was not manifested as relevant in the 1998 review. The present increase in the number of studies on symbiotic polychaetes appears to be almost fully restricted to the specialised academic world. Considering the current social perception of science and the fact that it is widely accepted that the aforementioned environmental changes are anthropogenically driven, a next logical step is that we have to be able to transfer our knowledge towards non-specialist audiences, including the general public. In other words, to contribute to the preservation of our seas, it is our duty to raise awareness of the potential ecological and economic impacts of these symbiotic associations and allow other eyes to enjoy the intrinsic beauty of symbiotic worms.

Acknowledgements

This paper is a contribution of D. Martin to the Research Projects MarSymbiOmics (CTM2013-43287-P), funded by Spanish 'Agencia Estatal de Investigación' (AEI), and PopCOmics (CTM2017-88080), funded by the AEI and the European Funds for Regional Development (FEDER), and to the Consolidated Research Group on Marine Benthic Ecology of the Generalitat de Catalunya (2017SGR378). The work of T.A. Britayev was supported by the research project 18-05-00459, funded by the Russian Scientific Foundation for Basic Science. We would also like to thank the anonymous reviewers and the editor for their constructive comments and criticisms, which resulted in greatly improving both the quality and the readability of this paper, as well as to Ned Deloach, Paul Human, Arne Nygren, Frederick Pleijel and O. V. Savinkin for kindly allowing us to reproduce their amazing pictures (credits are detailed in the respective figure legends).

References

Achari, G.P. 1974. On the polychaete *Gattyana deludens* Fauvel associated with the hermit crab *Diogenes diogenes* Herbst and *D. custos* Fabricius. *Journal of the Marine Biological Association of India* **16**, 839–842.

Agassiz, A. 1880. Reports on the results of dredging, under the supervision of Alexander Agassiz, in the Caribbean Sea in 1878–79, and along the Atlantic Coast of the United States during summer of 1880, by the U.S. Coast Survey Steamer 'Blake'. IX. Preliminary report on the Echini. *Bulletin of the Museum of Comparative Zoology at Harvard College* **8**, 69–84.

Aguado, M.T. & Glasby, C.J. 2015. Indo-Pacific Syllidae (Annelida, Phyllodocida) share an evolutionary history. *Systematics and Biodiversity* **13**, 369–385.

Aguado, M.T., Murray, A. & Hutchings, P.A. 2015. Syllidae (Annelida: Phyllodocida) from Lizard Island, Great Barrier Reef, Australia. *Zootaxa* **4019**, 35–60.

Aguado, M.T., Nygren, A. & Rouse, G.W. 2013. Two apparently unrelated groups of symbiotic annelids, Nautiliniellidae and Calamyzidae (Phyllodocida, Annelida), are a clade of derived chrysopetalid polychaetes. *Cladistics* **29**, 610–628.

Aguado, M.T. & Rouse, G.W. 2011. Nautiliniellidae (Annelida) from Costa Rican cold seeps and a western Pacific hydrothermal vent, with description of four new species. *Systematics and Biodiversity* **9**, 109–131.

Alderslade, P. 1998. Revisionary systematics in the gorgonian family Isididae, with descriptions of numerous new taxa (Coelenterata: Octocorallia). *Records of the Western Australian Museum* **S55**, 1–359.

Álvarez-Campos, P., San Martín, G. & Aguado, M.T. 2013. A new species and new record of the commensal genus *Alcyonosyllis* Glasby & Watson, 2001 and a new species of *Parahaplosyllis* Hartmann-Schröder, 1990, (Annelida: Syllidae: Syllinae) from Philippines Islands. *Zootaxa* **3734**, 159–168.

Amaral, A.C.Z. & Morgado, E.H. 1997. *Stratiodrilus* (Annelida: Polychaeta: Histriobdellidae) associated with a freshwater decapod, with the description of a new species. *Proceedings of the Biological Society of Washington* **110**, 471–475.

Amaral, A.C.Z. & Nonato, E. 1982. *Anelídeos poliquetas da costa brasileira. 3. Aphroditidae e Polynoidae. Conselho Nacional de Desenvolvimento Científico e Tecnológico.* Brasília, 46 pp.

Amato, J.F.R. 2001. A new species of *Stratiodrilus* (Polychaeta, Histriobdellidae) from freshwater crayfishes of southern Brazil. *Iheringia. Sèrie Zoologia* **90**, 37–44.

Andrews, E. A. 1891. A commensal annelid. *American Naturalist* **24**, 25–35.

Anker, A., Murina, G.V., Lira, C., Vera Caripe, J.A., Palmer, A.R. & Jeng, M.S. 2005. Macrofauna associated with echiuran burrows: A review with new observations of the innkeeper worm, *Ochetostoma erythrogrammon* Leuckart and Rüppel, in Venezuela. *Zoological Studies* **44**, 157–190.

Arwidsson, I. 1932. *Calamyzas amphictenicola*, ein ektoparasitischer Verwandter der Sylliden. *Zoologiska Bidrag fran Uppsala* **14**, 153–218.

Augener, H. 1922. Über litorale polychäten von Westindien. *Sitzungsberichte der Gesellschaft naturforschender Freunde* **1922**, 38–53.

Augener, H. 1926. Ueber das Vorkommen von *Spirorbis* Röhren an Einsiedlerkrebsen. *Zoologischer Anzeiger* **69**, 202–204.

Baird, W. 1865. Contributions toward a monograph of the species of Annelides belonging to the Aphroditacea, containing a list of the known species, and a description of some new species contained in the national collection of the British Museum. *Journal of the Linnean Society of London* **8**, 172–202.

Bałuk, W. & Radwański, A. 1997. The micropolychaete *Josephella commensalis* sp. n. commensal to the scleractinian coral *Tarbellastraea reussiana* (Milne-Edwards & Haime, 1850) from the Korytnica Clays (Middle Miocene; Holy Cross Mountains, Central Poland). *Acta Geologica Polonica* **47**, 211–224.

Barnich, R., Beuck, L. & Freiwald, A. 2013. Scale worms (Polychaeta: Aphroditiformia) associated with cold-water corals in the eastern Gulf of Mexico. *Journal of the Marine Biological Association of the United Kingdom* **93**, 2129–2143.

Barnich, R. & Fiege, D. 2000. Revision of the Mediterranean species of *Harmothoe* Kinberg, 1856 and *Lagisca* Malmgren, 1865 (Polychaeta: Polynoidae: Polynoinae) with descriptions of new genus and new species. *Journal of Natural History* **34**, 1889–1938.

Barnich, R. & Fiege, D. 2010. On the distinction of *Harmothoe globifera* (G.O. Sars, 1873) and some other easily confused polynoids in the NE Atlantic, with the description of a new species of *Acanthicolepis* Norman in McIntosh, 1900 (Polychaeta, Polynoidae). *Zootaxa* **2525**, 1–18.

Barnich, R., Gambi, M.C. & Fiege, D. 2012a. Revision of the genus *Polyeunoa* McIntosh, 1885 (Polychaeta, Polynoidae). *Zootaxa* **3523**, 25–38.

Barnich, R., Orensanz, J.M. & Fiege, D. 2012b. Remarks on some scale worms (Polychaeta, Polynoidae) from the Southwest Atlantic with notes on the genus *Eucranta* Malmgren, 1866, and description of a new *Harmothoe* species. *Marine Biodiversity* **42**, 395–410.

Becker, E.L., Cordes, E.E., Macko, S.A., Lee, R.W. & Fisher, C.R. 2013. Using stable isotope compositions of animal tissues to infer trophic interactions in Gulf of Mexico lower slope seep communities. *PLoS ONE* **8**, e74459.

Ben-Tzvi, O., Einbinder, S. & Brokovich, E. 2006. A beneficial association between a polychaete worm and a scleractinian coral? *Coral Reefs* **25**, 98–98.

Benham, W.B. 1916. Report on the Polychaeta obtained by the F.I.S. 'Endeavour' on the coasts of New South Wales, Victoria, Tasmania and South Australia. Part 2. *Fisheries. Zoological (and Biological) Results of the Fishing Experiments Carried out by F.I.S. 'Endeavour', 1909–1914* **4**, 127–169.

Bergsma, G.S. 2009. Tube-dwelling coral symbionts induce significant morphological change in *Montipora*. *Symbiosis* **49**, 143–150.

Bertrán, C., Vargas, L. & Quijón, P. 2005. Infestation of *Polydora rickettsi* (Polychaeta: Spionidae) in shells of *Crepidula fecunda* (Mollusca: Calyptraeidae) in relation to intertidal exposure at Yaldad Bay, Chiloé, Chile. *Scientia Marina* **69**, 99–103.

Bick, A. 2001. The morphology and ecology of *Dipolydora armata* (Polychaeta, Spionidae) from the western Mediterranean Sea. *Acta Zoologica* **82**, 177–187.

Blake, J.A. 1993. New genera and species of deep-sea polychaetes of the family Nautiliniellidae from the Gulf of Mexico and the Eastern Pacific. *Proceedings of the Biological Society of Washington* **106**, 147–157.

Blake, J.A., Böggemann, M., Desbruyères, D., Hourdez, S., Kudenov, J.D., Miura, T., Pleijel, F. & Segonzac, M. 2006. Annelida, polychaeta. In: *Handbook of the Deep-Sea Hydrothermal Vent Fauna*, D. Desbruyères et al. (eds). Brest: IFREMER, 55 pp.

Blake, J.A. & Evans, J.W. 1973. *Polydora* and related genera as borers in mollusc shells and other calcareous substrates. *The Veliger* **15**, 235–249.

Blake, J.A. & Woodwick, K.H. 1971. A reviev of the genus *Boccardia* Carazzi (Polychaeta: Spionidae) with description of two new species. *Bulletin of the Southern California Academy of Sciences* **70**, 31–42.

Bleidorn, C., Podsiadlowski, L., Zhong, M., Eeckhaut, I., Hartmann, S., Halanych, K. M. & Tiedemann, R. 2009. On the phylogenetic position of Myzostomida: can 77 genes get it wrong? *BMC Evolutionary Biology* **9**, 150.

Bo, M., Rouse, G.W., Martin, D. & Bavestrello, G. 2014. A myzostomid endoparasitic in black corals. *Coral Reefs* **33**, 273 only.

Bock, G., Fiege, D. & Barnich, R. 2010. Revision of *Hermadion* Kinberg, 1856, with a redescription of *Hermadion magalhaensi* Kinberg, 1856, *Adyte hyalina* (G.O. Sars, 1873) n. comb. and *Neopolynoe acanellae* (Verrill, 1881) n. comb. (Polychaeta: Polynoidae). *Zootaxa* **2554**, 45–61.

Boletzky, S. & von Dohle, W. 1967. Observations sur un capitellide (*Capitella hermaphodita* sp. n.) et d'autres polychetes habitant la ponte de *Loligo vulgaris*. *Vie et Milieu* **18**, 79–98.

Braby, C.E., Rouse, G.W., Johnson, S.B., Jones, W.J. & Vrijenhoek, R.C. 2007. Bathymetric and temporal variation among *Osedax* boneworms and associated megafauna on whale-falls in Monterey Bay, California. *Deep Sea Research Part I: Oceanographic Research Papers* **54**, 1773–1791.

Brasier, M. & Lindsay, J. 1998. A billion years of environmental stability and the emergence of eukaryotes: new data from northern Australia. *Geology* **26**, 555–558.

Brightwell, L.R. 1951. Some experiments with the common hermit crab (*Eupagurus bernhardus* Linn.) and transparent univalve shells. *Proceedings of the Zoological Society of London* **121**, 279–283.

Britayev, T.A. 1991. Life cycle of the symbiotic scale-worm *Arctonoe vittata* (Polychaeta: Polynoidae). In *Systematics, Biology and Morphology of World Polychaeta. Proceedings of the Second International Polychaeta Conference*, Copenhagen, Denmark: Zoological Museum. Ophelia Suppl. 5, 305–312.

Britayev, T.A. & Antokhina, T.I. 2012. Symbiotic polychaetes of from the Bay of Nhatrang Bay, Vietnam. In *Benthic fauna of the Bay of Nhatrang, Southern Vietnam*, T.A. Britayev & D.S. Pavlov (eds). Moscow: KMK Scientific Press Ltd., 11–54.

Britayev, T.A., Antokhina, T.I. & Marin, I.N. 2015. A scaleworm (Polychaeta: Polynoidae) living with corals. *Marine Biodiversity* **45**, 627–628.

Britayev, T.A., Beksheneva, L.F., Deart, Y.V. & Mekhova, E.S. 2016. Structure and variability of symbiotic assemblages associated with feather stars (Crinoidea: Comatulida) *Himerometra robustipinna*. *Oceanology* **56**, 666–674.

Britayev, T.A., Bratova, O.A. & Dgebuadze, P.Y. 2013. Symbiotic assemblage associated with the tropical sea urchin, *Salmacis bicolor* (Echinoidea: Temnopleuridae) in the An Thoi archipelago, Vietnam. *Symbiosis* **61**, 155–161.

Britayev, T.A., Doignon, G. & Eeckahaut, I. 1999. Symbiotic polychaetes from Papua New Guinea associated with echinoderms, with descriptions of three new species. *Cahiers de Biologie Marine* **40**, 359–374.

Britayev, T.A. & Fauchald, K. 2005. New species of symbiotic scaleworms *Asterophilia* (Polychaeta, Polynoidae) from Vietnam. *Invertebrate Zoology* **2**, 15–22.

Britayev, T.A., Gil, J., Altuna, A., Calvo, M. & Martin, D. 2014. New symbiotic associations involving polynoids (Polychaeta, Polynoidae) from Atlantic waters, with redescriptions of *Parahololepidella greeffi* (Augener, 1918) and *Gorgoniapolynoe caeciliae* (Fauvel, 1913). *Memoirs of Museum Victoria* **71**, 27–43.

Britayev, T.A., Krylova, E.M., Aksyuk, T.S. & Cosel, R. 2003a. Association of Atlantic hydrothermal mytilids of the genus *Bathymodiolus* spp. (Mollusca: Mytilidae) with the Polychaeta *Branchipolynoe* aff. *seepensis* (Polychaeta: Polynoidae): Commensalism or parasitism? *Doklady Biological Sciences* **391**, 371–374.

Britayev, T.A., Krylova, E.M., Martin, D., von Cosel, R., Aksiuk, T.S. & Martín, D. 2003b. Symbiont – host interraction in the association of the scaleworm *Branchipolynoe* aff. *seepensis* (Polychaeta: Polynoidae) with the hydrothermal mussel, *Bathymodiolus* spp. (Bivalvia: Mytilidae). *InterRidge News* **12**, 13–16.

Britayev, T.A. & Lyskin, S.A. 2002. Feeding of the symbiotic polychaete *Gastrolepidia clavigera* (Polynoidae) and its interactions with its hosts. *Doklady Biological Sciences* **385**, 352–356.

Britayev, T.A. & Martin, D. 2005. Scale-worms (Polychaeta, Polynoidae) associated with chaetopterid worms (Polychaeta, Chaetopteridae), with description of a new genus and species. *Journal of Natural History* **39**, 4081–4099.

Britayev, T.A. & Martin, D. 2016. Chaetopteridae Audouin & Milne Edwards, 1833. In *Handbook of Zoology: A Natural History of the Phyla of the Animal Kingdom*, A. Schmidt-Rhaesa (ed.). Berlin: De Gruyter. 1–17.

Britayev, T.A., Martin, D., Krylova, E.M., von Cosel, R. & Aksiuk, E.S. 2007. Life-history traits of the symbiotic scale-worm *Branchipolynoe seepensis* and its relationships with host mussels of the genus *Bathymodiolus* from hydrothermal vents. *Marine Ecology: An Evolutionary Perspective* **28**, 36–48.

Britayev, T.A. & Mekhova, E.S. 2014. Do symbiotic polychaetes migrate from host to host? *Memoirs of Museum Victoria* **71**, 21–25.

Britayev, T.A. & San Martín, G. 2001. Description and life-history traits of a new species of *Proceraea* with larvae infecting *Abietinaria turgida* (Polychaeta, Syllidae and Hydrozoa, Sertulariidae). *Ophelia* **54**, 105–113.

Britayev, T.A. & Smurov, A.V. 1985. The structure of a population of symbionts and related biological features, *Arctonoe vittata* (Polychaeta, Polynoidae) taken as an example. [In Russian]. *Zhurnal obschey Biologii* **46**, 355–366.

Britayev, T.A. & Zamyshliak, E.A. 1996. Association of the commensal scaleworm *Gastrolepidia clavigera* (Polychaeta: Polynoidae) with holothurians near the coast of South Vietnam. *Ophelia* **45**, 175–190.

Buhl-Mortensen, L. & Mortensen, P.B. 2004. Symbiosis in deep-water corals. *Symbiosis* **37**, 33–61.

Buhl-Mortensen, L. & Mortensen, P.B. 2005. Distribution and diversity of species associated with deep-sea gorgonian corals off Atlantic Canada. In: *Cold-Water Corals and Ecosystems*, A. Freiwald & J.M. Roberts (eds). Berlin: Springer, 849–879.

Cairns, S.D. 2006. Studies on western Atlantic Octocorallia (Coelenterata: Anthozoa). Part 6: The genera *Primnoella* Gray, 1858; *Thouarella* Gray, 1870; *Dasystenella* Versluys, 1906. *Proceedings of the Biological Society of Washington* **119**, 161–194.

Cairns, S.D. 2009. Review of Octocorallia (Cnidaria: Anthozoa) from Hawaii and adjacent seamounts. Part 2: Genera *Paracalyptrophora* Kinoshita, 1908; *Candidella* Bayer, 1954; and *Calyptrophora* Gray, 1866. *Pacific Science* **63**, 413–448.

Cairns, S.D. 2011. A revision of the Primnoidae (Octocorallia: Acyonacea) from the Aleutian Islands and Bering Sea. *Smithsonian Contributions to Zoology* **634**, 1–55.

Cairns, S.D. 2012. New Zealand Primnoidae (Anthozoa: Alcyonacea)-Part 1: Genera *Narella, Narelloides, Metanarella, Calyptrophora,* and *Helicoprimnoa. NIWA Biodiversity Memoirs* **126**, 1–100.

Cairns, S.D. & Bayer, F.M. 2008. A review of the Octocorallia (Cnidaria: Anthozoa) from Hawaiï and adjacent sseamounts: The Genus *Narella* Gray, 18701. *Pacific Science* **62**, 83–115.

Cairns, S.D. & Zibrowius, H. 1997. Cnidaria Anthozoa: azooxanthellate Scleractinia from the Philippine and Indonesian regions. In *Resultats des campagnes MUSORSTOM, vol. 16. Memoires du Museum national d'histoire naturelle, 172,* A. Crosnier & P. Bouchet (eds), 27–243.

Caullery, M. & Mesnil, F. 1896. Sur deus serpulidiens nouveaux (*Oriopsis mechnikowi* n. g., n. sp. et *Josephella marenzelleri* n.g., n. sp.). *Zoologischer Anzeiger* **19**, 482–486.

Campos, E., de Campos, A. & de León-González, J.A. 2009. Diversity and ecological remarks of ectocommensals and ectoparasites (Annelida, Crustacea, Mollusca) of echinoids (Echinoidea: Mellitidae) in the Sea of Cortez, Mexico. *Parasitology Research* **105**, 479–487.

Chevaldonné, P., Jollivet, D., Feldman, R.A., Desbruyères, D., Lutz, R.A. & Vrijennhoek, R.C. 1998. Commensal scale-worms of the genus *Branchipolynoe* (Polychaeta: Polynoidae) at deep-sea hydrothermal vents and cold seeps. *Cahiers de Biologie Marine* **39**, 347–350.

Chim, C.K., Ong, J.J.L. & Tan, K.S. 2013. An association between a hesionid polychaete and temnopleurid echinoids from Singapore. *Cahiers de Biologie Marine* **54**, 577–585.

Chisholm, J.R.M. & Kelley, R. 2001. Worms start the reefbuilding process. *Nature* **409**, 152 only.

Clark, R.B. 1956. *Capitella capitata* as a commensal, with a bibliography of parasitism and commensalism in the polychaetes. *Annals and Magazine of Natural History* **12**, 433–448.

Codreanu, R. & Mack-Fira, V. 1961. Sur un copépode, *Sunaristes paguri* Hesse 1867 et un Polychète, *Polydora ciliata* (Johnston) 1838, associés au pagure *Diogenes pugilator* (Roux) dans la Mer Noire et la Méditerranée. La notion de cryptotropisme. *Rapports de la Commission Internationale pour l'Exploration de la Mer Méditerranéene* **16**, 471–494.

Cuadras, J. & Pereira, F. 1977. Invertebrates associated with *Dardanus arrosor* (Anomura, Diogenidae). *Vie et Milieu* **27**, 301–310.

Culurgioni, J., D'Amico, V., Coluccia, E., Mulas, A. & Figus, V. 2006. Metazoan parasite fauna of conger eel *Conger conger* L. from Sardinian waters (Italy). *Ittiopatologia* **3**, 253–261.

Culver, C.S. & Kuris, A.M. 2000. The apparent eradication of a locally established introduced marine pest. *Biological Invasions* **2**, 245–253.

Culver, C.S. & Kuris, A.M. 2004. Susceptibility of California gastropods to an introduced South African sabellid polychaete, *Terebrasabella heterouncinata*. *Invertebrate Biology* **123**, 316–323.

Da Costa, E.M. 1778. *Historia Naturalis Testaceorum Britanniae or The British Conchology; Containing the Description and Other Particulars of Natural History of the Shells of Great Britain and Ireland*. Millan. London: White, Elmsley & Robson.

Dana, J.D. 1854. Catalogue and descriptions of Crustacea collected in California by Dr. John L. Le Conte. *Proceedings of the Academy of Natural Sciences of Philadelphia* **7**, 175–177.

Daudt, L.C.C. & Amato, J.F.R. 2007. Morphological variation of *Stratiodrilus circensis* (Polychaeta, Histriobdellidae) from a new host, *Aegla leptodactyla* (Crustacea, Anomura, Aeglidae) with identification of its type host species. *Zootaxa* **1450**, 57–62.

Dauer, D.M. 1991. Functional morphology and feeding behavior of *Polydora commensalis* (Polychaeta: Spionidae). *Ophelia Suppl.* **5**, 607–614.

David, A.A. & Williams, J.D. 2012. Morphology and natural history of the cryptogenic sponge associate *Polydora colonia* Moore, 1907 (Polychaeta: Spionidae). *Journal of Natural History* **46**, 1509–1528.

Day, J.H. 1963. The polychaete fauna of South Africa Part 8: new species and records from grab samples and dredgings. *Bulletin of the British Museum of Natural History* **10**, 381–445.

Day, R.L., Culver, C.S. & Kuris, A.M. 2000. The parasite *Terebrasabella heteroncinata* (Polychaeta) manipulates shell synthesis in *Haliotis rufescens*. *Journal of Shellfish Research* **19**, 507 only.

Dean, D. & Blake, J.A. 1966. Life history of *Boccardia hamata* on the east and west coasts of North America. *Biological Bulletin Marine Biological Laboratory, Woods Hole* **130**, 316–330.

De Assis, J.E., Bezerra, E.A.S., Brito, R.J., Gondim, A.I. & Christoffersen, M.L. 2012. An association between *Hesione picta* (Polychaeta: Hesionidae) and *Ophionereis reticulata* (Ophiuroidea: Ophionereididae) from the Brazilian coast. *Zoological Studies* **51**, 762–767.

de Bary, A. 1879. *Die Erscheinung der Symbiose: Vortrag gehalten auf der Versammlung Deutscher Naturforscher und Aerzte zu Cassel*. Strassburg: Trübner.

De Haan, W. 1841. Crustacea. In *Fauna japonica sive descriptio animalium, quae in itinere per Japoniam, jussu et auspiciis superiorum, qui summum in India Batavia Imperium tenent, suscepto, annis 1823–1830 collegit, notis, observationibus et adumbrationibus illustravit*, P.F. von Siebold (ed.), Leiden: Lugduni Batavorum, 243 pp, pls.1–55.

Delaroche, F. 1809. Suite du mémoire sur les espèces de poissons observées à Iviça. Observations sur quelques-uns des poissons indiqués dans le précédent tableau et descriptions des espèces nouvelles ou peu connues. *Annales du Muséum d'Histoire Naturelle de Paris* **13**, 313–361.

Di Camillo, C.J., Martin, D. & Britayev, T.A. 2011. Symbiotic association between *Solanderia secunda* (Cnidaria, Hydrozoa, Solanderiidae) and *Medioantenna variopinta* sp. nov. (Annelida, Polychaeta, Polynoidae) from North Sulawesi (Indonesia). *Helgoland Marine Research* **65**, 495–511.

Ditlevsen, H. 1917. *Annelids*. Copenhagen: I. H. Hagerup.

Dreyer, J., Miura, T. & van Dover, C.L. 2004. *Vesicomyicola trifulcatus*, a new genus and species of commensal polychaete (Annelida: Polychaeta: Nautilliniellidae) found in depp-sea clams from the Blake Ridge cold seep. *Proceedings of the Biological Society of Washington* **117**, 106–113.

Duarte, L.F.L. & Nalesso, R.C. 1996. The sponge *Zygomycale parishii* (Bowerbank) and its endobiotic fauna. *Estuarine, Coastal and Shelf Science* **42**, 139–151.

Eeckahaut, I. & Lanterbecq, D. 2005. Myzostomida: a review of the phylogeny and ultrastructure. *Hydrobiologia* **535/536**, 253–275.

Ehlers, E. 1864. *Die Borstenwürmer (Annelida Chaetopoda) nach systematischen und anatomischen Untersuchungen dargestellt.* Leipzig: Wilhelm Engelmann.

Eibye-Jacobsen, D. 1991. A revision of *Eumida* Malmgren, 1865 (Polychaeta: Phyllodocidae). *Steenstrupia* **17**, 81–140.

Eisig, H. 1906. *Ichthyotomus sanguinarius*, eine auf Aalen schmarotzende Annelide. *Fauna und Flora des Golfes von Neapel* **28**, 1–300.

Eliason, A. 1962. Undersökningar över öresund: XXXXI. Weitere untersuchungen über die polychaetenfauna des Örensunds. *Lunds Universitet. Arsskrift, Audelningen 2. Kungliga Fysiografiska Salskapetsi Lund. Handlinger* **58**, 1–98.

Emson, R.H., Young, C.M. & Paterson, G.L.J. 1993. A fire worm with a sheltered life: studies of *Benthoscolex cubanus* Hartman (Amphinomidae), an internal associate of the bathyal sea-urchin *Archeopneustes hystrix* (A. *Agassiz, 1880*). *Journal of Natural History* **27**, 1010–1028.

Fabricius, O. 1780. *Fauna Groenlandica, systematice sistens, Animalia Groenlandiae occidentalis hactenus indagata, quoad nomen specificum, triviale, vernaculumque synonyma auctorum plurium, descriptionem, locum, victum, generationem, mores, usum, capturamque singuli prout detegendi occasio fuit, maximaque parte secundum proprias observationes*, Hafniae Copenhagen: et Lipsiae.

Fage, L. 1936. Sur l'association d'un annélide polychète '*Lumbriconereis flabellicola*' n. sp. et d'un madrépore '*Flabellum pavoninum distinctum*' E. et H. *Comptes Rendues du XIIe Congrès International de Zoologie* **1**, 941–945.

Fage, L. & Legèndre, R. 1925. Sur une Annélide Polychète (*Iphitime cuenoti* Fauv.) commensale des crabes. *Bulletin de la Société Zoologique de France* **50**, 219–225.

Fage, L. & Legèndre, R. 1933. Les anélides polychètes du genre *Iphitime*. A propos d'une espèce nouvelle commensale des pagures, *Iphitime paguri*. *Bulletin de la Société Zoologique de France* **58**, 299–305.

Fauvel, P. 1914. Un Eunicien énigmatique *Iphitime cuenoti* n. sp. *Archives de Zoologie Expérimentale et Générale* **53**, 34–37.

Fauvel, P. 1939. Annélides Polychètes de l'Indochine recueillis par M. C. Dawidoff. *Commentators of the Pontifical Academy of Sciences* **3**, 243–368.

Feigenbaum, D. 1979. Predation on chaetoganths by typhloscolecid polychaetes, one explanation for headless specimens. *Journal of the Marine Biological Association of the United Kingdom* **59**, 631–633.

Fiege, D. & Barnich, R. 2009. Polynoidae (Annelida: Polychaeta) associated with cold-water coral reefs of the northeast Atlantic and the Mediterranean Sea. *Zoosymposia* **2**, 149–164.

Fiore, C.L. & Jutte, P.C. 2010. Characterization of macrofaunal assemblages associated with sponges and tunicates collected off the southeastern United States. *Invertebrate Biology* **129**, 105–120.

Fitzhugh, K. 1989. A systematic revision of the Sabellidae-Caobangiidae-Sabellongidae complex (Annelida: Polychaeta). *Bulletin of the American Museum of Natural History* **192**, 1–104.

Fitzhugh, K. 2006. The 'requirement of total evidence' and its role in phylogenetic systematics. *Biology and Philosophy* **21**, 309–351.

Fitzhugh, K. & Rouse, G.W. 1999. A remarkable new genus and species of fan worm (Polychaeta: Sbaellidae: Sabellinae) associated with marine gastropods. *Invertebrate Biology* **118**, 357–390.

Frey, R.W. 1987. Hermit crabs; neglected factors in taphonomy and paleoecology. *Palaios* **2**, 313–322.

Fromont, J. & Abdo, D.A. 2014. New species of *Haliclona* (Demospongiae: Haplosclerida: Chalinidae) from Western Australia. *Zootaxa* **3835**, 97–109.

Fujikura, K., Fujiwara, Y. & Kawato, M. 2006. A new species of *Osedax* (Annelida: Siboglinidae) associated with whale carcasses off Kyushu, Japan. *Zoological Science* **23**, 733–740.

Garberoglio, R.M. & Lazo, D.G. 2011. Post-mortem and symbiotic sabellid and serpulid-coral associations from the lower cretaceous of Argentina. *Revista Brasileira de Paelontologia* **14**, 215–228.

García-Hernández, J.E. & Hoeksema, B.W. 2017. Sponges as secondary hosts for Christmas tree worms at Curaçao. *Coral Reefs* **36**, 1243–1243.

Gardiner, S.L. 1976. Errant Polychaete annelids from North Carolina. *The Journal of the Elisha Mitchell Scientific Society* **91**, 77–220.

Genzano, G.N. & San Martín, G. 2002. Association between the polychaete *Procerastea halleziana* (Polychaeta: Syllidae: Autolytinae) and the hydroid *Tubularia crocea* (Cnidaria: Hydrozoa) from the Mar del Plata intertidal zone, Argentina. *Cahiers de Biologie Marine* **43**, 165–170.

Giard, A. 1893. Sur un type nouveau et aberrant de la famille des Sabellides, *Caobangia billeti*. *Comptes rendus des séances de la Société de Biologie* **5**, 473–476.

Giard, M.A. 1882. Sur un type synthétique d'annélide (*Anoplonereis herrmanni*) commensal des *Balanoglossus*. *Comptes Rendues hebdomadaires des séances de l'Academie des Sciences, Paris* **95**, 389–391.

Gilpin-Brown, J.B. 1969. Host-adoption in the commensal polychaete *Nereis fucata*. *Journal of the Marine Biological Association of the United Kingdom* **49**, 121–127.

Glasby, C.J. 1994. A new genus and species of polychaete, *Bollandia antipathicola* (Nereidoidea, Syllidae), from black coral. *Proceedings of the Biological Society of Washington* **107**, 615–621.

Glasby, C.J. & Aguado, M.T. 2009. A new species and new records of the anthozoan commensal genus *Alcyonosyllis* (Polychaeta: Syllidae). *The Beagle, Records of the Museums and Art Galleries of the Northern Territory* **25**, 55–63.

Glasby, C.J., Schroeder, P.C. & Aguado, M.T. 2012. Branching out: a remarkable new branching syllid (Annelida) living in a *Petrosia* sponge (Porifera: Demospongiae). *Zoological Journal of the Linnean Society* **164**, 481–497.

Glasby, C.J. & Watson, C. 2001. A new genus and species of Syllidae (Annelida: Polychaeta) commensal with octocorals. *The Beagle, Records of the Northern territory Museum of Arts and Sciences* **17**, 43–51.

Glover, A.G., Källström, B., Smith, C.R. & Dahlgren, T.G. 2005. World-wide whale worms? A new species of *Osedax* from the shallow north Atlantic. *Proceedings of the Royal Society B: Biological Sciences* **272**, 2587–2592.

Goerke, H. 1971. *Nereis fucata* (Polychaeta, Nereidae) als kommensale von *Eupagurus bernhardus* (Crustacea, Decapoda) Entwicklung einer population und verhalten der art. *Veröffentlichungen des Instituts für Meeresforschung in Bremerhaven* **13**, 79–81.

Goff, L.J. 1982. Symbiosis and parasitism: Another viewpoint. *BioScience* **32**, 255–256.

Goto, R. & Kato, M. 2011. Geographic mosaic of mutually exclusive dominance of obligate commensals in symbiotic communities associated with a burrowing echiuran worm. *Marine Biology* **159**, 319–330.

Gould, A.A. 1861. Descriptions of shells collected by the North Pacific exploring expedition. *Proceedings of the Boston Society of natural History* **8**, 14–40.

Grube, A.E. 1840. *Actinien, Echinodermen und Wurmen des Adriatischen und Mittelmeers nach eigenen sammlungen beschrieben*. Königsberg: J.H. Bon.

Grube, A.E. 1855. Beschreibungen neuer oder wenig bekannter Anneliden. *Archiv fur Naturgeschichte, Berlin* **21**, 81–136, pls.133–135.

Grube, A.E. 1862. Mittheilungen über die Serpulen, mit besonderer Berücksichtigung ihrer Deckel. *Jahres-Bericht der Schlesischen Gesellschaft für Vaterländische Cultur* **39**, 53–69.

Gustafson, R.G., Turner, R.D., Lutz, R.A. & Vrijenhoek, R.C. 1998. A new genus and five new species of mussels (Bivalvia: Mytilidae) from deep-sea sulfide/hydrocarbon seeps in the Gulf of Mexico. *Malacologia* **40**, 63–112.

Hand, C. 1975. Behaviour of some New Zealand sea anemones and their molluscan and crustacean hosts. *New Zealand Journal of Marine and Fresh Water Research* **9**, 509–527.

Hartman, O. 1942. The polychaetous Annelida. Report on the scientific results of the Atlantis Expedition to the West Indies under the joint auspices of the University of Havana and Harvard University. *Memorias de la Sociedad Cubana de Historia Natural* **16**, 89–104.

Hartman, O. 1947. Polychaetous annelids. Part VII. Capitellidae. *Allan Hancock Pacific Expeditions* **10**, 391–481.

Hartman, O. 1954. Marine annelids from the northern Marshall Islands. *United States Geological Survey Professional Papers* **260 Q**, 615–644.

Hartman, O. & Boss, K.J. 1965. *Antonbruunia viridis*, a new inquiline annelid with dwarf males, inhabiting a new species of pelecypod, *Lucinia fosteri* in the Mozambique channel. *Annals and Magazine of Natural History* **13**, 177–186.

Hartmann-Schröder, G. 1959. Zur Ökologie der Polychaeten des mangrove-estero-gebietes von El Salvador. *Beiträge zur Neotropischen Fauna* **1**, 69–183.

Hartmann-Schröder, G. 1960. Polychaeten aus dem Roten Meer. *Kieler Meeresforschungen* **16**, 69–125.

Hartmann-Schröder, G. 1978. Einige Sylliden-Arten (Polychaeta) von Hawaii und aus dem Karibischen Meer. *Mitteilungen aus dem Hamburgischen zoologischen Museum und Institut* **75**, 49–61.

Hartmann-Schröder, G. 1989. *Polynoe thouarellicola* n. sp. aus der Antarktis, assoziiert mit Hornkorallen, und Wiederbeschreibung von *Polynoe antarctica* Kinberg, 1858 (Polychaeta, Polynoidae). *Zoologischer Anzeiger* **3**, (4), 205–221.

Hartmann-Schröder, G. & Zibrowius, H. 1998. Polychaeta associated with Antipatharia (Cnidaria: Anthozoa): description of Polynoidae and Eunicidae. *Mitteilungen aus dem Hamburgischen zoologischen Museum und Institut* **95**, 29–44.

Hashimoto, J. & Okutani, T. 1994. Four new mytilid mussels associated with deepsea chemosynthetic communities around Japan. *Venus. Japanese Journal of Malacology* **53**, 61–83.

Haswell, W.A. 1900. On a new Histriobdellid. *Quarterly Journal of Microscopical Science* **43**, 299–335.

Hatfield, P.A. 1965. *Polydora commensalis* Andrews–Larval developpement and observations on adults. *Biological Bulletin Marine Biological Laboratory, Woods Hole* **128**, 356–368.

Hauenschild, C., Fischer, A. & Hofmann, D.K. 1968. Untersuchungen am pazifischen Palolowurm *Eunice viridis* (Polychaeta) in Samoa. *Helgoländer wissenschaftliche Meeresuntersuchungen* **18**, 254–295.

Heggøy, K.K., Schander, C. & Åkesson, B. 2007. The phylogeny of the annelid genus *Ophryotrocha* (Dorvilleidae). *Marine Biology Research* **3**, 412–420.

Hendrix, G.Y. 1975. A review of the genus *Phascolion* (Sipuncula) with the descriptions of two new species from the Western Atlantic. In *Proceedings of the International Symposium on the Biology of the Sipuncula and Echiura. Vol. 1*, M.E. Rice & M. Todorovic (eds). Beograd, Serbia: Naucno Delo, 117–137.

Hernández-Alcántara, P., Cruz-Prez, I.N. & Solís-Weiss, V. 2015. *Labrorostratus caribensis*, a new oenonid polychaete from the Grand Caribbean living in the body cavity of a nereidid, with emendation of the genus. *Zootaxa* **4948**, 127–139.

Hernández-Alcántara, P. & Solís-Weiss, V. 1998. Parasitism among polychaetes: a rare case illustrated by a new species: *Labrorostratus zaragozensis*, n. sp. (Oenonidae) found in the Gulf of California, Mexico. *Journal of Parasitology* **84**, 978–982.

Hoberg, M.K., McGee, S.G. & Feder, H.M. 1982. Polychaetes and amphipods as commensals with pagurids from the Alaska shelf. *Ophelia* **21**, 167–179.

Hoegh-Guldberg, O. 1999. Climate change, coral bleaching and the future of the world's coral reefs. *Marine and Freshwater Research* **50**, 839–866.

Hoeksema, B.W., Ten Hove, H. & Berumen, M.L. 2016. Christmas tree worms evade smothering by a coral-killing sponge in the Red Sea. *Marine Biodiversity* **46**, 15–16.

Hoeksema, B.W. & ten Hove, H.A. 2016. The invasive sun coral *Tubastraea coccinea* hosting a native Christmas tree worm at Curaçao, Dutch Caribbean. *Marine Biodiversity*, **47**, 59–65.

Hoeksema, B.W., van Beukelsom, M., Ten Hove, H.A., Ivanenko, V.N., Van der Meij, S.E.T. & van Moorsel, G. 2017. *Helioseris cucullata* as a host coral at St. Eustatius, Dutch Caribbean. *Marine Biodiversity* **47**, 71–78.

Høisæter, T. & Samuelsen, J. 2006. Taxonomic and biological notes on a species of *Iphitime* (Polychaeta, Eunicida) associated with *Pagurus prideaux* from Western Norway. *Marine Biology Research* **2**, 333–354.

Hong, J.-S., Lee, C.-L. & Sato, M. 2017. A review of three species of *Hesperonoe* (Annelida: Polynoidae) in Asia, with descriptions of two new species and a new record of *Hesperonoe hwanghaiensis* from Korea. *Journal of Natural History* **51**, 2925–2945.

Horton, T., Kroh, A., Bailly, N., Boury-Esnault, N., Brandão, S.N., Costello, M. J., Gofas, S., Hernandez, F., Mees, J., Paulay, G., Poore, G., Rosenberg, G., Stöhr, S., Decock, W., Dekeyzer, S., Vandepitte, L., Vanhoorne, B., Vranken, S., Adams, M. J., Adlard, R., Adriaens, P., Agatha, S., Ahn, K.J., Ahyong, S., Alvarez, B., Anderson, G., Angel, M., Arango, C., Artois, T., Atkinson, S., Barber, A., Bartsch, I., Bellan-Santini, D., Berta, A., Bieler, R., Błażewicz, M., Bock, P., Böttger-Schnack, R., Bouchet, P., Boyko, C.B., Bray, R., Bruce, N.L., Cairns, S., Campinas Bezerra, T.N., Cárdenas, P., Carstens, E., Cedhagen, T., Chan, B.K., Chan, T.Y., Cheng, L., Churchill, M., Coleman, C.O., Collins, A.G., Cordeiro, R., Crandall, K.A., Cribb, T., Dahdouh-Guebas, F., Daly, M., Daneliya, M., Dauvin, J.C., Davie, P., De Grave, S., de Mazancourt, V., Defaye, D., d'Hondt, J.L., Dijkstra, H., Dohrmann, M., Dolan, J., Downey, R., Drapun, I., Eisendle-Flöckner, U., Eitel, M., Encarnação, S.C.D., Epler, J., Ewers-Saucedo, C., Faber, M., Feist, S., Finn, J., Fišer, C., Fonseca, G., Fordyce, E., Foster, W., Frank, J.H., Fransen, C., Furuya, H., Galea, H., Garcia-Alvarez, O., Gasca, R., Gaviria-Melo, S., Gerken, S., Gheerardyn, H., Gibson, D., Gil, J., Gittenberger, A., Glasby, C., Glover, A., Gordon, D., Grabowski, M., Gravili, C. & Guerra-García, J.M. et al. 2017. World Register of Marine Species (WoRMS). WoRMS Editorial Board.

Huang, D.W., Fitzhugh, K. & Rouse, G.W. 2011. Inference of phylogenetic relationships within Fabriciidae (Sabellida, Annelida) using molecular and morphological data. *Cladistics* **27**, 356–379.

Imajima, M. 1966. The Syllidae (Polychaetous Annelids) from Japan. I. Exogoninae. *Publications of the Seto Marine Biological Laboratory* **13**, 385–404.

Imajima, M. 1997. Polychaetous annelids from Sagami Bay and Sagami Sea collected by the Emperor Showa of Japan and deposited at the Showa Memorial Institute, National Science Museum, Tokio. Families Polynoidae and Acoetidae. *National Science Museum Monographs* **13**, 1–131.

Imajima, M. 2003. Polychaetous Annelids from Sagami Bay and Sagami Sea Collected by the Emperor Showa of Japan and Deposited at the Showa Memorial Institute, National Science Museum, Tokyo II. Orders included within the Phyllodocida, Amphinomida, Spintherida and Eunicida. *National Science Museum Monographs* **23**, 1–221.

Intès, A. & Le Loeuff, P. 1975. Les annélides polychètes de cote d'Ivoire. I.-Polychètes errantes- compte rendu systématique. *Cahiers O.R.S.T.O.M., séries Océanographie* **13**, 267–321.

Jenkins, C.D., Ward, M.E., Turnipseed, M., Osterberg, J. & Van Dover, C.L. 2002. The digestive system of the hydrothermal vent polychaete *Galapagomystides aristata* (Phyllodocidae): evidence for hematophagy? *Invertebrate Biology* **121**, 243–254.

Jensen, A. & Frederiksen, R. 1992. The fauna associated with the bank-forming deepwater coral *Lophelia pertusa* (Scleractinaria) on the Faroe shelf. *Sarsia* **77**, 53–69.

Jensen, K.T. & Bender, K. 1973. Invertebrates associated with snail shells inhabited by *Pagurus bernhardus* (L.) (Decapoda). *Ophelia* **10**, 185–192.

Johnson, H.P. 1897. A preliminary account of the marine annelids of the Pacific coast, with descriptions of new species. *Proceedings of the California Academy of Sciences* **1**, 153–199.

Johnson, H.P. 1901. The Polychaeta of the Puget Sound region. *Proceedings of the Boston Society for Natural History* **29**, 381–437 pls. 319.

Johnson, J.Y. 1861. Description of a second species of *Acanthogorgia* (J. E. Gray) from Madeira. *Proceedings of the Zoological Society of London*, 296–298.

Jollivet, D., Comtet, T., Chevaldonné, P., Houdrez, S., Desbruyères, D. & Dixon, D.R. 1998. Unexpected relationship between dispersal strategies and speciation within the association *Bathymodiolus* (Bivalvia)–*Branchipolynoe* (Polychaeta) inferred from rDNA neutral ITS2 marker. *Cahiers de Biologie Marine* **39**, 359–362.

Jollivet, D., Empis, A., Baker, M.C., Houdrez, S., Comtet, T., Jouin-Toulmond, C., Desbruyères, D. & Tyler, P.A. 2000. Reproductive biology, sexual dimorphism, and population structure of the deep-sea hydrothermal vent scale-worm, *Branchipolynoe seepensis* (Polychaeta: Polynoidae). *Journal of the Marine Biological Association of the United Kingdom* **80**, 55–68.

Jones, M.L. 1974a. *Brandtika asiatica* new genus, new species, from Southeastern Asia and a redescription of *Monroika africana* (Monro) (Polychaeta: Sabellidae). *Proceedings of the Biological Society of Washington* **87**, 217–230.

Jones, M.L. 1974b. On the Caobangiidae, a new family of Polychaeta, with redescription of *Caobangia billeti* Giard. *Smithsonian Contributions to Zoology* **175**, 1–55.

Jones, W.J., Johnson, S.B., Rouse, G.W. & Vrijenhoek, R.C. 2008. Marine worms (genus *Osedax*) colonize cow bones. *Proceedings of the Royal Society B: Biological Sciences* **275**, 387–391.

Jourde, J., Sampaio, L., Barnich, R., Bonifácio, P., Labrune, C., Quintino, V. & Sauriau, P.-G. 2015. *Malmgrenia louiseae* sp. nov., a new scale worm species (Polychaeta: Polynoidae) from southern Europe with a key to European *Malmgrenia* species. *Journal of the Marine Biological Association of the United Kingdom* **95**, 947–952.

Kelaart, E.F. 1858. Description of new and little known species of Ceylon nudibranchiate molluscs, and zoophytes. *Journal of the Ceylon Branch of the Royal Asiatic Society* **3**, 84–139.

Keller, C. 1889. Die Spongienfauna des rothen Meeres (I. Hälfte). *Zeitschrift für wissenschaftliche Zoologie* **48**, 311–405.

Kiel, S. 2008. Parasitic polychaetes in the early Cretaceous hydrocarbon seep-restricted brachiopod *Peregrinella multicarinata*. *Journal of Paleontology* **82**, 1215–1217.

Knight-Jones, P. & Knight-Jones, E.W. 1977. Taxonomy and ecology of British Spirorbidae (Polychaeta). *Journal of the Marine Biological Association of the United Kingdom* **57**, 453–499.

Knudsen, J. 1944. A gephyrean, a polychaete and a bivalve (*Jousseaumielle concharum*, nov. sp.) living together (commensalistically) in the Indo-Malayan Seas. *Videnskabelige Meddelelser fra Dansk naturhistorisk Forening* **108**, 15–24.

Koch, H. 1846. Einige Worte zur Entwicklungsgeschichte von *Eunice*, mit einem Nachworte von A. Koelliker. *Neue Denkschriften der Allgemeinen Schweizerischen Gesellschaft fur die Gesammten Naturwissenschaften* **8**, 1–31.

Koehler, R. 1908. *Astéries, Ophiures et Echinides de l'expédition antarctique nationale écossaise. British Antarctic Expedition. Reports on the Scientific Investigations 2 (Biology)* **4**, 25–66 pl. IV–VIII.

Kolbasova, G.D. & Tzetlin, A.B. 2017. Developmental studies of the enigmatic worm *Caobangia billeti* Giard, 1893 (Annelida; Sabellidae), a symbiont of freshwater snails. *Journal of the Marine Biological Association of the United Kingdom* **97**, 1143–1153.

Koukouras, A., Russo, A., Voultsiadou-Koukoura, E., Arvanitidis, C. & Stefanidou, D. 1996. Macrofauna associated with sponge species of different morphology. *P.S.Z.N. I: Marine Ecology* **17**, 569–582.

Kumagai, N.H. & Aoki, M.N. 2003. Seasonal changes in the epifaunal community on the shallow-water gorgonian *Melithaea flabellifera. Journal of the Marine Biological Association of the United Kingdom* **83**, 1221–1222.

Kupriyanova, E.K. & Rouse, G.W. 2008. Yet another example of paraphyly in Annelida: molecular evidence that Sabellidae contains Serpulidae. *Molecular Phylogenetics and Evolution* **46**, 1174–1181.

Lamarck, J.B. 1816. Ordre second. Radiaires Échinodermes. *Histoire Naturelle des Animaux sans Vertèbres* **2**, 522–568.

Lamarck, J.B. 1818. *Histoire naturelle des Animaux sans Vertèbres, présentant les caractères généraux et particuliers de ces animaux, leur distribution, leurs classes, leurs familles, leurs genres, et la citation des principales espèces qui s'y rapportent; précédée d'une introduction offrant la détermination des caractères essentiels de l`animal, sa distinction du végétal et des autres corps naturels, enfin, l'exposition des principes fondamentaux de la zoologie.* Paris: Déterville & Verdière, Vol. **5**.

Lampert, K. 1883. Über einige neue Thalassemen. *Zeitschrift für wissenschaftliche Zoologie* **39**, 334–342.

Langerhans, P. 1880. Die Wurmfauna von Madeira. III. *Zeitschrift für wissenschaftliche Zoologie* **34**, 87–143.

Lanterbecq, D., Rouse, G.W. & Eeckhaut, I. 2009. Bodyplan diversification in crinoid-associated myzostomes (Myzostomida, Protostomia). *Invertebrate Biology* **128**, 283–301.

Lanterbecq, D., Rouse, G.W. & Eeckhaut, I. 2010. Evidence for cospeciation events in the host–symbiont system involving crinoids (Echinodermata) and their obligate associates, the myzostomids (Myzostomida, Annelida). *Molecular Phylogenetics and Evolution* **54**, 357–371.

Lattig, P. & Martin, D. 2009. A taxonomic revision of the genus *Haplosyllis* Langerhans, 1887 (Polychaeta: Syllidae: Syllinae). *Zootaxa* **2220**, 1–40.

Lattig, P. & Martin, D. 2011a. Sponge-associated *Haplosyllis* (Polychaeta: Syllidae: Syllinae) from the Caribbean Sea, with the description of four new species. *Scientia Marina* **75**, 733–758.

Lattig, P. & Martin, D. 2011b. Two new endosymbiotic species of *Haplosyllis* (Polychaeta: Syllidae) from the Indian Ocean and Red Sea, with new data on *H. djiboutiensis* from the Persian Gulf. *Italian Journal of Zoology* **78**, 112–123.

Lattig, P., Martin, D. & Aguado, M.T. 2010a. Four new species of *Haplosyllis* (Polychaeta: Syllidae: Syllinae) from Indonesia. *Journal of the Marine Biological Association of the UK* **90**, 789–798.

Lattig, P., Martin, D. & San Martín, G. 2010b. Syllinae (Syllidae: Polychaeta) from Australia. Part 4. The genus *Haplosyllis* Langerhans, 1879 (Polychaeta: Syllidae: Syllinae). *Zootaxa* **2552**, 1–36.

Lattig, P., Muñoz, I., Martin, D., Abelló, P. & Machordom, A. 2017. Comparative phylogeography of two symbiotic dorvilleid polychaetes (*Iphitime cuenoti* and *Ophryotrocha mediterranea*) with contrasting host and bathymetric patterns. *Zoological Journal of the Linnean Society* **179**, 1–22.

Laubier, L. 1960. Une nouvelle sous-espèce de Syllidien: *Haplosyllis depressa* Augener ssp. nov. *chamaeleon*, ectoparasite sur l'octocoralliaire *Muricea chamaeleon* Von Koch. *Vie et Milieu* **11**, 75–87.

Leach, W.E. 1815. *The Zoological Miscellany; Being Descriptions of New, or Interesting Animals.* London: R.P. Nodder, Vol. **2**.

Leach, W.E. 1816. A tabular view of the external characters of four classes of animals, which Linné arranged under Insecta; with the distribution of the genera composing three of these classes into orders, and descriptions of several new genera and species. *Transactions of the Linnean Society of London* **11**, 306–400.

Lee, J.W. & Rho, B.J. 1994. Systematic studies on Syllidae (Annelida, Polychaeta) from the South Sea and the East Sea in Korea. *The Korean Journal of Systematic Zoology* **10**, 131–144.

Lewis, J.B. 1998. Reproduction, larval development and functional relationships of the burrowing, spionid polychaete *Dipolydora armata* with the calcareous hydrozoan *Millepora complanata*. *Marine Biology* **130**, 651–662.

Light, W.J. 1970. *Polydora alloporis*, new species, a commensal spionid (Annelida, Polychaeta) from a hydrocoral off Central California. *Proceedings of the California Academy of Sciences (Fourth Series)* **37**, 459–472.

Lindner, A., Cairns, S.D. & Guzman, H.M. 2004. *Distichopora robusta* sp. nov., the first shallow-water stylasterid (Cnidaria: Hydrozoa: Stylasteridae) from the tropical eastern Pacific. *Journal of the Marine Biological Association of the UK* **84**, 943–947.

Linnaeus, C. 1758. Systema Naturae per Regna Tria Naturae, Secundum Classes, Ordines, Genera, Species, cum characteribus, differentiis, synonymis, locis. Tomus I. Editio *Decima, Reformata*. Stockholm: Laurentii Salvii.

Linnaeus, C. 1767. *Systema Naturae per Regna Tria Naturae, Editio Duodecima, Reformata, Tomus I, Pars II. Regnum Animale*. Stockholm: Laurentii Salvii, 533–1327 + 1–37.

Liu, P.J. & Hsieh, H.L. 2000. Burrow architecture of the spionid polychaete *Polydora villosa* in the corals *Montipora* and *Porites*. *Zological Studies* **39**, 47–54.

López, E., Britayev, T.A., Martin, D. & San Martín, G. 2001. New symbiotic associations involving Syllidae (Annelida: Polychaeta), with some taxonomic and biological remarks on *Pionosyllis magnifica* and *Syllis* cf. *armillaris*. *Journal of the Marine Biological Association of the United Kingdom* **81**, 399–409.

Lütken, C.F. 1871. Fortsatte kritiske org beskrivende Bidrag til Kundskab om Sostjernerne (Asteriderme). *Videnskabelige Meddelelser fra den naturhistoriske Forening i Kjøbenhavn* **1871**, 227–308.

Lützen, J. 1961. Sur une nouvelle espèce de Polychète, *Sphaerodoridium commensalis* n. gen., n. spec. (Polychaeta Errantia, famille des Sphaerodoridae) vivant en commensal de *Terebellides stroemi* Sars. *Cahiers de Biologie Marine* **2**, 409–416.

MacGinitie, G.E. 1939. The method of feeding of *Chaetopterus*. *Biological Bulletin Marine Biological Laboratory, Woods Hole* **77**, 115–118.

MacGinitie, G.E. & MacGinitie, N. 1968. *Natural History of Marine Animals*. New York: McGraw-Hill Book Co.

Mackie, A.S.Y. & Garwood, P.R. 1995. Annelida. In *Benthic Biodiversity in the Southern Irish Sea. BIOMOR Reports. Studies in Marine Biodiversity and Systematics*, A.S.Y. Mackie et al. (eds). Cardiff: National Museum and Galleries of Wales, 37–50.

Mackie, A.S.Y., Oliver, P.G. & Nygren, A. 2015. *Antonbruunia sociabilis* sp. nov (Annelida: Antonbruuniidae) associated with the chemosynthetic deep-sea bivalve *Thyasira scotiae* Oliver & Drewery, 2014, and a re-examination of the systematic affinities of Antonbruuniidae. *Zootaxa* **3995**, 20–36.

Magnino, G. & Gaino, E. 1998. *Haplosyllis spongicola* (Grübe) (Polychaeta, Syllidae) associated with two species of sponges from east Africa (Tanzania, Indian Ocean). *P.S.Z.N. I. Marine Ecology* **19**, 77–87.

Magnino, G., Pronzato, R., Sarà, A. & Gaino, E. 1999a. Fauna associated with the horny sponge *Anomoianthella lamella* Pulitzer-Finali & Pronzato, 1999 (Ianthellidae, Deponpongiae) from Papua-New Guinea. *Italian Journal of Zoology* **66**, 175–181.

Magnino, G., Sarà, A., Lancioni, T. & Gaino, E. 1999b. Endobionts of the coral reef sponge *Teonella swinhoei* (Porifera, Demospongiae). *Invertebrate Biology* **118**, 213–220.

Maldonado, M. & Young, C.M. 1998. Limits on the bathymetric distribution of keratose sponges: a field test in deep water. *Marine Ecology Progress Series* **174**, 123–139.

Malmgren, A.J. 1867. Annulata Polychaeta Spetsbergiae, Groelanlandiae, Islandiae et Scandinaviae hactenus cognita. Cum xiv. tabulis.

Martin, D., Aguado, M.T. & Britayev, T.A. 2009. Review of the symbiotic genus *Haplosyllides*, with description of a new species. *Zoological Science* **26**, 646–655.

Martin, D. & Britayev, T.A. 1998. Symbiotic polychaetes: review of known species. *Oceanography and Marine Biology: An Annual Review* **36**, 217–340.

Martin, D., Cuesta, J.A., Drake, P., Gil, J. & Pleijel, F. 2012. The symbiotic hesionid *Parasyllidea humesi* Pettibone, 1961 (Annelida: Polychaeta) hosted by *Scrobicularia plana* (da Costa, 1778) (Mollusca: Bivalvia: Semelidae) in European waters. *Organisms Diversity & Evolution* **12**, 145–153.

Martin, D., Marin, I. & Britayev, T.A. 2008. Features of the first known parasitic association between Syllidae (Annelida, Polychaeta) and crustaceans. *Organisms Diversity and Evolution* **8**, 279–281.

Martin, D., Meca, M.A., Gil, J., Drake, P. & Nygren, A. 2017a. Another brick in the wall: population dynamics of a symbiotic *Oxydromus* (Annelida, Hesionidae), described as a new species based on morphometry. *Contributions to Zoology* **86**, 181–211.

Martin, D., Núñez, J., Riera, R. & Gil, J. 2002. On the associations between *Haplosyllis* (Polychaeta, Syllidae) and gorgonians (Cnidaria, Octocorallaria), with a description of a new species. *Biological Journal of the Linnean Society* **77**, 455–477.

Martin, D., Nygren, A. & Cruz-Rivera, E. 2017b. *Proceraea exoryxae* sp. nov. (Annelida, Syllidae, Autolytinae), the first known polychaete miner tunneling into the tunic of an ascidian. *PeerJ* **5**, e3374, doi: 3310.7717/peerj.3374

Martin, D., Nygren, A., Hjelmstedt, P., Drake, P. & Gil, J. 2015. On the enigmatic symbiotic polychaete 'Parasyllidea' humesi Pettibone, 1961 (Hesionidae): taxonomy, phylogeny and behaviour. *Zoological Journal of the Linnean Society* **174**, 429–446.

Martinell, J. & Domènech, R. 2009. Commensalism in the fossil record: Eunicid polychaete bioerosion on Pliocene solitary corals. *Acta Palaeontologica Polonica* **54**, 143–154.

Martins, R., Magalhães, L., Peter, A., San Martín, G., Rodrigues, A.M. & Quintino, V. 2013. Diversity, distribution and ecology of the family Syllidae (Annelida) in the Portuguese coast (Western Iberian Peninsula). *Helgoland Marine Research* **67**, 775–788.

McDermott, J.J. 2001. Symbionts of the hermit crab *Pagurus longicarpus* Say 1817 (Decapoda: Anomura): new observations from New Jersey waters and a review of all known relationships. *Proceedings of the Biological Society of Washington* **114**, 624–639.

McIntosh, W.C. 1879. On a remarkably branched *Syllis* dredged by H.M.S. Challenger. *Journal of the Linnean Society of London* **14**, 720–724.

McIntosh, W.C. 1885. Report on the Annelida Polychaeta collected by the H.M.S. Challenger during the years 1873–1876. *Report on the Scientific Results of the Voyage of H.M.S. Challenger during the years 1872–76* **12**, 1–554.

Mekhova, E.S. & Britayev, T.A. 2015. Soft substrate crinoids (Crinoidea: Comatulida) and their macrosymbionts in Halong Bay (North Vietnam). *The Raffles Bulletin of Zoology* **63**, 438–445.

Micaletto, G., Gambi, M.C. & Cantone, G. 2002. A new record of the endosymbiont polychaete *Veneriserva* (Dorvilleidae), with description of a new sub-species, and relationships with its host *Laetmonice producta* (Polychaeta: Aphroditidae) in Southern Ocean waters (Antarctica). *Marine Biology* **141**, 691–698.

Miller, W. & Wolf, M. 2008. Crawling with worms: a look at two symbiotic relationships between polychaetes and urchins from the Bahamas. University of Oregon Libraries: Oregon Institute of Marine Biology Student Reports, 1–10. http://hdl.handle.net/1794/6875

Miranda, V.D.R. & Brasil, A.C.S. 2014. Two new species and a new record of scale-worms (Polychaeta) from Southwest Atlantic deep-sea coral mounds. *Zootaxa* **3856**, 211–226.

Misra, A. 1999. Polychaeta of West Bengal. *State Fauna Series 3: Fauna of West Bengal* Part 10 Zoological Survey of India, 125–225.

Miura, T. & Hashimoto, J. 1991. Two new branchiate scale-worms (Polynoidae: Polychaeta) from the hydrothermal vent of the Okinawa Trough and the volcanic seamount off Chichijima Island. *Proceedings of the Biological Society of Washington* **104**, 166–174.

Miura, T. & Hashimoto, J. 1996. Nautiliniellid Polychaetes living in the mantle cavity of bivalve mollusks from cold seeps and hydrothermal vents around Japan. *Publications of the Seto Marine Biological Laboratory* **37**, 257–274.

Miura, T. & Ohta, S. 1991. Two polychaete species from the deep-sea hydrothermal vent in the Middle Okinawa Trough. *Zoological Science* **8**, 383–387.

Miura, T. & Shirayama, Y. 1992. *Lumbineris flabellicola* (Fage, 1936), a lumbrinerid polychaete associated with a Japanese haermatypic coral. *Benthos Research* **43**, 23–27.

Mohammad, M.B.M. 1971. Intertidal polychaetes from Kuwait, Arabian Gulf, with descriptions of three new species. *Journal of Zoology, London* **163**, 285–303.

Molodtsova, T. & Budaeva, N. 2007. Modifications of *Corallium* morphology in black corals as an effect of associated fauna. *Bulletin of Marine Science* **81**, 469–480.

Moore, J.D., Juhasz, C.I., Robbins, T.T. & Grosholz, E.D. 2007. The introduced sabellid polychaete *Terebrasabella heterouncinata* in California: transmission, methods of control and survey for presence in native gastropod populations. *Journal of Shellfish Research* **26**, 869–876.

Moore, J.P. 1907. Descriptions of new species of spioniform annelids. *Proceedings of the Academy of Natural Sciences of Philadelphia* **59**, 195–207.

Morgado, E.H. & Tanaka, M.O. 2001. The macrofauna associated with the bryozoan *Schizoporella errata* (Walters) in southeastern Brazil. *Scientia Marina* **65**, 173–181.

Mortensen, P.B. 2001. Aquarium observations on the deep-water coral *Lophelia pertusa* (L., 1758) (Scleractinia) and selected associated invertebrates. *Ophelia* **54**, 83–104.

Mueller, C.E., Lundälv, T., Middelburg, J.J. & van Oevelen, D. 2013. The symbiosis between *Lophelia pertusa* and *Eunice norvegica* stimulates coral calcification and worm assimilation. *PLoS ONE* **8**, e58660.

Müller, O.F. 1788. Zoologica Danica seu Animalium Daniae et Norwegiae rariorum ac minus notoruum. Descritiones et Historia, Havniae.

Musco, L. & Giangrande, A. 2005. A new sponge-associated species, *Syllis mayeri* n. sp. (Polychaeta: Syllidae), with a discussion on the status of *S. armillaris* (Müller, 1776). *Scientia Marina* **69**, 467–474.

Naumann, M.S., Rix, L., Al-Horani, F.A. & Wild, C. 2016. Indications of mutual functional benefits within a polychaete-sponge association in the northern Red Sea. *Bulletin of Marine Science* **92**, 377–378.

Neo, M.L., Eckman, W., Vicentuan, K., Teo, S.L.-M. & Todd, P.A. 2015. The ecological significance of giant clams in coral reef ecosystems. *Biological Conservation* **181**, 111–123.

Nishi, E. & Tachikawa, H. 1998. A new species of scale worm *Hololepidella* (Polynoidae, Polychaeta) associated with an ophiuroid from Ogasawara Islands, Japan. *Natural History Research* **5**, 11–15.

Nishi, E. & Tachikawa, H. 1999. New record of a commensal scale worm *Medioantenna clavata* Imajima, 1997 (Polychaeta: Polynoidae), from Ogasawara Islands, Japan. *Natural History Research* **5**, 107–110.

Nogueira, J.M.M. 2000. Anelídeos poliquetas associados ao coral *Mussismilia hispida* (Verrill, 1868) em ilhas do litoral do Estado de São Paulo. Phyllodocida, Amphinomida, Eunicida, Spionida, Terebellida, Sabellida. PhD thesis, Universidade de São Paulo, São Paulo.

Norlinder, E., Nygren, A., Wiklund, H. & Pleijel, F. 2012. Phylogeny of scale-worms (Aphroditiformia, Annelida), assessed from 18SrRNA, 28SrRNA, 16SrRNA, mitochondrial cytochrome c oxidase subunit I (COI), and morphology. *Molecular Phylogenetics and Evolution* **65**, 490–500.

Núñez, J., Barnich, R., Santos, L. & Maggio, Y. 2011. Poliquetos escamosos (Annelida, Polychaeta) colectados en las campañas 'Fauna II, III, IV' (Proyecto 'Fauna Ibérica') y catálogo de las especies conocidas para el ámbito Íbero-balear. *Graellsia* **67**, 187–197.

Nutting, C.C. 1908. Descriptions of the Alcyonaria collected by the U.S. Bureau of Fisheries steamer Albatross in the vicinity of the Hawaiian Islands in 1902. *Proceedings of the United States National Museum* **34**, 543–601.

Nygren, A. & Pleijel, F. 2010. Redescription of *Imajimaea draculai*—a rare syllid polychaete associated with the sea pen *Funiculina quadrangularis*. *Journal of the Marine Biological Association of the United Kingdom* **90**, 1441–1448.

Nygren, A. & Pleijel, F. 2011. From one to ten in a single stroke–resolving the European *Eumida sanguinea* (Phyllodocidae, Annelida) species complex. *Molecular Phylogenetic Evolution* **58**, 132–141.

Okuda, S. 1937. Spioniform polychaetes from Japan. *Journal of the Faculty of Science, Hokkaido Imperial University, Series 6, Zoology* **5**, 217–254.

Okuda, S. 1950. Notes on some commensal polychaetes from Japan. *Annotations on Zoology, Japan* **24**, 49–53.

Oppelt, A., López Correa, M. & Rocha, C. 2017. Biogeochemical analysis of the calcification patterns of cold-water corals *Madrepora oculata* and *Lophelia pertusa* along contact surfaces with calcified tubes of the symbiotic polychaete *Eunice norvegica*: evaluation of a 'mucus' calcification hypothesis. *Deep Sea Research Part I: Oceanographic Research Papers* **127**, 90–104.

O'Reilly, M. 2016. Parasitic and commensal polychaetes (Fams. Arabellidae and Sphaerodoridae) and copepods (Fam. Saccopsidae) associated with lamella-worms (*Terebellides* spp.) in Scottish, and nearby, waters. *The Glasgow Naturalist* **26**, 61–70.

Orensky, L.D. & Williams, J.D. 2009. Morphology and ecology of a new sexually dimorphic species of *Polydora* (Polychaeta: Spionidae) associated with hermit crabs from Jamaica, West Indies. *Zoosymposia* **2**, 229–240.

Øresland, V. & Bray, R.A. 2005. Parasites and headless chaetognaths in the Indian Ocean. *Marine Biology* **147**, 725–734.

Øresland, V. & Pleijel, F. 1991. An ectoparasitic typhloscolecid polychaete on the chaetognath *Eukronia hamata* from the Antarctic Peninsula. *Marine Biology* **108**, 429–432.

Örsted, A.S. 1843. Annulatorum danicorum conspectus. Fasc. 1 Maricolae. Librariae Wahlianae, Hafniae.

Östergren, H. 1905. Zwei Koreanische Holothurien. *Archives de Zoologie Expérimentale et Générale* **21 Notes et Revue, 4ème Sèrie, 3ème Tome**, 199–206.

Pallas, P.S. 1766. Miscellanea Zoologica. Quibus novae imprimis atque obscurae animalium species describuntur et observationibus iconibusque illustrantur. Hagae Comitum apud Petrum van Cleef.

Pallas, P.S. 1776. Miscellanea zoologica quibus novae imprimis atque obscurae Animalium species describuntur et observationibus iconibusque illustrantur. Hague Comitum.

Pallas, P.S. 1788. Marina varia nova et rariora. *Nova Acta Acad. Imp. Sci. Petropolitanea* **2**, 229–249.

Paola, A., San Martín, G. & Martin, D. 2006. A new species of *Haplosyllis* Langerhans, 1879 (Polychaeta: Syllidae: Syllinae) from Argentina. *Proceedings of the Biological Society of Washington* **119**, 346–354.

Paresque, K., Fukuda, M.V. & Nogueira, J.M.M. 2016. *Branchiosyllis, Haplosyllis, Opisthosyllis* and *Trypanosyllis* (Annelida: Syllidae) from Brazil, with the description of two new species. *PLoS ONE* **11**, e0153442.

Paresque, K. & Nogueira, J.M.M. 2014. The genus Haplosyllis Langerhans, 1879 (Polychaeta: Syllidae) from northeastern Brazil, with descriptions of two new species. *Marine Biology Research* **10**, 554–576.

Paris, J. 1955. Commensalisme et parasitisme chez les annélides polychètes. *Vie et Milieu* **6**, 525–536.

Park, T., Lee, S. & Kim, W. 2016. New record of commensal scale worms, *Arctonoe vittata* (Grube, 1855) and *Hyperhalosydna striata* (Kinberg, 1856) (Polychaeta: Polynoidae) from Korean waters. *Journal of Species Research* **5**, 517–529.

Parmentier, E. & Michet, L. 2013. Boundary limits in symbiosis forms. *Symbiosis* **60**, 1–5.

Parulekar, A.H. 1969. *Neoaiptasia commensali*, gen. et. sp. nov.: an actiniarian commensal of hermit crabs. *Journal of the Bombay Natural History Society* **66**, 57–62.

Pearse, A.S. 1932. Observations on the parasites and commensals found associated with crustaceans and fishes at Dry Tortugas. *Publications of the Carnegie Institution of Washington*, **435**, 103–115.

Petersen, M.E. & Britayev, T.A. 1997. A new genus and species of polynoid scaleworm commensal with *Chaetopterus appendiculatus* Grube from the Banda Sea (Annelida: Polychaeta), with a review of commensals of Chaetopteridae. *Bulletin of Marine Science* **60**, 261–276.

Pettibone, M.H. 1953. *Some Scale-Bearing Polychaetes of Puget Sound and Adjacent Waters*. Seattle, WA: University of Washington Press, 98 pp.

Pettibone, M.H. 1961. New species of polychaete worms from the Atlantic Ocean, with a revision of the Dorvilleidae. *Proceedings of the Biological Society of Washington* **74**, 167–186.

Pettibone, M.H. 1963. Marine polychaete worms of the New England region. *Part 1. Families Aphroditidae through Trochochaetidae. Bulletin of the United States National Museum* **227**, 1–356.

Pettibone, M.H. 1970. Polychaeta Errantia of the Siboga Expedition. Part IV. Some additional polychaetes of the Polynoidae, Hesionidae, Nereidae, Goniadidae, Eunicidae, and Onuphidae, selected as new species by the late Dr. Hermann Augener with remarks on other related species. In: *Siboga-Expeditie Uitkomsten op Zoologisch, Bonatisch, Oceanographisch en Geol ogisch gebied verzameld in Nederlandsch Oost-Indië 1899–1900*, M. Weber, L. F. De Beaufort & J. H. Stock (eds). Leiden: E.J. Brill, 199–270.

Pettibone, M.H. 1986a. A new scale worm commensal with deep-sea mussels in the seep-sites at the Florida Escarpment in the Eastern Gulf of Mexico (Polychaeta: Polynoidae: Branchipolynoidae). *Proceedings of the Biological Society of Washington* **99**, 444–451.

Pettibone, M.H. 1986b. Review of the Iphioninae (Polychaeta: Polynoidae) and revision of *Iphione cimex* Quatrefages, *Gattyana deludens* Fauvel, and *Harmothoe iphionelloides* Johnson (Harmothoinae). *Smithsonian Contributions to Zoology* **428**, 1–43.

Pleijel, F., Rouse, G.W. & Nygren, A. 2012. A revision of *Nereimyra* (Psamathini, Hesionidae, Aciculata, Annelida). *Zoological Journal of the Linnean Society* **164**, 36–51.

Plouviez, S., Daguin-Thiébaut, C., Hourdez, S. & Jollivet, D. 2008. Juvenile and adult scale worms *Branchipolynoe seepensis* in Lucky Strike hydrothermal vent mussels are genetically unrelated. *Aquatic Biology* **3**, 79–87.

Polgar, G., Nishi, E., Idris, I. & Glasby, C.J. 2015. Tropical polychaete community and reef dynamics: insights from a Malayan Sabellaria (Annelida: Sabellariidae) reef. *The Raffles Bulletin of Zoology* **63**, 401–417.

Potts, F.A. 1910. Polychaeta of the Indian ocean. Pt. 2. The Palmyridae, Aphroditidae, Polynoidae, Acoetidae and Sigalionidae. *Transactions of the Linnean Society of Zoology* **16**, 325–353.

Potts, F.A. 1912. A new type of parasitism in the Polychaeta. *Proceedings of the Cambridge Philosophical Society* **16**, 409–413.

Prestandrea, N. 1839. Descrizione di due nuovi Crustacei dei Mari di Messina. *Atti della Accademia Gioneia di Catania* **1**, 131–136.

Quiroga, E. & Sellanes, J. 2009. Two new polychaete species living in the mantle cavity of *Calyptogena gallardoi* (Bivalvia: Vesicomyidae) at a methane seep site off central Chile (~36°S). *Scientia Marina* **73**, 399–407.

Radashevsky, V.I. 1993. Revision of the genus *Polydora* and related genera from the North West Pacific (Polychaeta, Spionidae). *Publications of the Seto Marine Biological Laboratory* **36**, 1–60.

Radashevsky, V.I. & Hsieh, H.L. 2000. *Polydora* (Polychaeta: Spionidae) species from Taiwan. *Zoological Studies* **39**, 203–217.

Randall, R. & Eldredge, L. 1976. Skeletal modification by a polychaete annelid in some scleractinian corals. In *Coelenterate Ecology and Behavior*, G.O. Mackie (ed.). New York: Springer, 453–465.

Rathke, H. 1843. Beiträge zur Fauna Norwegens. *Nova Acta der Kaiserlichen Leopold-Carolin Deutschen Akademie der Naturforscher, Halle* **20**, 1–264.

Ravara, A. & Cunha, M.R. 2016. Two new species of scale worms (Polychaeta: Aphroditiformia) from deep-sea habitats in the Gulf of Cadiz (NE Atlantic). *Zootaxa* **4097**, 422–450.

Read, G. & Fauchald, K. 2017. *World Polychaeta database*. Online. http://www.marinespecies.org/polychaeta (accessed 12 May 2017)

Read, G.B. 2001. Unique branching worm found in New Zealand. *Biodiversity Update (NIWA)* **4**, 1 only.

Read, G.B. 2015. *Caobangia* Giard, 1893. In *World Polychaeta database*. Online. http://www.marinespecies. org/polychaeta/aphia.php?p=taxdetails&id=325071, G.B. Read & K. Fauchald (eds) (accessed 28 April 2015)

Reiss, H., Knäuper, S. & Kröncke, I. 2003. Invertebrate associations with gastropod shells inhabited by *Pagurus bernhardus* (Paguridae) – secondary hard substrate increasing biodiversity in North Sea soft-bottom communities. *Sarsia* **88**, 404–415.

Ribeiro, S.M., Omena, E.P. & Muricy, G. 2003. Macrofauna associated to *Mycale microsigmatosa* (Porifera, Demospongiae) in Rio de Janeiro State, SE Brazil. *Estuarine, Coastal and Shelf Science* **57**, 951–959.

Risso, A. 1826. *Histoire naturelle des principales productions de l'Europe méridionale et particulièrement de celles des environs de Nice et des Alpes Maritimes*. Paris: Levrault.

Roberts, J.M. 2005. Reef-aggregating behaviour by symbiotic eunicid polychaetes from cold-water corals: do worms assemble reefs? *Journal of the Marine Biological Association of the United Kingdom* **85**, 813–819.

Rolando, L. 1822 (1821 volume). Description d'un animal nouveau qui appartient à la classe des echinodermes. *Memorie della Reale Accademia delle Scienze di Torino* **26**, 539–556.

Rossi, M.M. 1984. A new genus and species of iphitimid parasitic in an aphroditid (Polychaeta), with an emendation of the family Iphitimidae. *Bulletin of the Southern California Academy of Sciences* **83**, 163–166.

Rouse, G.W., Goffredi, S.K. & Vrijenhoek, R.C. 2004. *Osedax*: bone-eating marine worms with dwarf males. *Science* **305**, 668–671.

Rouse, G.W., Wilson, N., Goffredi, S.K., Johnson, S.B., Smart, T., Widmer, C., Young, C.M. & Vrijenhoek, R.C. 2009. Spawning and development in *Osedax* boneworms (Siboglinidae, Annelida). *Marine Biology* **156**, 395–405.

Rouse, G.W., Worsaae, K., Johnson, S.B., Jones, W.J. & Vrijenhoek, R.C. 2008. Acquisition of dwarf male 'harems' by recently settled females of *Osedax roseus* n. sp. (Siboglinidae; Annelida). *Biological Bulletin* **214**, 67–82.

Rowley, S. 2008. A critical evaluation of the symbiotic association between tropical tube-dwelling Polychaetes and their Hermatypic coral hosts, with a focus on *Spirobranchus giganteus* (Pallas, 1766). *The Plymouth Student Scientist* **1**, 335–353.

Rzhavsky, A.V., Kupriyanova, E.K. & Sikorski, A.V. 2013. Two new species of serpulid polychaetes (Annelida) from the Barents Sea. *Fauna norvegica* **32**, 27–38.

Sakai, K. 1986. On *Upogebia narutensis*, a new thalassinid (Decapoda, Crustacea), from Japan. *Researches on Crustacea, Carcinological Society of Japan* **15**, 23–28.

Salazar-Vallejo, S., González, N.E. & Salazar-Silva, P. 2015. *Lepidasthenia loboi* sp. n. from Puerto Madryn, Argentina (Polychaeta, Polynoidae). *ZooKeys* **546**, 21–37.

Samuelsen, T.J. 1970. The biology of six species of Anomura (Crustacea, Decapoda) from Raunefjorden, Western Norway. *Sarsia* **45**, 23–52.

San Martín, G., Álvarez-Campos, P. & Aguado, M.T. 2013. The genus *Branchiosyllis* Ehlers, 1887 (Annelida, Syllidae, Syllinae) from off the American coasts, with the description of a new species from Venezuela. *Pan-american Journal of Aquatic Sciences* **8**, 166–179.

San Martín, G., Hutchings, P. & Aguado, M.T. 2010. Syllinae (Polychaeta: Syllidae) from Australia. Part 3. Genera *Alcyonosyllis*, Genus A, *Parahaplosyllis*, and *Trypanosyllis* (*Trypanobia*). *Zootaxa* **2493**, 35–48.

San Martín, G. & López, E. 2002. New species of *Autolytus* Grube, 1850, *Paraprocerastea* San Martín and Alós, 1989, and *Sphaerosyllis* Claparède, 1863 (Syllidae, Polychaeta) from the Iberian Peninsula. *Sarsia* **87**, 135–143.

San Martín, G. & Nishi, E. 2003. A new species of *Alcyonosyllis* Glasby and Watson, 2001 (Polychaeta: Syllidae: Syllinae) from Shimoda, Japan commensal with the gorgonian *Melithaea flabellifera*. *Zoological Science* **20**, 371–375.

Sardá, R., Avila, C. & Paul, V.J. 2002. An association between a syllid polychaete, *Haplosyllis basticola* n. sp., and the sponge *Ianthella basta*. In: Guam Biodiversity. *Micronesica* **34**, 165–175.

Sars, M. 1835. *Beskrivelser og Iagttagelser over nogle moerkelige eller nye i Havet ved den Bergenske Kyst levende Dyr af Polypernes, Acalephernes, Radiaternes, Annelidernes og Molluskernes classer, med en kort Oversigt over de hidtil af Forfatteren sammesteds fundne Arter og deres Forekommen.* Bergen: Thorstein Hallegers Forlag hos Chr. Dahl, R. S.

Sato, M., Uchida, H., Itani, G. & Yamashita, H. 2001. Taxonomy and life history of the scale worm *Hesperonoe hwanghaiensis* (Polychaeta: Polynoidae), newly recorded in Japan, with special reference to commensalism to a burrowing shrimp, *Upogebia major*. *Zoological Science* **18**, 981–991.

Sato-Okoshi, W. 1998. Three species of polydorids (polychaeta, Spionidae) from Japan. *Species Diversity* **3**, 277–288.

Sato-Okoshi, W. 1999. Polydorid species (Polychaeta: spionidae) in Japan, with descriptions of morphology, ecology and burrow structure. 1. Boring species. *Journal of the Marine Biological Association of the United Kingdom* **79**, 831–848.

Savigny, J. 1816. *Mémoires sur les animaux sans vertèbres.* Paris: G. Dufour.

Schembri, P.J. & Jaccarini, V. 1978. Some aspects of the ecology of the echiuran worm *Bonellia viridis* and associated infauna. *Marine Biology* **47**, 55–61.

Schiaparelli, S., Alvaro, M.C. & Barnich, R. 2011. Polynoid polychaetes living in the gut of irregular sea urchins: a first case of inquilinism in the Southern Ocean. *Antarctic Science* **23**, 144–151.

Schiaparelli, S., Alvaro, M.C., Bohn, J. & Albertelli, G. 2010. 'Hitchhiker' polynoid polychaetes in cold deep waters and their potential influence on benthic soft bottom food webs. *Antarctic Science* **22**, 399–407.

Schmarda, L.K. 1861. *Neue wirbellose Thiere beobachtet und gesammelt auf einer Reise un die Erdr 1853 bis 1857. Erster Band (zweite halfte) Turbellarian, Rotatorien un Anneliden.* Leipzig: Wilhelm Engelmann.

Schmidt, O. 1870. *Grundzüge einer Spongien-Fauna des atlantischen Gebietes.* Leipzig: Wilhelm Engelmann.

Schoot, R.V.D., Scott, C.M., ten Hove, H.A. & Hoeksema, B.W. 2016. Christmas tree worms as epibionts of giant clams at Koh Tao, Gulf of Thailand. *Marine Biodiversity* **46**, 751–752.

Schroeder, P.C., Aguado, M.T., Malpartida, A. & Glasby, C.J. 2017. New observations on reproduction in the branching polychaetes *Ramisyllis multicaudata* and *Syllis ramosa* (Annelida: Syllidae: Syllinae). *Journal of the Marine Biological Association of the United Kingdom* **97**, 1167–1175.

Sellanes, J. & Krylova, E. 2005. A new species of *Calyptogena* (Bivalvia: Vesicomyidae) from a recently discovered methane seepage area off Concepcion Bay, Chile (~ 36° S). *Journal of the Marine Biological Association of the United Kingdom* **85**, 969–976.

Serpetti, N., Taylor, M.L., Brennan, D., Green, D.H., Rogers, A.D., Paterson, G.L.J. & Narayanaswamy, B.E. 2017. Ecological adaptations and commensal evolution of the Polynoidae (Polychaeta) in the Southwest Indian Ocean Ridge: a phylogenetic approach. *Deep Sea Research Part II: Topical Studies in Oceanography* **137**, 273–281.

Shields, M.A., Glover, A.G. & Wiklund, H. 2013. Polynoid polychaetes of the Mid-Atlantic Ridge and a new holothurian association. *Marine Biology Research* **9**, 547–553.

Simon, C.A., Kaiser, H., Booth, A.J. & Britz, P.J. 2002. The effect of diet and live host presence on the growth and reproduction of *Terebrasabella heterouncinata* (Polychaeta: Sabellidae). *Invertebrate Reproduction and Development* **41**, 277–286.

Simon, C.A., Kaiser, H. & Britz, P.J. 2004. Infestation of the abalone, *Haliotis midae*, by the sabellid, *Terebrasabella heterouncinata*, under intensive culture conditions, and the influence of infestation on abalone growth. *Aquaculture* **232**, 29–40.

Simon, C.A., Kaiser, H. & Britz, P.J. 2005. The life history responses of the abalone pest, *Terebrasabella heterouncinata*, under natural and aquaculture conditions. *Marine Biology* **147**, 135–144.

Simon, C.A., San Martín, G. & Robinson, G. 2014. Two new species of *Syllis* (Polychaeta: Syllidae) from South Africa, one of them viviparous, with remarks on larval development and vivipary. *Journal of the Marine Biological Association of the United Kingdom* **94**, 729–746.

Smyth, M.J. 1989. Bioerosion of gastropod shells: with emphasis on effects of coralline algal cover and shell microstructure. *Coral Reefs* **8**, 119–125.

Smyth, M.J. 1990. Incidence of boring organisms in gastropod shells on reefs around Guam. *Bulletin of Marine Science* **46**, 432–449.

Spooner, G.M., Wilson, D.P. & Trebble, N. 1957. Phylum Annelida. In: *Plymouth Marine Fauna*, D. P. Wilson (ed). Plymouth, UK: The Marine Biological Association of the U.K., 109–149.

Stachowitsch, M. 1977. The hermit crab microbiocoenosis—the role of mobile secondary hard bottom elements in a north Adriatic benthic community. In *Biology of Benthic Organisms. Proceedings of the 11th European Symposium on Marine Biology*, B.F. Keegan et al. (eds). Oxford: Pergamon Press, 549–558.

Stachowitsch, M. 1980. The epibiotic, Galway, Ireland and endolithic species associated with the gastropod shells inhabited by the hermit crabs *Paguristes oculatus* and *Pagurus cuanensis*. *Marine Ecology* **1**, 73–101.

Steiner, T.M. & Amaral, A.C.Z. 1999. The family Histriobdellidae (Annelida, Polychaeta) including descriptions of two new species from Brazil and a new genus. *Contributions to Zoology* **68**, 95–108.

Stewart, H.L., Holbrook, S.J., Schmitt, R.J. & Brooks, A.J. 2006. Symbiotic crabs maintain coral health by clearing sediments. *Coral Reefs* **25**, 609–615.

Stiller, M. 1996. Verbreitung und Lebensweise der Aphroditiden und Polynoiden (Polychaeta) im östlichen Weddellmeer und im Lazarevmeer (Antarktis). *Berichte zur Polarforschung* **185**, 1–200.

Stock, J.H. 1986. Cases of hyperassociation in the Copepoda (Herphyllobiidae and Nereicolidae). *Systematics and Parasitology* **8**, 71–81.

Stofel, C.B., Canton, G.C., Antunes, L.A.S. & Eutrópio, F.J. 2008. Fauna associated with the sponge *Cliona varians* (Porífera, Desmoespongiae). *Natureza on line* **6**, 16–18.

Storm, V. 1879. Bidrag til Kundskab om Trondhjemsfjordens Fauna. *Kongelige Norske videnskabers selskabs skrifter* **1879**, 9–36.

Studer, T. 1883. Verzeichniss der wahrend der Reise S.M.S. Gazelle an der weskuste von Afrika, Ascension und dem Cap der Guten Hoffnung Gesammelten Crustaceen. Berlin. *Abhandlungen der Klasse Preussischen Akademie der Wissenschaften* **1882**, 1–3.

Summers, M. & Rouse, G. 2014. Phylogeny of Myzostomida (Annelida) and their relationships with echinoderm hosts. *BCM Evolutionary Biology* **14**, 170 only.

Sun, R. & Yang, D.J. 2004. Annelida, Polychaeta II. Nereidida (= Nereimorpha). Nereididae, Syllidae, Hesionidae, Pilargidae, Nephtydae. In *Fauna Sinica. Invertebrata*, C. Huo & G. Zhao (eds). Beijing: China Science Press, 550 pp.

Swainson, W. 1823. The characters of several rare and undescribed shells. *The Philosophical Magazine and Journal* **61**, 375–378.

ten Hove, H.A. 1994. Serpulidae (Annelida: Polychaeta) from the Seychelles and Amirante Islands. In *Oceanic Reefs of the Seychelles. Cruise Reports Neth. Indian Ocean Programm II*, J.V.D. Land (ed.). Leiden: National Museum of Natural History, 107–116.

ten Hove, H.A. & Kupriyanova, E.K. 2009. Taxonomy of Serpulidae (Annelida, Polychaeta): The state of affairs. *Zootaxa* **2036**, 1–126.

Terrana, L. & Eeckahaut, I. 2017. Taxonomic description and 3D modelling of a new species of myzostomid (Annelida, Myzostomida) associated with black corals from Madagascar. *Zootaxa* **4244**, 277–295.

Thompson, J.A. & Henderson, W.D. 1905. On the other Alcyonaria. In *Report to the Government of Ceylon on the Pearl OysterFisheries in the Gulf of Manaar*, W.A. Herdmann (ed.). London: The Royal Society, 269–328.

Tokaji, H., Nakahara, K. & Goshima, S. 2014. Host switching improves survival rate of the symbiotic polychaete *Arctonoe vittata*. *Plankton and Benthos Research* **9**, 189–196.

Treadwell, A.L. 1909. *Haplosyllis cephalata* as an ectoparasite. *Bulletin of the American Museum of Natural History* **26**, 359–360.

Treadwell, A.L. 1924. Polychaetous annelids collected by the Barbados-Antigua Expedition from the University of Iowa in 1918. *University of Iowa, Studies in Natural History* **10**, 3–23.

Troncoso, N., Moreira, J. & Troncoso, J.S. 2000. *Tellimya phascolionis* (Dautzenberg & Fisher, 1925) (Bivalvia, Montacutidae) and other fauna associated with the sipunculid *Phascolion strombi* (Montagu, 1804) in the Ría de Aldán (Galicia, NW Península Ibérica). *Argonauta* **14**, 59–66.

Tzetlin, A.B. 1985. *Asetocalamyzas laonicola* gen. et sp. n., a new ectoparasitic polychaete from the White Sea. *Zoologicheski Zhurnal* **64**, 269–298.

Tzetlin, A.B. & Britayev, T.A. 1985. A new species of Spionidae (Polychaeta) with asexual reproduction associated with sponges. *Zoologica Scripta* **14**, 177–181.

Uchida, H. 2004. Hesionidae (Annelida, Polychaeta) from Japan. I. *Kuroshio Biosphere* **1**, 27–92.

Uriz, M.J., Agell, G., Blanquer, A., Turon, X. & Casamayor, E.O. 2012. Endosymbiotic calcifying bacteria: a new cue to the origin of calcification in metazoa? *Evolution* **66**, 2993–2999.

Uschakov, P.V. 1962. Polychaetous annelids of the fam. Phyllodocidae and Aphroditidae from the Antarctic and Subantarctic. [In Russian]. In: Biological results of the Soviet Antarctic Expedition (1955–1958), 1. *Isseldovaniya Fauni Moorei* **1**, 129–189.

Uschakov, P.V. & Wu, B.L. 1959. The polychaetous annelids of the families Phyllodocidae and Aphroditidae from the Yellow Sea. [In Russian]. *Archives de l'Institut d'Oceanologie Sinica* **1**, 1–40.

Utinomi, H. 1956. On the so-called 'Umi-Utiwa' a peculiar flabellate gorgonacean, with notes on a syllidean polychaete commensal. *Publications of the Seto Marine Biological Laboratory* **5**, 243–250.

Verrill, A.E. 1900. Additions to the Turbellaria, Nemertina, and Annelida of the Bermudas, with revisions of some New England genera and species. *Transactions of the Connecticut Academy of Arts and Sciences* **10**, 595–671.

Versluys, J. 1906. Die Gorgoniden der Siboga Expedition II. *Die Primnoidae. Siboga- Expeditie Monographie* **XIII**, 1–178.

Vicentuan-Cabaitan, K., Neo, M.L., Eckman, W., Teo, S.L. & Todd, P.A. 2014. Giant clam shells host a multitude of epibionts. *Bulletin of Marine Science* **90**, 795–796.

von Cosel, R., Comtet, T. & Krylova, E.M. 1999. *Bathymodiolus* (Bivalvia: Mytilidae) from hydrothermal vents on the Azores triple junction and the Logatchev hydrothermal field, Mid-Atlantic ridge. *The Veliger* **42**, 218–248.

von Cosel, R., Métivier, B. & Hashimoto, J. 1994. Three new species of *Bathymodiolus* (Bivalvia: Mytilidae) from hydrothermal vents in the Lau basin and the North Fiji basin, Western Pacific, and the Snake Pit area, Mid-Atlantic Ridge. *Veliger* **37**, 374–374.

Vortsepneva, E.V., Tzetlin, A.B., Purschke, G., Mugue, N., Hass-Cordes, E. & Zhadan, A.E. 2008. The parasitic polychaete known as *Asetocalamyzas laonicola* (Calamyzidae) is in fact the dwarf male of the spionid *Scolelepis laonicola* (comb. nov.). *Invertebrate Biology* **127**, 403–416.

Vortsepneva, E.V., Zhadan, A.E. & Tzetlin, A.B. 2006. Spermiogenesis and sperm ultrastructure of *Asetocalamyzas laonicola* Tzetlin 1985 (Polychaeta), an ectoparasite of the large spionid *Scolelepis* cf. *matsugae* Sikorsfi, 1994, from the White Sea. In: Scientific advances in Polychatete research. *Scientia Marina* **70**, 343–350.

Vrijenhoek, R.C., Johnson, S.B. & Rouse, G.C. 2008. Bone-eating *Osedax* females and their 'harems' of dwarf males are recruited from a common larval pool. *Molecular Ecology* **17**, 4535–4544.

Vrijenhoek, R.C., Johnson, S.B. & Rouse, G.C. 2009. A remarkable diversity of bone-eating worms (*Osedax*; Siboglinidae; Annelida). *BMC Biology* **7**, 74 only.

Walker, S.E. 1988. Taphonomic significance of hermit crabs (Anomura: Paguridea): epifaunal hermit crab–infaunal gastropod example. *Palaeogeography, Palaeoclimatology, Palaeoecology* **63**, 45–71.

Walker, S.E. & Carlton, J.T. 1995. Taphonomic losses become taphonomic gains: an experimental approach using the rocky shore gastropod, *Tegula funebralis*. *Palaeogeography, Palaeoclimatology, Palaeoecology* **114**, 197–217.

Ward, M.E., Shields, J.D. & van Dover, C.L. 2004. Parasitism in species of *Bathymodiolus* (Bivalvia: Mytilidae) mussels from deep-sea seep and hydrothermal vents. *Diseases of Marine Organisms* **62**, 1–16.

Ware, S. 1975. British Lower Greensand Serpulidae. *Palaeontology* **18**, 93–116.

Watling, L., France, S.C., Pante, E. & Simpson, A. 2011. Chapter 2. Biology of deep-water octocorals. In *Advances in Marine Biology*, M. Lesser (ed). London: Elsevier Ltd., 41–244.

Watson, C. & Faulwetter, S. 2017. Stylet jaws of Chrysopetalidae (Annelida). *Journal of Natural History* **51**, 2863–2924.

Webster, H.E. 1884. Annelida from Bermuda collected by G. Brown Goode. *Bulletin of the United States National Museum* **25**, 307–327.

Whiteman, N.K. 2008. Between a whale bone and the deep blue sea: the provenance of dwarf males in whale bone-eating tubeworms. *Molecular Ecology* **17**, 4395–4397.

Whorff, J. 1991. Commensals associated with *Xenophora* (*Onustus*) *longleyi* Bartsch (Mollusca: Gastropoda) in the Gulf of Mexico and Caribbean Sea. *The Veliger* **34**, 32–37.

Wielgus, J., Glassom, D., Ben-Shaprut, O. & Chadwick-Furman, N. 2002. An aberrant growth form of Red Sea corals caused by polychaete infestations. *Coral Reefs* **21**, 315–316.

Wielgus, J., Glassom, D. & Chadwick, N.E. 2006. Patterns of polychaete worm infestation of stony corals in the northern Red Sea and relationships to water chemistry. *Bulletin of Marine Science* **78**, 377–388.

Wielgus, J. & Levy, O. 2006. Differences in photosynthetic activity between coral sections infested and not infested by boring spionid polychaetes. *Journal of the Marine Biological Association of the United Kingdom* **86**, 727–728.

Wiklund, H., Altamira, I.V., Glover, A.G., Smith, C.R., Baco, A.R. & Dahlgren, T.G. 2012. Systematics and biodiversity of *Ophryotrocha* (Annelida, Dorvilleidae) with descriptions of six new species from deep-sea whale-fall and wood-fall habitats in the north-east Pacific. *Systematics and Biodiversity* **10**, 243–259.

Wiklund, H., Nygren, A., Pleijel, F. & Sundberg, P. 2005. Phylogeny of Aphroditiformia (Polychaeta) based on molecular and morphological data. *Molecular Phylogenetics and Evolution* **37**, 494–502.

Willette, D.A., Iñiguez, A.R., Kupriyanova, E.K., Starger, C.J., Varman, T., Toha, A. H., Maralit, B.A. & Barber, P.H. 2015. Christmas tree worms of Indo-Pacific coral reefs: Untangling the *Spirobranchus corniculatus* (Grube, 1862) complex. *Coral Reefs* **34**, 899–904.

Willey, A. 1902. XII. Polychaeta. In *Report on the Collections of Natural History Made in the Antarctic Regions during the Voyage of the "Southern Cross"*, B. Sharpe & J. Bell (eds). London: British Museum, 262–283, plates 241–246.

Williams, J.D. 2000. A new species of *Polydora* (Polychaeta: Spionidae) from the Indo-West Pacific and first record of a host hermit crab egg predation by a commensal polydorid worm. *Zoological Journal of the Linnean Society of London* **129**, 537–548.

Williams, J.D. 2001a. *Polydora* and related genera associated with hermit crabs from the Indo-West Pacific (Polychaeta: Spionidae), with descriptions of two new species and a second polydorid egg predator of hermit crabs. *Pacific Science. University of Hawai'i Press* **55**, 429–465.

Williams, J.D. 2001b. Reproduction and larval development of *Polydora robi* (Polychaeta: Spionidae), an obligate commensal of hermit crabs from Philippines. *Invertebrate Biology* **120**, 237–247.

Williams, J.D. 2002. The ecology and feeding biology of two *Polydora* species (Polychaeta: Spionidae) found to ingest the embryos of host hermit crabs (Anomura: Decapoda) from the Philippines. *Journal of Zoology, London* **257**, 339–351.

Williams, J.D. 2004. Reproduction and morphology of *Polydorella* (Polychaeta: Spionidae), including the description of a new species from the Philippines. *Journal of Natural History* **38**, 1339–1358.

Williams, J.D. & McDermott, J.J. 2004. Hermit crab biocoenoses: a worldwide review of the diversity and natural history of hermit crab associates. *Journal of Experimental Marine Biology and Ecology* **305**, 1–128.

Williams, J.D. & Radashevsky, V.I. 1999. Morphology, ecology, and reproduction of a new *Polydora* species from the east coast of North America. *Ophelia* **51**, 115–127.

Wiren, A. 1886. *Haematocleptes terebellidis* nouvelle annelide parasite de la famille des euniciens. *Bihang till Kungliga Svenska Vetenskaps-Akademiens Handlingar* **11**, 1–10.

Wirtz, P., de Melo, G. & de Grave, S. 2009. Symbioses of decapod crustaceans along the coast of Espírito Santo, Brazil. *Marine Biodiversity Records* **2**, e162.

Wisshak, M. & Neumann, C. 2006. A symbiotic association of a boring polychaete and an echinoid from the Late Cretaceous of Germany. *Acta Palaeontologica Polonica* **51**, 589–597.

Woodwick, K.H. 1963. Comparisson of *Boccardia columbiana* Berkeley and *Boccardia proboscidea* Hartman (Annelida, Polychaeta). *Bulletin of the Southern California Academy of Sciences* **62**, 132–139.

Yarrell, W. 1832. On the anatomy of the conger eel and on the differences between the conger and freshwater eels. *Proceedings of the Committee of Science and Correspondence of the Zoological Society of London* **1830–31**, 158–159.

Zapalski, M.K. 2011. Is absence of proof a proof of absence? Comments on commensalism. *Paleogeography, Paleoclimatology, Paleoecology* **302**, 484–488.

Zeidberg, L.D., Isaac, G., Widmer, C.L., Neumeister, H. & Gilly, W.F. 2011. Egg capsule hatch rate and incubation duration of the California market squid, *Doryteuthis* (= *Loligo*) *opalescens*: insights from laboratory manipulations. *Marine Ecology* **32**, 468–479.

Zibrowius, H. 1980. Les scléractiniaires de la Méditerranée et de l'Atlantique nord-oriental. *Mémoires de l'Institut Océanographique de Monaco* **11**, 1–227.

Zibrowius, H., Southward, E.C. & Day, J.H. 1975. New observations on a little-known species of *Lumbrineris* (Polychaeta) living on various Cnidarians, with notes on its recent and fossil Scleractinian hosts. *Journal of the Marine Biological Association of the United Kingdom* **55**, 83–108.

Author Index

Note: Page numbers in **boldface** denote complete articles.

Gent, P. R., 29
Gentil, F., 329, 350
Gentile, J. H., 312, 349
Genzano, G.N., 399
Georgiades, E. T., 325, 331, 344
Gera, A., 262
Gerber, E. P., 28
Gerber, L., 45, 94
Gerber, L. R., 87, 94
Gerdes, D., 115, 126, 129, 134, 157
Gerlach, F., 169
Gerlachm, F., 169
German, C. R., 112
Gerringa, L. J. A., 109
Gerrodette, T., 78, 79, 89
Gersonde, R., 19
Gervais, F., 12
Getter, C. D., 327, 328, 329, 330, 332, 333, 335, 339, 340,
    341, 347
Geuens, E., 169
Gezgin, T., 341
Ghalambor, C. K., 176, 177
Ghigliotti, L., 164
Gialanella, C., 252, 253, 254, 256
Giammanco, S., 239, 240
Gianakouras, D., 33
Giangrande, A., 238, 245, 249, 257, 258, 263, 265, 267,
    268, 285, 286, 287, 288, 289, 290, 391
Giard, M.A., 375, 376, 408
Gibbons, D. W., 132
Gibbons, J., 93
Gibbons, M., 19
Gibson, J. A. E., 185
Giese, A., 160
Giese, A. C., 126, 131, 145, 160, 182
Giese, M., 316
Gil, J., 390, 418, 420, 426
Gilbert, F., 342
Gilbert, N., 106, 132
Gilbert, R., 3, 28
Gilfillan, E. S., 338, 342, 343
Gili, J. -M., 114, 115, 116, 126, 129, 131, 142, 155, 157
Gili, J. –M., 142
Gill, P., 94
Gillanders, B. M., 257
Gille, S., 19
Gille, S. T., 3, 29, 40
Gillon, D., 337
Gilmer, R. W., 23
Gilmour, J. P., 269, 321
Gilpin-Brown, J.B., 377
Gilson, J., 29
Gimenez, I., 265
Gimino, G., 115
Ginsburg, D. T., 122, 124, 130
Giordano, D., 106, 145, 167, 169
Girardin, M., 137
Giraud, G., 255, 260, 282
Giulivi, C. F., 25
Glandon, H. L., 187
Glas, M. S., 238, 239, 251, 256, 257, 259, 269
Glasby, C.J., 388, 389, 391, 392, 417, 418

Gleitz, M., 22
Gloersen, P., 22, 27, 28
Glover, A.G., 387
Glover, D. M., 5, 35, 39, 40
Glowacki, P., 116, 160
Glynn, P. W., 350
Gnanadesikan, A., 2, 34, 42, 43, 184
Goarant, A., 16
Gobler, C. J., 186
Godfrey, J. S., 3
Goericke, R., 20
Goerke, H., 377
Goff, L.J., 371
Goffredi, S.K., 387
Goffredo, S., 256
Goldenberg, S. U., 238
Goldfarb, B., 89
Goldsworthy, P. M., 174, 323
Goldsworthy, S. D., 316
Gomes, V., 145, 175
Gomez, A., 145, 188, 189
Gomez, E., 186
Gomez, O., 327
Gon, O., 19, 141
González, A. F., 17
Gonzalez, C., 327, 328, 329, 330, 332, 333, 335, 339,
    340, 347
Gonzalez, E., 342
González, K., 170
Gonzalez, M., 188
González, N.E., 382
González, O., 175
González-Aravena, M., 170
Gonzalez-Bernat, M. J., 188
Gonzalez-Cabrera, P. J., 182
Gonzalvo, J., 77
Good, S. A., 29
Gooday, A. J., 178
Goodbody-Gringley, G., 337
Goodman, K. S., 327
Goodsell, P., 74
Goodwin, C., 245, 254, 263, 283, 284
Goosse, H., 106
Gorb, S. N., 186
Gorbunov, M. Y., 12
Gordon, A. L., 25
Gordon, L. I., 22
Gordon, R., 25
Gordon, S., 3, 106
Gore, D. J., 129, 133, 136, 171
Gormley, A., 90
Gormley, A. M., 79, 90, 91
Gorny, M., 116, 129, 133, 136
Goss, C., 14, 25, 31, 32, 40
Gosselin, J. -F., 78
Goto, R., 382
Gottini, V., 240
Gouezo, M., 262
Gould, A.A., 412
Gould, W. J., 29
Gouretski, V. V., 29
Gowlett-Holmes, K., 327

Subject Index

Note: Page numbers in **boldface** denote tables.